T0140064

Lecture Notes in Computer Science 13353

More information about this series at https://link.springer.com/bookseries/558

Derek Groen · Clélia de Mulatier ·
Maciej Paszynski · Valeria V. Krzhizhanovskaya ·
Jack J. Dongarra · Peter M. A. Sloot (Eds.)

Computational Science – ICCS 2022

22nd International Conference
London, UK, June 21–23, 2022
Proceedings, Part IV

Editors
Derek Groen
Brunel University London
London, UK

Maciej Paszynski
AGH University of Science and Technology
Krakow, Poland

Jack J. Dongarra
University of Tennessee at Knoxville
Knoxville, TN, USA

Clélia de Mulatier
University of Amsterdam
Amsterdam, The Netherlands

Valeria V. Krzhizhanovskaya
University of Amsterdam
Amsterdam, The Netherlands

Peter M. A. Sloot
University of Amsterdam
Amsterdam, The Netherlands

ISSN 0302-9743 ISSN 1611-3349 (electronic)
Lecture Notes in Computer Science
ISBN 978-3-031-08759-2 ISBN 978-3-031-08760-8 (eBook)
https://doi.org/10.1007/978-3-031-08760-8

This Springer imprint is published by the registered company Springer Nature Switzerland AG
The registered company address is: Gewerbestrasse 11, 6330 Cham, Switzerland

Preface

Welcome to the 22nd annual International Conference on Computational Science (ICCS 2022 - https://www.iccs-meeting.org/iccs2022/), held during 21–23 June, 2022, at Brunel University London, UK. After more than two years of a pandemic that has changed so much of our world and daily lives, this edition marks our return to a – partially – in-person event. Those who were not yet able to join us in London had the option to participate online, as all conference sessions were streamed.

Although the challenges of such a hybrid format are manifold, we have tried our best to keep the ICCS community as dynamic, creative, and productive as always. We are proud to present the proceedings you are reading as a result of that.

Standing on the River Thames in southeast England, at the head of a 50-mile (80 km) estuary down to the North Sea, London is the capital and largest city of England and the UK. With a rich history spanning back to Roman times, modern London is one of the world's global cities, having a prominent role in areas ranging from arts and entertainment to commerce, finance, and education. London is the biggest urban economy in Europe and one of the major financial centres in the world. It also features Europe's largest concentration of higher education institutions.

ICCS 2022 was jointly organized by Brunel University London, the University of Amsterdam, NTU Singapore, and the University of Tennessee.

Brunel University London is a public research university located in the Uxbridge area of London. It was founded in 1966 and named after the Victorian engineer Isambard Kingdom Brunel, who managed to design and build a 214m long suspension bridge in Bristol back in 1831. Brunel is well-known for its excellent Engineering and Computer Science Departments, and its campus houses a dedicated conference centre (the Hamilton Centre) which was used to host ICCS. It is also one of the few universities to host a full-length athletics track, which has been used both for practice purposes by athletes such as Usain Bolt for the 2012 Olympics and for graduation ceremonies.

The International Conference on Computational Science is an annual conference that brings together researchers and scientists from mathematics and computer science as basic computing disciplines, as well as researchers from various application areas who are pioneering computational methods in sciences such as physics, chemistry, life sciences, engineering, arts, and humanitarian fields, to discuss problems and solutions in the area, identify new issues, and shape future directions for research.

Since its inception in 2001, ICCS has attracted increasing numbers of attendees and higher-quality papers, and this year – in spite of the ongoing pandemic—was not an exception, with over 300 registered participants. The proceedings series has become a primary intellectual resource for computational science researchers, defining and advancing the state of the art in this field.

The theme for 2022, "The Computational Planet," highlights the role of computational science in tackling the current challenges of the all-important quest for sustainable development. This conference aimed to be a unique event focusing on recent developments in scalable scientific algorithms, advanced software tools, computational grids, advanced numerical methods, and novel application areas. These innovative novel models, algorithms, and tools drive new science through efficient application in physical systems, computational and systems biology, environmental systems, finance, and other areas.

ICCS is well-known for its excellent lineup of keynote speakers. The keynotes for 2022 were as follows:

- Robert Axtell, George Mason University, USA
- Peter Coveney, University College London, UK
- Thomas Engels, Technische Universität Berlin, Germany
- Neil Ferguson, Imperial College London, UK
- Giulia Galli, University of Chicago, USA
- Rebecca Wade, Heidelberg Institute for Theoretical Studies, Germany

This year we had 474 submissions (169 submissions to the main track and 305 to the thematic tracks). In the main track, 55 full papers were accepted (32%), and in the thematic tracks, 120 full papers (39%). A higher acceptance rate in the thematic tracks is explained by the nature of these, where track organizers personally invite many experts in a particular field to participate in their sessions.

ICCS relies strongly on our thematic track organizers' vital contributions to attract high-quality papers in many subject areas. We would like to thank all committee members from the main and thematic tracks for their contribution to ensure a high standard for the accepted papers. We would also like to thank Springer, Elsevier, and Intellegibilis for their support. Finally, we appreciate all the local organizing committee members for their hard work to prepare for this conference.

We are proud to note that ICCS is an A-rank conference in the CORE classification.

We wish you good health in these troubled times and look forward to meeting you at the next conference, whether virtually or in-person.

June 2022

Derek Groen
Clélia de Mulatier
Maciej Paszynski
Valeria V. Krzhizhanovskaya
Jack J. Dongarra
Peter M. A. Sloot

Organization

General Chair

Valeria Krzhizhanovskaya	University of Amsterdam, The Netherlands

Main Track Chair

Clélia de Mulatier	University of Amsterdam, The Netherlands

Thematic Tracks Chair

Maciej Paszynski	AGH University of Science and Technology, Poland

Scientific Chairs

Peter M. A. Sloot	University of Amsterdam, The Netherlands | Complexity Institute NTU, Singapore
Jack Dongarra	University of Tennessee, USA

Local Organizing Committee

Chair

Derek Groen	Brunel University London, UK

Members

Simon Taylor	Brunel University London, UK
Anastasia Anagnostou	Brunel University London, UK
Diana Suleimenova	Brunel University London, UK
Xiaohui Liu	Brunel University London, UK
Zidong Wang	Brunel University London, UK
Steven Sam	Brunel University London, UK
Alireza Jahani	Brunel University London, UK
Yani Xue	Brunel University London, UK
Nadine Aburumman	Brunel University London, UK
Katie Mintram	Brunel University London, UK
Arindam Saha	Brunel University London, UK
Nura Abubakar	Brunel University London, UK

Thematic Tracks and Organizers

Advances in High-Performance Computational Earth Sciences: Applications and Frameworks – IHPCES

Takashi Shimokawabe	University of Tokyo, Japan
Kohei Fujita	University of Tokyo, Japan
Dominik Bartuschat	Friedrich-Alexander-Universität Erlangen-Nürnberg, Germany

Artificial Intelligence and High-Performance Computing for Advanced Simulations – AIHPC4AS

Maciej Paszynski	AGH University of Science and Technology, Poland

Biomedical and Bioinformatics Challenges for Computer Science – BBC

Mario Cannataro	Università Magna Graecia di Catanzaro, Italy
Giuseppe Agapito	Università Magna Graecia di Catanzaro, Italy
Mauro Castelli	Universidade Nova de Lisboa, Portugal
Riccardo Dondi	University of Bergamo, Italy
Rodrigo Weber dos Santos	Universidade Federal de Juiz de Fora, Brazil
Italo Zoppis	Università degli Studi di Milano-Bicocca, Italy

Computational Collective Intelligence – CCI

Marcin Maleszka	Wroclaw University of Science and Technology, Poland
Ngoc Thanh Nguyen	Wroclaw University of Science and Technology, Poland
Dosam Hwang	Yeungnam University, South Korea

Computational Health – CompHealth

Sergey Kovalchuk	ITMO University, Russia
Stefan Thurner	Medical University of Vienna, Austria
Georgiy Bobashev	RTI International, USA
Jude Hemanth	Karunya University, India
Anastasia Angelopoulou	University of Westminster, UK

Computational Optimization, Modelling, and Simulation – COMS

Xin-She Yang	Middlesex University London, UK
Leifur Leifsson	Purdue University, USA
Slawomir Koziel	Reykjavik University, Iceland

Computer Graphics, Image Processing, and Artificial Intelligence – CGIPAI

Andres Iglesias	Universidad de Cantabria, Spain

Machine Learning and Data Assimilation for Dynamical Systems – MLDADS

Rossella Arcucci	Imperial College London, UK

Multiscale Modelling and Simulation – MMS

Derek Groen	Brunel University London, UK
Diana Suleimenova	Brunel University London, UK
Bartosz Bosak	Poznan Supercomputing and Networking Center, Poland
Gabor Závodszky	University of Amsterdam, The Netherlands
Stefano Casarin	Houston Methodist Research Institute, USA
Ulf D. Schiller	Clemson University, USA
Wouter Edeling	Centrum Wiskunde & Informatica, The Netherlands

Quantum Computing – QCW

Katarzyna Rycerz	AGH University of Science and Technology, Poland
Marian Bubak	Sano Centre for Computational Medicine and AGH University of Science and Technology, Poland | University of Amsterdam, The Netherlands

Simulations of Flow and Transport: Modeling, Algorithms, and Computation – SOFTMAC

Shuyu Sun	King Abdullah University of Science and Technology, Saudi Arabia
Jingfa Li	Beijing Institute of Petrochemical Technology, China
James Liu	Colorado State University, USA

Smart Systems: Bringing Together Computer Vision, Sensor Networks, and Machine Learning – SmartSys

Pedro Cardoso	University of Algarve, Portugal
João Rodrigues	University of Algarve, Portugal
Jânio Monteiro	University of Algarve, Portugal
Roberto Lam	University of Algarve, Portugal

Software Engineering for Computational Science – SE4Science

Jeffrey Carver	University of Alabama, USA
Caroline Jay	University of Manchester, UK
Yochannah Yehudi	University of Manchester, UK
Neil Chue Hong	University of Edinburgh, UK

Solving Problems with Uncertainty – SPU

Vassil Alexandrov	Hartree Centre - STFC, UK
Aneta Karaivanova	Institute for Parallel Processing, Bulgarian Academy of Sciences, Bulgaria

Teaching Computational Science – WTCS

Angela Shiflet	Wofford College, USA
Nia Alexandrov	Hartree Centre - STFC, UK

Uncertainty Quantification for Computational Models – UNEQUIvOCAL

Wouter Edeling	Centrum Wiskunde & Informatica, The Netherlands
Anna Nikishova	SISSA, Italy

Reviewers

Tesfamariam Mulugeta Abuhay
Jaime Afonso Martins
Giuseppe Agapito
Shahbaz Ahmad
Elisabete Alberdi
Luis Alexandre
Nia Alexandrov
Vassil Alexandrov
Julen Alvarez-Aramberri
Domingos Alves
Sergey Alyaev
Anastasia Anagnostou
Anastasia Angelopoulou
Samuel Aning
Hideo Aochi
Rossella Arcucci
Costin Badica
Bartosz Balis
Daniel Balouek-Thomert
Krzysztof Banaś
Dariusz Barbucha
João Barroso
Valeria Bartsch
Dominik Bartuschat
Pouria Behnodfaur
Jörn Behrens
Adrian Bekasiewicz
Gebrail Bekdas
Mehmet Ali Belen
Stefano Beretta
Benjamin Berkels
Daniel Berrar
Georgiy Bobashev
Marcel Boersma
Tomasz Boiński
Carlos Bordons
Bartosz Bosak
Giuseppe Brandi
Lars Braubach
Marian Bubak

Jérémy Buisson
Aleksander Byrski
Cristiano Cabrita
Xing Cai
Barbara Calabrese
Nurullah Calik
Almudena Campuzano
Mario Cannataro
Pedro Cardoso
Alberto Carrassi
Alfonso Carriazo
Jeffrey Carver
Stefano Casarin
Manuel Castañón-Puga
Mauro Castelli
Nicholas Chancellor
Ehtzaz Chaudhry
Thierry Chaussalet
Sibo Cheng
Siew Ann Cheong
Andrei Chernykh
Lock-Yue Chew
Su-Fong Chien
Marta Chinnici
Amine Chohra
Neil Chue Hong
Svetlana Chuprina
Paola Cinnella
Noélia Correia
Adriano Cortes
Ana Cortes
Enrique Costa-Montenegro
David Coster
Carlos Cotta
Helene Coullon
Daan Crommelin
Attila Csikasz-Nagy
Javier Cuenca
António Cunha
Pawel Czarnul
Lisandro D. Dalcin
Bhaskar Dasgupta
Clélia de Mulatier
Charlotte Debus
Javier Delserlorente
Pasquale De-Luca
Quanling Deng
Vasily Desnitsky
Mittal Dhruv
Eric Dignum
Riccardo Dondi
Rafal Drezewski
Hans du Buf
Vitor Duarte
Richard Dwight
Wouter Edeling
Nasir Eisty
Kareem El-Safty
Nahid Emad
Gökhan Ertaylan
Roberto R. Expósito
Fangxin Fang
Antonino Fiannaca
Christos Filelis-Papadopoulos
Pawel Foszner
Piotr Frąckiewicz
Martin Frank
Alberto Freitas
Ruy Freitas Reis
Karl Frinkle
Kohei Fujita
Takeshi Fukaya
Wlodzimierz Funika
Takashi Furumura
Ernst Fusch
Leszek Gajecki
Ardelio Galletti
Marco Gallieri
Teresa Galvão
Akemi Galvez-Tomida
Maria Ganzha
Luis Garcia-Castillo
Bartłomiej Gardas
Delia Garijo
Frédéric Gava
Piotr Gawron
Bernhard Geiger
Alex Gerbessiotis
Philippe Giabbanelli
Konstantinos Giannoutakis

Adam Glos
Ivo Goncalves
Alexandrino Gonçalves
Jorge González-Domínguez
Yuriy Gorbachev
Pawel Gorecki
Markus Götz
Michael Gowanlock
George Gravvanis
Derek Groen
Lutz Gross
Lluis Guasch
Pedro Guerreiro
Tobias Guggemos
Xiaohu Guo
Manish Gupta
Piotr Gurgul
Zulfiqar Habib
Mohamed Hamada
Yue Hao
Habibollah Haron
Ali Hashemian
Carina Haupt
Claire Heaney
Alexander Heinecke
Jude Hemanth
Marcin Hernes
Bogumila Hnatkowska
Maximilian Höb
Jori Hoencamp
Rolf Hoffmann
Wladyslaw Homenda
Tzung-Pei Hong
Muhammad Hussain
Dosam Hwang
Mauro Iacono
David Iclanzan
Andres Iglesias
Mirjana Ivanovic
Takeshi Iwashita
Alireza Jahani
Peter Janků
Jiri Jaros
Agnieszka Jastrzebska
Caroline Jay
Piotr Jedrzejowicz
Gordan Jezic
Zhong Jin
David Johnson
Guido Juckeland
Piotr Kalita
Drona Kandhai
Epaminondas Kapetanios
Aneta Karaivanova
Artur Karczmarczyk
Takahiro Katagiri
Timo Kehrer
Christoph Kessler
Loo Chu Kiong
Harald Koestler
Ivana Kolingerova
Georgy Kopanitsa
Pavankumar Koratikere
Triston Kosloske
Sotiris Kotsiantis
Remous-Aris Koutsiamanis
Sergey Kovalchuk
Slawomir Koziel
Dariusz Krol
Marek Krótkiewicz
Valeria Krzhizhanovskaya
Marek Kubalcík
Sebastian Kuckuk
Eileen Kuehn
Michael Kuhn
Tomasz Kulpa
Julian Martin Kunkel
Krzysztof Kurowski
Marcin Kuta
Panagiotis Kyziropoulos
Roberto Lam
Anna-Lena Lamprecht
Kun-Chan Lan
Rubin Landau
Leon Lang
Johannes Langguth
Leifur Leifsson
Kenneth Leiter
Florin Leon
Vasiliy Leonenko

Jean-Hugues Lestang
Jake Lever
Andrew Lewis
Jingfa Li
Way Soong Lim
Denis Mayr Lima Martins
James Liu
Zhao Liu
Hong Liu
Che Liu
Yen-Chen Liu
Hui Liu
Marcelo Lobosco
Doina Logafatu
Marcin Los
Stephane Louise
Frederic Loulergue
Paul Lu
Stefan Luding
Laura Lyman
Lukasz Madej
Luca Magri
Peyman Mahouti
Marcin Maleszka
Bernadetta Maleszka
Alexander Malyshev
Livia Marcellino
Tomas Margalef
Tiziana Margaria
Svetozar Margenov
Osni Marques
Carmen Marquez
Paula Martins
Pawel Matuszyk
Valerie Maxville
Wagner Meira Jr.
Roderick Melnik
Pedro Mendes Guerreiro
Ivan Merelli
Lyudmila Mihaylova
Marianna Milano
Jaroslaw Miszczak
Janio Monteiro
Fernando Monteiro
Andrew Moore
Eugénia Moreira Bernardino
Anabela Moreira Bernardino
Peter Mueller
Ignacio Muga
Khan Muhammad
Daichi Mukunoki
Vivek Muniraj
Judit Munoz-Matute
Hiromichi Nagao
Jethro Nagawakar
Kengo Nakajima
Grzegorz J. Nalepa
Yves Nanfack
Pratik Nayak
Philipp Neumann
David Chek-Ling Ngo
Ngoc Thanh Nguyen
Nancy Nichols
Sinan Melih Nigdeli
Anna Nikishova
Hitoshi Nishizawa
Algirdas Noreika
Manuel Núñez
Frederike Oetker
Schenk Olaf
Javier Omella
Boon-Yaik Ooi
Eneko Osaba
Aziz Ouaarab
Raymond Padmos
Nikela Papadopoulou
Marcin Paprzycki
David Pardo
Diego Paredesconcha
Anna Paszynska
Maciej Paszynski
Ebo Peerbooms
Sara Perez-Carabaza
Dana Petcu
Serge Petiton
Frank Phillipson
Eugenio Piasini
Juan C. Pichel
Anna Pietrenko-Dabrowska
Laércio L. Pilla

Armando Pinho
Yuri Pirola
Mihail Popov
Cristina Portales
Roland Potthast
Małgorzata Przybyła-Kasperek
Ela Pustulka-Hunt
Vladimir Puzyrev
Rick Quax
Cesar Quilodran-Casas
Enrique S. Quintana-Orti
Issam Rais
Andrianirina Rakotoharisoa
Raul Ramirez
Celia Ramos
Vishwas Rao
Kurunathan Ratnavelu
Lukasz Rauch
Robin Richardson
Miguel Ridao
Heike Riel
Sophie Robert
Joao Rodrigues
Daniel Rodriguez
Albert Romkes
Debraj Roy
Katarzyna Rycerz
Emmanuelle Saillard
Ozlem Salehi
Tarith Samson
Alberto Sanchez
Ayşin Sancı
Gabriele Santin
Vinicius Santos-Silva
Allah Bux Sargano
Robert Schaefer
Ulf D. Schiller
Bertil Schmidt
Martin Schreiber
Gabriela Schütz
Franciszek Seredynski
Marzia Settino
Mostafa Shahriari
Zhendan Shang
Angela Shiflet
Takashi Shimokawabe
Alexander Shukhman
Marcin Sieniek
Nazareen Sikkandar-Basha
Robert Sinkovits
Mateusz Sitko
Haozhen Situ
Leszek Siwik
Renata Słota
Oskar Slowik
Grażyna Ślusarczyk
Sucha Smanchat
Maciej Smołka
Thiago Sobral
Isabel Sofia Brito
Piotr Sowiński
Robert Speck
Christian Spieker
Michał Staniszewski
Robert Staszewski
Steve Stevenson
Tomasz Stopa
Achim Streit
Barbara Strug
Patricia Suarez
Dante Suarez
Diana Suleimenova
Shuyu Sun
Martin Swain
Jerzy Świątek
Piotr Szczepaniak
Edward Szczerbicki
Tadeusz Szuba
Ryszard Tadeusiewicz
Daisuke Takahashi
Osamu Tatebe
Carlos Tavares Calafate
Kasim Tersic
Jannis Teunissen
Mau Luen Tham
Stefan Thurner
Nestor Tiglao
T. O. Ting
Alfredo Tirado-Ramos
Pawel Topa

Bogdan Trawinski
Jan Treur
Leonardo Trujillo
Paolo Trunfio
Hassan Ugail
Eirik Valseth
Casper van Elteren
Ben van Werkhoven
Vítor Vasconcelos
Alexandra Vatyan
Colin C. Venters
Milana Vuckovic
Shuangbu Wang
Jianwu Wang
Peng Wang
Katarzyna Wasielewska
Jaroslaw Watrobski
Rodrigo Weber dos Santos
Mei Wen
Lars Wienbrandt
Iza Wierzbowska
Maciej Woźniak
Dunhui Xiao
Huilin Xing
Yani Xue
Abuzer Yakaryilmaz
Xin-She Yang
Dongwei Ye
Yochannah Yehudi
Lihua You
Drago Žagar
Constantin-Bala Zamfirescu
Gabor Závodszky
Jian-Jun Zhang
Yao Zhang
Wenbin Zhang
Haoxi Zhang
Jinghui Zhong
Sotirios Ziavras
Zoltan Zimboras
Italo Zoppis
Chiara Zucco
Pavel Zun
Simon Portegies Zwart
Karol Życzkowski

Contents – Part IV

Smart Systems: Bringing Together Computer Vision, Sensor Networks and Machine Learning

Uncertainty Quantification for Computational Models

Machine Learning and Data Assimilation for Dynamical Systems (Continued)

Machine Learning-Based Scheduling and Resources Allocation in Distributed Computing

Victor Toporkov(✉), Dmitry Yemelyanov, and Artem Bulkhak

National Research University "MPEI", Moscow, Russia
{ToporkovVV,YemelyanovDM,BulkhakAN}@mpei.ru

Abstract. In this work we study a promising approach for efficient online scheduling of job-flows in high performance and distributed parallel computing. The majority of job-flow optimization approaches, including backfilling and microscheduling, require apriori knowledge of a full job queue to make the optimization decisions. In a more general scenario when user jobs are submitted individually, the resources selection and allocation should be performed immediately in the online mode. In this work we consider a neural network prototype model trained to perform online optimization decisions based on a known optimal solution. For this purpose, we designed MLAK algorithm which implements 0–1 knapsack problem based on the apriori unknown utility function. In a dedicated simulation experiments with different utility functions MLAK provides resources selection efficiency comparable to a classical greedy algorithm.

Keywords: Resource · Scheduling · Online · Knapsack · Optimization · Neural network · Machine learning

1 Introduction and Related Works

Modern high-performance distributed computing systems (HPCS), including Grid, cloud and hybrid infrastructures provide access to large amounts of resources [1, 2]. These resources typically include computing nodes, network channels, software tools and data storages, required to execute parallel jobs submitted by HPCS users.

Most HPCS and cloud solutions have requirements to provide a certain quality of services (QoS) for users' applications scheduling, execution and monitoring. Correspondingly, QoS constraints usually include a set of requirements for a coordinated resources co-allocation [3–5], as well as a number of time and cost criteria and restrictions, such as deadline, response time, total execution cost, etc. [2–7].

Some of the most important efficiency indicators of a distributed computational environment include both system resources utilization level and users' jobs time and cost execution criteria [2–4].

HPCS organization and support bring certain economical expenses: purchase and installation of machinery equipment, power supplies, user support, maintenance works,

D. Groen et al. (Eds.): ICCS 2022, LNCS 13353, pp. 3–16, 2022.
https://doi.org/10.1007/978-3-031-08760-8_1

security, etc. Thus, HPCS users and service providers usually interact in economic terms, and the resources are provided for a certain payment. In such conditions, resource management and job scheduling based on the economic models is considered as an efficient way to coordinate contradictory preferences of computing system participants and stakeholders [3–7].

A metascheduler or a metabroker are considered as intermediate links between the users, local resource management and job batch processing systems [3, 4, 7, 8]. They define uniform rules for resources distribution and ensure the overall scheduling efficiency.

The most straightforward way to schedule a job-flow is by using the First-Come-First-Served (FCFS) procedure. FCFS executes jobs one by one in an order of arrival. Backfilling procedure [4, 9] makes use of advanced resources reservations in order to prevent starvation of jobs with a relatively large resource request requirements. Microscheduling [4, 5, 10] approach may be added to backfilling to affect global scheduling efficiency by choosing the appropriate secondary optimization criteria.

Online scheduling, on the other hand, requires HPCS scheduler to make resources allocation and optimization decisions immediately when jobs are submitted. One possible online scheduling strategy is to perform locally efficient resources selection for each job. However, in this case the global scheduling efficiency may be degraded. CoP microscheduling strategy [4] implements a set of heuristic rules to optimize job-flow execution time based on the resource's properties: performance, cost, utilization level, etc.

The main contribution of this paper is a machine learning-based approach which can be trained on efficient scheduling results to perform online scheduling based on secondary properties of the resources. To achieve this goal, an artificial neural network was designed in combination with a dynamic programming method. We consider a general 0–1 knapsack scheduling model and evaluate algorithms efficiency in a dedicated simulation experiment.

The paper is organized as follows. Section 2 presents a general problem statement and the corresponding machine learning model. Section 3 contains description of the proposed algorithms and neural network training details. Section 4 provides simulation details, results, and analysis. Finally, Sect. 5 summarizes the paper results.

2 Problem Statement

2.1 Online Resources Selection and Knapsack Problem

The 0–1 knapsack problem is fundamental for optimization of resources selection and allocation. The classic 0–1 knapsack problem operates with a set of N items having two properties: weight w_i and utility u_i. The general problem is to select a subset of items which maximizes total utility with a restriction C on the total weight:

$$\sum\nolimits_{i=1}^{N} x_i u_i \rightarrow \max, \tag{1.1}$$

$$\sum\nolimits_{i=1}^{N} x_i w_i \leq C, \tag{1.2}$$

where x_i - is a decision variable determining whether to select item i ($x_i = 1$) or not ($x_i = 0$) for the knapsack.

This problem definition (1.1), (1.2) fits the economic scheduling model with available computing resources having cost c_i (weight) and performance p_i (utility) properties. Many scheduling algorithms and approaches implement exact or approximate *knapsack* solutions for the resources' selection step [4, 11–14]. Sometimes the job scheduling problem may require additional constraints, for example, to limit the number n of items in the knapsack [12, 13] or to select items of different subtypes [14].

The most straightforward exact solution for the knapsack problem can be achieved with a brute force algorithm. However, with increasing N and C in (1.1), (1.2) its application eventually requires inadequately large computational costs. *Dynamic programming (DP)* algorithms can provide exact integer solution with a pseudo-polynomial computational complexity of $O(N * C)$ or $O(n * N * C)$ depending on the problem constraints. Dynamic programming algorithms usually rely on recurrent calculation schemes optimizing additive criteria (1.1) when iterating through the available items. For example, the following recurrent scheme can be used to solve the problem (1.1), (1.2):

$$\begin{aligned} f_i(c) = &\max\{f_{i-1}(c), f_{i-1}(c - w_i) + u_i\}, \\ &i = 1, .., N, c = 1, .., C, \end{aligned} \tag{1.3}$$

where $f_i(c)$ defines the maximum criterion (1.1) value allocated out of first i items with a total weight limit c.

When recurrent calculation (1.3) is finished, $f_N(C)$ will contain the problem solution.

Approximate solution can be obtained with more computationally efficient *greedy algorithms*. Greedy algorithms for the knapsack problem usually use a single heuristic function to estimate the items' importance for the knapsack in terms of their weight w_i and utility u_i ratio. Thus, the most common greedy solution for problem (1.1), (1.2) decreasingly arranges items by their u_i/w_i ratio and successively selects them into the knapsack up to the weight limit.

This greedy solution usually provides a satisfactory (1.1) optimization for an adequate computational complexity estimated as $O(N * LogN)$.

Most modern scheduling solutions in one way or another implement these algorithms or their modifications. For example, backfilling scheduling procedure defines additional rules for the *job queue* execution order and is able to minimize the overall queue completion time (a *makespan*). Once the execution priority is defined, each parallel job is scheduled independently based on the problem similar to (1.1), (1.2). One important requirement for the backfilling makespan optimization efficiency is that the job queue composition must be known in advance. The *backfilling* core idea implies execution of relatively small jobs from the back of the queue on the currently idle and waiting resources.

However, in a more general scenario the user jobs are submitted individually, and the resources selection and allocation should be performed immediately in the *online* mode. Thus, our main goal is to schedule user jobs independently in a way to optimize global scheduling criteria, for example average jobs' finish time or a makespan.

Similar ideas underlie the so-called microsheduling approaches, including CoP and PeST [4, 10]. They implement *heuristic* rules of how the resources should be selected for a job based on their meta-parameters and properties: utilization level, performance, local schedules, etc.

2.2 Machine Learning Model

Currently relevant is the topic of using machine learning methods to perform combinatorial optimization tasks, including the knapsack problem (1.1), (1.2) [15–17]. For example, [16] introduces a detailed research of a heuristic knapsack solver based on neural networks and deep learning. The neural solver was successfully tested on instances with up to 200 items and provided near optimal solutions (generally better compared to the greedy algorithm) in scenarios with a correlation between the items' utilities and weights.

In [17], a new class of recurrent neural networks is proposed to compute an optimal or provably good solutions for the knapsack problem. The paper considers a question of a network size theoretically sufficient to find solutions of provable quality for the Knapsack Problem. Additionally, the proposed approach can be generalized to other combinatorial optimization problems, including various Shortest Path problems, the Longest Common Subsequence problem, and the Traveling Salesperson problem.

In the current work we consider a more specific job *scheduling* problem based on a machine learning model. An efficient scheduling plan which minimizes makespan of a whole job-flow can be used to train an artificial neural network (ANN) to schedule each job individually (online) with a similar result. However, the job-flow scheduling plan provides only the efficient resources selections for each job (knapsack result), but not the corresponding utility values of the selected resources. Thus, for the training procedure we can use only secondary meta-parameters and properties of the resources. These typically include resources' cost, utilization level, performance attributes, average downtime, time distance to the neighbor reservations, etc. [4].

The more factors and properties of the efficient *reference* solution are considered the more accurate solution could be achieved *online*. Besides, online scheduling imposes additional restrictions on a priori knowledge of the computing environment composition and condition. The exact values of the resources' properties and utility function may be inaccurate or unknown.

In a more general and formal way, *the main task is to design a model, which will solve (predict solution of) 0–1 knapsack problem with a priori unknown utility* u_i *values based only on a set of secondary resource's properties*. Thus, to generalize this task we will use more complex knapsack model interpretation with items having four numeric properties a_i, b_i, d_i, g_i in an addition to the weight w_i. Utility values u_i will be calculated for each resource as a function F_{val} of properties a_i, b_i, d_i, g_i. This function will be used to calculate the optimal knapsack solution (by using a dynamic programming algorithm). Based on this solution the machine learning model will be trained to select resources based only on the input properties a_i, b_i, d_i, g_i, thus, simulating the online scheduling procedure.

In this paper, the following utility functions F_{val} will serve as examples of hidden conditions for selecting items in a knapsack:

$$F_{val} = a + b + d - g, \tag{2.1}$$

$$F_{val} = a * b + d * g^2, \quad (2.2)$$

$$F_{val} = \sin(a + b) + \cos d + g^2, \quad (2.3)$$

$$F_{val} = a + \lg(b + d) * g, \quad (2.4)$$

$$F_{val} = a * \lg b + d * e^{\frac{g}{10}}, \quad (2.5)$$

where a, b, d, g are knapsack item's properties in addition to the weight. The given functions contain almost the entire mathematical complexity spectrum in order to investigate at the testing stage how the function complexity affects the algorithm's accuracy and efficiency.

3 Algorithms Implementation

3.1 Artificial Neural Network Design and Training

An artificial neural network (ANN) can be represented as a sequence of layers that can compute multiple transformations to return a result. As the design of the network structure is mostly based on an empirical approach, we performed a consistent design and research of neural network architectures for the knapsack problem.

Firstly, we are faced with the task of classifying an action x_i with a certain item: whether to put it in a knapsack or not. Generally, classification tasks are solved with the decision tree models. However, unlike in a classic problem of individual elements classification, the items in a knapsack invest into a common property: their total weight should not exceed the constraint (1.2). Thus, it is infeasible to classify the elements separately, the model should accept and process everything at once. So, most suitable topology for such a classification problem is a fully connected multilayer neural network (multilayer perceptron).

To implement this model, the Python programming language was used with the Tensorflow framework and the Keras library [18]. Keras has a wide functionality for design artificial neural networks of diverse types.

After selecting the general structure, it is necessary to experimentally select the network parameters. These include: the number of layers, the number of neurons in layers, the neurons activation function, the quality criterion, the optimization algorithm.

First, we used binary cross-entropy as the most suitable loss function for predicting a set of dependent output values.

Next, we designed and tested a set of small candidate models to decide on other meta-parameters (see Table 1).

Table 1. ANN training results for 5-elements knapsack

Configuration number	Activation function	Optimizer	Number of layers	Neurons in hidden layers	Training set	Train/Test accuracy
1	sigmoid	SGD	1	35	10000	0.79/0.80
2	**sigmoid**	**Adam**	1	35	10000	0.87/0.86
3	relu	SGD	1	35	10000	0.51/0.49
4	relu	Adam	1	35	10000	0.77/0.78
5	sigmoid	Adam	5	35	10000	0.89/0.89
6	sigmoid	Adam	5	35	100000	0.93/0.94
7	**sigmoid**	**Adam**	**5**	**70**	**500000**	**0.96/0.96**
8	sigmoid	Adam	5	200	100000	0.97/0.94
9	sigmoid	Adam	9	35	100000	0.89/0.89
10	sigmoid	Adam	9	90	100000	0.94/0.93
11	relu	Adam	9	90	100000	0.69/0.69

From the initial training results (see Table 1), we can make the following conclusions:

1) a pair of Sigmoid activation function and Adam optimizer showed the best result in terms of the accuracy criteria;
2) increase in a number of ANN layers requires a larger size of the training set to achieve a higher accuracy;
3) the achieved 0.96 accuracy shows that ANN is able to solve knapsack problem fairly well given the right number of layers and the size of the training set.

Figure 1 shows how accuracy and loss values were improved on the validation set during the training of the best ANN configuration (number 7) from Table 1. The smoothness and linearity of the graphs indicates the adequacy of the selected parameters and the possibility of stopping at using 150–200 training epochs.

Next, we consistently increased the dimension of the knapsack problem and estimated how different hidden utility functions affect the ANN accuracy.

The training set was obtained as a dynamic programming-based exact solution for a randomly generated knapsack problem. The items' properties and the weight constraint were generated randomly to achieve the required features: 1) representativeness – a data set selected from a larger statistical population should adequately reproduce a large group according to any studied characteristic or property; 2) consistency – contradictory data in the training sample will lead to a low quality of network training.

The ANN input data is a training sample consisting of the knapsack element properties vectors a_i, b_i, d_i, g_i and the normalized weights vector $w_i' = w_i/C$. Weights normalization allows us to generalize the weight constraint in (1.2) to $C=1$ for any input sample. Vector y_answer_i of the correct selection is used for the loss function calculation and

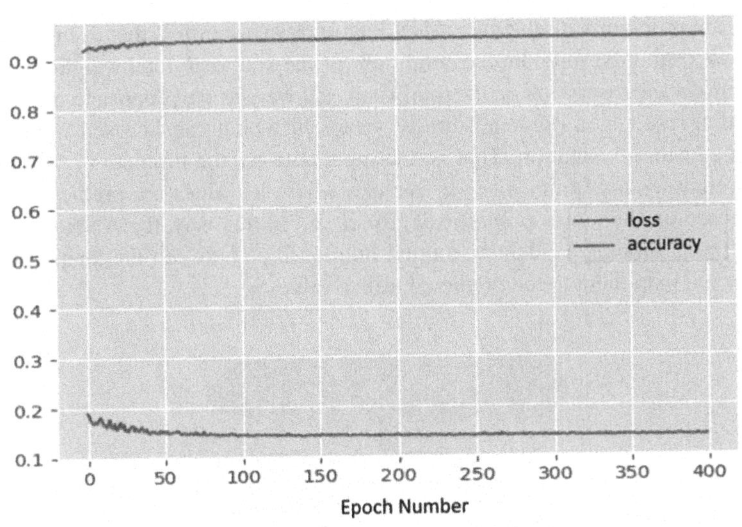

Fig. 1. ANN validation loss and accuracy for 5-element knapsack problem (configuration 7 from Table 1)

backpropagation step. The correct solution was obtained using a dynamic programming algorithm with explicit use of the hidden utility function.

The training and testing results for a 20-elements knapsack are presented in Table 2. As a main result, ANN was able to solve knapsack problem equally successfully for all the considered hidden functions (2.1)–(2.5) by using only the properties a_i, b_i, d_i, g_i of the knapsack items.

Table 2. Training results for 20-elements knapsack

Hidden utility function	Number of layers	Train/Test accuracy
$a + b + d - g$	14	0.94/0.93
$a * b + d * g^2$	17	0.92/0.91
$\sin(a + b) + \cos d + g^2$	14	0.93/0.92
$a + \lg(b + d) * g$	14	0.93/0.92
$a * \lg b + d * e^{\frac{g}{10}}$	14	0.92/0.91

3.2 MLAK Algorithm

While training a neural network, it is impossible to operate with formal mathematical concepts, in particular those defined for the knapsack problem (1.1), (1.2). The training relies on a set of pre-prepared examples of an optimal selection of the knapsack items.

The main problem with the pure ANN knapsack prediction is that even with a high accuracy we cannot be sure that the condition for the knapsack total weight is fulfilled.

To consider the restriction on the total knapsack weight, we propose to use the ANN classification result as a predicted utility vector h_i which can be used in a separate algorithmic knapsack solution. That is, the input data for the problem (1.1), (1.2) will contain weight w_i and utility $u_i = u_i^{'}$ vectors, where $u_i^{'}$ values are predicted for each element based on the item's properties a_i, b_i, d_i, g_i. In this way, the ANN will operate as a conversion module to identify mutual relationships between the knapsack items' properties and map them to the predicted utility values $u_i^{'}$.

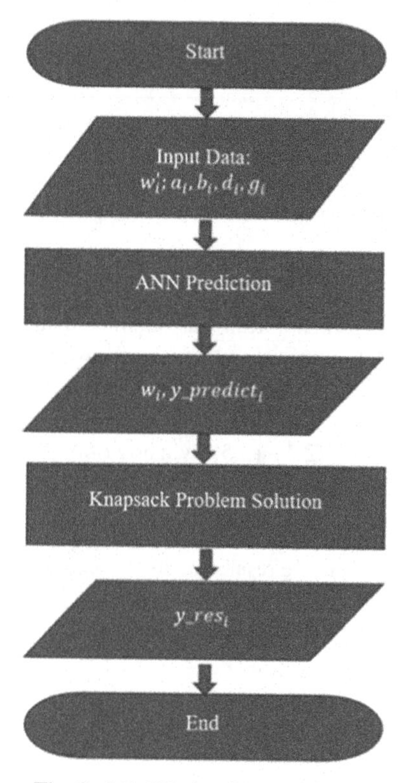

Fig. 2. MLAK algorithm flowchart

Figure 2 shows the flowchart of the proposed composite Machine Learning-based Algorithm for the Knapsack problem (hereinafter MLAK). It consists of ANN conversion module and a dynamic programming-based algorithm to provide the final solution for problem (1.1), (1.2) with an unknown, but predicted utility values and a constraint on the total weight.

The artificial neural network input for items $i = 1...n$:

- a_i, b_i, d_i, g_i – vectors of the properties;
- w_i – vector of the items' weights;

- $w_i^{'} = w_i/C$ – normalized vector of the weights;
- y_answer_i – an optimal selection result calculated by the dynamic programming method with use of a hidden utility function;
- $y\_predict_i$ – items selection predicted by the neural network;
- y_res_i – the final MLAK items selection.

```
Total Weight Limit: 159
Weight: [16, 26, 38, 42, 50, 62, 62, 67, 69, 74]
Utility: [28, 5, 182, 29, 100, 144, 152, 174, 18, 165]
True_Result:            [0, 0, 1, 0, 1, 0, 0, 1, 0, 0]
MLAK:                   [0, 0, 1, 0, 1, 0, 0, 1, 0, 0]
Greedy_Algorithm:       [1, 1, 1, 0, 0, 0, 0, 0, 0, 0]
ANN:                [0.26897, 0.01923, 0.99658, 0.02732, 0.42965,
0.24950, 0.61944, 0.72854, 0.00025, 0.09794]
```

Fig. 3. ANN and MLAK prediction comparison

Figure 3 presents sample data to demonstrate ANN and MLAK prediction differences with only a single utility property. In this example of a problem (1.1), (1.2) weight limit $C = 159$ is stated in the first line, while weight and utility vectors are presented on the following lines of Fig. 3. No hidden function is used, utility values are directly used as input data for ANN.

The exact solution (*True_result*) of x_i values was obtained with a dynamic programming procedure (1.3); approximate solution was provided by a greedy algorithm.

ANN result is a prediction vector of the x_i values. Based on this prediction, items (3, 7, 8) should be selected for the knapsack (indicated in red squares in Fig. 3), which is not the exact solution.

MLAK algorithm used ANN prediction vector as new utility values $u_i^{'}$ for the knapsack items. The dynamic programming procedure (1.3) was performed over $u_i^{'}$ and w_i values in (1.1), (1.2) with the result equal to the exact solution.

4 Simulation Experiment

4.1 Simulation Environment

We evaluate MLAK efficiency based on a comparison with classical knapsack algorithms (including dynamic programming and greedy implementation), as well as with a pure ANN knapsack implementation. ANN results were additionally modified to comply with the weight restriction: selected items with the smallest prediction confidence were removed one by one until the restriction is satisfied.

For this comparison we used the same (2.1)–(2.5) hidden utility functions F_{val}. Thus, MLAK and ANN received a_i, b_i, d_i, g_i properties as an input, while Dynamic programming (DP) and Greedy implementations used $u_i = F_{val}$ calculated utility functions to solve the knapsack problem.

Additionally, we implemented a Random selection algorithm to evaluate MLAK and ANN efficiency in the interval between the optimal solution provided by DP and a completely random result.

All the considered algorithms were given a set of the same 1000 knapsack problems as input. We consider two main efficiency indicators for each algorithm's solution:

- The resuting knapsack total utility and its relation to the DP result (average utility);
- The resulting accuracy as element-wise comparison with the DP result for all the experiments.

Both efficiency criteria are based on DP algorithm as it provides an optimal integer solution based on the known utility function.

Accuracy criterion will have 100% value when knapsack solution is identical to the DP in all 1000 experiments. Thus, low accuracy value does not necessarily mean low algorithm efficiency, as the same or comparable knapsack utilities sometimes may be achieved by selecting different combinations of items. Accuracy parameter shows how often the resulting solution matches the optimal one by performing similar optimization operations.

Therefore, an average total utility should serve as a main comparison criterion for problem (1.1), (1.2).

Besides, we measure and compare average working times. All the considered algorithms were implemented using Python language. Working time was observed on desktop PC with Core i5 and 8Gb RAM. MLAK working time includes both internal ANN and DP algorithms execution (see Fig. 2).

4.2 Simulation Results and Analysis

Simulation results collected over 1000 independent knapsack problems with 20 items are presented in Tables 3–7. Each table corresponds to a single hidden function (2.1)–(2.5).

Firstly, the results show that Greedy algorithm provided almost optimal average utility: nearly 99% compared to DP. This result is expected for 20 elements with randomly and uniformly generated properties and utility values. 50–85% accuracy shows that it is usually possible to achieve comparable optimization results with different items selected. Different utility functions are mostly affecting the accuracy difference (50–85% interval) as they provide varying diversity in the resulting utility values of the knapsack items. Low diversity usually leads to a higher accuracy values, as there are less options to achieve an efficient solution.

MLAK and ANN optimization efficiency is generally comparable to the Greedy implementation. Relative difference by the average utility between Greedy and MLAK is less than 1% for functions (2.1), (2.3), (2.4), and (2.6). ANN provides similar results with less than 1% lower utility compared to MLAK.

For functions (2.2) and (2.5) the relative difference with DP reaches 3%, which may be explained by much larger absolute values of the utility functions obtained from the same set of the randomly generated input properties (see column Average Utility in Tables 3–7). ANN prediction works less efficient when relations between the properties include multiplication and exponentiation operations.

However, even this less than 3% optimization loss (in the worst case observed scenarios) is rather small and reasonable when compared to the random selection result with more than 40% difference from DP solution. Besides, in this comparison DP and Greedy performed knapsack optimization using the actual utilities calculated from the hidden functions, while Random shows average results with no optimization.

When compared to each other, MLAK provides a slightly better average resulting utility and noticeably higher accuracy than ANN. Pure ANN usually generates quite efficient solutions which degrade when the weight constraint is applied. So, MLAK is one of the efficient ways to apply weight constraint over the pure ANN knapsack prediction.

In terms of the actual working time MLAK is inferior to all the other considered algorithms. Obviously, the strong difference in execution time between ANN-based and traditional algorithms is due to the nature and complexity of artificial neural networks, which are required to replicate hidden utility functions. For the considered 20-items knapsack problem MLAK prediction time of 0.05 s may seem quite insignificant, but larger problems will require increase in the ANN structure, training sample size, time and calculation efforts for the training.

Table 3. Function 2.1 optimization results

Algorithm	Average utility	Average utility, %	Accuracy, %	Average working time, s
Greedy	2175	98,7	48,0	0.00004
MLAK	2163	98,2	39,9	0.04206
ANN	2117	96,1	25,9	0.02363
Random	1270	57,6	0,6	0.00005
Greedy	2175	98,7	48,0	0.00004

Table 4. Function 2.2 optimization results

Algorithm	Average utility	Average utility, %	Accuracy, %	Average working time, s
DP	25.2*10^5	100,0	100	0.00609
Greedy	25*10^5	99,2	59,2	0.00004
MLAK	24.6*10^5	97,7	36,6	0.04145
ANN	23.9*10^5	94,9	21,6	0.02385
Random	13.5*10^3	53,5	1,4	0.00005

Table 5. Function 2.3 optimization results

Algorithm	Average utility	Average utility, %	Accuracy, %	Average working time, s
DP	154057	100,0%	100%	0.00743
Greedy	153127	99,4%	64,1%	0.00004
MLAK	151601	98,4%	42,5%	0.04531
ANN	148806	96,6%	22,6%	0.02380
Random	79732	51,7%	1,2%	0.00005

Table 6. Function 2.4 optimization results

Algorithm	Average utility	Average utility, %	Accuracy, %	Average working time, s
DP	3371	100	100	0.00734
Greedy	3339	99,0	49,6	0.00004
MLAK	3326	98,7	40,8	0.04556
ANN	3245	96,3	22,6	0.02400
Random	2024	60,0	1,4	0.00005

Table 7. Function 2.5 optimization results

Algorithm	Average utility	Average utility, %	Accuracy, %	Average working time, s
DP	6049*10^6	100,0	100	0.00823
Greedy	6036*10^6	99,8	85,9	0.00004
MLAK	5849*10^6	96,7	41,2	0.04784
ANN	5864*10^6	96,9	24,1	0.02391
Random	2407*10^6	39,8	1,0	0.00005

5 Conclusion

The paper introduced a promising machine learning-based approach for online scheduling and resources allocation. A generalized model for knapsack problem solution based on hidden (unknown) utility functions was proposed and simulated. The main design and practical development stages of the artificial neural network were presented and considered. Additional optimization step was proposed to apply weight constraint over the neural network prediction.

As a main result, the proposed algorithm MLAK showed the knapsack optimization efficiency comparable to a classical greedy implementation for five different hidden utility functions covering a wide spectrum of mathematical complexity. The importance of this result is that MLAK, unlike greedy algorithm, did not directly used the hidden utility functions of the elements. Instead, it was pre-trained on a set of optimal solutions for randomly generated knapsack problems.

Future work will concern problems of the algorithm scalability and more practical online job-flow scheduling implementations.

Acknowledgments. This work was supported by the Russian Science Foundation (project no. 22–21-00372).

References

1. Bharathi, S., Chervenak, A.L., Deelman, E., Mehta, G., Su, M., Vahi, K.: Characterization of scientific workflows. In: Proceedings of 2008 Third Workshop on Workflows in Support of Large-Scale Science, pp. 1–10 (2008)
2. Rodriguez, M.A., Buyya, R.: Scheduling dynamic workloads in multi-tenant scientific workflow as a service platforms. Futur. Gener. Comput. Syst. **79**(P2), 739–750 (2018)
3. Kurowski, K., Nabrzyski, J., Oleksiak, A., Weglarz, J.: Multicriteria aspects of grid resource management. In: Nabrzyski, J., Schopf, J.M., Weglarz J. (eds.) Grid Resource Management. State of the Art and Future Trends, pp. 271–293. Kluwer Academic Publishers (2003)
4. Toporkov, V., Yemelyanov, D.: Heuristic rules for coordinated resources allocation and optimization in distributed computing. In: Rodrigues, J.M.F., et al. (eds.) ICCS 2019. LNCS, vol. 11538, pp. 395–408. Springer, Cham (2019). https://doi.org/10.1007/978-3-030-22744-9_31
5. Toporkov, V., Yemelyanov, D., Toporkova, A.: Coordinated global and private job-flow scheduling in grid virtual organizations. J. Simulation Modelling Practice and Theory **107.** Elsevier (2021)
6. Sukhoroslov, O., Nazarenko, A., Aleksandrov, R.: An experimental study of scheduling algorithms for many-task applications. J. Supercomputing **75**, 7857–7871 (2019)
7. Samimi, P., Teimouri, Y., Mukhtar, M.: A combinatorial double auction resource allocation model in cloud computing. J. Inform. Sci. **357**(C), 201–216 (2016)
8. Rodero, I., Villegas, D., Bobroff, N., Liu, Y., Fong, L., Sadjadi, S.: Enabling interoperability among grid meta-schedulers. J. Grid Comput. **11**(2), 311–336 (2013)
9. Shmueli, E., Feitelson, D.G.: Backfilling with lookahead to optimize the packing of parallel jobs. J. Parallel Distrib. Comput. **65**(9), 1090–1107 (2005)
10. Khemka, B., Machovec, D., Blandin, C., Siegel, H.J., Hariri, S., Louri, A., Tunc, C., Fargo, F., Maciejewski, A.A.: Resource management in heterogeneous parallel computing environments with soft and hard deadlines. In: Proceedings of 11th Metaheuristics International Conference (MIC 2015) (2015)
11. Netto, M.A.S., Buyya, R.: A flexible resource co-allocation model based on advance reservations with rescheduling support. In: Technical Report, GRIDSTR-2007–2017, Grid Computing and Distributed Systems Laboratory. The University of Melbourne, Australia (2007)
12. Toporkov, V., Toporkova, A., Yemelyanov, D.: Slot co-allocation optimization in distributed computing with heterogeneous resources. In: Del Ser, J., Osaba, E., Bilbao, M.N., Sanchez-Medina, J.J., Vecchio, M., Yang, X.-S. (eds.) IDC 2018. SCI, vol. 798, pp. 40–49. Springer, Cham (2018). https://doi.org/10.1007/978-3-319-99626-4_4
13. Toporkov, V., Yemelyanov, D.: Optimization of resources selection for jobs scheduling in heterogeneous distributed computing environments. In: Shi, Y., Fu, H., Tian, Y., Krzhizhanovskaya, V.V., Lees, M.H., Dongarra, J., Sloot, P.M.A. (eds.) ICCS 2018. LNCS, vol. 10861, pp. 574–583. Springer, Cham (2018). https://doi.org/10.1007/978-3-319-93701-4_45

14. Toporkov, V., Yemelyanov, D.: Scheduling optimization in heterogeneous computing environments with resources of different types. In: Zamojski, W., Mazurkiewicz, J., Sugier, J., Walkowiak, T., Kacprzyk, J. (eds.) DepCoS-RELCOMEX 2021. AISC, vol. 1389, pp. 447–456. Springer, Cham (2021). https://doi.org/10.1007/978-3-030-76773-0_43
15. Xu, S., Panwar, S.S., Kodialam, M.S., Lakshman, T.V.: Deep neural network approximated dynamic programming for combinatorial optimization. In: Proceedings of the AAAI Conference on Artificial Intelligence, pp. 1684–1691 (2020)
16. Nomer, H.A.A., Alnowibet, K.A., Elsayed, A., Mohamed, A.W.: Neural knapsack: A neural network based solver for the knapsack problem. In: Proceedings of the IEEE Access, vol. 8, pp. 224200–224210 (2020)
17. Hertrich, C., Skutella, M.: Provably good solutions to the knapsack problem via neural networks of bounded size. In: Proceedings of the AAAI Conference on Artificial Intelligence 35(9), pp. 7685–7693 (2021)
18. Chollet, F.: Xception: Deep learning with depthwise separable convolutions. In: Proceedings of the IEEE Conference on Computer Vision and Pattern Recognition (CVPR), 2017, pp. 1800–1807 (2017)

Time Series Attention Based Transformer Neural Turing Machines for Diachronic Graph Embedding in Cyber Threat Intelligence

Binghua Song[1,2], Rong Chen[3], Baoxu Liu[1,2](✉), Zhengwei Jiang[1,2], and Xuren Wang[3]

[1] Institute of Information Engineering, Chinese Academy of Sciences, Beijing, China
{songbinghua,liubaoxu,jiangzhengwei}@iie.ac.cn
[2] School of Cyber Security, University of Chinese Academy of Sciences, Beijing, China
[3] Information Engineering College, Capital Normal University, Beijing, China
{2201002025,wangxuren}@cnu.edu.cn

Abstract. The cyber threats are often found to threaten individuals, organizations and countries at different levels and evolve continuously over time. Cyber Threat Intelligence (CTI) is an effective approach to solve cyber security problems. However, existing processes are considered inherent responses to known threats. CTI experts recommend proactively checking for emerging threats in existing knowledge. In addition, most researches focus on static snapshots of the CTI knowledge graph, while ignoring the temporal dynamics. To this end, we create a novel framework TSA-TNTM (Time Series Attention based Transformer Neural Turing Machines) for diachronic graph embedding framework, which uses time series self-attention mechanism to capture the non-linearly evolving entity representations over time. We demonstrate significantly improved performance over various approaches. A series of benchmark experiments illustrate that TSA-TNTM could generate higher quality than the state-of-the-art word embedding models in tasks pertaining to semantic analogy, clustering, threat classification and proactively identify emerging threats in CTI fields.

Keywords: Threat intelligence · Dynamic knowledge graph · Time series attention · Graph embedding

1 Introduction

1.1 Backgrounds

Due to the dynamically rise of the cyber attacks such as the attack against GitHub lasted 72 h in 2015 [20], cyber security is becoming a vital research area to our society. Attackers exploit vulnerabilities to attack systems, for example,

D. Groen et al. (Eds.): ICCS 2022, LNCS 13353, pp. 17–30, 2022.
https://doi.org/10.1007/978-3-031-08760-8_2

a hacker can exploit SQL-related injections to illegally break into the system and obtain user information. Cyber Threat Intelligence (CTI) based Knowledge Graphs (KGs) is an effective approach to reveal and predict the latent attack behaviors. Especially, KGs that represent structural relations between entities have become an increasingly popular research direction towards exploration and prediction of CTI. Most of advances in research based on knowledge graph focus on Knowledge Representation Learning (KRL) or Knowledge Graph Embedding (KGE) by mapping entities and relations into low-dimensional vectors, while capturing their static semantic meanings [3, 17].

The existing neural network models have achieved good performance in processing graph data. For example, the graph convolution network (GCN [14]) uses the CNN [21] mode to process the first-order, second-order or even more higher-order neighbors aggregation on nodes in graphs. Convolution algorithm extracts the similar features of nodes and graph topological features, Recurrent Neural Networks (RNN [35]) such as LSTM [10] can extract the input information at the current moment and the state information at the previous moment through a gating mechanism.

However, these algorithms can easily cause gradients disappearing or exploding, which make it difficult to extract features depend on the long time distances. In the field of threat intelligence, the span of cyber security behaviors is usually very long. For example, the data set [26] which we used in this paper has a very large time span from 1996 to 2010. Many hacker terms and semantics have changed with the development of new IT technology over time. So, these current existing models face a lot of challenges.

1.2 Contribution

Recently, there has been a growing research interest for researches in temporal KG embeddings, where the edges of a KG are also endowed with information about the time period(s) in which the relationship is considered valid. Many temporal knowledge graph models have been proposed such as RE-NET [13], Know-Evolve [30], ATiSE [34], JODIE [15], DGNN [32], GC-LSTM [4], D-GEF [26]. However, each of these models cannot adequately capture temporal and spatial information about the network and lack sufficient capability of spatio-temporal dependent feature extraction. It is difficult to efficiently realize the joint extraction of temporal and spatial features.

To this end, we leverage the existing NTM (Neural Turing Machines) [8] memory enhancement neural network model. The controller of existing NTM models mostly use traditional graph neural network methods such as GCN and RNN. Our TSA-TNTM model replace the controller with Transformer [31] model, and use the write head selectively write to memory block and use the read head to read from it again. In addition, we further propose a time series-based attention mechanism, focus on the time and space features of dynamic knowledge graph. Overall, our contributions are as follows:

1) We propose a time series attention based differentiable neural Turing machine model for dynamic CTI Knowledge Graph so as to promote the processing

seed and accuracy. We use the transformer model for the component of controller rather than LSTM to capture unlimited long distance spatio-temporal dependencies, and further modify the structure of transformer such as temporal position encoding (TPE) to improve the ability to capture the topological dynamics of the graph.

2) We propose an adaptive multi-head self-attention mechanism based on time series database which can speed up the spatio-temporal embeddings and prediction according to the different scenarios which focus on disparate concerns of attack behaviors in each snapshot.
3) Based on a real hacker forum data set across about 23 years, we made a series of experiments and achieved the remarkable performance over other related approaches.

2 Related Work

Many researchers considered the study of time series or dynamic knowledge graph. However, these studies are always focused on **general** fields, and they are mostly **static** knowledge graph.

2.1 General Static and Time-Based KG Models

In general fields, the most fundamental and widely used model is TransE [2], which treats the relationship in knowledge graph as some kind of translation vector between entities. However, TransE only works well for the one-to-one relationship. TransH [33] regards a relation as a translation operation on a hyperplane. TransR [17] projects the entities from the entity space to the corresponding relational space, and then transforms the projected entities, TransD [11] considers the diversity of entities and relationships, constructing a dynamic mapping matrix for each entity-relation pair.

The following combination of Trans family models and time series solved the problems of embedding entities and relationships of data at different times in low-dimensional vector space. Jin W et al. [13] proposed Recurrent Event Network (RE-NET), which employed a recurrent event encoder to encode past facts, and uses a neighborhood aggregator to model the connection of facts at the same timestamp in order to infer future facts. Trivedi R et al. [30] presented Know-Evolve, which learned non-linearly evolving entity representations over time. The occurrence of a fact is modeled as a multivariate point process. A García-Durán et al. [6] proposed a method to utilize recurrent neural networks to learn time-aware representations of relation types. Tingsong J et al. [12] proposed a time-aware KG embedding model using temporal order information among facts, specifically, using temporal consistency information as constraints to incorporate the happening time of facts. Julien L et al. [16] focused on the task of predicting time validity for unannotated edges. Shib S D et al. [5] proposed HyTE, which explicitly incorporates time in the entity relation space by associating each timestamp with a corresponding hyperplane. Xu C et al. [34]

proposed ATiSE which incorporates time information into entity or relation representations by using additive time series decomposition. Goel R et al. [7] built novel models for temporal KG completion through equipping static models with a diachronic entity embedding function.

These models can only handle **static** information and cannot handle complex **dynamic** graphs.

2.2 General Fields Dynamic KG Models

Due to the limitations of the static KG algorithms, many researchers began to study **dynamic** KGC techniques. Rakshit T et al. [29] proposed a model named DyRep, which is a latent mediation process bridging two observed dynamics on the network. Srijan K et al. [15] proposed JODIE, a coupled recurrent neural network model which employed two recurrent neural networks to update the embedding. Ma Y et al. [32] proposed DGNN, which could model the dynamic information as the graph evolving. Manessi F et al. [19] combined LSTM and GCN to learn long -short term dependencies together with graph structure, so that jointly exploiting structured data and temporal information. Chen J et al. [4] proposed GC-LSTM, a GCN embedded LSTM network for dynamic link prediction. Maheshwari A et al. [18] proposed a novel adversarial algorithm DyGAN to learn representation of dynamic networks, who leverage generative adversarial networks and recurrent networks to capture temporal and structural information.

These dynamic KG models are less studied and ignore **time** information and can not capture changes in semantic information over time.

2.3 Cyber Threat Intelligence Fields KG Models

Xiaokui S et al. [28] introduced threat intelligence computing as a graph computation problem, presenting a threat intelligence computing platform through the design and implementation of a domain-specific graph language. Samtani S et al. [26] created a novel Diachronic Graph Embedding Framework (D-GEF), which operated on a Graph of Words (GoW) representation of hacker forum text to generate word embeddings across time-spells. Rastogi N et al. [25] introduced an open-source malware ontology model named MALOnt, which allows the structured extraction of information and knowledge graph generation, especially for threat intelligence. Pingle A et al. [24] proposed a system to create semantic triples over cyber security text. Sarhan I et al. [27] presented Open-CyKG, which is constructed using an attention-based neural Open Information Extraction model to extract valuable cyber threat information from unstructured Advanced Persistent Threat (APT) reports.

In the field of cyber threat intelligence, there are fewer researches on knowledge graphs, especially with time information. Most of these studies focus on how to construct a knowledge graph, unable to predict potential threats, and ignore the changes in some threat terminology over time.

3 Method

3.1 Dynamic Graph Embedding Models

Most dynamic graph embedding models given a graph $G = G_1, G_2, \cdots, G_T$, where $G_T = (V_T, E_T)$, V is the node set, T is different timestamp and E is the edge set. Nodes have an edge if they are adjacent. Define mapping $f : V \rightarrow R^d$, function f preserves some proximity measure defined on the graph G, graph embedding is a time series of mappings $F = \{f_1, \cdots, f_T\}$ such that mapping f_T is a graph embedding for G_T and all mappings preserve the proximity measure for their respective graphs.

Current existing graph embedding methods cannot capture embedding shifts and changes in the temporal datasets mostly. Graph data is splitted into time-spells and the embeddings are created in each time-spell. However, each time-spell has a different semantic embedding space, which makes it difficult to directly compare the graph embeddings in different time-spells.

3.2 Attention Model

The attention model has become an important concept in neural networks. The Transformer model replaces convolution entirely with the attention mechanism and can be processed in parallel. There are many ways to calculate attention. We use the dot product formula, which is as shown:

$$Attention(Q, K, V) = softmax(\frac{QK^T}{\sqrt{d_k}})V \tag{1}$$

$$MultiHead(Q, K, V) = Concat(head_1, ..., head_h)W^Q \tag{2}$$

where Q, K, V represents the query, the key, and the value respectively, $\sqrt{d_k}$ is represented for normalization and $head_i = attention(QW_i^Q, KW_i^K, VW_i^V)$. Each position of output is the vector weighted average of all positions of the value, the weight is calculated by attention from all positions of the query and key.

Attention maps the query and key to a same high-dimensional space for calculating similarity, while multi-head attention allows the model to jointly attend to information from different representation sub-spaces at different positions [1]. For Example, the temporal and spatial features can be weighted in different head simultaneously.

3.3 TSA-TNTM Model

Based on the previous two steps, we will build our TSA-TNTM model. Firstly we describe the high-level structure of our model TSA-TNTM. The major novel component of the whole model is transformer-based controller, which can be utilized to capture unlimited spatio-temporal features for the dynamic CTI knowledge graph over time with high speed. The Neural Turing Machine (NTM) [8] had a similar structure, which extend the capabilities of neural networks by coupling them to external memory resources, which they can interact with each other by attention mechanism to read from and write to the memory block selectively according to different weights.

Our proposed model TSA-TNTM extend and modify the controller of NTM, which is equivalent to the CPU part of the computer. The extension of controller could be realized by graph neural network in the neural Turing machine [8]. We abandon the existing RNN [35], GCN [14] and other graph neural networks, adopt the prevalent transformer model, and use the attention mechanism to improve the dynamic time feature extraction capability and parallel processing capability of the threat intelligence knowledge graph.

In addition, we propose time series hybrid attention and realize seamless docking with memory in key-value mode. In time series hybrid attention, the three Tensors of Q (Query) value, K(Key) value, and V(Value) value are all from the same memory input. We calculate the dot product between Q value and K value firstly, and then divided by a scale $\sqrt{d_k}$ in order to prevent the result from being too large, where d_k is the dimension of a query and key tensor. Then use the softmax operation to normalize the result to a probability distribution, and multiply the result by the tensor V so that getting the weighted summation. This operation can be expressed as formula (1), where K and V establish a mapping with the key and value of the underlying time series database such as OpenTSDB [23] by parallel processing technology, so as to make the best of massive data processing of the underlying big data technology.

Finally, we introduce a positive and negative feedback mechanism. Memory is not only used for general data reading and writing, but also forms the input and output of the transformer. Among them, memory provides dynamic preprocessing capabilities based on attention for the input, and automatically learns the distribution and spatio-temporal characteristics of the graph data. This further reduces the burden on transformer, and improve the processing performance. In contrast to most models of working memory, our model is analogous to a Turing Machine or Von Neumann architecture but is differentiable end-to-end, allowing it to be efficiently trained with gradient descent. The model is as shown in Fig. 1.

As illustrated in Fig. 1, TSA-TNTM consist of a memory block followed by a controller block according to the architecture of Neural Turing Machines. Each block may contain multiple stacked layers of the same type. The memory block extract spatio-temporal features from the dynamic graph with unfixed-length through the multi-head self-attention aggregation. The details of this attention approach will be described in the next section. In this memory block, there are two ways to ingest the graph data: one way is to employee the online method to capture the topological evolution and node interactions of the dynamic graph continuously, the other way is to employee the offline method to capture the long-time historical temporal and topological evolution of multiple graphs. During the real time ingestion of dynamic graph, we can also use multi-head self-attention to extract multi-grained long span features to store into the time series database

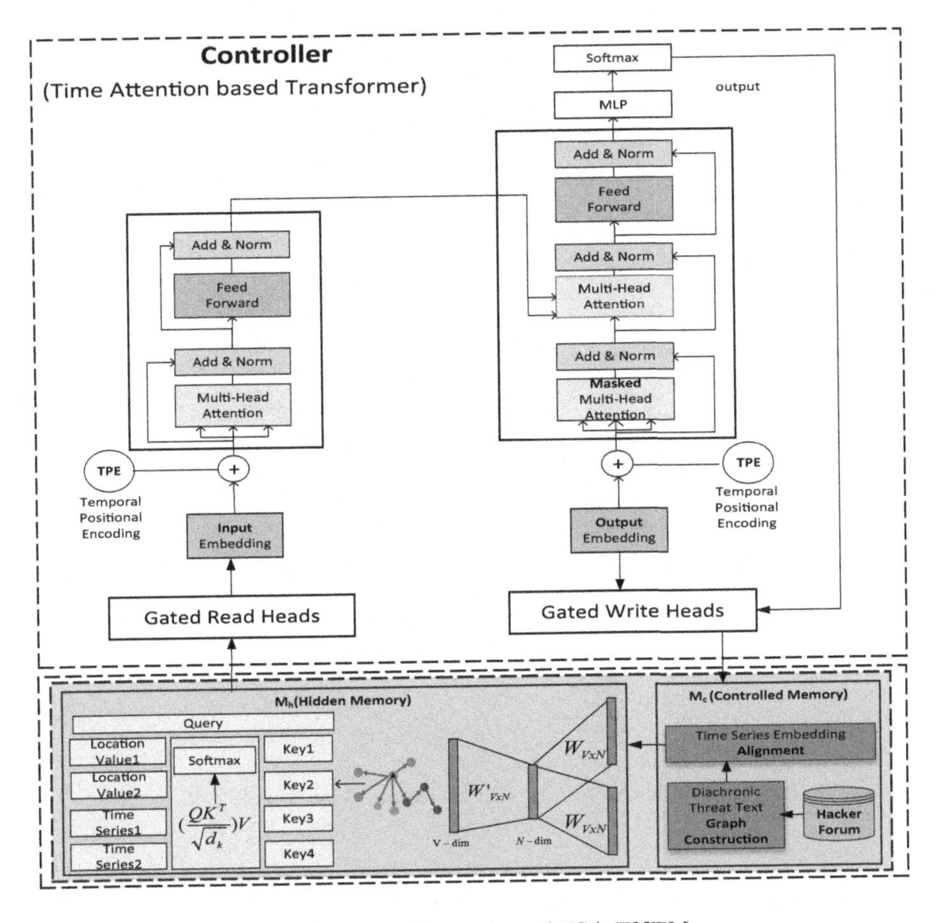

Fig. 1. Simple illustration of TSA-TNTM.

so as to deal with the long-distance correlation such as a new threat intelligence first appear in 1996 and changed representation in 1998, 2001, 2012 respectively.

The processing flow of the TSA-TNTM are as follows:

Firstly, Memory Block Construction. The main function of Memory Block is collecting CTI threat intelligence corpus in real time and dynamically constructing a knowledge graph. The Memory module takes the acquisition of the dynamic knowledge graph as its main function, uses the attention mechanism to optimize the storage performance of the Memory Block. The Graph Construction process in Memory Block is as follows:

1) Collect threat intelligence corpus text, preprocess and store it into time series database such as OpenTSDB [23]. In this paper, the hacker forum data we used is one of the corpus, we would consider to add other sources later.

2) Split data into time-spell and construct Threat Text Graph. According to the order of words in the sentence, build two adjacent words in accordance with first-order and second-order neighbors, adopt the CBOW model to learn

the weight of word embedding, construct the time information before and after the word as a separate entity structure and add it to the edge set list uniformly. The Fig. 2 is illustration of the constructed threat text graph in the time-spell 2019. From the graph, we can see that hackers exploit remote administration tools (RATs), email hacks, denial of service (DOS), and others to attack the system.

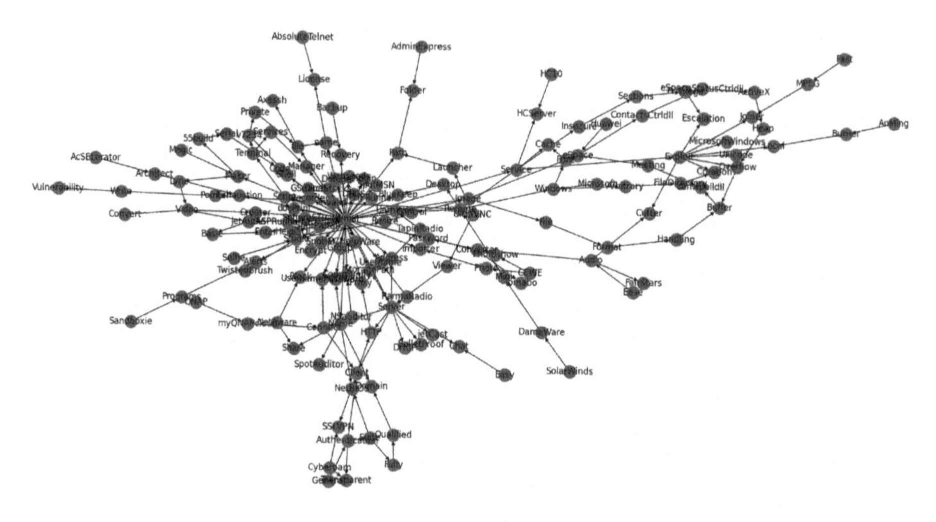

Fig. 2. The diachronic threat text graph in time-spell 2019.

3) Embedding alignment. Embedding spaces across time-spells are aligned while retaining cosine similarities by optimizing the following objective function:

$$R^{(t)} = \underset{Q^TQ=I}{\arg\min} ||W^{(t)}Q - W^{(t+1)}||_F \tag{3}$$

where $||\cdot||_F$ denotes the Forbnius norm.

4) Read the dynamic edge set list in real time and construct the graph in real time.
5) Calculate the attention score of the new input side information, and dynamically modify the weights. The related details can be found in Formula (1) and Formula (2).
6) Realize the initial embedding of the spatio-temporal features of the graph according to the attention weight, and adjust the relevant value in $L1$ and $L2$ norm.

Secondly, Controller Block Graph Embedding and Computation. Controller Block selectively read the word embedding vector of Memory Block through Gated Read Head, encoded by the Encoder, further extracts and processes spatio-temporal features, and outputs processing results through the Decoder, and then stores the results into Memory Block.

In the TSA-TNTM model, we divide the memory block into two parts: the **controller memory** M_c and the **hidden memory** M_h. The M_c is controlled by the controller block and store the controller's result into time series database for persistent store. The M_h is not memory by the controller and is connected with the M_c. The hidden memory saves the accumulated content into the controller, the procedure of read and write in time t is:

$$M_c(t) = h(M_c(t-1), w(t-1), c(t)) \tag{4}$$

$$M_{\mathrm{h}}(t) = aM_h(t-1) + bM_c(t) \tag{5}$$

$$r(t) = W_r(t)M_h(t) \tag{6}$$

where M_c is updated at time $\mathrm{t}-1$ by the write head $w(t-1)$ and generate the output of M_c at time t and update the M_h which is used to generate read head $r(t)$ at time t. $c(t)$ is the output of controller, h is the function of update the controlled memory and the write weight in order to realize the function of erasion and addition. The a and b is hybrid weight of the scalar and read head reads from the $M_h(t)$.

Based on the above analysis, it can be seen that TSA-TNTM expands the attention mechanism from the controller as a neural network to memory, thereby realizing the lightweight attention of the entire Turing model from storage to computing. Meanwhile, it realize unlimited spatio-temporal feature extraction by the expansion of memory.

4 Experiment and Evaluation

4.1 Dataset

We adopt the real hacker forum data set from D-GEF model [26] to complete the experiment in this paper. In order to analysis the temporal changes about hacker threats, the author collected a large and long-standing international hacker forum. According to the descriptions, collection procedures resulted in data set with 32766 threats posts made by 8429 hackers between January 1, 1996 and July 10, 2019 (across 23 years), but actually we only find 6,767 posts in their publicly opened data at the GitHub repository. The statistics of the data set are summarized into Table 1. The detail descriptions of hacker forum data set in this table can be viewed from the D-GEF model [26].

4.2 Evaluation Method and Computational Setup

We evaluate our model performances by using well-established metrics of accuracy, precision, recall, and F1-score (i.e., F-measure) as following:

$$\mathrm{Accuracy} = \frac{\mathrm{TP}+\mathrm{TN}}{\mathrm{TP}+\mathrm{TN}+\mathrm{FP}+\mathrm{FN}} \tag{7}$$

Table 1. Statistics for hacker forum dataset.

Threat type	Target platform	Date range	Exploit numbers	Total
Remote	Windows	03/23/2009–07/05/2019	1,418	1,864
	Linux	06/24/2000–07/02/2019	446	
Local	Windows	09/28/2004–06/20/2019	1,818	2,429
	Linux	01/01/1996–07/02/2019	611	
Total	–	01/01/1996–07/05/2019	4,293	4,293

$$\text{Precision} = \frac{\text{TP}}{\text{TP} + \text{FP}} \tag{8}$$

$$\text{Recall} = \frac{\text{TP}}{\text{TP} + \text{FN}} \tag{9}$$

$$\text{F1-score} = \frac{2 * \text{Precision} * \text{Recall}}{\text{Precision} + \text{Recall}} \tag{10}$$

where the True Positive (TP) means the percent of correctly classified normal samples; the False Positive (FP) means the percent of classifying abnormal samples as normal samples; the True Negative (TN) means the percent of classifying abnormal samples as abnormal samples; the False Negative (FN) means the percent of classifying normal samples as abnormal samples.

We use a mobile workstation notebook equipped with an Intel® Core i7-8750H@2.20GHz processor, 32 GB of RAM, and an NVIDIA® Quadro P2000 Graphical Processing Unit (GPU). The experiments use Python language, version 3.9. All word embedding approaches were implemented using the Genism package and Open Network Embedding (OpenNE) package. We use scikit-learn to process performance metrics and statistical tests. All of the experiments were trained over 50 epochs, and generated 128 dimensions embeddings.

4.3 Benchmark Methods

We select prevailing word and graph embedding approaches to compare with our TSA-TNTM model. In order to compare with D-GEF [26] in the same data set, considering almost no model predict the emerging threat in threat intelligence, our TSA-TNTM model just compare the classification of attack type and platform with six prevailing word embedding models and four famous graph embedding models.

Although the two classification tasks are based on the past facts, they contain temporal information about the threat facts, the subsequent happened facts are also the their future. So, the classification experiments also check the ability of prediction about the future **threat type** (local or remote attack) and **platform** (attack from Linux or windows). The happened facts are also the labels for prior facts. The methods include classic **word embedding** models and **graph embedding** methods.

Table 2. Summary of results from **Attack Type** classification experiments.

Category	Method	Accuracy	Precision	Recall	F1	AUC
TF-IDF	X^2	$87.6 \pm 1.3\%$	$88.1 \pm 1.2\%$	$77.6 \pm 1.2\%$	$80.6 \pm 1.5\%$	0.862
word2vec	SGNS	$84.6 \pm 1.3\%$	$82.6 \pm 1.3\%$	$74.6 \pm 1.3\%$	$76.1 \pm 2.3\%$	0.899
	CBOW	$83.1 \pm 1.3\%$	$81.6 \pm 1.3\%$	$70.6 \pm 1.3\%$	$73.6 \pm 1.3\%$	0.868
FastText	SGNS	$85.6 \pm 1.3\%$	$83.6 \pm 1.3\%$	$76.5 \pm 1.3\%$	$78.2 \pm 2.1\%$	0.901
	CBOW	$83.1 \pm 1.3\%$	$81.6 \pm 1.3\%$	$70.6 \pm 1.3\%$	$73.6 \pm 1.3\%$	0.868
Doc2Vec	DM	$82.5 \pm 1.2\%$	$79.5 \pm 1.3\%$	$68.5 \pm 1.5\%$	$71.6 \pm 1.8\%$	0.845
	DBOW	$85.7 \pm 1.2\%$	$81.8 \pm 1.3\%$	$74.6 \pm 1.3\%$	$76.9 \pm 1.5\%$	0.898
Graph factorization	GF	$87.6 \pm 1.3\%$	$85.5 \pm 1.3\%$	$79.5 \pm 1.5\%$	$81.2 \pm 2.1\%$	0.915
	HOPE	$84.5 \pm 1.3\%$	$82.1 \pm 1.2\%$	$72.7 \pm 1.2\%$	$75.6 \pm 1.3\%$	0.881
	GraRep	$86.5 \pm 1.2\%$	$83.8 \pm 1.3\%$	$77.6 \pm 1.3\%$	$80.6 \pm 1.3\%$	0.912
Random walk	DeepWalk	$88.6 \pm 1.3\%$	$86.6 \pm 1.3\%$	$80.5 \pm 1.3\%$	$82.2 \pm 2.1\%$	0.925
	node2vec	$87.9 \pm 1.5\%$	$85.6 \pm 2.1\%$	$80.6 \pm 1.3\%$	$82.2 \pm 2.3\%$	0.921
AutoEncoder	SDNE	$83.2 \pm 1.3\%$	$81.3 \pm 1.2\%$	$71.5 \pm 1.2\%$	$74.7 \pm 3.3\%$	0.872
EdgeReconstruction	LINE	$86.1 \pm 1.3\%$	$83.3 \pm 1.2\%$	$77.8 \pm 3.2\%$	$78.5 \pm 3.5\%$	0.912
TSA-TNTM	NTM	$91.7 \pm 1.5\%$	$92.5 \pm 2.6\%$	$87.6 \pm 2.3\%$	$86.6 \pm 2.5\%$	0.961

We conduct two experiments, which include:

1) **experiment 1: attack type (Local vs Remote)** classification benchmark. The experiment 1 is an attack type classification benchmark of the hacker forum data set from January 1, 1996 to July 10, 2019, which contains the attack type information for the past 23 years period. It is a typical use case to check word and graph embedding and prediction capability for dynamic graph across time-spells. In this experiment, we will use the above 10 benchmark methods and our TSA-TNTM model to compare the attack type classification prediction results by using the metrics of accuracy, precision, F1-score and AUC.

2) **experiment 2: platform (Linux vs Windows)** classification benchmark. The experiment 2 is a platform classification benchmark about the hacker forum dataset from January 1, 1996 and July 10, 2019, which contains the platform information for the past 23 years period. In this experiment, we will use the above ten benchmark methods and our TSA-TNTM model to compare the platform classification prediction results by using the metrics of accuracy, precision, F1-score and AUC.

4.4 Experiment Results

We report the accuracy, precision, recall, and F1 scores with a confidence interval for each algorithm. Besides our TSA-TNTM model, results across both classification tasks indicate that the random walk-based methods outperform the competing graph and word embedding approaches. In terms of F1 scores, the DeepWalk model has achieved a score of 82.6%. However, our TSA-TNTM model achieves the best scores in all metrics both experiment 1 and experiment 2. The experiment 1 results are as shown in Table 2 and the experiment 2 results are as shown in Table 3.

Table 3. Summary of results from **Attack Platform** experiments.

Category	Method	Accuracy	Precision	Recall	F1	AUC
TF-IDF	X^2	84.6 ± 1.3%	88.1 ± 1.2%	77.6 ± 1.2%	80.6 ± 1.5%	0.916
word2vec	SGNS	85.6 ± 1.3%	82.5 ± 1.5%	74.6 ± 1.3%	76.1 ± 2.3%	0.925
	CBOW	80.1 ± 1.3%	81.6 ± 1.3%	70.5 ± 1.3%	73.7 ± 1.2%	0.896
FastText	SGNS	85.6 ± 1.3%	83.5 ± 1.3%	76.5 ± 1.3%	78.2 ± 2.1%	0.922
	CBOW	79.5 ± 1.2%	80.7 ± 1.3%	70.6 ± 1.2%	73.5 ± 1.3%	0.882
Doc2Vec	DM	80.6 ± 1.3%	79.5 ± 1.3%	68.5 ± 1.3%	71.5 ± 2.1%	0.881
	DBOW	85.6 ± 1.3%	81.8 ± 1.3%	74.5 ± 1.3%	76.9 ± 1.3%	0.921
Graph factorization	GF	86.6 ± 1.3%	85.1 ± 1.3%	79.5 ± 1.3%	81.2 ± 2.1%	0.936
	HOPE	84.5 ± 1.3%	82.1 ± 1.3%	72.6 ± 1.2%	75.5 ± 1.3%	0.922
	GraRep	86.5 ± 1.3%	83.6 ± 1.3%	78.5 ± 1.3%	80.7 ± 1.3%	0.937
Random walk	DeepWalk	87.6 ± 1.3%	86.6 ± 1.3%	80.5 ± 1.3%	82.5 ± 2.1%	0.937
	node2vec	87.9 ± 1.3%	85.6 ± 1.3%	80.6 ± 1.3%	82.5 ± 1.3%	0.938
AutoEncoder	SDNE	83.5 ± 1.3%	81.3 ± 1.2%	71.5 ± 1.2%	74.7 ± 3.3%	0.916
EdgeReconstruction	LINE	86.1 ± 1.3%	83.3 ± 2.2%	77.6 ± 3.2%	79.5 ± 3.2%	0.938
TSA-TNTM	NTM	91.6 ± 1.2%	92.6 ± 2.1%	86.5 ± 2.3%	85.5 ± 2.3%	0.959

5 Conclusion

In this paper, we propose a model named TSA-TNTM based on time series attention. Our model can capture the temporal and spatial features assisted by time series database which breaking the long time dependencies limitation of previous models. Meanwhile, we propose an adaptive multi head self attention mechanism based on time series big data, which can promote the speed of temporal prediction across long timescales. Our model achieved the best results among the mentioned methods based on the data set of a real hacker forum.

In the future work, we consider improving our model on more larger datasets and more data sources such as the National Vulnerability Database (NVD) [22] which gather and store many Common Vulnerabilities and Exposures (CVEs) across long time span so as to find more unseen latent threats and predict emerging threats in the future. In addition, we also introduce the Hawkes process [9] for modeling sequential discrete events occurring in continuous time where the time intervals between neighboring events may not be identical. Thus, we can build a enhanced **continuous** time series based attention mechanism to further promote our TSA-TNTM model.

Acknowledgements. This research is supported by Key Laboratory of Network Assessment Technology, Chinese Academy of Sciences and Beijing Key Laboratory of Network Security and Protection Technology. We thank the anonymous reviewers for their insightful comments on the paper.

References

1. Bahdanau, D., Cho, K., Bengio, Y.: Neural machine translation by jointly learning to align and translate. arXiv preprint arXiv:1409.0473 (2014)
2. Bordes, A., Usunier, N., Garcia-Duran, A., Weston, J., Yakhnenko, O.: Translating embeddings for modeling multi-relational data. In: Advances in Neural Information Processing Systems, vol. 26 (2013)
3. Cao, Z., Xu, Q., Yang, Z., Cao, X., Huang, Q.: Dual quaternion knowledge graph embeddings. In: Proceedings of the AAAI Conference on Artificial Intelligence, vol. 35, pp. 6894–6902 (2021)
4. Chen, J., Wang, X., Xu, X.: GC-LSTM: graph convolution embedded LSTM for dynamic link prediction. arXiv preprint arXiv:1812.04206 (2018)
5. Dasgupta, S.S., Ray, S.N., Talukdar, P.: HyTE: hyperplane-based temporally aware knowledge graph embedding. In: Proceedings of the 2018 Conference on Empirical Methods in Natural Language Processing, pp. 2001–2011 (2018)
6. García-Durán, A., Dumančić, S., Niepert, M.: Learning sequence encoders for temporal knowledge graph completion. arXiv preprint arXiv:1809.03202 (2018)
7. Goel, R., Kazemi, S.M., Brubaker, M., Poupart, P.: Diachronic embedding for temporal knowledge graph completion. In: Proceedings of the AAAI Conference on Artificial Intelligence, vol. 34, pp. 3988–3995 (2020)
8. Graves, A., Wayne, G., Danihelka, I.: Neural turing machines. arXiv preprint arXiv:1410.5401 (2014)
9. Han, Z., Ma, Y., Wang, Y., Günnemann, S., Tresp, V.: Graph Hawkes neural network for forecasting on temporal knowledge graphs. arXiv preprint arXiv:2003.13432 (2020)
10. Hochreiter, S., Schmidhuber, J.: Long short-term memory. Neural Comput. **9**(8), 1735–1780 (1997)
11. Ji, G., He, S., Xu, L., Liu, K., Zhao, J.: Knowledge graph embedding via dynamic mapping matrix. In: Proceedings of the 53rd Annual Meeting of the Association for Computational Linguistics and the 7th International Joint Conference on Natural Language Processing (volume 1: Long papers), pp. 687–696 (2015)
12. Jiang, T., et al.: Towards time-aware knowledge graph completion. In: Proceedings of COLING 2016, the 26th International Conference on Computational Linguistics: Technical Papers, pp. 1715–1724 (2016)
13. Jin, W., Qu, M., Jin, X., Ren, X.: Recurrent event network: autoregressive structure inference over temporal knowledge graphs. arXiv preprint arXiv:1904.05530 (2019)
14. Kipf, T.N., Welling, M.: Semi-supervised classification with graph convolutional networks. arXiv preprint arXiv:1609.02907 (2016)
15. Kumar, S., Zhang, X., Leskovec, J.: Predicting dynamic embedding trajectory in temporal interaction networks. In: Proceedings of the 25th ACM SIGKDD International Conference on Knowledge Discovery & Data Mining, pp. 1269–1278 (2019)
16. Leblay, J., Chekol, M.W., Liu, X.: Towards temporal knowledge graph embeddings with arbitrary time precision. In: Proceedings of the 29th ACM International Conference on Information & Knowledge Management, pp. 685–694 (2020)
17. Lin, Y., Liu, Z., Sun, M., Liu, Y., Zhu, X.: Learning entity and relation embeddings for knowledge graph completion. In: Twenty-Ninth AAAI Conference on Artificial Intelligence (2015)
18. Maheshwari, A., Goyal, A., Hanawal, M.K., Ramakrishnan, G.: DynGAN: generative adversarial networks for dynamic network embedding. In: Graph Representation Learning Workshop at NeurIPS (2019)

19. Manessi, F., Rozza, A., Manzo, M.: Dynamic graph convolutional networks. Pattern Recogn. **97**, 107000 (2020)
20. Nestor, M.: GitHub has been under a continuous DDoS attack in the last 72 hours (2015)
21. Niepert, M., Ahmed, M., Kutzkov, K.: Learning convolutional neural networks for graphs. In: International Conference on Machine Learning, pp. 2014–2023. PMLR (2016)
22. NIST: National vulnerability database (2018). https://nvd.nist.gov/
23. openTSDB: OpenTSDB. http://opentsdb.net/
24. Pingle, A., Piplai, A., Mittal, S., Joshi, A., Holt, J., Zak, R.: Relext: relation extraction using deep learning approaches for cybersecurity knowledge graph improvement. In: Proceedings of the 2019 IEEE/ACM International Conference on Advances in Social Networks Analysis and Mining, pp. 879–886 (2019)
25. Rastogi, N., Dutta, S., Zaki, M.J., Gittens, A., Aggarwal, C.: MALOnt: an ontology for malware threat intelligence. In: Wang, G., Ciptadi, A., Ahmadzadeh, A. (eds.) MLHat 2020. CCIS, vol. 1271, pp. 28–44. Springer, Cham (2020). https://doi.org/10.1007/978-3-030-59621-7_2
26. Samtani, S., Zhu, H., Chen, H.: Proactively identifying emerging hacker threats from the dark web: a diachronic graph embedding framework (D-GEF). ACM Trans. Priv. Secur. (TOPS) **23**(4), 1–33 (2020)
27. Sarhan, I., Spruit, M.: Open-CYKG: an open cyber threat intelligence knowledge graph. Knowl.-Based Syst. **233**, 107524 (2021)
28. Shu, X., et al.: Threat intelligence computing. In: Proceedings of the 2018 ACM SIGSAC Conference on Computer and Communications Security, pp. 1883–1898 (2018)
29. Trivedi, R., Farajtabar, M., Biswal, P., Zha, H.: DyRep: learning representations over dynamic graphs. In: International Conference on Learning Representations (2019)
30. Trivedi, R., Farajtabar, M., Wang, Y., Dai, H., Zha, H., Song, L.: Know-evolve: deep reasoning in temporal knowledge graphs. arXiv preprint arXiv:1705.05742 (2017)
31. Vaswani, A., et al.: Attention is all you need. In: Advances in Neural Information Processing Systems, vol. 30 (2017)
32. Wang, J., Song, G., Wu, Y., Wang, L.: Streaming graph neural networks via continual learning. In: Proceedings of the 29th ACM International Conference on Information & Knowledge Management, pp. 1515–1524 (2020)
33. Wang, Z., Zhang, J., Feng, J., Chen, Z.: Knowledge graph embedding by translating on hyperplanes. In: Proceedings of the AAAI Conference on Artificial Intelligence, vol. 28 (2014)
34. Xu, C., Nayyeri, M., Alkhoury, F., Yazdi, H.S., Lehmann, J.: Temporal knowledge graph embedding model based on additive time series decomposition. arXiv preprint arXiv:1911.07893 (2019)
35. Zaremba, W., Sutskever, I., Vinyals, O.: Recurrent neural network regularization. arXiv preprint arXiv:1409.2329 (2014)

Reduced Order Surrogate Modelling and Latent Assimilation for Dynamical Systems

Sibo Cheng[1(✉)], César Quilodrán-Casas[1,2(✉)], and Rossella Arcucci[1,2]

[1] Data Science Institute, Imperial College London, London, UK
{sibo.cheng,c.quilodran,r.arcucci}@imperial.ac.uk
[2] Department of Earth Science and Engineering, Imperial College London, London, UK

Abstract. For high-dimensional dynamical systems, running high-fidelity physical simulations can be computationally expensive. Much research effort has been devoted to develop efficient algorithms which can predict the dynamics in a low-dimensional reduced space. In this paper, we developed a modular approach which makes use of different reduced-order modelling for data compression. Machine learning methods are then carried out in the reduced space to learn the dynamics of the physical systems. Furthermore, with the help of data assimilation, the proposed modular approach can also incorporate observations to perform real-time corrections with a low computational cost. In the present work, we applied this modular approach to the forecasting of wildfire, air pollution and fluid dynamics. Using the machine learning surrogate model instead of physics-based simulations will speed up the forecast process towards a real-time solution while keeping the prediction accuracy. The data-driven algorithm schemes introduced in this work can be easily applied/extended to other dynamical systems.

Keywords: Reduced order modelling · Deep learning · Data assimilation · Adversarial neural networks

1 Introduction

Reduced-order modelling is a reduced dimensionality model surrogate of an existing system in a reduced-order space. Reduced-order modelling (ROM), in combination with machine learning (ML) algorithms is of increasing interest of research in engineering and environmental science. This approach improves the computational efficiency for high-dimensional systems. Since forecasting the full physical space is computationally costly, much effort has been given to develop ML-based surrogate models in the pre-trained reduced-order space.

In recent years, the algorithm schemes which combine ROM and ML surrogate models have been applied in a variety of engineering problems, including computational fluid dynamics [5,30], numerical weather prediction [24] and

S. Cheng and C. Quilodrán-Casas—Equal contribution to this work.

D. Groen et al. (Eds.): ICCS 2022, LNCS 13353, pp. 31–44, 2022.
https://doi.org/10.1007/978-3-031-08760-8_3

nuclear science [28], among others, to speed up computational models without losing the resolution and accuracy of the original model. Typically, the first stage consists of reducing the dimension of the problem by compression methods such as Principal Component Analysis (PCA), autoencoder (AE), or a combination of both [28,29]. Solutions from the original computational models (known as snapshots) are then projected onto the lower-dimensional space, and the resulting snapshot coefficients are interpolated in some way, to approximate the evolution of the model.

In order to incorporate real-times observations, data assimilation (DA), originally developed in meteorological science, is a reference method for system updating and monitoring. Some recent researches [2,5,10,25] have focused on combining DA algorithms and ROMs so that the system correction/adjusting can be performed with a low computational cost. Adversarial training and Generative Adversarial Networks (GAN), introduced by [17], has also been used with ROM. Data-driven modelling of nonlinear fluid flows incorporating adversarial networks has been successfully studied previously [6]. GANs are also being used to capture physics of molecular dynamics [38] and have potential to aid in the modeling and simulation of turbulence [23].

The aim of this work is to create general workflows in order to tackle different applications combining DA and ML approaches. In the present work, we propose a modular approach, which combines ROM, ML surrogate models, and DA for complex dynamic systems with applications in computational fluid dynamic (CFD), wildfire spread and air pollution forecasting. The algorithms described in this work can be easily applied/extended to other dynamical systems. Numerical results in both applications show that the proposed approach is capable of real-time predictions, yielding an accurate result and considerable speed-up when compared to the computational time of simulations.

The paper is structured as follows: Sect. 2 presents the modular approach and its components. The applications are shown in Sect. 3 and 4. And finally, Discussions and Conclusions are presented in Sect. 5.

2 Components of the Modular Approach

The modular approach shown in this paper can be summarised by Fig. 1. The state model ($\mathbf{u}_t$) is compressed using ROMs approaches such as PCA, AE or a combination of both, followed by a ML-based forecast in the reduced space. This forecast is then corrected using DA, incorporating real-time observations ($\mathbf{v}_t$). This is an iterative process that can be used to improve the starting point of the next time-level forecast, thus improving its accuracy [3].

2.1 Reduced Order Modelling

In this section, we introduce two types of ROMs, namely the PCA and the convolutional autoencoder (CAE).

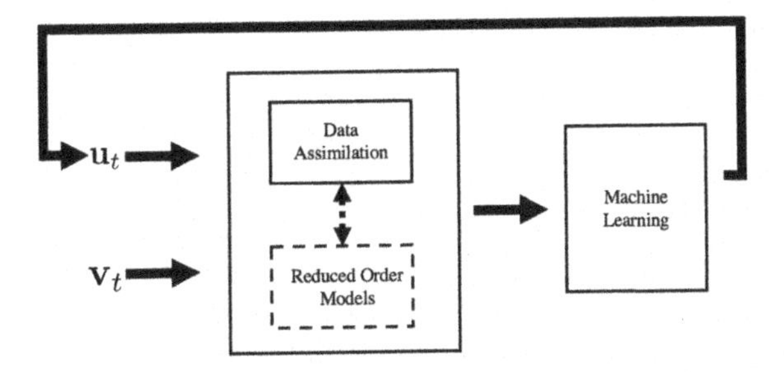

Fig. 1. Flowchart of the modular approach, modified using figure 1 of [4]

2.1.1 Principle Component Analysis

Principal component analysis, also known as Karhunen-Loève transform or Hotelling transform, is a reference method of ROM via an orthogonal linear projection. This approach has been widely applied in dynamical systems [35] with snapshots at different time steps. Applications can be found in a large range of engineering problems, including numerical weather prediction [21], hydrology [10] or nuclear engineering [16]. More precisely, a set of n_u simulated or observed fields $\{\mathbf{u}_{t_0, t_1, ..t_{n_u-1}}\}$ at different time are flattened and combined vertically to a matrix,

$$\mathbf{U} = \left[\mathbf{u}_{t_0}\middle|\mathbf{u}_{t_1}\middle|...\middle|\mathbf{u}_{t_{n_u-1}}\right]. \tag{1}$$

The principle components are then extracted by computing the empirical covariance matrix, that is,

$$\mathbf{C_u} = \frac{1}{n_u - 1}\mathbf{U}\mathbf{U}^T = \mathbf{L_U}\mathbf{D_U}\mathbf{L_U}^T, \tag{2}$$

where each column of $\mathbf{L_U}$ which represents an eigenvector of $\mathbf{C_u}$ and $\mathbf{D_U}$ is the associated eigenvalue diagonal matrix. The dynamical field $\mathbf{u}_t$ can then be compressed to

$$\tilde{\mathbf{u}}_t = {\mathbf{L}_{\mathbf{U},q}}^T \mathbf{u}_t, \tag{3}$$

where $\tilde{\mathbf{u}}_t$ denotes the compressed state vectors and q is the truncation parameter and the reconstruction to the full physical space reads

$$\mathbf{u}_t^{\text{PCA}} = \mathbf{L}_{\mathbf{U},q}\tilde{\mathbf{u}}_t = \mathbf{L}_{\mathbf{U},q}{\mathbf{L}_{\mathbf{U},q}}^T \mathbf{u}_t. \tag{4}$$

2.1.2 Convolutional Autoencoder

PCA, by design, is a linear method for ROM. In the last several years, much effort has been given to dimension reduction for chaotic dynamical systems via deep learning (DL) AEs [15]. Typically, an AE is an unsupervised neural network (NN) which consists two parts: an encoder E which maps the input variables to latent (i.e., compressed) vectors and a decoder D which performs reconstructions from the low-dimensional latent space to the full physical space. These processes can be summarised as:

$$\tilde{\mathbf{u}}_t = E(\mathbf{u}_t) \quad \text{and} \quad \mathbf{u}_t^{\text{AE}} = D(\tilde{\mathbf{u}}_t). \tag{5}$$

It is found that employing convolutional layers in AEs is helpful to i) reduce the number of parameters for high-dimensional systems, and ii) take into account local spatial patterns for structured data (e.g., images and times series) [18]. Following this idea, CAE was developed [18,32] where both the encoder E and the decoder D consist of a series of convolutional layers.

In general, the encoder and the decoder of an AE should be trained jointly, for instance, with the Mean square error (MSE) or the Mean absolute error (MAE) loss function of reconstruction accuracy, i.e.,

$$J_{\text{MSE}}(E, D) = \sum_{j=0}^{n_u-1} \frac{||\mathbf{u}_{t_j} - D \circ E(\mathbf{u}_{t_j})||^2}{n_u}, \quad J_{\text{MAE}}(E, D) = \sum_{j=0}^{n_u-1} \frac{||\mathbf{u}_{t_j} - D \circ E(\mathbf{u}_{t_j})||}{n_u}. \tag{6}$$

2.2 Machine Learning for Surrogate Models

2.2.1 Recurrent Neural Network

Long short-term memory networks neural networks, introduced in [19], is a variant of recurrent neural network (RNN), capable of dealing long term dependency, and vanishing gradient problems that traditional RNN could not handle. One of the components of our modular approach in the present work, is the sequence-to-sequence long short-term memory networks (LSTM). The LSTM learns the dynamics in the latent space from compressed training data.

LSTM can be unidirectional or bidirectional. Recently developed Bidirectional LSTM (BDLSTM) [33] differs from the unidirectional ones, as the latter can capture the forward and backward temporal dependencies in spatiotemporal data [12,20,26,31] . LSTMs are widely recognised as one of the most effective sequential models for times series predictions in engineering problems [27,37].

The LSTM network comprises three gates: input ($\mathbf{i}_{t_k}$), forget ($\mathbf{f}_{t_k}$), and output ($\mathbf{o}_{t_k}$); a block input, a single cell $\mathbf{c}_{t_k}$, and an output activation function. This network is recurrently connected back to the input and the three gates. Due to the gated structured and the forget state, the LSTM is an effective and scalable model that can deal with long-term dependencies [19]. The vector equations for a LSTM layer are:

$$\begin{aligned}
\mathbf{i}_{t_k} &= \phi(\mathbf{W}_{xi}\mathbf{u}_{t_k} + \mathbf{W}_{Hi}\mathbf{H}_{t_{k-1}} + \mathbf{b}_i) \\
\mathbf{f}_{t_k} &= \phi(\mathbf{W}_{xf}\mathbf{u}_{t_k} + \mathbf{W}_{Hf}\mathbf{H}_{t_{k-1}} + \mathbf{b}_f) \\
\mathbf{o}_{t_k} &= \phi(\mathbf{W}_{xo}\mathbf{u}_{t_k} + \mathbf{W}_{Ho}\mathbf{H}_{t_{k-1}} + \mathbf{b}_o) \\
\mathbf{c}_{t_k} &= \mathbf{f}_{t_k} \circ \mathbf{c}_{t_{k-1}} + \mathbf{i}_{t_k} \circ \tanh(\mathbf{W}_{xc}\mathbf{u}_{t_k} + \mathbf{W}_{Hc}\mathbf{H}_{t_{k-1}} + \mathbf{b}_c) \\
\mathbf{H}_{t_k} &= \mathbf{o}_{t_k} \circ \tanh(\mathbf{c}_{t_k})
\end{aligned} \tag{7}$$

where ϕ is the sigmoid function, $\mathbf{W}$ are the weights, $\mathbf{b}_{i,f,o,c}$ are the biases for the input, forget, output gate and the cell, respectively, $\mathbf{x}_{t_k}$ is the layer input, $\mathbf{H}_{t_k}$ is the layer output and $\circ$ denotes the entry-wise multiplication of two vectors. This is the output of a unidirectional LSTM.

For a BDLSTM, the output layer generates an output vector $\mathbf{u}_{t_k}$:

$$\mathbf{u}_{t_k} = \psi(\overrightarrow{\mathbf{H}_{t_k}}, \overleftarrow{\mathbf{H}_{t_k}}) \tag{8}$$

where ψ is a concatenating function that combines the two output sequences, forwards and backwards, denoted by a right and left arrow, respectively.

2.2.2 Adversarial Network

The work of [17] introduced the idea of adversarial training and adversarial losses which can also be applied to supervised scenarios and have advanced the state-of-the-art in many fields over the past years. Additionally, robustness may be achieved by detecting and rejecting adversarial examples by using adversarial training [34]. GAN are a network trained adversarially. The basic idea of GAN is to simultaneously train a discriminator and a generator, where the discriminator aims to distinguish between real samples and generated samples. By learning and matching the distribution that fits the training data $\mathbf{x}$, the aim is that new samples, sampled from the matched distribution formed by the generator, will produce, or generate, 'realistic' features from the latent vector $\mathbf{z}$

The GAN is composed by a discriminator network ($\mathcal{D}$) and a generator network ($\mathcal{G}$) The GAN losses, binary cross-entropy, therefore, can be written as:

$$\begin{aligned} \mathcal{L}_{\mathcal{D}}^{adv} &= -\sum log\mathcal{D}(\mathbf{x})) + log(1 - \mathcal{D}(\mathcal{G}(\mathbf{z})) \\ \mathcal{L}_{\mathcal{G}}^{adv} &= -\sum log(\mathcal{D}(\mathcal{G}(z)) \end{aligned} \tag{9}$$

This idea can be developed further if we consider similar elements of the adversarial training of GAN and applied to other domains, e.g. time-series, extreme events detection, adversarial attacks, among others.

2.3 Data Assimilation

Data assimilation algorithms aim to estimate the state variable $\mathbf{u}$ relying on a prior approximation $\mathbf{u}_b$ (also known as the background state) and a vector of observed states $\mathbf{v}$. The theoretical value of the state vector is denoted by $\mathbf{u}_{\text{true}}$, so called the true state, which is out of reach in real engineering problems. Both the background and the observation vectors are supposed to be noisy in DA, characterised by the associated error covariance matrices $\mathbf{B}$ and $\mathbf{R}$, respectively, i.e.,

$$\mathbf{B} = Cov(\epsilon_b, \epsilon_b), \quad \mathbf{R} = Cov(\epsilon_o, \epsilon_o), \tag{10}$$

with the prior errors ϵ_b and ϵ_o defined as:

$$\epsilon_b = \mathbf{u}_b - \mathbf{u}_{\text{true}} \quad \epsilon_o = \mathcal{H}(\mathbf{u}_{\text{true}}) - \mathbf{v}. \tag{11}$$

Since the true states are out of reach in real applications, the covariance matrices $\mathbf{B}$ and $\mathbf{R}$ are often approximated though statistical estimations [7,14]. The $\mathcal{H}$ function in Eq. 11 is called the transformation operator, which maps the state variables to the observable quantities. $\mathcal{H}(\mathbf{u}_{\text{true}})$ is also known as the model equivalent of observations.

By minimizing a cost function J defined as

$$\begin{aligned} J(\mathbf{u}) &= \frac{1}{2}(\mathbf{u}-\mathbf{u}_b)^T\mathbf{B}^{-1}(\mathbf{u}-\mathbf{u}_b) + \frac{1}{2}(\mathbf{v}-\mathcal{H}(\mathbf{v}))^T\mathbf{R}^{-1}(\mathbf{v}-\mathcal{H}(\mathbf{v})) \\ &= J_b(\mathbf{u}) + J_o(\mathbf{u}) \end{aligned} \tag{12}$$

DA approaches attempt to find an optimally weighted *a priori* analysis state,

$$\mathbf{u}_a = \underset{\mathbf{u}}{\text{argmin}}\Big(J(\mathbf{u})\Big). \tag{13}$$

The $\mathbf{B}$ and the $\mathbf{R}$ matrices, determining the weights of background and observation information (as shown in Eq. 12), is crucial in DA algorithms [11,36]. When $\mathcal{H}$ can be approximated by a linear function H and the error covariances B and R are well specified, Eq. 12 can be solved via Best Linear Unbiased Estimator (BLUE) [7]:

$$\mathbf{u}_a = \mathbf{u}_b + K(\mathbf{v} - \mathbf{H}\mathbf{u}_b) \tag{14}$$

where $\mathbf{K}$ denotes the Kalman gain matrix,

$$\mathbf{K} = \mathbf{B}\mathbf{H}^T(\mathbf{H}\mathbf{B}\mathbf{H}^T + \mathbf{R})^{-1}. \tag{15}$$

The optimisation of Eq. 12 often involves gradient descent algorithms (such as "L-BFGS-B") and adjoint-based numerical techniques. In the proposed modular approach of the present paper, we aim to perform DA in the low-dimensional latent space to reduce the computational cost, enabling a real-time model updating. The latent assimilation (LA) approach was first introduced in the work of [2] for CO_2 spread modeling. A generalised Latent Assimilation algorithm was proposed in the recent work of [9]. The observation quantities $\mathbf{v}_t$ are first preprocessed to fit the space of the state variables $\mathbf{u}_t$, i.e.,

$$\tilde{\mathbf{v}}_t = E(\mathbf{v}_t). \tag{16}$$

As a consequence, the transformation operator becomes the identity function in the latent space, leading to the loss function of LA:

$$\tilde{J}(\tilde{\mathbf{u}}_t) = \frac{1}{2}(\tilde{\mathbf{u}}_t - \tilde{\mathbf{u}}_{t,b})^T\mathbf{B}^{-1}(\tilde{\mathbf{u}}_t - \tilde{\mathbf{u}}_{t,b}) + \frac{1}{2}(\tilde{\mathbf{u}}_t - \tilde{\mathbf{v}}_t)^T\mathbf{R}^{-1}(\tilde{\mathbf{u}}_t - \tilde{\mathbf{v}}_t), \tag{17}$$

where the latent background state $\tilde{\mathbf{u}}_{t,b}$ is issued from the RNN predictions as mentioned in Sect. 2.2.1. The analysis state,

$$\mathbf{u}_{t,a} = \underset{\mathbf{u}_t}{\text{argmin}}\Big(\tilde{J}(\tilde{\mathbf{u}}_t)\Big), \tag{18}$$

can then replace the background prediction $\tilde{\mathbf{u}}_{t,b}$, which can be used as the starting-point for the next-level prediction in ML algorithms.

3 Application to Wildfires

The first application is real-time forecasting of wildfire dynamics. Wildfires have increasing attention recently in fire safety science world-widely, and it is an extremely challenging task due to the complexities of the physical models and the geographical features. Real-time forecasting of wildfire dynamics which raises increasing attention recently in fire safety science world-widely, is extremely challenging due to the complexities of the physical models and the number of geographical features. Running physics-based simulations for large-scale wildfires can be computationally difficult, if not infeasible. We applied the proposed modular approach for fire forecasting in near real-time, which combines reduced-order modelling, recurrent neural networks RNN, data assimilation DA and error covariance tuning. More precisely, based on snapshots of dynamical fire simulations, we first construct a low-dimensional latent space via proper orthogonal decomposition or convolution AE. A LSTM is then used to build sequence-to-sequence predictions following the simulation results projected/encoded in the reduced space. In order to adjust the prediction of burned areas, latent DA coupled with an error covariance tuning algorithm is performed with the help of daily observed satellite wildfire images as observation data. The proposed method was tested on two recent large fire events in California, namely the Buck fire and the Pier fire, both taking place in 2017 as illustrated in Fig. 2.

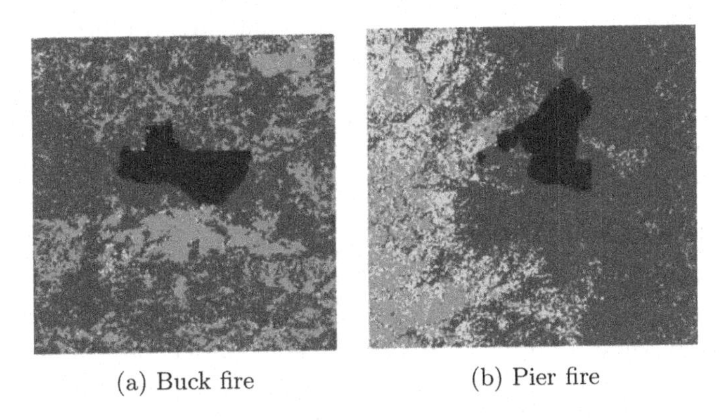

(a) Buck fire (b) Pier fire

Fig. 2. Observed burned area of the Buck fire and the Pier fire at the first day

We first employed an operational cellular automata (CA) fire spread model [1] to generate the training dataset for ROM and RNN surrogate modelling. This CA model is a probabilistic simulator which takes into account a number of local geophysical features, such as vegetation density (see Fig. 2) and ground elevation. Once the latent space is acquired, the ML-based surrogate model is then trained using the results of stochastic CA simulations in the corresponding area of fire events. With a much shorter online execution time, the so-obtained data-driven model provides similar results as physics-based CA simulations (stochastic) in the sense that the mean and the standard deviation of CA-CA and CA-LSTM differences are similar as shown in Fig. 3 for the Pier fire. In fact, the ROM- and ML-based approach run roughly 1000 times faster than the original CA model as shown in Fig. 3(b).

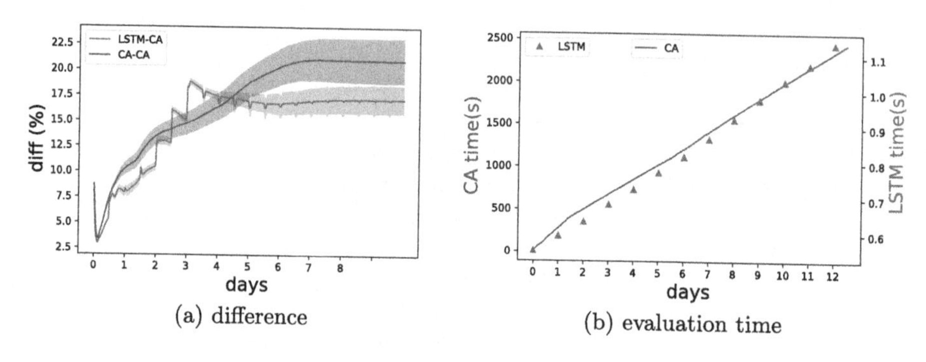

(a) difference

(b) evaluation time

Fig. 3. Model difference and computational time

Each step in CA-LSTM predictions is roughly equivalent to 30 min in real time while the satellite observations are of daily basis. The latter is used to adjust the fire prediction consistently since the actual fire spread also depends heavily on other factors such as real-time climate or human interactions which are not included in the CA modelling. The evolution of the averaged relative root mean square error (R-RMSE) is shown in Fig. 4. The numerical results show that, with the help of error covariance tuning [8,14], DA manages to improve the model prediction accuracy in both fire events.

4 Application to Computational Fluid Dynamics and Air Pollution

Similar to the wildfire problem, we also present an general workflow to generate and improve the forecast of model surrogates of CFD simulations using deep learning, and most specifically adversarial training. This adversarial approach aims to reduce the divergence of the forecasts from the underlying physical model. Our two-step method, similar to the wildfire application, integrates a PCA AE with adversarial LSTM networks. Once the reduced-order model (ROM) of the CFD solution is obtained via PCA, an adversarial autoencoder (AAE) is used on the principal components time series. Subsequently, a LSTM model is adversarially trained, named adversarial LSTM (ALSTM), on the latent space produced by the principal component adversarial autoencoder (PC-AAE) to make forecasts. Here we show, that the application of adversarial training improves the rollout of the latent space predictions.

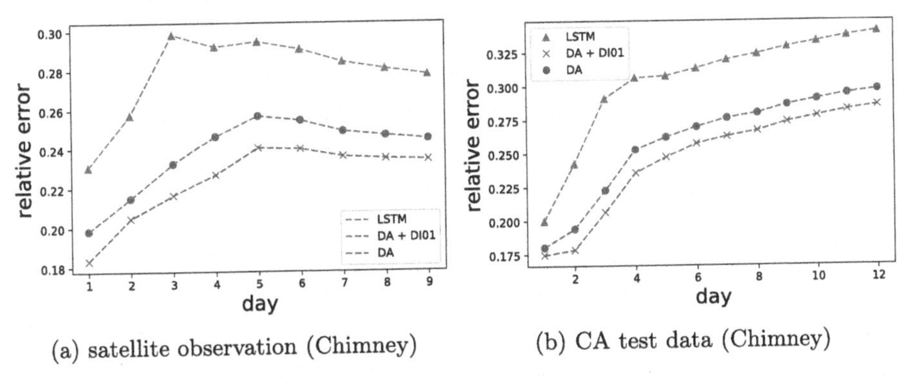

(a) satellite observation (Chimney) (b) CA test data (Chimney)

Fig. 4. Relative error of pure LSTM and assimilated prediction compared to satellite observations

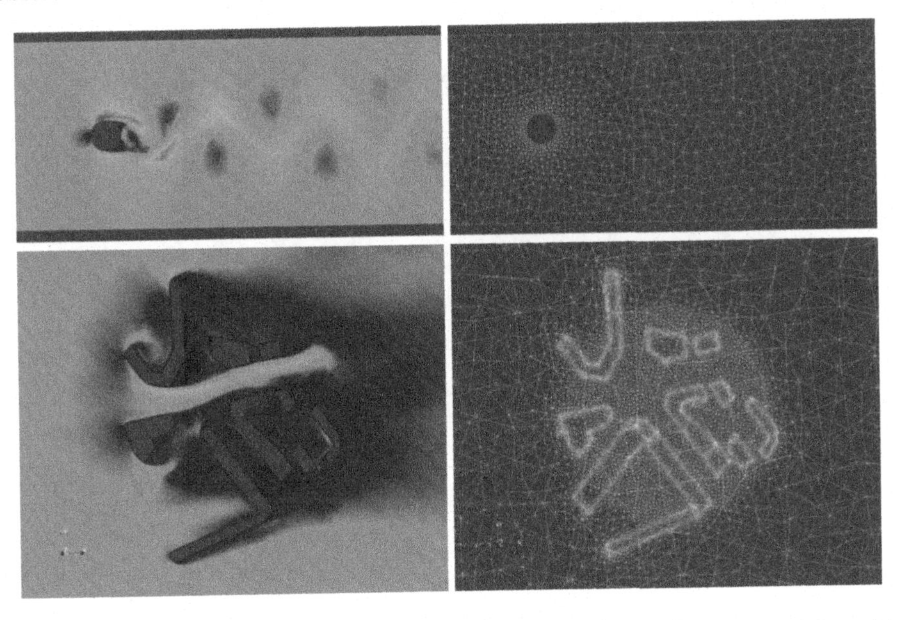

Fig. 5. Top: Flow past the cylinder - 2D CFD simulation. Bottom: 3D Urban Air pollution CFD simulation. Right: Unstructured meshes of both domains

Different case studies are shown in Fig. 5:

- FPC: the 2D case describes a typical flow past the cylinder CFD, in which a cylinder placed in a channel at right angle to the oncoming fluid making the steady-state symmetrical flow unstable. This simulation has a Reynolds number (Re) of 2,300 with $m = 5,166$ nodes and $n = 1,000$ time-steps.
- 3DAirPollution: The 3D case is a realistic case including 14 buildings representing a real urban area located near Elephant and Castle, South London,

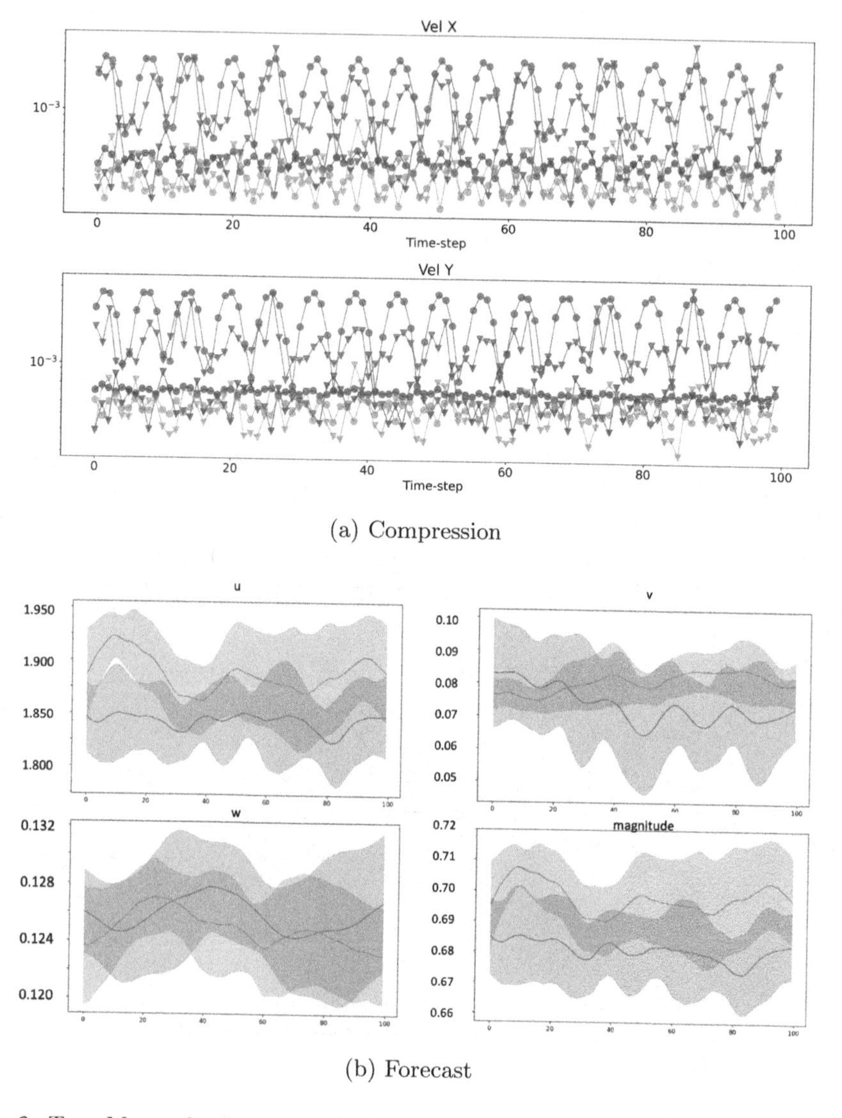

(a) Compression

(b) Forecast

Fig. 6. Top: Mean absolute error (MAE) of the different compression methods on the velocities (ms^{-1}) of the flow past the cylinder domain. Red, magenta, and cyan represents a compression to 8, 4, and 2 dimensions, respectively. Circle and triangle markers are a PC and a PC-AAE compression, respectively. Bottom: Ensemble, of 50 different starting points, of the MAE of the forecast, for 100 time-levels, of the velocities (ms^{-1}) of the 3D Urban Air Pollution domain. Blue is the original test data, and orange is the prediction obtained with the PC-AAE and the ALSTM. (Color figure online)

UK. The 3D case (720 m × 676 m × 250 m) is composed of an unstructured mesh including $m = 100,040$ nodes per dimension and $n = 1,000$ time-steps.

The two-dimensional (2D) CFD case study was performed using Thetis [22] and the three-dimensional (3D) CFD simulations were carried out using Fluidity [13]. For these domains, the framework was trained and validated on the first 1000 time-steps of the simulation, and tested on the following 500 time-steps.

A PCA as applied to a 2-dimensional velocity field (ms^{-1}) in Flow past the cylinder and likewise to the velocities of the 3D model. The full-rank PCs were used as input for the AAE and divided in 3 different experiments named LS_τ and compared to the corresponding reconstruction $\mathbf{x}_\tau$ with $\tau = \{2, 4, 8\}$ PCs. The results of the mean absolute error using the different dimension reduction approaches are shown in Fig. 6a for the flow past the cylinder case. The AAE outperforms a simple truncation of the PCs in both domains.

In terms of forecasting, our framework generalises well on unseen data Fig. 6b. This is because of the Gaussian latent space obtained with the adversarial AE constraints the further predictions and forces the predictions back into the distribution. Furthermore, the adversarial training of the LSTM learns how to stay within the distribution data. Followed by the training of an adversarial LSTM, we can assess the forecasts using our workflow. An ensemble of 50 different starting points from the test dataset were used to be forecasted in time for 100 time-levels. The ensemble of Mean Absolute Errors results are based on a dimension reduction 8 dimensions in the latent space of the AAE, which is a compression of 5 orders of magnitude. The error percentage of the means of these forecasts is 5% in the test dataset.

5 Conclusions

In the present paper, we introduced a ROM- and ML-based modular approach for efficient predictions of high-dimensional dynamical systems. In addition, this method can also incorporate real-time observations for model correction/adjusting with a low computational cost. A variety of ROM and RNN approaches can be included in the algorithm scheme regarding different applications. The replacement of the physics-based simulation/resolution by these models will speed up the forecast process towards a real-time solution. And, the application of adversarial training could potentially produce more physically realistic scenarios. We showed the strength of the proposed method in predicting wildfire spread and air pollution diffusion in this work. Furthermore, this framework is data-agnostic and could be applied to different physical models when enough data is available.

Acknowledgements. This research is funded by the Leverhulme Centre for Wildfires, Environment and Society through the Leverhulme Trust, grant number RC-2018-023. This work is partially supported by the EP/T000414/1 PREdictive Modelling with QuantIfication of UncERtainty for MultiphasE Systems (PREMIERE) and by the EPSRC grant EP/T003189/1 Health assessment across biological length scales for personal pollution exposure and its mitigation (INHALE).

References

1. Alexandridis, A., Vakalis, D., Siettos, C., Bafas, G.: A cellular automata model for forest fire spread prediction: the case of the wildfire that swept through Spetses Island in 1990. Appl. Math. Comput. **204**(1), 191–201 (2008)
2. Amendola, M., et al.: Data assimilation in the latent space of a neural network (2020)
3. Asch, M., Bocquet, M., Nodet, M.: Data assimilation: methods, algorithms, and applications, vol. 11. SIAM (2016)
4. Buizza, C., et al.: Data learning: integrating data assimilation and machine learning. J. Comput. Sci. **58**, 101525 (2022)
5. Casas, C.Q., Arcucci, R., Wu, P., Pain, C., Guo, Y.K.: A reduced order deep data assimilation model. Physica D **412**, 132615 (2020)
6. Cheng, M., Fang, F., Pain, C.C., Navon, I.: Data-driven modelling of nonlinear spatio-temporal fluid flows using a deep convolutional generative adversarial network. Comput. Meth. Appl. Mech. Eng. **365**, 113000 (2020)
7. Cheng, S., Argaud, J.P., Iooss, B., Lucor, D., Ponçot, A.: Background error covariance iterative updating with invariant observation measures for data assimilation. Stoch. Environ. Res. Risk Assess. **33**(11), 2033–2051 (2019)
8. Cheng, S., Argaud, J.-P., Iooss, B., Lucor, D., Ponçot, A.: Error covariance tuning in variational data assimilation: application to an operating hydrological model. Stoch. Env. Res. Risk Assess. **35**(5), 1019–1038 (2020). https://doi.org/10.1007/s00477-020-01933-7
9. Cheng, S., et al.: Generalised latent assimilation in heterogeneous reduced spaces with machine learning surrogate models. arXiv preprint arXiv:2204.03497 (2022)
10. Cheng, S., Lucor, D., Argaud, J.P.: Observation data compression for variational assimilation of dynamical systems. J. Comput. Sci. **53**, 101405 (2021)
11. Cheng, S., Qiu, M.: Observation error covariance specification in dynamical systems for data assimilation using recurrent neural networks. Neural Comput. Appl., 1–19 (2021). https://doi.org/10.1007/s00521-021-06739-4
12. Cui, Z., Ke, R., Pu, Z., Wang, Y.: Stacked bidirectional and unidirectional LSTM recurrent neural network for forecasting network-wide traffic state with missing values. Transp. Res. Part C Emerg. Technol. **118**, 102674 (2020)
13. Davies, D.R., Wilson, C.R., Kramer, S.C.: Fluidity: a fully unstructured anisotropic adaptive mesh computational modeling framework for geodynamics. Geochem. Geophys. Geosyst. **12**(6) (2011)
14. Desroziers, G., Ivanov, S.: Diagnosis and adaptive tuning of observation-error parameters in a variational assimilation. Q. J. R. Meteorol. Soc. **127**(574), 1433–1452 (2001)
15. Dong, G., Liao, G., Liu, H., Kuang, G.: A review of the autoencoder and its variants: a comparative perspective from target recognition in synthetic-aperture radar images. IEEE Geosci. Remote Sens. Mag. **6**(3), 44–68 (2018)
16. Gong, H., Cheng, S., Chen, Z., Li, Q.: Data-enabled physics-informed machine learning for reduced-order modeling digital twin: application to nuclear reactor physics. Nucl. Sci. Eng. **196**, 668–693 (2022)
17. Goodfellow, I., et al.: Generative adversarial nets. In: Advances in Neural Information Processing Systems, pp. 2672–2680 (2014)
18. Hinton, G.E., Salakhutdinov, R.R.: Reducing the dimensionality of data with neural networks. Science **313**(5786), 504–507 (2006)

19. Hochreiter, S., Schmidhuber, J.: Long short-term memory. Neural Comput. **9**(8), 1735–1780 (1997)
20. Huang, Z., Xu, W., Yu, K.: Bidirectional LSTM-CRF models for sequence tagging. arXiv preprint arXiv:1508.01991 (2015)
21. Jaruszewicz, M., Mandziuk, J.: Application of PCA method to weather prediction task. In: Proceedings of the 9th International Conference on Neural Information Processing, 2002, ICONIP 2002, vol. 5, pp. 2359–2363. IEEE (2002)
22. Kärnä, T., Kramer, S.C., Mitchell, L., Ham, D.A., Piggott, M.D., Baptista, A.M.: Thetis coastal ocean model: discontinuous Galerkin discretization for the three-dimensional hydrostatic equations. Geosci. Model Dev. **11**(11), 4359–4382 (2018)
23. Kim, B., Azevedo, V.C., Thuerey, N., Kim, T., Gross, M., Solenthaler, B.: Deep fluids: a generative network for parameterized fluid simulations. In: Computer Graphics Forum, vol. 38, pp. 59–70. Wiley Online Library (2019)
24. Knol, D., de Leeuw, F., Meirink, J.F., Krzhizhanovskaya, V.V.: Deep learning for solar irradiance nowcasting: a comparison of a recurrent neural network and two traditional methods. In: Paszynski, M., Kranzlmüller, D., Krzhizhanovskaya, V.V., Dongarra, J.J., Sloot, P.M.A. (eds.) ICCS 2021. LNCS, vol. 12746, pp. 309–322. Springer, Cham (2021). https://doi.org/10.1007/978-3-030-77977-1_24
25. Liu, C., et al.: EnKF data-driven reduced order assimilation system. Eng. Anal. Boundary Elem. **139**, 46–55 (2022)
26. Liu, G., Guo, J.: Bidirectional LSTM with attention mechanism and convolutional layer for text classification. Neurocomputing **337**, 325–338 (2019)
27. Nakamura, T., Fukami, K., Hasegawa, K., Nabae, Y., Fukagata, K.: Convolutional neural network and long short-term memory based reduced order surrogate for minimal turbulent channel flow. Phys. Fluids **33**(2), 025116 (2021)
28. Phillips, T.R.F., Heaney, C.E., Smith, P.N., Pain, C.C.: An autoencoder-based reduced-order model for eigenvalue problems with application to neutron diffusion. Int. J. Numer. Meth. Eng. **122**(15), 3780–3811 (2021)
29. Quilodrán Casas, C., Arcucci, R., Guo, Y.: Urban air pollution forecasts generated from latent space representations. In: ICLR 2020 Workshop on Integration of Deep Neural Models and Differential Equations (2020)
30. Quilodrán-Casas, C., Arcucci, R., Mottet, L., Guo, Y., Pain, C.: Adversarial autoencoders and adversarial LSTM for improved forecasts of urban air pollution simulations. Published as a Workshop Paper at ICLR 2021 SimDL Workshop (2021)
31. Quilodrán-Casas, C., Silva, V.L., Arcucci, R., Heaney, C.E., Guo, Y., Pain, C.C.: Digital twins based on bidirectional LSTM and GAN for modelling the COVID-19 pandemic. Neurocomputing **470**, 11–28 (2022)
32. Rawat, W., Wang, Z.: Deep convolutional neural networks for image classification: a comprehensive review. Neural Comput. **29**(9), 2352–2449 (2017)
33. Schuster, M., Paliwal, K.K.: Bidirectional recurrent neural networks. IEEE Trans. Signal Process. **45**(11), 2673–2681 (1997)
34. Shafahi, A., et al.: Adversarial training for free! In: Advances in Neural Information Processing Systems, pp. 3358–3369 (2019)
35. Sirovich, L.: Turbulence and the dynamics of coherent structures. II. Symmetries and transformations. Q. Appl. Math. **45**(3), 573–582 (1987)
36. Tandeo, P., et al.: A review of innovation-based methods to jointly estimate model and observation error covariance matrices in ensemble data assimilation. Mon. Weather Rev. **148**(10), 3973–3994 (2020)

37. Tekin, S.F., Karaahmetoglu, O., Ilhan, F., Balaban, I., Kozat, S.S.: Spatio-temporal weather forecasting and attention mechanism on convolutional LSTMs. arXiv preprint arXiv:2102.00696 (2021)
38. Wu, H., Mardt, A., Pasquali, L., Noe, F.: Deep generative Markov state models. arXiv preprint arXiv:1805.07601 (2018)

A GPU-Based Algorithm for Environmental Data Filtering

Pasquale De Luca[1], Ardelio Galletti[2], and Livia Marcellino[2](✉)

[1] International PhD Programme/UNESCO Chair "Environment, Resources and Sustainable Development", Department of Science and Technology, Parthenope University of Naples, Centro Direzionale, Isola C4, (80143) Naples, Italy
pasquale.deluca@uniparthenope.it

[2] Department of Science and Technology, Parthenope University of Naples, Centro Direzionale, Isola C4, (80143) Naples, Italy
{ardelio.galletti,livia.marcellino}@uniparthenope.it

Abstract. Nowadays, the Machine Learning (ML) approach is needful to many research fields. Among these, the Environmental Science (ES) which involves a large amount of data to be processed and collected. On the other hand, in order to provide a reliable output, those data information must be assimilated. Since this process requires a large execution time when the input dataset is very huge, here we propose a parallel GPU algorithm based on a curve fitting method, to filter the starting dataset, by exploiting the computational power of the CUDA tool. The innovative aspect of the proposed procedure can be used in several application fields. Our experiments show the achieved results in terms of performance.

Keywords: Machine learning · Curve fitting · Filtering · GPU parallel algorithm · HPC

1 Introduction

In recent years, the Machine Learning approach is becoming very helpful for environmental data analysis, e.g. weather prediction, air pollution quality analysis or earthquakes. The large amount of available data can be successfully used to define, classify, monitor and predict the ecosystem conditions in which we live [1]. However, data are known to have errors of a random nature because, very often the acquisition tools can provide false measurements or with missing data. This implies that mathematical models used to analyze them can provide unacceptable results.

To overcome this problem Data Assimilation (DA) plays a key role in ML methods. DA is a standard practice, which combines mathematical models and detected measures, and it is heavily employed in numerical weather prediction [2]. Lately its application is becoming widespread in many other areas of climate, atmosphere, ocean, and environment modeling; i.e. in all circumstances where one intends to estimate the state of a large dynamical system based on limited information. In

D. Groen et al. (Eds.): ICCS 2022, LNCS 13353, pp. 45–52, 2022.
https://doi.org/10.1007/978-3-031-08760-8_4

this context, here we deal with a smoothing method based on local least-squares polynomial approximation to filter initial data, known as Savitzky-Golay (SG) filter [3]. This filter provides to fit a polynomial to a set of sampled values, in order to smooth noisy data while maintaining the shape and height of peaks. The innovative aspect of the proposed procedure can be used in several applications with satisfactory results: for example for image analysis, signals electrocardiograms processing and for environmental measurements filtering [4–6]. Nevertheless, the large amount of data to be processed requires several waiting hours and this represents a problem that must be solved. High-Performance Computing (HPC) offers a powerful tool to overcome this issue through parallel strategies, suitably designed to be applied in several application fields [7–9].

In this work, we present a novel parallel algorithm, for Graphics Processing Units (GPUs) environment, appropriately designed to efficiently perform the SG filter. Our implementation exploits the computational power of the Compute Unified Device Architecture (CUDA) framework [10], together with the cuBLAS and cuSOLVER libraries, in order to achieve an appreciable gain of performance in dealing advanced mathematical algebraic operations.

Then the rest of the paper is organized as follows. Section 2 recalls the mathematical model related to the SG filter. In Sect. 3, the GPU-CUDA parallel approach and the related algorithm are described. The experiments discussed in Sect. 4 confirm the efficiency of the proposed implementation in terms of performance. Finally, conclusions in Sect. 5 close the paper.

2 Numerical Model Details

In this section we recall some mathematical preliminaries about the model implemented. This allows us to describe a pseudo-algorithm which is the basis of the GPU-parallel implementation, described in next section, we propose. The discussion follows scheme and main notations presented in [11]. In the following, we limit the discussion to the basic information to design the GPU-parallel implementation and recommend the reader to see papers [3,11,12] for further details. The model we consider is the Savitky-Golay filter, which consists in fact in applying a least square fitting procedure to the entries values lying in moving windows of the input signal.

Let us begin by denoting by: $x[n], (n = \dots, -2, -1, 0, -1, 2, \dots)$ the entries of the input, and set two nonnegative integer values ML and MR. Then, for all value $x[i]$ to be filtered, let consider the window:

$$\mathbf{x}_i = \big(x[i - ML], \dots, x[i], \dots, x[i + MR]\big) \tag{1}$$

centered at $x[i]$ and including ML values "to the left" and MR values "to the right". $\mathbf{x}_i$, that is: In the SG filter, to get the filtered value $y[i]$, firstly we find the polynomial:

$$p_N(n) = \sum_{k=0}^{N} a_k n^k \quad -ML \le n \le MR \tag{2}$$

of degree N that minimizes, in the least square sense, its distance from values in $\mathbf{x}_i$, i.e. the quantity: $\varepsilon_N = \sum_{k=-ML}^{MR} \left(p_N(k) - x[i+k]\right)^2$. Then we set $y[i] = p_N(0) = a_0$., as the value the polynomial takes at $n = 0$. We observe that p_N is the least square polynomial approximating points $(i, x[i])$, (for $i = -ML, \ldots, MR$) and it can be proved that its coefficients a_k solve the normal equations linear system:

$$\mathcal{A}^T \mathcal{A} a = \mathcal{A}^T \mathbf{x}_i^T, \quad \text{where} \quad \mathcal{A} = \left(i^j \right)_{i=-ML,\ldots,MR}^{j=0,\ldots,N}. \tag{3}$$

It follows that a_0 and all other coefficients of p_N depend on (are linear combination of) $\mathbf{x}_i$. However, by rearranging (3) we get: $a = (\mathcal{A}^T\mathcal{A})^{-1}\mathcal{A}^T\mathbf{x}_i^T$ the 0-th row of the pseudo-inverse matrix:

$$H = (\mathcal{A}^T\mathcal{A})^{-1}\mathcal{A}^T, \tag{4}$$

whose entries do not depend on $\mathbf{x}_i$. In other words each filtered value $y[i]$ is a linear combination of values in $\mathbf{x}_i$ with coefficients in the 0-th row of H, that can be pre-computed once, independently from the filtered value we need, i.e.:

$$y[i] = \sum_{k=-ML}^{MR} h_{0,k} \cdot x[i+k] \tag{5}$$

Previous discussion allows us to introduce the Pseudo-algorithm 1, which summarizes the main steps needed to solve the numerical problem.

Algorithm 1. Sequential pseudo-algorithm

```
Input:        x, ML, MR, N.      Output:      y
1: build  A                  % as in (3)
2: build  H                  % as in (4)
3: extract  H0               % the 0-th row of H
4: for i ∈ Z do
5:    extract x_i            % from the input x
6:    compute y[i]           % as in (5)
7: end for
```

3 Parallel Approach and GPU Algorithm

Observing the significant results achieved in [13–15], due to large amount of produced data from scientific community, we have chosen to develop a GPU-based parallel implementation, equipped by a suitable Domain Decomposition (DD) with overlapping, already proposed in [16], for most modern GPU architecture. From now on, in this section, to describe the parallel algorithm, we set: $ML = MR = M$ and assume the input signal x to have finite size s, i.e., there is no likelihood of confusion by setting: $x = \big(x[0], x[1], \ldots, x[s-1]\big)$. We observe that to apply the algorithm to all moving windows, included the edge ones, we also need to increase the size of the input with a *zero-padding* procedure, which consists of extending the original input signal with artificial M zeros at the left and M zeros at the right boundaries, as follows:

$$x^M = (0, \dots, 0, x[0], \dots, x[s-1], 0, \dots, 0). \tag{6}$$

The overall parallel scheme is shown in the Pseudo-algorithm 2.

Algorithm 2. Parallel pseudo-algorithm

```
Input:      x, M N .   Output:      y
 1: build x^M as in (6)              % zero-padding of x
    % step 1: Pseudo-inverse building
 2: build A, as in (3) by using the cuSOLVER routine
 3: build H, as in (4) by using the cuBLAS routine
 4: extract  H_0                     % the 0-th row of H
    % step 2: Domain Decomposition with overlapping
 5: for each t_i do
 6:     S_i ← x_i
 7: end for
    % step 3: Computation of final output
 8: compute y as in (7), by using the cuBLAS routine
```

The main steps of the Pseudo-algorithm 2 are described below:

STEP 1 - *Pseudo-inverse building.*
This step is based on the QR factorization of a matrix A, which is suitably efficient for computing the pseudo-inverse H when A is symmetric. To this aim, the algorithm exploits the potential of the cuSOLVER and cuBLAS libraries [17,18] for numerical linear algebra operations on GPU. In particular, we firstly build matrix $\mathcal{A}$ by using the `cusolverDnDgeqrf` cuSOLVER routine, then to build of H, we observe that it holds: $H = R^{-1}Q^T$, where Q and R come from the QR factorization of $\mathcal{A}$. Then, we use the `cublasDgetriBatched` cuBLAS routine to compute R^{-1} and the `cublasDgemm` CUBLAS routine, to get the matrix-matrix multiplication H.

STEP 2 - *Domain Decomposition with overlapping.*
Here, we decompose the problem by building a matrix: $\mathcal{S} = [\mathbf{x}_0, \mathbf{x}_1, \dots, \mathbf{x}_{s-1}]^T$ whose rows are the $\mathbf{x}_i$ defined in (1). According to SIMT paradigm, which is used by the CUDA configuration, we use: s threads, $t_0, t_1, \dots, t_{s-1}$, and each of them copies the $i-th$ moving window from x^M to $i-th$ row of $\mathcal{S}$, in a fully parallel way. To this aim, we implemented a CUDA kernel which takes advantage of the large number of processing units of the GPU environment.

STEP 3 - *Computation of final output.*
The final output y can be regarded as the matrix-vector product

$$y = S \cdot H_0^T. \tag{7}$$

This operation is performed by using the `cublasDgemv` CUBLAS routine.

4 Performance Analysis

In this section, some experimental results show and confirm the efficiency of proposed software. Following, the hardware specifications where the GPU algorithm has been implemented and tested, are listed: 1 x CPU Intel i7 860 with 4 cores, 2 threads per core, 2.80 Ghz, 8 GB of RAM; 1 x GPU NVIDIA Quadro K5000 with 1536 CUDA Cores, 706 MHz Core GPU Clock, and 4 GB 256-bit, GDDR5 memory configuration. Thanks to CUDA framework, Algorithm 2 exploits the overall GPU's parallel computational power. The evaluation of a parallel algorithm is covered by different metrics, and here: we preliminarily show a execution time comparison in order to highlight the gain of performance obtained, then we focus on the spent time for each computational expensive operation listed in Pseudo-algorithm 2. Finally, last subsection shows a example output of the parallel algorithm applied to a set of environmental data.

Execution Time Comparison. Table 1 exhibits an execution time comparison among the GPU implementation and the native function *sgolayfilt* in MATLAB, by varying the input data sample dimension. We underline the *sgolayfilt* function runs in a parallel way by using 4 cores. Input data come from the environmental dataset related to the last London Atmospheric Emissions Inventory (LAEI) for year 2019. The area covered by the LAEI includes Greater London, as well as areas outside Greater London up to the M25 motorway (see [19]). These emissions have been used to estimate ground level concentrations of key pollutants NO_x, NO_2, PM_{10} and $PM_{2.5}$ across Greater London for year 2019, using an atmospheric dispersion model. In this test we used $M = 2$ and $N = 3$. The CUDA configuration is static and set to 512×1024 block per threads. We highlight a significant time reduction of GPU algorithm with respect to multi-core execution. The achieved speed-up is closely linked to accelerated operations computed in parallel way by giving a strong impact of time reduction. Despite the accomplished good performance, we must stop to increase the input data sample due to small memory size adopted by the CPU. Conversely, the large dimension and accurate management of GPU memory together, allows us to introduce a great increment of the input data sample in order to process big data sample.

Table 1. Execution times: MATLAB vs. GPU.

	Time (ms)	
s	Multi-core	GPU
3.4×10^1	5.92	2.96
4.0×10^3	6.59	3.54
4.6×10^4	7.01	3.45
7.3×10^4	7.50	3.50
3.9×10^5	7.90	3.51

CUDA Kernels Analysis. An additional performance analysis acts as a support for previous experimental result. More precisely, for each single kernel by varying the input sample dimension, we show the time consumption for each CUDA kernel which computes the operations of most expensive procedure of Pseudo-algorithm 2. Table 2 shows time values for each CUDA kernel by positively confirming the efficiency of the proposed software. Specifically, a good work-load balancing among CUDA threads is performed. Transfer and copy times are avoided to underline the gain of performance for each kernel.

Table 2. Time execution analysis for each CUDA kernel in milliseconds.

	Execution time (ms)				
Operation	3.4×10^1	4.0×10^3	4.6×10^4	7.3×10^4	3.9×10^5
QR	0.11	0.11	0.11	0.11	0.11
Pseudo-inverse	0.30	0.32	0.31	0.30	0.31
S computation	0.12	0.12	0.13	0.07	0.28
Polynomial evaluation	0.11	0.12	0.20	0.26	0.92
Overall time	**0.68**	**0.69**	**0.78**	**0.84**	**1.73**

Environmental Data Testing. Last plot, Fig. 1, aims to show how the algorithm performs when applied to data of environmental nature. In particular, this qualitative test is referred to the Greater London mean PM_{10} particulate matter arising from the dataset in [19]. Input data is plotted in blue while the SG filtered output is the red line. We recall that results sharply overlap with the ones obtained by means of the MATLAB routine `sgolayfilt`. We remark that, unlike the MATLAB procedure, our implementation is able to manage input data, of very large size, in a reasonable time.

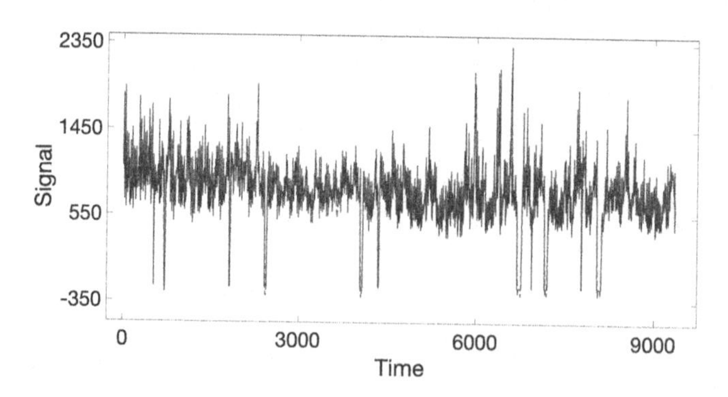

Fig. 1. Application of the SG filter to PM_{10}. Input signal is reported in blue, filtered output of the Savitzky-Golay filter in red. (Color figure online)

5 Conclusion

In this work, we presented a novel GPU-parallel algorithm, based on the SG filter, to approximate by a polynomial sampled data values. This procedure can be seen as a pre-processing step to correctly assimilate data within the Machine Learning approach in order to provide a reliable output and without affecting execution times. Our experiments showed very good performance even with real environmental data. Future work could consider a further application of our implementation to the ECG denoising field.

References

1. Kanevski, M.: Machine Learning for Spatial Environmental Data: Theory, Applications, and Software. EPFL press, Lausanne (2009)
2. Cuomo, S., Galletti, A., Giunta, G., Marcellino, L.: Numerical effects of the gaussian recursive filters in solving linear systems in the 3dvar case study. Numer. Math. Theory Methods Appl. **10**(3), 520–540 (2017)
3. Savitzky, A., Golay, M.J.: Smoothing and differentiation of data by simplified least squares procedures. Anal. Chem. **36**(8), 1627–1639 (1964)
4. Liu, Y., Dang, B., Li, Y., Lin, H., Ma, H.: Applications of Savitzky-Golay filter for seismic random noise reduction. Acta Geophysica **64**(1), 101–124 (2015). https://doi.org/10.1515/acgeo-2015-0062
5. Chen, J., Jönsson, P., Tamura, M., Gu, Z., Matsushita, B., Eklundh, L.: A simple method for reconstructing a high-quality NDVI time-series data set based on the Savitzky-Golay filter. Rem. Sens. Environ. **91**(3–4), 332–344 (2004)
6. Rodrigues, J., Barros, S., Santos, N.: FULMAR: follow-up lightcurves multitool assisting radial velocities. In: Posters from the TESS Science Conference II (TSC2), p. 45 (2021)
7. D'Amore, L., Casaburi, D., Galletti, A., Marcellino, L., Murli, A.: Integration of emerging computer technologies for an efficient image sequences analysis. Integr. Comput.-Aided Eng. **18**(4), 365–378 (2011)
8. De Luca, Pasquale, Galletti, Ardelio, Giunta, Giulio, Marcellino, Livia: Accelerated Gaussian convolution in a data assimilation scenario. In: Krzhizhanovskaya, V.V., et al. (eds.) ICCS 2020. LNCS, vol. 12142, pp. 199–211. Springer, Cham (2020). https://doi.org/10.1007/978-3-030-50433-5_16
9. De Luca, P., Galletti, A., Marcellino, L.: Parallel solvers comparison for an inverse problem in fractional calculus. In: 2020 Proceeding of 9th International Conference on Theory and Practice in Modern Computing (TPMC 2020) (2020)
10. https://developer.nvidia.com/cuda-zone
11. Schafer, R.W.: What is a Savitzky-Golay filter? [lecture notes]. IEEE Signal Process. Mag. **28**(4), 111–117 (2011)
12. Luo, J., Ying, K., Bai, J.: Savitzky-Golay smoothing and differentiation filter for even number data. Signal Process. **85**(7), 1429–1434 (2005)
13. Cuomo, S., De Michele, P., Galletti, A., Marcellino, L.: A GPU parallel implementation of the local principal component analysis overcomplete method for DW image denoising. In: 2016 IEEE Symposium on Computers and Communication (ISCC), pp. 26–31. IEEE (2016)

14. Cuomo, S., De Michele, P., Galletti, A., Marcellino, L.: A GPU-parallel algorithm for ECG signal denoising based on the NLM method. In: 2016 30th International Conference on Advanced Information Networking and Applications Workshops (WAINA), pp. 35–39. IEEE (2016)
15. De Luca, P., Galletti, A., Giunta, G., Marcellino, L.: Recursive filter based GPU algorithms in a Data Assimilation scenario. J. Comput. Sci. **53**, 101339 (2021)
16. De Luca, P., Galletti, A., Marcellino, L.: A Gaussian recursive filter parallel implementation with overlapping. In: 2019 15th International Conference on Signal-Image Technology & Internet-Based Systems (SITIS), pp. 641–648 (2019)
17. https://docs.nvidia.com/cuda/cusolver/index.html
18. https://docs.nvidia.com/cuda/cublas/index.html
19. https://data.london.gov.uk/

Multiscale Modelling and Simulation

Multipoint Meshless FD Schemes Applied to Nonlinear and Multiscale Analysis

Irena Jaworska(✉)

Cracow University of Technology, Krakow, Poland
irena@cce.pk.edu.pl

Abstract. The paper presents computational schemes of the multipoint meshless method – the numerical modeling tool that allows accurate and effective solving of boundary value problems. The main advantage of the multipoint general version is its generality – the basic relations of derivatives from the unknown function depend on the domain discretization only and are independent of the type of problem being solved. This feature allows to divide the multipoint computational strategy into two stages and is advantageous from the calculation efficiency point of view. The multipoint method algorithms applied to such engineering problems as numerical homogenization of heterogeneous materials and nonlinear analysis are developed and briefly presented. The paper is illustrated by several examples of the multipoint numerical analysis.

Keywords: Meshless FDM · Higher order approximation · Multipoint method · Homogenization · Nonlinear analysis · Elastic-plastic problem

1 Introduction

Besides the most commonly applied method of computational analysis – the Finite Element Method, the alternative methods, such as various meshless methods are more and more developed [1] contemporary tools for analysis of engineering problems. The paper introduces computational schemes of the application of the recently developed higher order multipoint meshless finite difference method [2, 3] to the boundary value problems. The new multipoint approach (higher order extension of the meshless FDM (MFDM) [4]) is based on the arbitrary irregular meshes, the moving weighted least squares (MWLS) approximation, and the local or various global formulations of boundary value problems. The multipoint technique leads to greater flexibility when compared with the FEM, provides *p*-type solution quality improvement, and may be used to solve various types of engineering problems. Due to its characteristic features, the method application especially for the nonlinear and multiscale analyses needs specific solutions, which are presented in the paper.

The paper is organized as follows. Section 2 describes the idea of the multipoint meshless FDM and its two versions. The general version is especially useful to solve various engineering problems including nonlinear ones. The basic steps of the multipoint MFDM analysis are presented in this part. The algorithm of the method application

D. Groen et al. (Eds.): ICCS 2022, LNCS 13353, pp. 55–68, 2022.
https://doi.org/10.1007/978-3-031-08760-8_5

to the two-scale analysis of heterogeneous materials, based on the multipoint MFDM characteristic features, is discussed in detail in Sect. 3. The general multipoint strategy in the nonlinear analysis is outlined in Sect. 4. Finally, a short summary and some concluding remarks are given.

2 The Fundamentals of the New Multipoint Meshless FDM

The idea of the multipoint method as the higher order FDM (translated as the "Hermitian method" in the English version of Collatz's book) was proposed by L. Collatz [5] more than sixty years ago, and forgotten since then due to complex calculations, too difficult to apply without the modern computational techniques. The original multipoint concept has been modified and extended recently [3] to the new multipoint meshless FDM, which is based on an arbitrary irregular cloud of nodes, the MWLS approximation [4] instead of the polynomial interpolation, and the local or various global formulations of the boundary value problems. This allows to obtain a fully automatic high quality solution as well as an error estimation tool. Generalized and unified description of the multipoint method is presented in the articles [2, 3]. The concept of the multipoint approach is based on raising the approximation order of the unknown function by using a combination of its values together with a combination of additional degrees of freedom at all nodes of a stencil.

The idea of the approach is illustrated by derived multipoint difference operators of the various approximation orders of the following type

$$\sum_{j(i)} c_j u_j = \sum_{j(i)} \alpha_j f_j \tag{1}$$

Here, i – is the central node number of the stencil consisting of nodes $P_j, j = 1 \ldots m, f$ – the additional degrees of freedom (d.o.f), c_j and α_j – coefficients. In this formula, a combination of the additional d.o.f. values *at each node* of the stencil (MFD star) is used instead of the function value f at the *central node only*, as it is in the classic (M)FDM solution approach.

The analysis of the method effectivity was done. As is depicted in Fig. 1, to obtain the required level of solution error, the multipoint method needs a decreased number of nodes when compared with the standard MFDM. Additionally, the convergence rate of derivatives obtained by the multipoint method has the same order as it is for the solution itself (phenomena of "superconvergence"). It is very important for engineering problems due to problem formulation posed in terms of the derivatives and unknowns (stresses, strains, etc.).

2.1 The General Approach of the Multipoint Method – Versions

There are two basic versions of the multipoint method – the *specific* one, where the values f of the considered differential equation right-hand side are assumed as the additional d.o.f. (1), and the *general* approach – the unknown values of selected function derivatives are used as the additional d.o.f.

$$\sum_{j(i)} c_j u_j = \sum_{j(i)} \alpha_j u_j^{(k)} \tag{2}$$

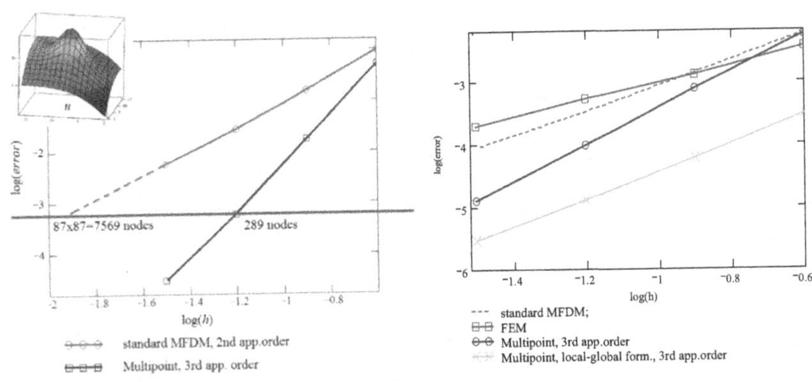

Fig. 1. Solution convergence for the multipoint method, the standard MFDM, and the FEM (Poisson's BVP with the different right-hand side)

Although the specific formulation is simpler and easier in implementation, its application is more restricted, mainly to the linear boundary value problem (BVP) posed in the strong formulation. The general formulation [2] is more complex but it can be used for all types of the BVP, including nonlinear ones, and all types of problem formulation – strong (local), weak (global), and mixed (global-local, MLPG). The global relation between the additional d.o.f. (unknown k-th order derivatives) and the basic ones.

$$\mathbf{U}^{(k)} = \mathbf{AU} \tag{3}$$

which follows from the multipoint general FD operator (2), is the key feature of the general version. Here $\mathbf{A}$ – is the coefficient matrix.

Having found such dependencies for any derivative emerging in the considered BVP, any differential equation can be approximated. In this way, the general multipoint approach can be considered as a tool for solving a wide class of engineering problems.

2.2 The Basic Steps of the Multipoint MFDM Analysis

The basic multipoint MFDM algorithm (described in detail in [2, 3]) is based on the meshless FDM scheme [4] and consists of the several steps (Fig. 2), which are listed below with some comments:

Selection of the Appropriate Boundary Value Problem Formulation:
local (strong), global (weak), or global-local (e.g. MLPG [6]). Global formulation may be posed in the problem domain Ω as the following variational principle

$$b(v,\ u) = l(v), \qquad \forall\ v \in V \tag{4}$$

as well as minimization of the energy functional given in the general form.

In the weak formulation (4), the trial u (searched problem solution) and test v function may be the same – it is the Bubnov-Galerkin approach, or different from each other – the Petrov-Galerkin one. In the last case, the multipoint method may be used also with various mixed global-local problem formulations. The MLPG5 (Meshless Local Petrov-Galerkin) global-local formulation [6], due to the use Heaviside-type test function may be computationally more efficient than the other formulations.

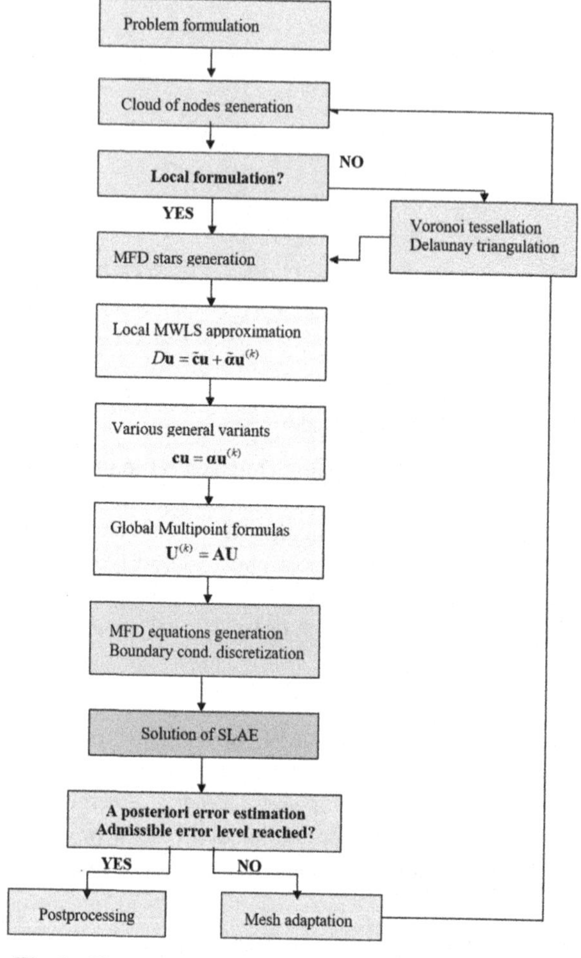

Fig. 2. The flow chart of the multipoint MFDM algorithm

Domain Discretization

- nodes generation (e.g. by applying Netgen [8], or the Liszka's type generator [7]);
- domain partition by Voronoi tessellation and Delaunay triangulation;
- domain topology determination (neighborhoods).

Only a cloud of nodes is needed in the case of local problem formulation. For global or global-local formulation, the integration and, consequently, the integration cells, like Voronoi polygons (subdomains assigned to individual nodes) [9] or Delaunay triangles, are needed.

Optimal MFD Star (Stencil) Generation

The MFD star (stencil) is the basis of the MWLS approximation and can be generated by using various criteria (Fig. 3), like the nearest distance (the simplest one), the cross (the optimal one), or the Voronoi neighbors criterion (best, but more complex) [4, 9].

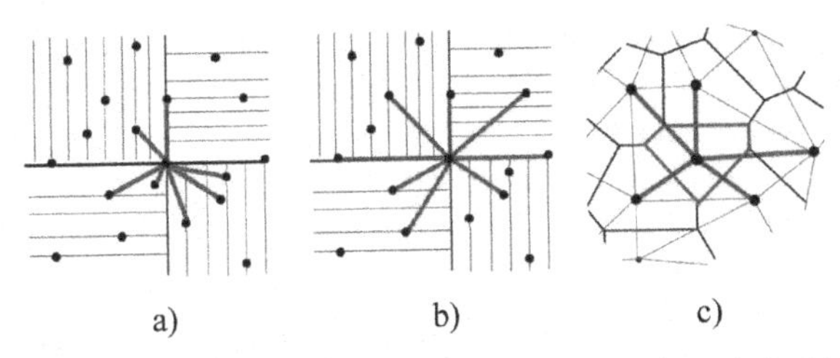

Fig. 3. Stencil criteria in 2D: a) nearest distance, b) cross criterion, c) the Voronoi neighbors

Local Approximation by Moving Weighted Least Squares (MWLS) Method
The important part of the multipoint MFDM algorithm is the computation of the FD operators using MWLS approximation. Construction of the local (based on the stencil) function approximation is performed by assuming appropriate d.o.f. at stencil nodes. The Taylor polynomial of degree p is considered in the MFDM [3, 4] as the basis function. Minimization of the weighted error functional $\partial J / \partial\{D\mathbf{u}\} = \mathbf{0}$ yields at each node the local multipoint MFD formulas for a set of derivatives $D\mathbf{u} = D\mathbf{u}_i = \left\{u_i, u_i^{'}, u_i^{''}, \ldots u_i^{(p)}\right\}$ and finally for the local basic operators (2).

Generation of the Multipoint Difference Formulas
Simultaneous equations $\mathbf{Cu} = \alpha\mathbf{u}^{(k)}$ are generated in the whole domain Ω afterward.

For the nonlinear problems or problems posed in global formulation, the approximation of unknown function u and its derivatives are provided by the *general* multipoint technique only. Several variants of the general multipoint approach may be used [2]. Finally, the solution provides the relation (3) formula for each $k = 1,\ldots n$ derivative.

Generation of the MFD Simultaneous Equations
Having found the FD formula (3) in the whole domain for all derivatives emerging in the considered BVP, these relations are applied to the given problem. After such discretization PDE depends on the primary unknowns u only.

Solution of the System of Linear Algebraic Equations (SLAE)
Error Estimation
In general, the error e of the obtained solution u is evaluated by the exact (analytical) one u^{E} (benchmark tests only): $e = u^E - u$; or in the case when it is unknown, by applying an improved solution u^H as the reference one $e^H = u^H - u$.

Due to the high quality of the multipoint result, it can be used to calculate the reference solution needed for a posteriori error estimation [12]. The p-type error analysis can be performed by comparison of the multipoint higher order solution with the one obtained by the second order MFDM. The error analysis may be applied for two purposes: an examination of the solution quality and, if required, the mesh refinement. The mesh can be easily locally modified by moving existing or entering new nodes, which is simpler in the meshless methods than in the FEM.

Postprocessing of the Results
At the postprocessing stage, after the nodal values of the displacement are obtained, the strains and stresses are evaluated. The advantage of the MFDM-based methods

is that difference formulas are generated at once for the full set of derivatives $\mathbf{Du}$, and the characteristic feature of the multipoint general case is the relation between the unknown function and its derivative (3). All derivative operators depend on the domain discretization only. Therefore, the stress and strain fields can be easily calculated by using coefficient matrices (3).

3 Multipoint Approach in the Two-Scale Analysis of Heterogeneous Materials

The higher order multipoint approach may be applied at the macro as well as the micro levels in the procedure of homogenization [11, 13] (Fig. 4). At the macro level, the heterogeneous material with a periodic distribution of inclusions was considered. The values of the effective material constants were determined for a single representative volumetric element RVE with defined material parameters for the matrix and inclusions.

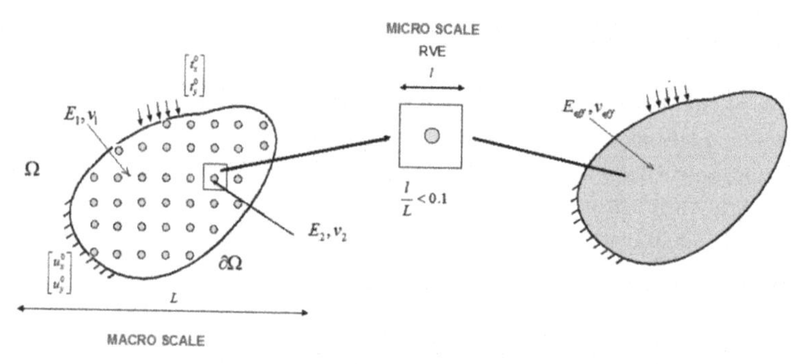

Fig. 4. Formulation of the homogenization problem

At the macro level, the problem for a heterogeneous or porous material may be posed in both weak and strong formulation, instead at the micro RVE level, the variational form (4) of the linear elasticity problem was assumed. In general, the homogenization solution approach based on two separate scales consists of the following steps [10]:

Discretization of the Domain

- initial mesh generation for both – the macro and microstructures;
- selection of the points at the macroscale (typically the Gauss points), for which the effective parameters will be evaluated at the microscale;
- determination of type and size of the RVE.

Microscale Analysis

- solution of the BVP at the RVE;
- computation of the effective material tensor $\mathbf{C}_{\text{eff}}$;
- verification of the multiscale analysis;
- error estimation.

Transfer of the Homogenized Effective Parameters to Gauss Points at the Macrostructure
Macroscale problem solution.
Postprocessing and error estimation.

The multipoint approach to computational homogenization was developed and described in detail in the article [11] taking into account the characteristic features of the method. The homogenization algorithm is presented in Fig. 5 with some comments added below.

Steps 3–7 (Fig. 5) regard both the macro and micro levels but have different details:

- At the macrostructure level, the selected *formulation* depends on the solved problem. At the micro level – the variational formulation (global or global-local) was assumed instead.
- On the *discretization* stage, the quality of the mesh (density, distribution of nodes) for the macrostructure is not so important and mostly depends on the geometry of the

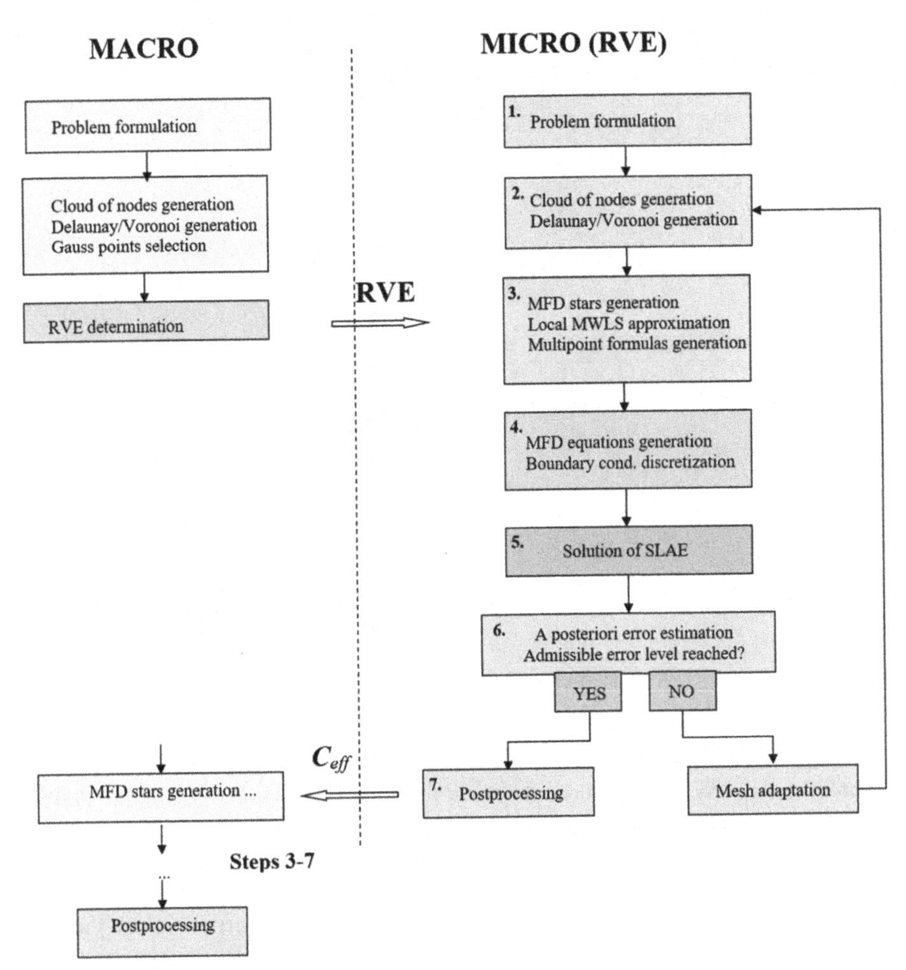

Fig. 5. The numerical homogenization algorithm

domain and solved problem. The RVE mesh may be generated either independently on the inclusion distribution or may be adjusted to them. In the latter case, a better result is expected, but such influence is not very significant for sufficiently fine mesh [11].

- *Selection and generation of the stencils* (MFD stars) for the local approximation is very important in the RVE domain of the multiscale problem. The stencils should be adjusted to the inclusion distribution [11]. It is significant due to the oscillations phenomenon occurring near the inclusion boundary in another case (Fig. 6). Their amplitude was significantly influenced by the Gauss point number used for numerical integration. To generate the optimal stencils adjusted to the inclusion, the nearest criterion (the simplest one) or the Voronoi neighbors (more complex) stencil generation have to be applied. Although the asymmetrical stencils cause a slightly worse approximation quality, they allow obtaining good results at the inclusion boundary.

  Inside the domain – at the main matrix of heterogeneous material or subdomains of inclusions, the best choice is to use the cross criterion due to its simplicity.

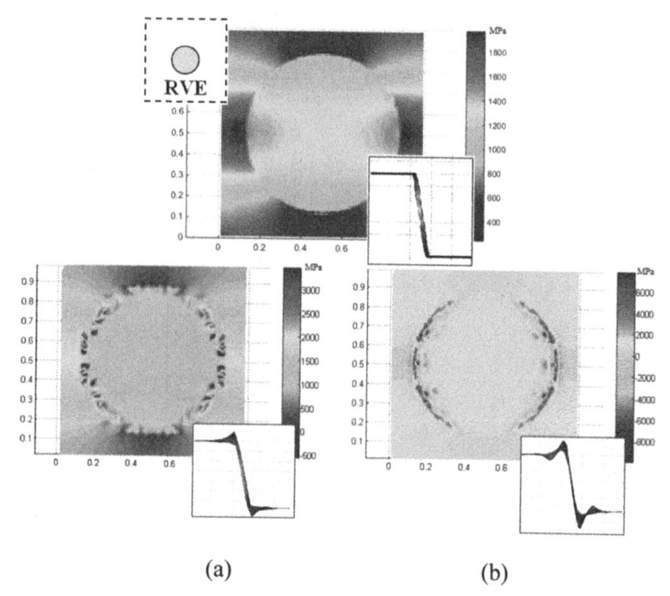

Fig. 6. Stress σ_{xx}: stencils adapted to the inclusion; stencils do not correspond to the inclusion with 1(a) and 3(b) Gauss point.

- In the case of local formulation (macrostructure) only trial function u has to be *approximated* by the MWLS method. In the weak formulation (the microscale), the Petrov-Galerkin approach (the trial u and test v function are different) was applied. The trial function u and its derivatives were approximated by the multipoint meshless FDM. The test function v and its derivatives at Gauss points can be calculated by the MWLS approximation based on the same stencils as u, or by simple interpolation on the integration subdomain, e.g. Delaunay triangle. In the preliminary tests, the type and order of approximation of test function v do not influence on results.

- Various versions of the general multipoint case can be applied for *difference formulas generation* [2]. The best way due to the necessity of adjusting the stencil to the inclusion at the RVE is version 4-"XY", where $u_{,x}$ as well as $u_{,y}$ derivatives are used as the additional d.o.f. in the Multipoint FD operator.
- For the local problem formulation (in the macroscale only), the *system of difference equations is generated* by the collocation technique. The *numerical integration* is additionally required in the case of the global formulation. Simultaneous algebraic equations are generated directly from the variational principle (5) by aggregation from each integration cell (Delaunay triangle or Voronoi polygon, which could be different from the test or trial function subdomains) and taking into account the multipoint FD operators.
- At the *postprocessing* stage the strain and stress volume averages can be evaluated and the effective values of the material parameter tensor $\boldsymbol{C}_{eff}$ are calculated.

The preliminary comparison of the effective material constants calculated by the MFDM algorithm, with the ones obtained by the FEM [17] as presented in Fig. 7, shows a more stable solution process in the case of the MFDM.

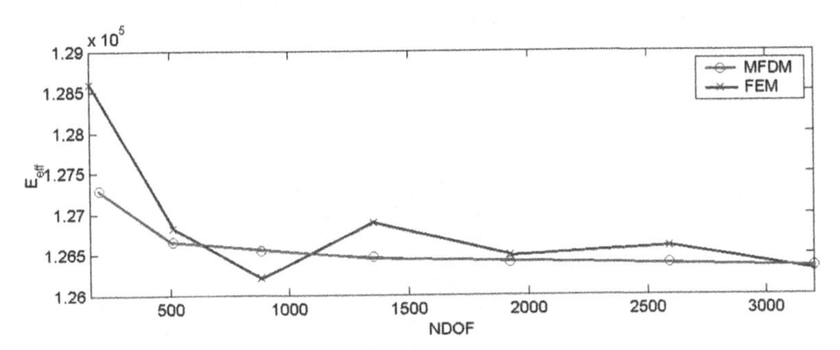

Fig. 7. Convergence of the effective Young's modulus obtained by the FEM and MDFM

4 The Nonlinear Analysis by the Multipoint MFDM

The multipoint MFDM has also been developed for the analysis of geometrically and physically nonlinear problems [14]. The strategy of the multipoint method for nonlinear analysis consists of the same parts as in the case of the linear BVP described above and several additional steps related to nonlinearity. The whole algorithm may be divided into two stages.

The initial one concerns the calculation of the primary derivative relations of type (3) in the whole domain. The higher order and mixed derivative dependencies (3) can be obtained from the first order derivatives by the formulae composition approach as is shown in the following examples

$$\mathbf{u}_x = \mathbf{A}\mathbf{u}, \quad \mathbf{u}_y = \mathbf{B}\mathbf{u} \quad \Rightarrow \quad \mathbf{u}_{xx} = \mathbf{A}^2\mathbf{u}, \quad \mathbf{u}_{xy} = \mathbf{A}\mathbf{B}\mathbf{u}, \quad \mathbf{u}_{yy} = \mathbf{B}^2\mathbf{u}, \quad \mathbf{u}_{xxy} = \mathbf{A}^2\mathbf{B}\mathbf{u}$$

The process of the calculation of the primary derivatives in the general multipoint approach is quite complex and computationally demanding [2], but it does not depend on the analyzed problem. Moreover, the relations (3) have to be computed only once for the assumed domain and discretization. This first part of the algorithm can be implemented in C++ or Fortran environment to obtain the coefficient matrices optimally.

Having found the global multipoint coefficient matrices for derivatives, the relations (3) can be applied to the given BVP in the second part of the algorithm. This part, strictly related to the analyzed nonlinear problem, may be easily computed in, e.g. Matlab environment. Due to its matrix-oriented programming language and extended graphic possibilities, Matlab is a perfect tool for this purpose. Besides the optimization of the computational time, this partitioning is especially useful for engineering tasks, where problem formulation may be changed but the discretization remains the same.

The nonlinear algorithm is based on the key features of the MFDM and especially of the higher order multipoint method. First of all, it is the independence of the FD operators from the problem formulation. Using this feature, the nonlinear problem formulation PDE may be easily rewritten in terms of unknown function only.

The Multipoint meshless FDM solution algorithm is presented in the flowchart (Fig. 8).

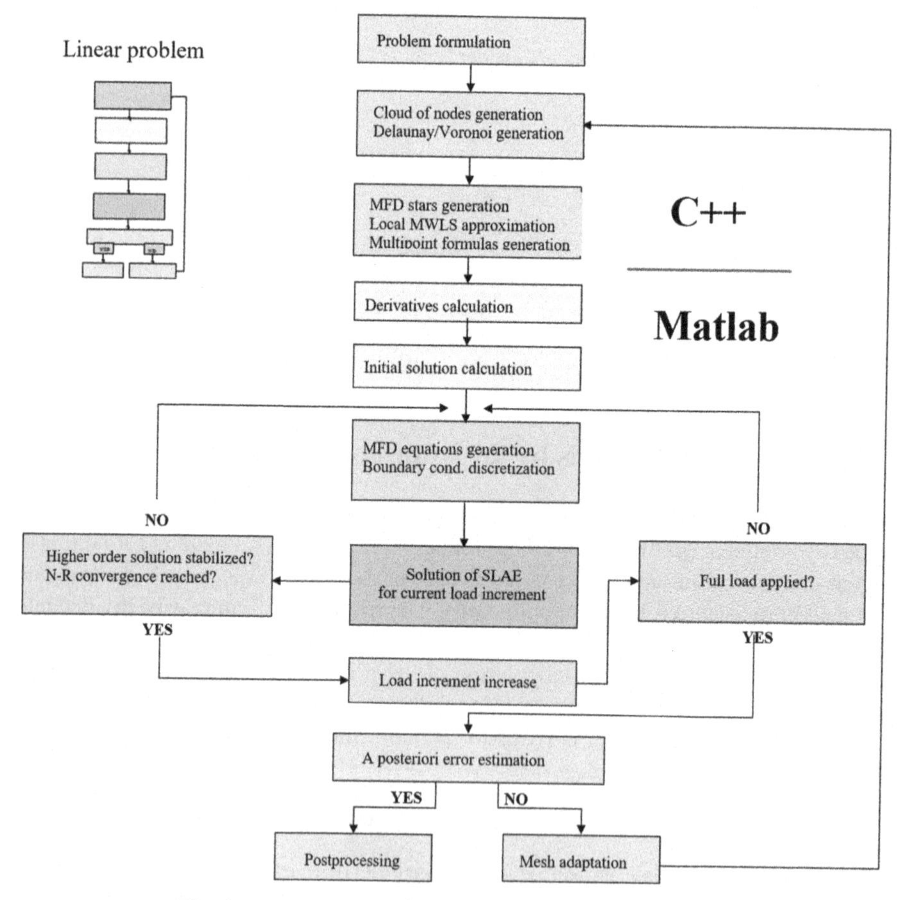

Fig. 8. The flow chart of the nonlinear analysis algorithm

4.1 Geometrically Nonlinear Problems

Obtained by the multipoint general approach simultaneous nonlinear algebraic equations can be solved by using an iterative technique. The most commonly applied procedure for this purpose is the Newton–Raphson method. The Jacobian tangent matrix on each iteration step has to be evaluated by the formula

$$\frac{\partial F_i}{\partial u_j} = \frac{\partial F_i}{\partial u_k}\frac{\partial u_k}{\partial u_j} + \frac{\partial F_i}{(\partial u_x)_k}\frac{(\partial u_x)_k}{\partial u_j} + \frac{\partial F_i}{\left(\partial u_y\right)_k}\frac{\left(\partial u_y\right)_k}{\partial u_j} + \ldots$$

The first part of the Jacobian matrix $\left(\partial F_i \middle/ \partial u_j^{(s)},\ \ s = 0, 1, \ldots\right)$ may be analytically calculated, e.g. by using the symbolic differentiation, while the second part (the derivatives $\partial u_j^{(s)} \middle/ \partial u_i$,) is evaluated numerically only once for the discretization. Appropriate multipoint difference operators (3) generated for the set of the partial derivatives can be used here. For example, if $\mathbf{u}_x = \mathbf{Au}$, $\mathbf{u}_y = \mathbf{Bu}$, then $\frac{\partial \mathbf{u}}{\partial \mathbf{u}} = \mathbf{I}$, $\frac{\partial \mathbf{u}_x}{\partial \mathbf{u}} = \mathbf{A}$, $\frac{\partial \mathbf{u}_y}{\partial \mathbf{u}} = \mathbf{B}$, where $\mathbf{A}$ and $\mathbf{B}$ – are the FD coefficient matrices, $\mathbf{I}$ is the identity matrix. In this way, the calculation time can be significantly reduced.

Several simple benchmarks and nonlinear engineering problems, including the deflections of the ideal membrane and analysis of the large deflection of plates using the von Karman plate theory, were tested. Two iterative approaches were used – the simple one using the staggered scheme (4th order von Karman problem), and the Newton-Raphson (NR) iterative scheme. The ideal membrane deflections problem was solved by the incremental-iterative approach and the NR method. Despite a fully random irregular cloud of nodes assumed, a high quality almost axially symmetric solution was obtained (Fig. 9). The 14 pressure increments and only 73 iteration steps were needed to reach the required level of the error (assumed 10^{-12}) by the 3rd order general multipoint approach. It is a very encouraging result. The standard MFDM (using relaxation modification of NR algorithm [15]) requires twice more iteration steps to reach the appropriate level of error.

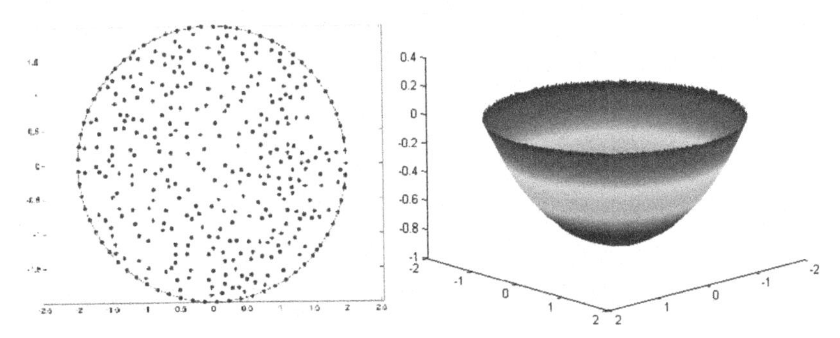

Fig. 9. Random irregular mesh and numerical solution for ideal membrane deflection

4.2 Physically Nonlinear Problems

In the case of physically nonlinear problems in elastic-plastic analysis, the iterative procedure may lead to increased errors in the neighborhood of the elastic-plastic interface, and the incremental-iterative approach is especially useful. The multipoint operators can be calculated only once at the initially assumed discretization, or the solution may be improved by using the adaptation process [16, 17] based on the error estimation near the elastic-plastic boundary.

The oscillation problem similar to the situation in the homogenization analysis was expected to occur in the physically nonlinear analysis along the elastic-plastic boundary. However, the preliminary numerical results do not demonstrate the oscillations. This can be due to the different types of problem formulation (the weak one in the numerical homogenization), might be related to the type of discontinuity or other discretization factors, and needs further investigation.

The additional part of the nonlinear strategy in the case of elastic-plastic analysis is presented in the scheme (Fig. 10).

The physically nonlinear de Saint-Venant's problem of a prismatic bar elastic-plastic torsion has been analyzed by the multipoint MFDM and outlined in the paper [14]. The boundary between elastic and plastic parts was determined by using the Nadai "roof" approach [18]. The problem was solved by the Newton-Raphson method. In Fig. 11 (b) the elastic-plastic boundaries for various increasing loads θ of square-shape cross-section bar are presented. The Prandtl stress function of the torsional problem is depicted in Fig. 11(a).

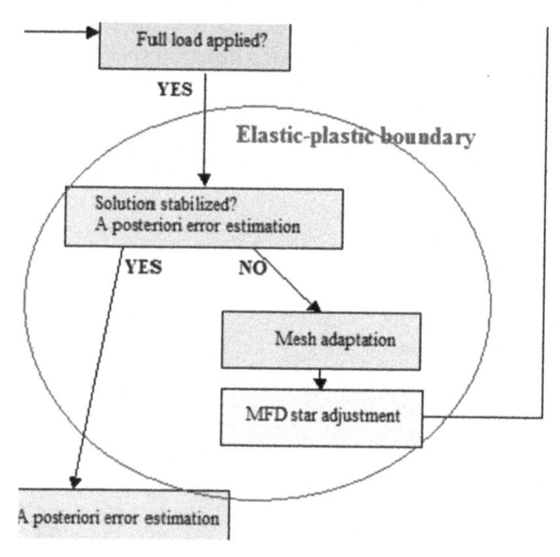

Fig. 10. An additional part of the multipoint nonlinear strategy for elastic-plastic analysis

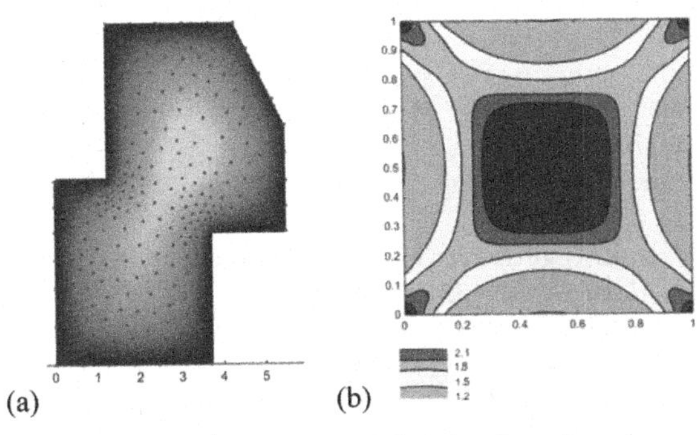

Fig. 11. Prismatic bar torsion: (a) the Prandtl stress function for polygon cross-section; (b) elasto-plastic boundaries of the square cross-section for loads $\theta = 1.2, 1.5, 1.8$ and 2.1

5 Final Remarks

The paper introduces the application schemes of the new higher order multipoint meshless method for more demanding engineering problems such as numerical homogenization and nonlinear analysis. The appropriate algorithms were developed, characteristic features of the method application for particular types of engineering problems were examined, and necessary procedures were written and tested.

Like the other methods, the multipoint meshless approach has its advantages and disadvantages. They were examined, tested, and analyzed. The following characteristic features belong to the advantages of the multipoint method application to the analyzed problems:

- the possibility of using various formulations of BVP (weak, strong, mixed);
- modification and adaptation of whole mesh or only part of them without difficulties;
- higher order approximation and improved precision of the obtained solution on the arbitrarily irregular meshes, which allows for decreased number of nodes;
- the difference operators for the particular derivatives are generated besides the solution at once without additional computational cost;
- the multipoint FD operators depend on the domain discretization rather than on the analyzed specific problem;
- convergence rate of solution derivatives has the same order as it is for the solution itself (phenomena of "superconvergence");
- the general approach of the multipoint MFDM allows for the analysis of various problems including nonlinear ones;
- in nonlinear analysis – computationally more efficient updating and evaluation of the tangent matrices needed in iteration approach;
- the multipoint solution may be used as the improved reference solution instead of the true analytical one for a posteriori error estimation.

In subsequent research, it is planned to combine both the nonlinear analysis and heterogeneous material homogenization and apply it for numerical modeling in e.g. concrete mechanics.

References

1. Chen, J.-S., Hillman, M., Chi, S.-W.: Meshfree methods: progress made after 20 years. J Eng Mech **143**(4), 04017001 (2017)
2. Jaworska, I.: On the ill-conditioning in the new higher order multipoint method. Comput. Math. Appl. **66**(3), 238–249 (2013)
3. Jaworska, I., Orkisz, J.: Higher order multipoint method – from Collatz to meshless FDM. Eng. Anal. Bound. Elem. **50**, 341–351 (2015)
4. Orkisz, J.: Finite Difference Method (part III). In: Handbook of Computational Solid Mechanics, Springer-Verlag, 336–432 (1998)
5. Collatz, L.: Numerische Behandlung von Differential-gleichungen, Springer-Verlag (1955)
6. Atluri, S.N.: The Meshless Method (MLPG) for Domain & Bie Discretizations (2004)
7. Liszka, T.: An automatic grid generation in flat domain. Mech. i Komp **4**, 181–186 (1981)
8. Schöberl, J.: NETGEN An Advancing Front 2D/3D-mesh generator based on abstract rules. Comput. Vis. Sci. **1**, 41–52 (1997)
9. Preparata, F.P., Shamos, M.I.: Computational geometry. Springer-Verlag (1985)
10. Jaworska, I., Milewski, S.: On two-scale analysis of heterogeneous materials by means of the meshless finite difference method. Int. J. Multiscale Comput. Eng. **14**(2), 113–134 (2016)
11. Jaworska, I.: Higher order multipoint meshless FDM for two-scale analysis of heterogeneous materials. Int. J. Multiscale Comput. Eng. **17**(3), 239–260 (2019)
12. Jaworska, I., Orkisz, J.: Estimation of a posteriori computational error by the higher order multipoint meshless FDM. Comput. Inform. **36**, 1447–1466 (2017)
13. Geers, M.G.D., Kouznetsova, V.G., Brekelmans, W.A.M.: Multi-scale first-order and second-order computational homogenizastion of microstructures towards continua. Int. J. Multiscale Comput. Eng. **1**(4), 371–385 (2003)
14. Jaworska, I., Orkisz, J.: On nonlinear analysis by the multipoint meshless FDM. Eng. Anal. Bound. Elem. **92**, 231–243 (2018)
15. Milewski, S., Orkisz, J.: In search of optimal acceleration approach to iterative solution methods of simultaneous algebraic equations. Comput Math Appl **68**(3), 101–117 (2014)
16. Demkowicz, L., Rachowicz, W., Devloo, P.: A fully automatic hp-adaptivity. J. Sci. Comput. **17**(1–4), 117–142 (2002)
17. Oleksy, M., Cecot, W.: Application of HP-adaptive finite element method to two-scale computation. Arch. Comput. Methods Eng. **22**(1), 105–134 (2015)
18. Nadai, A.: Der Beginn des Fliesvorganges in einem tordierten Stab, ZAMM3, 442-454 (1923)

Automating and Scaling Task-Level Parallelism of Tightly Coupled Models via Code Generation

Mehdi Roozmeh(✉) and Ivan Kondov

Steinbuch Centre for Computing, Karlsruhe Institute of Technology,
Hermann-von-Helmholtz-Platz 1, 76344 Eggenstein-Leopoldshafen, Germany
{mehdi.roozmeh,ivan.kondov}@kit.edu

Abstract. Tightly coupled task-based multiscale models do not scale when implemented using a traditional workflow management system. This is because the fine-grained task parallelism of such applications cannot be exploited efficiently due to scheduling and communication overheads. Existing tools and frameworks allow implementing efficient task-level parallelism, however with high programming effort. On the other hand, Dask and Parsl are Python libraries for low-effort up-scaling of task-parallel applications but still require considerable programming effort and do not equally provide functions for optimal task scheduling. By extending the wfGenes tool with new generators and a static task graph scheduler, we enhance Dask and Parsl to tackle these deficiencies and to generate optimized input for these systems from a simple application description and enable rapid design of scalable task-parallel multiscale applications relying on thorough graph analysis and automatic code generation. The performance of the generated code has been analyzed by using random task graphs with up to 10,000 nodes and executed on thousands of CPU cores. The approach implemented in wfGenes is beneficial for improving the usability and increasing the exploitation of existing tools, and for increasing productivity of multiscale modeling scientists.

Keywords: Scientific workflow · Tightly coupled model · Task-based parallelism · Code generation · Scalability · Productivity

1 Introduction

Exascale computing has a high potential for multiscale simulation in computational nanoscience. Due to the limited physical scalability and the large number of instances of sub-models, as well as the complexity of the couplings between sub-models of different scales, exploiting exascale computing in this domain is still a challenge. Particularly difficult is the rapid and scalable design of novel *tightly coupled* multiscale applications for which the domain scientists urgently need tools that facilitate the rapid integration of sub-models while maintaining high scalability and efficiency of the produced applications.

For *loosely coupled* multiscale applications, scientific workflows and workflow management systems (WMSs) have been established solutions. Thereby, the

D. Groen et al. (Eds.): ICCS 2022, LNCS 13353, pp. 69–82, 2022.
https://doi.org/10.1007/978-3-031-08760-8_6

underlying models are usually wrapped by scripts or Python functions and so integrated as nodes and tasks into a workflow, while the couplings are represented by dataflow links. The workflow nodes of such applications can be executed as separate jobs on one or more high performance computing (HPC) clusters.

Recently, we have demonstrated [16] that workflows for different WMSs can be automatically generated from a single abstract description, WConfig, and have provided a proof-of-concept implementation in the wfGenes tool [11]. The WConfig input file, written in JSON or YAML format, is a simple description of a scientific workflow, that specifies in an arbitrary order all functions with their input parameters and returned objects by unique global names. The wfGenes tool has been first employed in the settings of a collaborative project [1] in which the participants use two different WMSs, FireWorks [10] and SimStack [2], to perform multiscale simulation in nanoscience. This often requires defining a set of simulation workflows in two different languages simultaneously. Previously, we have shown how to generate a workflow to compute the adsorption free energy in catalysis for these two WMSs by using wfGenes [16]. Such a workflow is prototypical for *loosely coupled* applications, where the number of data dependencies is small in relation to the average execution time of single tasks. However, there is another class of multiscale models [6,8] in which the ratio between the number of data dependencies and the average task execution time is very high. These *tightly coupled* models are usually implemented in a program running as a single job on an HPC cluster in order to minimize the task scheduling overheads and the times for data transfers between dependent tasks.

In previous work [8], we implemented and optimized a tightly coupled multiscale model describing charge and exciton transfer in organic electronics. Thereby, we used the Python language to integrate electronic structure codes, such as Turbomole and NWChem, and the mpi4py package [7] to schedule the tasks and data transfers. We found that the use of Python greatly facilitated the adoption of the Message Passing Interface (MPI) through mpi4py as well as the integration of simulation codes in the domain of computational materials science that are provided as Python application programming interfaces. Nevertheless, we found that this approach had several disadvantages: i) The development effort was high due to lack of specific semantics for modeling application's parallelism. ii) The task graph and the task execution order had to be produced manually before starting the application. iii) The lack of domain-specific semantics prevented code reuse in other applications. The developer had to repeat the cumbersome procedure "from scratch" to design new task-parallel applications using that approach. More recently, Dask [15] and Parsl [3,4] have provided semantics for writing implicitly parallel applications in Python by decorating particular objects. While being powerful for rapid design of parallel applications in Python, Parsl and Dask have not been designed as typical WMSs from their outset.

In this work, we extend wfGenes to generate Python input code for Parsl and Dask starting from an existing workflow description. This is extremely beneficial for the use cases where the workflow model is not available as a graph and/or the developer is familiar neither with these tools nor with Python. We find that the newly integrated task graph scheduler works with two different executor strategies of Dask and Parsl, lazy evaluation and immediate execution, respectively, and enables maximum level of parallelism while preserving the functionality of generated code. In the next Sect. 2, we provide an overview of related work. In Sect. 3, we provide some implementation insights into the new features, the generators for Dask and Parsl, and the static task graph scheduler. We employ the thus extended wfGenes tool to generate executable Python code from task graphs representative for a tightly coupled application and measure the parallel performance in Sect. 4. In Sect. 5 we summarize the paper.

2 Related Work

Numerous frameworks, tools and WMSs can be employed in task-based multiscale computing. In principle, dedicated environments, such as the domain-specific Multiscale Modeling and Simulation Language (MMSL) and the Multiscale Coupling Library and Environment (MUSCLE) can be adopted for any applications of multiscale modeling and computing applications, in particular for solving tightly coupled problems (see Ref. [18] and the references therein). In this work, we pursue a more general concept allowing to use the same tools also in a high-throughput computing context [17] and to address additional domain-specific requirements from computational nanoscience and virtual materials design. Therefore, we identify task-based parallel computing as a common concept in this more general context of usage.

Task-based parallel applications can be implemented in many different ways. The most preferred approach to do this in HPC is based on established standards such as MPI and OpenMP. The MPI standard [12] is only available for C and Fortran, and for other languages available via third-party libraries, e.g. Boost [5] for C++ and mpi4py [7] for Python. Although MPI provides a powerful interface allowing to implement any kind of parallelism, there are no specific definitions to support task-based parallelism directly. Starting from version 3, OpenMP [14] provides support for task-based parallelism based on compiler directives. However, OpenMP only supports shared-memory platforms and C, C++ and Fortran languages and requires support by the corresponding compiler.

There are domain-specific languages for writing task-based parallel computing applications. Swift [19] is a domain-specific language enabling concurrent programming to exploit task and data parallelism implicitly rather than describing the workflow as a static directed acyclic graph. A Swift compiler translates the workflow for different target execution backends. For instance, Swift/T [20] provides a Swift compiler to translate code for the dataflow engine Turbine [21] that uses the asynchronous dynamic load balancer (ADLB) based essentially on MPI. Another domain-specific language, Skywriting [13] also provides semantics for task parallelism and for handling dataflow.

Starting from version 3.2, the Python standard library provides the `concurrent` package to facilitate task parallelism through an abstract interface allowing asynchronous execution of the same Python code on different backends, e.g. implemented in the `multiprocessing` and `mpi4py` [7] packages. However, these can be regarded as execution engines for tasks that have to be scheduled according to their data dependencies and resource requirements.

3 Implementation

3.1 wfGenes Architecture

The wfGenes implementation has been described in detail in previous work [16]. For a better understanding of the new features introduced below, we will here briefly outline the basic stages that are shown in Fig. 1. First, the model designer authors an abstract workflow description (WConfig) in that all details about the concrete backend system implementation are left out. After that, the WConfig is validated and analyzed, and a task graph of the application is created including the dependencies describing the control flow and the dataflow. In addition, wfGenes extracts various properties of a directed acyclic graph relying on multilevel analysis in the following two steps. First, a thorough dependency analysis is performed using join operations over the input/output lists of the nodes. Second, the depth-level–breadth structure of the graph, discussed in Sect. 4, is measured by counting the dependent partitions of the graph, the depth levels, and total number of parallel tasks in each depth level, called breadth. Finally, a workflow in the language of the target backend system is automatically generated and validated against a schema provided from the target backend system. In our previous work [16], generators for two different backend systems, FireWorks [10] and SimStack [2], have been implemented as a proof of concept.

3.2 Task-Level Parallelism

The graph analysis stage, shown in Fig. 1, includes a node-level global dependency analysis and a task-level local dependency analysis that enable graph-aware code generation in wfGenes [16]. Combining the results of these two analyses allows automated optimization of the granularity of the workload towards increasing the degree of parallelism of the generated input code. This is carried out through a transformation of the workflow graph as depicted in Fig. 2. The original WConfig describes two logical nodes, A and B, that are strictly sequential, each containing a group of tasks that can be scheduled in parallel, as shown in Fig. 2a. Now, if there is no need to schedule Node A and Node B on different resources, for example on different computing clusters or in different batch jobs, then Tasks 4 and 5 can be scheduled much earlier, as shown in Fig. 2b. In this way, the degree of parallelism can be increased by automatically replacing the node-level with task-level granularity without having to modify the original user input in WConfig. The maximum degree of parallelism that can be realized in Fig. 2a is two while it is three in Fig. 2b. It is noted that the order of execution, shown in Fig. 2b as rows from top to bottom, is for the maximum degree of parallelism that can be achieved if sufficient computing resources are provided.

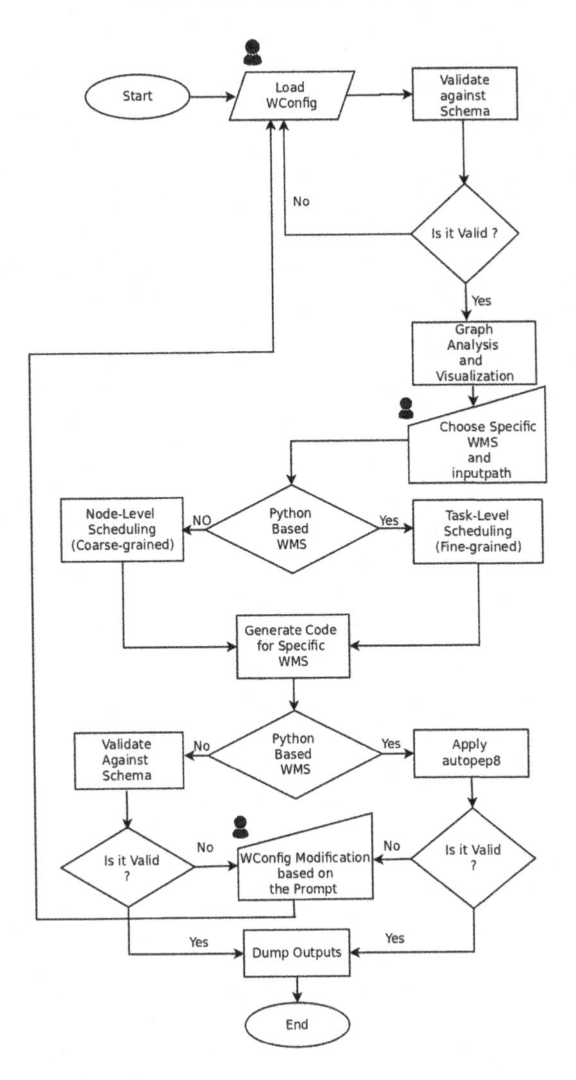

Fig. 1. Flowchart of wfGenes process stages

3.3 Task Scheduling

Parsl and Dask resemble each other in many different aspects, however, the differences become more apparent with the task scheduling. A proper scheduling strategy for any task-parallel application relies on the task graph. Both Dask and Parsl are equipped with mechanisms to extract the task graph of the application model from decorated Python code. Although these decorators have different syntax, they both enable an underlying scheduler to perform static analysis and to direct the execution toward parallel computing. While the stages of task graph generation and the task scheduling are logically distinct, they are not well

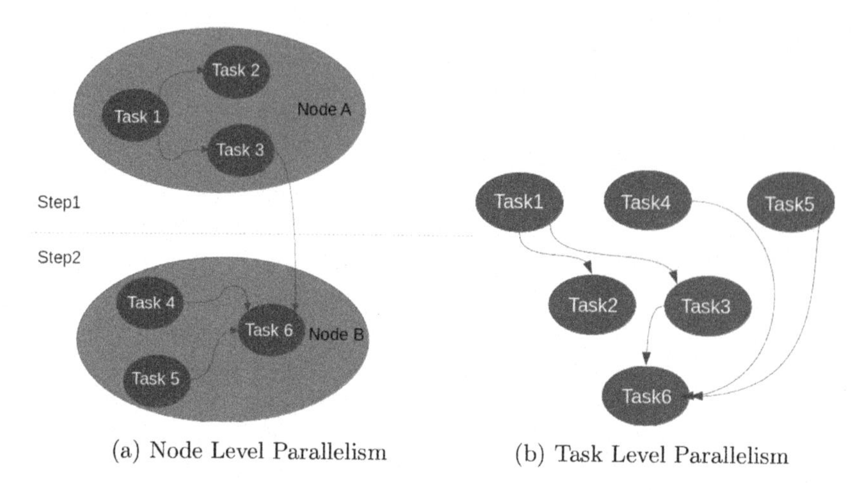

(a) Node Level Parallelism (b) Task Level Parallelism

Fig. 2. Transforming the workflow granularity in order to exploit the maximum possible parallelism.

separated in time. In particular, Parsl schedules tasks *on the fly*, i.e. during the scheduling and execution stages. In other words, Parsl's executor performs immediate evaluation of the delayed execution objects (futures) when the `result()` method is called and blocks further computation until all tasks within the call are completed before it proceeds with tasks defined further in the python code after this call. This behavior may lead to unwanted idle workers since the scheduler cannot launch any tasks after this barrier before synchronization. On the other hand, Dask performs lazy evaluation of deferred execution objects after constructing the relevant portion of the task graph by applying the `compute()` method to these objects. This strategy is problematic for computations with task graphs that evolve at run time, i.e. dynamic workflows. In particular, Dask lazy evaluation objects cannot be used in loop boundaries or in conditional statements. These problems are resolved in wfGenes by ordering the task calls in a way i) to avoid unnecessary barriers when using Parsl and ii) to obtain results of delayed execution objects when they are needed to dynamically branch the data flow when using Dask. This is done in the generation phase using the task graph produced by the newly integrated task graph scheduler.

3.4 Task Graph Scheduler

To circumvent these scheduling and dependency handling issues, we have extended wfGenes with an independent task graph scheduler that is switched before the code generator, as is shown in Fig. 1. The task ordering and grouping technique implemented in the task scheduler enables Parsl's executor to launch the maximum number of independent tasks in parallel and avoids unnecessary barriers during execution.

Building the task graph prior to code scheduler allows exploiting fully the task parallelism. In the case of Python code generated for Dask, the wfGenes' task scheduler is not necessary since Dask extracts the dependency information from the Python code with no additional effort from developer's side. Here, we use Dask's first generation scheduler which is based on lazy evaluation of the task graph prior to execution. The second generation of Dask executors uses immediate execution of callables that resembles the strategy for Parsl used in this work. Nevertheless, for execution of dynamic workflows, the two tools offer no low-effort solution since the user must implement explicit barriers to assure synchronization.

Figure 3 shows the execution patterns of two different versions of Python code produced by wfGenes for Parsl with disabled and enabled wfGenes' task graph scheduler. In the case of disabled scheduler (Fig. 3a) almost all tasks are executed sequentially. With enabling the scheduler (Fig. 3b) independent tasks are run in parallel. Although there is some overhead leading to increased pending times, the total running time is about twice shorter. This demonstrates the benefit from wfGenes' task graph scheduler in terms of improved parallel speedup when code for Parsl is generated.

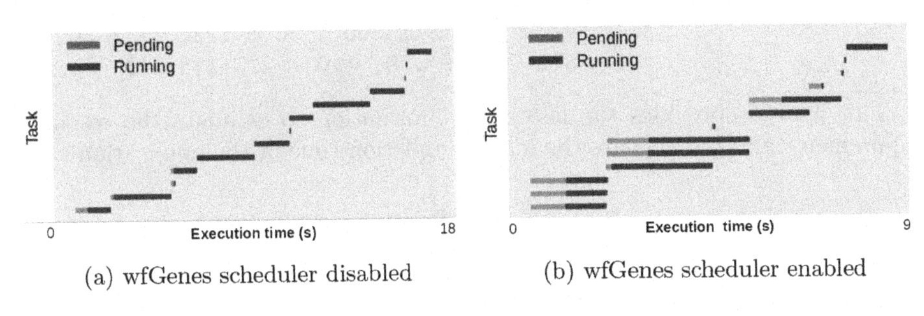

(a) wfGenes scheduler disabled

(b) wfGenes scheduler enabled

Fig. 3. Parallel execution with Parsl of code generated from a task graph including 13 tasks with maximum concurrency of three tasks with wfGenes' task graph scheduler turned off and on. For better visualization, Fig. 3b is zoomed by a factor of two.

4 Use Case

In this section, we address a use case of a tightly coupled multiscale model that is realized as a fine-granular task-parallel application. The up-scaling of such a multiscale application on HPC clusters is a highly non-trivial scheduling problem as it has been shown for computing charge carrier transport properties in organic semiconductors [8]. The scheduling strategy in Ref. [8] has been implemented with huge effort using the state-of-the-art techniques and tools. Although this application can now be implemented with significantly less effort using Dask and Parsl, that have become available in the mean time [3,4,15], we are showing here an approach with a much better usability, in that the Python code for Dask and Parsl does not have to be manually written but can be generated from a simple workflow description.

Here we are not going to repeat the specific application in Ref. [8]. Instead, we use several randomly generated task graphs that have use-case specific characteristics relevant for the performance and scalability of the task-based multiscale application. These characteristics are the total number of tasks, the number of parallel tasks and the task execution times. Such an approach allows us to generalize the results of our experiments beyond the domains of the available application use cases by varying these parameters. To this end, a random graph generator [9] is used to produce random task graphs that are then converted into valid WConfig workflow descriptions, that in turn are used to generate input for FireWorks, Dask and Parsl.

In order to simulate the execution time, we call a sleep function in every task. As we use random graphs, the ideal completion time depends on the structure of the graph and the maximum number of independent nodes at each depth level. The depth levels are the points in time when one or more nodes are scheduled for execution. The breadth at a depth level is the maximum number of nodes that can be scheduled in parallel. wfGenes inspects the WConfig and reports the graph structure and the breadth at each depth level. For example, the graph used in the measurements has the following depth-level–breadth structure:

```
depth_level-breadth: {'1': 2952, '2': 3890, '3': 1881,
                      '4': 769, '5': 300', '6': 124, '7': 50,
                      '8': 23, '9': 9, '10': 2, '11': 1}
```

This analysis provides the necessary information to estimate the resource requirements and to calculate the ideal completion time of the application

$$T_{\text{total}} = T_{\text{task}} \sum_{d=1}^{D} \lceil \frac{B_d}{P} \rceil \tag{1}$$

where D is the maximum depth level of the graph, B_d is the breadth, i.e. the maximum number of parallel tasks at depth level d, P is the number of available processing elements, that is here the number of CPU cores, and T_{task} is the average task execution time. For calculating the parallel speedup we need the time for running the same task graph sequentially. To this end, we have used T_{task} multiplied by the total number of tasks. This estimated time is in a good agreement with the measured sequential execution time.

4.1 Measurement Results

A task graph with 10,000 nodes has been produced and transformed to WConfig as described in the previous section. Afterwards, separate inputs for FireWorks, Parsl and Dask have been generated. The workflows have been executed on different number of processor cores on a node of the HPC systems bwUniCluster (with two Intel Xeon Gold 6230 processors) and HoreKa (with two Intel Xeon Platinum 8368 processors). For every run, the parallel speedup has been calculated by dividing the measured total running time by the total time of sequential execution, i.e. on one worker.

In Fig. 4 the speedup on all 80 hardware threads of a bwUniCluster compute node is shown for different task execution times. Dask shows an almost constant speedup of about 60 for all task execution times due to minimum monitoring effort on parallel workers. In contrast to Dask, FireWorks has a very low speedup of around 7 for the task graph with execution time of 1 s. The speedup is improved with increasing the execution time. This can be explained with the communication overheads due to queries to a remote MongoDB database performed by the FireWorks executor. Very short task execution times become comparable to the times of these queries and the latter limit the scalability of parallel execution with the FireWorks executor. With sufficiently long task execution times, FireWorks' speedup is comparable to that of Dask and Parsl, as shown in Fig. 4.

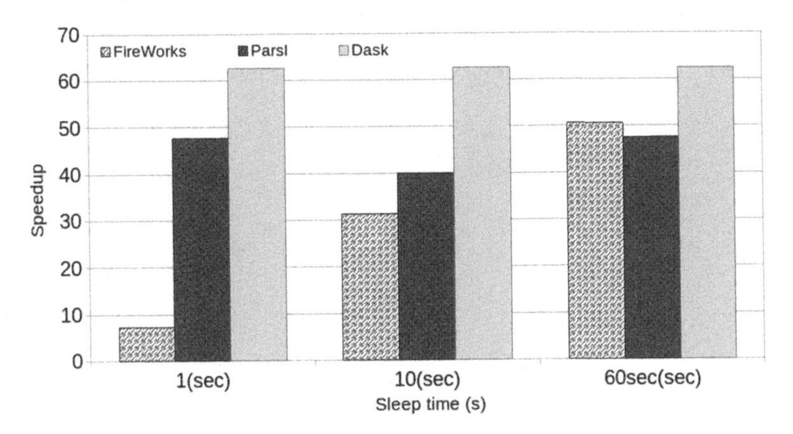

Fig. 4. Parallel speedup of FireWorks, Parsl and Dask for varying task execution times using 80 workers

Furthermore, in Fig. 5 we consider the speedup with varying the number of workers. In addition to *balanced* task graphs, in which the task execution times of all tasks are either 1 or 10 s, we consider *unbalanced* task graphs in which 10% of the tasks have long execution times and 90% have short execution times. Mixing two types of tasks with different execution times in similar ratios is prototypical for multiscale task-parallel applications for computing charge carrier transport properties in organic semiconductors [8]. In Figs. 5a and 5b, we show the performance for a balanced task graph with 10,000 tasks for 1 and 10 s task execution times, respectively, by measuring the total running times on 1 up to 32 HoreKa compute nodes using 128 workers per compute node. In addition, we measure the performance for unbalanced task graphs: one in which 10% and 90% of the tasks run 10 s and 1 s each, respectively (Fig. 5c) and one in which 10% and 90% of the tasks run 100 s and 10 s each, respectively (Fig. 5d).

In all measurements, the running time decreases with the number of workers due to parallel execution of independent tasks. After a certain number of workers, e.g. 1000 in Fig. 5a, no more time gain can be observed because no more tasks are available for parallel execution. The same limit holds also for the ideal time shown in Fig. 5 calculated using Eq. (1). In the cases of balanced workloads, Dask exhibits a slightly better performance than Parsl and a very similar scaling with the number of workers running on up to 32 nodes. In the case of a balanced workload with long task execution time shown in Fig. 5b, the measured running times with both Dask and Parsl are virtually the same as the ideal time.

With unbalanced workloads, Parsl shows better performance and overall speedup with respect to Dask. The measured time with Dask for the unbalanced workload in Fig. 5c does not improve any more already with 100 workers and gives rise to a flat plateau-like dependence that is due to Dask's executor implementation. Dask's executor cannot scale as good as Parsl due to a local limit of the SLURM workload manager not allowing more than 64 concurrently queued jobs. Therefore, no measurement with Dask is available in Fig. 5d. This problem is circumvented in the case of Parsl by submitting a single job allocating several nodes that can be efficiently utilized by Parsl's executor, as depicted in

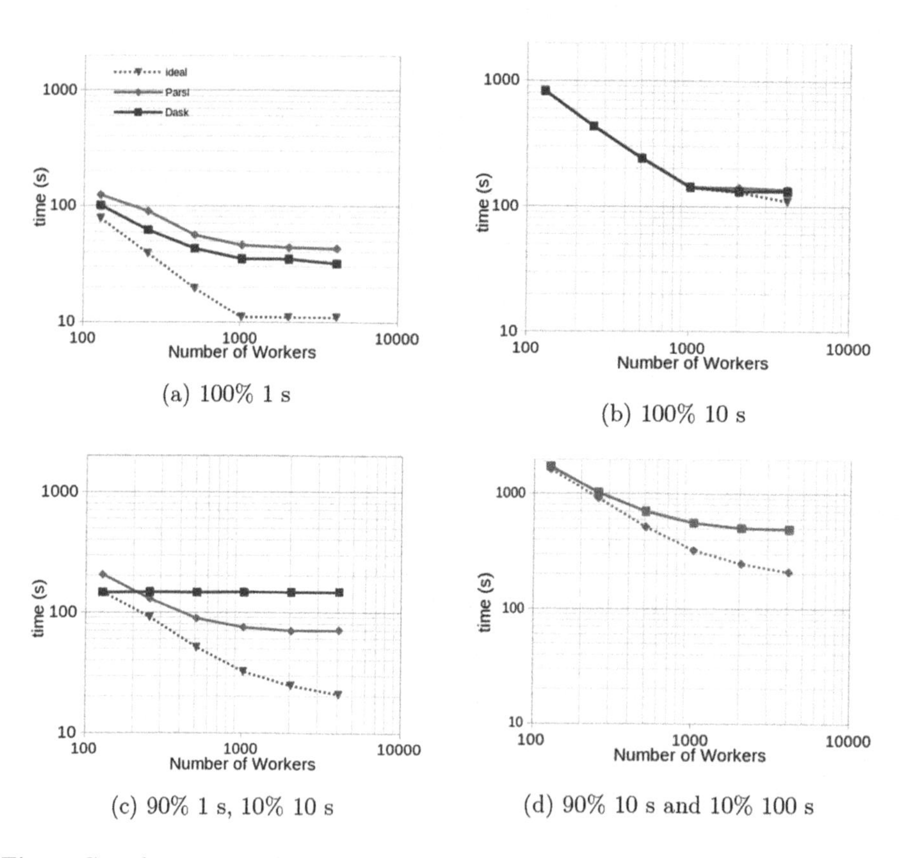

(a) 100% 1 s

(b) 100% 10 s

(c) 90% 1 s, 10% 10 s

(d) 90% 10 s and 10% 100 s

Fig. 5. Completion time of a balanced (a, b) and an unbalanced workload (c, d). The ideal time is calculated by Eq. (1).

Fig. 5d. The workers in Fig. 5d are uniformly distributed over up to 256 HoreKa compute nodes while in Figs. 5a, 5b and 5c the scaling is performed on up to 64 nodes.

Figure 6 depicts the efficiency (%) of parallelization that is here defined as the ratio of the measured total running time with the ideal time defined in Eq. (1). The best parallel efficiency with both Dask and Parsl is achieved for balanced workloads with long task execution times (10 s) since the communication latency has less impact on the overall performance. Furthermore, Dask has overall better efficiency than Parsl that is more pronounced for the workload with short task execution times (1 s).

With unbalanced workloads, the efficiency is generally reduced presumably due to a load imbalance that cannot be mitigated at run time due to the lack of dynamic load balancing. Communication latency times and scheduling overheads do not seem to be the only reasons for the reduced performance. This can be seen in the measurement of the same unbalanced task graph with ten times larger task execution times (in Figs. 6, blue dotted line) for which the parallel efficiency is improved by up to 25% but still 30% lower than the efficiency of the balanced case with the long task execution time. Strikingly, the code for the Parsl executor generated with wfGenes task graph scheduler shows higher efficiency with unbalanced workloads. In contrast, Dask shows poor scalability with the number of workers for the unbalanced workload.

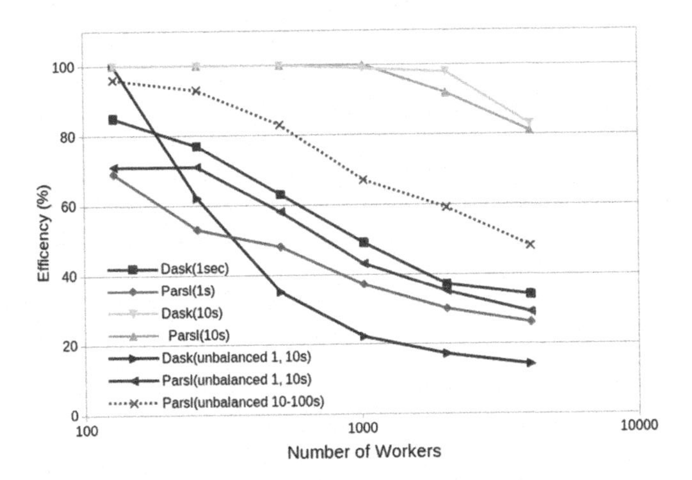

Fig. 6. Parallel efficiency with Dask and Parsl for task graphs with variations of the task execution times (Color figure online)

A similar measurement comparing FireWorks, Dask and Parsl has been performed in a recent study [3]. However, every workload used in Ref. [3] includes independent tasks with equal execution times. Here, we investigate the effect of dependencies, using a random task graph, and of the imbalance on the performance and scalability. In addition, the task execution ordering in the generated code for Parsl has been automatically optimized employing the wfGenes task graph scheduler. In the most simple case of a balanced workload our results are in agreement with Ref. [3]. In the case of Parsl, the `HighThroughputExecutor` over SLURM has been used offering the high throughput execution model for up to 4000 nodes. In the case of Dask, the SLURM scheduler shows the best performance characteristics for these use cases.

5 Conclusion

Tightly coupled multiscale models often exhibit task-based parallelism with fine granularity. On the other hand, the complexity of the models and of the available tools limits developer's productivity in the design of novel applications and deployment on high performance computing resources. We have addressed these issues by extending the wfGenes tool. To eliminate the coding effort for application developers, we have added support for Parsl and Dask into the generator stage of wfGenes. In order to optimize the scheduling in both Dask and Parsl, we have integrated a task graph scheduler to wfGenes that allows optimal ordering of the Python function calls in the generated Python code, such that enables maximum task parallelism.

We demonstrate the benefits of our approach by generating valid and efficient Python code for Parsl and Dask and executed the produced code on up to 256 computing nodes on an HPC cluster. The performance measurements have shown that the application executed with Dask has better performance and parallel scaling for all balanced workloads whereby all tasks have the same duration. In contrast, Parsl outperforms for unbalanced workloads, in which 10% of the tasks have ten times longer duration than the rest 90% of the tasks. It must be noted that the code for Parsl has been generated employing the wfGenes' internal task graph scheduler. Additionally, it has been shown that the parallelism of the application cannot be sufficiently exploited without wfGenes' task graph scheduler. Our measurements with the FireWorks executor suggest that the performance penalties introduced by the MongoDB queries, necessary for the operation of the executor, are only acceptable for long task execution times (larger than 60 s). Therefore, scaling applications with task graphs including more than 10,000 nodes is not recommended for short running tasks using FireWorks. The measurements show that the right choice from Dask, Parsl and FireWorks largely depends on the resource requirements profile of the specific workload, even for the same application.

Our approach is not limited to the demonstrated use case. Rather it can be employed in any use case where a formal description of a tightly coupled simulation is available but the implementation in a concrete WMS or parallelization with Dask and Parsl is not otherwise feasible. For example, by combining the built-in functions `FOREACH` and `MERGE` in WConfig, common dataflow patterns such as map-reduce can be formally described and through wfGenes translated to workflows for different target WMSs, e.g. FireWorks and SimStack, or into Parsl and Dask inputs. In use cases with extensive number of tightly coupled tasks, the end user can benefit from the high performance and scalability of the generated code without having the burden to code the application in the syntax and semantics of the target system. In future work, we will enable embedding of Parsl and Dask scripts into FireWorks workflows, and will exploit useful features, such as memoization and checkpointing in Parsl.

Acknowledgment. The authors gratefully acknowledge support by the GRK 2450. This work was partially performed on the HoreKa supercomputer funded by the Ministry of Science, Research and the Arts Baden-Württemberg and by the Federal Ministry of Education and Research. The authors acknowledge support by the state of Baden-Württemberg through bwHPC.

References

1. Research Training Group 2450 – Tailored Scale-Bridging Approaches to Computational Nanoscience. https://www.compnano.kit.edu/
2. SimStack. The Boost for Computer Aided-Design of Advanced Materials. https://www.simstack.de/
3. Babuji, Y., et al.: Parsl: pervasive parallel programming in python. In: Proceedings of the 28th International Symposium on High-Performance Parallel and Distributed Computing (HPDC 2019), pp. 25–36. ACM Press, Phoenix (2019). https://doi.org/10.1145/3307681.3325400
4. Babuji, Y., et al.: Scalable parallel programming in python with Parsl. In: Proceedings of the Practice and Experience in Advanced Research Computing on Rise of the Machines (Learning) (PEARC 2019), pp. 1–8. ACM Press, Chicago (2019). https://doi.org/10.1145/3332186.3332231
5. Boost. Boost C++ Libraries (2020). http://www.boost.org/. Accessed 18 Dec 2020
6. Borgdorff, J., et al.: A distributed multiscale computation of a tightly coupled model using the multiscale modeling language. Proc. Comput. Sci. **9**, 596–605 (2012). https://doi.org/10.1016/j.procs.2012.04.064
7. Dalcín, L., Paz, R., Storti, M., D'Elía, J.: MPI for Python: performance improvements and MPI-2 extensions. J. Parallel Distrib. Comput. **68**(5), 655–662 (2008). https://doi.org/10.1016/j.jpdc.2007.09.005
8. Friederich, P., Strunk, T., Wenzel, W., Kondov, I.: Multiscale simulation of organic electronics via smart scheduling of quantum mechanics computations. Proc. Comput. Sci. **80**, 1244–1254 (2016). https://doi.org/10.1016/j.procs.2016.05.495
9. Haghighi, S.: Pyrgg: Python random graph generator. J. Open Source Softw. **2**(17) (2017). https://doi.org/10.21105/joss.00331
10. Jain, A., et al.: FireWorks: a dynamic workflow system designed for high-throughput applications. Concurr. Comput. Pract. Exp. **27**(17), 5037–5059 (2015). https://doi.org/10.1002/cpe.3505
11. Roozmeh, M.: wfGenes. https://git.scc.kit.edu/th7356/wfgenes. Accessed 14 Jan 2022
12. Message Passing Interface Forum. MPI: A Message-Passing Interface Standard, Version 3.1. Tech. rep., High-Performance Computing Center Stuttgart, Stuttgart, Germany (2015). https://www.mpi-forum.org/docs/mpi-3.1/
13. Murray, D.G., Hand, S.: Scripting the cloud with Skywriting. In: HotCloud'10: Proceedings of the 2nd USENIX Workshop on Hot Topics in Cloud Computing. USENIX (2010). https://www.usenix.org/legacy/events/hotcloud10/tech/full_papers/Murray.pdf

14. OpenMP Architecture Review Board. OpenMP application program interface version 3.0 (2008). http://www.openmp.org/mp-documents/spec30.pdf. Accessed 18 Dec 2020
15. Rocklin, M.: Dask: parallel computation with blocked algorithms and task scheduling. In: Python in Science Conference, Austin, pp. 126–132. (2015). https://doi.org/10.25080/Majora-7b98e3ed-013
16. Roozmeh, M., Kondov, I.: Workflow generation with wfGenes. In: IEEE/ACM Workflows in Support of Large-Scale Science (WORKS), pp. 9–16. Institute of Electrical and Electronics Engineers (IEEE) (2020). https://doi.org/10.1109/WORKS51914.2020.00007
17. Schaarschmidt, J., et al.: Workflow engineering in materials design within the BATTERY 2030 + Project. Advanced Energy Materials, p. 2102638 (2021). https://doi.org/10.1002/aenm.202102638
18. Veen, L.E., Hoekstra, A.G.: Easing multiscale model design and coupling with MUSCLE 3. In: Krzhizhanovskaya, V.V., et al. (eds.) ICCS 2020. LNCS, vol. 12142, pp. 425–438. Springer, Cham (2020). https://doi.org/10.1007/978-3-030-50433-5_33
19. Wilde, M., Hategan, M., Wozniak, J.M., Clifford, B., Katz, D.S., Foster, I.: Swift: a language for distributed parallel scripting. Parallel Comput. **37**(9), 633–652 (2011). https://doi.org/10.1016/j.parco.2011.05.005
20. Wozniak, J.M., Armstrong, T.G., Wilde, M., Katz, D.S., Lusk, E., Foster, I.T.: Swift/T: large-scale application composition via distributed-memory dataflow processing. In: 2013 13th IEEE/ACM International Symposium on Cluster, Cloud, and Grid Computing, pp. 95–102. IEEE, Delft (2013). https://doi.org/10.1109/CCGrid.2013.99
21. Wozniak, J.M., et al.: Turbine: a distributed-memory dataflow engine for high performance many-task applications. Fundam. Inf. **128**(3), 337–366 (2013). https://doi.org/10.3233/FI-2013-949

An Agent-Based Forced Displacement Simulation: A Case Study of the Tigray Crisis

Diana Suleimenova[1](✉), William Low[2], and Derek Groen[1,3]

[1] Department of Computer Science, Brunel University London, London, UK
diana.suleimenova@brunel.ac.uk
[2] Save the Children, London, UK
[3] Centre for Computational Science, University College London, London, UK

Abstract. Agent-based models (ABM) simulate individual, micro-level decision making to predict macro-scale emergent behaviour patterns. In this paper, we use ABM for forced displacement to predict the distribution of refugees fleeing from northern Ethiopia to Sudan. Since Ethiopia has more than 950,000 internally displaced persons (IDPs) and is home to 96,000 Eritrean refugees in four camps situated in the Tigray region, we model refugees, IDPs and Eritrean refugees. It is the first time we attempt such integration, but we believe it is important because IDPs and Eritrean refugees could become refugees fleeing to Sudan. To provide more accurate predictions, we review and revise the key assumptions in the Flee simulation code that underpin the model, and draw on new information from data collection activities. Our initial simulation predicts more than 75% of the movements of forced migrants correctly in absolute terms with the average relative error of 0.45. Finally, we aim to forecast movement patterns, destination preferences among displaced populations and emerging trends for destinations in Sudan.

Keywords: Agent-based modelling · Simulation · Forced displacement

1 Introduction

Agent-based modelling (ABM) is a computational approach that provides an opportunity to model complex systems. It can explicitly model social interactions and networks emerging from it. Its popularity is in part due to the decentralised nature of the approach, which allows a heterogeneous mix of many agents to act and interact autonomously, in turn leading to emergent behaviours in the system at higher levels.

The initial concept of ABM was introduced in the late 1940s.s. However, it has become popular in the 1990s s as the use of ABM required computational advancements [1]. Today, ABM is widely applied to various research disciplines, such as biology, business, economics, social sciences, and technology, and practical areas including infrastructure, civilisation, terrorism, military and crowd

D. Groen et al. (Eds.): ICCS 2022, LNCS 13353, pp. 83–89, 2022.
https://doi.org/10.1007/978-3-031-08760-8_7

modelling. There are also newly emerging application domains, such as cyber-security and the social factors of climate change [2].

In this paper, we concentrate on the domain of crowds involving human movement patterns and evacuation modelling. Schelling [3] was one of the first to represent people as agents and their social behaviour as agent interactions. Only after two decades, the idea of modelling human behaviour and their movement patterns in society and geography was broadened in literature [4]. More recently, human movement modelling has expanded using various ABM software tools, as well as variations in source code languages, model developments, and their level of scalability [5]. Although ABM models problems ranging from small-scale behavioural dynamics to large scale migration simulations [6], it is becoming particularly prominent for population displacement studies [7,8]. It is already used in a wide range of refugee-related settings, such as disaster-driven migration, which incorporate changes in climate and demographics [9].

Modelling forced displacement requires an understanding of the type of migrants, methods structuring available data, modelling approach and measures of uncertainty associated with data [10]. It is also crucial to consider the course of movement of refugees and internally displaced persons (IDPs), including when they decide to leave, where they choose to flee and whether to stay or flee further from the first destination choice [11]. Hence, ABMs could be applied interactively to assist governments, organisations and non-governmental organisations (NGOs) in predicting when and where the forced displacement are likely to arrive [12], and which camps are most likely to become occupied in the short term.

The rest of the paper is structured as follows. In Sect. 2, we introduce an agent-based simulation development approach (SDA) that allows construction and execution of forced displacement simulations. We create a forced displacement model for the Tigray region in Ethiopia, which has an ongoing conflict crisis since the beginning of November 2020, using the proposed approach (Sect. 3). In Sect. 4, we discuss the obtained simulation results, validate the prediction against the UNHCR data and visualise forecast simulation results for camps in Sudan. Finally, we conclude our paper with final remarks and future work in Sect. 5.

2 Simulation Development Approach Using Flee

Suleimenova et al. [13] proposed a generalised and automated SDA to predict the distribution of incoming forced displacement, who fled because of war, persecution and/or political instability, across destination camps in neighbouring countries. The proposed approach includes six main phases, namely situation selection, data collection, model construction, model refinement, simulation execution and analysis (see Fig. 1).

In the first phase, we select a country and time period of a specific conflict, which result in large scale forced displacement. In the second phase, we obtain relevant data to the conflict from four data sources: the Armed Conflict Location and Event Data Project (ACLED, http://acleddata.com), the OpenStreetMap platform (https://openstreetmap.org), the United Nations High Commissioner

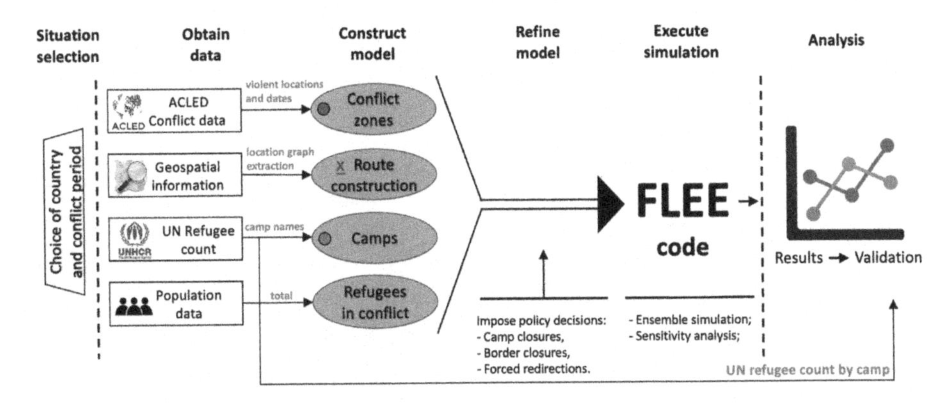

Fig. 1. A simulation development approach to predict the distribution of incoming forced population across destination camps.

for Refugees (UNHCR, http://data2.unhcr.org/en/situations) and population data. Third, we construct our initial model using these data sets and create, among other things, a network-based ABM model. Once we have built the initial model, we refine it as part of the fourth phase. Here, we manually extract information on camp and border closures, as well as forced redirection. The fifth phase involves the main simulation, which we run to predict or forecast, given a total number of forced population in the conflict, the distribution of displaced people across the individual camps. We run our simulations using the Flee simulation code (https://github.com/djgroen/flee). Importantly, Flee has the algorithmic assumptions (i.e. ruleset) for forced displacement simulations where each simulation step represents one day (see [13,14] for a detailed description of assumptions). Once the simulations have been completed, we analyse and validate the prediction results against the full UNHCR forced displacement numbers and/or visualise the forecast results as part of the sixth phase.

To demonstrate the use of SDA with Flee, we construct a model based on the conflict instance of the Tigray region and execute simulations to predict the distribution of incoming refugees to the neighbouring country of Sudan, as well as forecast future trends to efficiently allocate camp resources.

3 Forced Displacement in Tigray

The conflict in the Tigray region, which has about six million people, started on 4th November 2020, when Prime Minister Abiy Ahmed ordered a military offence against regional forces, namely Tigray People's Liberation Front (TPLF). The conflict is thought to have killed thousands and displaced more than 950,000 people internally according to UNHCR estimates, and of which about 50,000 people are displaced to Sudan. The majority have crossed at Hamdayet and Lugdi border points while the rest have crossed at Abderafi and Wad Al Mahi border points. Most of these refugees have been transferred away from the border points to the Um Rakuba and Village 8 settlement sites, and some refugees still remain at the border point waiting for other family members who may have gotten lost along the way, as well as in hope to return home soon.

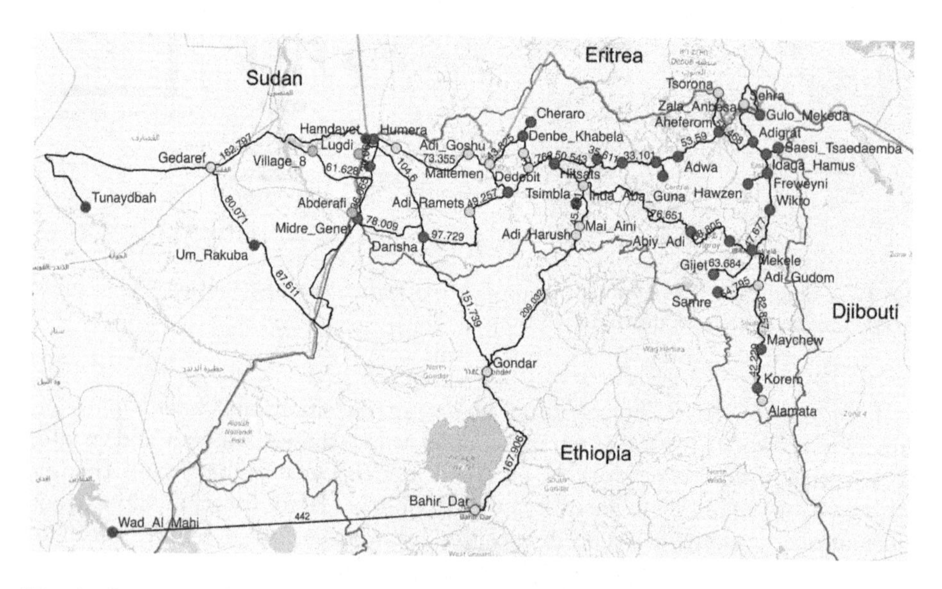

Fig. 2. Overview of geographic network model for the Tigray region with potential conflict zones marked in red, camps in dark green, forced redirected locations in light green and other settlements in yellow. Interconnecting routes are given in a simplified straight-line representation including their length in kilometres. (Color figure online)

Ethiopia is also home to 178,000 Eritrean refugees, including 96,000 mainly accommodated in four refugee camps in the western part of the Tigray region. However, their conditions have been unknown due to the communications blackout. But, in recent days, many Eritrean refugees have reportedly fled to safer locations within Tigray and other regions in Ethiopia. In addition, it has been reported that some Eritrean refugees are among the more recent arrivals into Sudan. To have better insight and to define the ruleset for migration prediction and conflict forecast for this simulation instance, we have tried to follow up every reliable source including reports from UNHCR, UNOCHA, ACLED, etc. in addition to the latest news and updates from news agencies like the Associated Press (AP), Reuters, and BBC.

We created an initial model for the Tigray region using ACLED, OpenStreetMaps, UNHCR and other sources (see Fig. 2). It includes 30 conflict locations, 18 towns, four camps in Sudan, which are Hamdayet, Um Rakuba, Wad Al Mahi and Tunaydbah, and three forwarding hubs, namely Lugdi, Village 8 and Abderafi. The latter locations are the main Sudanese border crossing points for displaced people emerging from Tigray. The two towns (Tsorona and Sehra) in Eritrea are hypothetically added to the model because UNHCR suggests that there are people arriving in Eritrea along the border with Ethiopia according to unverified reports.

When a conflict erupts, Flee predicts the resulting movement of displaced population along with a simplified travel network, and how many arrive at the camps. The prediction is based on assumptions that are summarised as follows. First, our simulation begins with 100,000 displaced persons and 96,000 Eritrean refugees residing in the Tigray region on the 4th November 2020. It does not include Ethiopian refugees in Amhara, Afar and Benishangul-Gumuz regions, as well as in Djibouti or Eritrea because we do not know their numbers at this time. Second, we model both refugees and IDPs, as IDPs could become refugees if the conflict intensifies. It is our first effort in modelling both types of forced migrants. Third, displaced persons travel on foot, at a speed of up to 40 kilometres per day. Moreover, persons may leave conflict regions two days before a conflict erupts due to the spread of danger warnings from nearby locations or families. Furthermore, any day of conflict in any Tigrayan settlement will create 10,000 displacements in our forecast. We validated this very simple heuristic with the displacement data in November, and found that it is 90% accurate on average over that period. Finally, distance to neighbouring locations is important in the choice of next destination for people, but a bit less so than in the original simulations by Suleimenova et al. [13], to reduce the chance that people keep going back and forth between very nearby settlements.

4 Results and Analysis

We run our simulation for 146 days, from the nominal start date of the crisis, November 4th 2020, until March 29th 2021. We obtained the total number of refugees arriving in Sudan from UNHCR and validate the simulation results to gauge model performance. Subsequently, we introduce conflicts at random locations in the selected region of Tigray, at a frequency determined by the chosen level of intensity. To cover a range of potential future conflict patterns, we define 18 conflict scenarios based on five potential woredas (i.e. districts) of Tigray and three levels of conflict intensity. Specifically, we create future conflicts by placing randomized major conflict events in the 100 day forecasting period which starts on March 30th 2021. The three intensity levels are defined as follows: Low with 5 events, Medium with 10 conflict events and High with 15 conflict events. Each conflict event is assumed to take place in one location and to last anywhere between 2 and 20 days. The motivation behind this is to examine the importance of location and/or intensity of conflicts driving the number of arrivals in Sudan.

We illustrate our simulation results in comparison to the UNHCR data till the simulation day 146 in Fig. 3, where the refugee population growth is underpredicted in three camps while overpredicted in Wad Al Mahi. The large mismatch in the Hamdayet, Um Rakuba and Tunaydbah camps is primarily due to their far proximity from conflict locations. In comparison, Wad Al Mahi is the furthest camp but the number of arrivals is only 700 refugees according to data.

After day 146, we forecast the distribution of incoming refugees in Sudan for additional 100 days and observe a drastic increase in the numbers of refugees for Hamdayet and Um Rakuba due to their close proximity to conflicts.

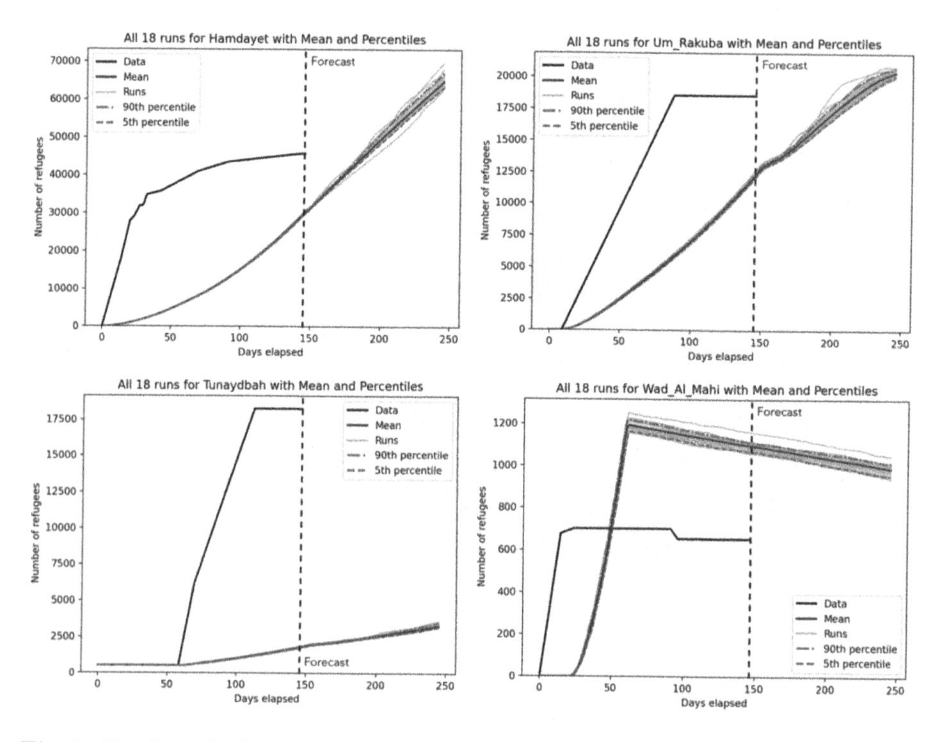

Fig. 3. Number of refugees for four camps as forecasted by forced displacement simulation for the Tigray conflict. The UNHCR data (black line) is represented for 146 days till the availability of the actual data.

5 Conclusion

This is our first attempt at forecasting arrivals in the Tigray conflict in North Ethiopia. Many aspects of our model are not mature yet, limiting the conclusions we can draw from this paper. However, we did identify that people displaced from conflicts further away tend to go to Eritrea, Djibouti, or simply other parts of Tigray in the short term (but may arrive in Sudan gradually over the longer term). Moreover, individual events in nearby towns, such as Humera, trigger an immediate spike in arrivals on our simulation. Events further afield do not trigger a spike in arrivals but do have a more gradual boosting effect on the number of arrivals over time. In future work, we will vastly increase the number of simulations, and use averaging techniques to make more precise forecasts with quantified uncertainty ranges. We will also review our IDP model, and aim to include realistic assumptions or any specific pressing questions that we should tackle with future forecasts.

Acknowledgements. This work was supported by the HiDALGO project, which has received funding from the European Union Horizon 2020 research and innovation programme under grant agreement No 824115. The authors are grateful to Save The Children and Achut Manandhar for his constructive suggestions.

References

1. Arora, P., Malhotra, P., Lakhmani, I., Srivastava, J.: Agent based software development. Int. J. Sci. Res. Eng. Technol. **6**(4) (2017)
2. Allan, R.: Survey of agent-based modelling and simulation tools. Science and Technology Facilities Council (2010). http://purl.org/net/epubs/manifestation/5601/DLTR-2010-007.pdf
3. Schelling, T.C.: Dynamic models of segregation. J. Math. Sociol. **1**, 143–186 (1971)
4. Epstein, J.M., Axtell, R.: Growing Artificial Societies: Social Science from the Bottom Up. MIT Press, Cambridge (1996)
5. Abar, S., Theodoropoulos, G.K., Lemarinier, P., O'Hare, G.M.P.: Agent based modelling and simulation tools: a review of the state-of-art software. Comput. Sci. Rev. **24**, 13–33 (2017)
6. Macal, C.M., North, M.J.: Tutorial on agent-based modelling and simulation. J. Simul. **4**(3), 151–162 (2010)
7. Castle, C., Crooks, A.: Principles and concepts of agent-based modelling for developing geospatial simulations. Centre for Advanced Spatial Analysis, University College London (2006)
8. Crooks, A., Castle, C., Batty, M.: Key challenges in agent-based modelling for geo-spatial simulation. Comput. Environ. Urban Syst. **32**(6), 417–430 (2008)
9. Entwisle, B., et al.: Climate shocks and migration: an agent-based modeling approach. Popul. Environ. **38**(1), 47–71 (2016)
10. Raymer, J., Smith, P.W.F.: Editorial: modelling migration flows. J. Roy. Stat. Soc. **173**(4), 703–705 (2010)
11. Hébert, G.A., Perez, L., Harati, S.: An agent-based model to identify migration pathways of refugees: the case of Syria. In: Perez, L., Kim, E.-K., Sengupta, R. (eds.) Agent-Based Models and Complexity Science in the Age of Geospatial Big Data. AGIS, pp. 45–58. Springer, Cham (2018). https://doi.org/10.1007/978-3-319-65993-0_4
12. Estrada, L.E.P., Groen, D., Ramirez-Marquez, J.E.: A serious video game to support decision making on refugee aid deployment policy. Procedia Comput. Sci. **108**, 205–214 (2017)
13. Suleimenova, D., Bell, D., Groen, D.: A generalized simulation development approach for predicting refugee destinations. Sci. Rep. **7**, 13377 (2017)
14. Groen, D.: Simulating refugee movements: where would you go? Procedia Comput. Sci. **80**, 2251–2255 (2016)

Quantum Computing

Practical Solving of Discrete Logarithm Problem over Prime Fields Using Quantum Annealing

Michał Wroński(✉)

Military University of Technology, Kaliskiego Str. 2, Warsaw, Poland
michal.wronski@wat.edu.pl

Abstract. This paper investigates how to reduce discrete logarithm problem over prime fields to the QUBO problem to obtain as few logical qubits as possible. We show different methods of reduction of discrete logarithm problem over prime fields to the QUBO problem. In the best case, if n is the bitlength of a characteristic of the prime field $\mathbb{F}_p$, there are required approximately $2n^2$ logical qubits for such reduction. We present practical attacks on discrete logarithm problem over the 4-bit prime field $\mathbb{F}_{11}$, over 5-bit prime field $\mathbb{F}_{23}$ and over 6-bit prime field $\mathbb{F}_{59}$. We solved these problems using D-Wave Advantage QPU. It is worth noting that, according to our knowledge, until now, no one has made a practical attack on discrete logarithm over the prime field using quantum methods.

Keywords: Discrete logarithm problem · D-Wave Advantage · Quantum annealing

1 Introduction

Shor's quantum algorithm for factorization and discrete logarithm computation [10] is one of the essential researches in modern cryptology. Since then, there have been many efforts to build a general-purpose quantum computer that solves real-world cryptographic problems. Unfortunately, till now, such powerful general-purpose quantum computers do not exist. On the other hand, quantum annealing is an approach that takes more and more popularity. The most powerful computer using quantum annealing technology is the D-Wave Advantage computer. One of the most exciting applications of quantum annealing to cryptography is transforming the factorization algorithm into the QUBO problem and then solving this problem using the D-Wave computer [7].

Moreover, the newest D-Wave computers have much more physical qubits than general-purpose quantum computers. It is believed that this approach also may be helpful, primarily until large general-purpose quantum computers will

This work was supported by the Military University of Technology's University Research Grant No. 858/2021: "Mathematical methods and models in computer science and physics.".

D. Groen et al. (Eds.): ICCS 2022, LNCS 13353, pp. 93–106, 2022.
https://doi.org/10.1007/978-3-031-08760-8_8

exist. It seems that, in some cases, D-Wave computers may be used to solve cryptographic problems, which cannot be solved nowadays by general-purpose quantum computers.

This paper shows how to transform discrete logarithm problem (DLP) over prime fields to the QUBO problem. We consider different approaches to such transformation, aiming to obtain the smallest possible number of logical qubits. The best method allows one to convert discrete logarithm problem to the QUBO problem using approximately $2n^2$ logical qubits.

Our contribution is:

- presenting different methods of reduction of discrete logarithm problem to the QUBO problem, where the best method requires approximately $2n^2$ logical qubits for such reduction;
- presenting practical attacks on discrete logarithm problem over the 4-bit prime field $\mathbb{F}_{11}$, over 5-bit prime field $\mathbb{F}_{23}$ and over 6-bit prime field $\mathbb{F}_{59}$ using D-Wave Advantage QPU.

It is worth noting that, according to our knowledge, until now, no one has made a practical attack on discrete logarithm over the prime field using quantum methods.

2 Quantum Annealing and Cryptography

Shor's quantum algorithm for factorization and discrete logarithm began the race to construct a quantum computer to solve real-world cryptographic problems. Nowadays, the two approaches of quantum computing for cryptography are the most popular.

The first approach is quantum annealing, used in D-Wave computers. The second approach is general-purpose quantum computing. The important thing is that the first approach has limited applications, where mainly QUBO and Ising problems may be solved using such quantum computers.

QUBO (Quadratic Unconstrained Binary Optimization) [3] is a significant problem with many real-world applications. One can express the QUBO model by the following optimization problem:

$$\min_{x \in \{0,1\}^n} x^T Q x, \tag{1}$$

where Q is an $N \times N$ upper-diagonal matrix of real weights, x is a vector of binary variables. Moreover, diagonal terms $Q_{i,i}$ are linear coefficients, and the nonzero off-diagonal terms are quadratic coefficients $Q_{i,j}$.

QUBO problem may also be viewed as a problem of minimizing the function

$$f(x) = \sum_i Q_{i,i} x_i + \sum_{i<j} Q_{i,j} x_i x_j. \tag{2}$$

Let us note, that the QUBO problem is a special case of the BQM (Binary Quadratic Model) problem, where BQM may be given as

$$\sum_{i=1} a_i v_i + \sum_{i<j} b_{i,j} v_i v_j + c, \tag{3}$$

where a_i and $b_{i,j}$ are real numbers and $v_i \in \{-1, +1\}$ or $\{0, 1\}$. The transformation of the QUBO problem to the BQM problem is straightforward and what we need to do is to forget the constant c appearing in the BQM problem.

What is essential from the cryptological point of view, many problems may be translated to the QUBO problem. The most exciting example of such transformation is integer factorization. It is worth noting that the quantum factorization record had belonged to the D-Wave computer for some time. Using transformation of integer factorization to the QUBO problem, Dridi and Alghassi [4] factorized integer 200, 099, which result was later beaten by Jiang et al. [7], and by Wang et al. [12], who factorized 20-bit integer 1, 028, 171. It is worth noting that quantum annealing was also used to find relations in the index calculus method for elliptic curves where using D-Wave Leap and hybrid sampler, elliptic curve discrete logarithm problem over the 8-bit prime field has been solved [13].

On the other hand, general-purpose quantum computers have limited resources. The most powerful Intel, Google, and IBM quantum computers have 49, 72, and 127 qubits, respectively [5,6,11]. It means that the resources of general quantum computers are nowadays too small to solve real-world cryptographic problems.

The D-Wave computers using quantum annealing are developing rapidly and have many more qubits than a few years before. The most potent quantum annealing computer, D-Wave Advantage [2], has 5, 760 working qubits. This quantum annealer allows solving general problems with up to 1, 000, 000 variables and dense problems with 20, 000 variables. A detailed description of how the D-Wave computer works may be found in [3].

3 Methods of Transformation of Discrete Logarithm Problem to the QUBO Problem

This section will present different approaches to transforming the discrete logarithm problem to the BQM problem, which problem may be easily transformed to the QUBO problem by removing the constant appearing in the given BQM problem.

We begin by defining discrete logarithm problem

$$g^y = h, \tag{4}$$

in the prime field $\mathbb{F}_p$, where $g, h \in \mathbb{F}_p^*$ and $y \in \{1, \ldots, Ord(g)\}$. This problem is equivalent to

$$g^y \equiv h (mod\ p), \tag{5}$$

for integers $g, h \in \{1, \ldots, p-1\}, y \in \{1, \ldots, Ord(g)\}$.

Let m be the bitlength of $Ord(g)$. We begin by making the following transformation. Let us note that y may be written using m bits and if $y = 2^{m-1}u_m + \cdots + 2u_2 + u_1$, where $u_1, \ldots, u_m$ are binary variables, then

$$g^y = g^{2^{m-1}u_m + \cdots + 2u_2 + u_1} = g^{2^{m-1}u_m} \cdots g^{2u_2} g^{u_1}, \tag{6}$$

It is worth noting that writing $y = 2^{m-1}u_m + \cdots + 2u_2 + u_1$ allows to obtain $y > Ord(g)$, but because we operate in a cyclic group, one can always get the result from $\{1, \ldots, Ord(g)\}$ computing $y \mod Ord(g)$.

Let us also note that

$$g^{2^{i-1}u_i} = \begin{cases} 1, u_i = 0, \\ g^{2^{i-1}}, u_i = 1, \end{cases} \tag{7}$$

which is equivalent to

$$g^{2^{i-1}u_i} = 1 + u_i \left(g^{2^{i-1}} - 1 \right). \tag{8}$$

Now we use the observation above to define different transformation approaches of discrete logarithm problem over prime fields to the BQM and thus equivalent QUBO problem.

3.1 Solving Modular Equations

To clarify our approach, we will begin with a simple example of the transformation of the modular equations to the BQM problem. We will show how it works considering linear modular equations. It is worth noting that the problem given by Eq. (9) is the discrete logarithm problem in the additive group of field $\mathbb{F}_p$.

Let us consider the equation

$$ax \equiv b(mod\ p). \tag{9}$$

Because $x \in \{0, \ldots, p-1\}$, we can rewrite the Eq. (9) as

$$a\left(u_1 + 2u_2 + \cdots + 2^{n-2}u_{n-1} + (p - 2^{n-1} + 1)u_n\right) \equiv b(mod\ p), \tag{10}$$

because $x = u_1 + 2u_2 + \cdots + 2^{n-2}u_{n-1} + (p - 2^{n-1} + 1)u_n$ for binary variables $u_1, \ldots, u_n$, and therefore $x \in \{0, \ldots, p-1\}$ [1].

Now one should rewrite this equation as

$$\begin{aligned} &u_1(a\ mod\ p) + u_1(2a\ mod\ p) + \cdots + u_{n-1}(2^{n-2}a\ mod\ p) \\ &+u_n((p - 2^{n-1} + 1)a\ mod\ p) + (-b\ mod\ p) - kp = 0, \end{aligned} \tag{11}$$

where k is some integer. Let us note that after such reduction, all monomials appearing in the equation above (instead of $-kp$) are positive. It means that one can bound k, because $(a\ mod\ p) + (2a\ mod\ p) + \cdots + (2^{n-2}a\ mod\ p) + ((p - 2^{n-1} + 1)a\ mod\ p) + ((-b)\ mod\ p) \geq kp$. What is more, we can find a general bound on k, because every monomial coefficient is from the set $\{0, \ldots, p-1\}$.

Because we have $n+1$ monomials, we can find that $(p-1)(n+1) \geq kp$, which means that $k \leq \frac{(n+1)(p-1)}{p} < n+1$, so $k \leq n$ and finally k may be written using $l = \lfloor \log_2 n \rfloor + 1$ binary variables, similarly as x was written before. This idea may be found, for example, in [1].

Now we can rewrite the equation above

$$\begin{aligned} &f = u_1(a \ mod \ p) + u_1(2a \ mod \ p) + \dots + u_{n-1}(2^{n-2}a \ mod \ p) \\ &+u_n((p-2^{n-1}+1)a \ mod \ p) + (-b \ mod \ p) - (k_1 + 2k_2 + \dots + k_{l-1}(2^{l-2}) \\ &+k_l(n-2^{l-1}+1))p = 0. \end{aligned} \tag{12}$$

Finally, one should find the minimal energy of f^2 (this energy should be equal to 0), where f^2 will be indeed in the BQM form.

3.2 Transformation of Discrete Logarithm Problem to the QUBO Problem - Brutal Approach

Let us note that using Eqs. (6) and (8), one obtains the following equation

$$\begin{aligned} g^y &= \left(1 + u_m\left(g^{2^{m-1}} - 1\right)\right) \cdot \dots \cdot \left(1 + u_2\left(g^2 - 1\right)\right)\left(1 + u_1\left(g - 1\right)\right) \\ &= \left(1 + u_m\left((g^{2^{m-1}} - 1) mod \ p\right)\right) \dots \left(1 + u_2\left((g^2 - 1) mod \ p\right)\right) \\ &\cdot \left(1 + u_1\left((g-1) mod \ p\right)\right). \end{aligned} \tag{13}$$

We can see that g^y can be represented as the polynomial of degree m of m Boolean variables. We will show how to linearize this polynomial. Let us note that linearization may be performed in the following way.

If $m = 1$, then $1 + u_1(g-1)$ and it is indeed linear polynomial.

If $m = 2$, then $f = (1 + u_1(g-1))\left(1 + u_2(g^2-1)\right) = 1 + u_1(g-1) + u_2(g^2-1) + u_1u_2(g-1)\left(g^2-1\right)$. The variable u_1u_2 may be substituted by an auxiliary variable $v_1 = u_1u_2$. The penalty will be added later. So one can see that $f = 1 + u_1(g-1) + u_2(g^2-1) + v_1(g-1)\left(g^2-1\right)$ and is in linear form.

We can keep on such a procedure, and finally, one obtains the linear polynomial of $2^m - 1$ variables.

Having polynomial f in linear form, now we should transform modular equation $f \equiv h(mod \ p)$ to the equation over integers

$$(f-h) \ mod \ p - kp = 0, \tag{14}$$

where $k \in \mathbb{Z}$ and for every polynomial f, operation $f \ mod \ p$ is equivalent to the reduction of all of the coefficients of polynomial f modulo p.

If one wants to solve Eq. (14) searching for minimal energy of optimization problem, it is necessary to square Eq. (14), obtaining in result polynomial F and Eq. (15).

$$F = \left((f-h) \ mod \ p - kp\right)^2 = 0. \tag{15}$$

Let us note that k is bounded by the maximal number of monomials appearing in the polynomial $(f-h) \ mod \ p$, which is equal to 2^m. Finally, $k_{max} \leq$

$\lfloor \frac{2^m(p-1)}{p} \rfloor < 2^m$ and bitlength of k is equal to m at most. Moreover, to obtain proper energy, we have to add penalties to the function F, according to Eq. (16), obtaining $F_{Pen} = F + Pen$, where Pen are penalties obtained during linearization. Polynomial F_{Pen} has minimal energy equal to 0 (our BQM problem is constructed so that minimal energy is equal to 0 because in our BQM problem appears constant energy offset). If one removes this constant energy offset c, one obtains a problem in the QUBO form, but its minimal energy will equal $-c$.

The DLP transformation to the QUBO problem using a brutal approach requires, in general, $2^m + m - 1$ variables for m-bit order of an element g. The number of variables in such a case does not depend on the bitlength of p. Let us note that this exponential growth makes that the presented method may be applied only for small prime fields $\mathbb{F}_p$.

It is crucial that one can obtain the BQM problem in two little different ways:

1. one can at first linearize each equation f_i to obtain the linearized equation f_{Lin_i}, then compute the sum $F_{Pen} = \sum_{i=1}^{w} f_{Lin_i}^2 + Pen$, where Pen denotes penalties obtained during linearization; polynomial F_{Pen} is in such a case in BQM form;
2. one can at first compute the sum $F = \sum_{i=1}^{w} f_i^2$, and then make of quadratization of the polynomial F, obtaining F_{Quadr}, finally obtaining polynomial $F_{Pen} = F_{Quadr} + Pen$ in BQM form, where Pen denotes penalties obtained during quadratization.

In our methods of reducing discrete logarithm problem to the BQM problem, the first method simply allows us to compute the maximal number of required variables in the resulting BQM problem and thus in the equivalent QUBO problem. The second method allows one often to obtain a smaller number of variables than the first method. Even though, in practical experiments, often the second method was used.

However, in the case of solving linear modular equations, it is unnecessary to reduce high-order terms, while transforming discrete logarithm problem to the BQM problem, it is necessary to reduce 2-local terms while using approach 1 of obtaining BQM problem, and 4-local terms and 3-local terms (for any $w \geq 2$ one can make a similar reduction from $w+1$-local term to w-local term) while using approach 2 of obtaining BQM problem. Now we will show how the resulting 3-local terms may be reduced to 2-local terms. Let us note that each penalty monomial of the form $x_i x_j x_l$ will be transformed, according to [7], in the following way

$$x_i x_j x_l \rightarrow u_k x_l + 2(x_i x_j - 2u_k(x_i + x_j) + 3u_k). \tag{16}$$

It means that

$$x_i x_j x_l = u_k x_l = u_k x_l + 2(x_i x_j - 2u_k(x_i + x_j) + 3u_k), \tag{17}$$

if $x_i x_j = u_k$ and

$$x_i x_j x_l < x_l u_k + 2(x_i x_j - 2u_k(x_i + x_j) + 3u_k), \tag{18}$$

if $x_i x_j \neq u_k$.

It results in that the term $x_i x_j x_l$ may be transformed to quadratic form by replacing $x_i x_j$ with u_k plus a constraint, given by penalty term:

$$min(x_i x_j x_l) = min\left(x_l u_k + 2(x_i x_j - 2u_k(x_i + x_j) + 3u_k)\right). \qquad (19)$$

3.3 Transformation of Discrete Logarithm Problem to the QUBO Problem Using Approximately $2n^2$ Logical Qubits - Efficient Approach

In this section, we will transform the discrete logarithm problem to the QUBO problem using a regular binary tree of maximal height for decomposition. It is possible to obtain an equivalent QUBO problem using approximately $2n^2$ logical qubits in such a case, where n is the bitlength of a characteristic of the prime field $\mathbb{F}_p$. Let us define for every $i = \overline{1,m}$ equality $x_i = g^{2^{i-1}u_i}$. The scheme of such a regular binary tree of maximal height for general m (the bitlength of $Ord(g)$) used for decomposition of discrete logarithm problem to the QUBO problem is presented in Fig. 1.

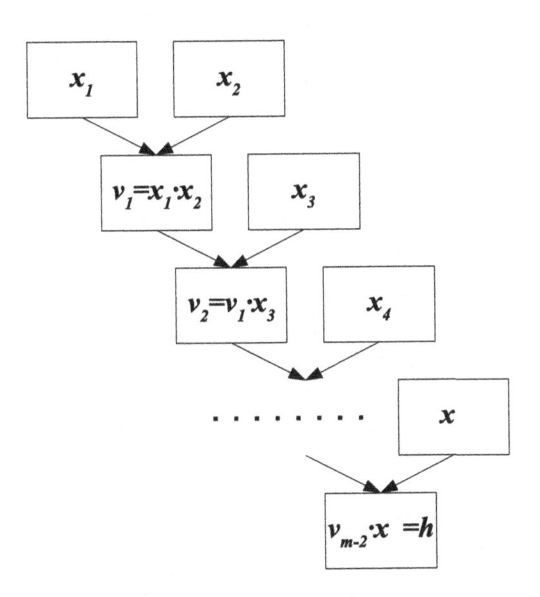

Fig. 1. The scheme of decomposition of discrete logarithm problem using the efficient approach.

It is worth noting that other decomposition methods are possible, as a method using a binary balanced tree. Unfortunately, in such a case, the expected number of logical variables of equivalent QUBO problem is equal to approximately $\frac{n^3}{2}$. We, therefore, do not describe here this method.

At first, let us note that using the problem decomposition scheme presented in Fig. 1, in every step, we can create a new variable $v_i = v_{i-1}x_{i+1}$, which is equivalent to $v_i \equiv v_{i-1}x_{i+1}(mod\ p)$. It is also easy to show that the total number of new variables v_i will be equal to $m-2$, because $x_1, \ldots, x_m$ are leaves of the binary tree with $m-1$ inner nodes, where each inner node is equivalent to some auxiliary variable. However, the root is not equivalent to any auxiliary variable, but it is equivalent to $v_{m-2}x_m \equiv h(mod\ p)$, so the number of auxiliary variables v_i is equal to $m-2$.

What is more, each of the equations for $v_1, v_2, v_{m-2}, \ldots, v_{m-3}, v_{m-2}, v_{m-2}x_m$ need to be transformed to the equation over integers.

$$\begin{cases} f_1 = (v_1 - x_1x_2)\ mod\ p - k_1p = 0, \\ f_2 = (v_2 - v_1x_3)\ mod\ p - k_2p = 0, \\ \ldots \\ f_{m-3} = (v_{m-3} - v_{m-4}x_{m-2})\ mod\ p - k_{m-3}p = 0, \\ f_{m-2} = (v_{m-2} - v_{m-3}x_{m-1})\ mod\ p - k_{m-2}p = 0, \\ f_{m-1} = (h - v_{m-2}x_m)\ mod\ p - k_{m-1}p = 0. \end{cases} \tag{20}$$

We will precisely count how many auxiliary variables are necessary. Let us note that for variables $v_1, \ldots, v_{m-2}$ it is necessary n bits to represent v_i, because $v_i \in \{1, \ldots, p-1\}$ and at most $n+1$ bits to represent k_i, because $(v_i - v_{i-1}x_{i+1})\ mod\ p$, according to Eq. (8) is equivalent to

$$\left(v_i - v_{i-1} - v_{i-1}u_{i+1}\left(g^{2^i}+1\right)\right) mod\ p. \tag{21}$$

Let us note that v_i is limited by its definition by $p-1$. Using the binary representation of $-v_{i-1}$ and making reduction modulo p, one obtains polynomial of binary variables and coefficients from interval $\{0, \ldots, p-1\}$. It means that maximal value of polynomial (21) is equal to $(2n+1)(p-1)$ and therefore, $kp \leq (2n+1)(p-1)$, which means that $k \leq \frac{(2n+1)(p-1)}{p} < 2n+1$, so $k \leq 2n$ and the bitlength of k is equal to $\lfloor \log_2(2n) \rfloor + 1$ at most.

Additionally, for every $i = \overline{1, m-2}$, during linearization of $(v_i - v_{i-1}x_{i+1})\ mod\ p$ it is necessary to linearize terms appearing in $v_{i-1}x_{i+1}$, which requires n variables, because, according to Eq. (8), $x_{i+1} = 1 + u_{i+1}\left(g^{2^i}-1\right)$ has two monomials but depends on only one variable.

Let us denote as $f_{lin_1}, \ldots, f_{lin_{n-1}}$ polynomials $f_1, \ldots, f_{m-1}$ after linearization. Then the final polynomial F in BQM form is equal to

$$F_{Pen} = (f_{lin_1})^2 + \cdots + (f_{lin_{m-1}})^2 + Pen, \tag{22}$$

where Pen are penalties obtained during linearization and minimal energy of F_{Pen} is equal to 0.

So the total number of variables is equal:

- for $x_1, \ldots, x_m$ - it is required to have m binary variables,
- for $v_1, \ldots, v_{m-2}$ - it is required to have $(m-2)n$ binary variables,
- for $k_1, \ldots, k_{m-1}$ - it is required to have $(m-1)\left(\lfloor \log_2(2n) \rfloor + 1\right)$ binary variables,
- for auxiliary variables obtained during linearization of each polynomial $f_1, \ldots f_{m-1}$ it is required $(m-1)n$ variables,

Finally, obtained BQM (and thus equivalent QUBO) problem requires $m+(m-2)n+(m-1)\left(\lfloor \log_2(2n) \rfloor + 1\right)+(m-1)n = 2mn+2m-3n+(m-1)\lfloor \log_2(2n) \rfloor - 1$, which is approximately equal to $2mn$ variables. If we also assume that $m \approx n$ (what is true if the given generator is the generator of the multiplicative subgroup of field $\mathbb{F}_p$), then the total number of variables is equal to approximately $2n^2$.

3.4 Mixed Approach

In the mixed approach, the crucial observation is that, especially for small prime fields, the brutal approach is more efficient than approaches using binary tree decomposition. The key idea is to use both methods to obtain the problem using fewer logical qubits.

In the first step, we multiply k first terms $x_1, x_2, \ldots, x_k$, as same as in the brutal approach, obtaining equation $(v_1 - x_1 \cdot x_2 \cdots x_k) \mod p - k_1 p = 0$. We use the decomposition method of discrete logarithm problem using a regular binary tree of maximal height in all following steps. The scheme of such decomposition is presented in Fig. 2.

If choosing the number of variables we like to multiply in the first step is made carefully, the final QUBO problem may consist of a smaller number of variables than any other presented approach.

4 Experiments

We experimented with different approaches and solvers: classical, hybrid, and quantum. However, we aimed to solve discrete logarithm problem over prime fields using quantum solver for D-Wave Advantage QPU. In every case, as characteristic of n-bit prime field $\mathbb{F}_p$ we chose the biggest n-bit prime, for which $d = \frac{p-1}{2}$ is prime. We also chose as generators elements of order equal to d. The most notable result of solving discrete logarithm problems over the prime field using D-Wave Advantage QPU was solving discrete logarithm problem over the 4-bit prime field $\mathbb{F}_{11}$ using the efficient approach, 5-bit prime field $\mathbb{F}_{23}$ using the mixed approach and over 6-bit prime field $\mathbb{F}_{59}$ using the brutal approach.

4.1 Solving Discrete Logarithm Problem over $\mathbb{F}_{11}$ Using Efficient Approach

In the case of the prime field $\mathbb{F}_{11}$, the smallest QUBO problem can be obtained using the brutal approach, but we wanted to show the application of the efficient approach for solving discrete logarithm problem.

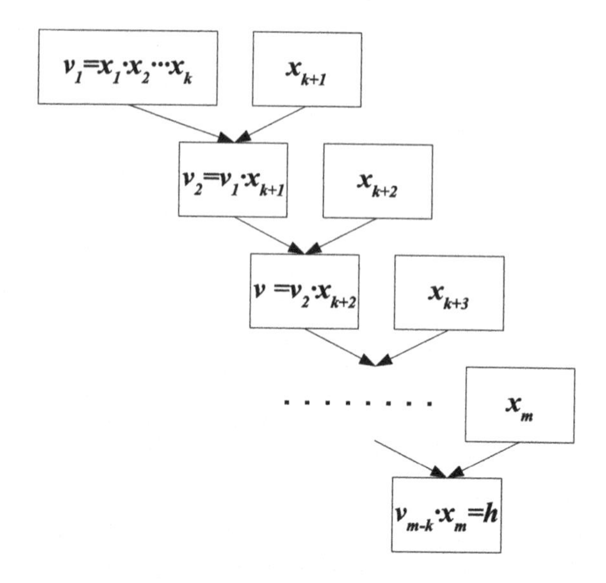

Fig. 2. The scheme of decomposition of discrete logarithm problem using the mixed approach.

In this experiment, we solved the discrete logarithm problem over a field $\mathbb{F}_{11}$ which is the 4-bit prime field. The given generator was 4, and the order of 4 in $\mathbb{F}_{11}$ is 5. We solved the following discrete logarithm problem:

$$4^y \equiv 9(mod\ 11). \tag{23}$$

Obtained QUBO problem consisted of 18 logical qubits. The problem has been embedded in the D-Wave Advantage computer, in the Pegasus topology, using 36 physical qubits.

Unfortunately, we could not quantumly solve discrete logarithm problems over larger fields using an efficient approach.

4.2 Solving Discrete Logarithm Problem over $\mathbb{F}_{23}$ Using Mixed Approach

However, in the case of the prime field $\mathbb{F}_{23}$, the smallest QUBO problem can be obtained using the brutal approach, we wanted to show the application of a mixed approach for solving discrete logarithm problem.

In this experiment, we solved the discrete logarithm problem over a field $\mathbb{F}_{23}$ which is the 5-bit prime field. The given generator was 2, and the order of 2 in $\mathbb{F}_{23}$ is 11. We solved the following discrete logarithm problem (with $m = 4$ and $k = 3$):

$$2^y \equiv 13(mod\ 23). \tag{24}$$

Obtained QUBO problem consisted of 32 logical qubits. The problem has been embedded in the D-Wave Advantage computer, in the Pegasus topology, using 75 physical qubits.

Unfortunately, we were not able to quantumly solve discrete logarithm problem over $\mathbb{F}_{59}$ using a mixed approach. In such a case, the obtained QUBO problem consisted of 41 logical qubits, but after embedding in the D-Wave Advantage computer, the final problem required 130 physical qubits. We made in such a case several experiments, but we did not obtain the proper minimal energy.

4.3 Solving Discrete Logarithm Problem over $\mathbb{F}_{59}$ Using Brutal Approach

In the last experiment, we solved the discrete logarithm problem over a field $\mathbb{F}_{59}$ which is the 6-bit prime field. The given generator was 4, and the order of 4 in $\mathbb{F}_{59}$ is 29. We solved the following discrete logarithm problem:

$$4^y \equiv 27(mod\ 59). \tag{25}$$

Obtained QUBO problem consisted of 30 logical qubits. The problem has been embedded in the D-Wave Advantage computer, in the Pegasus topology, using 79 physical qubits.

4.4 Experiments Summary

Unfortunately, we could not solve discrete logarithm problems over prime fields of a more significant length than 6 using quantum solver and D-Wave Advantage QPU.

We provide parameters of solution of discrete logarithm problem using D-Wave Advantage QPU over $\mathbb{F}_{11}$ using the efficient approach, $\mathbb{F}_{23}$ using the mixed approach, and $\mathbb{F}_{59}$ using the brutal approach.

Table 1. D-Wave Advantage solver parameters used in solving QUBO problems equivalent to the discrete logarithm problems.

Parameter	Value
Name (chip ID)	Advantage_system4.1
Qubits	5,760
Topology	Pegasus
Number of reads	10,000
Annealing time (μs)	20
Anneal schedule	[[0,0],[20,1]]
H gain schedule	[[0,0],[20,1]]
Programming thermalization (μs)	1000

Table 2. Parameters used in solving QUBO problem equivalent to the problem of finding discrete logarithm over $\mathbb{F}_{11}$ using the efficient approach, $\mathbb{F}_{23}$ using the mixed approach and $\mathbb{F}_{59}$ using the brutal approach.

Parameter	DLP over $\mathbb{F}_{11}$	DLP over $\mathbb{F}_{23}$	DLP over $\mathbb{F}_{59}$
Number of source variables	18	32	30
Number of target variables	36	75	81
Max chain length	4	5	4
Chain strength	8,258.58	20,000.00	2,025,065.28
QPU access time (μs)	959,248.6	1,324,651	913,450.6
QPU programming time (μs)	8,448.6	8,451	8,450.6
QPU sampling time (μs)	950,800	1,316,200	905,000
Total post processing time (μs)	11,209	17,384	17,023
Post processing overhead time (μs)	4,252	3,244	2,341

Figure 3 shows how different QUBO problems were embedded on the D-Wave Advantage computer.

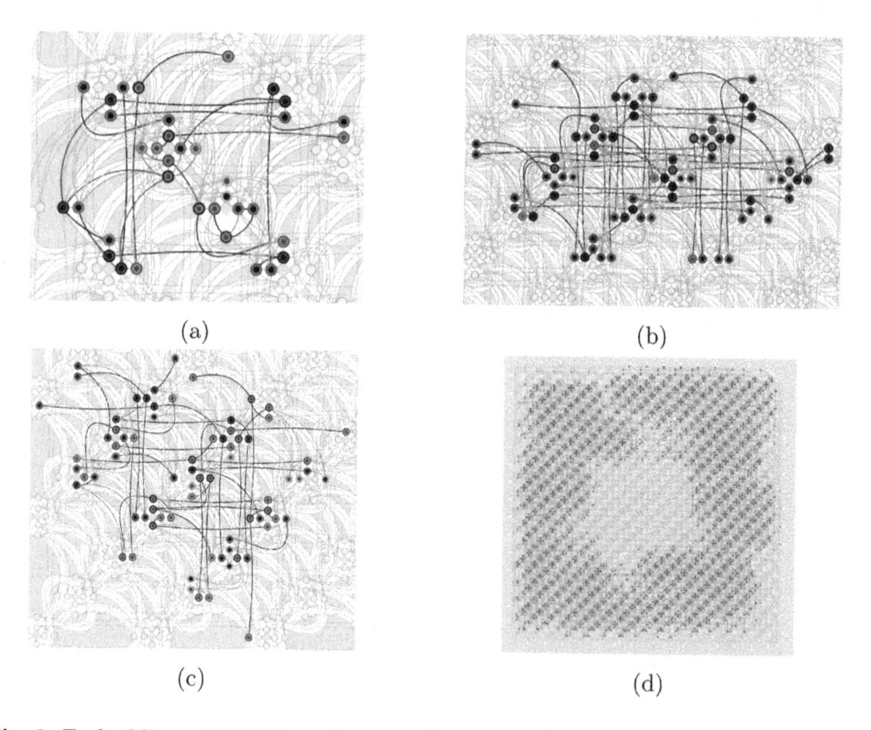

Fig. 3. Embedding of QUBO problems equivalent to discrete logarithm problems over the prime field $\mathbb{F}_{11}$ using efficient approach (Fig. 3a), prime field $\mathbb{F}_{59}$ using brutal approach (Fig. 3b), prime field $\mathbb{F}_{23}$ using mixed approach (Fig. 3c), and prime field $\mathbb{F}_{65267}$ using mixed approach (Fig. 3d).

We also checked the limitations of proposed methods using the D-Wave Advantage computer. We prepared a QUBO problem equivalent to a discrete logarithm problem over the 16-bit prime field $\mathbb{F}_{65267}$.

We used the mixed approach to try to solve the following discrete logarithm problem:

$$4^y \equiv 64643(mod\ 65267). \tag{26}$$

The equivalent QUBO problem consisted of 444 variables. We embedded this problem into the D-Wave Advantage QPU using 3,724 physical qubits. This embedding is presented in Fig. 3d.

According to our experiments, it looks that discrete logarithm problem over 16-bit prime field probably could not be properly solved. The biggest problem in such a case is a large number of variables and long chains, where maximal chain length is equal to 20. It is also interesting that discrete logarithm problem over 17-bit prime, consisting of 537 logical qubits, could not be embedded into D-Wave Advantage QPU.

The same QUBO problems, which have been solved using quantum annealing, we also have solved using a classical CPU solver. In such a case, solving QUBO problems equivalent to discrete logarithm problems over the prime field $\mathbb{F}_{11}$, $\mathbb{F}_{23}$, and $\mathbb{F}_{59}$, took 0.050 s, 0.065 s, and 0.057 s, respectively.

The other observation is that using the brutal approach and classical CPU solver, we solved the QUBO problem equivalent to a discrete logarithm problem over a 10-bit prime in just 4.274 s. It is very interesting because this QUBO problem consists of 521 variables.

It looks that in the case of solving the QUBO problem, very important is the problem definition. The number of variables is also significant, but the number of connections between variables is often more critical. What is more, the QUBO problem obtained using a brutal approach and classical CPU solver may be easily solved even for the 521 variables problem. Unfortunately, the brutal approach is naturally limited by an exponential number of variables required to construct an equivalent QUBO problem.

5 Further Works and Conclusion

In this paper, we presented methods of transformation of discrete logarithm problem over prime fields to the QUBO problem. We showed different approaches to such transformation. The best methods allow one to obtain equivalent QUBO problem using approximately $2n^2$ variables (logical qubits). It is worth noting that in the case of factorization, in general, known methods allow transforming factorization problem to the QUBO problem using approximately $\frac{n^2}{4}$ variables if n is bitlength of integer N that one wants to factorize.

The main result of the paper is a practical experiment, where discrete logarithm problem over $\mathbb{F}_{59}$ has been solved using D-Wave Advantage QPU. Even though it is a small problem, according to our knowledge, no one has reported a solution of discrete logarithm problem over prime fields using quantum methods until now.

Because the expected asymptotic time of solving the QUBO problem, even knowing the number of variables, is now unknown, it is hard to estimate the time in which presented QUBO problems could be solved. There are some expectations that for n variables QUBO problem, the minimal energy using quantum annealing should be found in $O\left(e^{\sqrt{n}}\right)$ [9], but more research in this area should be done. What is more, the presented methods should not outperform Shor's polynomial-time algorithm for large prime fields.

More research should be done to solve more significant problems using our methods. What is more, it seems that the presented methods may also be applied using quantum superconducting computers. Such researches have been done in the case of factorization problem [8], and it seems that transformation of discrete logarithm problem to find the ground state of the Hamiltonian and to solve then such problem using a superconducting quantum computer will also be possible.

References

1. Chen, Y.A., Gao, X.S., Yuan, C.M.: Quantum algorithm for optimization and polynomial system solving over finite field and application to cryptanalysis. arXiv preprint arXiv:1802.03856 (2018)
2. D-WAVE, T.Q.C.C.: The d-wave advantage system: An overview. Technical report (2020)
3. D-WAVE, T.Q.C.C.: Getting started with the d-wave system. User manual (2020)
4. Dridi, R., Alghassi, H.: Prime factorization using quantum annealing and computational algebraic geometry. Sci. Rep. **7**, 43048 (2017)
5. Greene, T.: Google reclaims quantum computer crown with 72 qubit processor (2018)
6. Intel. The future of quantum computing is counted in qubits (2018)
7. Jiang, S., Britt, K.A., McCaskey, A.J., Humble, T.S., Kais, S.: Quantum annealing for prime factorization. Sci. Rep. **8**(1), 1–9 (2018)
8. Karamlou, A.H., Simon, W.A., Katabarwa, A., Scholten, T.L., Peropadre, B., Cao, Y.: Analyzing the performance of variational quantum factoring on a superconducting quantum processor. NPJ Quant. Inf. **7**(1), 1–6 (2021)
9. Mukherjee, S., Chakrabarti, B.K.: Multivariable optimization: quantum annealing and computation. Eur. Phys. J. Sp. Top. **224**(1), 17–24 (2015). https://doi.org/10.1140/epjst/e2015-02339-y
10. Shor, P.W.: Algorithms for quantum computation: discrete logarithms and factoring. In: Proceedings 35th Annual Symposium on Foundations of Computer Science, pp. 124–134. IEEE (1994)
11. Sparkes, M.: A new quantum leader? (2021)
12. Wang, B., Hu, F., Yao, H., Wang, C.: prime factorization algorithm based on parameter optimization of ising model. Sci. Rep. **10**(1), 1–10 (2020)
13. Wroński, M.: Index calculus method for solving elliptic curve discrete logarithm problem using quantum annealing. In: Paszynski, M., Kranzlmüller, D., Krzhizhanovskaya, V.V., Dongarra, J.J., Sloot, P.M.A. (eds.) ICCS 2021. LNCS, vol. 12747, pp. 149–155. Springer, Cham (2021). https://doi.org/10.1007/978-3-030-77980-1_12

Quantum-Classical Solution Methods for Binary Compressive Sensing Problems

Robert S. Wezeman[1(✉)], Irina Chiscop[1], Laura Anitori[2], and Wim van Rossum[2]

[1] Department of Cyber Security and Robustness, The Netherlands Organisation for Applied Scientific Research, The Hague, Netherlands
{robert.wezeman,irina.chiscop}@tno.nl

[2] Department of Radar Technology, The Netherlands Organisation for Applied Scientific Research, The Hague, Netherlands
{laura.anitori,wim.vanrossum}@tno.nl

Abstract. Compressive sensing is a signal processing technique used to acquire and reconstruct sparse signals using significantly fewer measurement samples. Compressive sensing requires finding the most sparse solution to an underdetermined linear system, which is an NP-hard problem and as a consequence in practise is only solved approximately. In our work we restrict ourselves to the compressive sensing problem for the case of binary signals. For that case we have defined an equivalent formulation in terms of a quadratic binary optimisation (QUBO) problem, which we solve using classical and (hybrid-)quantum computing solving techniques based on quantum annealing. Phase transition diagrams show that this approach significantly improves the number of problem types that can be successfully reconstructed when compared to a more conventional $\mathcal{L}_1$ optimisation method. A challenge that remain is how to select optimal penalty parameters in the QUBO formulation as was shown can heavily impact the quality of the solution.

Keywords: Binary compressive sensing · Quadratic unconstrained binary optimisation · Quantum annealing

1 Introduction

The Nyquist sampling theorem states that to be able to perfectly reconstruct a signal, it has to be sampled at a rate that is at least twice the highest frequency in the original signal, the *Nyquist rate* [28]. Unfortunately for many applications, this Nyquist rate is too high for sampling to be feasible in practise. Let $\boldsymbol{x} \in \mathbb{R}^N$ be a real N-dimensional signal and $\{\boldsymbol{\Psi}_i\}_{i=1}^N$ be a set of basis vectors. We can write a signal as $\boldsymbol{x} = \sum_{i=1}^N s_i \boldsymbol{\Psi}_i$. The signal is said to be K-sparse in the basis Ψ if the N-dimensional vector $\boldsymbol{s}$ with coefficients s_i has at most K nonzero elements. Compressive sensing (CS) is a signal processing technique that makes it possible to acquire and reconstruct a signal more efficiently, given that the signal has

D. Groen et al. (Eds.): ICCS 2022, LNCS 13353, pp. 107–121, 2022.
https://doi.org/10.1007/978-3-031-08760-8_9

a sparse representation ($K \ll N$) in some basis [10]. CS has a large number of applications in several fields ranging from (medical) imaging, communication systems and pattern recognition, to speech and sound processing [26].

The task of compressive sensing is equivalent to finding the most sparse solution of an underdetermined linear system. Consider that problem, recovering the most sparse vector $\boldsymbol{x} \in \mathbb{R}^N$ from a undersampled set of $M \ll N$ measurements $\boldsymbol{y} \in \mathbb{R}^M$. These vectors are related by what is called the measurement matrix $A \in \mathbb{R}^{M \times N}$ which in this article has its elements drawn independent and identically distributed random from some distribution on $\mathbb{R}$. The reconstruction problem is then described as follows

$$\min_x \|\boldsymbol{x}\|_0 \quad \text{subject to} \quad A\boldsymbol{x} = \boldsymbol{y}, \tag{1}$$

where $\|\cdot\|_0$ is the $\mathcal{L}_0$-norm, the number of non-zero elements of a vector. To guarantee a unique solution to this problem, additional constraints on the measurement matrix are required, such as the Restricted Isometry Property (RIP) [15,16]. Given that this problem can be solved uniquely, compressive sensing then enables a much shorter signal acquisition time together with reduced amounts of data, because the needed number of samples M can be much smaller than what would be required according to the Nyquist rate. Unfortunately, this problem does not have a closed form solution and furthermore, it is an NP-complete problem and thus hard to solve having combinatorial complexity [7,10].

Summarised, compressed sensing based signal acquisition allows a significant decrease in the sampling rate of sparse signals, but in return requires a hard optimisation problem to be solved. Algorithms used to solve the sparse reconstruction problem tend to be very slow and rely heavily on the speed at which matrix-vector multiplications can be done for the measurement matrix A. An overview of different computational techniques for solving the sparse reconstruction problem is given in [30]. One of the major algorithmic approaches relaxes the original problem by replacing the combinatorial $\mathcal{L}_0$ norm on $\boldsymbol{x}$ by the $\mathcal{L}_1$-norm, the sum of the magnitudes of a vector $\boldsymbol{x}$. The obtained problem, given by Eq. (2), becomes a convex optimisation problem which can then be solved using standard convex optimisation routines.

$$\min_x \|\boldsymbol{x}\|_1 \quad \text{subject to} \quad A\boldsymbol{x} = \boldsymbol{y}. \tag{2}$$

Candès and Romberg proof that $\mathcal{L}_1$ optimisation obtains the exact solution to Eq. (1) under the constraint that the number of samples $M > C\mu^2(A)K \log N$, where $\mu(A) = \sqrt{N} \max_{k,j} |A_{k,j}|$, K is the sparsity of $\boldsymbol{x}$ and C is some constant [14]. The required computation time, even in the case when convex optimisation routines are used, can still remain an obstacle for real-time applications and is limited by the available computational resources. This is due to the iterative nature of the algorithms, in which each iteration is closely related to the corresponding processing step in conventional processing. Therefore, when many iterations are required this results in a significant increase in computations.

Meanwhile, a new computing paradigm, quantum computing, is quickly approaching us. It is expected that quantum computers will be able to solve specific problems faster than the current generation of classical computer are capable of. In [11] thirteen applications for radar and sonar information processing have been identified, among which compressive sensing, that can possibly be improved using quantum computing.

In this work, we make first steps towards a quantum computing approach for compressive sensing. We simplify the problem by only considering binary valued signals and formulate that problem as a quadratic unconstrained binary optimisation (QUBO) problem. Binary sparse signal recovery is relevant for applications such as event detection in wireless sensor networks [29], group testing [27] and spectrum hole detection for cognitive radios [21]. Using quantum annealing techniques to solve binary CS has been studied before by Ayanzadeh et al. [8,9], however, only limited number of numerical results have there been given. We extend upon their work by solving the binary CS problem using both classical techniques as well as annealing based (hybrid-)quantum computing techniques. The obtained results are compared with a more conventional $\mathcal{L}_1$ optimisation approach by looking at phase transition diagrams.

The structure of this paper is as follows. In Sect. 2, we describe the considered problem of binary compressive sensing and formulate it as a QUBO problem. In Sect. 3, we describe solution methods and we describe how to create phase transition diagrams which can be used to quantitative compare different solution methods. In Sect. 4, the numerical results are described and we end with Sect. 5, where conclusions are given together with directions for further research.

2 Problem Formulation

In this paper we focus on the original problem of recovering binary signals from a limited number of measurements, which can be formulated as follows:

$$\min_{x \in \{0,1\}} \|\boldsymbol{x}\|_0 \quad \text{subject to} \quad \|A\boldsymbol{x} - \boldsymbol{y}\|_2 = 0. \tag{3}$$

This problem is known to be NP-complete [9], however it can be tackled using quantum annealers. This type of quantum devices require either an Ising or Quadratic Unconstrained Binary Optimisation (QUBO) problem formulation [5]. As shown in [9] Eq. (3) can be translated to the following QUBO:

$$\min_{\boldsymbol{x} \in \{0,1\}^N} \gamma \|\boldsymbol{x}\|_0 + \|A\boldsymbol{x} - \boldsymbol{y}\|_2^2. \tag{4}$$

After expanding the norms, this translates to:

$$\min_{\boldsymbol{x} \in \{0,1\}^N} \quad \sum_i \Big(\gamma + \sum_l A_{li}[-2y_l + A_{li}]\Big) x_i + \sum_{i,j;i<j} \Big(2 \sum_k A_{ki} A_{kj}\Big) x_i x_j. \tag{5}$$

We note that the objective in Eq. (5) represents the same optimization problem as in Eq. (4) but its optimal objective value is smaller by $\|\boldsymbol{y}\|_2^2$, since this constant term does not get taken into account in the QUBO. This formulation, which applies a penalty parameter γ to the $\mathcal{L}_0$ norm, is the only one we encountered in literature [9]. However, one can also opt to multiply a penalty parameter λ with the constraint term:

$$\min_{\boldsymbol{x}\in\{0,1\}^N} \|\boldsymbol{x}\|_0 + \lambda \|A\boldsymbol{x} - \boldsymbol{y}\|_2^2. \tag{6}$$

which is then equivalent to:

$$\min_{\boldsymbol{x}\in\{0,1\}^N} \sum_i \Big(1 + \lambda \sum_l A_{li}[-2y_l + A_{li}]\Big)x_i + \sum_{i,j;i<j} \Big(2\lambda \sum_k A_{ki}A_{kj}\Big)x_i x_j. \tag{7}$$

From a theoretical point of view, the two formulations in Eq. (5) and Eq. (7) are equivalent when $\gamma = \frac{1}{\lambda}$. When γ and λ are set to appropriate values, both should lead to the optimal value of $\boldsymbol{x}$. However, quantum annealing-based solvers are sensitive to small variations in the QUBO values, due to the finite precision available [4]. This means that depending on the problem size (number of measurements M and size of signal N), and on the entries of matrix A, one formulation may results in better solution quality than the other. From now on we identify the formulation in Eq. (5) as QUBO type 1 and the formulation in Eq. (7) as QUBO type 2.

3 Solution Methods and Evaluation

For solving the binary CS problem we consider a classical approach for $\mathcal{L}_1$-minimisation and several quantum-based and hybrid solution methods which address $\mathcal{L}_0$-minimisation. When solving usual combinatorial problems, one would look at the optimality gap or lower and upper bounds as metrics to describe the quality of the solutions found. However, in compressive sensing, phase transition diagrams [18] are widely used to indicate whether a certain algorithm is successful in recovering the signal under certain conditions. In this section we describe all the solution methods used and elaborate on the use of phase diagrams in CS applications.

3.1 $\mathcal{L}_1$-Minimisation

As already described in Sect. 1, signal recovery can also be tackled by solving a convex $\mathcal{L}_1$ minimisation problem, which is equivalent under certain strict conditions on the number of measurements taken. Alternating Direction Method of Multipliers (ADMM) is a technique that has proven itself to be suitable for solving distributed convex optimisation problems and in particular large-scale problems arising in compressive sensing [13]. In this paper we focus on a set of publicly available solvers [32] implemented in Matlab that apply ADMM [31] to solve the following problem:

$$\min_{\boldsymbol{x}\in\mathbb{R}^N} \|\boldsymbol{x}\|_1 + (1/2\rho) \|A\boldsymbol{x} - \boldsymbol{y}\|_2^2. \tag{8}$$

where $\rho = \frac{0.01}{\|\boldsymbol{y}\|_\infty}$ and $\|\cdot\|_\infty$ is the infinity norm, the element with the largest absolute value of a vector. We have chosen this solver as a benchmark for $\mathcal{L}_1$-based approaches.

3.2 $\mathcal{L}_0$-Minimisation

The QUBO formulations given by Eq. (5) and Eq. (7) corresponding to the $\mathcal{L}_0$-minimisation binary CS problem, were solved using a classical commercial solver and built-in algorithms of the Ocean tool suite provided by D-Wave Systems [1]. In the following paragraphs, we provide some background on each of these algorithms.

Gurobi Optimiser is a commercial optimisation software library for solving mixed-integer linear and quadratic optimisation problems.[1] It is widely used for large-scale applications in different industries [6]. We have implemented this solver using the *gurobipy* Python package. To ensure a reasonable computation time, we have limited the Gurobi running time per problem to 30 s.

Simulated Annealing (SA) is a stochastic meta-heuristic [17], which emerged as a local search method that can escape from being trapped in local optima. There are two stochastic steps in simulated annealing. First, a solution s' is chosen based from the set of neighbors of the current solution s according to some given distribution (each neighbor usually has the same probability). Then, the chosen solution is accepted with probability $p(s, s', c) = e^{\min\{c(s)-c(s'),0\}/c}$, where c is a positive control parameter which decreases with increasing number of iterations and converges to 0. The performance of the algorithm relies on the cooling schedule, i.e., which specifies the initial and final values of the control parameter c together with a decrement function. In our numerical experiments, we employed the simulated annealing algorithm provided by D-Wave Systems with default parameters. Further details concerning the annealing (cooling) schedule and parameters can be found in the official software documentation [2]. Similar to Gurobi, we have imposed a limit of 30 s to the simulated annealing process on each problem.

Quantum Annealing is a meta-heuristic which utilizes quantum fluctuations [22] and quantum computation by adiabatic evolution [20] in solving a particular type of optimisation problems, and is currently implemented in the quantum devices produced by D-Wave Systems. The evolution of a quantum state on D-Wave's Quantum Processing Unit (QPU) is described by a time-dependent Hamiltonian, composed of initial Hamiltonian H_0, whose ground state is easy to create, and final Hamiltonian H_1, whose ground state encodes the solution of the problem at hand:

$$H(t) = \left(1 - \frac{t}{T}\right) H_0 + \frac{t}{T} H_1. \tag{9}$$

The system in Eq. (9) is initialized in the ground state of the initial Hamiltonian, i.e. $H(0) = H_0$. The adiabatic theorem states that if the system evolves according to the Schrödinger equation, and the minimum spectral gap of $H(t)$ is not zero, then for time T large enough, $H(T)$ will converge to the ground state of H_1, which encodes the solution of the problem. The D-Wave quantum annealer accepts

[1] https://www.gurobi.com/products/gurobi-optimizer/.

as input H_1 either as an Ising Hamiltonian, or as its equivalent formulation, the QUBO. We remark that although Eq. (9) suggests that the annealing is performed linearly, in practice D-Wave uses a non-linear annealing schedule. Additionally, the chosen formulation needs to be embedded on the hardware. The current architecture of D-Wave's Advantage 4.1 quantum computing system contains approximately 5000 qubits, with a total number of 35.000 qubit couplers (each qubit is connected to 15 other qubits) [24]. When the problem size is too large, such an embedding is impossible and so the quantum annealing process cannot be applied to the entire problem. In this scenario, hybrid algorithms need to be employed, in which the problem is first decomposed into smaller sub-problems, the each of these being solved with quantum annealing and at the end, the complete solution vector is reconstructed from all sub-sample solutions. In this paper, we use the term 'quantum annealing' only for situations in which the entire problem can be passed on to the D-Wave QPU, without any need for decomposition.

QBsolv is a hybrid solver introduced by D-Wave to tackle very large optimisation problems [12]. It combines a problem decomposition step with classical search algorithms and quantum annealing to find a global optimal solution. At first, an initial solution for the entire problem is obtained using classical Tabu search. Based on this solution, the problem is split into sub-problems of smaller size by ordering the problem variables according to their impact on the objective function. Then, each of these sub-problems is solved using quantum annealing, and the current solution is updated with the corresponding bits from the sub-problem vector. Then, a new Tabu search is applied starting from the current solution, and the algorithm loop is repeated until a maximum number of iterations is reached. QBsolv makes use of the different numerical precision in classical algorithms and quantum annealing. Whilst solving the sub-problems with quantum annealing on D-Wave is affected by limited precision, the classical Tabu search can solve the full problem QUBO in the standard format of IEEE double-precision (64-bit) floating-point values [12]. In this paper we have used the built-in implementation of the D-Wave Ocean tool suite[2] with default parameters.

Hybrid Solvers are offered by D-Wave Systems to enable solving arbitrarily large optimisation problems [3]. There are two types of resources available: cloud-based hybrid solvers (also known as Leap hybrid solvers) and the *dwave-hybrid* Python framework, which allows the creation of custom hybrid workflows. We focused on the first category, since according to D-Wave's software documentation [3], these solvers implement state-of-the-art classical algorithms together with intelligent allocation of the QPU to sections of the problem where it is most beneficial. The classical components herein utilize quantum queries to the D-Wave QPU to guide their search of the larger solution space. Whilst the generic structure of Leap's hybrid solvers is described in a technical report [25], no details concerning the implemented algorithms (which classical

[2] https://github.com/dwavesystems/qbsolv/.

meta-heuristics are implemented, how they utilize the solution output from the QPU etc.) are disclosed. We therefore remain cautious when interpreting the results obtained with the LeapHybridSampler, the hybrid solver that was selected for our problem.

3.3 Phase Transition Diagrams

Maleki and Donoho [23] have developed a framework that can be used to compare quantitatively the properties of different reconstruction algorithms. Performance is measured by what is called the undersampling-sparsity trade-off, and can be visualised by a phase transition curve. The idea is to find the "breakdown point", the point where an algorithm can still successfully reconstruct a sparse solution provided K is smaller than a certain definite fraction of N.

We define $\delta = \frac{M}{N}$ to be a measure of problem indeterminacy and define $\rho = \frac{K}{M}$ to be a measure for the sparsity of the problem. The difficulty of a problem instance can be visualised by its point in the two-dimensional phase space $(\delta, \rho) \in [0,1] \times [0,1]$. In Fig. 1, examples are shown how a typical phase transition diagram looks, in this case for different problem sizes, using the $\mathcal{L}_1$ optimisation recovery approach as described in Sect. 3.1. The hardest problems are in the top left of this plane, trying to reconstruct a not so sparse signal with relatively few measurements. The different colors represent the probability for successful reconstruction of the sparsest solution. This figure illustrates the typically behaviour of reconstruction algorithms, below a certain threshold the algorithms works well, while above that threshold reconstruction fails. Typically the transition zone becomes narrow and better defined for larger problem sizes and depends on both the reconstruction algorithm but also the type of measurement matrix [23].

In this paper we will only consider Gaussian random measurement matrices. A Gaussian random measurement matrix can be generated by sampling its elements a_{ij} independent and identically distributed from the standard normal distribution with mean zero and standard deviation one. This type of matrix is commonly used in CS as it is known to satisfy the RIP condition with high probability [15] and has its phase transition shape well studied [19].

For a given reconstruction algorithm, for a fixed problem size N, the construction of phase transition diagrams is as follows:

1. Create a grid by varying both parameters δ and ρ in $(0, 1)$.
2. For each combination (δ, ρ) we calculate $M = \lceil \delta N \rceil$ and $K = \lceil \delta \rho N \rceil$.
3. At each combination (δ, ρ) we create L problem instances $(A, \boldsymbol{x}, \boldsymbol{y})$ and obtain L algorithm outputs $\boldsymbol{x}_{solve}$.
4. For each problem instance we declare it as a successful reconstruction if

$$\frac{\|\boldsymbol{x} - \boldsymbol{x}_{solve}\|_2}{\|\boldsymbol{x}\|_2} \leq tol,$$

for some tolerance parameter tol, where $\|\cdot\|_2$ is the standard Euclidean norm.

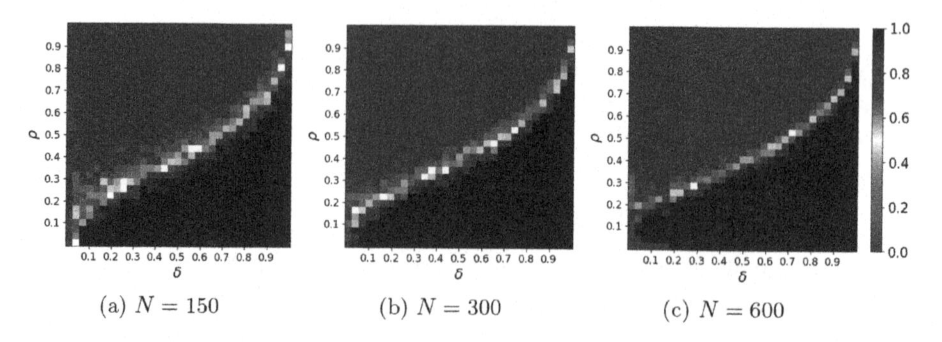

(a) $N = 150$ (b) $N = 300$ (c) $N = 600$

Fig. 1. Phase transition plots generated with Matlab-$\mathcal{L}_1$ solver for Gaussian random measurement matrices and 20 problem instances per step. The color indicate the probability that reconstruction for $(\delta = \frac{M}{N}, \rho = \frac{K}{M})$ instances is successful.

4 Results

In this section we present an overview of the various numerical experiments that we performed. We studied different aspects of binary compressing sensing problems focusing on two aspects: the QUBO penalty parameter (γ or λ) and the QUBO formulation. Once suitable choices have been identified for these parameters, we also performed a comparison of different classical and (hybrid-)quantum solvers. All phase transition plots in this chapter are generated with an error tolerance value $tol = 0.1$ and a grid step size of 33 for both δ and ρ.

4.1 Impact of Penalty Parameter

When using a QUBO-based approach to solve the binary compressive sensing, we need to find suitable values for the penalty parameter γ (or λ). To this end, we created phase diagrams for signal size $N = 300$, using simulated annealing for $\gamma \in \{1, 5, 10, 15, 20, 25, 30, 35, 40\}$. We have chosen simulated annealing due to limited computation time available on the D-Wave QPU. Figure 2 shows all the resulting diagrams. From this figure we observe that the optimal choice of γ ranges between 15 and 25, as these values result in the largest success regions. It is also worthwhile to notice that in the case of $\gamma \in \{1, 5, 10\}$ the Simulated Annealing algorithm achieves successful signal recovery for $\delta > 0.7$, independent of the sparsity of the signal. The phase transition diagrams obtained for $\gamma = 5$, or $\gamma = 10$ look counter-intuitive. In these examples, if the algorithm is successful in the $\delta, \rho \in [0, 0.1] \times [0, 0.1]$ where the signal is sparse and the measurements are very few, it should theoretically also be capable of recovering signals with the same sparsity when more measurements are done, i.e. in regions $\delta, \rho \in [0, 0.1] \times [0.1, 0.4]$. This phenomenon could be explained by the time-out limit of 30s set on the simulated annealing solver, which may struggle to distinguish between many 'good' solutions in the search space. The variation in obtained phase transition suggests that γ is dependent not only on δ and ρ (or equivalently, M, N, K) but also on the entries of the matrix A. The optimal value of γ obtained using simulated annealing seems to be a suitable choice for QBsolv as well. Figure 3a shows the phase transition diagram obtained with QBsolv and $\gamma = 20$. The region of successful signal recovery is quite large, especially when compared to any of the simulated annealing phase transition diagrams.

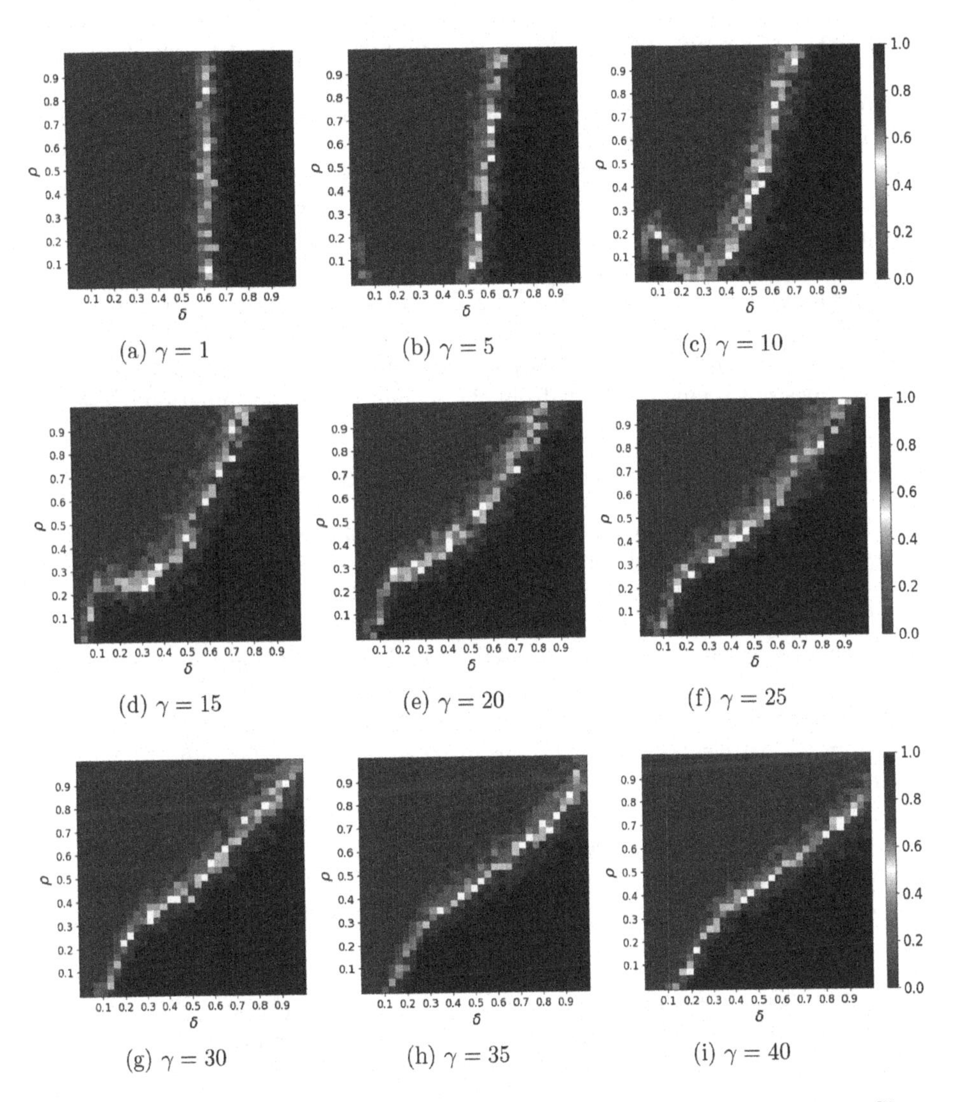

Fig. 2. Phase transition diagrams generated for $N = 300$ using QUBO type 1, Simulated Annealing and 20 problem instances per step. The color indicate the probability that reconstruction for $(\delta = \frac{M}{N}, \rho = \frac{K}{M})$ instances is successful.

4.2 Evaluation of Different QUBO Formulations

In this subsection we investigate whether the second QUBO formulation (QUBO type 2) presented in Sect. 3 exhibits a different and eventually improved performance in comparison to the QUBO type 1 employed so far. Recall that the two QUBO formulations presented are theoretically equivalent, when $\gamma = \frac{1}{\lambda}$. Figure 3b and Fig. 3c show phase transition diagrams obtained with QBsolv and QUBO type 2 for two different λ values. The QUBO type 2 formulation with $\lambda = 0.1$ delivers a phase transition which is very similar to the QUBO type 1 formulation with $\gamma = 20$. The size of the successful recovery region is essentially identical, with slight trade-off in a few points. On the other hand, the QUBO type 2 formulation with $\lambda = 10$ results in a very different phase transition, which suggests that under this parameter setting, the QBsolv performance is mostly dependent on the number of measurements taken and not the sparsity rate. The same experiment was performed using simulated annealing. We see that equivalent formulations give the same results. QUBO type 2 with $\lambda = 0.1$ in Fig. 3e yields similar phase transition as QUBO type 1 with $\gamma = 10$ in Fig. 2c; the same holds for Fig. 3f and Fig. 2a). Based on experiments, the simulated annealing sampler does not seem to benefit from the second QUBO formulation.

4.3 Comparison of Classical and Quantum Solvers

To compare the performance of different algorithms we decided to solve the binary compressive sensing problem with all the solvers described in Sect. 3. Since we had limited computation time on the D-Wave QPU, we chose smaller signal sizes, with $N = 20$ with $N = 40$, and performed only one trial at each step of the phase diagram. We also opted for Qubo formulation 2 with $\lambda = 10$ as parameter, since it seemed to yield a phase transition which was not dependent on the sparsity rate. The results of these experiments can be visualized in Fig. 4. For $N = 20$ we clearly see that Gurobi, together with the hybrid-quantum solvers, QBsolv and D-Wave hybrid, achieve the best performance having the largest region of successful signal recovery. The Matlab-$\mathcal{L}_1$ solver displays the expected behaviour, for both values of N which were considered. Both simulated annealing and the quantum annealing on the D-Wave QPU, with some exception, can recover signals successfully if enough measurements, i.e., $M > 0.6N$ are performed. For $N = 40$, we notice a decrease in the performance of all methods. In particular, quantum annealing suffers from the lack of custom parameter tuning. In this case, it is Gurobi that achieves the largest success region, followed by QBsolv and DWave-hybrid which perform worse in the upper-left corner of the diagram, where there are very few measurements taken.

Finally, we note that all results presented so far may heavily be influenced by the lack of custom parameter optimization. It is likely that each QUBO-based solver has a different optimum QUBO formulation and penalty value that can depend on the specific measurement matrix. However, due to the lack of computation resources, we have not been able to perform such parameter tuning.

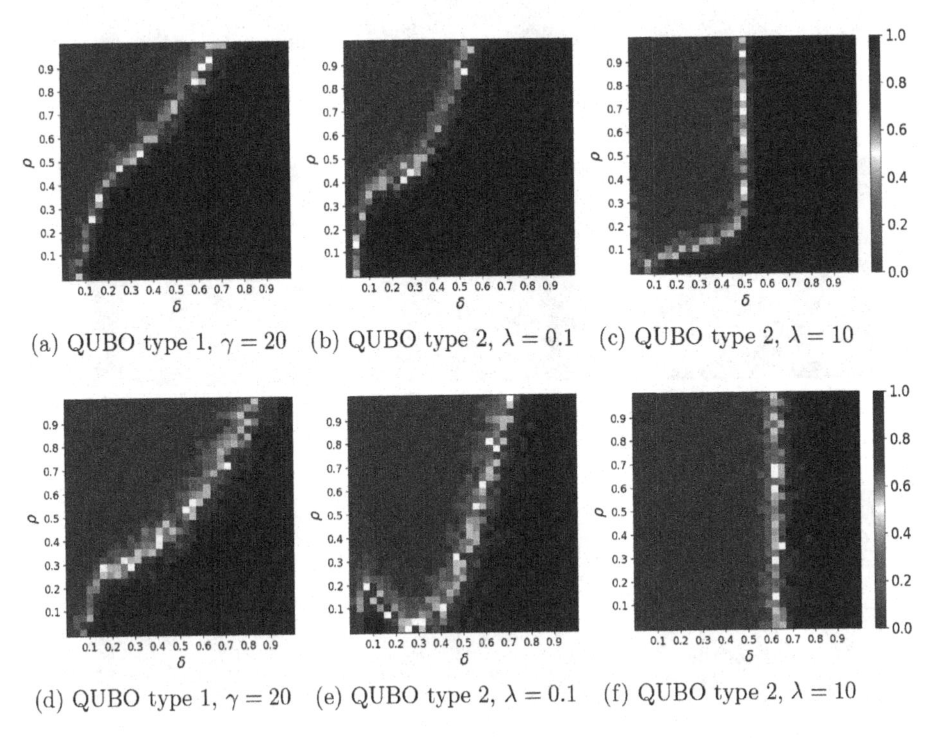

(a) QUBO type 1, $\gamma = 20$ (b) QUBO type 2, $\lambda = 0.1$ (c) QUBO type 2, $\lambda = 10$

(d) QUBO type 1, $\gamma = 20$ (e) QUBO type 2, $\lambda = 0.1$ (f) QUBO type 2, $\lambda = 10$

Fig. 3. Phase diagrams with QBsolv (a, b and c) and simulated annealing (d, e and f), $N = 300$, and 20 problems per step. The color indicate the probability that reconstruction for ($\delta = \frac{M}{N}, \rho = \frac{K}{M}$) instances is successful.

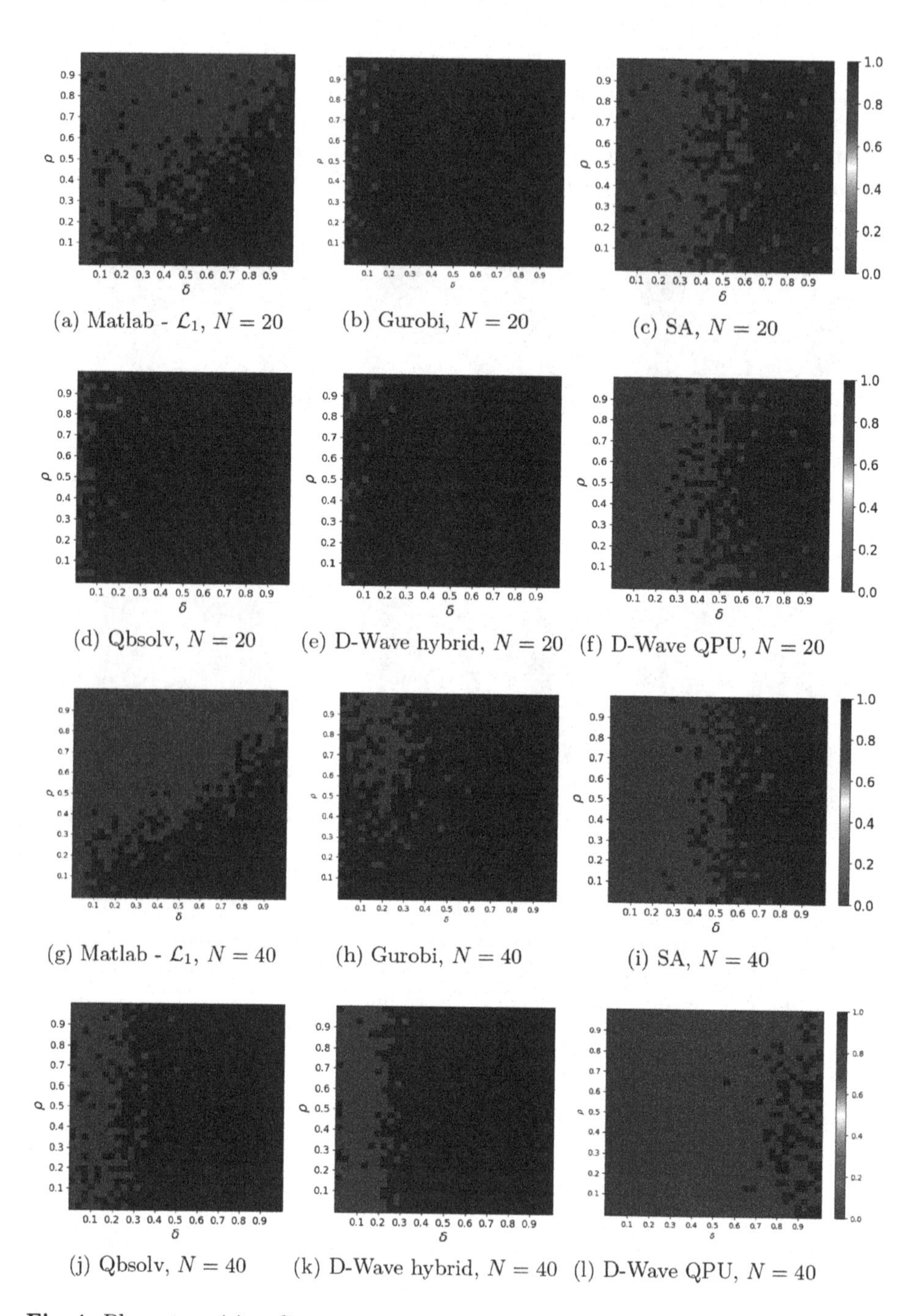

(a) Matlab - $\mathcal{L}_1$, $N = 20$ (b) Gurobi, $N = 20$ (c) SA, $N = 20$

(d) Qbsolv, $N = 20$ (e) D-Wave hybrid, $N = 20$ (f) D-Wave QPU, $N = 20$

(g) Matlab - $\mathcal{L}_1$, $N = 40$ (h) Gurobi, $N = 40$ (i) SA, $N = 40$

(j) Qbsolv, $N = 40$ (k) D-Wave hybrid, $N = 40$ (l) D-Wave QPU, $N = 40$

Fig. 4. Phase transition diagrams generated $N = 20$ and $N = 40$ using 1 problem instance per step. QUBO problems have been solved with QUBO type 2 with $\lambda = 10$. The color indicate the probability that reconstruction for ($\delta = \frac{M}{N}, \rho = \frac{K}{M}$) instances is successful.

5 Conclusions

In this paper we tackled the binary compressive sensing problem and provide the first numerical results using a real quantum device. Different classical and (hybrid-)quantum QUBO solvers have been compared quantitatively to a classical $\mathcal{L}_1$-based approach by calculating phase-transition diagrams.

Based on the results presented in Sect. 4, we conclude that using a QUBO approach for solving the binary compressive sensing problem, the resulting phase-transition diagrams are significantly improved from the classical $\mathcal{L}_1$-based approach. For the Gaussian measurement matrices considered, a clear phase transition could be identified for all classical and (hybrid)-quantum solvers. In particular, for hybrid approaches such as QBsolv and D-Wave hybrid the phase transition was found to only depend on the undersampling rate, and not the sparsity of the signal. This is a positive result as the number of measurement taken is in principle a controllable quantity in signal recovery experiments.

For the small problem instances considered we see that the Gurobi solver slightly outperforms the hybrid-quantum annealing approaches. The hope is, once hardware grows and larger problems can be embedded on the quantum chip, that quantum annealing becomes a option when problem sizes become so large that they are unfeasible to solve using classical approaches such as Gurobi. Nevertheless our work shows the high potential of QUBO-based formulations compared to $\mathcal{L}_1$ minimization. The QUBOs presented in this paper can also be adapted to handle real or complex-valued through the appropriate usage of slack variables.

Currently, we do not have a full understanding yet on how to optimally select the different parameters (QUBO type, penalty value) for each solver. We expect that the choice of value for the penalty parameters γ or λ when using the QUBO formulation depends on ρ, δ and also the entries of the measurement matrix A. Hence, the only way to eventually infer an expression of γ or λ is to design a structured grid search, and evaluate a large set of values for each problem instance considered at each point in the phase diagram. This is one of the aspects we consider worthwhile for further investigation.

Acknowledgements. This work was supported by the Dutch Ministry of Defense under Grant V2104.

References

1. D-Wave Ocean Software Documentation. https://docs.ocean.dwavesys.com/en/stable/. Accessed 11 Dec 2021
2. D-Wave Ocean Software Documentation: Simulated Annealing Sampler. https://docs.ocean.dwavesys.com/en/stable/docs_neal/reference/sampler.html. Accessed 08 Jun 2022
3. D-Wave Problem Solving Handbook: Using Hybrid Solvers. https://docs.dwavesys.com/docs/latest/handbook_hybrid.html

4. D-Wave System online documentation: QPU-specific characteristics. https://docs.dwavesys.com/docs/latest/handbook_qpu.html. Accessed 08 Jun 2022
5. D-Wave System online documentation: what is quantum annealing? https://docs.dwavesys.com/docs/latest/c_gs_2.html#getting-started-qa. Accessed 14 Jan 2022
6. State of Mathematical Optimization Report. Technical report, Gurobi Optimization (2021)
7. Anitori, L.: Compressive sensing and fast simulations, applications to radar detection. Ph.D. thesis, TU Delft (2012)
8. Ayanzadeh, R., Halem, M., Finin, T.: An ensemble approach for compressive sensing with quantum. arXiv e-prints arXiv:2006.04682, June 2020
9. Ayanzadeh, R., Mousavi, S., Halem, M., Finin, T.: Quantum annealing based binary compressive sensing with matrix uncertainty. arXiv e-prints arXiv:1901.00088, December 2018
10. Baraniuk, R.G.: Compressive sensing [lecture notes]. IEEE Signal Process. Mag. **24**(4), 118–121 (2007)
11. Bontekoe, T.H., Neumann, N.M.P., Phillipson, F., Wezeman, R.S.: Quantum computing for radar and sonar information processing (2021, unpublished)
12. Booth, M., Reinhardt, S.P., Roy, A.: Partitioning optimization problems for hybrid classical/quantum execution. Technical report, D-Wave: The Quantum Computing Company, October 2018
13. Boyd, S., Parikh, N., Chu, E., Peleato, B., Eckstein, J.: Distributed optimization and statistical learning via the alternating direction method of multipliers. Found. Trends Mach. Learn. **3**(1), 1–122 (2011)
14. Candès, E., Romberg, J.: Sparsity and incoherence in compressive sampling. Inverse Probl. **23**(3), 969–985 (2007)
15. Candès, E.J., Romberg, J.K., Tao, T.: Stable signal recovery from incomplete and inaccurate measurements. Commun. Pure Appl. Math. **59**, 1207–1223 (2005)
16. Candès, E.J.: The restricted isometry property and its implications for compressed sensing. C.R. Math. **346**(9), 589–592 (2008)
17. Dekkers, A., Aarts, E.: Global optimization and simulated annealing. Math. Program. **50**(1–3), 367–393 (1991)
18. Donoho, D., Tanner, J.: Observed universality of phase transitions in high-dimensional geometry, with implications for modern data analysis and signal processing. Philos. Trans. R. Soc. A Math. Phys. Eng. Sci. **367**(1906), 4273–4293 (2009)
19. Donoho, D.: Compressed sensing. IEEE Trans. Inf. Theory **52**(4), 1289–1306 (2006)
20. Farhi, E., Goldstone, J., Gutmann, S., Sipser, M.: Quantum computation by adiabatic evolution. arXiv:quant-ph/0001106v1 (2000)
21. Hayashi, K., Nagahara, M., Tanaka, T.: A user's guide to compressed sensing for communications systems. IEICE Trans. Commun. **E96.B**(3), 685–712 (2013)
22. Kadowaki, T., Nishimori, H.: Quantum annealing in the transverse Ising model. Phys. Rev. E **58**, 5355–5363 (1998)
23. Maleki, A., Donoho, D.L.: Optimally tuned iterative reconstruction algorithms for compressed sensing. IEEE J. Sel. Top. Signal Process. **4**(2), 330–341 (2010)
24. McGeoch, C., Farré, P.: The Advantage System: Performance Update. Technical report, D-Wave: The Quantum Computing Company, October 2021
25. McGeoch, C., Farré, P., Bernoudy, W.: D-Wave Hybrid Solver Service + Advantage: Technology Update. Technical report, D-Wave: The Quantum Computing Company, September 2020
26. Rani, M., Dhok, S.B., Deshmukh, R.B.: A systematic review of compressive sensing: concepts, implementations and applications. IEEE Access **6**, 4875–4894 (2018)

27. Romanov, E., Ordentlich, O.: On compressed sensing of binary signals for the unsourced random access channel. Entropy **23**(5), 605 (2021)
28. Shannon, C.: Communication in the presence of noise. Proc. IRE **37**(1), 10–21 (1949)
29. Shirvanimoghaddam, M., Li, Y., Vucetic, B., Yuan, J., Zhang, P.: Binary compressive sensing via analog fountain coding. IEEE Trans. Signal Process. **63**(24), 6540–6552 (2015)
30. Tropp, J.A., Wright, S.J.: Computational methods for sparse solution of linear inverse problems. Proc. IEEE **98**(6), 948–958 (2010)
31. Yang, J., Zhang, Y.: Alternating direction algorithms for $\mathcal{L}_1$-problems in compressive sensing. SIAM J. Sci. Comput. **33**, 250–278 (2011)
32. Zhang, Y.: User's Guide for YALL1: Your ALgorithms for L1 Optimization. Technical report, Rice University, Houston, Texas (2009)

Studying the Cost of n-qubit Toffoli Gates

Francisco Orts(✉), Gloria Ortega, and Ester M. Garzón

Supercomputing -Algorithm Research Group, Informatics Department,
University of Almería, Almería, Spain
{francisco.orts,gloriaortega,gmartin}@ual.es
https://hpca.ual.es/

Abstract. There are several Toffoli gate designs for quantum computers in the literature. Each of these designs is focused on a specific technology or on optimising one or several metrics (T-count, number of qubits, etc.), and therefore has its advantages and disadvantages. While there is some consensus in the state of the art on the best implementations for the Toffoli gate, scaling this gate for use with three or more control qubits is not trivial. In this paper, we analyse the known techniques for constructing an n-qubit Toffoli gate, as well as the existing state-of-the-art designs for the 2-qubit version, which is an indispensable building block for the larger gates. In particular, we are interested in a construction of the temporary logical-AND gate with more than two control qubits. This gate is widely used in the literature due to the T-count and qubit reduction it provides. However, its use with more than two control qubits has not been analysed in detail in any work. The resulting information is offered in the form of comparative tables that will facilitate its consultation for researchers and people interested in the subject, so that they can easily choose the design that best suits their interests. As part of this work, the studied implementations have been reproduced and tested on both quantum simulators and real quantum devices.

Keywords: Quantum circuits · Toffoli gate · n-qubit Toffoli gate

1 Introduction

The circuit paradigm is the most widely used paradigm for programming a quantum computer [11]. This paradigm consists of using quantum gates (unitary operations) to manipulate the information contained in qubits. In the classical world, it is possible to limit the number of existing logic gates. For example, there are only two classical gates that act on a bit: the NOT gate and the identity. However, in the quantum case, there are infinitely many gates that can act on 1 qubit, given their special nature [4]. Although it is impossible to have a universal set of gates that generate the rest of the infinite quantum gates (although there is a set that allows us to approximate them), there are certain gates that are well known in the community and widely used due to the operation they perform. This is the case of the Toffoli gate, which, given three values c_1, c_2, and t, performs the

D. Groen et al. (Eds.): ICCS 2022, LNCS 13353, pp. 122–128, 2022.
https://doi.org/10.1007/978-3-031-08760-8_10

operation $t \oplus c_1c_2$. Using the correct notation, the operation can be expressed as $Toffoli(|c_1\rangle|c_2\rangle|t\rangle) = |c_1\rangle|c_2\rangle|t \oplus c_1c_2\rangle$. By convention, c_1 and c_2 are usually called control qubits, and t target qubit. A simple example of the usefulness of this operation is to operate on $t = 0$ and considering only the standard bases as possible values for c_1 and c_2. In such a case, the result c_1c_2 coincides with the classical AND operation. The Toffoli gate is useful in operations as varied as adders [12], cryptography [17], image processing [13], etc.

When designing a quantum circuit, one should try to make it as small as possible in order to optimise the use of resources. In fact, a small circuit is a very valuable resource even if it has no quantum properties, since it can be used as part of major circuits [15]. However, it is not always easy to measure the cost of a circuit in order to determine if it is "smaller" than other. Two metrics are particularly important in today's NISQ devices: the number of ancilla qubits, and the so-called T-count. Regarding the ancilla qubits, it is necessary to minimise them as current quantum devices have a low number of qubits, [10]. The second metric, the T-count, is the number of T-gates used by a circuit. NISQ devices are very sensitive to external and internal noise. The use of T-gates allows the use of error detection and correction codes to reduce the effects of noise. However, the cost of the T-gate is much higher than the cost of other gates (in the order of 100 times more), so it is important and necessary to keep the number of T-gates small [14].

In this paper, we focus on studying the use of the Toffoli gate using more than two control qubits, formally labeled as n-qubit Toffoli gates for the general n case (being n the number of involved qubits). As will be demonstrated later, implementing Toffoli gates with n control qubits will require the use of Toffoli gates with two control qubits (that is, the normal Toffoli gate). Therefore, it is necessary to study the implementations of the Toffoli gate available in the state of the art. It is worth mentioning that some implementations are focused on a particular technology. For example, there are Toffoli gate designs focused exclusively on reducing the number of controlled gates required for its implementation [8]. These gates have this objective as in linear optics is only possible to implement controlled quantum gates probabilistically. In this work we will not consider gates dependent on the physical technologies of the quantum computer, but we will consider those focused on reducing the T-count and the number of ancilla qubits.

2 Decomposition of n-qubit Controlled Gates

Controlled operations are operations that are only executed when all control qubits are set to one. Let be U a unitary operation that acts on a single qubit. For every unitary operation U there exist unitary operators A, B and C such that $ABC = I, U = e^{i\alpha} A \times B \times C$, being α some overall phase factor [1]. Note that, in case the qubit control is $|0\rangle$, only the operations $ABC = I$ will be performed. In case it is $|1\rangle$, the operation $e^{i\alpha} A \times B \times C = U$ will be computed [11]. If we now focus on the case of a 2-qubit controlled operation U, the approach is different.

It is necessary to find an operator V such that $U = V^2$. For instance, in the case of the Toffoli gate the operation U would be X. Then, if V is defined as $(1-i)(I+iX)/2$, we see how the equality $X = ((1-i)(I+iX)/2)^2$ is satisfied.

Note that for the construction of a gate with n control qubits, gates with $n-1$ control qubits are used. Therefore, methodologies are proposed to design gates controlled by n qubits using iterative processes based on gates controlled by 1 or 2 qubits, using auxiliary qubits to store intermediate results. Nielsen and Chuag [11] presented a procedure to implement $C^n(U)$ gates consisting of computing the product $c_1c_2...c_n$ using Toffoli gates. The idea is to first compute c_1c_2 on an auxiliary qubit, then to compute the product of this value with c_3 on another auxiliary qubit, and so on. The operation will therefore require $n-1$ auxiliary qubits. Finally, to avoid rubbish outputs the circuit must be reversed.

Since order does not matter in a product, He et al. proposed to parallelise the computation $c_1c_2...c_n$ so that the result can be obtained more quickly [6]. This product is computed as follows:

1. Step 1: $p_1 = c_1c_2$ is computed and stored in the first ancilla qubit. At the same time, $p_2 = c_3c_4$ is also computed and stored in the second ancilla qubit. Since both operations does not share any qubit, they can be computed in parallel.
2. Step 2: $p_3 = p_1p_2$ is computed.
3. Step 3: $p_4 = p_3c_5$ is computed.
4. Step 4: p_4 is "copied" into the target qubit.
5. Step 5−7: The circuit is reverted using Bennett's garbage removal scheme [3].

Barenco et al. proposed another scheme to build a multiple control Toffoli gate [2]. This work takes into account factors such as the possibility of working directly with negated controls and error correction. The main advantage obtained is that it manages to implement the gate using one less qubit. On the downside, the cost is much higher than previous schemes, going from needing $n-1$ Toffoli gates with two control qubits to needing $4n-8$.

Although we have found other schemes in the literature, they contain one or more gates that are theoretically defined but whose implementation is not addressed. This is why we do not include them in this paper. As a summary of this section, the information on the analysed schemes is compiled in Table 1.

Table 1. Existing schemes to build n-qubit Toffoli gates. The number of ancilla qubits may be increased due to the implementation of the 2-qubit Toffoli gates.

Design	Number of Toffoli gates	Delay	Ancilla qubits
Nielsen and Chuang [11]	$n-1$	$O(n)$	$n-1$
He et al. [6]	$n-1$	$O(Log(n))$	$n-1$
Barenco et al. [9]	$4n-8$	$O(n)$	$n-2$

3 Implementations of the Toffoli Gate

In the previous subsection, has been demonstrated the need to use 2-qubit Toffoli gates in order to build the n-qubit versions. It is therefore essential to consider existing versions of the 2-qubit Toffoli gate. There are designs in the literature that allow an approximation of the result. That is, they do not guarantee the correct result even in a noise-free device. This kind of gates has not been included in this work.

A Toffoli gate design has already been presented in the previous section. To make this design effective, it is necessary to specify an implementation for the controlled-V gate, such that $V^2 = X$ is satisfied. Amy et al. proposed a design for this gate, consisting of two CNOT gates, two Hadamard gates, two T gates, and one T' gate [1]. The T-count of this version of the V gate (as well as the V' gate) is 3. Since the Toffoli gate contains three V gates, its T-count is 9. It also requires 3 qubits for its implementation. Although this may seem obvious, some later implementations use extra qubits for their implementation. This is why it is important to keep this count.

From the previous Toffoli gate, some operations can be reorganised and simplified, as explained in Amy et al. [1]. This new version reduces the T-count by 2 with respect to the previous design, keeping the same number of qubits. Likewise, the number of CNOT and Hadamard gates is also reduced.

Jones proposed a new implementation [7] which allows the T-count to be reduced to 4. At the cost, however, of using an ancilla qubit to hold the intermediate result before being transferred to the target qubit. This Toffoli gate is based on an implementation of the iX gate proposed by Selinger [16], which has a T-count of 4. Jones' Toffoli gate has no T gates beyond those contained in Selinger's iX gate, so its quantum cost is also 4. Jones' contribution is not only limited to this reduction of the T-count, as explained below.

In quantum computing, reversing qubits after use to make them available for future operations is considered a good practice. In fact, in NISQ devices, it is a necessity due to the limited number of available qubits. After applying a Toffoli gate, two possibilities can occur:

- The value contained in the target qubit must be retained, since it will be one of the values measured and used at the end of the circuit.
- The value contained in the target qubit is only used temporarily or as an auxiliary operation and, once used, is no longer needed.

In the latter case, the operation performed by the Toffoli gate must be reversed. For this, it is necessary to apply a second Toffoli gate, which will then be focused only on returning the qubits involved to the value they had prior to the application of the first Toffoli gate [3]. Note that this operation will increase the total T-count (but not the number of qubits).

Returning to Jones' circuit, the c_1c_2 operation is computed using the iX gate and a S' gate. Subsequently, the operation $t \oplus c_1c_2$ is performed on the target qubit, and the c_1c_2 operation contained in the ancilla qubit is uncomputed.

However, instead of applying the inverse circuit to reverse the operation, Jones resorts to a measure-and-fixup approach. Specifically, a Hadamard gate is applied to then measure and classically control the obtained value to correct the phase (which will be 1 if the measurement is 0, and $(-1)^{q_0 q_1}$ if it is 1). The cost of applying the inverse circuit would be, in terms of T-count, 4 (and the global T-count 8). Using this approximation, the T-count is 0 (and the global T-count 4).

Despite the obvious improvements it offers, the Toffoli gate proposed by Jones has two points to bear in mind. First, it uses an auxiliary qubit, i.e., one more qubit than the other implementations. Second, although the c_1c_2 operation is reversed, the same is not true for the $t \oplus c_1c_2$ operation. Aware of these points, Gidney [5] separated Jones' Toffoli gate into two parts. A first gate, called temporary logical-AND gate, computes the c_1c_2 operation on an ancilla qubit prepared in state $\frac{1}{\sqrt{2}}(|0\rangle + e^{\frac{i\pi}{4}}|1\rangle)$. And a second gate whose function is to reverse the operation carried out by the previous one using the described measure-and-fixup approach. By this simple idea of delaying the uncomputation of the AND operation until such time as it is no longer needed, Gidney saves the extra qubit used by the previous circuit, and reverses the gate operations altogether.

As a summary, the information on the possible implementations of the Toffoli gate is compiled in Table 2.

Table 2. Comparison in terms of T-count excluding uncomputation, T-count including uncomputation, and number of ancilla qubits between the best implementations of the Toffoli gate.

Design	T-count	T-count with uncomputation	Ancilla inputs
Amy et al. (1). [1]	9	18	0
Amy et al. (2). [1]	7	14	0
Jones [7]	4	4	1
Gidney [5]	4	4	0

4 Analysis and Comparison

To test the correct operation of the circuits and to reinforce the metrics used, all the gates have been tested and measured using the IBM Q Experience platform, with the circuits being written in Python and tested on real devices and, when not possible due to circuit size, on a simulator. The measurement methodology described in Orts et al. [12].

The possible implementations of m-qubit Toffoli gates arise from the combination of the schemes listed in Table 1 with the Toffoli gate implementations studied in Table 2. Table 3 shows the obtained metrics for each scheme-design combination, in terms of ancilla inputs and T-count. Since this information is trivial, the delay is not indicated in the table. For the sake of clarity, we again clarify that the fastest scheme is the He et al. one (logarithmic order) [7], followed by the Nielsen and Chuang scheme (linear order) [11], and finally the Barenco scheme (linear order, but slower than the Nielsen scheme because it performs a higher number of operations, as discussed in Section 2) [2].

Table 3. Comparison in terms of T-count (including uncomputation) and ancilla inputs of m-qubit Toffoli gates created according to existing schemes using the most efficient Toffoli gates (of 2 control qubits) available in the literature. It is important to note that, although the Nielsen and Chuang and He et al. schemes share the same values, the Nielsen scheme has a linear delay and the He et al. scheme has a logarithmic delay. Regardless of the chosen scheme, the most appropriate uncomputation technique is adopted for each gate.

Design	Nielsen and Chuang [11]		He et al. [6]		Barenco [2]	
	T-count	Ancilla inputs	T-count	Ancilla inputs	T-count	Ancilla inputs
Amy et al. (1) [1]	$18n - 18$	$n - 1$	$18n - 18$	$n - 1$	$64n - 144$	$n - 2$
Amy et al. (2) [1]	$14n - 14$	$n - 1$	$14n - 14$	$n - 1$	$56n - 112$	$n - 2$
Jones [7]	$4n - 4$	$2n - 2$	$4n - 4$	$2n - 2$	$16n - 32$	$2n - 3$
Gidney [5]	$4n - 4$	$n - 1$	$4n - 4$	$n - 1$	$16n - 32$	$n - 2$

In terms of T-count, the best implementations are given by the Nielsen and Chuan [11] and He et al. [7] schemes using the Jones gates or the Gidney temporary, with a final value of $4n - 4$. Since the logical-AND temporary gate allows to reduce the number of ancilla inputs (it needs $n - 1$ versus $2n - 2$ using the Jones gate), it becomes the best option in these terms. Note also that the number of $n - 1$ ancilla qubits is the best possible for these two schemes, and is also achieved by the designs of Amy et al. However, using Barenco's scheme, the number of ancilla inputs needed can be reduced by one. At the cost, however, of a large increase in the T-count. The best value obtained in this scheme is again obtained using the temporary, achieving a T-count of $16n - 32$ with $n - 2$ ancilla inputs. That is, a $12n - 28$ increase in the T-count to save a single qubit.

5 Conclusions

In this paper we provide a review of the existing techniques for constructing n-qubit Toffoli gates. These gates has been analized (and reproduced) in terms of noise tolerance and ancilla qubits. Their design is provided using only Clifford+T gates. Moreover, a revision on the state-of-the-art reversible Toffoli gates has been carried out. Appropriate metrics have been considered for the measurement and comparison of quantum gates and circuits. The analysis has been carried out with two essential goals in mind: first, to find all the procedures to build a n-qubit Toffoli gate, and second to dispose of the best implementations of the Toffoli gates to, in combination with the mentioned procedures, to build the most optimized n-qubit Toffoli gate possible in terms of T-count and number of ancilla qubits. Special attention has been paid to the temporary logical-AND gate, whose design to act with more than two control qubits had not yet been analyzed in the literature. Finally, the possibilities has been compared in such terms, highlighting the advantages and the drawbacks of each candidate.

References

1. Amy, M., Maslov, D., Mosca, M., Roetteler, M.: A meet-in-the-middle algorithm for fast synthesis of depth-optimal quantum circuits. IEEE Trans. Comput.-Aided Des. Integr. Circ. Syst. **32**(6), 818–830 (2013)
2. Barenco, A., et al.: Elementary gates for quantum computation. Phys. Rev. A **52**(5), 3457 (1995)
3. Bennett, C.: Logical reversibility of computation. IBM J. Res. Dev. **17**(6), 525–532 (1973)
4. Bernhardt, C.: Quantum Computing for Everyone. MIT Press, Cambridge (2019)
5. Gidney, C.: Halving the cost of quantum addition. Quantum **2**, 74 (2018)
6. He, Y., Luo, M., Zhang, E., Wang, H., Wang, X.: Decompositions of n-qubit Toffoli gates with linear circuit complexity. Int. J. Theor. Phys. **56**(7), 2350–2361 (2017)
7. Jones, C.: Low-overhead constructions for the fault-tolerant Toffoli gate. Phys. Rev. A **87**(2), 022328 (2013)
8. Lanyon, B., et al.: Simplifying quantum logic using higher-dimensional Hilbert spaces. Nat. Phys. **5**(2), 134–140 (2009)
9. Maslov, D., Dueck, G., Miller, D., Negrevergne, C.: Quantum circuit simplification and level compaction. IEEE Trans. Comput.-Aided Des. Integr. Circ. Syst. **27**(3), 436–444 (2008)
10. Mohammadi, M., Eshghi, M.: On figures of merit in reversible and quantum logic designs. Quant. Inf. Process. **8**(4), 297–318 (2009)
11. Nielsen, M., Chuang, I.: Quantum computation and quantum information (2002)
12. Orts, F., Ortega, G., Combarro, E.F., Garzón, E.M.: A review on reversible quantum adders. J. Netw. Comput. Appl. **170**, 102810 (2020)
13. Orts, F., Ortega, G., Cucura, A.C., Filatovas, E., Garzón, E.M.: Optimal fault-tolerant quantum comparators for image binarization. J. Supercomput. **77**(8), 8433–8444 (2021)
14. Orts, F., Ortega, G., Garzon, E.M.: Efficient reversible quantum design of sign-magnitude to two's complement converters. Quant. Inf. Comput. **20**(9–10), 747–765 (2020)
15. Pérez-Salinas, A., Cervera-Lierta, A., Gil-Fuster, E., Latorre, J.: Data re-uploading for a universal quantum classifier. Quantum **4**, 226 (2020)
16. Selinger, P.: Quantum circuits of t-depth one. Phys. Rev. A **87**(4), 042302 (2013)
17. Zhou, R., Wu, Q., Zhang, M., Shen, C.: Quantum image encryption and decryption algorithms based on quantum image geometric transformations. Int. J. Theor. Phys. **52**(6), 1802–1817 (2013)

Reducing Memory Requirements of Quantum Optimal Control

Sri Hari Krishna Narayanan[1], Thomas Propson[2(✉)], Marcelo Bongarti[3], Jan Hückelheim[1], and Paul Hovland[1]

[1] Argonne National Laboratory, Lemont, IL 60439, USA
{snarayan,jhueckelheim,hovland}@anl.gov
[2] University of Chicago, Chicago, IL, USA
tcpropson@uchicago.edu
[3] Weierstrass Institute for Applied Analysis and Stochastics, Berlin, Germany
bongarti@wias-berlin.de

Abstract. Quantum optimal control problems are typically solved by gradient-based algorithms such as GRAPE, which suffer from exponential growth in storage with increasing number of qubits and linear growth in memory requirements with increasing number of time steps. These memory requirements are a barrier for simulating large models or long time spans. We have created a nonstandard automatic differentiation technique that can compute gradients needed by GRAPE by exploiting the fact that the inverse of a unitary matrix is its conjugate transpose. Our approach significantly reduces the memory requirements for GRAPE, at the cost of a reasonable amount of recomputation. We present benchmark results based on an implementation in JAX.

Keywords: Quantum · Autodiff · Memory

1 Introduction

Quantum computing is computing using quantum-mechanical phenomena, such as superposition and entanglement. It holds the promise of being able to efficiently solve problems that classical computers practically cannot. In quantum computing, quantum algorithms are often expressed by using a quantum circuit model, in which a computation is a sequence of quantum gates. Quantum gates are the building blocks of quantum circuits and operate on a small number of

This material is based upon work supported by the U.S. Department of Energy, Office of Science, Office of Advanced Scientific Computing Research, under the Accelerated Research in Quantum Computing and Applied Mathematics programs, under contract DE-AC02-06CH11357, and by the National Science Foundation Mathematical Sciences Graduate Internship. We gratefully acknowledge the computing resources provided on Bebop and Swing, a high-performance computing cluster operated by the Laboratory Computing Resource Center at Argonne National Laboratory.

D. Groen et al. (Eds.): ICCS 2022, LNCS 13353, pp. 129–142, 2022.
https://doi.org/10.1007/978-3-031-08760-8_11

qubits, similar to how classical logic gates operate on a small number of bits in conventional digital circuits.

Practitioners of quantum computing must map the logical quantum gates onto the physical quantum devices that implement quantum gates through a process called *quantum control.* The goal of quantum control is to actively manipulate dynamical processes at the atomic or molecular scale, typically by means of external electromagnetic fields. The objective of *quantum optimal control* (QOC) is to devise and implement shapes of pulses of external fields or sequences of such pulses that reach a given task in a quantum system in the best way possible.

We follow the QOC model presented in [20]. Given an intrinsic Hamiltonian H_0, an initial state $|\psi_0\rangle$, and a set of control operators $H_1, H_2, \ldots H_m$, one seeks to determine, for a sequence of time steps $t_0, t_1, \ldots, t_N$, a set of control fields $u_{k,j}$ such that

$$\mathbb{H}_j = H_0 + \sum_{k=1}^{m} u_{k,j} H_k \tag{1}$$

$$U_j = e^{-i\mathbb{H}_j(t_j - t_{j-1})} \tag{2}$$

$$K_j = U_j U_{j-1} U_{j-2} \ldots U_1 U_0 \tag{3}$$

$$|\psi_j\rangle = K_j |\psi_0\rangle. \tag{4}$$

An important observation is that the dimensions of K_j and U_j are $2^q \times 2^q$, where q is the number of qubits in the system. One possible objective is to minimize the trace distance between K_N and a target quantum gate K_T:

$$F_0 = 1 - |\mathrm{Tr}(K_T^\dagger K_N)/D|^2, \tag{5}$$

where D is the Hilbert space dimension. The complete QOC formulation includes secondary objectives and additional constraints One way to address this formulation is by adding to the objective function weighted penalty terms representing constraint violation: $\min_{u_{k,j}} \left(\sum_{i=0}^{2} w_i F_i + \sum_{i=3}^{6} w_i G_i \right)$, where

$$F_1 = 1 - \frac{1}{n} \sum_j |\mathrm{Tr}(K_T^\dagger K_j)/D|^2 \tag{6}$$

$$F_2 = 1 - \frac{1}{n} \sum_j |\langle \psi_T | \psi_j \rangle|^2 \tag{7}$$

$$G_3 = 1 - |\langle \psi_T | \psi_n \rangle|^2 \tag{8}$$

$$G_4 = |u|^2 \tag{9}$$

$$G_5 = \sum_{k,j} |u_{j,k} - u_{k,j-1}|^2 \tag{10}$$

$$G_6 = \sum_j |\langle \psi_F | \psi_j \rangle|^2 \tag{11}$$

and ψ_F is a forbidden state.

Algorithm 1. Pseudocode for the GRAPE algorithm.

Guess initial controls $u_{k,j}$.
repeat
  Starting from H_0, calculate
    $\rho_j = U_j U_{j-1} \dots U_1 H_0 U_1^\dagger \dots U_{j-1}^\dagger U_j^\dagger$.
  Starting from $\lambda_N = K_T$, calculate
    $\lambda_j = U_{j+1}^\dagger \dots U_N^\dagger K_T U_N \dots U_j$.
  Evaluate $\frac{\partial \rho_j \lambda_j}{\partial u_{k,j}}$
  Update the $m \times N$ control amplitudes:
    $u_{j,k} \longrightarrow u_{j,k} + \epsilon \frac{\partial \rho_j \lambda_j}{\partial u_{k,j}}$
until $\mathrm{Tr}(K_T^\dagger K_N) <$ threshold
return $u_{j,k}$

QOC can be solved by several algorithms, including the gradient ascent pulse engineering (GRAPE) algorithm [17]. A basic version of GRAPE is shown in Algorithm 1. The derivatives $\frac{\partial \rho_j \lambda_j}{\partial u_{k,j}}$ required by GRAPE can be calculated by hand coding or finite differences. Recently, these values have been calculated efficiently by automatic differentiation (AD or autodiff) [20].

AD is a technique for transforming algorithms that compute some mathematical function into algorithms that compute the derivatives of that function [3,12,21]. AD works by differentiating the functions intrinsic to a given programming language (`sin()`, `cos()`, `+`, `-`, etc.) and combining the partial derivatives using the chain rule of differential calculus. The associativity of the chain rule leads to two main methods of combining partial derivatives. The forward mode combines partial derivatives starting with the independent variables and propagating forward to the dependent variables. The reverse mode combines partial derivatives starting with the dependent variables and propagating back to the independent variables. It is particularly attractive in the case of scalar functions, where a gradient of arbitrary length can be computed at a fixed multiple of the operations count of the function.

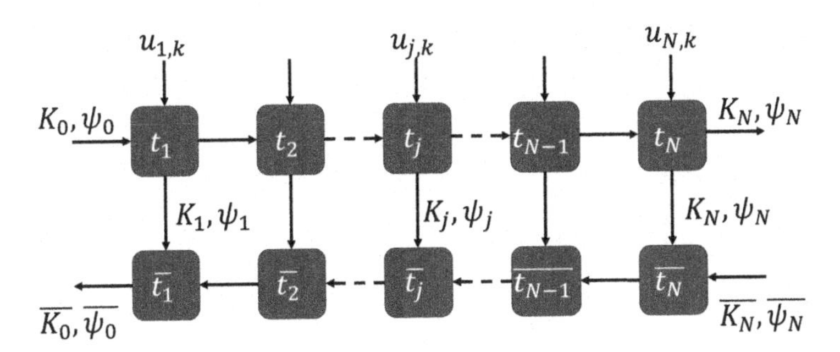

Fig. 1. Reverse-mode gradient computation for QOC. Each forward time step, j, computes K_j, U_j, and ψ_j and stores them in memory. The reverse sweep starts at time step N. Each reverse time step j uses the previously stored K_j, U_j, and ψ_j.

The reverse mode of AD is appropriate for QOC because of the large number ($m \times N$) of inputs and the small number of outputs (the cost function(s)). As shown in Fig. 1, standard reverse-mode AD stores the results of intermediate time steps (K_j, U_j, ψ_j) in order to compute $\frac{\partial \rho_j \lambda_j}{\partial u_{k,j}}$. This implies that reverse-mode AD requires additional memory that is exponentially proportional to q. Current QOC simulations therefore are limited in both the number of qubits that can be simulated and the number of time steps in the simulation. In this work we explore the suitability of *checkpointing* as well as *unitary matrix reversibility* to overcome this additional memory requirement.

The rest of the paper is organized as follows. Section 2 discusses related work. Section 3 presents our approach to reducing the memory requirements of QOC. Section 4 details the QOC implementation in JAX, and the evaluation of this approach is presented in Sect. 5. Section 6 concludes the paper and discusses future work.

2 Related Work

QOC has been implemented in several packages, such as the Quantum Toolbox in Python (QuTIP) [15,16]. In addition to GRAPE, QOC can be solved by using the chopped random basis (CRAB) algorithm [6,9]. The problem is formulated as the extremization of a multivariable function, which can be numerically approached with a suitable method such as steepest descent or conjugate gradient. If computing the gradient is expensive, CRAB can instead use a derivative-free optimization algorithm. In [20], the AD capabilities of TensorFlow are used to compute gradients for QOC.

Checkpointing is a well-established approach in AD to reduce the memory requirements of reverse-mode AD [10,18]. In short, checkpointing techniques trade recomputation for storing intermediate states; see Sect. 3.1 for more details. For time-stepping codes, such as QOC, checkpointing strategies can range from simple periodic schemes [12], through binomial checkpointing schemes [11] that minimize recomputation subject to memory constraints, to multilevel checkpointing schemes [2,23] that store checkpoints to a multilevel storage hierarchy. Checkpointing schemes have also been adapted to deep neural networks [4,7,14,22] and combined with checkpoint compression [8,19].

3 Reducing Memory Requirements

We explore three approaches to reduce the memory required to compute the derivatives for QOC.

3.1 Approach 1: Checkpointing

Checkpointing schemes reduce the memory requirements of reverse-mode AD by recomputing certain intermediate states instead of storing them. These schemes checkpoint the inputs of selected time steps in a *plain-forward* sweep. To compute the gradient, a stored checkpoint is read, followed by a forward sweep and a reverse sweep for an appropriate number of time steps. Figure 2 illustrates periodic checkpointing for a computation of 10 time steps and 5 checkpoints. In the case of QOC with N time steps and periodic checkpointing interval C, one must store $O(C + \frac{N}{C})$ matrices of size $2^q \times 2^q$.

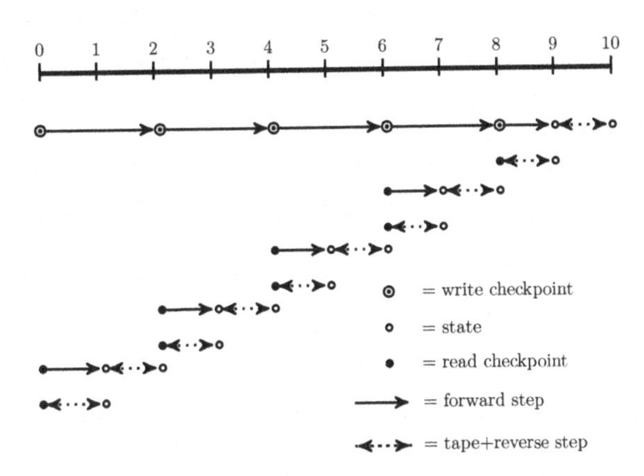

Fig. 2. Periodic checkpointing schedule for $N = 10$ time steps and 5 checkpoints ($C = 2$).

3.2 Approach 2: Reversibility of Unitary Matrices

The second approach is to exploit the property of unitary matrices that the inverse of a unitary matrices is its conjugate transpose.

$$U^\dagger U = U U^\dagger \tag{12}$$

$$U^\dagger = U^{-1} \tag{13}$$

Computing the inverse by exploiting the reversibility property of unitary matrices is an exact and inexpensive process. The use of the inverse allows us to compute K_{j-1} from K_j and ψ_{j-1} from ψ_j.

$$K_j = U_j U_{j-1} U_{j-2} \dots U_1 U_0 \tag{14}$$

$$K_{j-1} = U_j^\dagger K_j \tag{15}$$

$$\psi_{j-1} = \psi_0 K_{j-1} \tag{16}$$

Thus, one does not have to store any of the K_j matrices required to compute the adjoint of a time step. This approach reduces the memory requirement by half.

More importantly, using reversibility can unlock a further reduction in the memory requirements by not storing U_j, but rather using only the $u_{j,k}$ control values to recompute U_j. As a result, no intermediate computations need to be stored, and therefore the only additional requirements are to store the derivatives of the function with respect to the controls and other variables, thus basically doubling the memory requirements relative to the function itself.

3.3 Approach 3: Periodic Checkpointing Plus Reversibility

The reversibility property of unitary matrices is exact only in real arithmetic. A floating-point implementation may incur roundoff errors. Therefore, Eq. 15 might not hold exactly, especially for large numbers of time steps. That is,

$$K_{j-1} \approx U_j^\dagger K_j \tag{17}$$

$$\psi_{j-1} \approx \psi_0 K_{j-1} \tag{18}$$

in floating-point arithmetic. Because K_{j-1} is computed each time step, the error continues to grow as the computation proceeds in the reverse sweep.

To mitigate this effect, we can combine the two approaches, checkpointing every C time steps and, during the reverse pass, instead of computing forward from these checkpoints, computing backward from the checkpoints by exploiting reversibility. Thus, floating-point errors in Eq. 15 are incurred over a maximum of C time steps, and we reduce the number of matrices of size $2^q \times 2^q$ stored from $O(C + \frac{N}{C})$ to $O(\frac{N}{C})$.

Table 1. Overview of the object sizes, number of object instances stored for the forward computation, and number of additional instances that need to be stored for store-all, checkpointing, reversibility, and checkpointing plus reversibility. The total memory size in the last row is the product of the object size and the number of instances.

Variable	Size	Forward	Store	Checkpoint	Revert	Rev + Ckp
$u_{k,j}$	1	Nm	+0	+0	+0	+0
H	$2^q \cdot 2^q$	m	+0	+0	+0	+0
$\mathbb{H}_j$	$2^q \cdot 2^q$	1	+0	+0	+0	+0
U_j	$2^q \cdot 2^q$	1	$+N$	$+C$	+0	+0
K_j	$2^q \cdot 2^q$	1	$+N$	$+\frac{N}{C} + C$	+0	$+\frac{N}{C}$
ψ_j	2^q	1	$+N$	$+\frac{N}{C} + C$	+0	$+\frac{N}{C}$
Mem ($\mathcal{O}$)		$2^{2q}m + Nm$	$+2^{2q}N$	$+2^{2q}\left(\frac{N}{C} + C\right)$	+0	$+2^{2q}\frac{N}{C}$

3.4 Analysis of Memory Requirements

Table 1 summarizes the memory requirements for the forward pass of the function evaluation as well as the added cost of the various strategies for computing the gradient. Conventional AD, which stores the intermediate states U_j, K_j, and ψ_j at every time step, incurs an additional storage cost proportional to the number of time steps *times* the size of the $2^q \times 2^q$ matrices. Periodic checkpointing reduces the number of matrices stored to $\frac{N}{C} + C$. The checkpointing interval that minimizes this cost occurs when $\frac{\partial}{\partial C}\left(\frac{N}{C} + C\right) = \frac{-N}{C^2} + 1 = 0$, or $C = \sqrt{N}$. Exploiting reversibility enables one to compute U_j from $u_{j,k}$ and H_k and K_{j-1} from K_j and U_j, resulting in essentially zero additional memory requirements, beyond those required to store the derivatives themselves. Combining reversibility with periodic checkpointing eliminates the number of copies of U_j and K_j to be stored from $\frac{N}{C} + C$ to $\frac{N}{C}$.

4 Implementation

As an initial step we have ported to the JAX machine learning framework [5] a version of QOC that was previously implemented in TensorFlow. JAX provides a NumPy-style interface and supports execution on CPU systems as well as GPU and TPU (tensor processing unit) accelerators, with built-in automatic differentiation and just-in-time compilation capability. JAX supports checkpointing through the use of the `jax.checkpoint` decorator and allows custom derivatives to be created for functions using the `custom_jvp` decorator for forward mode and the `custom_vjp` decorator for reverse mode. To enable our work, we have contributed `jax.scipy.linalg.expm` to JAX to perform the matrix exponentiation operation using Padé approximation and to compute its derivatives. This code is now part of standard JAX releases.

Our approach requires us to perform checkpointing or use custom derivatives only for the Python function that implements Eqs. 1–4 for a single time step j or a loop over a block of time steps. Standard AD can be used as before for the rest of the code. By implication, our approach does not change for different objective functions.

We show here the implementation of the periodic checkpointing plus reversibility approach and direct the reader to our open source implementation for further details [1]. Listing 1 shows the primal code that computes a set of time steps. The function `evolve_step` computes Eqs. 1–4.

Listing 1. Simplified code showing a loop to simulate QOC for N time steps

```
def evolve_step_loop(start, stop, cost_eval_step, dt, states, K,
                     control_eval_times, controls):
    for step in range(start,stop):
        # Evolve the states and K to the next time step.
        time = step * dt
        states, K  = evolve_step(dt, states, K, time,
                                          control_eval_times, controls)
    return states, K
```

Listing 2 is a convenience wrapper with for the primal code. We decorate the wrapper with `jax.custom_vjp` to inform JAX that we will provide custom derivatives for it. User-provided custom derivatives for a JAX function consist of a forward sweep and a reverse sweep. The forward sweep must store all the information required to compute the derivatives in the reverse sweep. Listing 3 is the provided forward sweep. Here, as indicated in Table 1, we are storing the K matrix and the state vector. Note that this form of storage is effectively a checkpoint, even though it does not use `jax.checkpoint`.

Listing 2. A wrapper to evolve_step_loop(), which will have derivatives provided by the user.

```
@jax.custom_vjp
def evolve_step_loop_custom(start, stop, cost_eval_step, dt, states,
                            K, control_eval_times, controls):
    states, K = evolve_step_loop(start, stop, cost_eval_step, dt,
                                 states,K,control_eval_times,
                                 controls)
    return states, K
```

Listing 3. Forward sweep of the user-provided derivatives.

```
def evolve_loop_custom_fwd(start, stop,cost_eval_step, dt, states, K,
                           control_eval_times, controls):
    states, K = _evaluate_schroedinger_discrete_loop_inner(
                           start, stop, cost_eval_step, dt, states, K,
                           control_eval_times, controls)
    #Here we store the final state and K for use in the backward pass
    return (states,K), (start, stop, cost_eval_step, dt, states,
           K, control_eval_times,controls)
```

Listing 4 is a user-provided reverse sweep. It starts by restoring the values passed to it by the forward sweep. While looping over time steps in reverse order, it recomputes Eqs. 1–2. It then computes Eqs. 13, 15, and 16 to retrieve K_{j-1} and ψ_{j-1}, which are then used to compute the adjoints for the time step. The code to compute the adjoint of the time step was obtained by the source transformation AD tool Tapenade [13].

5 Experimental Results

We compared standard AD, periodic checkpointing, and full reversibility or periodic checkpointing with reversibility, as appropriate. We conducted our experiments on a cluster where each compute node was connected to 8 NVIDIA A100 40 GB GPUs. Each node contained 1 TB DDR4 memory and 320 GB GPU memory. We validated the output of the checkpointing and reversibility approaches against the standard approach implemented using JAX. We used the JAX memory profiling capability in conjunction with GO pprof to measure the memory needs for each case. We conducted three sets of experiments to evaluate the approaches, varying the number of qubits, the number of time steps, or the checkpoint period.

Listing 4. User-provided reverse sweep that exploits reversibility.

```
def evolve_loop_custom_bwd(res,g_prod):
  #Restore all the values stored in the forward sweep
  start, stop, cost_eval_step, dt, states,
        K, control_eval_times, controls = res
  _M2_C1 = 0.5
  controlsb = jnp.zeros(controls.shape, states.dtype)
  #Go backwards in time steps
  for i in range(stop-1,start-1,-1):
    #Reapply controls to compute a step unitary matrix
    time = i * dt
    t1 = time + dt * _M2_C1
    x = t1
    xs = control_eval_times
    ys = controls
    index = jnp.argmax(x <= xs)
    y = ys[index - 1] + (((ys[index] - ys[index - 1]) /
        (xs[index] - xs[index - 1])) * (x - xs[index - 1]))
    controls_ = y
    hamiltonian_ = (SYSTEM_HAMILTONIAN
                + controls_[0] * CONTROL_0
                + jnp.conjugate(controls_[0]) * CONTROL_0_DAGGER
                + controls_[1] * CONTROL_1
                + jnp.conjugate(controls_[1]) * CONTROL_1_DAGGER)
    a1 = -1j * hamiltonian_
    magnus = dt * a1
    step_unitary, f_expm_grad = jax.vjp(jax.scipy.linalg.expm, (magnus),
    has_aux=False)
    #Exploit reversibility of unitary matrix
    #and calculate previous state and K
    step_unitary_inv=jnp.conj(jnp.transpose(step_unitary))
    states=(jnp.matmul(step_unitary_inv,states))
    K=(jnp.matmul(step_unitary_inv,K))
    _, f_matmul = jax.vjp(jnp.matmul,step_unitary, states)
    _, f_matmul_K = jax.vjp(jnp.matmul,step_unitary, K)
    #Go backwards for the timestep
    step_unitaryb,Kb=f_matmul_K(g_prod[1])
    step_unitaryb,statesb=f_matmul(g_prod[0])
    magnusb = f_expm_grad(step_unitaryb)
    a1b=dt*magnusb[0]
    hamiltonian_b = jnp.conjugate(-1j)*a1b
    controls1b=jnp.array((jnp.sum(jnp.conjugate(CONTROL_0)*hamiltonian_b) +
      jnp.conjugate(jnp.sum(jnp.conjugate(CONTROL_0_DAGGER)*hamiltonian_b)),
      jnp.sum(jnp.conjugate(CONTROL_1)*hamiltonian_b) +
      jnp.conjugate(jnp.sum(jnp.conjugate(CONTROL_1_DAGGER)*hamiltonian_b))),
      dtype=hamiltonian_b.dtype)
    tempb = (x-control_eval_times[index-1])*controls1b/
      (control_eval_times[index]-control_eval_times[index-1])
    controlsb=jax.ops.index_update(controlsb,
      jax.ops.index[index-1],controlsb[index-1]+controls1b - tempb)
    controlsb=jax.ops.index_update(controlsb,
      jax.ops.index[index],controlsb[index]+tempb)
    g_prod=statesb,Kb
  return (0.0,0.0,0.0,0.0,statesb,Kb,0.0,-1*controlsb)

evolve_loop_custom.defvjp(evolve_loop_custom_fwd, evolve_loop_custom_bwd)
```

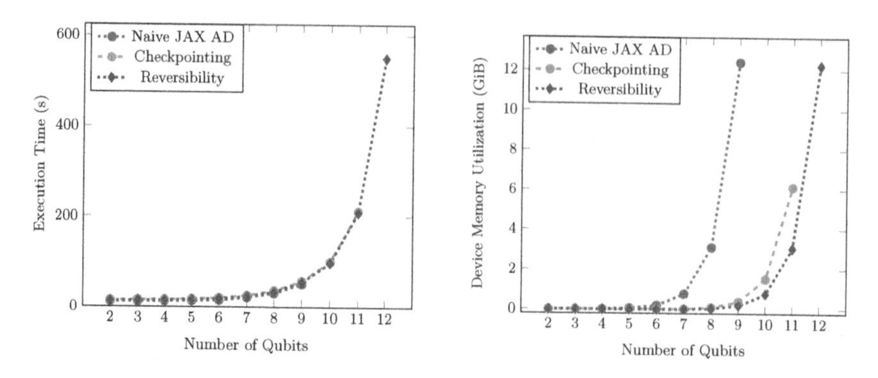

Fig. 3. Comparison of execution time and device memory requirements for standard AD, periodic checkpointing, and full reversibility with increasing number of qubits. The QOC simulation consisted of 100 time steps with a checkpoint period of 10.

5.1 Vary Qubits

We first varied the number of qubits q, keeping the number of time steps fixed at 100 and the checkpoint period fixed at $C = \sqrt{N} = 10$. Figure 3 shows the memory consumed by standard AD, periodic checkpointing, and full reversibility. One can see that the device memory requirements for the standard approach are highest whereas the requirements for reversibility are lowest, although all three grow exponentially as a function of q, as predicted by the analysis in Sect. 3. Furthermore, we note that the standard approach can be executed for a maximum of 9 qubits and runs out of available device memory on the 10th qubit. The periodic checkpointing approach can be run for 11 qubits and runs out of available device memory on the 12th. The full reversibility approach can be run for 12 qubits and exceeds available device memory on the 13th. Figure 3 (left) also shows the execution time for the various approaches. The times are similar for the cases that can be executed before running out of memory. As expected, the time grows exponentially as a function of q.

5.2 Vary Time Steps

Next we fixed the number of qubits at $q = 8$ or $q = 9$ and varied the number of time steps, N. For periodic checkpointing we used the optimal checkpoint period, $C = \sqrt{N}$. We expect the time to be roughly linear in N and independent of C because every U_j and K_j must be computed once during the forward pass and one more time on the reverse pass. We expect periodic checkpointing and full reversibility to be slower than standard AD because they both trade some amount of recomputation for reduced storage requirements. We expect full reversibility to be somewhat faster than periodic checkpointing alone because periodic checkpointing must compute forward from the checkpoint, storing intermediate K_j along the way, while full reversibility skips the second forward pass and is able to restore K_j during the reverse pass directly from the controls

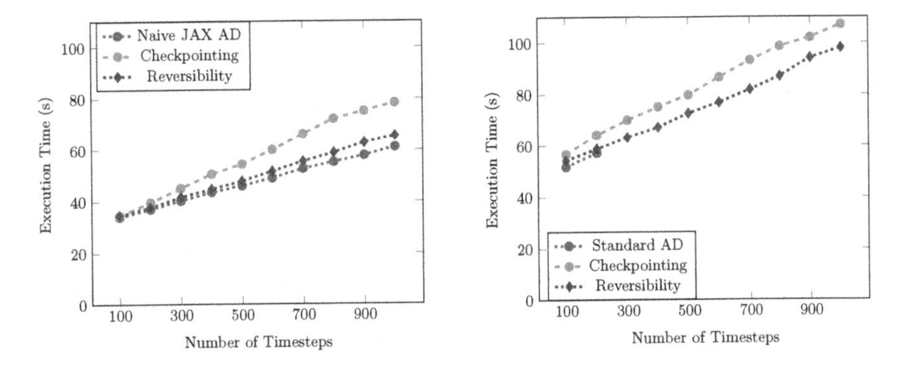

Fig. 4. Comparison of the execution time for standard AD, periodic checkpointing, and periodic reversibility approaches with increasing number of time steps. The QOC simulation consisted of 8 (left) or 9 (right) qubits. The checkpoint period was chosen to be the square root of the number of time steps.

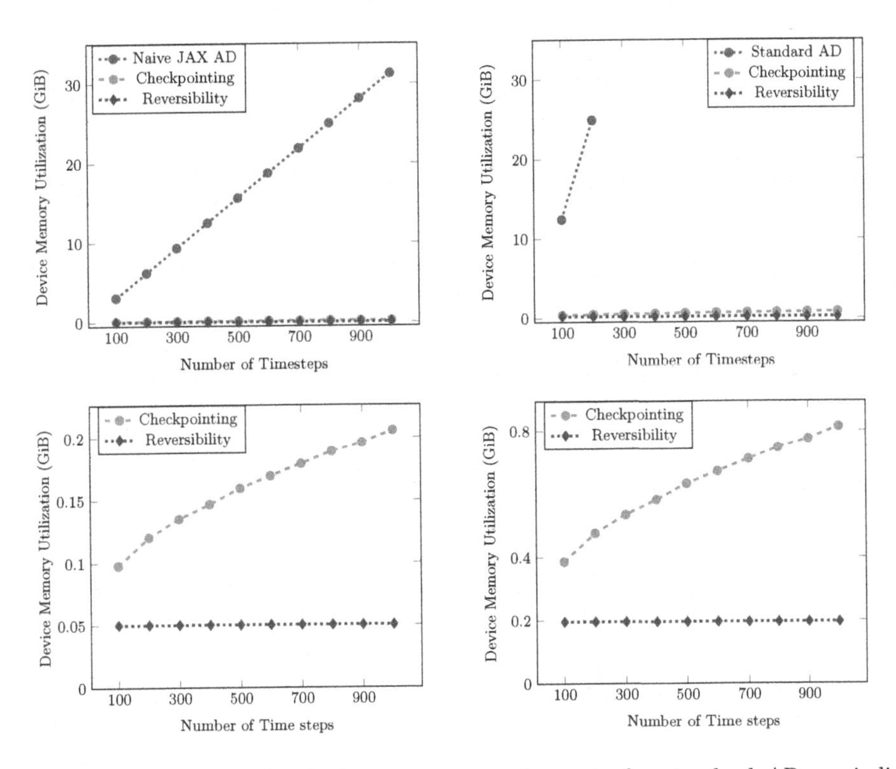

Fig. 5. Comparison of the device memory requirements for standard AD, periodic checkpointing, and periodic reversibility approaches with increasing number of time steps. The QOC simulation consisted of 8 (left) or 9 (right) qubits. The checkpoint period was chosen to be the square root of the number of time steps. Top row shows all three approaches; bottom row omits standard AD.

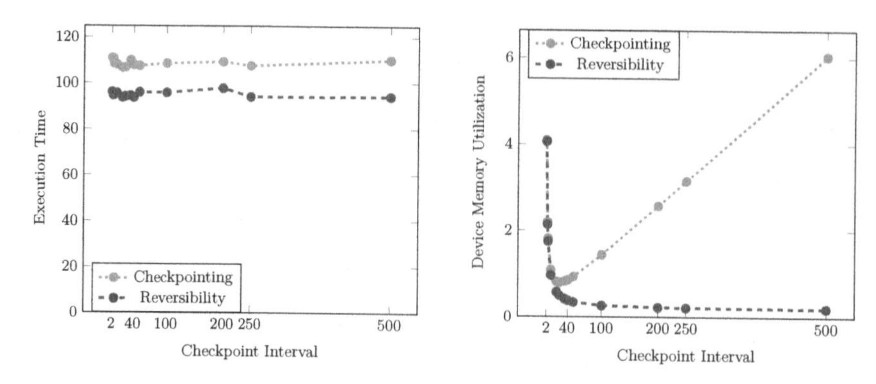

Fig. 6. Comparison of the execution time and device memory requirements for periodic checkpointing and checkpointing plus reversibility approaches with increasing number of time steps. The QOC simulation consisted of 1,000 time steps and 9 qubits.

and K_{j+1}. Figure 4 is consistent with these expectations, although standard AD quickly runs out of memory for the case $q = 9$.

Based on the analysis in Sect. 3, we expect the memory requirements of standard AD to be linear in the number of time steps and the memory requirements of full reversibility to be independent of the number of time steps. We expect the memory requirements of periodic checkpointing to vary as a function of $\frac{N}{C} + C$; since C is chosen to be $C = \sqrt{N}$, the memory should vary as a function of $\sqrt{N}$. Figure 5 clearly shows the linear dependence of standard AD and independence of full reversibility on the number of time steps. The memory requirements for periodic checkpointing are also consistent with expectations.

5.3 Vary Checkpointing Period

We examined the dependence of execution time and memory requirements on the checkpointing period, C, keeping the number of qubits fixed at $q = 9$ and the number of time steps fixed at $N = 1,000$. We expect the time to be roughly independent of C because every U_j and K_j must be computed once during the forward pass and one more time on the reverse pass. We expect periodic checkpointing with reversibility to be somewhat faster than periodic checkpointing alone because periodic checkpointing must compute forward from the checkpoint, storing intermediate K_j along the way, while periodic checkpointing with reversibility skips the second forward pass and is able to restore K_j during the reverse pass directly from the controls and K_{j+1}. The timing results in Fig. 6 (left) are consistent with these expectations.

We expect the memory requirements of periodic checkpointing with reversibility to vary as a function of $\frac{N}{C}$ or, since N is constant, as a function of $\frac{1}{C}$. We expect the memory requirements of periodic checkpointing alone to vary as a function of $\frac{N}{C} + C$, with a minimum at $C = \sqrt{N} \approx 32$. Again, the memory utilization results in Fig. 6 (right) are consistent with these expectations.

6 Conclusion and Future Work

We have implemented a version of quantum optimal control (QOC) using the JAX framework. We have compared standard automatic differentiation (AD), periodic checkpointing, and reversibility—a nonstandard AD approach that recognizes that the inverse of a unitary matrix is its conjugate transpose. Checkpointing and reversibility are both superior to standard AD. The reversibility approach, however, allows more qubits to be simulated when the number of time steps is large. Recognizing that reversibility (Eq. 15) is precise in real arithmetic but is not precise in floating-point arithmetic, we demonstrated that reversibility can be combined with periodic checkpointing, reducing memory requirements relative to periodic checkpointing alone while ensuring that roundoff errors are not accumulated over a period of more than C time steps.

In the future, we will study methods to estimate the amount of roundoff error as a function of C in order to choose a period that minimizes memory requirements while incurring acceptable roundoff errors. We will investigate applying lossy compression to the checkpoints and compare the trade-offs in storage and accuracy between periodic checkpointing with lossy compression and periodic checkpointing with reversibility. Moreover, we will combine periodic checkpointing, lossy compression, and reversibility to enable QOC to be applied to even larger numbers of qubits and time steps.

References

1. (2022). https://github.com/sriharikrishna/qoc
2. Aupy, G., Herrmann, J., Hovland, P., Robert, Y.: Optimal multistage algorithm for adjoint computation. SIAM J. Sci. Comput. **38**(3), C232–C255 (2016)
3. Baydin, A.G., Pearlmutter, B.A., Radul, A.A., Siskind, J.M.: Automatic differentiation in machine learning: a survey. J. Mach. Learn. Res. **18**(153), 1–43 (2018). http://jmlr.org/papers/v18/17-468.html
4. Beaumont, O., Herrmann, J., Pallez, G., Shilova, A.: Optimal memory-aware backpropagation of deep join networks. Phil. Trans. R. Soc. A **378**(2166), 20190049 (2020)
5. Bradbury, J., et al.: JAX: composable transformations of Python+NumPy programs (2018). http://github.com/google/jax
6. Caneva, T., Calarco, T., Montangero, S.: Chopped random-basis quantum optimization. Phys. Rev. A **84**, 022326 (2011). https://doi.org/10.1103/PhysRevA.84.022326
7. Chen, T., Xu, B., Zhang, C., Guestrin, C.: Training deep nets with sublinear memory cost. arXiv preprint arXiv:1604.06174 (2016)
8. Cyr, E.C., Shadid, J., Wildey, T.: Towards efficient backward-in-time adjoint computations using data compression techniques. Comput. Meth. Appl. Mech. Eng. **288**, 24–44 (2015)
9. Doria, P., Calarco, T., Montangero, S.: Optimal control technique for many-body quantum dynamics. Phys. Rev. Lett. **106**, 190501 (2011). https://doi.org/10.1103/PhysRevLett.106.190501
10. Griewank, A.: Achieving logarithmic growth of temporal and spatial complexity in reverse automatic differentiation. Optim. Methods Softw. **1**(1), 35–54 (1992)

11. Griewank, A., Walther, A.: Algorithm 799: Revolve: an implementation of checkpointing for the reverse or adjoint mode of computational differentiation. ACM Trans. Math. Softw. **26**(1), 19–45 (2000). https://doi.org/10.1145/347837.347846
12. Griewank, A., Walther, A.: Evaluating Derivatives: Principles and Techniques of Algorithmic Differentiation. No. 105 in Other Titles in Applied Mathematics, 2nd edn. SIAM, Philadelphia (2008). http://bookstore.siam.org/ot105/
13. Hascoet, L., Pascual, V.: The Tapenade automatic differentiation tool: principles, model, and specification. ACM Trans. Math. Softw. (TOMS) **39**(3), 1–43 (2013)
14. Jain, P., et al.: Checkmate: breaking the memory wall with optimal tensor rematerialization. In: Proceedings of Machine Learning and Systems, vol. 2, pp. 497–511 (2020)
15. Johansson, J., Nation, P., Nori, F.: QuTiP: an open-source python framework for the dynamics of open quantum systems. Comput. Phys. Commun. **183**(8), 1760–1772 (2012). https://doi.org/10.1016/j.cpc.2012.02.021
16. Johansson, J., Nation, P., Nori, F.: QuTiP 2: a Python framework for the dynamics of open quantum systems. Comput. Phys. Commun. **184**(4), 1234–1240 (2013). https://doi.org/10.1016/j.cpc.2012.11.019
17. Khaneja, N., Brockett, R., Glaser, S.J.: Time optimal control in spin systems. Phys. Rev. A **63**, 032308 (2001). https://doi.org/10.1103/PhysRevA.63.032308
18. Kubota, K.: A Fortran77 preprocessor for reverse mode automatic differentiation with recursive checkpointing. Optim. Meth. Softw. **10**(2), 319–335 (1998). https://doi.org/10.1080/10556789808805717
19. Kukreja, N., Hückelheim, J., Louboutin, M., Washbourne, J., Kelly, P.H., Gorman, G.J.: Lossy checkpoint compression in full waveform inversion. Geoscientific Model Development Discussions, pp. 1–26 (2020)
20. Leung, N., Abdelhafez, M., Koch, J., Schuster, D.: Speedup for quantum optimal control from automatic differentiation based on graphics processing units. Phys. Rev. A **95**, 042318 (2017). https://doi.org/10.1103/PhysRevA.95.042318
21. Naumann, U.: The Art of Differentiating Computer Programs. Society for Industrial and Applied Mathematics (2011). https://doi.org/10.1137/1.9781611972078
22. Rajbhandari, S., Ruwase, O., Rasley, J., Smith, S., He, Y.: ZeRO-infinity: breaking the GPU memory wall for extreme scale deep learning. In: Proceedings of the International Conference for High Performance Computing, Networking, Storage and Analysis, SC 2021. Association for Computing Machinery, New York (2021). https://doi.org/10.1145/3458817.3476205
23. Schanen, M., Marin, O., Zhang, H., Anitescu, M.: Asynchronous two-level checkpointing scheme for large-scale adjoints in the spectral-element solver Nek5000. Procedia Comput. Sci. **80**(C), 1147–1158 (2016). https://doi.org/10.1016/j.procs.2016.05.444

Quantum Annealing and Algebraic Attack on Speck Cipher

Elżbieta Burek(✉) and Michał Wroński(✉)

Military University of Technology, Kaliskiego Street 2, Warsaw, Poland
{elzbieta.burek,michal.wronski}@wat.edu.pl

Abstract. Algebraic attacks using quantum annealing are a new idea of cryptanalysis. This paper shows how to obtain a QUBO problem equivalent to the algebraic attack on the Speck cipher, using as small a number of logical variables as possible. The main idea of minimizing the number of variables in the algebraic attack on this ARX cipher was appropriate cipher partition and insertion of additional variables. Using such an idea, in the case of the most popular variants: Speck-128/128 and Speck-128/256, the equivalent QUBO problem has 19,311 and 33,721 logical variables, which is more efficient than the same attack on AES cipher, where for AES-128 and AES-256, an equivalent QUBO problem consist of 29,770 and 72,597 logical variables, respectively. It is an open question if this kind of attack may overtake, in some cases, brutal or Grover's attack.

Keywords: Cryptanalysis · Algebraic attacks · Speck · D-Wave advantage · Quantum annealing

1 Introduction

Quantum computing has allowed the development of new approaches to computational problems that classical computers cannot cope with. One such problem in cryptanalysis of block ciphers is solving large systems of multivariate polynomial equations during algebraic attacks. In general, the idea of algebraic attacks is based on two steps: the first is to represent the cipher as a system of multivariate polynomial equations, and the second is to solve the created system.

In [3] Burek et al. showed how to transform obtained system of multivariate equations into the QUBO problem.

QUBO (Quadratic Unconstrained Binary Optimization) is a combinatorial optimization problem in which the cost function $f(x)$ is defined on an n-dimensional binary vector space $\mathbb{B}^n$ onto $\mathbb{R}$, as follows: $QUBO: \ min\ f(x) = x^t Q x$, where x is a vector of binary variables and Q is an upper diagonal matrix of real weights.

This work was supported by the Military University of Technology's University Research Grant No. 858/2021: "Mathematical methods and models in computer science and physics".

D. Groen et al. (Eds.): ICCS 2022, LNCS 13353, pp. 143–149, 2022.
https://doi.org/10.1007/978-3-031-08760-8_12

Since the variables are binary, $x_i^2 = x_i$ holds, and the cost function can be represented as: $QUBO: \ min\ f(x) = \sum_i Q_{i,i}x_i + \sum_{i<j} Q_{i,j}x_ix_j$.

It is worth noting that algebraic attacks on symmetric ciphers using general-purpose quantum computing have been studied in [4,6], where variants of the HHL [7] algorithm has been used.

The contribution presented in this paper is the presentation of the application of an algebraic attack using quantum annealing on the Speck cipher. We focused on obtaining equivalent QUBO problem using as small variables as possible. The main idea of minimizing the number of variables in the algebraic attack on the Speck cipher was appropriate ciphers partition and insertion of additional variables. In the case of the most popular variants: Speck128/128 and Speck128/256, we obtained the equivalent QUBO problem consisting of 19,311 and 33,721 logical variables. According to our experiments, applying quantum annealing to the algebraic attacks on Speck should be much more efficient than the same attack on AES cipher, where in the case of the algebraic attack on AES-128 and AES-256, an equivalent QUBO problem consist of 29,770 and 72,597 logical variables respectively. It is an open question if this kind of attack may overtake, in some cases, brutal or Grover's attacks. However, assuming that complexity of solving of QUBO problem consisting of N variables requires $O\left(e^{\sqrt{N}}\right)$ elementary operations [8], one can obtain an attack faster than the brute force on Speck-128/256 consisting of 31 of 34 rounds, which is better than the best known classical attack on this cipher variant, which works for 25 rounds.

2 Algebraic Attack on Speck Using Quantum Annealing

This section will present the method of representing the Speck cipher using multivariate polynomial equations to obtain a system of multivariate polynomial equations with as few monomials as possible, which consequently allows obtaining a problem in the QUBO form with as few binary variables as possible.

2.1 Speck Cipher

The Speck cipher is a family of lightweight block ciphers of the ARX type, presented in [2] as highly-optimized block ciphers intended for software and hardware implementations.

An instance of the Speck cipher will be designated, according to [2], as Speck$2n/mn$, where $2n$ is the length of the input block, n is the word length, and mn is the key length. The Speck$2n/mn$ cipher uses the n-bit word operations, as bitwise xor, addition modulo 2^n and right and left rotations.

The general structure of the Speck$2n/mn$ cipher is shown in Fig. 1, where T denotes the number of rounds, $\oplus$ denotes the bitwise xor operation, $\boxplus$ denotes the addition modulo 2^n, and $\gg \alpha$ and $\ll \beta$ denote a right rotation by α and left rotation by β bits, respectively.

The round function of the encryption algorithm of the Speck$2n/mn$ cipher is a map $R: GF(2)^n \times GF(2)^n \rightarrow GF(2)^n \times GF(2)^n$, where $GF(q)$ is Galois field with q elements, defined as follows: $R(x_{i+1}, y_{i+1}) = (((x_i \gg \alpha) + y_i) \oplus k_i, (y_i \ll \beta) \oplus ((x_i \gg \alpha) + y_i) \oplus k_i)$, where x_i and y_i is, respectively, the left and the right n-bit word of the input block of i round, k_i is the round key and i is the number of the round.

The round key generation algorithm uses the round function. The key is divided into m n-bit words, where the least significant n bits are the round key of the first round, and the next n-bit words are successive l_i words: $K = [l_{m-2}, \dots, l_0, k_0]$, where $l_i, k_0 \in GF(2)^n$. The words l_i and the round keys k_i are determined as: $l_{i+m-1} = ((l_i \gg \alpha) + k_i) \oplus i$ and $k_{i+1} = (k_i \ll \beta) \oplus l_{i+m-1}$.

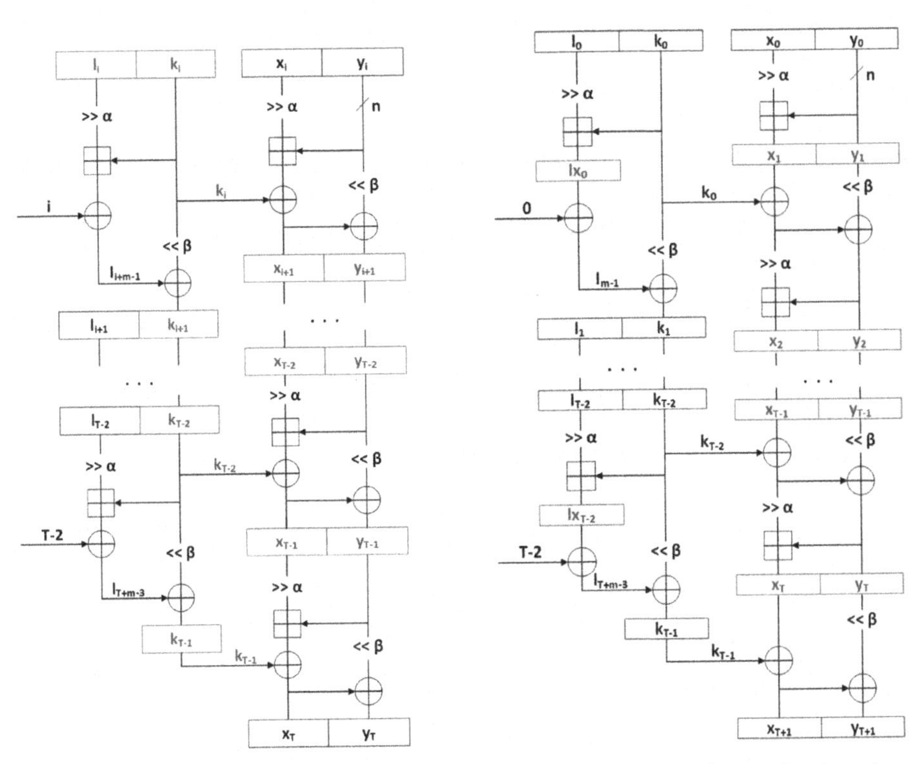

(a) Split of Speck$2n/mn$ cipher according to cipher documentation.

(b) Split of Speck$2n/mn$ cipher by using additional variables.

Fig. 1. Structure of the Speck$2n/mn$ cipher for the presented approaches. (Color figure online)

2.2 Efficient Approach to Generating Multivariate Polynomial Equations

In the approach to generating multivariate polynomial equations, where the range of the round was held by the Speck algorithm documentation, the additional binary variables have been introduced for intermediate states between rounds and round keys. The number of additional binary variables for intermediate states is $(T-1)2n$, and for round keys $(T-1)n$, as presented in red in Fig. 1a. The multivariate polynomials were generated over $GF(2)^n$, separately for the left and right words, each state between rounds, and each round key. Finally, the degree of the left word polynomial equations is $2n+1$, so the number of binary variables in the QUBO problem will be very large.

In our approach to generating multivariate polynomial equations, the range of the round was changed. Figure 1b shows in red how the additional intermediate variables were introduced. Additional binary variables were introduced for round keys and intermediate states, which were introduced after the addition modulo 2^n in the encryption algorithm and the round key generation algorithm. Since there is no key addition in the first round in Fig. 1b, the bits of the words x_1 and y_1 are also known: $x_1 = (x_0 \gg \alpha) + y_0$ and $y_1 = y_0$.

The number of additional binary variables for the round keys is $(T-1)n$, for intermediate states in the encryption algorithm, it is $(T-1)2n$, and for intermediate states lx_i, in the round key generation algorithm, it is $2n$.

The xor operation of a_j and b_j bits may be written as $a_j \oplus b_j = a_j + b_j - 2a_jb_j$, therefore for the n-bit a and b words is executed as: $a \oplus b = \sum_{j=0}^{n-1} 2^j(a_j + b_j - 2a_jb_j)$. Since in this approach addition modulo 2^n is executed after the xor operation, then: $(a+b)\ mod\ 2^n = \sum_{j=0}^{n-1} 2^j(a_j + b_j) - c \cdot 2^n$, where the bit c is the carry bit of sum.

In this approach, the equation representing the left-word of one round of the encryption algorithm, except the last round, takes the following form: $x_{i+1} = (((x_i \oplus k_i) \gg \alpha) + y_{i+1})\ mod\ 2^n$, and after performing all operations it is form as:

$$\sum_{j=0}^{n-1} 2^j((x_i)_{(j+\alpha) mod\ n} + (k_i)_{(j+\alpha) mod\ n} - 2(x_i)_{(j+\alpha) mod\ n}(k_i)_{(j+\alpha) mod\ n} + (y_{i+1})_j - (x_{i+1})_j) - c \cdot 2^n = 0. \tag{1}$$

The equation representing the right-word of one round of the encryption algorithm, except the last round, takes the following form: $y_{i+1} = ((y_i \ll \beta) \oplus x_i \oplus k_i)$, which can be finally converted to:

$$\sum_{j=0}^{n-1} 2^j((y_i)_{(j-\beta) mod\ n} + (x_i)_j - 2(y_i)_{(j-\beta) mod\ n}(x_i)_j + (k_i)_j + - 2(k_i)_j(y_i)_{(j-\beta) mod\ n} - 2(k_i)_j(x_i)_j + 4(k_i)_j(y_i)_{(j-\beta) mod\ n}(x_i)_j - (y_{i+1})_j) = 0. \tag{2}$$

Similarly, the last round of the encryption algorithm can be represented by the following equations: $x_{T+1} = (x_T \oplus k_{T-1})$, for the left word, which is equivalent to:

$$\sum_{j=0}^{n-1} 2^j((x_T)_j + (k_{T-1})_j - 2(x_T)_j(k_{T-1})_j - (x_{T+1})_j) = 0, \tag{3}$$

and for the right word: $y_{T+1} = (y_T \lll \beta) \oplus x_{T+1}$, which is equivalent to:

$$\sum_{j=0}^{n-1} 2^j ((y_T)_{(j-\beta) mod\ n} + (x_{T+1})_j - 2(y_T)_{(j-\beta) mod\ n}(x_{T+1})_j - (y_{T+1})_j) = 0. \quad (4)$$

Two multivariate polynomial equations also represent each round of the round key generation algorithm. The first equation associates the binary variables of the word l_i with the binary variables of the word lx_i, and the second equation relates the binary variables of the word lx_i with the binary variables of the k_i and k_{i+1} round keys.

The equation defining the lx_i word is as follows: $lx_i = ((l_i \ggg \alpha) + k_i)\ mod\ 2^n$, which can be converted to the form:

$$\sum_{j=0}^{n-1} 2^j ((l_i)_{(j+\alpha) mod\ n} + (k_i)_j - (lx_i)_j) - c \cdot 2^n = 0. \quad (5)$$

The equation defining the k_{i+1} round key has the following form: $k_{i+1} = ((lx_i \oplus i) \oplus (k_i \lll \beta))$, which is equivalent to:

$$\begin{aligned}\sum_{j=0}^{n-1} 2^j ((lx_i)_j + (i)_j - 2(lx_i)_j(i)_j + (k_i)_{(j-\beta) mod\ n} - 2(lx_i)_j(k_i)_{(j-\beta) mod\ n} + \\ - 2(i)_j(k_i)_{(j-\beta) mod\ n} + 4(lx_i)_j(i)_j(k_i)_{(j-\beta) mod\ n}) = 0,\end{aligned} \quad (6)$$

where $(i)_j$ is the j-th bit of known constant i. The degree of the polynomial in Eq. (6) is 3. However, such a degree occurs only in the monomial with the i constant bit, so the monomial will have the degree 2 if the constant bit is 1, otherwise, the monomial will vanish.

In the proposed approach to generating multivariate polynomial equations representing the Speck$2n/mn$ cipher, the degree of polynomials is constant and does not depend on the length of the input block. T-round Speck$2n/mn$ cipher can be represented by the system of: $T-1$ polynomials of degree 2, of the form as in Eq. (1), $T-1$ polynomials of degree 3, of the form as in Eq. (2), one polynomial of degree 2, of the form as in Eq. (3) and one polynomial of degree 2, of the form as in Eq. (4) for the encryption algorithm. Additional, for the round key generation algorithm: $T-1$ polynomials of degree 1, of the form as in Eq. (5) and $T-1$ polynomials of degree 2, of the form as in Eq. (6).

3 Transformation of Algebraic Attacks on Speck Using Quantum Annealing

Cryptanalysis of Speck algorithm has been widely described, see [1,5,10]. This section will describe the results of the transformation of algebraic attacks using quantum annealing on Speck.

We used the transformation method of algebraic attacks to the QUBO problem presented in [3].

It is worth presenting the following observation, which can be found in [9]. First, let us note that there are many different variants of the Speck cipher. Each variant has a different block size ($2n$) and key length (mn). If $2n \geq mn$, there is approximately one proper key for each pair of plaintext - ciphertext. Things are getting different if $2n < mn$. In such a case, having only one pair of plaintext - ciphertext, for each pair, approximately 2^{mn-2n} keys will be proper for this pair, but only one will be proper for all other pairs. If one wants to find the key used for encryption with high probability, in such a case, there are required $\lceil \frac{m}{2} \rceil$ plaintext - ciphertext pairs.

For each variant of the Speck cipher, we computed the number of variables of the equivalent QUBO problem. Let us note that for variants in which block length is smaller than key length, we used 2 pairs of plaintext - ciphertext, and therefore, in such cases, such QUBO problem is constructed from two smaller QUBO problems - each problem for each pair. Unfortunately, our final QUBO problem must consist of two smaller. We most frequently obtain some proper solution using quantum annealing, but not all solutions. It means that we cannot solve such systems independently for each pair.

Table 1. Results of transformation of the system of multivariate quadratic equations describing the AES and Speck ciphers to the QUBO problem.

Cipher variant	Number of rounds	Number of variables	Cipher variant	Number of rounds	Number of variables
AES-128	10	29,770	Speck96/96	28	12,418
AES-192	12	62,153	Speck64/128	27	13,711
AES-256	14	72597	Speck128/128	32	19,311
Speck32/64	22	5,789	Speck96/144	29	21,716
Speck48/72	22	8,470	Speck128/192	33	32,659
Speck48/96	23	8,884	Speck128/256	34	33,721
Speck64/96	26	13,167			

Because QUBO problem, equivalent to the algebraic attack on Speck cipher, in general, consists of less number of variables than analogic QUBO problem in the case of AES cipher with the same block size and key size (see Table 1), we conclude, that Speck cipher is easier to break using quantum annealing. However, it is hard to speculate if this attack can outperform brute force or Grover's attack. The computational complexity of solving the QUBO problem using quantum annealing still requires much research. However, it is claimed that such complexity depends mostly on the number of variables. The precise time complexity of solving the QUBO problem using quantum annealing has not been computed yet. Using heuristics, it is possible to estimate the expected time of solving QUBO problem consisting of N variables as $O\left(e^{\sqrt{N}}\right)$ [8]. Unfortunately, using current quantum annealers, it is impossible to break any variant of Speck cipher in practice.

4 Conclusion

This paper presents the transformation of the algebraic attack on the Speck cipher to the QUBO problem. We showed how to obtain the smallest possible number of variables for a QUBO problem. To obtain such a small number of variables, we proposed a novel way of describing the algebraic structure of each of the algorithms.

The computational complexity of solving the QUBO problem using quantum annealing has not been fully studied yet, and much more research in this area is required.

Further works should be more research on the computational complexity of solving algebraic attacks on cryptographic algorithms using quantum annealing and applying the presented method to other symmetric algorithms.

References

1. Abed, F., List, E., Lucks, S., Wenzel, J.: Differential cryptanalysis of round-reduced SIMON and SPECK. In: Cid, C., Rechberger, C. (eds.) FSE 2014. LNCS, vol. 8540, pp. 525–545. Springer, Heidelberg (2015). https://doi.org/10.1007/978-3-662-46706-0_27
2. Beaulieu, R., Shors, D., Smith, J., Treatman-Clark, S., Weeks, B., Wingers, L.: The Simon and speck families of lightweight block ciphers. Cryptology eprint archive (2013)
3. Burek, E., Wroński, M., Mańk, K., Misztal, M.: Algebraic attacks on block ciphers using quantum annealing. IEEE Trans. Emerg. Top. Comput. (in press)
4. Chen, Y.-A., Gao, X.-S.: Quantum algorithm for Boolean equation solving and quantum algebraic attack on cryptosystems. J. Syst. Sci. Complex. **35**, 1–40 (2021)
5. Dwivedi, A.D., Morawiecki, P., Srivastava, G.: Differential cryptanalysis of round-reduced speck suitable for internet of things devices. IEEE Access **7**, 16476–16486 (2019)
6. Gao, J., Li, H., Wang, B., Li, X.: Quantum security of AES-128 under HHL algorithm. Quantum Inf. Comput. **22**(3&4), 0209–0240 (2022)
7. Harrow, A.W., Hassidim, A., Lloyd, S.: Quantum algorithm for linear systems of equations. Phys. Rev. Lett. **103**(15), 150502 (2009)
8. Mukherjee, S., Chakrabarti, B.K.: Multivariable optimization: quantum annealing and computation. Eur. Phys. J. Spec. Top. **224**(1), 17–24 (2015). https://doi.org/10.1140/epjst/e2015-02339-y
9. Pakhomchik, A.I., Voloshinov, V.V., Vinokur, V.M., Lesovik, G.B.: Converting of Boolean expression to linear equations, inequalities and QUBO penalties for cryptanalysis. Algorithms **15**(2), 33 (2022)
10. Song, L., Huang, Z., Yang, Q.: Automatic differential analysis of ARX block ciphers with application to SPECK and LEA. In: Liu, J.K., Steinfeld, R. (eds.) ACISP 2016. LNCS, vol. 9723, pp. 379–394. Springer, Cham (2016). https://doi.org/10.1007/978-3-319-40367-0_24

Benchmarking D-Wave Quantum Annealers: Spectral Gap Scaling of Maximum Cardinality Matching Problems

Cameron Robert McLeod and Michele Sasdelli(✉)

Australian Institute for Machine Learning, The University of Adelaide, Adelaide, Australia
cameron.mcleod@student.adelaide.edu.au,
michele.sasdelli@adelaide.edu.au

Abstract. Quantum computing, in particular Quantum Annealing (QA), provides a theoretically promising alternative to classical methods for solving combinatorially difficult optimization problems. In particular, QA is suitable for problems that can be formulated as a Quadratic Unconstrained Binary Optimization (QUBO) problem, such as SAT, graph colouring and travelling salesman. With commercially available QA hardware, like that offered by D-Wave Systems (D-Wave), reaching scales capable of tackling real world problems, it is timely to assess and benchmark the performance of this current generation of hardware. This paper empirically investigates the performance of D-Wave's 2000Q (2048 qubits) and Advantage (5640 qubits) quantum annealers in solving a specific instance of the maximum cardinality matching problem, building on the results of a prior paper that investigated the performance of earlier QA hardware from D-Wave. We find that the Advantage quantum annealer is able to produce optimal solutions to larger problem instances than the 2000Q. We further consider the problem's structure and its implications for suitability to QA by utilising the Landau-Zener formula to explore the potential scaling of the diabatic transition probability. We propose a method to investigate the behaviour of minimum energy gaps for scalable problems deployed to quantum annealers. We find that the minimum energy gap for our target QA problem does not scale favourably. This behaviour raises questions as to the suitability of this problem for benchmarking QA hardware, as it potentially lacks the nuance required to identify meaningful performance improvements between generations.

Keywords: D-Wave · Quantum annealing · Maximum matching · Landau-Zener

1 Introduction

The promise of quantum computing has been a tantalising prospect ever since the concept of utilising quantum behaviour to perform computation was first proposed, most notably in the early 1980s s with the work of Paul Benioff [2–4] and Richard Feynman [10]. Since the inception of this idea there has been significant effort invested, and subsequent advances in hardware, however, it remains a speculative area with no

D. Groen et al. (Eds.): ICCS 2022, LNCS 13353, pp. 150–163, 2022.
https://doi.org/10.1007/978-3-031-08760-8_13

clear indication if the promise of quantum computing will be realised [8] or if classical computing will reign supreme.

There are a number of competing approaches to quantum computing including Quantum Gate Array (QGA) [18], One-way Quantum Computers (OQC) [20], more exotic and theoretical Topological Quantum Computers (TQC) [14], Adiabatic Quantum Computers (AQC) [1] and Quantum Annealing (QA) approaches [9] and many others such as those covered in [19]. While the QGA model is arguably the most investigated, QA has shown recent promise with commercially available hardware reaching scales, i.e. number of qubits, with the potential of tackling useful real world problems [13,26]. QA as described in [24], inhabits a regime that is intermediate between the idealised assumptions of universal AQC and unavoidable experimental compromises.

The contribution of this work is mainly twofold. We first extend the work of [23] to the latest QA hardware from D-Wave and contrast the performance between quantum annealers. We then consider the structure of the specific problem we have utilised to assess the quantum annealer performance. It is known that as the energy gap between two energy states of a quantum system decreases, the greater the likelihood the system will 'jump' from one to the other [28]. We utilise the Landau-Zener formula to calculate the expected diabatic transition probabilities for increasing problem sizes. We estimate the scaling of the problem's minimum energy gap, that is the minimum energy gap between the ground state and the first excited state, and discuss the implications for suitability to benchmark QA. We hope that this work forms a building block in the ongoing assessment and benchmarking of quantum annealer performance.

2 Quantum Annealing on D-Wave

The QA implementation used by D-Wave is of the form shown in Eq. 1 where $\hat{\sigma}_x^{(i)}$ and $\hat{\sigma}_z^{(i)}$ are Pauli matrices operating on a qubit, q_i, and h_i and $J_{i,j}$ are the qubit biases and coupling strengths.

$$\mathcal{H}_{ising} = -\frac{A(s)}{2}\left(\sum_i \hat{\sigma}_x^{(i)}\right) + \frac{B(s)}{2}\left(\sum_i h_i \hat{\sigma}_z^{(i)} + \sum_{i>j} J_{i,j}\hat{\sigma}_z^{(i)}\hat{\sigma}_z^{(j)}\right) \tag{1}$$

Using D-Wave terminology [17] the initial Hamiltonian is called the tunnelling Hamiltonian and is in its lowest energy state when all qubits are in a superposition state of 0 and 1. The final Hamiltonian, also called the problem Hamiltonian, encodes the spin problem to be optimised, Eq. 2, by extending it from discrete spins to quantum states. The ground state of the problem Hamiltonian corresponds to the solution of the Ising problem. Initially the quantum annealer starts in the lowest energy state of the tunnelling Hamiltonian and slowly introduces the problem Hamiltonian, i.e. the annealing cycle begins at $s = 0$ with $A(s) \gg B(s)$ and ends at $s = 1$ with $A(s) \ll B(s)$.

$$\mathrm{E}_{ising}(s) = \sum_i h_i s_i + \sum_{i>j} J_{i,j} s_i s_j \tag{2}$$

Once the annealing cycle is completed, $s = 1$, the $\hat{\sigma}_z^{(i)}$ can be replaced by classical spin variables, $s_i = \pm 1$. The energy of the system is then described as per Eq. 2 with the

s_i's corresponding to the solution for the target problem. If the system has remained in the ground state then the corresponding values of the s_i's represent the optimal solution to the target problem.

2.1 QUBO Formulation

QA requires that Hamiltonians be written as the quantum version of the Ising spin glass [21]. Ising spin glasses often go by the name of Quadratic Unconstrained Binary Optimisation (QUBO) problems [12,16]. A general QUBO model is expressed as an optimisation problem, as shown in Eq. 3, where x is a vector of binary decision variables and Q is a square matrix:

$$minimise/maximise\ y = x^T Qx. \tag{3}$$

Quadratic penalties are added to the objective function to impose constraints on the problem. These penalties are constructed such that their contribution is zero for feasible solutions and some positive amount for infeasible solutions.

Consider the maximum cardinality matching problem, whose goal is, for some graph, G, to find a matching containing as many edges as possible such that each vertex is adjacent to at most one of the selected edges. The maximum matching problem, for a graph, $G = (V, E)$, can be represented as shown in Eq. 4.

$$\text{Maximise} \sum_{e \in E} x_e \quad s.t. \quad \forall v \in V, \sum_{e \in E(v)} x_e \leq 1 \tag{4}$$

where $x_e \in \{0, 1\}$ and $E(v)$ denotes the set of edges which have v as an endpoint.

Once the target problem has been formulated as a QUBO, it is then able to be implemented and run on a quantum annealer. To implement the QUBO on the quantum annealer it is first converted into an equivalent Ising problem, which requires the mapping of binary variables, $x_i \in \{0, 1\}$, to spin variables, $s_i \in \{-1, 1\}$ via the relation $s_i = 2x_i - 1$. These spin variables are then mapped to physical qubits on the QPU. Many of these implementation details are handled by the API associated with the quantum annealer as is the case with submitting problems to D-Wave's quantum annealers.

3 Maximum Cardinality Matching

The goal of the maximum matching problem is to find, for some graph, a matching containing as many edges as possible, that is, a maximum cardinality subset of the edges, such that each vertex is adjacent to at most one edge of the subset. More formally the problem can be stated as for some (undirected graph), $G = (V, E)$, the maximum matching problem asks for $M \subseteq E$ such that $\forall e, e' \in M^2, e \neq e'$ we have that $e \cap e' = \emptyset$ and such that $|M|$ is maximum [15].

3.1 Graph Family

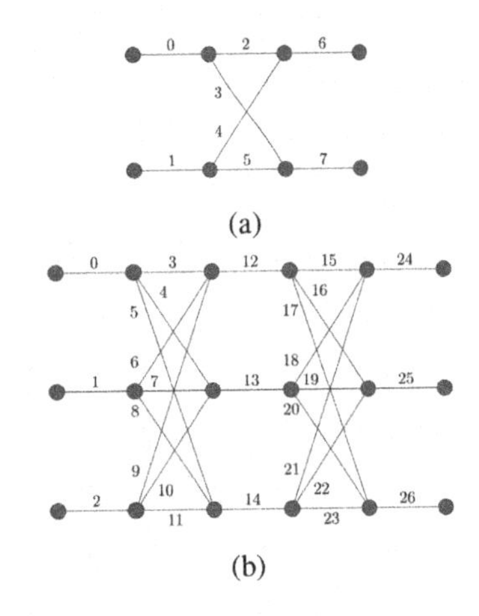

Fig. 1. Schemes representing the G_1(a) and G_2(b) graphs.

As per [23] a specific family of graphs are utilised for the application of the maximum matching problem. This graph family, G_n, consists of alternating layers of sparsely and densely connected nodes where the number of rows are equal to $(n+1)$ and the number of layers equal to $(2n+1)$. The total number of edges for a specific G_n graph is $(n+1)^3$ and the number of nodes is $2(n+1)^2$. As an example, the G_1 and G_2 graphs are shown in Fig. 1. A rigorous definition of this graph family can be found in [22].

There are two main reasons for the selection of this family of graphs. Firstly, it is trivial to see that a maximum matching for G_n consists of all the edges in the sparsely connected layers. It is therefore easy to check solutions returned by a quantum annealer. Secondly, the work presented in [22] shows that the expected number of iterations required by a class of annealing algorithms to reach a maximum matching is in $O(\exp(n))$. It is these two properties that make this graph family an interesting testing ground for quantum annealers.

3.2 QUBO Formulation

To formulate the maximum matching problem as a QUBO we must incorporate the constraints of Eq. 4 into the objective function. The optimisation problem is looking to incentivise the inclusion of all edges, this corresponds to the diagonal terms in Q. The constraint that selected edges are not adjacent to other selected edges corresponds to the off-diagonal terms in Q. An example QUBO formulation for a G_1 graph is shown in Eq. 5, where we have chosen -1, being a minimisation problem, to represent the value of an included edge and a penalty term of P for adjacent edges. A vector of $[1,1,0,0,0,0,1,1]$ minimises the value of Eq. 5, pending a sufficiently large P. Given an arbitrary graph, the QUBO formulation to find a maximum matching would require that the ratio of off-diagonal to diagonal elements be greater than 1. Naturally the penalty of violating a constraint must outweigh the benefit of incorporating another edge into the solution. In our formulation we opt for a value of $P = 2$. This value was chosen to avoid excessive scaling of biases and coupling values, as at implementation time on the annealer D-Wave re-scales h_i and $J_{i,j}$ values between [−2,2] and [−1,1] respectively. The choice of P is itself an area that warrants further investigation and an optimal choice will likely depend on the specific problem structure.

We refer to the formulation shown in Eq. 5 as the 'general' formulation and the formulation in [23] as the 'prior' formulation from here out. This distinction is due to [23] utilising $1 + 2|E|$ and $-2|E|$ for their diagonal and off-diagonal values, respectively, in Q. We note that the prior formulation is valid for the utilised G_n graphs but that it is not necessarily true for general graphs given $\frac{2|E|}{1+2|E|} < 1$. We provide performance comparisons for both QUBO formulations, to provide continuity in the performance benchmark dataset and also to investigate any impacts on solution quality when a general formulation is used.

$$\text{Min} y = \begin{bmatrix} x_0 \\ x_1 \\ x_2 \\ x_3 \\ x_4 \\ x_5 \\ x_6 \\ x_7 \end{bmatrix}^\top \begin{bmatrix} -1 & 0 & P & P & 0 & 0 & 0 & 0 \\ 0 & -1 & 0 & 0 & P & P & 0 & 0 \\ 0 & 0 & -1 & P & P & 0 & P & 0 \\ 0 & 0 & 0 & -1 & 0 & P & 0 & P \\ 0 & 0 & 0 & 0 & -1 & P & P & 0 \\ 0 & 0 & 0 & 0 & 0 & -1 & 0 & P \\ 0 & 0 & 0 & 0 & 0 & 0 & -1 & 0 \\ 0 & 0 & 0 & 0 & 0 & 0 & 0 & -1 \end{bmatrix} \begin{bmatrix} x_0 \\ x_1 \\ x_2 \\ x_3 \\ x_4 \\ x_5 \\ x_6 \\ x_7 \end{bmatrix} \quad (5)$$

4 Improved QPU Topology

One of the key changes in hardware from D-Wave has been the improvement in the physical QPU architecture (or topology). D-Wave refers to the two most recent architectures as Chimera (present in the 2000Q and older models) and Pegasus (present in the Advantage) [6]. The Chimera architecture consists of recurring $K_{4,4}$, bipartite graph unit cells which are coupled together and each qubit is physically coupled to 6 other qubits. The Pegasus architecture, present in the Advantage quantum annealer, increases the interconnection density with each qubit being connected to 15 other qubits. This increase in qubit connectivity has positive implications with respect to chain lengths and minor embedding.

When looking to solve an arbitrary problem on a D-Wave quantum annealer the problem's structure may not naturally match the annealer's topology, as such the problem needs to be 'embedded'. Physical qubits need to be linked, or coupled, to form 'chains' each representing a logical variable of original problem [5]. This introduces the issue of chain breaks in the solutions found by the quantum annealer. Resolving these chain breaks to find a consensus solution is typically achieved via empirical approaches, such as majority voting by the qubits in the broken chain.

5 Quantum Annealing Experiments

To conduct our experiments we made use of D-Wave's open-source SDK, Ocean, and the quantum annealers 'DW_2000Q_6', 'Advantage_system1.1' and 'Advantage_system4.1'. Where we refer to 'Advantage' it is a reference to both the 1.1 and 4.1 variants, where we refer to a specific annealer the version number is specified. For all experiments we utilised, unless specified otherwise, the default annealing duration of $20\mu s$ and 10,000 samples (annealing cycles). Similar to [23] we note that both shorter and longer annealing durations had negligible impacts, although a more nuanced analysis could be of future benefit. We also utilised the default chain strength, which is calculated using D-Wave's 'uniform torque compensation'.

6 Results Overview

Experimental results on the newest Advantage architecture are shown in Table 1. Up to the G3 graph, the quantum annealer is able to obtain the optimal solution. It is not guaranteed that the solutions returned will be valid for the original problem i.e. they may violate the problem constraints. As noted in [23] a post processing step would be required when using QA to ensure results produce valid matchings. Checking the validity of solutions is trivial and it is an insightful metric to assess the performance of QA hardware. Table 2 provides a summary of the number of valid solutions returned out of the 10,000 annealing cycles. An insight of these results is that our general QUBO formulation produces significantly more valid solutions. We postulate that the quantity of valid solutions is directly correlated to the ratio between diagonal and off-diagonal terms in our Q matrix. We also note that both the 2000Q and Advantage quantum annealers produce valid results for the G_3 graph, whereas the 2X failed to produce a single valid matching, demonstrating an improvement in the hardware's capability, albeit minor.

Table 1. Results advantage - general and prior QUBO formulations (all samples)

	Best	Worst	Mean	Med.	St. dev.	Best	Worst	Mean	Med.	St. dev.
	2000Q (General)					2000Q (Prior)				
G1	−4	−1	−4.0	−4	0.2	−68	−36	−67.9	−68	1.1
G2	−9	−4	−7.9	−8	0.8	−495	−279	−451.5	−442	37.8
G3	−16	−5	−11.4	−12	1.3	−2064	−1046	−1563.2	−1553	108.0
G4	−21	−5	−16.1	−16	2.0	−5526	−3279	−4579.6	−4527	292.1
	Advantage1.1 (General)					Advantage1.1 (Prior)				
G1	−4	−2	−4.0	−4	0.2	−68	−52	−68.0	−68	0.6
G2	−9	−2	−7.1	−7	1.0	−495	−227	−428.0	−441	43.3
G3	−16	−3	−11.2	−11	1.6	−2064	−1042	−1592.8	−1554	127.5
G4	−22	−1	−13.9	−14	3.0	−5527	−2528	−4281.0	−4279	387.6
G5	−29	6	−13.6	−14	4.7	−12137	−4800	−9373.3	−9545	960.2
	Advantage4.1 (General)					Advantage4.1 (Prior)				
G1	−4	−1	−4.0	−4	0.1	−68	−53	−68.0	−68	0.2
G2	−9	−2	−7.1	−7	1.0	−495	−276	−431.0	−441	42.0
G3	−16	−2	−10.9	−11	1.7	−2064	−915	−1560.6	−1553	127.4
G4	−22	−2	−15.2	−15	2.6	−5776	−2779	−4428.6	−4525	347.4
G5	−29	3	−17.8	−18	4.0	−12568	−6093	−10258.4	−10407	845.6

We make the following observations from our empirical tests:

- The 2000Q and Advantage were able to produce the optimal solution up to G_3, where as the 2X was only able to produce the optimal solution up to G_2
- There is general improvement (a better lower bound) on the worst solution returned

- There is general improvement in the mean and median solutions returned with the newer annealers, with the exception of Advantage on G_4
- The Advantage was able to return valid, but non-optimal, solutions with our general formulation up to and including G_5

While embeddings for G_6 and G_7 were identified on Advantage further analysis was not conducted given the decline in solution quality beyond the G_4 and G_5 graphs.

Table 2. Comparison of valid solutions returned by quantum annealers

(a) Valid solutions returned by 2000Q

Graph	Num. of valid sols.	
	Prior	General
G1	9959	9992
G2	4418	9561
G3	21	5575
G4	0	1516

(b) Valid solutions returned by Advantage1.1

Graph	Num. of valid sols.	
	Prior	General
G1	9987	10000
G2	2864	8509
G3	50	3511
G4	0	123
G5	0	2

(c) Valid solutions returned by Advantage4.1

Graph	Num. of valid sols.	
	Prior	General
G1	9998	9997
G2	3062	8821
G3	59	3255
G4	1	456
G5	0	7

The general formulation produced a greater number of valid solutions, as shown in Table 2. The newer Advantage architecture produced a reduced number of valid solutions compared to 2000Q. For example we obtained 1516 valid solutions on the 2000Q versus 123 on Advantage1.1 and 456 on Advantage4.1 using our general formulation for the G_4 graph. Even though Advantage returned fewer valid solutions, it still produced a better overall result, achieving a solution of -22 versus the -21 obtained on the 2000Q. The driver of this unexpected result is not clear, although Advantage4.1 closes this gap slightly. Potentially the more complex QPU architecture of Advantage compromises with an increased variability in results (noise), noting generally higher standard deviation values. A key benefit of Advantage is an increased connection density over the previous Chimera architecture, leading to easier identification of minor embeddings and both reduced total number and overall length of chains required.

Table 3 shows the number of returned samples, for each graph size, that contain at least one chain break, keeping in mind that chain breaks were resolved by majority voting. It was noted during testing that the number of samples that contained chain breaks, from a batch of 10,000, was rather variable. In particular, for G_3 the number of samples containing chain breaks would sometimes be higher for sample batches on the 2000Q and other times be higher for Advantage. This overlap in chain break occurrence is not entirely unexpected given that both the 2000Q and Advantage require a similar number of chains for the embedding, 60

Table 3. Number of chain breaks for each graph size on different formulations and architectures. G_5 does not fit on 2000Q.

Graph	# of Chain breaks			
	2000Q		Advantage1.1	
	Prior	General	Prior	General
G1	13	4	0	0
G2	208	21	93	31
G3	440	594	739	431
G4	3648	3603	1583	1503
G5	–	–	5859	5862

and 50 respectively. However, at the scale of G_4 the benefits are clear, with Advantage typically suffering from only half as many chain breaks as the 2000Q.

7 Energy Gaps

As noted in Sect. 1 QA generalises AQC and inhabits a regime that is intermediate between the idealised assumptions of universal AQC and unavoidable experimental compromises. As such we are unable to rely on the adiabatic theorem to ensure that an evolving quantum system will remain in its ground state. For non-adiabatic systems, as the energy gap between two states of a quantum system decreases, the likelihood that the system will 'jump', or more precisely undergo a diabatic transition, from one to state to another [28] increases. This possibility of diabatic transition is directly related to the likelihood that the system obtains the global solution.

Being able to determine if a problem suffers, or will suffer at larger scales, from a reducing gap between the first and second eigenvalues would be informative as to its suitability to QA. However, determining the minimum gap is not easy, in fact being able to calculate this gap would generally imply that the problem could be solved directly.

Given the limitations of calculating eigenstates, and corresponding eigenvalues, for the G_2 and beyond, we produce a number of interim graphs between the G_1 and G_2. These graphs attempt to replicate the inherent structure of this family of graphs, while providing a number of interim data points to assess the relationship between the minimum energy gap and number of variables.

7.1 Landau-Zener Formula

To investigate the scaling of the minimum energy gap we utilise the Landau-Zener (LZ) formula, which gives the probability of a diabatic transition between two energy states [28]. The LZ formula is intended for two-state quantum systems and employs several approximations. Even with these limitations the formula provides a useful tool to gain insight into how diabatic transitions may behave for varying graph sizes. The LZ formula along with supporting calculations are shown in Eqs. 6, 7 and 8 [25].

$$P_D = e^{-2\pi\Gamma} \tag{6}$$

$$\Gamma = \frac{a^2/\hbar}{\left|\frac{\partial}{\partial t}(E_2 - E_1)\right|} = \frac{a^2/\hbar}{\left|\frac{dq}{dt}\frac{\partial}{\partial q}(E_2 - E_1)\right|} = \frac{a^2}{\hbar|\alpha|} \tag{7}$$

$$\Delta E = E_2(t) - E_1(t) \equiv \alpha t \tag{8}$$

The quantity a is half the minimum energy gap, in our case the minimum energy gap between the first and second eigenstate. For the calculation of α in Eq. 8 we assume a linear change in the gap between eigenvalues as defined in Eq. 9.

$$\alpha = \frac{\text{Initial Gap - Minimum Gap}}{\text{Time to Minimum Gap}} \tag{9}$$

A decreasing minimum energy gap a drives the exponent in Eq. 6 to zero and hence the probability of a diabatic transition, P_D, to 1.

7.2 Modified Graph Family

The calculation of eigenvalues quickly becomes infeasible as the size of the system increases. Given that the graph family we have been utilising, G_n, scales in accordance with $(n+1)^3$ we subsequently require a matrix of size $2^{(n+1)^3} \times 2^{(n+1)^3}$ to represent the quantum system. As a result, even calculating eigenstates for the G_2 (27 variables) was beyond the capability of our consumer hardware. Given this limitation we constructed a number of 'interim' graphs, between G_1 and G_2, that increase in size while preserving key structural elements such as sparse outer layers and a denser connecting layer.

The graphs we utilise to investigate energy gaps are shown in Fig. 2 and range in size from 8 to 22 variables (edges). For each of these graphs the optimal solution to the maximum matching problem is still to select all the edges in the sparse outer layers.

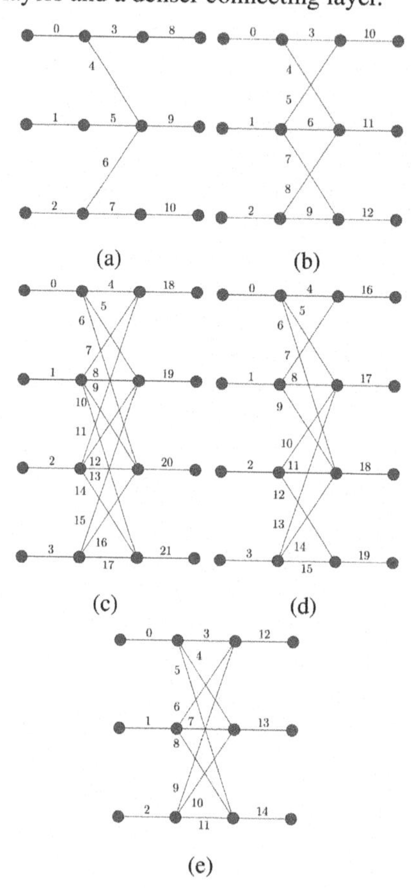

Fig. 2. (a), (b), (c), (d) and (e) show the interim graphs utilised to investigate the minimum energy gap

7.3 Eigenvalue Calculation

To determine the minimum energy gap between the ground state and first excited state we calculate the eigenvalues using the following process.

Construct Initial Hamiltonian: As per Eq. 1 we form the initial Hamiltonian which is the sum of Pauli x matrices, $\hat{\sigma}_x$, see Eqs. 10 and 11.

$$H_i = \sum_i \hat{\sigma}_x^{(i)} \tag{10}$$

$$\hat{\sigma}_x^{(i)} \equiv \mathbb{I}_1 \otimes \cdots \otimes \mathbb{I}_{i-1} \otimes \hat{\sigma}_x^{(i)} \otimes \mathbb{I}_{i+1} \otimes \cdots \otimes \mathbb{I}_n \tag{11}$$

Obtain Biases and Couplers and Scale: To obtain the relevant biases and couplers for the particular graph we first form the corresponding Q matrix using the form shown in Eq. 5, noting we again use a value of $P = 2$ for off-diagonal values. Using D-Wave's Ocean libraries the matrix is then converted into a binary quadratic model and subsequently an Ising model. The linear biases, h_i, and quadratic couplers, J_i, are then extracted from this Ising model.

Construct Final Hamiltonian: As per Eq. 1 we form the final Hamiltonian which is the sum of Pauli z matrices, $\hat{\sigma}_z$, over both the biases, h_i, and couplers, $J_{i,j}$, see Eqs. 12 and 13.

$$H_f = \sum_i h_i \hat{\sigma}_z^{(i)} + \sum_{i>j} J_{i,j} \hat{\sigma}_z^{(i)} \hat{\sigma}_z^{(j)} \tag{12}$$

$$\begin{aligned} \hat{\sigma}_z^{(i)} \hat{\sigma}_z^{(j)} \equiv \mathbb{I}_1 \otimes \cdots \otimes \mathbb{I}_{i-1} \otimes \hat{\sigma}_z^{(i)} \otimes \mathbb{I}_{i+1} \otimes \\ \cdots \otimes \mathbb{I}_{j-1} \otimes \hat{\sigma}_z^{(j)} \otimes \mathbb{I}_{j+1} \otimes \cdots \otimes \mathbb{I}_n \end{aligned} \tag{13}$$

Define the System Evolution: The evolution of the system occurs as per Eq. 14, where t_f is the annealing duration, nominally being $20\mu s$ and $t \in [0, t_f]$. The value of s is then used to determine the corresponding $A(s)$ and $B(s)$ (units of GHz) as per the annealing schedule [7].

$$H(s) = A(s)H_i + B(s)H_f \tag{14}$$

Calculate Eigenvalues: The first (ground state) and second (first excited state) eigenvalues of H, from Eq. 14, are calculated as a function of s. With these eigenvalues the initial gap, when $s = 0$, can be calculated and also the value of s that corresponds to the smallest gap. With these values the value of α can be calculated using Eq. 9 and the fact that the time to the minimum energy gap can be calculated by $t = st_f$.

7.4 Energy Gap Results and Expected Diabatic Transition Probability

In Fig. 3 we plot the minimum energy gap between the first and second eigenvalue with a fitted exponential trendline. While the range of graph sizes utilised is narrow, the data suggest that it is plausible that the minimum energy gap decreases exponentially with graph size. If these results are indicative of the scaling of the minimum energy gap for larger G_n graphs we would expect that the probability of diabatic transitions during annealing would increase dramatically.

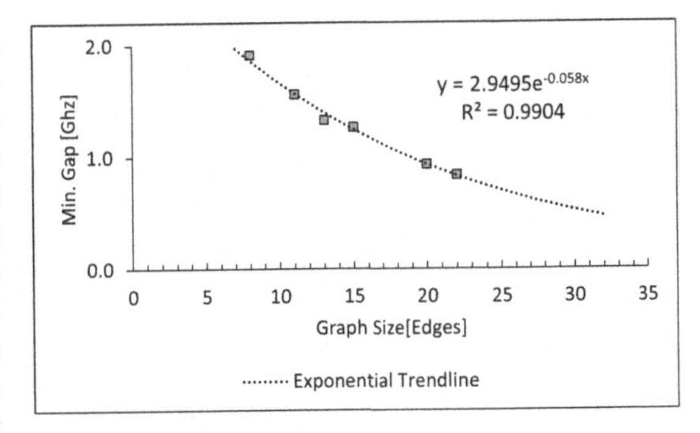

Fig. 3. Minimum energy gaps for varying graph sizes (with exponential trendline)

The other factor that impacts the diabatic transition probability is the time until the minimum energy gap, with this value showing up in the denominator of Eq. 9. This time varies slowly as a function of graph size on our interim graphs, but it is insufficient to offset an exponentially decreasing minimum energy gap. We also assume a constant initial energy gap of 19.6 GHz, which is based on the average initial energy gap for the interim graphs, with the initial gap ranging from 19.5 GHz to 19.7 GHz.

We now extrapolate the quantities obtained from the eigenvalues of the Hamiltonian for larger graphs. Using the LZ formula on the extrapolated quantities we calculate the expected diabatic transition probability for an annealing cycle. This procedure spares us from the need of simulating the quantum annealing process by integrating the Schrödinger equation.

The increase in diabatic transition probability corresponds to a decreasing likelihood that an annealing cycle won't be affected by a diabatic transition. Figure 4 plots the expected number of annealing cycles, out of 10,000 cycles (the default value), where a diabatic transition does not occur, based on our energy gap modelling (grey squares), versus the number of optimal solutions we actually obtained from our empirical tests on Advantage1.1 (black squares). We specifically compare to the results obtained from Advantage using our general formulation and an annealing duration of $20\mu s$.

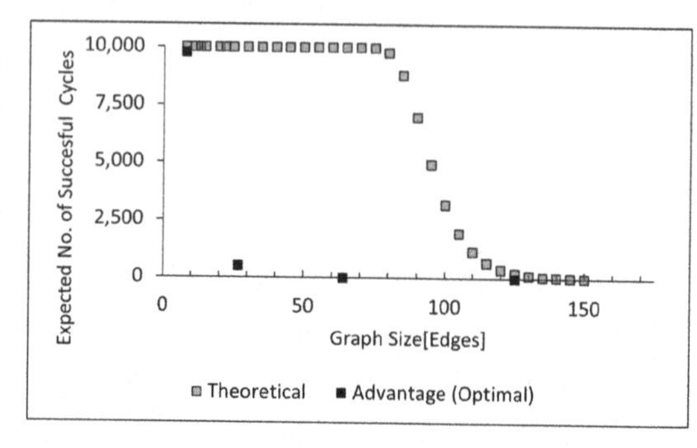

Fig. 4. Theoretical expected no. of successful annealing cycles (out of 10,000) versus number of optimal solutions returned from Advantage1.1

The results of the Advantage architecture are compatible with the theoretical result, once noise is taken into account. Between G_1 (8 edges) and G_2 (27 edges) there is a significant drop in the number of optimal solutions obtained and by G_3 only a single optimal solution is returned out of 10,000 annealing cycles and none for G_4. Our theoretical model indicates that obtaining optimal solutions, even for future hardware, is highly unlikely for G_5 (216 edges) and beyond.

7.5 Energy Gap Analysis Limitations

We acknowledge that in our probing of the behaviour of the minimum energy gap we have made a number of assumptions, namely that our fabricated graphs are representative of the original graph family, that the minimum energy gap and time to minimum energy gap follow an exponential and logarithmic relationship respectively and that the results we obtain can be extrapolated. Further to this, the LZ formula is based on a two-state quantum system with additional assumptions that ignore external impacts. Our systems are clearly multi-state, with degenerate energy levels occurring from the first excited state, and as such diabatic transitions may occur between any states of the system. As such, our analysis is not intended to be a precise calculation of theoretical behaviour, but it is meant to capture the possible scaling of the problem, to provide initial insights into how the diabatic transition probability scales for large graphs and to give an indication of the quantum complexity of our problem, and thus the suitability of such a problem as a benchmark for QA hardware.

There are likely a number of additional factors, as discussed in [11], that further contribute to imperfect annealing, such as thermal excitation, unintended qubit couplings, errors in couplings and biases within the implemented Hamiltonian and decoherence in energy eigenstates. It is not unexpected that our empirical results fall short of our theoretical model. However, we do believe that even given the limitations and assumptions of our approach that our analysis lends credence to the possibility that larger instances of our target problem may become intractable for future generations of QA hardware. At the very least, the minimum energy gap of benchmarking problems should be considered when assessing performance of quantum annealers to ensure performance improvements can be adequately discerned.

8 Discussion of Future Work

The empirical tests we have conducted on the latest QA hardware from D-Wave are an extension of prior work from [23]. It is likely that performance could be improved by further considering other annealing variables. For instance, custom annealing schedules, chain strengths, chain break resolution methods and annealing offsets [27] (adjustment of annealing path per qubit) could be optimised for improved results. We leave these optimisations for future work where the impacts of such optimisations can be contrasted to our implementation to assess the benefits of parameter tuning.

Secondly, our investigation into energy gap scaling utilised fabricated graphs up to a size of 22 variables (edges). It would be insightful to extend the analysis up to the G_2 graph, as it would provide a direct point of comparison to the original graph family and assist with validating the scaling behaviour of the maximum matching problem on the G_n graph family.

9 Conclusion

We have extended the prior work of [23] and the optimisation problem of maximum cardinality matching to the latest QA hardware from D-Wave, the 2000Q and Advantage quantum annealers. In doing so we propose an improved version of the QUBO formulation of this problem, specifically we modify the ratio of diagonal to off-diagonal terms of the Q matrix so that the formulation is applicable to arbitrary graph structures. We find that this formulation produces significantly more valid solutions. With the increased number of qubits and improved topology we also obtain empirical results for larger problem instances, although solution quality at these scales remains notably poor. We do note that the minor embeddings produced for these problems are far more compact on the newer Pegasus topology.

Our testing shows improvements in the capability of QA hardware to solve combinatorially difficult optimisation problems. Our results demonstrate a general improvement in the lower bound of solution quality and importantly we observe that the 2000Q and Advantage are capable of producing the optimal solution for G_3, whereas [23] were unable to do so using the 2X quantum annealer.

Following from our empirical tests we probe the behaviour of the minimum energy gap of our target problem by formulating a number of 'interim' graphs, that maintain key structural elements. We numerically calculate the first and second eigenvalues over an evolutionary path that replicates the behaviour of the Advantage1.1 quantum annealer. These eigenvalues are then used to calculate the minimum energy gaps that occur during annealing. To further investigate the minimum energy gap behaviour for larger graphs we assume that the minimum energy gap decreases exponentially, which our data support. We combine this with the LZ formula to calculate the probability of a diabatic transition during an annealing cycle. We observe that even with this theoretical approach, the increasing likelihood of diabatic transitions as graph sizes increase suggests that optimal solutions for this problem on graph sizes of G_5 and above are unlikely to be achieved in the near term.

Ultimately our empirical results show improvement in the capability of QA hardware with the 2000Q and Advantage being able to solve the maximum cardinality matching problem for a G_3, which escaped previous generations. However, our further investigations into energy gaps and diabatic transitions raises questions about the suitability of maximum cardinality matching problem on this graph family as a benchmarking tool for QA hardware, as it potentially lacks the nuance required to identify meaningful performance improvements between generations of QA hardware. We hope that the methods developed in our work can be used to guide the design of better benchmarking problems for QA hardware.

Acknowledgments. We would like to thank Professor Tat-Jun Chin for his support, expertise and feedback that was invaluable in producing this work.

References

1. Albash, T., Lidar, D.A.: Adiabatic quantum computation. Rev. Mod. Phys. **90**(1) (2018). https://doi.org/10.1103/revmodphys.90.015002
2. Benioff, P.: The computer as a physical system: a microscopic quantum mechanical Hamiltonian model of computers as represented by turing machines. J. Statist. Phys. **22**(5), 563–591 (1980)
3. Benioff, P.: Quantum mechanical Hamiltonian models of turing machines. J. Statist. Phys. **29**(3), 515–546 (1982)
4. Benioff, P.: Quantum mechanical models of turing machines that dissipate no energy. Phys. Rev. Lett. **48**(23), 1581–1585 (1982)
5. Cai, J., Macready, W.G., Roy, A.: A practical heuristic for finding graph minors (2014)
6. D-Wave Systems Inc. Programming the D-Wave QPU: Parameters for Beginners. 14–1045A-A
7. D-Wave Systems Inc. QPU-Specific Anneal Schedules - D-Wave Systems. https://docs.dwavesys.com/docs/latest/doc_physical_properties.html. Accessed 5 Feb 2022
8. Dyakonov, M.: When will useful quantum computers be constructed? Not in the foreseeable future, this physicist argues. Here's why: the case against: quantum computing. IEEE Spectrum **56**(3), 24–29 (2019)
9. Farhi, E., Goldstone, J., Gutmann, S., Sipser, M.: Quantum computation by adiabatic evolution (2000)
10. Feynman, R.P.: Simulating physics with computers. Int. J. Theor. Phys. **21**(6–7), 467–488 (1982)
11. Gardas, B., Dziarmaga, J., Zurek, W.H., Zwolak, M.: Defects in quantum computers. Sci. Rep. **8**(1), 4539 (2018)
12. Glover, F., Kochenberger, G., Du, Y.: Quantum bridge analytics i: a tutorial on formulating and using qubo models. 4OR **17**(4), 335–371 (2019)

13. King, A.D., et al.: Scaling advantage over path-integral Monte Carlo in quantum simulation of geometrically frustrated magnets. Nat. Commun. **12**(1), 1113 (2021)
14. Kitaev, A.: Fault-tolerant quantum computation by Anyons. Ann. Phys. **303**(1), 2–30 (2003)
15. Korte, B.: Combinatorial Optimization Theory and Algorithms. Algorithms and Combinatorics, 5th edn, vol. 21, Springer, Heidelberg (2012). https://doi.org/10.1007/3-540-29297-7
16. Lucas, A.: Ising formulations of many np problems. Front. Phys. **2** (2014)
17. Morita, S., Nishimori, H.: Mathematical foundation of quantum annealing. J. Math. Phys. **49**(12), 125210 (2008)
18. Nielsen, M.A., Chuang, I.L.: Programmable quantum gate arrays. Phys. Rev. Lett. **79**(2), 321–324 (1997)
19. Nimbe, P., Weyori, B.A., Adekoya, A.F.: Models in quantum computing: a systematic review. Quant. Inf. Process. **20**(2), 1–61 (2021). https://doi.org/10.1007/s11128-021-03021-3
20. Raussendorf, R., Briegel, H.J.: A one-way quantum computer. Phys. Rev. Lett. **86**(22), 5188–5191 (2001)
21. Santoro, G.E., Martonak, R., Tosatti, E., Car, R.: Theory of quantum annealing of an ising spin glass. Science (Am. Assoc. Advan. Sci.) **295**(5564), 2427–2430 (2002)
22. Sasaki, G., Hajek, B.: The time complexity of maximum matching by simulated annealing. J. ACM **35**(2), 387–403 (1988)
23. Vert, D., Sirdey, R., Louise, S.: Revisiting old combinatorial Beasts in the quantum age: quantum annealing versus maximal matching. In: Krzhizhanovskaya, V.V., et al.: (eds.) ICCS 2020. LNCS, vol. 12142, pp. 473–487. Springer, Cham (2020). https://doi.org/10.1007/978-3-030-50433-5_37
24. Vinci, W., Lidar, D.A.: Non-stoquastic hamiltonians in quantum annealing via geometric phases. NPJ Quant. Inf. **3**(1), 1–6 (2017)
25. Wittig, C.: The Landau-Zener formula. J. Phys. Chem. B **109**(17), 8428–8430 (2005)
26. Yarkoni, S., et al.: Quantum shuttle: traffic navigation with quantum computing (2020)
27. Yarkoni, S., Wang, H., Plaat, A., Bäck, T.: Boosting quantum annealing performance using evolution strategies for annealing offsets tuning. In: Feld, S., Linnhoff-Popien, C. (eds.) QTOP 2019. LNCS, vol. 11413, pp. 157–168. Springer, Cham (2019). https://doi.org/10.1007/978-3-030-14082-3_14
28. Zener, C.: Non-adiabatic crossing of energy levels. Proc Roy. Soc. London. Seri. A Contain. Pap. Math. Phys. Char. **137**(833), 696–702 (1932)

Challenges and Future Directions in the Implementation of Quantum Authentication Protocols

Juliet McLeod[1], Ritajit Majumdar[2], and Sanchari Das[1](✉)

[1] University of Denver, Denver, CO 80208, USA
Sanchari.Das@du.edu
[2] Indian Statistical Institute, Kolkata, West Bengal 700108, India

Abstract. Quantum computing is a powerful concept in the technological world that is critically valued in information security due to its enhanced computation powers. Researchers have developed algorithms that allow quantum computers to hack into information security concepts that were previously considered difficult, if not impossible, including asymmetric key cryptography and elliptic curve cryptography. Studies have been done to focus on improving security protocols through quantum computing to counter these vulnerabilities. One such focus is on the topic of quantum authentication (QA). However, while several QA protocols have been theorized, only a few have been implemented and further tested. Among the protocols, we selected and implemented five quantum authentication protocols to determine their feasibility in a real-world setting. In this late-breaking work, we discuss the difficulties and obstacles developers might face while implementing authentication protocols that use quantum computing.

Keywords: Quantum Computing · Authentication · User studies · Quantum authentication

1 Introduction

Authentication plays a critical role to protect user data and online user presence. Several researchers are focusing on improving these authentication technologies while adding advance computing strategies, one of which is *Quantum Computing* [20]. As we move into the realm of quantum computing, we must consider authentication in a quantum sphere as well. Researchers have postulated protocols to implement quantum authentication, with varying degrees of difficulty both in execution and in implementation [2]. However, there are no analyses of the difficulties in the actual implementation of these protocols from the developer perspective due to limited hardware capabilities. Additionally, there has been no reporting of any user studies to test the adaptability of these protocols from the user perspective in a real-world environment [16].

D. Groen et al. (Eds.): ICCS 2022, LNCS 13353, pp. 164–170, 2022.
https://doi.org/10.1007/978-3-031-08760-8_14

In this late-breaking work, we report on our implementation of five quantum authentication protocols. This work contributes by detailing the technical difficulties in implementing and applying quantum authentication protocols with currently available infrastructure. We identified four primary obstacles in programming quantum authentication protocols. First, quantum computing is currently implemented for interaction only on a local computer. Second, there are no publicly available quantum communication channels. Third, quantum key distribution is difficult to realize or simulate. Fourth, classical computers are unable to read or store a quantum state. We plan to extend this study by presenting these protocols to users and determining how well users can understand and use these quantum authentication protocols.

2 Related Work

Due to the lack of noise-free, general-purpose, large-scale quantum computers, most QA protocols are proposed theoretically. However, a general lack in user-studies has been evident from the literature review done by Majumdar and Das [16]. Most of the general users of such quantum technology are not aware of quantum mechanics, and quantum computing and the technological feasibility is often questionable. For example, McCaskey et al. discusses how users are often not familiar with the technological implementation, codes, or device implemented in quantum computation [17]. One can argue that a fully working quantum computer is most likely decades away. However, we want to point out that working Quantum Random Number Generators (QNRG) [10] are already available, and many laboratories are trying to implement the Quantum Cryptography protocols [3].

3 Method

We began by identifying QA protocols for the implementation purposes. Here, we kept protocols that included QKD along with QA or protocols that verified user identity through a trusted third-party. While conducting this search, we identified 17 protocols that fit our criteria. These protocols included Barnum et al. [2], Curty & Santos [4], Dan et al. [5], Das et al. [6], Hong et al. [11], Hwang et al. [12], Kiktenko et al. [13], Lee et al. [14], Ljunggren et al. [15], Shi et al. [20], Wang et al. [21], Zawadzki [22], Zhang et al. [19], Zhang et al. [23], Zhao et al. [24], Zhu et al. [25], and Zuning & Sheng [26].

After the initial search, we considered the pre-shared key requirements of these protocols, which varied among the set of quantum entangled-state pairs, classical keys, classical sets of bits, and knowledge and registration with a trusted third-party. Thereafter, we focused on the number of transmissions between the two or three parties, since larger numbers of transmissions are more resource-intensive and more prone to errors. We also analyzed the type of quantum channel involved, which varied between maximally or non-maximally entangled-state pairs, squeezed state pairs, and GHZ states. Given the nature of communication through these channels, we evaluated whether the protocol involved more than two participants. Finally, we considered whether the protocol required any additional technical implementation, such as quantum encryption or a pseudorandom number generator.

Based on our filtering mechanism, we implemented five protocols. First, Shi et al. requires a pre-shared entangled-state key pair. It communicates solely over a quantum channel but requires multiple transmissions [20]. Second, Zawadzki uses a pre-shared entangled-state pair and a pre-shared classical key. It completes few transmissions over an unprotected classical channel [22]. Third, Dan et al. utilizes a pre-shared entangled-state pair and pre-shared classical user IDs. It uses a quantum channel and has many transmissions [5]. Fourth, Hong et al. requires a pre-shared classical key, but only needs single-photon messages for authentication. It uses a quantum channel and has multiple transmissions [11]. Finally, Das et al. uses pre-shared classical user IDs. It uses both a protected quantum and an unprotected classical channel. The quantum channel has few transmissions, but the classical channel has multiple [6]. We implemented the five above-mentioned quantum authentication protocols in Python. We used a popular Python quantum computing package called Qiskit to simulate quantum computing on a classical computer [1]. The characteristics of each of these protocols are summarized in Table 1.

Table 1. Keys and channels used by each quantum authentication protocol.

	Shi et al.	Zawadzki	Dan et al.	Hong et al.	Das et al.
Entangled-state pair	X	X	X		
Classical key		X	X	X	X
Quantum channel	X		X	X	X
Classical channel		X			X

4 Results: Challenges in Implementation

4.1 Challenges with the Implementation of Quantum Authentication Protocols

First, Qiskit is good at simulating local quantum devices. This is sufficient for quantum computations, but causes difficulty when attempting to execute quantum teleportation. In quantum teleportation, two subjects, Alice and Bob, are assumed to share one qubit each of the entangled pair while being separated spatially. It is not possible to simulate this spatial separation in Qiskit. The entire quantum circuit needs to be developed on a local quantum device, which somewhat defeats the purpose of teleportation.

Second, there are currently no publicly-available quantum channels. It is hard to conceptualize sending a qubit to another entity; it is even harder to implement.

Qiskit addresses this problem by avoiding it, as there is no way to save a qubit directly since quantum memory (QRAM) is not available yet. A workaround that Qiskit provides is storing the statevector snapshot. Nevertheless, in reality it is not possible to obtain the statevector from a quantum circuit. This becomes a problem when combined with Qiskit's suggested means of teleportation, since both the entangled pair and the teleported qubit reside on the same quantum device. This limitation of Qiskit largely diminishes the protocols' integrity and security since now the relevant parties and the eavesdropper are on the same system. Therefore, while this allows simulation of the protocol, it is far from the ideal scenario.

Third, pre-shared key distribution is a problem. All of the QA protocols we identified require the two participants to possess a shared quantum pair, usually entangled. However, there are limitations in place that hinder this assumption from happening. There are a large number of theoretical quantum key distribution protocols. Many of these protocols currently achieve perfect security, but require a quantum communication channel.

Finally, a classical computer measures a qubit, and obtains a classical value, i.e., the state in which the qubit collapsed. Many QA protocols require distinguishing among the Bell states. Bell states are a set of four orthogonal states that correspond to different rotations of the qubit and represent a simple entangled state. This entanglement is sufficient to guarantee security in a variety of ways, but requires the ability to measure in Bell bases. Qiskit allows measurement in computational bases only, and therefore we needed to add relevant rotations to each qubit to compensate for this limitation.

4.2 Impact of Challenges with the Implementation

More than 80% of the protocols we initially surveyed required a pre-shared quantum key, including all five of the protocols we implemented. This is a prerequisite for each protocol. In order to start the protocol, the participants must already have a key shared. Unfortunately, based on current technology, there is no way to generate an entangled quantum key and distribute it to two or more participants. This is a result of a lack of quantum communication channels. It is possible to generate an entangled quantum key on a single computer, but it is currently impossible to transfer one of the qubits in that key to another computer. This inability causes any protocol that requires a prerequisite shared quantum key to function inadequately. The protocols we implemented that require a shared quantum key include Shi et al., Zawadzki, and Dan et al. [5,20,22].

Second, the lack of quantum storage creates interesting issues with quantum protocols. The inability to store quantum states requires the programmer to measure them before storing their values. This removes the uncertainty in their values. Measuring the values early creates a unique problem. When a participant measures a qubit, they must choose a basis to measure it in. If they choose the wrong basis, their results could be inaccurate. This concept is important to the security of quantum message channels, since an eavesdropper will not know the correct basis and will obtain inaccurate results with high probability if they try to measure intercepted qubits.

Third, the lack of quantum storage prohibits measuring quantum changes. One common practice in quantum authentication is measuring Bell states. As stated before, measurement in Bell basis is not supported in Qiskit or with current quantum devices. However, knowing which Bell state a qubit is in is important for security. The workaround to this is to incorporate additional rotations prior to measurement. This issue with Bell state measurement occurs in Shi et al., Dan et al., and Hong et al. [5,11,20].

5 Future Study Design: Vision

In our study design, the participants will first take a pre-screening survey to determine their technical ability regarding computers, authentication, and quantum topics. These technical questions are taken from the SEBIS questionnaire by Egelman et al. [8,9] and the expert evaluation survey by Rajivan et al. [18]. We will select eligible participants to include a diverse set of technical experience, as technical expertise could be a critical factor in evaluating the effectiveness of the feasibility of QA for an in-lab study. Participants will use think-aloud while they execute the selected simulation of the protocols motivated by the study design of Das et al. [7]. We also plan to implement the QA for regular accounts which our users can implement in their daily life after the first phase of this experiment, then conduct a timeline analysis to see how the participants' continued usage is impacted. There are technical infeasibility issues for this extension which this study emphasizes but we plan to overcome those through this research and by starting with simulated accounts for the initial phase of the study. Along these lines, recent Quantum Networks[1] can be utilized instead of Qiskit to emulate real-world quantum communication.

6 Conclusion

Quantum computing and quantum authentication are becoming critical in the information security domain due to their computational power and secure identification capabilities. However, less is known about the implementation of the QA protocols, particularly from the feasibility perspective. In this paper, to explore and learn further on this, we report on our investigation of the challenges of implementing user authentication protocols that utilize quantum computing. First, we implemented five QA protocols in Python using the Qiskit library. While programming these protocols, we identified significant difficulties in implementing these or similar QA protocols. These difficulties include a lack of quantum teleportation channels, issues with pre-shared key distribution, and a lack of quantum storage. Additionally, we discuss why these difficulties exist and why they are problematic for accurately testing quantum authentication

[1] https://www.quantum-network.com/.

protocols. After the implementation, in this late-breaking work, we also report on our future direction plan to continue this research by conducting user studies through a think-aloud protocol to test the efficacy of the five protocols.

Acknowledgement. We would like to acknowledge the Inclusive Security and Privacy-focused Innovative Research in Information Technology: InSPIRIT Research Lab at the University of Denver. We would also like to thank Nayana Das for their feedback on the Quantum Authentication protocols. Any opinions, findings, and conclusions or recommendations expressed in this material are solely those of the authors and do not necessarily reflect the views of the University of Denver or the Indian Statistical Institute.

References

1. Aleksandrowicz, G., et al.: Qiskit: an open-source framework for quantum computing (2021). https://doi.org/10.5281/zenodo.2573505
2. Barnum, H., Crepeau, C., Gottesman, D., Smith, A., Tapp, A.: Authentication of quantum messages. In: 43rd Annual IEEE Symposium on the Foundations of Computer Science, pp. 449–458 (2002). https://doi.org/10.1109/SFCS.2002.1181969
3. Bourennane, M., et al.: Experiments on long wavelength (1550 nm) "plug and play" quantum cryptography systems. Opt. Express **4**(10), 383–387 (1999)
4. Curty, M., Santos, D.J.: Quantum authentication of classical messages. Phys. Rev. A **64**, 062309 (2001). https://doi.org/10.1103/PhysRevA.64.062309
5. Dan, L., Chang-Xing, P., Dong-Xiao, Q., Nan, Z.: A new quantum secure direct communication scheme with authentication. Chin. Phys. Lett. **27**(5), 050306 (2010). https://doi.org/10.1088/0256-307X/27/5/050306
6. Das, N., Paul, G., Majumdar, R.: Quantum secure direct communication with mutual authentication using a single basis. Int. J. Theor. Phys. **60**, 4044–4065 (2021) (arXiv preprint arXiv:2101.03577) (2021). https://doi.org/10.1007/s10773-021-04952-4
7. Das, S., Dingman, A., Camp, L.J.: Why Johnny doesn't use two factor a two-phase usability study of the FIDO U2F security key. In: Meiklejohn, S., Sako, K. (eds.) FC 2018. LNCS, vol. 10957, pp. 160–179. Springer, Heidelberg (2018). https://doi.org/10.1007/978-3-662-58387-6_9
8. Egelman, S., Harbach, M., Peer, E.: Behavior ever follows intention? A validation of the security behavior intentions scale (Sebis). In: Proceedings of the 2016 CHI Conference on Human Factors in Computing Systems, pp. 5257–5261 (2016)
9. Egelman, S., Peer, E.: Scaling the security wall: developing a security behavior intentions scale (Sebis). In: Proceedings of the 33rd Annual ACM Conference on Human Factors in Computing Systems, pp. 2873–2882 (2015)
10. Herrero-Collantes, M., Garcia-Escartin, J.C.: Quantum random number generators. Rev. Mod. Phys. **89**(1), 015004 (2017)
11. Hong, C., Heo, J., Jang, J.G., Kwon, D.: Quantum identity authentication with single photon. Quantum Inf. Process. **16**(10), 1–20 (2017). https://doi.org/10.1007/s11128-017-1681-0
12. Hwang, T., Luo, Y.-P., Yang, C.-W., Lin, T.-H.: Quantum authencryption: one-step authenticated quantum secure direct communications for off-line communicants. Quantum Inf. Process. **13**(4), 925–933 (2013). https://doi.org/10.1007/s11128-013-0702-x

13. Kintenko, E., et al.: Lightweight authentication for quantum key distribution. IEEE Trans. Inf. Theory **66**(10), 6354–6368 (2020). https://doi.org/10.1109/TIT.2020.2989459
14. Lee, H., Lim, J., Yang, H.: Quantum direct communication with authentication. Phys. Rev. A **73**(4), 042305 (2006). https://doi.org/10.1103/PhysRevA.73.042305
15. Ljunggren, D., Bourennane, M., Karlsson, A.: Authority-based user authentication in quantum key distribution. Phys. Rev. A **62**(2), 022305 (2000). https://doi.org/10.1103/PhysRevA.62.022305
16. Majumdar, R., Das, S.: SOK: an evaluation of quantum authentication through systematic literature review. In: Proceedings of the Workshop on Usable Security and Privacy (USEC) (2021)
17. McCaskey, A., Dumitrescu, E., Liakh, D., Humble, T.: Hybrid programming for near-term quantum computing systems. In: 2018 IEEE International Conference on Rebooting Computing (ICRC), pp. 1–12. IEEE (2018)
18. Rajivan, P., Moriano, P., Kelley, T., Camp, L.J.: Factors in an end user security expertise instrument. Inf. Comput. Secur. **25** (2017)
19. Sheng, Z., Jian, W., Chao-Jing, T., Quan, Z.: A composed protocol of quantum identity authentication plus quantum key distribution based on squeezed states. Commun. Theor. Phys. **56**(2), 268–272 (2011). https://doi.org/10.1088/0253-6102/56/2/13
20. Shi, B.S., Li, J., Liu, J.M., Fan, X.F., Guo, G.C.: Quantum key distribution and quantum authentication based on entangled state. Phys. Lett. A **281**(2–3), 83–87 (2001). https://doi.org/10.1016/S0375-9601(01)00129-3
21. Wang, J., Zhang, Q., Jing Tang, C.: Multiparty simultaneous quantum identity authentication based on entanglement swapping. Chin. Phys. Lett. **23**(9), 2360–2363 (2006). https://doi.org/10.1088/0256-307X/23/9/004
22. Zawadzki, P.: Quantum identity authentication without entanglement. Quantum Inf. Process. **18**(1), 1–12 (2018). https://doi.org/10.1007/s11128-018-2124-2
23. Zhang, S., Chen, Z.-K., Shi, R.-H., Liang, F.-Y.: A novel quantum identity authentication based on Bell states. Int. J. Theor. Phys. **59**(1), 236–249 (2019). https://doi.org/10.1007/s10773-019-04319-w
24. Zhao, B., et al.: A novel NTT-based authentication scheme for 10-Ghz quantum key distribution systems. IEEE Trans. Industr. Electron. **63**(8), 5101–5108 (2016). https://doi.org/10.1109/TIE.2016.2552152
25. Zhu, H., Wang, L., Zhang, Y.: An efficient quantum identity authentication key agreement protocol without entanglement. Quantum Inf. Process. **19**(10), 1–14 (2020). https://doi.org/10.1007/s11128-020-02887-z
26. Zuning, C., Zheng, Q.: A “ping-pong” protocol with authentication. In: 5th IEEE Conference on Industrial Electronics and Applications, pp. 1805–1810 (2010). https://doi.org/10.1109/ICIEA.2010.5515357

Distributed Quantum Annealing on D-Wave for the Single Machine Total Weighted Tardiness Scheduling Problem

Wojciech Bożejko[1(✉)], Jarosław Pempera[1], Mariusz Uchroński[1,2], and Mieczysław Wodecki[3]

[1] Department of Control Systems and Mechatronics, Wrocław University of Science and Technology, Janiszewskiego 11-17, 50-372 Wrocław, Poland
{wojciech.bozejko,jaroslaw.pempera,mariusz.uchronski}@pwr.edu.pl

[2] Wroclaw Centre for Networking and Supercomputing, Wybrzeże Wyspiańskiego 27, 50-370 Wrocław, Poland

[3] Department of Telecommunications and Teleinformatics, Wrocław University of Science and Technology, Wybrzeże Wyspiańskiego 27, 50-370 Wrocław, Poland
mieczyslaw.wodecki@pwr.edu.pl

Abstract. In the work, we are proposing a new distributed quantum annealing method of algorithm construction for solving an NP-hard scheduling problem. A method of diversification of calculations has been proposed by dividing the space of feasible solutions and using the fact that the quantum annealer of the D-Wave machine is able to optimally solve (for now) small-size subproblems only. The proposed methodology was tested on a difficult instance of a single machine total weighted tardiness scheduling problem proposed by Lawler.

Keywords: Discrete optimization · Scheduling · Quantum algorithm · Quantum annealing

1 Introduction

The idea of quantum computing and computers emerged in the 1980s as a result of work of Richard Feynman, Paul Benioff and Yuri Manin. An approach, known as the gate model, expresses the interactions between qubits as quantum gates. Quantum gates, because of quantum physics, operate differently than classical electrical gates such as AND or OR. In a quantum computer with a gate model, there is no AND gate. Instead, there are Hadamard gates and Toffoli gates. Unlike many classical logic gates, quantum logic gates are reversible.

Instead of expressing the problem in terms of quantum gates, the user presents it as an optimization problem and the quantum annealing computer tries to find the best solution. D-Wave Systems makes quantum annealing computers available to the public today, proposing an approach to quantum computing that is admittedly limited to the use of quantum annealing, but that fits

D. Groen et al. (Eds.): ICCS 2022, LNCS 13353, pp. 171–178, 2022.
https://doi.org/10.1007/978-3-031-08760-8_15

well with the needs of the operations research discipline. This paper presents a method for solving a one-machine problem with a sum-of-cost criterion based on quantum annealing.

2 The Specificity of Quantum Annealers

Quantum annealing (QA) [8] is a hardware optimization method implemented through quantum fluctuations, instead of – as in simulated annealing – temperature fluctuations. It can be treated as metaheuristics, because the solutions achieved on a quantum computer designed to perform quantum annealing do not have a guarantee of optimality. This process is carried out by a special type of quantum computer searching for the minimum energy Ising Hamiltonian configuration, the ground states of which represent the optimal solution to the problem under consideration. The Ising model, traditionally used in mechanics, is formulated for variables denoting the directions of spins: 'up' and 'down' – which corresponds to the values of $+1$ and -1. The energy function is given by the formula: $E_{Ising}(s) = \sum_{i=1}^{N} h_i s_i + \sum_{i=1}^{N} \sum_{j=i+1}^{N} J_{i,j} s_i s_j$, where N is the number of qubits, $s_i \in \{+1, -1\}$, and the vector h and the matrix J are the Ising Hamiltonian coefficients. In practice, it is easier to adapt the problem under consideration to the binary quadratic model, BQM, traditionally considered in operations research, using binary variables: $E_{BQM}(v) = \sum_{i=1}^{N} a_i v_i + \sum_{i=1}^{N} \sum_{j=i+1}^{N} b_{i,j} v_i v_j + c$, where $v_i \in \{1, +1\}$ or $\{0, 1\}$ and $a_i, b_{i,j}, c$ are some real numbers depending on the instance of the problem being solved. Of course, the BQM model is a generalization of the Ising model.

In practice, on D-Wave machines, a constrained model is used to solve optimization problems, which can be formulated as follows: minimize an objective

$$\sum_{i} a_i x_i + \sum_{i<j} b_{ij} x_i x_j + c \tag{1}$$

subject to constrains

$$\sum_{i} a_i^{(c)} x_i + \sum_{i<j} b_{ij}^{(c)} x_i x_j + c^{(c)} \leq 0 \quad c = 1, \ldots, C_{ineq}, \tag{2}$$

$$\sum_{i} a_i^{(c)} x_i + \sum_{i<j} b_{ij}^{(c)} x_i x_j + c^{(c)} = 0 \quad c = 1, \ldots, C_{eq}, \tag{3}$$

where x_i, $i = 1, \ldots, n$ can be binary or integer variables, a_i, b_{ij} and c are real values and C_{ineq}, C_{eq} are the number of inequality and equality constraints respectively.

3 Single Machine Total Weighted Tardiness Problem

In the considered problem of tasks scheduling on a single machine with minimization of total delays costs (marked in the literature by $1||\sum w_i T_i$) we have a set of tasks that must be performed on one machine. Each task has an associated execution time, desired completion date, and the weight of the late penalty function. The order in which the tasks are performed should be determined, minimizing the sum of the costs of delays. It is one of the most studied, strongly NP-hard problems with total-cost objective functions. The first work on this topic, Rinnooy Kan et al. [10] has been published in the mid-1970s. Optimal algorithms, based on the dynamic programming or branch and bound methods, were published by: Congram et al. [5], and Wodecki [11]. These are mainly metaheuristics that have been widely used since the 1990s: tabu search of Bożejko et al. [3], ant colony optimization algorithm (Den Basten et al. [6]). Extensive reviews of the literature on scheduling problems with due dates was also presented by Adamu and Adewumi [1]. The literature also deals with problems with random execution times or completion dates (Rajba and Wodecki [9], Bożejko et al. [2]).

The single-machine problem of minimizing the sum of costs delays (*Total Weighted Tardiness Problem*, abbreviated AS TWT), marked in the literature by $1||\sum w_i T_i$, can be formulated as follows: the set of tasks is given $\mathcal{J} = \{1, 2, \ldots, n\}$. For the $i \in \mathcal{J}$ task, let us define: p_i – processing time, d_i – due date, and w_i – weight of the cost function for the task's tardiness. Each task must be performed on the machine, the following restrictions must be met: (a) the machine can perform at most one task at any given time, (b) task execution cannot be interrupted, (c) the task execution may begin at time zero.

Any solution to the TWT problem can be represented by the sequence $S_1, S_2, \ldots, S_n$ of times when tasks meet the constraints:

$$S_i + p_i \leq S_j \vee S_j + p_j \leq S_i,\ i \neq j,\ i, j = 1, 2, \ldots, n, \tag{4}$$

$$S_i \geq 0,\ i = 1, 2, \ldots, n \tag{5}$$

Solution $S_1, S_2, \ldots, S_n$ can be represented by the order of execution of tasks expressed by a permutation $\pi \in \Pi$ of elements of the set $\mathcal{J}$, where Π is the set of all such permutations.

Let $C_{\pi(i)} = S_{\pi(i)} + p_{\pi(i)}$ be completion time and $d_{\pi(i)}$ the *due date* of a task $\pi(i) \in \mathcal{J}$. Then $T_{\pi(i)} = \max\{0, C_{\pi(i)} - d_{\pi(i)}\}$ is a *tardiness*, and $f_{\pi(i)}(C_{\pi(i)}) = w_{\pi(i)} T_{\pi(i)}$ *tardiness weight*. For any permutation of $\pi \in \Pi$, penalty for tasks tardiness (solution cost) is

$$\mathcal{F}(\pi) = \sum_{i=1}^{n} f_{\pi(i)}(C_{\pi(i)}) = \sum_{i=1}^{n} w_{\pi(i)} T_{\pi(i)}. \tag{6}$$

In the considered problem, the optimal order $\pi^* \in \Pi$ in which tasks should be performed should be determined minimizing the total penalty. In the further part of the work, we present a new method for solving the TWT problem and an algorithm in which a quantum computer performing quantum annealing was used for calculations.

4 Solution Method

The idea of solving the problem is based on the divide and conquer method. It can be summarized as follows.

From n element set of tasks $\mathcal{J}$ find all k $(0 < k < n)$ element subsets. Next:

1. Choose one of the subsets. We denote it as $\mathcal{K}$. Let $C_{\mathcal{K}} = \sum_{z \in \mathcal{K}} p_z$ be the due date for $\mathcal{K}$.
2. Using the QA quantum annealing algorithm, determine the order of execution, starting from $C_{\mathcal{K}}$ for the remaining tasks, i.e. tasks from the set $\overline{\mathcal{K}} = \mathcal{J} \setminus \mathcal{K}$. Let $F(\overline{\mathcal{K}})$ be the cost of executing them.
3. Find a set of all element permutations from $\mathcal{K}$.
4. For the δ permutation, calculate the F_δ penalty for performing tasks in the order δ.
5. Calculate $F_{\min}(\mathcal{K})$, minimum after all permutations, from $F_\delta + F(\overline{\mathcal{K}})$.
6. Perform 1–5 consecutively for all k elementary subsets of the task set $\mathcal{J}$.

The minimum, after subsets, of the $F_{\min}(\mathcal{K})$ values determined in 5th point is a solution to the problem considered in the study.

In the further part of the work, we assume that the optimization algorithms run on a quantum computer determine optimal solutions. If not, then the solution to the problem discussed in this paper is suboptimal.

Let TWT (A, α) be a sub-problem of the TWT problem under consideration, in which $A \subseteq \mathcal{J}$ and α is the starting point for the tasks from the A set. With these markings, TWT $(\mathcal{J}, 0)$ is the problem under consideration in this paper.

For a natural number k $(0 < k < n)$, $L = \binom{n}{k}$ is the number of k elementary subsets (combinations) of the elements of the set $\mathcal{J}$. Let

$$\mathcal{A} = \{A_1, A_2, \ldots, A_L\}, \tag{7}$$

will be a set of such k elementary subsets. Come on, through

$$\overline{A}_i = \mathcal{J} \setminus A_i, \; A_i \in \mathcal{A}$$

we denote the completion of A_i in the task set $\mathcal{J}$.

We consider a subset of $A_i \in \mathcal{A}$. We assumed that $|A_i| = k$, $0 < k < n$. From the set A_i we can generate a $\tau = k!$ permutation. Let

$$\Phi^i = \{\pi_1^i, \pi_2^i, \ldots, \pi_\tau^i\} \tag{8}$$

will be a set of these permutations. Cost of executing tasks in the order π_j^i $(i = 1, 2, \ldots, L, \; j = 1, 2, \ldots, \tau)$

$$\mathcal{F}(\pi_j^i) = \sum_{l=1}^{k} \max\{0, C_{\pi_j^i(l)} - d_{\pi_j^i(l)}\} \cdot w_{\pi_j^i(l)}, \tag{9}$$

where the completion time of the task $\pi_j^i(l)$ is $C_{\pi_j^i(l)} = \sum_{s=1}^{l} p_{\pi_j^i(s)}$, for $l = 1, 2, \ldots, k$.

For the A_i set, $C(A_i) = \sum_{l \in A_i} p_l$ is the end of all tasks from A_i. Let $\overline{\pi}_i$ be the solution to the problem TWT $(\overline{A}_i, C(A_i))$ with the value of $\mathcal{F}(\overline{A}_i)$ determined by the algorithm QA $(\overline{A}_i, C(A_i))$ $(i = 1, 2, \ldots, L)$. Then the set

$$\Pi^i = \{(\pi_j^i, \overline{\pi}^i) : \pi_j^i \in \Phi^i, \ j = 1.2, \ldots, \tau\},$$

contains permutations of $\mathcal{J}$ tasks resulting from concatenation of π_j^i elements of A_i set with $\overline{\pi}^i$ elements of $\overline{A}_i$, which is a solution to the TWT problem determined by the Quantum Annealing QA$(\overline{A}_i, C(A_i))$. The cost of the permutation $(\pi_j^i, \overline{\pi}^i) \in \Pi^i$ is $\mathcal{F}(\pi_j^i, \overline{\pi}^i) = \mathcal{F}(\pi_j^i) + \mathcal{F}(\overline{A}_i)$. Then $\mathcal{F}_i^* = \min\{\mathcal{F}(\pi_j^i, \overline{\pi}^i) : j = 1, 2, \ldots, \tau\}$ is the value of the optimal[1] solutions for permutations from the set Π^i, i.e. permutations in which the first k positions include tasks from the set A_i and on the following $k+1, k+2 \ldots, n$ tasks from the set $\overline{A}_i$.

Remark 1. For the set of $\mathcal{J}$ tasks and a natural number $(k, \ 0 < k < n)$, if the QA algorithm $(\overline{A}_i)$ determines optimal solutions, then $\mathcal{F}^* = \min\{\mathcal{F}_i^* : i = 1, 2, \ldots, L\}$, is the optimal value of the solution of the TWT problem.

Algorithm 1 shows the Distributed Quantum Annealing (DQA) scheme for determining, on a quantum computer, a solution to the single machine total weighted tardiness problem.

Algorithm 1: Distributed Quantum Annealing DQA

Input : $\mathcal{J}$ – a set of tasks, p_i, d_i, w_i – task parameters,
k – number of elements of generated subsets;
Output: $\mathcal{F}^*$ – penalty function value (solution of TWT problem);

1 Generate $\mathcal{A}$ containing all k - elementary subsets of n - element set;
2 **for all** $A_i \in \mathcal{A}$ **do**
3   $\overline{A}_i = \mathcal{J} \setminus A_i$; $C(A_i) = \sum_{l \in A_i} p_l$;
4   Find the solution value $\mathcal{F}(\overline{A}_i)$ of TWT problem $(\overline{A}_i, C(A_i))$ using the QA;
5   According to (8) determine the set of permutations Φ^i elements from A_i;
6   **for all** $\pi_j^i \in \Phi^i$ **do**
7   calculate $\mathcal{F}(\pi_j^i)$ basis on (9);
8   $\mathcal{F}_i^* = \min\{\mathcal{F}(\pi_j^i) + \mathcal{F}(\overline{A}_i) : \pi_j^i \in \Phi^i\}$;
9 $\mathcal{F}^* = \min\{\mathcal{F}_i^* : i = 1, 2, \ldots, |\mathcal{A}|\}$;

Example 1. For $n = 8$ tasks $(\mathcal{J} = \{1, 2, 3, 4, 5, 6, 7, 8\})$ and $k = 3$, a set of 3-element subsets of $\mathcal{A}$ has $\binom{n}{k} = \binom{8}{5} = 56$ items. Using the quantum annealing algorithm, 56 sub-problems have to be solved. Each 3-element subset has $3! = 6$ strings. The total of such strings is $56 \cdot 6 = 336$.

[1] Assuming that quantum annealing will generate an optimal solution.

5 Implementation on DWave Quantum Annealer

The implementation of the algorithm uses a constrained quadratic model with a constraints available under Ocean Developer Tools for D-Wave Systems. Ocean software is a suite of tools D-Wave Systems provides on the D-Wave GitHub repository for solving hard problems with quantum computers.

For solving TWT problem we introduce integer variables S_i, T_i and binary variables $x_{i,j}$, $i \neq j$, $i, j \in \{1, \ldots, n\}$ which equals to 1 if job i precedes job j.

Our aim is to minimize:

$$\sum_i w_i T_i \tag{10}$$

Subject to constrains:

$$S_j - T_j + p_j - d_j \leq 0 \quad j = 1, \ldots, n, \tag{11}$$

$$-T_j \leq 0 \quad j = 1, \ldots, n, \tag{12}$$

$$S_k - S_j + (p_j - p_k)x_{jk} + 2(S_j - S_k)x_{jk} + p_k \leq 0 \quad j < k,\ j, k = 1, \ldots, n, \tag{13}$$

$$-S_j \leq -S_0 \quad j = 1, 2, \ldots, n. \tag{14}$$

6 Results

Computer experiments have been conducted in D-Wave Leap environment on `hybrid_constrained_quadratic_model_version1p` solver executed on a North America quantum annealer. Exact QPU (Quantum Processing Unit) processing time has been measured.

Case Study. An instance taken from the work of Lawler [7] has been used for an experiment for checking usefulness of the proposed methodology, due to the limited quantum machine time available per month. Data of the used instance is shown in the Table 1.

Table 1. Lawler test instance

i	1	2	3	4	5	6	7	8
p_i	121	79	147	83	130	102	96	88
d_i	260	266	269	336	337	400	683	719
w_i	1	1	1	1	1	1	1	1

The weight for all tasks have been set as $w_i = 1$, $i = 1, 2, \ldots, n$ (original instance is dedicated for the single machine problem with total tardiness cost function (without weights), however its difficulty's is significant (it is hard to solve). The optimal value is 755 (see Lawler [7]).

It is important to note, that it was impossible to achieve optimal solution by execution QA algorithm on the whole Lawler instance on D-Wave machine – percentage relative errors of the obtained solutions were very big, not less than 20.17% (regardless of the number of repetitions of the QA algorithm calls). However, as we can see above, it was possible to obtain optimal solution $(1, 2, 4, 6, 5, 7, 8, 3)$ in $i = 2$ iteration by the DQA method proposed in this work.

7 Conclusions

The study considers the single machine tasks scheduling problem with the criterion of minimizing the weighted sum of tardiness. A distributed quantum annealing DQA algorithm has been proposed. Currently, the possibilities of quantum annealers (they are only produced by D-Wave company) allow us for the optimal solution of very small instances, specifically for the problem under consideration, up to $n = 5$. The application of the proposed DQA approach made it possible to determine the optimal solution for a very difficult Lawler example for the number of tasks $n = 8$. The proposed methodology allows for optimal solving of similar problems (e.g. TSP) and larger sizes.

Acknowledgments. Calculations have been partially carried out using resources provided by Wroclaw Centre for Networking and Supercomputing (http://wcss.pl), grant No. 96.

References

1. Adamu, M.O., Adewumi, A.O.: A survey of single machine scheduling to minimize weighted number of tardy jobs. J. Ind. Manag. Optim. **10**, 219–241 (2014)
2. Bożejko, W., Rajba, P., Wodecki, M.: Stable scheduling of single machine with probabilistic parameters. Bull. Pol. Acad. Sci. Tech. Sci. **65**, 219–231 (2017)
3. Bożejki, W., Grabowski, J., Wodecki, M.: Block approach-tabu search algorithm for single machine total weighted tardiness problem. Comput. Industr. Eng. **50**, 1–14 (2006)
4. Chenga, T.C.E., Nga, C.T., Yuanab, J.J., Liua, Z.H.: Single machine scheduling to minimize total weighted tardiness. Eur. J. Oper. Res. **165**, 423–443 (2005)
5. Congram, R.K., Potts, C.N., van de Velde, S.L.: An iterated Dynasearch algorithm for the single-machine total weighted tardiness scheduling problem. INFORMS J. Comput. **14**, 52–67 (2002)
6. Den Basten, M., Stützle, T.: Ant colony optimization for the total weighted tardiness problem, precedings of PPSN-VI. Eur. J. Oper. Res. **1917**, 611–620 (2000)
7. Lawler, E.L.: A "Pseudopolynomial" algorithm for sequencing jobs to minimize total tardiness. Ann. Discrete Math. **1**, 331–342 (1977)
8. McGeoch, C.C.: Theory versus practice in annealing-based quantum computing. Theoret. Comput. Sci. **816**, 169–183 (2020)
9. Rajba, P., Wodecki, M.: Stability of scheduling with random processing times on one machine. Applicationes Mathematicae **39**, 169–183 (2012)

10. Rinnooy Kan, A.H.G., Lageweg, B.J., Lenstra, J.K.: Minimizing total costs in one-machine scheduling. Oper. Res. **25**, 908–927 (1975)
11. Wodecki, W.: A branch-and-bound parallel algorithm for single-machine total weighted tardiness problem. Int. J. Adv. Manuf. Technol. **37**, 996–1004 (2008)

Efficient Constructions for Simulating Multi Controlled Quantum Gates

Stefan Balauca(✉) and Andreea Arusoaie

Faculty of Computer Science, Al. I. Cuza University of Iaşi, 700506 Iaşi, Romania
stefann.balauca@gmail.com, andreea.arusoaie@uaic.ro

Abstract. Multi Controlled Gates, with Multi Controlled Toffoli as primary example are a building block for a lot of complex quantum algorithms in the domains of discrete arithmetic, cryptography, machine learning, and image processing. However, these gates cannot be physically implemented in quantum hardware and therefore they need to be decomposed into many smaller elementary gates. In this work we analyse previously proposed circuit constructions for MCT gates and describe 6 new methods for generating MCT circuits with efficient costs, less restrictions, and improved applicability.

Keywords: Multi-controlled Toffoli gate · Quantum circuit · Ancilla qubit · Efficient quantum algorithms

1 Introduction

Since the first ideas of quantum computing arose in the 1980s, various mathematicians, physicists, and computer scientists have put huge amounts of work in designing both quantum hardware and software. The applications they have found so far include number theory, cryptography, quantum physics, molecular bio-chemistry, search problems, and machine learning [13,18]. The algorithms proposed are, at least in theory, far more efficient than the classical ones [7,15], but very few of them have been actually tested on real hardware [17].

For the moment, quantum computing faces some practical problems, most of which are related to a physical phenomenon named decoherence. This effect means that a quantum system's state can collapse after some time because of interactions with its surroundings [4,5]. In order to reduce decoherence and create feasible quantum computers, we need to cool the system down to a few milli-Kelvins, which requires very expensive equipment [20]. On the other hand, the task of mathematicians and computer scientists is to design efficient and robust algorithms that will need less qubits, quantum gates, and time in order to be executed, since the effects of decoherence scale with each of these parameters.

In this work, we have focused on a specific, yet very important and representative quantum algorithm, the Multi-Controlled Toffoli Gate (MCT). These gates are the quantum equivalent of the logical AND operation, and are necessary in many quantum arithmetic and discrete computing algorithms. [2] has also

D. Groen et al. (Eds.): ICCS 2022, LNCS 13353, pp. 179–194, 2022.
https://doi.org/10.1007/978-3-031-08760-8_16

shown that MCT gates can be used to simulate any multi-controlled gate, which in turn are widely used in increasingly complex applications (Quantum Simulations [13], Machine Learning [11], Image Processing [19]). Efficient implementations of MCT gates are therefore essential in real near-term quantum computer applications.

The paper is structured as follows: In Sect. 2 we introduce the most important quantum computing concepts needed for understanding the contents of this work. In Sect. 3 we describe various implementation ideas previously proposed by the scientific community. In Sects. 4.1, 4.2, and 4.3 we expose and analyse the new constructions we propose. In Sect. 5 we describe the implementation of these constructions using Qiskit [9] and compare the complexities of our ideas with the ones of the existing ideas. We finish this paper by presenting our conclusions and some ideas of future improvements and research.

Notation: Throughout this work we will use the notation C^nX to represent a n-controlled Toffoli gate. In this context, $C^1X = CX$ is the basic CNOT gate and $C^2X = TOFF$ is the original Toffoli gate or 2-controlled NOT.

Note: All figures representing quantum circuits in this work have been realised using the graphical library Quantikz [10].

2 Basic Concepts

2.1 Circuit Costs

In order to understand what an efficient circuit means, we have to describe some cost functions that are usually used. We will refer to these cost functions when analysing the complexity of the circuit.

1. *Total Gate Count*: the total number of elementary gates in circuit;
2. *CX Gate Count*: the number of elementary 2-qubit gates (usually CX);
3. *Circuit Depth*: the number of time steps necessary for executing the circuit;
4. *CX Depth*: circuit depth after all the one-qubit gates have been removed;
5. *Circuit Volume*: a measure introduced by IBM [3] to better represent circuit performance, defined as: number of qubits $\times$ circuit depth.

For the most part of the paper, our focus will be towards the CX depth of the circuit, because it is proportional to the full circuit depth, but it is easier to generalise between different hardware implementations (most of them implement the CX gate directly, or some other two-qubit gate that can be easily transformed in CX, such as CZ: $CX = H \cdot CZ \cdot H$).

2.2 Ancilla Qubits

It is well known that the most efficient solutions many classical and quantum problems make use of extra memory to achieve a speed-up. However, in the case of Quantum Computing, due to its reversible nature, this speed-up comes

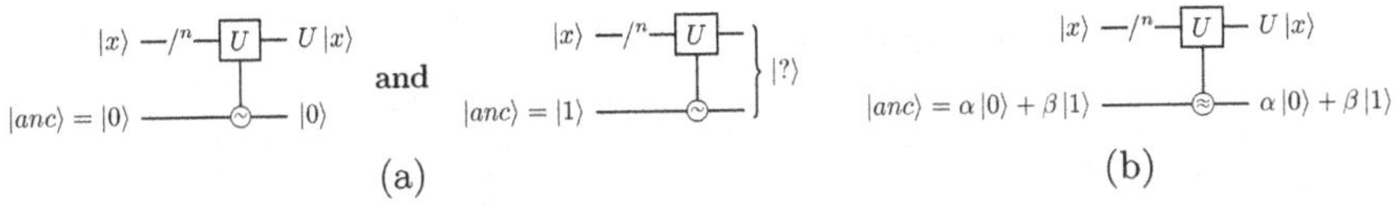

$|x\rangle$ —/^{n}— U — $U|x\rangle$ and $|x\rangle$ —/^{n}— U — $|?\rangle$ $|x\rangle$ —/^{n}— U — $U|x\rangle$

$|anc\rangle = |0\rangle$ — $|0\rangle$ $|anc\rangle = |1\rangle$ $|anc\rangle = \alpha|0\rangle + \beta|1\rangle$ — $\alpha|0\rangle + \beta|1\rangle$

(a) (b)

Fig. 1. Ancilla qubits notation: (a) clean, (b) dirty

with a cost: any auxiliary qubit used must return to its initial state after the computation is done, otherwise it will remain entangled to the other qubits, and therefore any operation applied to it may modify the overall state of the system.

In literature, the auxiliary qubits are called **ancilla**, and they can be of two types: *clean* ancillae and *dirty* ancillae.

Clean Ancilla: These are qubits that need to be in the state $|0\rangle$ before the computation is started and returned to this state afterwards. We will represent the usage of clean ancilla by a circle with a single tilde as in Fig. 1a in order to illustrate that only the $|0\rangle$ state is left unchanged.

Dirty Ancilla: These qubits may be in any state before the computation is started and must be returned to the exact same state afterwards. We will represent the usage of dirty ancilla by a circle with a double tilde (Fig. 1b) in order to illustrate that both fundamental states are left unchanged by the gate.

3 Existing Methods (Prior Work)

In this Section we introduce and present the evolution of quantum circuits for simulating multi-controlled gates starting from the first ideas developed simultaneously with the concept of quantum computing.

3.1 First Ideas

In the theoretical model of quantum computations, the reference work which analyses Multi Controlled Gates, with MCT as the primary example is [2]. The authors have shown how one can achieve quantum MCT gates with or without using auxiliary qubits. Some of the algorithms described there are implemented by [9] and we will briefly discuss them here.

No Ancilla: Even though Tomaso Toffoli had proven in 1980 that constructions of multi-controlled gates is impossible in the context of classical reversible computing without the use of auxiliary bits because of parity constraints [16], the ability of generating square roots of quantum gates makes this task possible in the context of quantum computing. However, the construction relies on exploring all the possible states of the control register of qubits in a Gray-code ordering and therefore needs an exponential number of gates. The cost of this construction makes it impractical for a number of control qubits greater than 5.

Since then, other papers have proposed variants of this circuit pretending to have a theoretical linear depth, while still using no ancilla [6,14]. However, the authors admit that the 2-qubit gates used are not easy to implement on real quantum hardware and would need further decomposition into many basic gates and the overall complexity increases.

Cascading Operations: The easiest and most intuitive way of implementing MCT gates is by using classical Boolean logic concepts. These ideas are firstly presented in [16], as classical circuits, and later in [2] in their quantum form. Since the implementations are based on Boolean logic, they require additional qubits to store partial calculation results. Figure 2 presents the base construction of these circuits by using Toffoli gates in both clean and dirty ancilla setups. It is easy to see how the construction generalises to any number n of control qubits at a cost of using $n - 2$ ancilla qubits and a linear number of total gates. The depth of the circuit is proportional to the number of gates because parallelism cannot be used (operations are done in a sequential order). The shape of the resulting circuits has given this construction the name 'V-chain'.

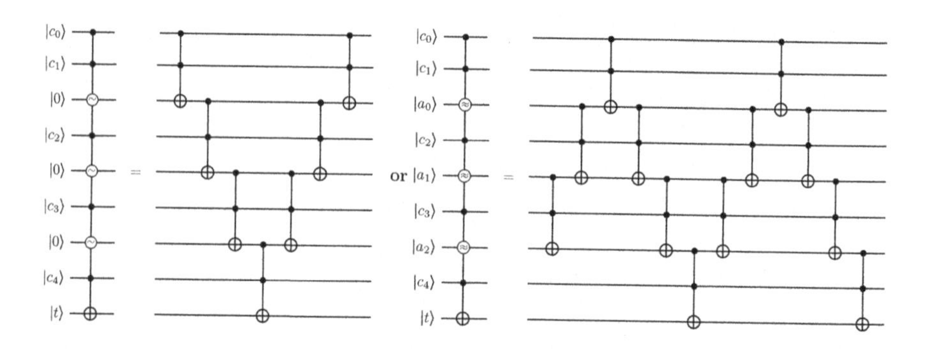

Fig. 2. Structure of the V-chain construction for the case of 5 control qubits and 3 ancilla qubits (left: using clean ancilla, right: using dirty ancilla)

A great use for the dirty ancilla V-chain construction is also presented in Lemma 7.3 from [2]: for creating an arbitrary large multi-controlled X circuit one can use a single ancilla qubit and three or four V-chain constructions with half the number of the initial controls, as illustrated in Fig. 3. The dirty ancilla qubits needed for the V-chain gates are provided by the other half of the circuit which does not participate in the current computation.

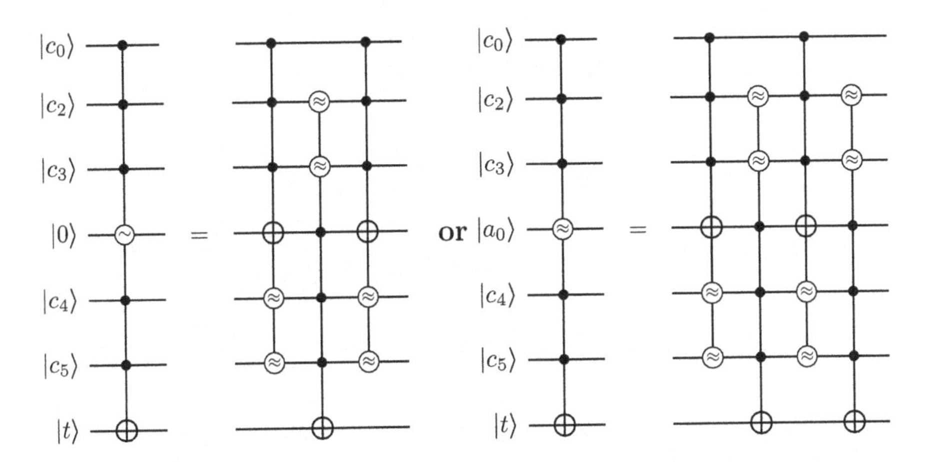

Fig. 3. One Ancilla Construction using dirty ancilla V-chain sub-circuits (left: one clean ancilla, right: one dirty ancilla)

Implemented by Qiskit: The three constructions described above are implemented in [9] with the names `noancilla`, `v-chain` and `recursive` respectively. However, the recursive implementation has a design flaw because the sub-circuits used are also `recursive` and not `v-chain` as they should be. This makes the overall gate-cost and depth complexities of the circuit to be quadratic rather than linear as presented in [2]. Because of the widely use of Qiskit in many Quantum Computing applications and as a learning environment, lots of students or even specialists might mistakenly believe that a linear construction for multi-controlled gates is not possible by using only one ancilla qubit. In order to prove that is not the case, we have implemented, tested and measured the complexities of these constructions using `v-chain` gates for the sub-circuits (more details in Sect. 5).

3.2 Logarithmic Depth Using Parallelism

In a more recent article [8], authors have described a construction that rearranges the gates of the clean ancilla V-chain circuit in order to achieve a logarithmic depth by making use of parallelism (see Fig. 10a for construction details).

3.3 Relative Phase and Partial Gate Cancelling

An important way of reducing the gate and depth cost of quantum circuits is the usage of gate constructions that have *almost* the same functionality as the gates they are trying to replicate in circumstances where the difference is either irrelevant or can be cancelled by another conjugated gate construction. The most common trick is to use relative phase gates: these gates are identical with their counterparts when taken in absolute value, but differ only by a relative phase applied over some of the states. In matrix form, this is equivalent to a multiplication with a diagonal matrix D whose diagonal entries are of the form $e^{i\varphi}$. We will use the prefix R to denote a relative phase gate, therefore RC^2X will represent a relative phase Toffoli gate.

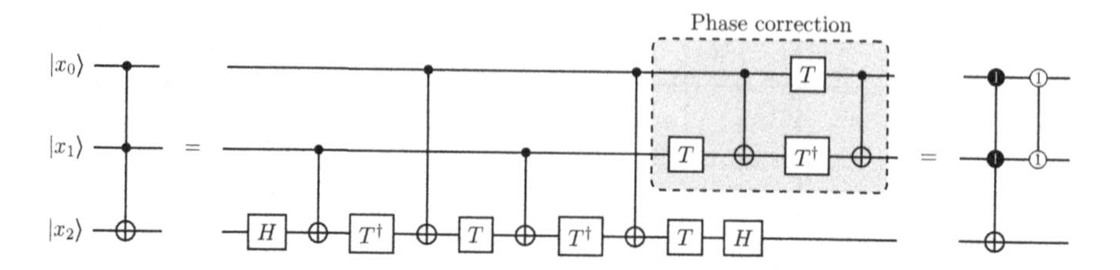

Fig. 4. Decomposition of a Toffoli gate into a Relative Phase Toffoli and a Phase Correction Gate. The circled 1s on the control qubits of the last circuit represent the relative phase. The black circles represent the relative phase induced by the first part of the circuit, and the white circles represent the corresponding phase correction.

Since the relative phase of a state is not observable, these kind of gates have the same effect as the original ones if measurement is applied right after the relative phase gate. However, in more complex circuits the relative phase may interfere with other gates applied and lead to undesired effects. An example of such a gate construction can be observed in Fig. 4.

An advantage of the diagonal matrix is that it can commute with many more operators than an ordinary unitary operator can, and therefore the relative phase may be cancelled by a conjugated phase applied somewhere later in the circuit. In particular, a relative phase applied on a qubit can be cancelled after applying a gate that has that qubit as a control, as shown in Sect. 3 of [12]. Moreover, depending on the circuit shape, a gate and it's inverse can be decomposed in pairs of gates that affect different numbers of qubits. If used in a circuit as presented in Fig. 6, we can further reduce the number of gates by eliminating the two corresponding sub-gates that are inverse to each other.

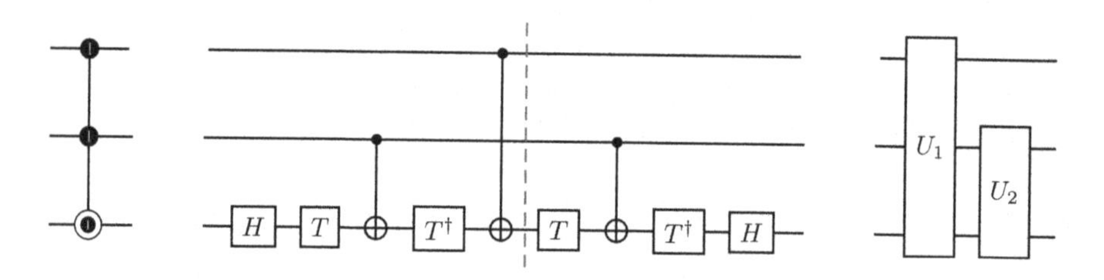

Fig. 5. Decomposition of a relative phase 2-controlled Toffoli gate [12]. The full construction represents the relative phase gate and the dashed red line marks the decomposition of this construction into U_1 and U_2. The gate U_1 represents the short version of the RC^2X gate and has a depth of 5 with 2 CX gates and 3 single-qubit gates.

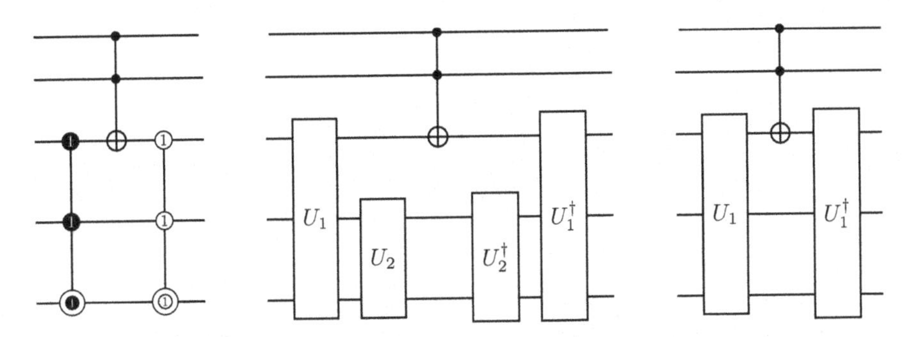

Fig. 6. Usage of conjugated short RC^2X gates in order to reduce the costs of the circuit. The first part of the figure presents the decomposition of the two conjugated RC^2X gates by using the formula $(AB)^\dagger = B^\dagger A^\dagger$, while the second part shows how to apply the gate reduction.

3.4 Using 3-Controlled Relative Phase Toffoli Gates

The main contribution of [12] was the introduction of 3-controlled relative phase Toffoli gates with the construction from Fig. 7 which is very cost-efficient, especially when the decomposition is used as in Fig. 6. In this case the cost of a single gate is of 4 CX gates and 6 one-qubit gates. Also, the author proposes a more efficient V-chain construction: all the Toffoli gates in the V-chain construction (Fig. 2) except for the ones at the bottom of the circuit can be replaced by 3-controlled relative phase gates. Moreover, when using dirty ancilla, we can actually use the short version of these gates for all but the two gates at the top. This method doubles the number of total control qubits in the circuit without the need to add more auxiliary qubits. The gate and depth cost of this circuit keep their linear complexities with respect to the number of control qubits n, but the constant factor decreases by 1/3 in the case of the dirty ancilla construction. For the rest of this paper we will refer to these circuits as 'V-chain-2'.

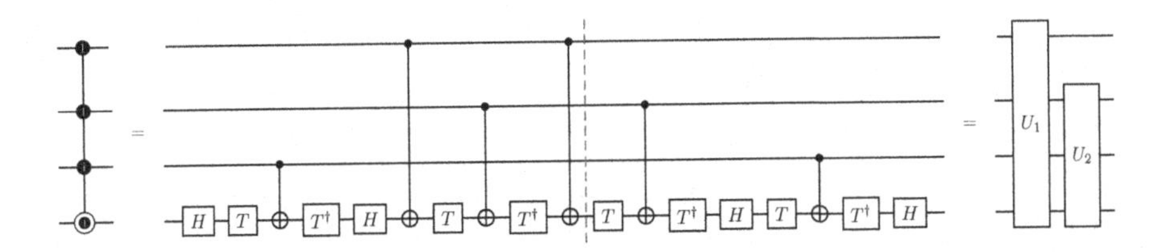

Fig. 7. Decomposition of a relative phase 3-controlled Toffoli gate (RC^3X) [12]. The construction represents the relative phase gate and the dashed red line marks the decomposition into U_1 and U_2. The gate U_1 represents the short version of the RC^3X gate and has a depth of 10 with 4 CX gates and 6 single-qubit gates.

4 Our Proposed Methods

4.1 One Ancilla Constructions

Starting from the V-chain-2 constructions, we have developed two constructions for the one-ancilla MCT circuits in Fig. 3 that are more efficient than those presented in [2] and [8].

One Clean Ancilla: In the first part of Fig. 3 we can observe that there is a V-chain gate repeated twice. This means that reducing the size of these gates will reduce the overall cost of the circuit. In this construction, we have to split the n control qubits into two groups of m_1 and m_2 qubits each, that will take turns in being control and dirty ancilla for the three V-chain sub-circuits (Fig. 8). Since each of these sub-circuits needs $m/2$ ancilla qubits for their m control qubits, we have two constraints for m_1 and m_2: $(m_1 - 3)/2 < m_2$ and $(m_2 - 2)/2 < m_1$. Since we want to optimise such that the value of m_1 is minimal, we will reach a configuration with $m_1 = \left\lceil \frac{n}{3} \right\rceil$ and $m_2 = \left\lfloor \frac{2n}{3} \right\rfloor$.

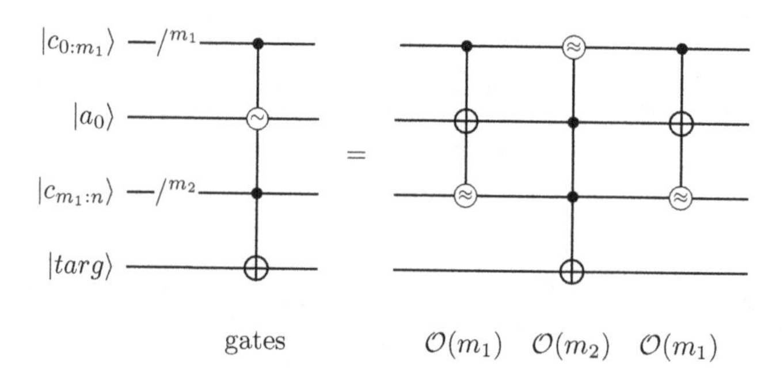

Fig. 8. One clean Ancilla construction with splitting the controls into two groups of m_1 and m_2 qubits respectively. The gate acting on the first group of m_1 qubits appears twice so we need to minimise its cost.

Calculating the cost of this construction we observe a decrease by a little more than 1/3 over the version proposed in [2] (Details in Sect. 5).

One Dirty Ancilla: When designing the circuit using only one dirty ancilla, there is only one improvement to make: replace the V-chain gates with V-chain2 ones. Since both the top and bottom part of the circuit contain 2 gates each, splitting the controls in two groups, m_1 and m_2 does not have any effect on the gate count or circuit depth. In this case we obtain again a decrease by 1/3 over the version in [2].

4.2 Depth Optimisation with Dirty Ancilla

Since we have observed that the depth of the V-chain-2 circuit that uses $n/2$ dirty ancilla is two times less than the depth of the one-dirty-ancilla construction, we started the search for a general k-dirty-ancilla construction where k varies between 1 and $n/2$. In order to achieve optimal depth in our circuit, we propose the construction in Fig. 9.

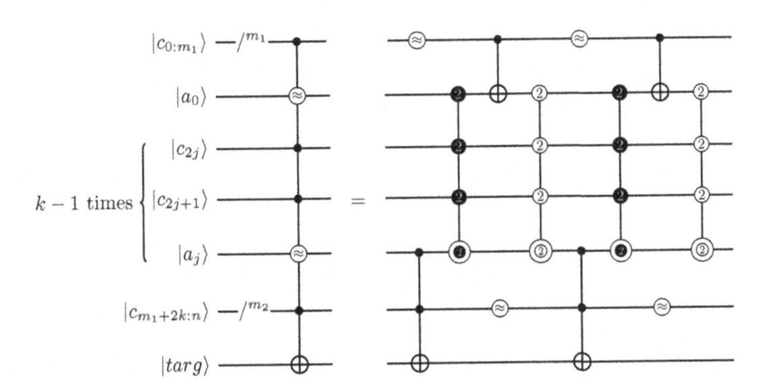

Fig. 9. Construction of a circuit simulating a MCT gate with k dirty ancilla qubits. We use $k-1$ of the ancilla qubits in the middle part of the circuit to store the results of $2 \times (k-1)$ pairs of short RC^3X gates. The two groups of m_1 and m_2 qubits at the top and bottom of the circuit alternate between being control qubits or dirty ancillae for the 4 V-chain-2 gates required. We have eliminated the line linking the ancilla qubits with their corresponding gate for clarity.

We mention that the authors of [1] have also proposed a dynamic programming approach to generate an optimal-depth circuit with an arbitrary number of dirty ancilla. However, they have not used the improved version V-chain-2, and therefore the costs of their circuits are higher (details in Sect. 5).

4.3 Depth Optimisation with Clean Ancilla

It should be clear by now from the discussion in Sect. 3 that clean ancilla qubits are more useful for reducing the circuit's gate count and depth. However, as stated in Sect. 1, they are a resource harder to obtain than the dirty ones, so we must always adapt our circuits to the amount of qubits available.

Logarithmic with Half Ancilla: The first idea to reduce the clean Ancilla count is to use relative phase 3-controlled gates as we have already done in the case of dirty ancilla. In order to maintain the logarithmic depth of the circuit, we use the same construction presented in [8], but replace all regular Toffoli gates with relative phase ones, except for the one acting on the target (Fig. 10).

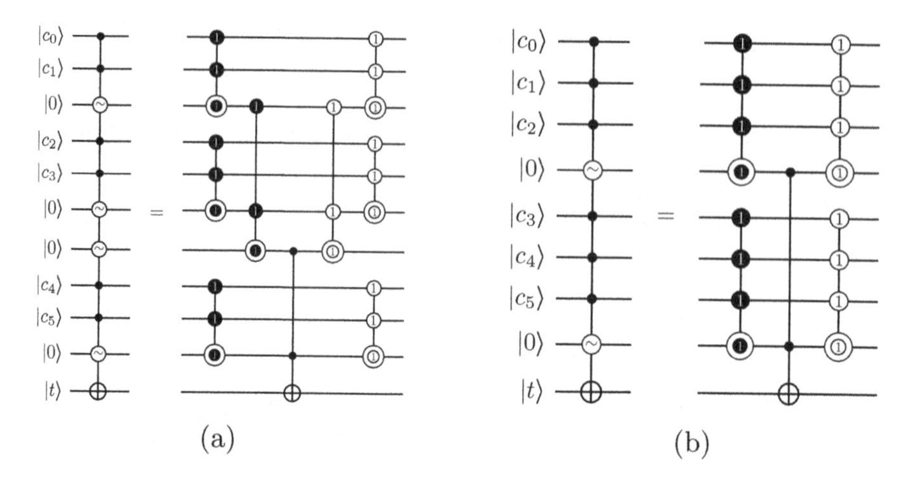

Fig. 10. Logarithmic depth constructions for $n = 6$ control qubits using (a) RC^2X gates and $n-2$ clean ancilla or (b) RC^3X gates and $n/2$ clean ancilla. The first construction keeps the same depth even if we increase the number of control qubits to $n = 8$ and add two more clean ancillae.

The circuit depth will now be

$$D_3(n) = 2\lfloor \log_3 n \rfloor \times \text{depth}(RC^3X) + \text{depth}(C^2X) = \mathcal{O}(\log n).$$

In comparison, when using 2-controlled relative phase gates, the depth is

$$D_2(n) = 2\lfloor \log_2 n \rfloor \times \text{depth}(RC^2X) + \text{depth}(C^2X) = \mathcal{O}(\log n).$$

We can observe that the constant factor of our construction is higher, since

$$\frac{D_3(n)}{D_2(n)} = \frac{\text{depth}(RC^3X)}{\text{depth}(RC^2X)} \cdot \frac{\log_3 n}{\log_2 n} = \frac{18}{9}\log_3 2 = \log_3 4 \approx 1.26.$$

Depending on the physical implementation of the quantum computer, future engineers should decide whether the $n/2$ additional clean auxiliary qubits required for a 1.3× decrease in depth represent a fair price or not.

Note: The volumes of the two circuits are almost equal, but the one we proposed is 5% smaller:

$$\frac{V_3(n)}{V_2(n)} = \frac{(n + n/2)D_3(n)}{(n + n - 1)D_2(n)} \approx \frac{3}{4}\log_3 4 \approx 0.95.$$

Lower cost with Logarithmic Ancilla: The next idea we present is one that reduces the depth and gate cost of the circuit in exchange for $\mathcal{O}(\log n)$ clean ancilla. The main idea is to recursively add layers to the circuit while using the already used control qubits as dirty ancilla before the restoring phase takes place. Using these qubits as dirty ancilla allows us to use the short version of V-chain-2 gates. The construction is detailed in Fig. 11.

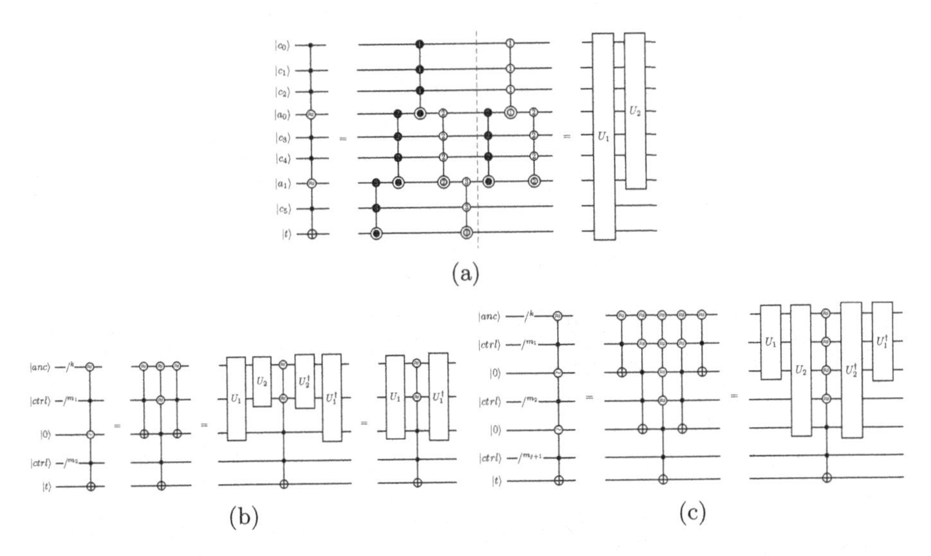

Fig. 11. Construction based on recursively appending layers of V-chain-2 gates. (a) Decomposition of V-chain-2 gate to generate its short form. (b) Usage of short V-chain-2 gates by conjugated gate cancelling: the U_2 and $U_2^\dagger$ gates cancel each other because the gate between them uses the qubits only as dirty ancilla, so it preserves their state. (c) The procedure of recursively adding another layer of short V-chain-2 gates (U_1 and $U_1^\dagger$): we continue this procedure with smaller gates until we have no more ancilla left. The outer pair of V-chain-2 gates should only use the other $j-1$ clean ancillae in the circuit and will therefore have $m_1 = 2j-1$ controls.

This construction allows us to use j clean ancilla for building a circuit with $n = \mathcal{O}(3^{j-1} \cdot j)$ control qubits, and because of the clever way of using the reduced V-chain gates the depth and gate count of the circuit are around 30% less than in the case of the construction with just one clean ancilla. When compared to the logarithmic depth construction which uses a linear number of clean ancilla, the gate count of this circuit is only higher with 25%, but the depth being linear is definitely a downside. However, in a situation where the number of available clean ancilla is low, this construction might still be preferred since with just 4 ancilla we could simulate circuits with $n \approx 80$ controls.

Sqrt Compromise: The logarithmic depth construction could be generalised such that we can use gates with any number of control qubits as the building block. Considering the size of the building block gate to be k controls we would need $2 \cdot \lfloor \log_k n \rfloor + 1 \doteq 2L+1$ layers of k-controlled V-chain-2 gates. However this would mean that we need some more auxiliary qubits for the first layer of gates (if we do not have these qubits we may split the first layer in two sub-layers with half of the controls acting as dirty ancilla for the other half). The number of auxiliary qubits needed will then be

$$\#_{clean} = \sum_{i=1}^{L-1} k^i = \frac{k^L - 1}{k - 1} \approx \frac{n}{k-1}.$$

The depth of the circuit can be calculated as

$$D_k(n) = (2L + 1) * \text{depth}(k\text{-V-chain-2}) = \mathcal{O}(kL).$$

In the case of $k \approx \sqrt{n}$, we have $L = 2$ and then $\#_{clean} \approx \dfrac{n}{k-1} \approx \sqrt{n}$. The depth of the circuit will be $D_{\sqrt{n}}(n) = \mathcal{O}(2\sqrt{n}) = \mathcal{O}(\sqrt{n})$.

Again, the values for k and L should be chosen depending on the availability of clean ancilla and the willingness to sacrifice the circuit's depth and therefore its execution time.

5 Implementation and Comparisons

We have implemented all the presented methods in Qiskit, the open-source Quantum Computing environment developed by IBM. In order to calculate the real costs of the implemented circuits, we have used Qiskit's `transpile(circuit)` function, which returns the equivalent circuit with only elementary gates (one-qubit gates and CNOTs). The experiments were done with `optimization_level=1` and we have used `FakeQasmSimulator` and `FakeManhattan` as backends. `FakeQasmSimulator` has a virtually infinite number of fully connected qubits, while `FakeManhattan` is the largest fake simulator available in Qiskit, with 65 qubits. `FakeManhattan` also needs additional SWAP gates in order to simulate full connectivity between qubits.

5.1 One Ancilla Constructions

In Fig. 12 we present a comparison of the circuit depth for the various implementations of one-ancilla MCT gates discussed. We also include the `noancilla` method here. We can observe in the figure that the two Qiskit implemented methods become very expensive even for small values of n. It is also clear that the clean ancilla methods have a smaller depth and our proposed methods represent an improvement over the existing ones, both in the ideal case (Fig. 12a) and in the real-life scenario (Fig. 12b). The circuits run on the Manhattan Quantum Computer have a 4–5 times higher depth, and the linear trends are not stable because of the extra SWAP gates required.

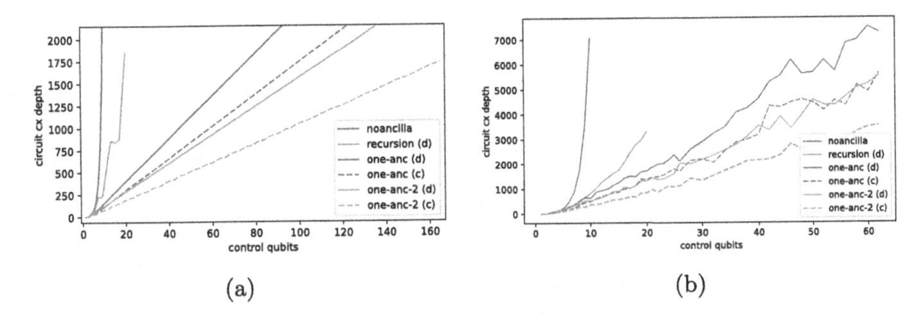

Fig. 12. Circuit depth of one-ancilla and no-ancilla circuits for MCT gates on a theoretical, fully connected quantum computer (a), and on the FakeManhattan computer with 65 qubits (b). The first two methods are implemented by Qiskit, but their costs are not linear. Our methods (pink) have better costs than the others. Dashed lines represent constructions which use clean ancilla.

5.2 Dirty Ancilla Constructions

In Fig. 13a we compare the depth of our circuit with the depth obtained by [1] and in Fig. 13b we compare the depth and total gate cost for our circuit. It can be concluded from these comparisons that if we want to minimise the depth of the circuit we can use roughly $k \approx n/3$ dirty ancilla qubits with no great improvement over the $n/2$ ancilla version, while for also minimising the total gate count we need $k \approx n/2$.

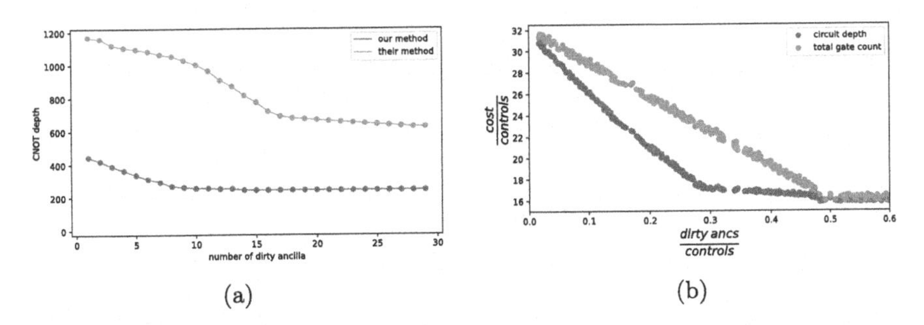

Fig. 13. Circuit costs of the circuit with varying number of dirty ancilla qubits. (a) Comparison between the CX depths of our method and the method proposed in [1] for a circuit with $n = 30$ control qubits. (b) Comparison between the total CX count and the CX depth of the circuit in our method. The horizontal axis shows the ratio between the number of ancilla and number of controls. The vertical axis shows the ratio between the cost and the number of controls. The data used was obtained for values of n between 10 and 100 so that the linear trends can be observed to be the same regardless of the value of n.

5.3 Clean Ancilla Constructions

In Fig. 14 we can observe a CX depth comparison between the methods that make use of clean ancillae. The V-chain-2 method is only shown for comparison, as it should never be used, since it can be replaced by its logarithmic depth equivalent. It is clear from this figure that our methods `one-anc-2` and `log-anc` have better depths than the method `one-anc-1` described in [2].

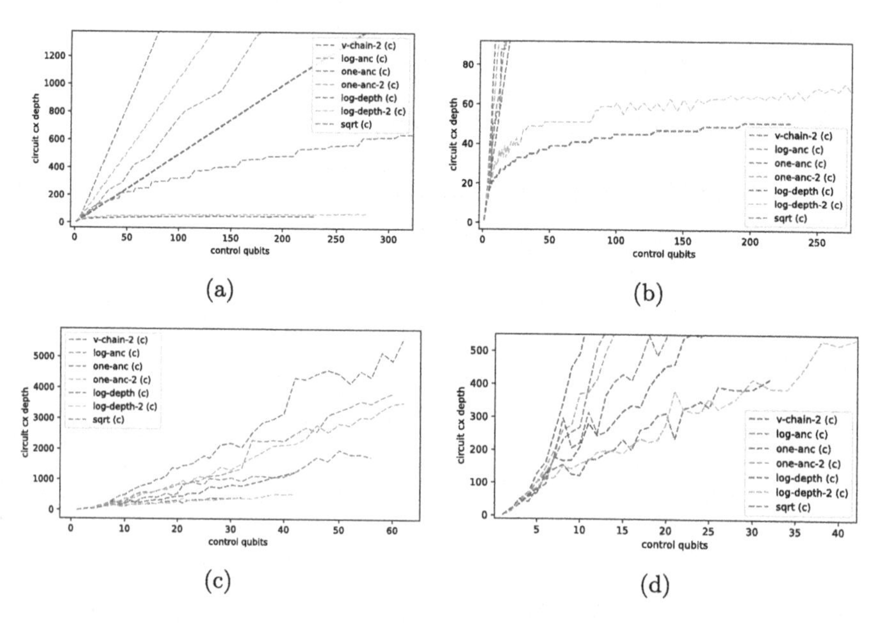

Fig. 14. Comparison between the CX depth of different implementations of MCT circuits with clean ancilla. (a) Comparison between the linear-depth constructions and the `sqrt` construction. (b) Detail of the logarithmic depth constructions: our method has a slightly higher depth, but uses only half as many ancilla. (c) Comparison of the costs on FakeManhattan. (d) Detail of logarithmic depth constructions on FakeManhattan: no noticeable difference between the two methods;

We can also conclude that the `sqrt` method proves its efficiency for higher values of n, where the circuit depth and ancilla need start to grow slower. When comparing the two logarithmic depth approaches, we can observe the theoretical prediction that our method `log-depth-2` comes at a halved ancilla cost and slightly increased depth, while on the Manhattan Quantum Computer there is no noticeable difference between its depth and that of the original method.

6 Conclusion and Future Work

In this work we have presented and analysed various methods for implementing Multi-Controlled Toffoli gates in Quantum Circuits and have proposed new, more efficient, and versatile constructions.

Since the Multi Controlled Gates are widely used in lots of Quantum applications we expect the ideas presented here to help reducing the costs of scalable Algorithms which will one day be run on fault-tolerant Quantum hardware.

While the constructions presented here might not be optimal in any given circumstances, we plan on designing an algorithm that will find the best circuit with respect to a given cost function and resource constraints.

We also plan to search for optimisations of other well-known and widely used Quantum Algorithms such as Quantum Fourier Transform, Arbitrary Permutations, and SAT Grover Oracles.

We invite future hardware engineers to choose the solution best suited for their physical implementation and encourage quantum software engineers to try and improve these constructions.

References

1. Baker, J.M., Duckering, C., Hoover, A., Chong, F.T.: Decomposing quantum generalized Toffoli with an arbitrary number of ancilla (2019)
2. Barenco, A., et al.: Elementary gates for quantum computation. Phys. Rev. A **52**(5), 3457–3467 (1995). https://doi.org/10.1103/PhysRevA.52.3457
3. Cross, A.W., Bishop, L.S., Sheldon, S., Nation, P.D., Gambetta, J.M.: Validating quantum computers using randomized model circuits. Phys. Rev. A **100**, 032328 (2019). https://doi.org/10.1103/PhysRevA.100.032328
4. Dirac, P.A.M.: The Principles of Quantum Mechanics. International Series of Monographs on Physics. Clarendon Press (1981)
5. DiVincenzo, D.P.: The physical implementation of quantum computation. Fortsch. Phys. **48**(9–11), 771–783 (2000)
6. Gidney, C.: Using quantum gates instead of ancilla bits (2015). https://algassert.com/circuits/2015/06/22/Using-Quantum-Gates-instead-of-Ancilla-Bits.html
7. Grover, L.K.: A fast quantum mechanical algorithm for database search. In: Proceedings of the Twenty-Eighth Annual ACM Symposium on Theory of Computing (STOC 1996), pp. 212–219. Association for Computing Machinery, New York (1996). https://doi.org/10.1145/237814.237866
8. He, Y., Luo, M.-X., Zhang, E., Wang, H.-K., Wang, X.-F.: Decompositions of n-qubit Toffoli gates with linear circuit complexity. Int. J. Theoret. Phys. **56**(7), 2350–2361 (2017). https://doi.org/10.1007/s10773-017-3389-4
9. Héctor Abraham, E.A.: Qiskit: an open-source framework for quantum computing (2019). https://doi.org/10.5281/zenodo.2562110
10. Kay, A.: Quantikz (2018). https://doi.org/10.17637/rh.7000520.v4
11. Krovi, H., Magniez, F., Ozols, M., Roland, J.: Quantum walks can find a marked element on any graph. Algorithmica **74**(2), 851–907 (2015). https://doi.org/10.1007/s00453-015-9979-8
12. Maslov, D.: Advantages of using relative-phase Toffoli gates with an application to multiple control Toffoli optimization. Phys. Rev. A **93**(2) (2016). https://doi.org/10.1103/PhysRevA.93.022311
13. Nielsen, M.A., Chuang, I.L.: Quantum Computation and Quantum Information: 10th Anniversary Edition. Cambridge University Press (2010). https://doi.org/10.1017/CBO9780511976667

14. Saeedi, M., Pedram, M.: Linear-depth quantum circuits for n-qubit Toffoli gates with no ancilla. Phys. Rev. A **87**, 062318 (2013). https://doi.org/10.1103/PhysRevA.87.062318
15. Shor, P.: Algorithms for quantum computation: discrete logarithms and factoring. In: Proceedings 35th Annual Symposium on Foundations of Computer Science, pp. 124–134 (1994). https://doi.org/10.1109/SFCS.1994.365700
16. Toffoli, T.: Reversible computing. In: de Bakker, J., van Leeuwen, J. (eds.) ICALP 1980. LNCS, vol. 85, pp. 632–644. Springer, Heidelberg (1980). https://doi.org/10.1007/3-540-10003-2_104
17. Vandersypen, L.M.K., Steffen, M., Breyta, G., Yannoni, C.S., Sherwood, M.H., Chuang, I.L.: Experimental realization of Shor's quantum factoring algorithm using nuclear magnetic resonance. Nature **414**(6866), 883–887 (2001). https://doi.org/10.1038/414883a
18. Yanofsky, N.S., Mannucci, M.A.: Quantum Computing for Computer Scientists. Cambridge University Press, Cambridge (2008). https://doi.org/10.1017/CBO9780511813887
19. Yao, X.W., Wang, H., et al.: Quantum image processing and its application to edge detection: theory and experiment. Phys. Rev. X **7**(3) (2017). https://doi.org/10.1103/PhysRevX.7.031041
20. Zu, H., Dai, W., de Waele, A.: Development of dilution refrigerators-a review. Cryogenics **121**, 103390 (2022). https://www.sciencedirect.com/science/article/pii/S001122752100148X

Experiment-Driven Quantum Error Reduction

Krzysztof Werner, Kamil Wereszczyński(✉), and Agnieszka Michalczuk

Department of Computer Graphics, Vision and Digital Systems,
Silesian University of Technology, Gliwice, Poland
kamil.wereszczynski@polsl.pl

Abstract. Error correction is wide and well elaborated area of quantum information theory. Those methods, however, demand additional resources, like quantum gates, qubits or time. We have observed, in statistical sense, that the qubit's error in real quantum computers, once calibrated doesn't change much until next one. Then being so, for quantum sampling based computations, one can determine the correction experimentally and use it until the next calibration, without a need of utilize additional resources. In this work we present the method of determining such a correction and applying it to practical quantum-sampling algorithms.

Quantum sampling is the method, which we deliberately decline to obtain one deterministic result of one-shot computation in. Instead of that, we provide a number of same experiments. Then we observe the probability distribution function (PDF) thus obtained, which is considered as the final result of computation. We have observed and experimentally proved in this work, that error of this probability distribution is correlated with the local quantum phase of qubits involved in computations. Hence we are able to create a *Phase Distortion Unraveling* (PDU) function for each qubit and for whole system as well, that depends on this phase. Briefly, the final result after correction is the sum of PDF and PDU.

Keywords: Quantum computing · Quantum error correction · Quantum sampling · Quantum information theory · NISQ era

1 Introduction

We introduce the novel method of experiment-based error correction in quantum sampling computation process: *Phase Distortion Unraveling* PDU. Initially, the *determined errors* set ε_d is obtained as the difference between the experimental results and expectation values for each input binary string d representing eigen

This publication was supported by the Department of Graphics, Computer Vision and Digital Systems, under statue research project (Rau6, 2022), Silesian University of Technology (Gliwice, Poland).

D. Groen et al. (Eds.): ICCS 2022, LNCS 13353, pp. 195–201, 2022.
https://doi.org/10.1007/978-3-031-08760-8_17

state $|x_d\rangle$. Then the *Phase Distortion Unraveling Function* is computed as the function interpolated by set of points (x_d, ε_d). Once obtained, it can be applied to the results of quantum sampling until the next physical calibration. In current work we present theoretical and experimental proofs of correctness of the PDU method: we will show that the PDU improves quantum calculation results statistically significant. Moreover, we will show that PDU is temporary stable, which we mean that the final error doesn't change significantly over time from the last physical calibration.

Our work is motivated by necessity of disposing the practically applicable error correction in the era of NISQ (Noisy, Intermediate Quantum) computers without requirement of involving extra qubits or gates. Due to limitation in QV (quantum volume), the methods that can correct errors in the evolution phase decreasing QV by grabbing qubits or gates simultaneously is not practically applicable PDU allows to proceed the computations utilizing whole QV and perform the correction procedure on the results. We are aware that PDU utility in beyond-NISQ era is problematic, however in the next ten-fifteen years it avails to proceed with experiments involving QV approaching the current maximum.

Current works in the area of error correction run towards of stabilizer [1] codes, the surface codes, cyclic codes and other less common. In the area of stabilizer codes two works of Nguyen and Kim [8,9] are interesting. In the first one, they shows the stabilizer codes generated from Hermitian self-orthogonal ones and in the second - based on the binary formalization. Lv et al. in [6] shows another conversion od quantum error codes: from quasi-cyclic to stabilizer codes. Ryan-Anderson et al. in [10] proposed the real-time method based on stabilized codes implemented on 10-qubits ion-trapped computer. Dymarsky and Shapere in [3] published the theoretical consideration about stabilizer codes in the perspective of CFT (Conformal Field Theory). Bravyi et al. in [2] describes the method of correction of coherent errors using surface codes. Litinski in [5] describe the interesting issue - he discuss the strategies of surface codes applied to different scale quantum computers.

There are methods of error mitigation [4,7], which is basically the process of creating the additional operator applied to the given gate, that represents the inverse of an error that has been determined experimentally beforehand. This methods utilize the notion of expectation value or gate or detector tomography. Our method treats the quantum circuit as a black box containing an algorithm for solution of one specific problem, through the quantum sampling procedure. It is problem-oriented (not gate-oriented) method, hence for each problem and quantum computer the PDU should be designated. On the one hand it is the limitation, but on the other hand it is much simpler then mitigation methods, since it relies on observing the difference between experimentally obtained and expected results so it doesn't need any complex computation.

In this work we use PDU on the Quantum Cosine Series Sampling operator [11], which generally describes the method of quantum computing based on interpretation of outputs eigen state appearance normalized histograms as the function that can be mapped into the sine-cosine Fourier series. The result is interpreted in the context of problem to be solved, like image processing.

2 Materials and Methods

Let's consider "one half" of first component of QCoSamp $\nu_1(x)$ where parameters $r_1 = s_1 = 0$. We will use it for determining an error in resulting PDF, therefore we will call it reference function, which we formally define as: The *reference function* is the function $\gamma : \mathbb{R} \cap [-1, 1] \longrightarrow [0, 1] \cap \mathbb{R}$ for which for some set $\{x_0, ..., x_K\}$ there are known expected values $\{\gamma(x_0), ..., \gamma(x_K)\}$, e.g. determined classically. For each argument x_k that we can encode, we conduct sufficient[1] number of experiments realizing the reference function $\gamma = \nu_1$. Because we measure the last qubit, the histogram of outputs has two beams for $|0\rangle$ and $|1\rangle$. We normalize it, take the value for $|0\rangle$ and we denote it $\tilde{\gamma}(x_k)$. In that manner we construct the values of a function $\tilde{\gamma}(x)$ which is unknown in general, however we have obtained its values for 2^X arguments x_k, from experiments. On the other hand we can say that, for each $x_k : \tilde{\gamma}(x_k) = \gamma(x_k) + \varepsilon_k$, where ε_k is considered as error. It is different for different k, most probably. Therefore we can say, similarly, that there exists a function $\varepsilon(x) = \tilde{\gamma}(x) - \gamma(x)$, which we don't know, but we know some of its values: $\varepsilon_k = \varepsilon(x_k)$. Hence we can interpolate this function and use as the correction for the next computations. This interpolated function we call *Phase Distortion Unraveling Function* PDU. The PDU function has to be used for the same setup of qubits it was determined for. It is because the PDU contains both the errors of specific qubits themselves, and the errors connected with relations between qubits.

Experimental Protocol. In the experiments we have proved that it is possible to once experimentally designated correction function basing on one component $\frac{1}{2}(1+cos(x))$, will improve significantly the results of subsequent $\frac{1}{2}(1+cos(nx+r))$ components, until next physical calibration of the system. All experiments were done using library Qiskit for Python. This library was developed by IBM, and allowed Us to run Our experiments on real quantum computers (backends). Backends available to us were IBMQ: Lima, Manila, Bogota, Belem, Quito and Santiago.

Temporal Stability Experiment. This experiment was aimed to examine if an error on IBMQ quantum computers is temporal stable. For this purpose we proceed with the same circuit in different times after declared calibration. Then we determine an error and compute the trendline over the time. For this experiment we run whole set of circuits - consisting of circuits for 16 different input values on a selected quantum computer, multiple times. After each run we were collecting actual values returned by quantum computer, along side expected results, and storing them in a file.

[1] In ideal world we should make statistical analysis what does the word "sufficient" means in the reality, however in real world we are limited to the offer of NISQ computers suppliers. Hence in this work we consider that 8.192 repetitions of the experiment, which is maximal number we can make on IBM Q Experience, is sufficient.

Significance of the PDU Method. As the second experiment we examined if the PDU procedure of error correction is statistically significant. For this purpose we determine the PDU function for the given quantum computer. Then we run the circuits representing the same $\frac{1}{2}(1 + cos(x))$ and different function $\frac{1}{2}(1 + cos(2x))$ for inputs from the range $[-\pi, \pi]$ with the resolution $\frac{1}{8}\pi$ since we had disposed 6 qubits computers. The experiments was run in different time after the process of physical calibration. We denoted the results, MSE (Mean Square Error) according to the result expected and the time of an experiment. After we have collected the data we computed the statistical measures like average, standard deviation and Pearson correlation coefficient for the whole dataset containing results from all computers and for each computer separately.

3 Results

Temporal Stability Experiment. For this experiment, we collected data from various quantum computers and circuits shown in Sect. 2. Data were collected between 05.2021 and 01.2022. Data we collected showed us general boundaries of quantum errors - especially quantum noise.

We observed that on some quantum computers error values were more concise - like IMBQ Manila, where 83% of measured error values differed no more than 20% from an average, and standard error deviation for error was 0.033. On others - like IBMQ Santiago - error values were spread out. Over 56% of error values exceeded the average by more then 20%. Hence calculating error correction factors needs to be measured/done every time quantum computer is being calibrated. During Our tests we discovered, that despite information provided by IBM regarding last calibration time for certain quantum computers - declared error values on gates changed once per day. We decided to run the same tests for hours since initial calibration - which set error values on gates. Finally, we proceed with the main experiment in this task: to prove that errors are on the stable level over time. We use Athens, Belem, Bogota, Lima, Manila, Quito and Santiago. The trendlines for each of them are shown on the Fig. 1 We see that 4 of computers has downward trend, which is quite surprising, two of them - stable and only one has the upward trend. However we expected stable trends, the difference of MSE over time doesn't exceed the threshold of 0.02 in the perspective of 20 h. Which we recognize as promising for our method stability and usability.

Significance of the PDU Method. On Fig. 2 there is shown example how PDU influences on the final result. We can see, that the correction is distinct visually. Numerically, in the cases shown there, we observed mean square error to drop from 0.0190 to 0.0016, and standard deviation from 0.1333 to 0.0401 - which was consistent with the other results we have obtained during our experiments.

The PDU process increase the value of Pearson correlation coefficient value in general (see Table 1), which means that the correlation between results corrected by PDU and reference expected ones is better then before this process.

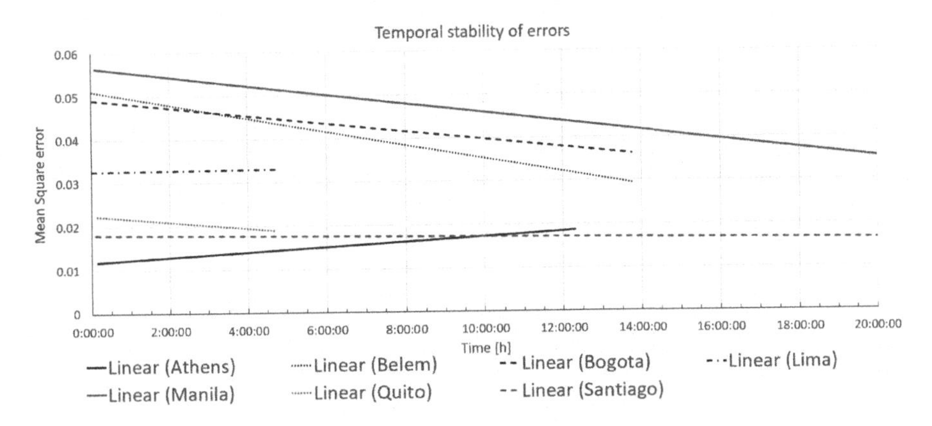

Fig. 1. Temporal stability trendlines for Athens, Belem, Bogota, Lima, Manila, Quito and Santiago computers in time period of 0–20 h after physical calibration

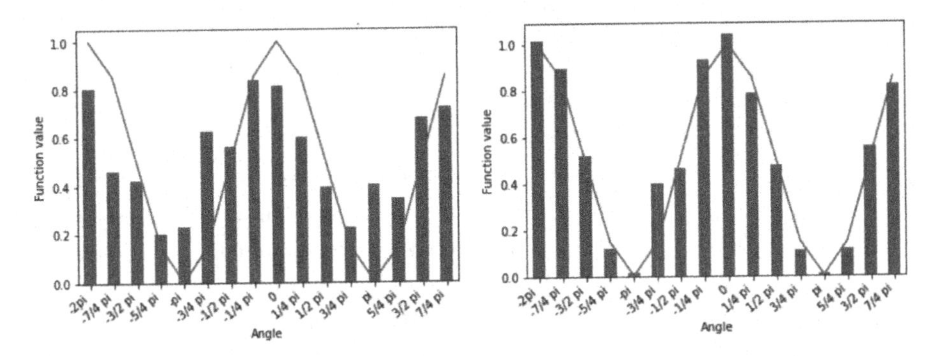

Fig. 2. Results for function $\frac{1}{2}(1 + cos(2x))$ collected on backend IBMQ Belem before (left) and after (right) applying error correction factors. Blue lines means the reference plot of the given function. Beans are the results obtained from quantum computer before (left) and after (right) PDU correction process. (Color figure online)

The increase is smallest for IBMQ Athens (0.0032), Quito (0.0298) and Santiago (0.099). For those computers that we disposed datasets consisting of over 1000 tuples the increase of Pearson correlation factor is obvious and is in the range 0.0728 (Lima) to 0.1438 (Manila). Overall factors are: 0.876 for the original results and 0.9707 after PDU, which gives 0.0947 increase. Moreover, the coefficients value are high, sometimes very close to 1 with small p-values, which means that observed dependence between reference and corrected values is not accidental. Which is experimental prove of statistical significance of results we obtained.

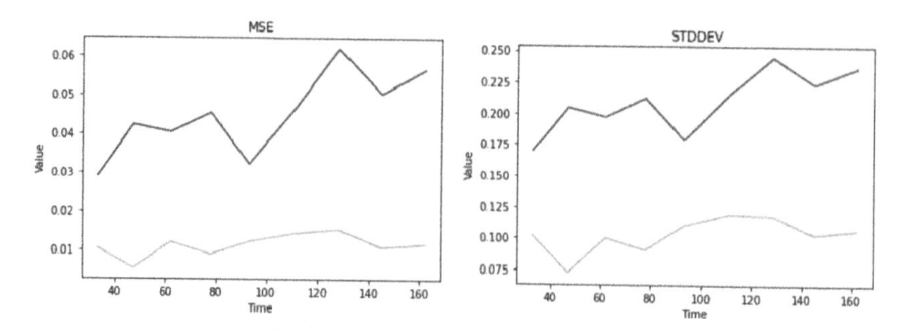

Fig. 3. Values of mean square error and standard deviation on error for functions $\frac{1+cos(x)}{4} + \frac{1+cos(2x)}{4}$ collected on IBMQ Lima. The red line shows the quantities (MSE, STDDEV) before PDU correction and blue ones- after PDU correction. (Color figure online)

Table 1. The correlation coefficients: Pearson product - moment, Kendall's τ and Spearmans ρ with p-values generated for examination of simplicity for result after PDU correction vs reference (ideal) results.

Computer name	Samples number	Pearson		Kendall's		Spearman	
		Score	p-value	τ score	p-value	ρ score	p-value
Athens	100	0.98	2.2e−16	0.91	2.2e−16	0.98	2.2e−16
Belem	1360	0.96	2.2e−16	0.85	2.2e−16	0.96	2.2e−16
Bogota	1856	0.97	2.2e−16	0.85	2.2e−16	0.96	2.2e−16
Lima	990	0.97	2.2e−16	0.84	2.2e−16	0.96	2.2e−16
Manila	1027	0.96	2.2e−16	0.84	2.2e−16	0.95	2.2e−16
Quito	154	0.98	2.2e−16	0.89	2.2e−16	0.97	2.2e−16
Santiago	519	0.99	2.2e−16	0.91	2.2e−16	0.98	2.2e−16

4 Conclusions

In this paper we have presented PDU, the new, practical method for quantum error correction suitable for quantum sampling protocols of quantum computing, made without any additional involvement of quantum gates or qubits, made after measurement. We show that our approach, laying in one time determination of PDU and applying it many times until the next physical calibration of the given computer is proper, due to temporal stability of errors and the correction process, which was proved in Sect. 3. Moreover, we have shown the single results of PDU correction on two real quantum computers (Fig. 2). Finally, we have presented that our method increase significantly (by ca. 10% overall) the level of Pearson correlation coefficient, which confirms experimentally effectiveness of our method (Fig. 3).

Surprisingly, from results obtained follows that temporal stability, meaning as the horizontal trend of error volatility over time, is less important for the increase of correlation, which was our initial assumption. Looking on the Fig. 1 and Table 1, we see that e.g., Santiago is temporal stability better then Manila, but increase is smaller. We can also observe that for computers that has higher increase (Belem, Bogota, Manila), the volatility trend is decreasing. Maybe there is some rule in this observation, but it need further investigations.

In conclusion we can say that PDU, the new method of quantum correction is suitable for use in the era of NISQ computers and we prove experimentally significant increase of correlation with reference values of the results corrected with PDU.

References

1. Aaronson, S., Gottesman, D.: Improved simulation of stabilizer circuits. Phys. Rev. A **70**(5), 052328 (2004). https://doi.org/10.1103/PhysRevA.70.052328, arXiv: quant-ph/0406196
2. Bravyi, S., Englbrecht, M., König, R., Peard, N.: Correcting coherent errors with surface codes. npj Quantum Inf. **4**(1), 55 (2018). https://doi.org/10.1038/s41534-018-0106-y
3. Dymarsky, A., Shapere, A.: Quantum stabilizer codes, lattices, and CFTs. J. High Energy Phys. **2021**(3), 160 (2021). https://doi.org/10.1007/JHEP03(2021)160
4. Endo, S., Benjamin, S.C., Li, Y.: Practical quantum error mitigation for near-future applications. Phys. Rev. X **8**(3), 031027 (2018) https://doi.org/10.1103/PhysRevX.8.031027
5. Litinski, D.: A game of surface codes: large-scale quantum computing with lattice surgery. Quantum **3**, 128 (2019). https://doi.org/10.22331/q-2019-03-05-128
6. Lv, J., Li, R., Wang, J.: An explicit construction of quantum stabilizer codes from quasi-cyclic codes. IEEE Commun. Lett. **24**(5), 1067–1071 (2020) https://doi.org/10.1109/LCOMM.2020.2974731, https://ieeexplore.ieee.org/document/9019839/
7. Maciejewski, F.B., Zimborás, Z., Oszmaniec, M.: Mitigation of readout noise in near-term quantum devices by classical post-processing based on detector tomography. Quantum **4**, 257 (2020). https://doi.org/10.22331/q-2020-04-24-257
8. Nguyen, D.M., Kim, S.: Quantum stabilizer codes construction from Hermitian self-orthogonal codes over GF(4). J. Commun. Netw. **20**(3), 309–315 (2018) https://doi.org/10.1109/JCN.2018.000043, https://ieeexplore.ieee.org/document/8437211/
9. Nguyen, D.M., Kim, S.: A novel construction for quantum stabilizer codes based on binary formalism. Int. J. Mod. Phys. B **34**(08), 2050059 (2020) https://doi.org/10.1142/S0217979220500599
10. Ryan-Anderson, C., et al.: Realization of real-time fault-tolerant quantum error correction. Phys. Rev. X **11**(4), 041058 (2021) https://doi.org/10.1103/PhysRevX.11.041058
11. Wereszczyński, K., Michalczuk, A., Pęszor, D., Paszkuta, M., Cyran, K., Polański, A.: Cosine series quantum sampling method with applications in signal and image processing. arXiv:2011.12738 [quant-ph], November 2020

Quantum Approaches for WCET-Related Optimization Problems

Gabriella Bettonte(✉), Valentin Gilbert, Daniel Vert, Stéphane Louise, and Renaud Sirdey

Université Paris-Saclay, CEA List, Gif-sur-Yvette, France
{gabriella.bettonte,valentin.gilbert,daniel.vert2,stephane.louise, renaud.sirdey}@cea.fr

Abstract. This paper explores the potential of quantum computing on a WCET(Worst-Case Execution Time (of a program).)-related combinatorial optimization problem applied to a set of several polynomial special cases. We consider the maximization problem of determining the most expensive path in a control flow graph. In these graphs, vertices represent blocks of code whose execution times are fixed and known in advance. We port the considered optimization problem to the quantum framework by expressing it as a QUBO. We then experimentally compare the performances in solving the problem of classic Simulated Annealing (SA), Quantum Annealing (QA), and Quantum Approximate Optimization Algorithm (QAOA). Our experiments suggest that QA represents a fast equivalent of simulated annealing. Indeed, we measured the approximation ratio on the results of QA and SA, showing that their performances are comparable, at least on our set of simplified problems.

Keywords: WCETs · Quantum computing · QUBO · Combinatorial optimization · Quantum annealing · QAOA

1 Introduction

The interest given to quantum computing primarily comes from the ability of qubits to store a superposition state that reflects all the possible inputs and outputs of a given algorithm until a measurement is performed. This property is called *quantum parallelism* and can, in certain cases [8,21], give a performance boost for quantum algorithms. However, the advantages of quantum computing do not come without caveats. Only some classes of problems can be solved by quantum computing, with a definite efficiency increase compared to classical computing [19]. One crucial research issue related to quantum computing is determining with precision which problems are better solved by means of quantum approaches [20].

This paper explores the potential of quantum computing by examining problems involved with determining the Worst-Case Execution Time (WCET) of a restricted set of programs. The problems arising in WCET evaluation cover a wide range of complexity classes, from undecidability in the general case to

D. Groen et al. (Eds.): ICCS 2022, LNCS 13353, pp. 202–217, 2022.
https://doi.org/10.1007/978-3-031-08760-8_18

NP-hardness and polynomial-time solvability in some restricted cases [25]. As such, WCET evaluation appears to provide a relevant playground to put the quantum computing promise to the test. The execution time of a program running on a machine depends on multiple factors, such as the initial system state, the hardware, and the input data. The validation of a real-time system requires knowing the WCET. However, the analysis of the exact WCET of a program is complicated by dynamic hardware mechanisms. A possible way to cope with the analysis complexity of WCET is by omitting such hardware mechanisms [15].

All the possible combinations of input data and execution paths need to be considered for computing the actual WCET of a program. This computation of an exact WCET is usually unfeasible with classic computers because the number of possible program paths can grow exponentially with the program size (notwithstanding decidability issues in the general case). Classical computing performs static analysis that approximates the WCET without actually executing the program. This approximation has to be an upper bound of the actual value of the worst-case execution time to assure the system's safeness. However, these WCET approximations should also not be overly pessimistic to avoid system over-dimensioning.

Yet, performing an exhaustive examination, even implicitly, of all possible program paths is a *sufficient* condition, even if not necessary, for understanding the worst-case scenario [15]. Thus, quantum computing could allow computing better worst-case-execution-time approximations, thanks to its promising computational power higher than classical computing. In a nutshell, the *program path analysis* method determines which sequence of instructions requires the highest execution time. The problem of determining a program's WCET generally is undecidable. Still, if there are no recursive function calls, dynamic structures, and unbounded loops in the program, it becomes decidable [11,17]. Those strong hypothesis are enforced to be true in the field of real-time embedded systems, which is the primary target of WCET research.

In this paper, as a first step, we focus on a simplified program model: we consider only programs with IF-ELSE conditions (i.e., programs in which statically bounded loops have been unrolled), assuming uninterrupted executions and independence from the hardware. We assume the execution time of an instruction to be a constant: each instruction leads to a cache miss, and all data have to be fetched from the main memory. It is worth noticing that all these hypotheses make the problem solvable in polynomial time. Nevertheless, considering the limitation of existing quantum hardware, we argue that they represent an interesting model to be transposed to the quantum framework for benchmarking and that quantum computing approaches should be also evaluated against polynomial problems [24] rather than only on NP-hard ones [28]: Performing well on the former is presumably necessary to perform well on the latter. Furthermore, distance to optimality is of course easier to measure when attempting to solve polynomial-time problems on quantum hardware.

In particular, we examined well-known approaches for the estimation of worst-case execution times [13,14,22,25]. Using Integer Linear Programming

(ILP) problems, we can naturally describe the structure of our problem and the set of possible program paths, reducing the issue of estimating the WCET of a program into an optimization problem. The sum of the execution time of the executed *basic blocks* gives the cost function. Our goal is to find the maximum of this function. It is possible to adapt this optimization problem to the quantum computing framework through the penalty function method. The cost function and the constraints of the original problem are represented using linear and quadratic terms. Thus, the problem can be reformulated as a QUBO (Quadratic Unconstrained Binary Optimization) problem [4,16] and solved by Quantum Approximate Optimization Algorithm (QAOA) and Quantum Annealing.

In this paper, we solve the problem with different methods: Quantum annealing on D-Wave machines, classical Simulated Annealing and Quantum Approximate Optimization Algorithm.The aim is to compare the performance of these methods (in term of optimization quality). Our tests suggest that D-Wave machines, in ideal cases, achieve the performances of classical Simulated Annealing while being much faster.

The paper is organized as follows. Section 2 provides the necessary preliminaries. Section 3 defines our problem and reformulates it in the form of a QUBO problem. Section 4 describes our evaluation parameters and the machines for the experiments. Section 5 collects the results of our experiments. Section 6 concludes the paper.

2 Combinatorial Optimization

Combinatorial optimization computes the maximum or minimum of a function over a discrete domain. A combinatorial optimization problem is expressed as:

$$z^* = \underset{z \in S}{\text{argmax}} \qquad f(z) \tag{1}$$

$$\begin{cases} g_l(z) = 0, & l = 1, \ldots, L \\ h_m(z) < 0 & m = 1, \ldots, M \end{cases} \tag{2}$$

where z is a discrete integer variables, $f(z)$ is the cost function, and $\mathcal{S}$ the set of decision variables satisfying the equality and inequality constraints given in (2) [26].

An integer linear programming (ILP) formulation is the mathematical formulation of an optimization problem in which variables are restricted to integer values and the constraints and cost function are linear [3]. ILP canonical form is expressed as:

$$\max\ c^T x \tag{3}$$

subject to

$$\begin{cases} Ax = b, \\ 0 \leq x, & x \in \mathbb{Z}^n \end{cases} \tag{4}$$

A Quadratic Unconstrained Binary Optimization (QUBO) problem is a combinatorial problem defined through an upper triangular matrix $Q \in \mathbb{R}^{N\text{x}N}$ and a vector x of binary variables. The goal of the optimization problem is to determine the vector of binary variables $\forall i, x_i \in \{0,1\}$ that minimizes (or maximizes) the objective function :

$$\sum_i Q_{ii}x_i + \sum_{i<j} Q_{ij}x_ix_j. \tag{5}$$

Using the penalty function method, we can rewrite any optimization problem into one without any constraints. For instance, given the equality constaint $g(z) = 0$, we can transform (1) into:

$$z^* = \underset{z}{\text{argmax}} \quad f(z) + \lambda g(z). \tag{6}$$

In this paper, we solve the QUBO problem to find the most expensive execution path by using quantum annealing (QA) and the Quantum Approximate Optimization Algorithm (QAOA). We compare the results with these of Classical Simulated Annealing (SA).

At this point, it is worth noting that any QUBO cost function can be transformed into a generalized 2D-Ising Hamiltonian with a simple transformation of variable $x_i = \frac{1+\sigma_i^z}{2}$ with $\sigma_i^z \in \{-1,1\}$:

$$\mathcal{H}_P = \sum_i^n h_i\sigma_i^z + \sum_{i<j}^n J_{ij}\sigma_i^z\sigma_j^z \tag{7}$$

2.1 Quantum Annealing

Quantum annealing is a computational process that relies on the adiabatic theorem to solve combinatorial optimization problems. As a principle, it implements a time-dependent Hamiltonian composed of an initial Hamiltonian $\mathcal{H}_0$ whose ground state is easy to calculate and a final one tied to the cost function of the optimization problem as seen in Eq. (7): $\mathcal{H}_{\text{Ising}}(t) = A(t)\mathcal{H}_0 + B(t)\mathcal{H}_P$ such that $\mathcal{H}_{\text{Ising}}(0) = \mathcal{H}_0$ and $\mathcal{H}_{\text{Ising}}(\tau) = \mathcal{H}_P$ where τ is the optimal annealing time.

We choose $\mathcal{H}_0 = -\sum_i^n \sigma_i^x$ whose ground state corresponds to an equal superposition of the states of the computational basis. The adiabatic theorem states that if the time evolution is slow enough (i.e., τ is large enough), then the (global) optimal solution can be obtained with high probability. In practice, the final result is conditioned by the size of the spectral gap, the evolution time, the environmental or intrinsic decoherence effects, and the size of the coherence domain of the qubits on the chip.

If the quantum annealing can reach a minimum energy configuration, then the associated state vector solves the equivalent QUBO problem. Reformulating combinatorial problems in QUBO form preserves the underlying structure of the objective function [7].

2.2 Quantum Approximate Optimization Algorithm

The Quantum approximate optimization algorithm (QAOA)[5] is a quantum-classical algorithm used to solve optimization problems. It can be seen as an ansatz for the simulation of the Adiabatic process on a gate-based quantum computer. The approximation of the process is done using a parameter p stating the number of steps for the simulation: A p-depth QAOA consists of alternatively apply the two unitary propagators associated with both Hamiltonians $U(\mathcal{H}_P, \gamma)$ and $U(\mathcal{H}_M, \beta)$ on the initial state $|s\rangle$ which is a uniform superposition of all states of the computational basis.

$$|\psi\rangle = U(\mathcal{H}_M, \beta_p)U(\mathcal{H}_P, \gamma_p)...U(\mathcal{H}_M, \beta_1)U(\mathcal{H}_P, \gamma_1)|s\rangle \tag{8}$$

Each classical optimization round is used to optimize angles $\beta_1..\beta_p$ and $\gamma_1..\gamma_p$ to maximize the expectation value obtained from the run of the quantum circuit.

3 Problem Designing: From Control Flow Graphs to QUBO

We consider a simple micro-architecture [14] such that each basic block B_i of the program takes a constant time c_i to execute. A basic block is defined as a sequence of consecutive instructions. The flow of control enters into the block at the beginning and leaves at the end, without halt or possibility of branching except in the end. Let variable x_i be the execution count of the basic block B_i and N be the number of basic blocks in the program. In this paper, we assume that $x_i \in \{0, 1\}$, meaning that each basic block could be executed only once. The total execution time of the program is given by the linear expression:

$$\text{Program execution time} = \sum_i^N c_i x_i. \tag{9}$$

The problem has intrinsic constraints: not every execution flow is possible. For instance, if there is an IF-ELSE condition in our code, only the block respecting that condition is executed. Thus, one part of the program is ignored, depending on the input data. The WCET is given by the cost of the most expensive flow in terms of execution time. Notice that, in our model, we consider that the solution is unique. Our goal is to find the sequence of variables $x_0, \cdots, x_{N-1}$ such that it maximizes the cost function. To clarify all this, let us consider an example of a program with an IF condition: concretely, we have a sequential path that at some point splits into two branches and then reconnects again to the main path (Fig. 1).

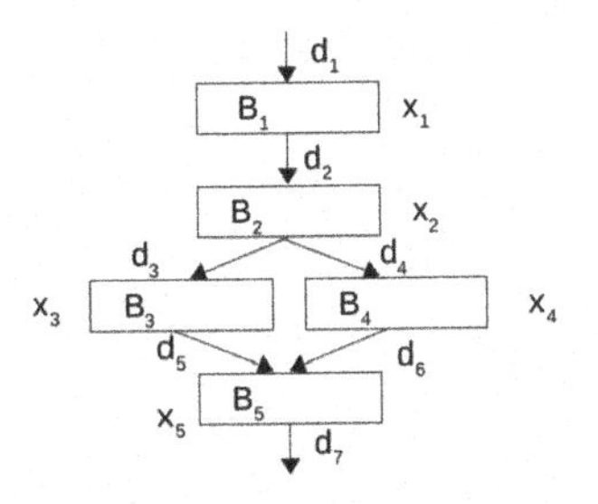

Fig. 1. Control flow diagram of a simple program structure

In this particular example we have only two possible paths of execution and the objective of our optimisation problem is to find the execution path that gives us the maximal cost:

$$\text{argmax} \quad (c_1x_1 + c_2x_2 + c_3x_3 + c_5x_5 \quad , \quad c_1x_1 + c_2x_2 + c_4x_4 + c_5x_5) \tag{10}$$

We call x the vector representing the execution path, with $x_i = 1$ if the corresponding basic block B_i is executed and $x_i = 0$ otherwise. The solution is:

$$x = \begin{cases} \begin{bmatrix} 1 \, 1 \, 1 \, 0 \, 1 \end{bmatrix} & \text{if } (c_1x_1 + c_2x_2 + c_3x_3 + c_5x_5 > c_1x_1 + c_2x_2 + c_4x_4 + c_5x_5) \\ \\ \begin{bmatrix} 1 \, 1 \, 0 \, 1 \, 1 \end{bmatrix} & \text{otherwise.} \end{cases}$$

In this example, the solution would be obtained simply considering the path that includes the most expensive branch of the IF condition. However, let us consider the constraints of this graph to compute the solution. The nodes of this graph are the blocks of code. The edges are the d_is, representing the action of *entering* or the *quitting* of a basic block. This graph represents the execution of a program, so we know for sure that the first and the last block have to be executed, so the edges entering and quitting these nodes will be equal to one. Only one branch of the condition will be executed (if, for instance, $d_3 = 1$, then $d_4 = 0$). For this example, the set of constraints to consider is:

$$\begin{cases} d_1 = 1, \\ x_1 = d_1 = d_2 \\ x_2 = d_2 = d_3 + d_4 \end{cases} \qquad \begin{cases} x_3 = d_3 = d_5 \\ x_4 = d_4 = d_6 \\ x_5 = d_5 + d_6 = d_7 \end{cases} \tag{11}$$

We transform this optimization problem to a QUBO moving the constraints to the cost function (penalty method). We omit the constant values in the cost function and we consider $x_i^2 = x_i, \forall i$ because $x_i \in \{0, 1\}$. Our optimization problem thus becomes:

$$\begin{aligned} &\text{argmax}_x(\textstyle\sum_i^5 c_i x_i - \lambda(1 - x_3 - x_4)^2) \\ &= \text{argmax}_x(\textstyle\sum_i^N c_i x_i - \lambda(1 - x_3^2 - x_4^2 + 2x_3x_4)) \\ &= \text{argmax}_x(c_1x_1^2 + c_2x_2^2 + (c_3 + \lambda)x_3^2 + (c_4 + \lambda)x_4^2 - 2\lambda x_3x_4 + c_5x_5^2)) \end{aligned} \tag{12}$$

Thus, for the considered problem, the matrix Q is:

$$Q = \begin{pmatrix} c_1 & 0 & 0 & 0 & 0 \\ 0 & c_2 & 0 & 0 & 0 \\ 0 & 0 & c_3 + \lambda & -2\lambda & 0 \\ 0 & 0 & 0 & c_4 + \lambda & 0 \\ 0 & 0 & 0 & 0 & c_5 \end{pmatrix} \tag{13}$$

where $c_1, \ldots, c_5 \in \mathbb{N}$ and $x = (x_1, x_2, x_3, x_4, x_5)^T$.

This paper focuses on a finite set of study cases to perform our experiments. The most basic program is not more than a linear sequence of instructions, thus with a deterministic and constant execution time, (Fig. 2(a)). We explored programs made by chains of consecutive IFs (Fig. 2(b)). This problem is interesting because, essentially, any program could be reduced to a loop with a condition in it. So, by unrolling loops, a program could be seen as a chain of conditions. Still, we want to underline that the problem is polynomial: the solution is easily founded by considering the most expensive branch at each IF.

Then we analyzed an expanded version by allowing the exclusive conditions to be more than two SWITCHes (Fig. 2(c)). Here again, the solution is given by the path that takes the most expensive branch at each IF condition, so the problem is still polynomially solvable. However, both examples give us the possibility of building an interesting benchmark for quantum computing.

Enlarging slightly more the focus on the targeted problem, we allowed the IFs blocks to have other nested IFs blocks inside them (Fig. 2(d)). This situation is slightly more complex than the previous ones because the paths need to be enumerated to find the actual worst-case path. All these cases of study are called *series-parallel graphs.*

The IFs chain and the SWITCH study case are easily generalizable from the matrix (13). The nested IF study case is more complex because it is not enough to choose the most expensive branch at each step. Thus, it is not possible to

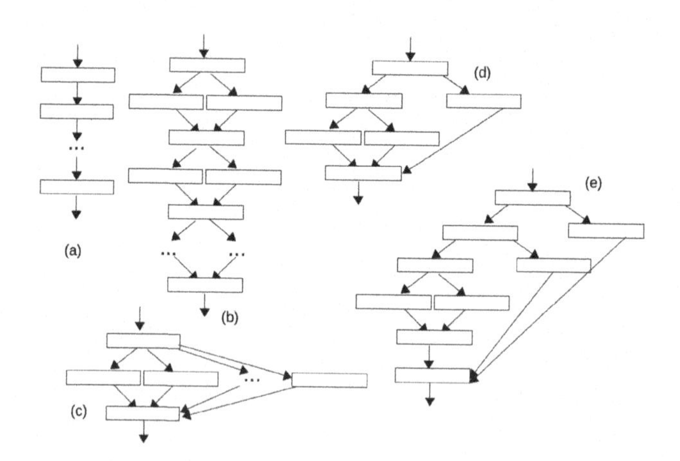

Fig. 2. Series-parallel graphs

determine a pattern. We develop explicitly here the Q matrix only for a particular example. The Q matrix for the graph shown in Fig. 2(d) is:

$$Q = \begin{pmatrix} c_0 & 0 & 0 & 0 & 0 & 0 \\ 0 & c_1 & -2\lambda & 2\lambda & 2\lambda & 0 \\ 0 & 0 & c_2 + 2\lambda & -2\lambda & -2\lambda & 0 \\ 0 & 0 & 0 & c_3 & -4\lambda & 0 \\ 0 & 0 & 0 & 0 & c_4 & 0 \\ 0 & 0 & 0 & 0 & 0 & c_5 \end{pmatrix}$$

3.1 Choice of the Lambda Value

Let $\{c_1, ...c_k\} \in C$ be the set of weights for each node of the graph. $\{x_1, ...x_k\}$ are boolean values defining if the basic block is executed. In the single IF case, the local cost function to the IF case is expressed as $(c_i + \lambda)x_i + (c_j + \lambda)x_j - 2\lambda x_i x_j$ with c_i, c_j the weigths of each path and x_i, x_j the boolean values. The factor λ is appropriate if it follows the set of conditions:

$$\begin{aligned} &\forall\ c_i, c_j \in \mathbb{R}^+ \text{we must have :} \\ c_i + \lambda > c_i + \lambda + c_j + \lambda - 2\lambda &\Leftrightarrow \lambda > c_j \\ c_j + \lambda > c_i + \lambda + c_j + \lambda - 2\lambda &\Leftrightarrow \lambda > c_i \end{aligned} \tag{14}$$

For simplicity, we use a single λ for the whole expression, even if it contains multiple IF cases. We also consider that λ is an integer as each graph weight is integer. Therefore, λ should be greater than each weight of the graph yielding:

$$\lambda = \max(C) + 1 \tag{15}$$

This choice of λ appears to behave well with the nested IFs and SWITCH too.

4 Benchmark Metrics and Computers

We compare the performances of SA, QA and QAOA while solving the optimisation problem of finding the most expensive execution path in the case tests we mentioned above: IF chains, SWITCH and nested IF. This section presents metrics, algorithms and quantum computers used for the benchmark.

4.1 Benchmark Metrics

Problems presented in this paper are toy problems in which optimal solutions can be found in polynomial time. However, their simplicity allows us to fully evaluate and understand the results of our experiments. Before executing the experiments, each of our optimization problems is transposed into their minimization form. To compare our results, we took into consideration two parameters:

- The approximation ratio (16). It represents the quality of the solution found compared to the whole energy landscape of the problem. E is the energy obtained from a single simulation, E_{min} is the energy of the optimal solution and E_{max} the energy of the worst solution.

$$r = \frac{E - E_{max}}{E_{min} - E_{max}} \tag{16}$$

  This parameter is interesting because, although the solution may not be the exact one, it could be very close.
- The optimal solution probability. It represents the proportion of optimal solutions (solutions having the energy E_{min}) obtained during the simulation.

4.2 Simulated Annealing

We configure the simulated annealing with an exponential decrease of temperature $T_1 = 0.95 * T_0$. In the following experiments, we set the thermal equilibrium at $T = 10^{-3}$. These iterations are fixed with the number of input variables n and may vary between $n^{0,5}$ to $n^{1,5}$. We consider SA running in a degraded mode where the number of iterations per temperature step is inferior to n. Each simulation is based on 100 runs of the SA to extract the mean of energy and the probability of getting the optimal solution.

4.3 Quantum Annealing with D-Wave Systems

Our experiments on D-Wave systems involve adiabatic quantum computing used to find the vector that minimizes the input cost function. At the moment, they represent the most advanced quantum machines having thousands of qubits. However, they require problems under the form of QUBOs and the topology of their chips limits their performance [23]. D-Wave systems used during our benchmarks are:

- **DW_2000Q_6** with 2048 qubits (2041 usable qubits) and 6 connections between each qubit (cf. Chimera topology).
- **Advantage_system4.1** with 5760 qubits (5627 usable qubits) and 15 connections between each qubit (cf. Pegasus topology).

For each experiment, results are computed with and without gauge inversion. The principle of a gauge inversion is to apply a Boolean inversion to the σ_i operators in our Hamiltonian. This technique preserves the optimal solution of the problem while limiting the effect of local biases of the qubits, as well as the machine accuracy errors [2]. Following the commonly used procedure (e.g. [1]), we randomly selected 10% of the physical qubits used as spin inversion for each instance. Each simulation is based on 1000 runs of D-Wave systems to extract the mean of energy and the optimal solution probability.

4.4 Simulation of QAOA

The simulation of QAOA is performed using the Qiskit library [6]. QAOA circuits are built from QUBO instances using penalty terms to express constraints. In our experiments, weights are specified with integers, hence $\gamma \in [0, 2\pi]$ and $\beta \in [0, \pi]$. We did not find patterns to perform interpolation optimization as in [27]. We followed the *parameter fixing strategy* [12] to set angles at p-depth. This method starts at $p = 1$ and randomly generates 100 pairs of angles γ_1 and β_1. Then, we run a local optimizer on each of these pairs. We used COBYLA, a gradient-free optimizer. We run 1024 times the QAOA circuit at each optimization step to sample the mean expectation value corresponding to γ and β angles. At the end of the 100 optimization loops, we get 100 optimized pairs of angles. We select γ_1^* and β_1^* such as they minimize the value of the cost function. This process is then repeated at $p = n$, initializing the problem with $\gamma_1^* ... \gamma_{n-1}^*$ and $\beta_1^* ... \beta_{n-1}^*$ and 100 pairs of angles γ_n and β_n. For the simulation of the QAOA we used the *aer_simulator*, which provides a good speed performance.

5 Experimental Results

To represent the cost of each basic block, we generated random integer, such that $c_0, ..., c_n \in \{1, \ldots, 50\}$, and used them as input for our experiments. Each problem is designed to have only a single optimal solution, meaning that each branch candidate for solution should have a different global cost. In Sects. 5.1 and 5.2, each data point is smoothed over 30 randomly generated instances.

5.1 IF Chains

Each IF chain (Fig. 2(a)) is composed of a succession of several IF conditions. This study case represents an ideal problem for the D-Wave as the corresponding QUBO matrix is sparse. Hence, the encoding on D-Wave systems does not require any duplication of qubits, neither for the 2000Q_6 nor the Advantage4.1 system. We start from one IF condition for this benchmark and grow the problem up to 10 IF conditions. A single block separates each IF condition block. We benchmark in Fig. 3 D-Wave systems against the resolution with classical simulated annealing. As the problem grows, QA seems to outperform SA progressively. For the IF chain, D-Wave system 2000_Q_6 systematically outperforms the Advantage_system4.1. Although the Advantage system is more recent than the 2000Q_6 system, it seems more sensitive to noise. We do not simulate QAOA for IF chains as it gives rather poor results compared to SA and D-Wave systems.

Figure 4 shows the QAOA energy landscape at $p = 1$ for a 3-if chain picked randomly. The heatmap exhibits many local minima, which impact the angle optimization of QAOA. This heatmap is more complex than heatmaps usually obtained for the well-studied Max-cut problem ([12],Fig. 5a). This complex energy landscape seems to be closely related to penalty terms that impact the whole energy landscape. Moreover, minimizing the mean energy does not always lead to an increased probability of getting the optimal solution.

5.2 SWITCH

SWITCHes presented in Fig. 2(b) are harder to solve for D-Wave system since the plurality of choices leads to a dense matrix and requires higher connectivity between qubits. The impact of the density on QAOA is lower thanks to SWAP gates. In this execution case, we notice that the D-Wave Advantage4.1 performs better than the D-Wave 2000Q_6 for SWITCH cases 3 and 4. This improvement is due to the qubit duplication occurring on D-Wave 2000Q_6 whereas there is no qubit duplication on the Advantage4.1 for these instances (Fig. 5e). On larger instances (from 10 to 15 SWITCHes), the advantage of the D-Wave Advantage4.1 provided by its number of connections is questionable. However, when the case is not ideal for D-Wave systems, the performances are poor against SA, even when SA is running under a degraded mode ($n^{0.5}$ iterations per temperature step).

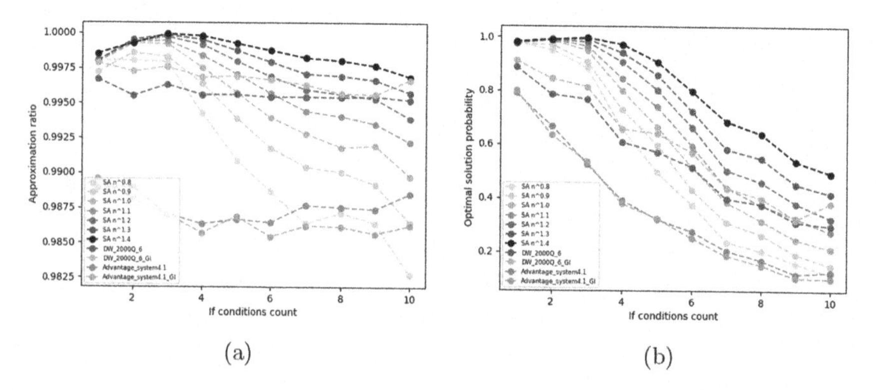

(a) (b)

Fig. 3. Benchmark of the QUBO resolution by D-Wave systems and SA for IF chains from 1 to 10 IF conditions. Each D-Wave simulation is done with and without GI (Gauge Inversion). n is the number of blocks in the if chain. SA number of iteration per temperature step is expressed according to n. a) Mean of approximation ratio. b) Mean of the probability to get the optimal solution.

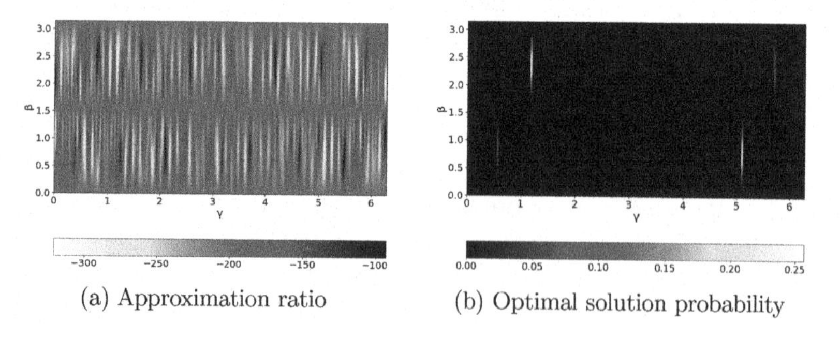

(a) Approximation ratio (b) Optimal solution probability

Fig. 4. QAOA heatmap at $p = 1$ for a random 3 IFs chain. Approximation ratio is computed from the mean of the expectation value at angles γ and β.

5.3 Nested IFs

Nested IFs cases increase the density of the QUBO matrix. Table 1 shows two cases of nested IFs. The results are smoothed over 30 costs generated randomly. We compare the results obtained with SA, QA, and QAOA simulator. As for SWITCH cases, nested IF cases are difficult to be solved for D-Wave systems.

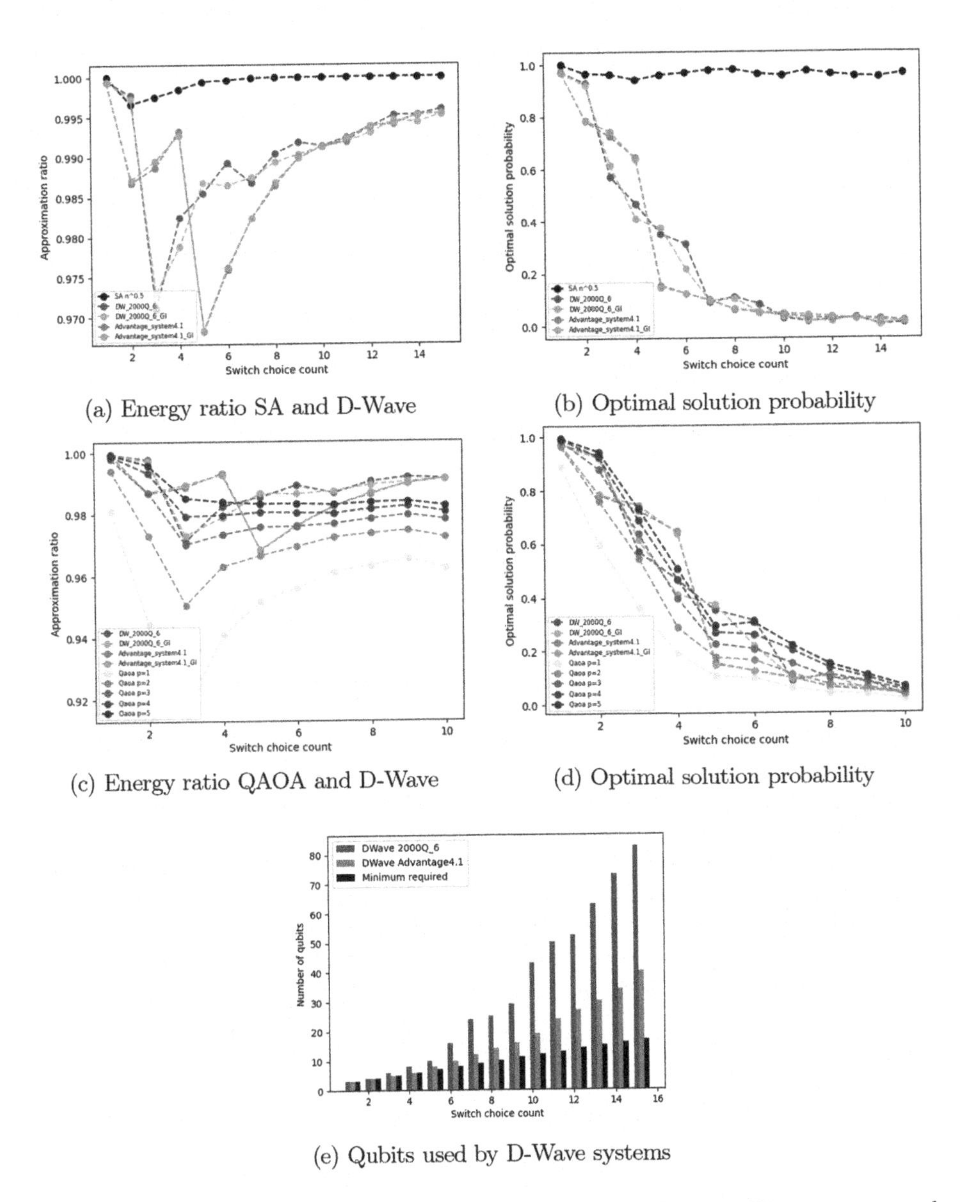

(a) Energy ratio SA and D-Wave

(b) Optimal solution probability

(c) Energy ratio QAOA and D-Wave

(d) Optimal solution probability

(e) Qubits used by D-Wave systems

Fig. 5. Comparison of switch instances solved with classical SA, D-Wave systems and QAOA simulator. (e) Mean of the number of qubits used by D-Wave systems to solve switch instances against the optimal number of qubits on ideal fully connected topology.

They result to be not competitive against simulated annealing. QAOA simulations provide results of a lower performance compared to the D-Wave systems. The results obtained with QAOA start to decrease at $p = 6$. This decrease may be due to the difficulty of COBYLA to optimize γ and β angles at this depth. As the optimization at depth p always starts from angles $\beta_1^*...\beta_{p-1}^*$ and $\gamma_1^*...\gamma_{p-1}^*$, the global optimization may be stuck at local minima imposed by the parameters found at $p-1$. The energy landscape of problems with penalty terms would deserves in-depth study to understand and find patterns of optimization of QAOA angles.

Table 1. Nested IFs simulation

Nested if results	Case 1 (see Fig. 2d)		Case 2 (see Fig. 2e)	
	Mean energy	Opt sol prob	Mean energy	Opt sol prob
SA $n^{0.5}$	0.985	0.66	0.999	0.934
SA $n^{0.6}$	0.986	0.69	0.999	0.923
SA $n^{0.7}$	0.989	0.76	0.999	0.956
D-Wave 2000Q	0.985	0.57	0.993	0.56
D-Wave Advantage	0.983	0.54	0.975	0.29
QAOA $p = 4$	0.958	0.44	0.909	0.10
QAOA $p = 5$	0.965	0.49	0.920	0.13
QAOA $p = 6$	0.926	0.40	0.855	0.06

5.4 Real Case

We provide a concrete application inspired by the bubble sort algorithm provided in the Mälardalen WCET research group [9][18]. The algorithm's goal is to sort an array of integers in ascending order. We consider an ideal scenario with an in-order, single-issue Arm processor similar to an M_0 with prefetched cache memory. Blocks composing the graph are built from the instructions obtained after the compilation of C code (using *gcc* Arm 8.3 with *-O2* level of optimization). The number of micro-instruction in each basic block defines its cost as given by our in-house WCET explorer with block identification and pipeline simulation. For the sake of simplicity, we limited to 3 the number of elements in the vector to sort. Table 2 shows the obtained result while solving the problem with SA and QA. From Table 2 we conclude that the 2000Q system performs quite well, even duplicating 4 qubits. The Advantage system, while duplicating only 1 qubits performs poorly. We may conclude that duplicating on Advantages machines downgrade drastically the performances of the machine.

Table 2. Real case simulation

Method	Mean energy	Opt sol prob	Qubits used	Max qubits duplication
SA $n^{0.5}$	0.944	0.58	/	/
SA n^1	0.972	0.79	/	/
SA $n^{1.1}$	0.990	0.91	/	/
D-Wave 2000Q_6	0.987	0.868	15	1
D-Wave Advantage4.1	0.901	0.325	12	1

6 Conclusion

In this paper, we explored the potential of quantum computing in dealing with the combinatorial optimization problem of finding the most expensive execution path in a graph on a restricted set of simple instances. Our work offers an example of using quantum computing in a new application field. The performances we obtained using D-Wave machines and the QAOA do not suggest that quantum computing is the silver bullet solution that will replace the classical computing solid and road-tested techniques. Still, we claim the results we obtained are encouraging enough to explore further the potential of quantum computing on the proposed problem (for instance, the backtracking quantum algorithm). Indeed, our experiments suggest that quantum computing gives us a fast equivalent of simulated annealing for ideal problems, such as the IFs chains.

We can draw some conclusions from the experience of this paper on D-Wave machines. Their topology is quite limiting, and it is not straightforward to adapt the considered optimization problem to it. We noticed that the Advantage4.1 machine outperforms the 2000Q machine when the considered problems involve qubit duplications on 2000Q device and not on the Advantage4.1 machine. It may be worth exploring the performances of these machines while solving problems that need more qubit duplications to be adapted to both topologies. The goal is to find parameters that show a priori which D-Wave machine is the most suitable to solve the considered problem. The chain of IFs is ideal for the topology of D-Wave systems, and their results are competitive with SA. Concretely, SA on 5600 variables at n^2 would need billions of evaluations of the cost function against millisecond runtime for D-Wave systems. However, we stress again that the problem is polynomially solvable using methods other than SA, and the interest is the possibility of building a benchmark for quantum computers.

We can as well draw some conclusions about QAOA. At $p = 5$ the performances of QAOA are almost the same as D-Wave. QAOA does not have the problem of qubit duplication since we can circumvent the chip's topology with SWAP gates. Still, we performed our experiments on a simulator and not on a physical machine as D-Wave systems. An idea to investigate is to restrict the search on the feasible subspace of the problem [10]. In our case, this would be a subspace where every solution preserves the flow constraint.

Another aspect that may be explored in the future is the behavior of our models considering the effect of cache memories. Additionally, our choice of λ has no guarantees of being optimal, even if it provides good results. As a perspective, it may be interesting to perform a pre-processing to find, through simulations, the optimal λ. However, the effort of finding the optimal λ could overcome its benefits.

Acknowledgements. The authors acknowledge financial support from the AIDAS project of the Forschungszentrum Julich and CEA.

References

1. Albash, T., Lidar, D.: Demonstration of a scaling advantage for a quantum annealer over simulated annealing. Phys. Rev. **X**, 8 (2018)
2. Boixo, S., Albash, T., Spedalieri, F.M., Chancellor, N., Lidar, D.A.: Experimental signature of programmable quantum annealing. Nat. Commun. **4**, 2067 (2013)
3. Hax, B., Magnanti, T.: Applied mathematical programming (1977)
4. Chang, C.C., Chen, C.C., Koerber, C., Humble, T.S., Ostrowski, J.: Integer programming from quantum annealing and open quantum systems (2020)
5. Farhi, E., Goldstone, J., Gutmann, S.: A quantum approximate optimization algorithm. arXiv preprint arXiv:1411.4028 (2014)
6. Gambetta, J., et al.: Qiskit: an open-source framework for quantum computing (2022)
7. Glover, F., Kochenberger, G., Du, Y.: A tutorial on formulating and using QUBO models. arXiv preprint arXiv:1811.11538 (2018)
8. Grover, L.K.: A fast quantum mechanical algorithm for database search (1996)
9. Gustafsson, J., Betts, A., Ermedahl, A., Lisper, B.: The Mälardalen WCET benchmarks - past, present and future. In: Proceedings of the 10th International Workshop on Worst-Case Execution Time Analysis (2010)
10. Hadfield, S., Wang, Z., O'Gorman, B., Rieffel, E., Venturelli, D., Biswas, R.: From the quantum approximate optimization algorithm to a quantum alternating operator ansatz. Algorithms **12**(2), 34 (2019)
11. Kligerman, E., Stoyenko, A.D.: Real-time Euclid: a language for reliable real-time systems. IEEE Trans. Softw. Eng. **SE-12**(9), 941–949 (1986)
12. Lee, X., Saito, Y., Cai, D., Asai, N.: Parameters fixing strategy for quantum approximate optimization algorithm. In: 2021 IEEE International Conference on Quantum Computing and Engineering (QCE). IEEE (2021)
13. Li, Y.-T., Malik, S., Wolfe, A.: Efficient microarchitecture modeling and path analysis for real-time software. In: Proceedings 16th IEEE Real-Time Systems Symposium, pp. 298–307 (1995)
14. Li, Y.-T., Malik, S., Wolfe, A.: Cache modeling for real-time software: beyond direct mapped instruction caches. In: 2011 IEEE 32nd Real-Time Systems Symposium, vol. 254 (1996)
15. Liu, J.-C., Lee, H.-J.: Deterministic upperbounds of the worst-case execution times of cached programs. In: 1994 Proceedings Real-Time Systems Symposium, pp. 182–191 (1994)
16. Lucas, A.: Ising formulations of many NP problems. Front. Phys. **2**, 5 (2014)
17. Puschner, P., Koza, C.: Calculating the maximum execution time of real-time programs. Real-Time Syst. **1**, 159–176 (1989)
18. M.W. Research Group: Wcet benchmarks. http://www.mrtc.mdh.se/projects/wcet/benchmarks.html

19. Rønnow, T.F., et al.: Defining and detecting quantum speedup. Science **345**(6195), 420–424 (2014)
20. Rønnow, T.F., et al.: Defining and detecting quantum speedup. Science **345**, 420–424 (2014)
21. Shor, P.W.: Polynomial-time algorithms for prime factorization and discrete logarithms on a quantum computer. SIAM J. Comput. **26**(5), 1484–1509 (1997)
22. Theiling, H., Ferdinand, C.: Combining abstract interpretation and ILP for microarchitecture modelling and program path analysis. In: Proceedings 19th IEEE Real-Time Systems Symposium (Cat. No. 98CB36279), pp. 144–153 (1998)
23. Vert, D., Sirdey, R., Louise, S.: On the limitations of the chimera graph topology in using analog quantum computers. In: Proceedings of the 16th ACM International Conference on Computing Frontiers, pp. 226–229. ACM (2019)
24. Vert, D., Sirdey, R., Louise, S.: Benchmarking quantum annealing against "hard" instances of the bipartite matching problem. SN Comput. Sci. **2**, 106 (2021)
25. Wilhelm, R., et al.: The worst-case execution-time problem-overview of methods and survey of tools. ACM Trans. Embed. Comput. Syst. (2008)
26. Zaman, M., Tanahashi, K., Tanaka, S.: Pyqubo: Python library for mapping combinatorial optimization problems to QUBO form. CoRR, abs/2103.01708 (2021)
27. Zhou, L., Wang, S.T., Choi, S., Pichler, H., Lukin, M.D.: Quantum approximate optimization algorithm: Performance, mechanism, and implementation on near-term devices. Phys. Rev. **X**, 10(2) (2020)
28. Salehi, Ö., Glos, A., Miszczak, J.A.: Unconstrained binary models of the travelling salesman problem variants for quantum optimization. Quantum Inf. Process. **21**(2), 1–30 (2022)

A First Attempt at Cryptanalyzing a (Toy) Block Cipher by Means of QAOA

Luca Phab[1], Stéphane Louise[2(✉)], and Renaud Sirdey[2]

[1] Université Paris-Saclay, Gif-sur-Yvette, France
luca.phab@tutanota.com
[2] Université Paris-Saclay, CEA, List, 91120 Palaiseau, France
{stephane.louise,renaud.sirdey}@cea.fr

Abstract. The discovery of quantum algorithms that may have an impact on cryptography is one of the main reasons of the rise of quantum computing. Currently, all quantum cryptanalysis techniques are purely theoretical and none of them can be executed on existing or near-term quantum devices. So, this paper investigates the capability of already existing quantum computers to attack a toy block cipher (namely the Heys cipher) using the Quantum Approximate Optimization Algorithm (QAOA). Starting from a known-plaintext key recovery problem, we transform it into an instance of the MAX-SAT problem. Then, we propose two ways to implement it in a QAOA circuit and we try to solve it using publicly available IBM Q Experience quantum computers. The results suggest that the limited number of qubits requires the use of exponential algorithms to achieve the transformation of our problem into a MAX-SAT instance and, despite encouraging simulation results, that the corresponding quantum circuit is too deep to work on nowadays (too-) noisy gate-based quantum computers.

Keywords: Cryptography · Quantum computing · QAOA · MAX-SAT · Block cipher

1 Introduction

Quantum computing offers a new paradigm that can solve certain problems much more efficiently than classical computing. At the same time, a large part of modern cryptography is precisely based on the difficulties to solve specific problems that are conjectured hard to solve with classical computing.

On one hand, Shor's algorithm [16] can solve factorization and discrete logarithm problems, that are of huge importance in cryptography. On the other hand, Grover's algorithm [8] can be utilized as a quantum brute-force algorithm that is much more efficient than the classical brute-force (but remains exponential). Hence, quantum computing may be a serious threat to cryptography and that is why a lot of research is conducted on quantum cryptanalysis. For a few years, we have seen an emerging field on quantum non-black box cryptanalysis [4–6,11–13], meaning that those approaches are cipher specific by exploiting

D. Groen et al. (Eds.): ICCS 2022, LNCS 13353, pp. 218–232, 2022.
https://doi.org/10.1007/978-3-031-08760-8_19

their internal structure. Yet all those works assume a large scale and noise-free quantum computer that does not currently exist.

In this context, this paper investigates (for the first time) whether Noisy Intermediate Scale Quantum (NISQ) machines may have an impact in cryptanalysis of symmetric ciphers. To that end, we will study an attack on a toy cipher, namely the Heys cipher, using the Quantum Approximate Optimization Algorithm (QAOA) that is expected to be less sensitive to decoherence due to its hybrid nature. The proposed attack can be summarized as follows: We first need a plaintext and the corresponding ciphertext. Then, we transform it into an instance of a combinatorial optimization problem. Finally, we execute the quantum algorithm to solve the instance and if an optimal solution is found then we can deduce a key that can encrypt the plaintext into the ciphertext.

This paper is organized as follows: In Sect. 2, we review QAOA. Section 3 provides some background on cryptography and describes the studied toy cipher and the attack steps. Then, in Sect. 4, we show how to build the quantum circuit and indicate its complexity according to the basis gates of the quantum hardware utilized. Finally, Sect. 5 details the implementation and the experimental results before concluding.

2 Quantum Approximate Optimization Algorithm

QAOA [7] is an algorithm created in 2014 by Farhi et al. that approximates solutions of combinatorial optimization problems. It is a quantum circuit whose parameters are optimized through a classical optimization algorithm. Its hybrid nature allows this algorithm to limit the depth of the quantum circuit, hence, it seems relevant as an algorithm of choice to solve our problem on NISQ devices.

It is based on a well-known quantum mechanics theorem, called the "adiabatic theorem", stating that a quantum system in a ground state for a hamiltonian will remain in a ground state for that hamiltonian if it changes slowly enough over time.

Let H be the hamiltonian such that:

$$H(t) = (1 - s(t)) \cdot H_D + s(t) \cdot H_P \tag{1}$$

where $s : [0, T] \rightarrow [0, 1]$ is a smooth function with $s(0) = 0$ and $s(T) = 1$, H_D is the "driver hamiltonian" that has an easy-to-build ground state and H_P is the "problem hamiltonian" whose ground states encode the optimal solutions of our problem. So, according to the adiabatic theorem, a quantum system in an easy-to-build ground state for H_D will evolve into a ground state for H_P, which encodes optimal solutions of the problem, after waiting a time T.

Let C be the cost function that we want to minimize and which takes n binary variables as input. The corresponding problem hamiltonian can be chosen as:

$$H_P = \sum_{x \in \{0,1\}^n} C(x) \left|x\right\rangle \left\langle x\right| \tag{2}$$

QAOA simulates the approximate time evolution of a quantum system for the hamiltonian H starting from one of its ground states, using n qubits.

Based on the Schrödinger equation and the Trotter-Suzuki formula [17], the time evolution of the quantum system $|\psi\rangle$ for the hamiltonian H is approximated by:

$$|\psi(t)\rangle \approx \prod_{j=1}^{p} \exp\left(-\frac{i\Delta t}{\hbar}(1 - s(j\Delta t))H_D\right) \cdot \exp\left(-\frac{i\Delta t}{\hbar} s(j\Delta t)H_P\right) |\psi(0)\rangle \quad (3)$$

whose precision as an approximation is improved by increasing p.

Let $U_{H_P}(\gamma) = \exp(-i\gamma H_P)$ and $U_{H_D}(\beta) = \exp(-i\beta H_D)$ be unitary operators, the QAOA circuit computes the following state:

$$|\vec{\beta}, \vec{\gamma}\rangle = U_{H_D}(\beta_{p-1})U_{H_P}(\gamma_{p-1}) \cdots U_{H_D}(\beta_0)U_{H_P}(\gamma_0) |\psi(0)\rangle \quad (4)$$

where $\vec{\beta} = (\beta_0, ..., \beta_{p-1}) \in [0, 2\pi[^p$ and $\vec{\gamma} = (\gamma_0, ..., \gamma_{p-1}) \in [0, 2\pi[^p$. We do not know the values of p, $\vec{\beta}$ and $\vec{\gamma}$ such that $|\vec{\beta}, \vec{\gamma}\rangle$ is a ground state for H_P so we try to find them empirically, at constant p, by minimizing the function:

$$\begin{aligned} F_p : [0, 2\pi[^p \times [0, 2\pi[^p &\rightarrow \mathbb{R} \\ (\vec{\beta}, \vec{\gamma}) &\mapsto \langle \vec{\beta}, \vec{\gamma}| H_P |\vec{\beta}, \vec{\gamma}\rangle \end{aligned} \quad (5)$$

using a classical optimizer. That function calculates the average of the eigenvalues of H_P weighted by the probability distribution of states of $|\vec{\beta}, \vec{\gamma}\rangle$ and its minimum is reached exactly when the superposition is in a ground state for H_P. It can be computed with the probability distribution that we can approximate by executing several times the quantum circuit and measuring the resulting final state:

$$\forall x \in \{0,1\}^n, \ \mathbb{P}_{\vec{\beta},\vec{\gamma}}(\text{measuring state } |x\rangle) \approx \frac{k_x}{k} \quad (6)$$

where k is the number of executions and k_x is the number of times that the state $|x\rangle$ was measured.

Finally, we just need to compute the quantum circuit with the parameters $\vec{\beta}^*$ and $\vec{\gamma}^*$, found by optimization, and then to measure the superposition to get a state that encodes a solution z where $C(z)$ is close to $\min_x C(x)$.

3 Problem Statement

3.1 Cryptography Background

Cryptography is all around us, especially when we use our communication devices, as smartphones or computers. It aims to ensure the security properties of an exchange, which are confidentiality, integrity, authentication and non-repudiation. To protect the confidentiality of a message, the sender utilizes a cipher that transforms (encrypts) the message (plaintext) into an unintelligible one (ciphertext), sends it on a channel, that can be public, to the recipient, who applies the inverse of the transformation (decrypts) to obtain the original message. An extra parameter (key) is needed for the encryption and decryption functions.

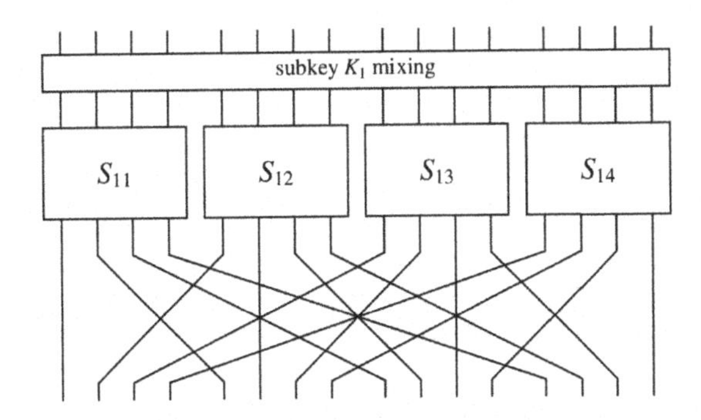

Fig. 1. First round of the Heys cipher.

In a symmetric cipher, as studied in the present work, the encryption key and the decryption key are computed from a secret shared between the sender and the recipient. Among them, a large part are iterated block ciphers, that encrypt a predefined fixed-size plaintext (block) into a ciphertext with the same size, using the repetition of several operations. Each iteration (round) needs a key (round key) derived from a master key. The number of rounds is a significant parameter of the cipher because increasing it generally improves its security but reduces its efficiency, which is also an important criterion. Doing a cryptanalysis on a reduced number of rounds is an usual way to study that type of ciphers, which is easier and that can give insights on its security.

In this paper, the proposed cryptanalysis is a known-plaintext attack, meaning that the attacker has access to a set of plaintext-ciphertext pairs, which is a realistic possibility. Indeed, that kind of attack was already utilized previously, during World War II for instance. Allies guessed the content (or partial content) of some messages (mainly weather forecasts) and utilized that to break the German cipher Enigma [19].

3.2 Heys Cipher

The Heys cipher [9] is an iterated block cipher created by Howard M. Heys for educational purposes to teach linear and differential cryptanalysis. It is a toy substitution-permutation network (SPN) with a block size of 16 bits and where each of its rounds is composed of:

- a mixing operation between the round key and the current block
- a substitution
- a permutation

The first round is described in Fig. 1. The cipher finishes with a new mixing operation which enables the decryption function to be as similar as the encryption function. That final step can be considered as a round because it requires the utilization of a round key. The mixing operation is commonly an exclusive or (xor). The cipher has the advantage to be particularly simple and has the same construction as a lot of ciphers in practical utilization currently or formerly, like e.g. DES or AES.

3.3 Principle of the Attack

After getting a plaintext and the corresponding ciphertext for which we want to find the key utilized for the encryption, we transform them into an instance of the SAT problem, namely a propositional formula, where the variables embody the bits of the sought key. If we had an oracle solving the SAT problem, we could utilize it to find an assignment satisfying the formula and thus deduce a key that can encrypt the plaintext into the ciphertext. But, as SAT is NP-complete, all known classical algorithms cannot solve all formula polynomially. In particular, formulas that come from cryptanalytic attacks can be expected to be difficult to solve (otherwise, the underlying primitive would have significant security issues). Among quantum algorithms, QAOA is a very promising one that could run on NISQ machines, so we convert the formula into an equivalent formula in Conjunctive Normal Form (CNF), which is an instance of the combinatorial optimization version of the SAT problem (MAX-SAT). As our formula is satisfiable by construction, an optimal solution is an assignment satisfying the formula, so the algorithm will act as the required oracle. It is important to notice that we need to find algorithms to convert a plaintext-ciphertext pair that do not add too many auxiliary variables, given that the number of qubits required depends on the number of binary variables of our problem and that the number of qubits of a quantum machines (real devices or simulators) is very limited.

4 Retrieving the Encryption Key from a Plaintext-Ciphertext Pair Using QAOA

4.1 Plaintext-Ciphertext Pair into Conjunctive Normal Form Formula (CNF) Conversion

The algorithm aim is to transform a plaintext and the corresponding ciphertext into a propositional formula where the variables embody the utilized cipher key bits (without adding any auxiliary variables). It is based on a theorem saying that a formula where we substitute variables with formulas remains a formula. So, we start with a simple formula and we substitute variables with formulas, as it goes along the cipher operations, in the current formula.

Let $(p_i)_{i\in\{1,\ldots,16\}}$ be the plaintext bits, $(c_i)_{i\in\{1,\ldots,16\}}$ the ciphertext bits and $(k_i)_{i\in\{1,\ldots,16\}}$ the key bits. The starting formula is:

$$\psi_{init} = \bigwedge_{i=1}^{16} (p_i \leftrightarrow x_i) \tag{7}$$

Then, for each operation, some variables of the current formula are replaced with formulas according to the operation type. We denote $(a_i)_{i\in\{1,\ldots,16\}}$ the input variables of an operation and $(b_i)_{i\in\{1,\ldots,16\}}$ the output variables.

J-Th Xor: For all $i \in \{1, ..., 16\}$, we have $b_i = a_i \oplus k_{16(j-1)+i}$, hence $a_i = b_i \oplus k_{16(j-1)+i}$. So, variables $(x_i)_{i\in\{1,\ldots,16\}}$ in the current formula are substituted with formulas $(\phi_{1,j,i})_{i\in\{1,\ldots,16\}}$ as follows:

$$\phi_{1,j,i} = y_i \oplus k_{16(j-1)+i} \equiv \left(\left(y_i \vee k_{16(j-1)+i}\right) \wedge \left(\neg y_i \vee \neg k_{16(j-1)+i}\right)\right) \tag{8}$$

Table 1. (a) Table representing the inputs and outputs of a random sbox on 2 bits. (b) Table representing the possible outputs of the sbox described in (a) where $a_1 = 1$ and the corresponding formulas.

(a)

input (a_2a_1)	00	01	10	11
output (b_2b_1)	10	00	01	11

(b)

output	formula
00	$\psi_{00} = (\neg z_1 \wedge \neg z_2)$
11	$\psi_{11} = (z_1 \wedge z_2)$

***S_{jk}* Substitution:** The S_{jk} substitution acts on bits of index from $4(k-1)+1$ to $4(k-1)+4$. The replacement formula $\phi_{2,j,i}$ where $i = 4(k-1)+h$, $h \in \{1, 2, 3, 4\}$ is a disjunction of conjunctions created from the outputs for which the bit of index i in input is equal to 1.

For example, suppose that we want to construct the replacement formula $\phi_{2,j,i}$ from the sbox[1] described in Table 1(a) for $i = 1$. The outputs where $a_1 = 1$ in input are "00" and "11". For each of those outputs, the corresponding formula, as shown in Table 1(b), is the conjunction of $(l_s)_{s\in\{1,2\}}$ such that:

$$l_s = \begin{cases} z_s & \text{if } b_s = 1 \\ \neg z_s & \text{otherwise} \end{cases} \tag{9}$$

That way, if $a_1 = 1$ then one of the formulas ψ_{00} or ψ_{11} must be True thus the replacement formula is:

$$\phi_{2,j,1} = \psi_{00} \vee \psi_{11} \tag{10}$$

***σ* Permutation:** For all $i \in \{1, ..., 16\}$, we have $b_i = a_{\sigma(i)}$, hence $a_i = b_{\sigma^{-1}(i)}$. So, variables $(z_i)_{i\in\{1,\ldots,16\}}$ are replaced with formulas $(\phi_{3,j,i})_{i\in\{1,\ldots,16\}}$ as follows:

$$\phi_{3,j,i} = x_{\sigma^{-1}(i)} \tag{11}$$

[1] A sbox (substitution box) is an algorithm component that compute a substitution in a cipher.

Finally, after substituting the variables for all cipher operations, all it is required is to replace $(x_i)_{i\in\{1,...,16\}}$ according to the value of the ciphertext bits to obtain a formula where the only variables embody the key bits.

It is a naive algorithm where the size of the output formula increases exponentially with the number of cipher rounds. That is due to the substitution operations where the linked replacement formulas contain many multiple copies of the same variable. However, with a number of rounds small enough, the size of the formulas remains acceptable.

Then, the formula must be in conjunctive normal form to be an instance of the MAX-SAT problem. There are two main algorithms to achieve that transformation. The first one [10] use the De Morgan laws but the output formula grows exponentially with the size of the input formula, while the second, called Tseitin transformation [18], is polynomial but add auxiliary variables. We chose the algorithm that does not increase the number of variables (and thus the number of qubits needed by the quantum devices) to be able to perform experiments on real hardware on our instances.

It is worth noting that we probably could transform the plaintext-ciphertext pair directly into a CNF by introducing auxiliary variables as in [14] where the output formula has a polynomial number of variables and clauses.

4.2 Solving the MAX-SAT Problem Using QAOA

Let φ be a propositional formula in conjunctive normal form and $P_\varphi = \{x_1, ..., x_n\}$ the set of the variables in φ such that:

$$\varphi = \bigwedge_{j=1}^{m} C_j = \bigwedge_{j=1}^{m} \left(\bigvee_{k=1}^{m_j} l_{j,k} \right) \tag{12}$$

where $l_{j,k} = x_{j,k}$ or $l_{j,k} = \neg x_{j,k}$. We denote $M = \max_{j\in\{1,...,m\}} m_j$ the maximum number of literals in a clause. An assignment of n variables is embodied by $z = (z_1, ..., z_n) \in \{0,1\}^n$ such that:

$$\forall l \in \{1, ..., n\},\ (z_l = 1 \Leftrightarrow x_l = True) \text{ and } (z_l = 0 \Leftrightarrow x_l = False) \tag{13}$$

The cost function associated with φ is defined as follows:

$$C(z) = -\sum_{j=1}^{m} C_j(z) \text{ where } C_j(z) = \begin{cases} 1 \text{ if } z \text{ satisfies } C_j \\ 0 \text{ otherwise} \end{cases} \tag{14}$$

and can be transformed into a problem hamiltonian such that:

$$H_P = \sum_{x\in\{0,1\}^n} C(x) |x\rangle \langle x| = -\sum_{j=1}^{m} \hat{C}_j \text{ where } \hat{C}_j = \sum_{x\in\{0,1\}^n} C_j(x) |x\rangle \langle x| \tag{15}$$

$$= -\sum_{j=1}^{m} (I - \frac{1}{2^{m_j}} \prod_{k=1}^{m_j} (I + \varepsilon_{j,k} \cdot \sigma_{j,k}^z)) \text{ where } \varepsilon_{j,k} = \begin{cases} -1 \text{ if } l_{j,k} = \neg x_{j,k} \\ 1 \quad \text{otherwise} \end{cases} \tag{16}$$

Using the definition of H_P in Eq. 15 and the Trotter-Suzuki formula, U_{H_P} can be written as the product:

$$U_{H_P}(\gamma) = \prod_{j=1}^{m} U_{-\hat{C}_j}(\gamma) \tag{17}$$

So, the quantum circuit of $U_{H_P}(\gamma)$ is composed of the quantum circuits of $U_{-\hat{C}_j}(\gamma)$ with $j \in \{1, ..., m\}$, in any order since the $\hat{C}_j$ commute. For all $j \in \{1, ..., m\}$, we can note that:

$$U_{-\hat{C}_j}(\gamma)\,|z\rangle = \exp\left(i\gamma\hat{C}_j\right)|z\rangle = \begin{cases} \exp(i\gamma)\,|z\rangle & \text{if } z \text{ satisfies } C_j \\ 1\,|z\rangle & \text{else} \end{cases} \tag{18}$$

But, the satisfiability of $C_j = \bigvee_{k=1}^{m_j} l_{j,k}$ only depends on variables $(x_{j,k})_{k\in\{1,...,m_j\}}$ so $U_{-\hat{C}_j}(\gamma)\,|z\rangle$ only depends on qubits $(z_{j,k})_{k\in\{1,...,m_j\}}$. Moreover, there is a unique assignment of $(z_{j,k})_{k\in\{1,...,m_j\}}$ such that $U_{-\hat{C}_j}(\gamma)\,|z\rangle = |z\rangle$, for all others, $U_{-\hat{C}_j}(\gamma)\,|z\rangle = \exp(i\gamma)\,|z\rangle$. Using those statements, we can model the quantum circuit of $U_{-\hat{C}_j}(\gamma)$ as shown in Fig. 2. To the best of our knowledge, the multi-controlled R_ϕ gate cannot be decomposed into 1-qubit and 2-qubit gates in a polynomial number of gates. So, the complexity of the circuit is:

$$\text{Comp}(U_{H_P}) = \mathcal{O}(m \cdot M \cdot 2^M) \tag{19}$$

$$\text{Depth}(U_{H_P}) = \mathcal{O}(m \cdot 2^M) \tag{20}$$

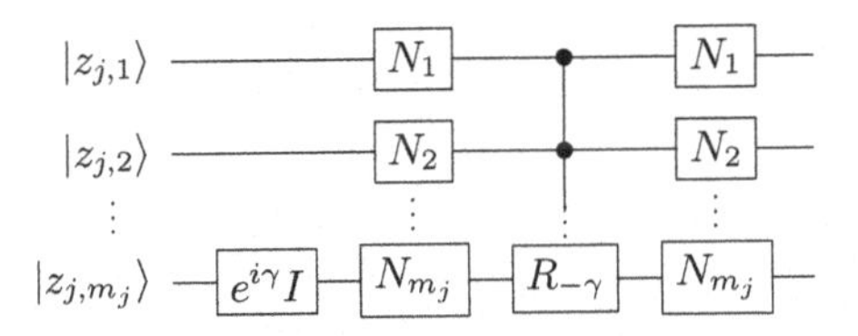

Fig. 2. Quantum circuit modelling $U_{-\hat{C}_j}(\gamma)$ where N_k is the X gate if $l_{j,k} = x_{j,k}$ and the identity gate if $l_{j,k} = \neg x_{j,k}$, for all $k \in \{1, ..., m_j\}$.

We can also use the definition of H_P in Eq. 16 using the Pauli operator σ^z. In that case, for all $I \subseteq \{1, ..., n\}$ with $|I| \le M$, there exists $a_I \in \mathbb{R}$, such that:

$$U_{H_P}(\gamma) = \exp\left(i\gamma \sum_I a_I \prod_{k\in I} \sigma_k^z\right) = \prod_I \exp\left(i\gamma a_I \prod_{k\in I} \sigma_k^z\right) \tag{21}$$

So, the quantum circuit of $U_{H_P}(\gamma)$ is composed of quantum circuits of $i\gamma a \prod_k \sigma_k^z$, in any order, which can be modelling by the quantum circuit in Fig. 3(a). Its complexity is:

$$\text{Comp}(U_{H_P}) = \mathcal{O}\left(\sum_{k=1}^{M} \binom{n}{k} \cdot (2k-1)\right) \tag{22}$$

$$\text{Depth}(U_{H_P}) = \text{Comp}(U_{H_P}) \tag{23}$$

The driver hamiltonian is often the same, namely:

$$H_D = -\sum_{j=1}^{n} \sigma_j^x \tag{24}$$

It is a well-known mixing operator for which a ground state is:

$$|+\rangle^{\otimes n} = H^{\otimes n} |0\rangle^{\otimes n} \tag{25}$$

The quantum circuit of $U_{H_D}(\beta) = \bigotimes_{j=1}^{n} R_x(-2\beta)$ is modelled in Fig. 3(b) and its complexity is:

$$\text{Comp}(U_{H_D}) = \mathcal{O}(n) \tag{26}$$

$$\text{Depth}(U_{H_D}) = \mathcal{O}(1) \tag{27}$$

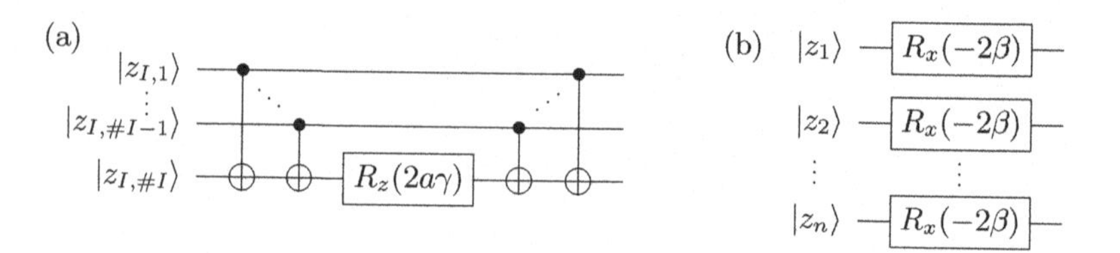

Fig. 3. Quantum circuits modelling (a) $i\gamma a \prod_k \sigma_k^z$ and (b) $U_{H_D}(\beta)$

5 Experimental Results

First, we generate random plaintexts and keys and we encrypt them using our implementation of the Heys cipher, on two over five rounds, to obtain a set of triplets (key, plaintext, ciphertext). Then, we transform each one into a Conjunctive Normal Form formula (CNF) using our implementation of propositional formula that can do some simplifications to limit the size of the output formula. To study the formulas for different number of variables, we replace in the formula some variables with the corresponding key bits.

Then, we utilize the SymPy library [3] to build the hamiltonian as a symbolic expression based on the formula with the Pauli operator σ^z in Eq. 16. The final step is to construct the corresponding quantum circuit with the Qiskit library [2] and to give it as input to our own implementation of QAOA. It utilizes the Qiskit Constraint Optimization BY Linear Approximation (COBYLA) optimizer [15] (that seems to be the best among the Qiskit optimizers for this work).

Before trying the algorithm on IBM Q Experience [1] real quantum devices, we have studied the theoretical performance of QAOA on our problem with the Qiskit Aer simulators. To estimate the performances, the success probability and the following approximation ratio was used as a metric:

$$r^* = \frac{F_p\left(\vec{\beta}_p^*, \vec{\gamma_p}^*\right)}{\min_x C(x)} \tag{28}$$

where $\vec{\beta}_p^*$ and $\vec{\gamma}_p^*$ are the optimal parameters found by the classical optimizer at step p.

5.1 Experimental Results on Simulators

For all instances, we utilize the "qasm_simulator" that returns only one state at the end of the circuit. So we approximate the value of F_p, as described in Eq. 6, by running then measuring 256 times the quantum circuit. For the smallest instances, we can also use the"statevector_simulator" that returns the entire superposition state so the value of F_p is computed exactly. The results from "qasm_simulator" are considered as "experimental" while the results from "statevector_simulator" are considered as "theoretical". After doing some experiments at $p \in \{1, 2, 3, 4\}$, we notice similarities between different instances or between the same instance at p and $p + 1$. As expected, we observe that the variations of F_p seem to correspond with the variations of the success probability. Besides, for all tested p and instances, for all parameters at step p leading to a low cost solution, there exists parameters at step $p + 1$ with similar first $2p$ components and also leading to a low cost solution. Finally, for all tested p and instances, a value close to the minimum of F_p is obtained when the values $\beta_{p,i}$ are both small and decreasing and $\gamma_{p,i}$ are both small and increasing with i.

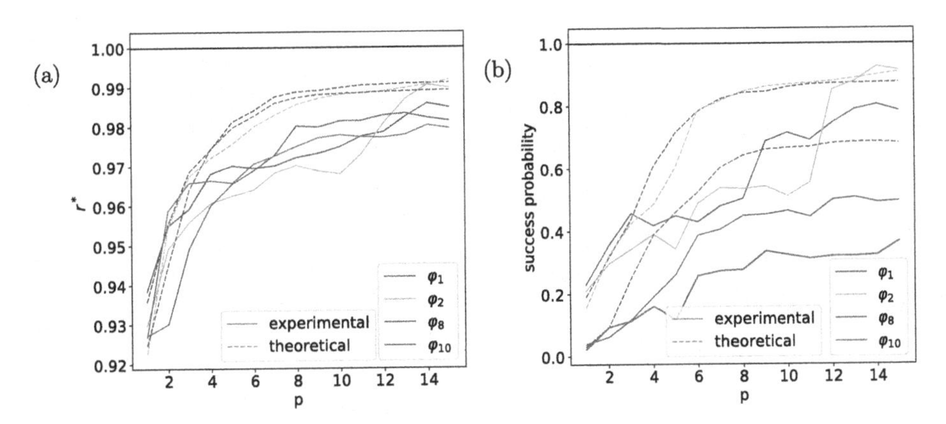

Fig. 4. Evolution of (a) r^* and (b) the success probability both as a function of p, using the first heuristic in Sect. 5.1, on formulas φ_1 (5 variables and 19 clauses), φ_2 (5 variables and 21 clauses), φ_8 (10 variables and 41 clauses) and φ_{10} (15 variables and 69 clauses).

Hence, instead of starting the classical optimizer with several random parameters (which needs a number of initializations growing with p and becoming quickly impracticable), we utilize heuristics to select the initial parameters at step $p+1$ given the optimal parameters found by the optimizer at step p.

The first studied heuristic keeps all components of $\vec{\beta}_p^*$ and $\vec{\gamma}_p^*$ and does a grid search to find the best last component of $\vec{\beta}_{p+1}^{init}$ and $\vec{\gamma}_{p+1}^{init}$, that is:

$$\vec{\beta}_{p+1}^{init} = \left(\beta_{p,0}^*, ..., \beta_{p,p-1}^*, \beta\right) \quad , \quad \vec{\gamma}_{p+1}^{init} = \left(\gamma_{p,0}^*, ..., \gamma_{p,p-1}^*, \gamma\right) \tag{29}$$

with $(\beta, \gamma) \in [0, 2\pi[\times[0, 2\pi[$.

The performances of QAOA using that strategy for formulas with a different number of variables and clauses are shown in Fig. 4. We clearly see that r^* and the success probability follow a logarithmic evolution with a big increase at the beginning and ends up stabilizing on a plateau. However, for the success probability, the height of the plateau seems to reduce with the increase of the number of variables and clauses. So, that strategy does not seem appropriate to solve our MAX-SAT problem when the number of variables increases, given that we absolutely need an optimal solution to deduce the key. In addition, it is important to notice that we have to execute QAOA using optimization as many times as the number of grid points, for each p.

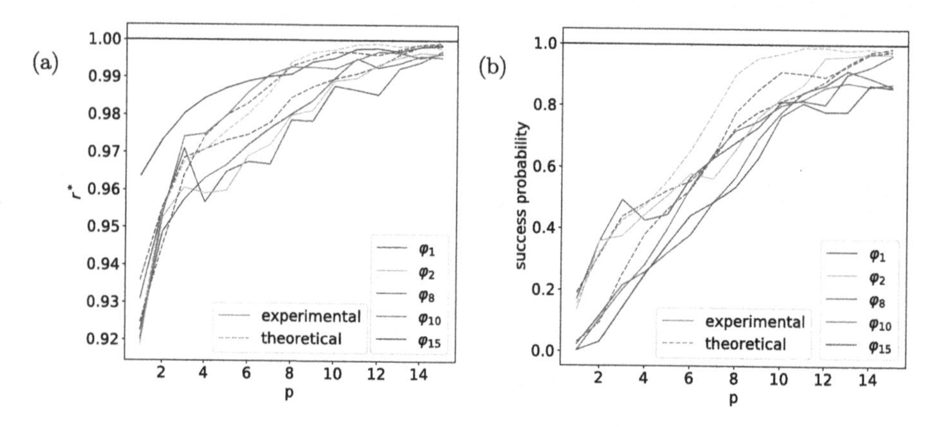

Fig. 5. Evolution of (a) r^* and (b) the success probability both as a function of p, using the INTERP heuristic, on formulas φ_1 (5 variables and 19 clauses), φ_2 (5 variables and 21 clauses), φ_8 (10 variables and 41 clauses), φ_{10} (15 variables and 69 clauses) and φ_{15} (29 variables and 164 clauses).

The second heuristic, called INTERP [20], is based on the regularity in the evolution of the components of certain optimal parameters, as mentioned previously. It consists of doing a linear interpolation on the components of $\vec{\beta}_p^*$ and $\vec{\gamma}_p^*$ to get the initial parameters on $p+1$ components. That strategy seems to be efficient for the MAX-CUT problem where the optimal parameters have a similar shape with those of MAX-SAT. The components of the initial parameters are:

$$\beta_{p+1,i}^{init} = \frac{i}{p}\beta_{p,i-1}^* + \frac{p-i}{p}\beta_{p,i}^* \tag{30}$$

$$\gamma_{p+1,i}^{init} = \frac{i}{p}\gamma_{p,i-1}^* + \frac{p-i}{p}\gamma_{p,i}^* \tag{31}$$

with $\beta_{p,-1}^* = \beta_{p,p}^* = 0 = \gamma_{p,-1}^* = \gamma_{p,p}^*$ and starting with $\vec{\beta}_1^{init} = (0.5)$ and $\vec{\gamma}_1^{init} = (0.5)$. By using that heuristic, we observe in Fig. 5(a) that the evolution of r^* runs through three steps. Indeed, we see first a large increase, that becomes moderate before reaching a stabilization near to 1. The height of the plateau does not seem to be affected by the number of variables, contrary to the previous strategy. The general shape of the curves for the success probability is different, as shown in Fig. 5(b). All curves seem to increase in a more or less linear way before reaching a plateau, which is always high, no matter what the number of variables and clauses of the instance are. That strategy seems to be particularly interesting for our MAX-SAT problem, especially as we need to perform QAOA with optimization only once for each p.

Its main drawback is that we need to ensure an adequate shape of $\vec{\beta}_p^*$ and $\vec{\gamma}_p^*$ at each p. If one of the $\beta_{p,i}^*$ or $\gamma_{p,i}^*$ is not in the continuity of the others then it could impair the efficiency of the strategy, as shown in Fig. 6. Indeed, during one of the two executions, at $p = 10$, β_9 is too high and thus r^* and the success probability collapse from that p.

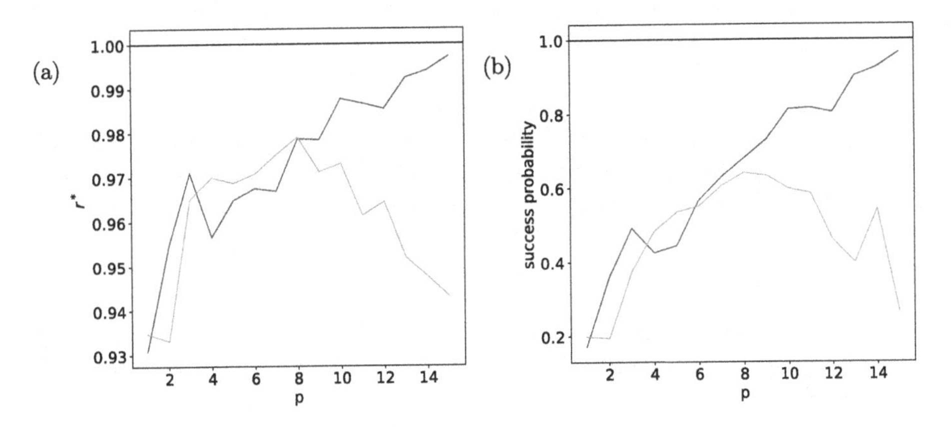

Fig. 6. Comparison of (a) r^* and (b) the success probability both as a function of p between two executions of QAOA using the INTERP heuristic on the same formula φ_1 (5 variables and 19 clauses).

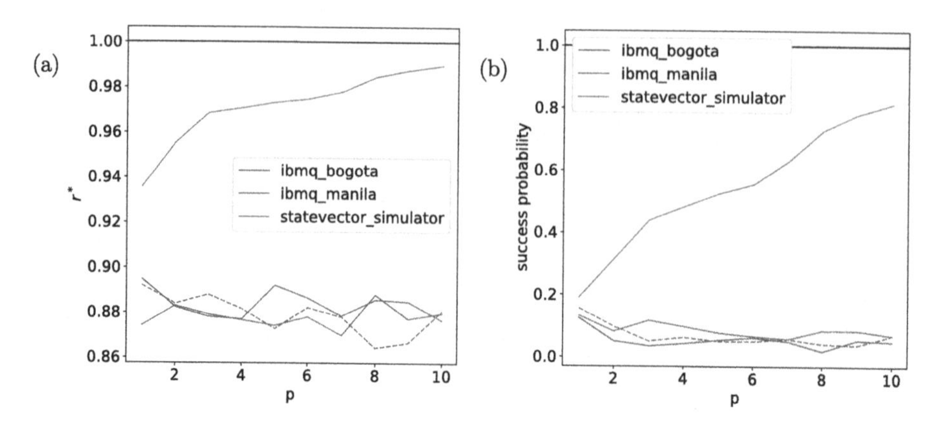

Fig. 7. Comparison of (a) r^* and (b) the success probability both as a function of p between executions of QAOA using INTERP (solid line) or INTERP$_{th}$ (dashed line) heuristics for the same formula on real quantum computers "ibmq_bogota" and "ibmq_manila" (5 qubits and quantum volume of 32) and on a statevector simulator.

5.2 Experimental Results on Real Quantum Hardware

We try to solve our problem on the smallest studied instance φ_1 (5 variables with 19 clauses), using the INTERP heuristic until $p = 10$, with real quantum computers (5 qubits and quantum volume of 32) from IBM Q Experience. As with simulators, we run the quantum circuit then measure 256 times the final quantum state every time we need to approximate F_p.

Given that the noise from errors and decoherence of the quantum computer can be potentially significant, there is a high probability that the optimal parameters $\vec{\beta}_p^*$ and $\vec{\gamma}_p^*$, found by the optimizer at each p, have not an adequate shape for the INTERP heuristic. So, we also conduct another experiment where we try to reduce the impact of this INTERP drawback by starting the optimization with the optimal parameters found with the statevector simulator as initial parameters for each p (that we call "INTERP$_{th}$").

The results of all these experiments are shown in Fig. 7. We observe, at $p = 1$ and for all tested quantum computers, that the success probability reaches between 13% and 16% instead of the theoretical 20% of the statevector simulator, with optimal parameters close to the expected ones. But when $p \geq 2$, the success probability remains more or less constant and does not exceed 10%. The approximation ratio r^* also remains constant and low (around 0.88) for all p. It can be explained by the fact that the quantum computers utilized to execute QAOA have a small quantum volume (32 at most) and the quantum circuit has a depth of $1 + 27p$ gates which is already high considering the precision of each individual gate. Notice that the optimal parameters found at each p do not have the expected shape, except for $p = 1$.

6 Discussion and Perspectives

This paper was a first attempt to estimate the capability of already existing quantum computers on solving a straightforward cryptanalytic problem. First of all, as the number of qubits of NISQ devices that we had access to was very low, we were constrained to utilize algorithms that do not add any auxiliary variables to transform a plaintext-ciphertext pair into a Conjunctive Normal Form formula (CNF). But, currently and as far as we know, such polynomial algorithms do not exist. Moreover, our modelling of the MAX-SAT problem into a quantum circuit is not polynomial as long as we do not know how to decompose the multi-controlled R_ϕ gate polynomially into 1-qubit and 2-qubit basic gates of the quantum hardware.

We have studied the performances of QAOA using the INTERP heuristic and our simulation results show that it seems to be efficient to solve our instances with a success probability that increases linearly until it reaches a high plateau, at least at low p. However, we were not able to confirm that on real quantum hardware given that the resulting quantum circuits were too deep from $p \geq 2$, leading to a hardware-induced noise which is too significant to have reliable results.

Those encouraging results on simulators of QAOA using INTERP are promising but should be treated cautiously. Indeed, all the experiments were realized on few specific small instances of the MAX-SAT problem so we of course cannot affirm that it will work as well on larger instances or on all instances of that problem. Furthermore, the probable link between the number of optimal solutions and r^* or the success probability was not taken into account in the present work.

To summarize, the proposed attack appears not to be feasible, at short and medium term, because of the complexity of the transformation of the plaintext-ciphertext pair into the quantum circuit utilized in QAOA and because of the fact that publicly available quantum computers are not allowing deep enough quantum circuits.

More exploration of QAOA capabilities to solve efficiently some instances of the MAX-SAT problem on a simulator in an attempt at determining which specific instances are "QAOA-friendly" is a relevant perspective. Furthermore, we can conjecture that the INTERP heuristic is sensitive to the noise given that the optimal parameters found by the optimizer need to have a specific shape, so it would be interesting to design other heuristics more resistant to such noise.

References

1. Ibm q experience. https://quantum-computing.ibm.com
2. Qiskit source code. https://github.com/QISKit/
3. Sympy source code. https://github.com/sympy
4. Bonnetain, X., Hosoyamada, A., Naya-Plasencia, M., Sasaki, Yu., Schrottenloher, A.: Quantum attacks without superposition queries: the offline Simon's algorithm. In: Galbraith, S.D., Moriai, S. (eds.) ASIACRYPT 2019. LNCS, vol. 11921, pp. 552–583. Springer, Cham (2019). https://doi.org/10.1007/978-3-030-34578-5_20

5. Bonnetain, X., Naya-Plasencia, M.: Hidden shift quantum cryptanalysis and implications. In: Peyrin, T., Galbraith, S. (eds.) ASIACRYPT 2018. LNCS, vol. 11272, pp. 560–592. Springer, Cham (2018). https://doi.org/10.1007/978-3-030-03326-2_19
6. Bonnetain, X., Naya-Plasencia, M., Schrottenloher, A.: Quantum security analysis of AES. IACR Trans. Symm. Cryptol. **2019**(2), 55–93 (2019)
7. Farhi, E., Goldstone, J., Gutmann, S.: A quantum approximate optimization algorithm. arXiv preprint arXiv:1411.4028 (2014)
8. Grover, L.K.: A fast quantum mechanical algorithm for database search. In: Proceedings of the Twenty-Eighth Annual ACM Symposium on Theory of Computing, pp. 212–219 (1996)
9. Heys, H.M.: A tutorial on linear and differential cryptanalysis. Cryptologia **26**(3), 189–221 (2002)
10. Jukna, S., et al.: Boolean Function Complexity: Advances and Frontiers, vol. 5. Springer, Heidelberg (2012)
11. Kaplan, M., Leurent, G., Leverrier, A., Naya-Plasencia, M.: Quantum differential and linear cryptanalysis. arXiv preprint arXiv:1510.05836 (2015)
12. Kuwakado, H., Morii, M.: Quantum distinguisher between the 3-round Feistel cipher and the random permutation. In: 2010 IEEE International Symposium on Information Theory, pp. 2682–2685. IEEE (2010)
13. Kuwakado, H., Morii, M.: Security on the quantum-type even-Mansour cipher. In: 2012 International Symposium on Information Theory and its Applications, pp. 312–316. IEEE (2012)
14. Massacci, F., Marraro, L.: Logical cryptanalysis as a sat problem. J. Autom. Reason. **24**(1), 165–203 (2000)
15. Powell, M.J.: A direct search optimization method that models the objective and constraint functions by linear interpolation. In: Gomez, S., Hennart, J.P. (eds.) Advances in Optimization and Numerical Analysis. Mathematics and Its Applications, vol. 275, pp. 51–67 (1994). Springer, Dordrecht. https://doi.org/10.1007/978-94-015-8330-5_4
16. Shor, P.W.: Algorithms for quantum computation: discrete logarithms and factoring. In: Proceedings 35th Annual Symposium on Foundations of Computer Science, pp. 124–134. IEEE (1994)
17. Sun, Y., Zhang, J.Y., Byrd, M.S., Wu, L.A.: Adiabatic quantum simulation using trotterization. arXiv preprint arXiv:1805.11568 (2018)
18. Tseitin, G.S.: On the complexity of derivation in propositional calculus. In: Siekmann, J.H., Wrightson, G. (eds.) Automation of Reasoning. Symbolic Computation, pp. 466–483. Springer, Heidelberg (1983). https://doi.org/10.1007/978-3-642-81955-1_28
19. Welchman, G.: The Hut Six Story: Breaking the Enigma Codes. M. & M, Baldwin (1997)
20. Zhou, L., Wang, S.T., Choi, S., Pichler, H., Lukin, M.D.: Quantum approximate optimization algorithm: Performance, mechanism, and implementation on near-term devices. Phys. Rev. X **10**(2), 021067 (2020)

Scheduling with Multiple Dispatch Rules: A Quantum Computing Approach

Poojith U. Rao(✉) and Balwinder Sodhi

Indian Institute of Technology, Ropar, India
{poojith.19csz0006,sodhi}@iitrpr.ac.in

Abstract. Updating the set of Multiple Dispatch Rules (MDRs) for scheduling of machines in a Flexible Manufacturing System (FMS) is computationally intensive. It becomes a major bottleneck when these rules have to be updated in real-time in response to changes in the manufacturing environment. Machine Learning (ML) based solutions for this problem are considered to be state-of-the-art. However, their accuracy and correctness depend on the availability of high-quality training data. To address the shortcomings of the ML-based approaches, we propose a novel Quadratic Unconstrained Binary Optimization (QUBO) formulation for the MDR scheduling problem. A novel aspect of our formulation is that it can be efficiently solved on a quantum annealer. We solve the proposed formulation on a production quantum annealer from D-Wave and compare the results with single dispatch rule based baseline model.

Keywords: Quantum annealing · Multiple dispatch rules · Flexible manufacturing system

1 Introduction

Scheduling in manufacturing systems refers to the process of allocating a common set of production resources such as machines to different jobs simultaneously. The aim of a scheduling problem is to allocate the resources such that an overall objective is optimized. Some of the common objectives are optimizing make-span, total tardiness, mean lateness, mean flow-time and so on. The Job Shop Scheduling Problem (JSSP) aims to calculate the start time of each operation on machines in order to optimize the overall objective. This problem is constrained such that each operation of a job is scheduled only once and when all preceding operations in the job are completed. Here each machine can process only one job at a time [11]. Flexible JSSP (F-JSSP) is an extension of JSSP where each operation can be scheduled on any machine from a given set of machines [2]. JSSP is proven to be an NP-hard problem [3] and F-JSSP, which is its extension, is also NP-hard making it a very difficult combinatorial optimization problem.

Occurrences of unforeseen events in the job shop environment may void the applicability of a scheduling decision. Mitigating the effects of such events necessitates real-time corrective action. Real-time Scheduling (RTS) of operations is

D. Groen et al. (Eds.): ICCS 2022, LNCS 13353, pp. 233–246, 2022.
https://doi.org/10.1007/978-3-031-08760-8_20

a prominent approach to address this issue. The RTS in manufacturing systems has to take decisions in real-time based on the changes in the job shop environment. One category of such decisions is about allotting different Dispatch Rules (DRs) to machines so that the overall scheduling objective is satisfied. In a typical manufacturing shop scenario, the time taken for an operation on a machine is in the order of minutes. Any RTS approach which can decide the DRs within those few minutes is generally desirable.

1.1 Motivation

The main challenge with the approaches that aim to find a (global) optimal job schedule is that they are inefficient and often impractical for real-time scheduling. This is because the job schedule needs to be recomputed quickly in response to changes in the job shop environment. That is, a machine in a shop will not be able to deploy the next job until the optimal schedule is recomputed. Since a practical shop has to handle a large number of jobs and machines, such approaches do not scale well for real-time scheduling in such an environment.

A DR based approach eases the requirement for immediate re-computation of the job schedules for the machines, whenever there is a change in the environment. In the case of DR-based scheduling, each machine on the shop is assigned a DR which the machine will use to select the next job in real-time. The machine can continue to use a given DR for job scheduling, while a "higher-layer" computes an updated DR that reflects the new reality of the shop's environment. Since the DRs immediately react to dynamic events in the environment, they achieve the best time efficiency. Though the dynamic DRs may fail to guarantee a global minimum, they are good for RTS because of zero delay in job scheduling in the machine. This has motivated us, like many other researchers, to apply different techniques to find the optimal set of DRs to drive real-time job scheduling in machine shops.

There are various approaches available today that can perform the RTS decisions of dispatch rules. Predominant among them is based on machine learning methods [11] because they provide approximate optimal answers within a quick response time. However, the downside of such methods is the significant amount of time and computing resources they consume to build the predictive models that they rely on. Secondly, they also require a sufficiently large volume of relevant training data, which may not always be available.

To address these problems, we propose a quantum algorithmic approach. Quantum algorithms are known to offer exponential speedup over their classical counterparts for specific problems. With the availability of systems such as D-Wave's Quantum annealer, the quantum annealing approach has become a promising option for solving complex combinatorial optimization problems. In order to take advantage of such quantum platforms, we have developed a novel quantum formulation for the F-JSSP scheduling *with dispatch rules*.

2 Related Work

JSSP has been attempted on a quantum annealer by [8]. They show how the problem can be split into small optimization problems such that it can be solved on a limited quantum hardware capacity. Similarly, [2] explore how a parallel flexible JSSP can be solved on a quantum annealer. They formulate the problem as a QUBO problem along with various variable pruning techniques to solve bigger scheduling problems on quantum annealers. These works indicate the quantum annealer approach is indeed promising for solving the scheduling problems.

Apart from quantum algorithms, a sizeable body of classical work is available in the literature where different algorithms and heuristics are applied for finding the optimal dispatch rules. The new job insertion problem in dynamic flexible job shop scheduling problem has also been addressed in the literature [7,9,14]. [9] develop a deep Q-learning method with double DQN and soft weight update to tackle this problem and to learn the most suitable action (e.g., dispatch rule) at each rescheduling point. [7] propose a new Karmarkar-Karp heuristic and combine it with a genetic algorithm for solving the dynamic JSSP. They show that their proposed methodologies generate excellent results. [14] propose a reinforcement learning-based MDRs selection mechanism to tackle this problem. They determine the system state by a two-level self-organizing map and use the Q-learning algorithm as a reinforcement learning agent. [6] take a random-forest based approach for learning the dispatch rules to minimize the total tardiness. They compare their method with other decision-tree-based algorithms and show their approach is effective in terms of extracting scheduling insights.

3 Solution Architecture

The overall architecture for the proposed quantum formulation is shown in Fig. 1. The input to the FMS is job requests and the DRs for each machine. The FMS outputs the finished products by scheduling the machines using the assigned dispatch rules. When determining the scheduling decisions, our approach considers the jobs received during fixed time intervals T (e.g., every 60 min). T is treated as a hyper-parameter in our approach, which can be estimated using suitable heuristics (a naive one would be just to fix the value of T to the average job completion time).

For the jobs arriving during every T interval, the DR estimator calculates the best possible DRs. The mathematical models for estimating DRs are developed in the following section.

4 Problem Formulation

A job shop has a set of n jobs $J = \{j_1, j_2, ..., j_n\}$ which need to be executed on k machines $M = \{m_1, m_2, ..., m_k\}$. Each job j_i has a sequence of operations $O_i = \{o_{i,1}, o_{i,2}, ...\}$ which need to be executed in a specific order. The objective is to assign a dispatch rule $r \in R$ for each machine such that overall tardiness is minimized. The DR assignment problem is framed as a QUBO problem such that we may use quantum annealing to solve it.

4.1 Example Scenario Used for Testing and Describing Our Approach

We use a modification of the example FMS used by a well-known work by Montazeri to describe and test our approach [10]. It consists of 3 machine families (F1, F2, F3), three load/unload stations (L1, L2, L3) and a sufficient Work in Process (WiP) buffer. Machine family F1 and F2 consist of 2 machines each (F11, F12 and F21, F22), and we assume zero loading time for material carriers (e.g., conveyor belt). Each machine in the manufacturing system is connected by a conveyor belt. The belt can carry a maximum of 3 products at any given time. The time taken by the conveyor to transport intermediate products between the machines is shown in the Fig. 1. The conveyor belt is modelled as a machine family C with three machines C1, C2, C3. As it has 3 machines, a maximum of 3 products can be carried at anytime. The processing time for these machines is equal to the time needed for the product to be transported on the conveyor belt. There are five different products that need to be manufactured by the FMS. The routing and timing for each of them are given in Table 1. The DRs considered in this work are SIO, SPT, SRPT, SDT, SMT, SIO, LPT, LRPT, LDT, LMT. The description of these rules can be found in Table 2. We would like to note that our approach works in general scenarios similar to the above.

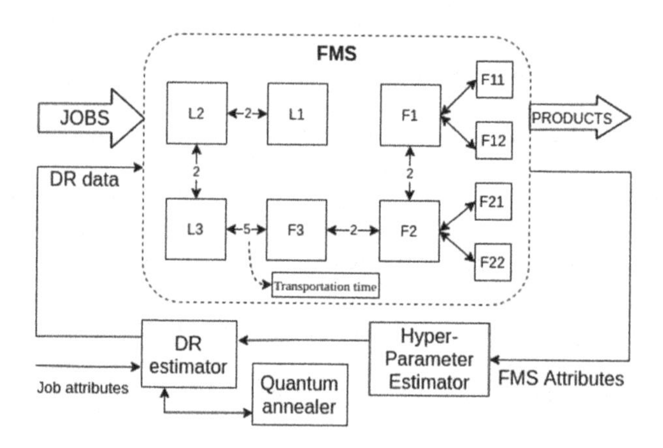

Fig. 1. Proposed solution architecture

4.2 Proposed QUBO Formulation for the Problem

In the following subsections, we describe the crucial elements of the proposed QUBO formulation for the F-JSSP problem that also takes into consideration the dispatch rules in an RTS scenario.

Table 1. Routing and timing information of products/jobs

Job	Operations	Timing (minutes)
P1	L2, F2, L3, F1, L3, F2, L2	2, 11, 10, 20, 3, 14, 2
P2	L2, F2, L3, F1, L3, F2, L2	2, 10, 10, 24, 3, 10, 2
P3	L2, F2, L2, F1, L3, F2, L3	2, 15, 2, 30, 10, 21, 3
P4	L2, F2, L2, F1, L3, F2, L3	2, 12, 2, 26, 10, 13, 3
P5	L2, F2, L3, F1, L3, F1, L3	8, 16, 5, 25, 5, 22, 10

Table 2. Description of different dispatch rules

DR	Description
SIO	Select the job with the shortest imminent operation time
SPT	Select the job with the shortest processing time
SRPT	Select the job with the shortest remaining processing time
SDT	Select the job with the smallest ratio obtained by dividing the processing time of the imminent operation by the total processing time
SMT	Select the job with the smallest value obtained by multiplying the processing time of the imminent operation by the total processing time
LIO	Select the job with the largest imminent operation time
LPT	Select the job with the largest processing time
LRPT	Select the job with the largest remaining processing time
LDT	Select the job with the largest ratio obtained by dividing the processing time of the imminent operation by the total processing time
LMT	Select the job with the largest value obtained by multiplying the processing time of the imminent operation by the total processing time

Decision Variables. Two sets of binary decision variables are used for representing different decisions needed for solving the problem. The first set of binary variables $x_{j,o,m,t}$ represent the decision whether the oth operation of jth job must be scheduled on a machine m at time t. The second set of binary variables $x_{m,r}$ represent if the dispatch rule r is assigned for machine m.

$$x_{j,o,m,t} = \begin{cases} 1 & \text{if the operation o of job j must be scheduled on machine m at time t} \\ 0 & \text{otherwise} \end{cases} \tag{1}$$

$$x_{m,r} = \begin{cases} 1 & \text{if machine m must be assigned rule r} \\ 0 & \text{otherwise} \end{cases} \tag{2}$$

The constraints and objective functions that we formulate using these decisions variables will be expressed as Hamiltonians in the following subsections [13]. Further, to understand the background of choosing the specific terms in an objective function or constraint, please refer to the literature on QUBO formulation [4].

Operation Scheduling. A job comprises of a sequence of operations needed to be performed in order to manufacture a product (see example in Table 1). A product requires each operation to be performed (and thus scheduled) only once. This constraint is expressed as the Hamiltonian in Eq. 3

$$H_1 = \sum_{j \in J} \sum_{o \in O} \left[\sum_{m,t} x_{j,o,m,t} - 1 \right]^2 \tag{3}$$

Preference Order. Every job has a predefined order in which the operations need to be carried out. In our example from Table 1, the product P1 needs the operations to be scheduled in this order: L2 → F2 → L3 → F1 → L3 → F2 → L2. If this order is violated, the schedule generated will not be a valid one. In order to force the scheduler to follow the defined order, a penalty is added for each violation.

Let $x_1 = x_{j,o_1,m_1,t_1}$ and $x_2 = x_{j,o_2,m_2,t_2}$ represent scheduling decisions for two consecutive operations of the same job j at time t_1 and t_2. Let p represent the processing time of operation o_1. In a valid schedule, the operation o_1 must be scheduled before o_2. o_1 and o_2 need to be scheduled (i.e.,for them $x_1 = 1$ and $x_2 = 1$) such that t_2 must be greater than $t_1 + p$ (i.e., o_2 can be scheduled only after o_1 is completed). All combinations of x_1 and x_2 that do not satisfy the condition are violations and must be penalized. The set A in Eq. 4 contains all such combinations and the penalty term is shown in Eq. 5.

$$A = [(x_{j,o,m,t_1}, x_{j,o+1,m,t_2}) \forall t_2 < t_1 + p\ ] \tag{4}$$

$$H_2 = \sum_{x_1, x_2 \in A} x_1 x_2 \tag{5}$$

Machine Scheduling. Each machine can handle only one operation at a time. Schedules resulting in overlapping operations on a machine must be avoided. To avoid all such conflicting (i.e., those having overlapped schedule) operations a penalty term is added.

Let $x_1 = x_{j_1,o_1,m,t_1}$ and $x_2 = x_{j_2,o_2,m,t_2}$ represent scheduling decisions for two operations on the same machine m at time t_1 and t_2 respectively. Let p be the processing time of operation o_1. In a valid schedule, the operations o_1 and o_2 must be scheduled such that they do not overlap each other. If both o_1 and o_2 need to be scheduled i.e. $x_1 = 1$ and $x_2 = 1$, then t_2 must not lie in the range t_1 to $t_1 + p$. All combinations of x_1 and x_2 that do not satisfy the condition are violations and must be penalized. The set B in Eq. 6 contains all such combinations and the penalty term is shown in Eq. 7.

$$B = [(x_{j_1,o_1,m,t_1}, x_{j_2,o_2,m,t_2}) \forall t1 \leq t_2 \leq t_1 + p\ ] \tag{6}$$

$$H_3 = \sum_{x_1,x_2 \in B} x_1 x_2 \tag{7}$$

Squeezed Scheduling. In the case of DR based scheduling, the DRs select the operation to be scheduled for each machine. When the current operation is completed, the scheduler picks the next operation from the queue based on the DR and schedules it immediately. That is, when an operation O is scheduled at time t on machine M, then the next operation is scheduled at $t + p + 1$ if at all available where p is the processing time of the operation O.

To encourage the solutions that have operations scheduled one after the other whenever possible, a reward of -1 is given when the time gap between two operations scheduled on the same machine is zero. Let $C_{m,t'}$ represent the set of all operations which can be scheduled on machine m at time t', then the Eq. 9 assigns the above-mentioned reward.

$$C_{m,t'} = [x_{j,o,m,t'} \text{ Where, } t' = t + p + 1] \tag{8}$$

$$H_4 = \sum_{j,o,m,t} \left[\sum_{y \in C_{m,t'}} -x_{j,o,m,t} \times y \right] \tag{9}$$

Tardiness Minimization. The objective we have considered in our problem formulation is to minimize the total tardiness. In order to accomplish this, the penalty function we have considered is given in Eq. 10. It is formulated as a sum of the product of decision variables denoting the last operation of each job and the delay incurred for that particular job. The delay is calculated as the difference between the job completion time (i.e., scheduled time + processing time p) and the deadline d.

$$H_5 = \sum_{j,t} x_{j,o,m,t} \times \max(0, (t + p - d)) \tag{10}$$

Rule Assignment. The objective is to assign each machine a DR such that the overall cost function is optimized. This implies that each machine must be assigned a single DR. This constraint is expressed as:

$$H_6 = \sum_{m \in M} \left[\sum_{r \in R} x_{m,r} - 1 \right]^2 \tag{11}$$

Rule Selection. Each rule selects a unique next operation from a queue of operations waiting to be scheduled on the machine based on its definition. For a sequence of operations that are scheduled on a machine immediately one after the other, the dispatch rule which imitates the same must be assigned to the machine. At each instance of time when an operation is scheduled on the machine, we assign a weight to all the rules based on the operation it chooses to be scheduled. The rule which would have picked the same operation that was scheduled is given assigned a reward of -1.

If an operation with processing time p is scheduled on machine m at time t, then let $C_{m,t+p+1}$ represent all operations that can be scheduled once the current operation is completed. In other words, $C_{m,t+p+1}$ contains all the operations waiting to be scheduled on the machine from which the DRs have to select the next operation from. We define function $d_{m,t+p+1}(r, x, y)$ to assign the penalty to decision variables. For each operation in $C_{m,t+p+1}$ if a rule r can select the operation o once the current operation is completed, then a reward of -1 is assigned is assigned to the operation. The function $d_{m,t}(r, x, y)$ is defined as in Eq. 12. The Hamiltonian that assigns the weights based on the above conditions is given in Eq. 13. It can be noted that 13 is not in the QUBO formulation and cannot be solved directly.

$$d_{m,t}(r,x,y) = \begin{cases} -1 & \text{if } y \text{ is the next operation selected from the list of operations in } C_{m,t} \text{ by rule r after operation } x \text{ is completed} \\ 0 & \text{otherwise} \end{cases} \tag{12}$$

$$H_7 = \sum_{j,o,m,t} x_{j,o,m,t} \times \left[\sum_{r \in R} x_{m,r} \left[\sum_{y \in C_{m,t+p+1}} y d_{m,t+p+1}(r, x_{j,o,m,t}, y) \right] \right] \tag{13}$$

We solve the energy functions in two sequential phases. In the first phase, we calculate the best-squeezed schedule using the energy function H_{sch} described in Eq. 14. This gives us the optimal schedule that can be simulated using dispatch rules.

$$H_{sch} = H_1 + H_2 + H_3 + H_4 + H_5 \tag{14}$$

The schedule calculated is used to find the best set of dispatch rules using the Hamiltonian described in Sect. 4.2. With the schedule known, it becomes possible to accurately find the waiting set of operations in each machine at any given time i.e. the $C_{m,t}$ can be accurately calculated. It represents the operations whose preceding operation was completed and are yet to be scheduled at time t

on machine m. The schedule returned from the first phase is used to replace the variables $x_{j,o,m,t}$ and y in the Eq. 13. The term H_7, thus becomes:

$$\hat{H}_7 = \sum_{j,o,m,t} S[x_{j,o,m,t}] \quad \times \\ \left[\sum_{r \in R} x_{m,r} \left[\sum_{y \in C_{m,t+p+1}} S[y] d_{m,t+p+1}(r, x_{j,o,m,t}, y)\right]\right] \tag{15}$$

As the equation now is in the quadratic form, the best set of DRs that produce a schedule similar to S is evaluated by solving the energy function in Eq. 16.

$$H_r = H_6 + \hat{H}_7 \tag{16}$$

4.3 Variable Pruning

The complexity of solving the problem increases with the number of variables. Reducing the number of variables reduces the time needed to solve the problem and helps the model find a solution near the global minimum. The maximum number of variables needed are $|J| \times |O| \times |M| \times |T| + |M| \times |R|$. We apply three heuristics for reducing the number of variables, as described next:

Ignoring Operation-Incompatible Machines. An operation can be scheduled only on machines that can process them. For example, the operation $F1$ of job $J1$ can only be scheduled on machines that belong to the machine family $F1$: $F11$ *and* $F12$. Removing all incompatible machine-operation decision variables significantly reduces the number of redundant variables.

Suitable Value for Operation Time-Span. Timespan T is a major factor that hugely varies the total number of decision variables. Choosing a large T increases the complexity of the problem as the model will have a large number of decision variables to choose from for each operation. On the other hand, reducing T's value can make the problem unsolvable because performing all scheduled operations requires a certain minimum duration. Thus, a balanced value of T is chosen heuristically.

Time-Span Bounds for an Operation. The time when an operation can be scheduled has certain lower and upper bounds depending on its position in the job's operation sequence. The decision variables that lead to those schedules that violate the operation's time span bounds can be safely removed. For example, the time span during which an operation can be scheduled depends on the processing time needed by the operations preceding and following it. Thus, the earliest time at which an operation can be scheduled is the sum of processing times of all its preceding operations. Similarly, the latest time before which the operation must be scheduled is the sum of its processing time and that of the operations following it.

5 Experiments and Results

We performed several experiments to validate the correctness and efficacy of our approach. The important ones, along with their results, are described below.

Table 3. Tardiness of different of MDRs

Jobs	Products	DL	Tardiness										
	P1-P2-P3-P4-P5		QUANT	SIO	SPT	SRPT	SDT	SMT	LIO	LPT	LRPT	LDT	LMT
J1	2-2-2-2-2	200	**42**	61	75	90	94	90	124	256	170	137	199
J2	10-0-0-0-0	200	**40**	45	**40**	45	45	45	126	**40**	126	126	126
J3	0-10-0-0-0	200	**22**	37	23	37	37	37	41	23	38	41	41
J4	0-0-10-0-0	250	**9**	15	**9**	15	15	15	13	**9**	13	13	13
J5	0-0-0-10-0	220	**13**	33	**13**	33	33	33	**13**	**13**	19	**13**	**13**
J6	5-5-0-0-0	200	**35**	**35**	37	37	39	42	76	**35**	90	76	99
J7	5-0-0-5-0	200	**6**	11	26	26	45	26	15	130	98	19	183

Table 4. Energy vs tardiness of various dispatch rules

Rule	Min. tardiness	Min. energy
QUANT	**42**	**−658.16**
SIO	61	−656.56
SPT	75	−655.32
SRPT	90	−654.13
SDT	94	−652.7
SMT	90	−654.12
LIO	124	−650.61
LPT	256	−636.35
LRPT	170	−647
LDT	137	−649.87
LMT	199	−643.52

5.1 Tardiness Comparison of the Proposed Approach with Single Dispatch Rules (SDR)

For validating the proposed approach, we compared the tardiness of different job groups on the FMS mentioned in the case study (see Table 1). Each job group consists of 10 random jobs and a deadline for completion. All jobs are submitted to the FMS at time $t = 0$, and the tardiness is calculated for the schedules generated by the proposed approach and those generated by SDR approach. The results are shown in the Table 3.

The solution to the proposed approach was solved on a real quantum annealer provided by D-Wave. Please note that the performance of these devices is affected by ambient noise. The answers calculated by these devices are only near-optimal solutions. Thus, the solution obtained for our problem formulation is not the best schedule but a sample near it. The numbers in Table 3 indicates that just within 2–3 min, the annealer is able to calculate the set of dispatch rules.

5.2 Validating the Correctness of Overall Energy Function

The solution to the scheduling problem is the combination of decision variables having minimum value to the energy function in Eq. 14. We calculate the energy value of the solution produced by different SDRs and compare it with the value obtained by using the proposed method. Our results in Table 4 show that the overall energy values are positively correlated with the tardiness. That is, the schedule with the least tardiness has the least energy value.

In order to verify the correctness of the Hamiltonian H_r (Eq. 16) used for finding the best dispatch rules, we initialized all the machines with pre-determined dispatch rules. We first calculated the schedule generated by these rules by simulating the machines in the job shop environment. Later, we calculated back the best set of dispatch rules for this schedule using the Hamiltonian H_r. As expected, the pre-determined set of rules from which the schedule was generated had the minimum energy value. This validates the correctness of our hypothesis and approach.

6 Discussion

The number of binary variables in the formulation increase with the number of machines, jobs, operations and the time-span parameters. As an example for a $15 \times 15 \times 15$ (Jobs $\times$ Operations $\times$ Machines) instance with time span of 250, the number of binary variables is around thirty five thousand. On the contrary the largest quantum annealer today has only 5000 qubits with 15 way connectivity. Ultimately, the linear coefficients in QUBO are mapped to qubit biases and quadratic coefficients to coupler strengths. With limited connectivity of qubits, QUBO formulations having large number of quadratic terms need much larger number of qubits to embed the problem on the annealer. Hence large problems such as ours cannot be directly solved on these devices without classical heuristics directly on annealers.

In this work we tested our QUBO formulation using the Leap's hybrid solvers for finding the best set of dispatch rules. These solvers implement state-of-the-art classical algorithms such as Tabu search together with intelligent allocation of the quantum computer to parts of the problem where it benefits most. While this enables them to accommodate very large problems, it also reduces their ability to find near optimal solutions.

We solved the above QUBO problem on public datasets [12,15] available for $6 \times 6 \times 6$, $8 \times 8 \times 8$, $10 \times 10 \times 10$ and $15 \times 15 \times 15$ scenarios using Leap's hybrid solver and compared it with the baseline models as described above. The deadline for each job was set to 1.2 times the optimal makespan mentioned in the dataset for each job. The *time_limit* parameter of the hybrid solver was kept at a large value such that no further improvement in the solution quality was observed on further increment. The results obtained showed that the Leap's hybrid solver was not able to sample solutions with minimum energy even with multiple attempts. The baseline models and the Leap hybrid solver found better solutions in equal number of cases. But in all cases the QUBO formulation was able to determine the best solution accurately based on the energy value.

We observed that the energy value of the annealer solution (non optimal) was in some case around 5–10% higher than the energy value of the best answer known to us (through baseline models). This showed us two things 1) Leap's hybrid solver cannot sample solutions nearer to the true optimal for our problem formulation. This is also consistent with their claim on the website [1] 2) The QUBO formulation is able to detect the best answer in each case.

As the constraints in the problem are encoded as penalties, the coefficients of each term determine the energy characteristics of the total Hamiltonian. In order to ensure the constraints are always satisfied, the coefficients for the penalty terms are kept high. As the solutions returned by leap is sub optimal, the gap (ratio or difference) between the coefficient values of penalties and objective function plays a significant role. When it is small, the solutions has many constraint violations and in case the gap is high, the objective function value is not satisfactorily optimized.

Instead of formulating such complex Hamiltonians with many constraints another approach is to use more sophisticated heuristic algorithms such as Quantum Alternating Operator Ansatz (QAOA). QAOA is a quantum classical hybrid meta-heuristic framework proposed by [5] for performing approximate optimization on gate-based quantum computers. It is well suited for scenarios such as JSSP where the feasible solution space is much smaller, such as optimization where solutions have to satisfy multiple hard constraints. But a state-of-art gate-based quantum computer currently available can support less than 100 qubits.

7 Conclusion and Future Work

Finding the best set of Multiple Dispatch Rules (MDRs) needed for real-time scheduling in a flexible manufacturing system scenario is an important scheduling problem. We have formulated it as a Quantum Unconstrained Binary Optimization (QUBO) problem and solved it on a quantum annealer. The results obtained from quantum annealer have the least tardiness compared with the schedules produced by Single Dispatch Rules (SDRs). The strong correlation between energy values of different schedules with the tardiness validates that the proposed method can find MDRs that produce schedules with minimum tardiness.

The annealers provided by D-Wave currently cannot obtain the global minimum to the energy functions. This puts a limit on the quality of solutions obtained for our formulations. The quality and the time needed to calculate solutions will improve with advancements in quantum technology.

Future research should focus on how the FMS attributes can be used to find good hyperparameter settings such as length of time span. To further check the robustness of the approach, the following types of experiments can be conducted: 1) Finding the overall tardiness value when large job groups are split into small subgroups and solved sequentially. 2) Checking performance of the approach when the length of time span is varied and 3) Comparison of the quantum method with other classical methods. 4) Performance when run directly on quantum hardware.

References

1. D-Wave. https://docs.dwavesys.com/docs/latest/handbook_hybrid.html
2. Denkena, B., Schinkel, F., Pirnay, J., Wilmsmeier, S.: Quantum algorithms for process parallel flexible job shop scheduling. CIRP J. Manuf. Sci. Technol. **33**, 100–114 (2021)
3. Garey, M.R., Johnson, D.S., Sethi, R.: The complexity of flowshop and jobshop scheduling. Math. Oper. Res. **1**(2), 117–129 (1976)
4. Glover, F., Kochenberger, G., Du, Y.: A tutorial on formulating and using QUBO models (2019)
5. Hadfield, S., Wang, Z., O'Gorman, B., Rieffel, E.G., Venturelli, D., Biswas, R.: From the quantum approximate optimization algorithm to a quantum alternating operator ansatz. Algorithms **12**(2), 34 (2019)
6. Jun, S., Lee, S., Chun, H.: Learning dispatching rules using random forest in flexible job shop scheduling problems. Int. J. Prod. Res. **57**(10), 3290–3310 (2019)
7. Kundakcı, N., Kulak, O.: Hybrid genetic algorithms for minimizing makespan in dynamic job shop scheduling problem. Comput. Ind. Eng. **96**, 31–51 (2016)
8. Kurowski, K., Węglarz, J., Subocz, M., Różycki, R., Waligóra, G.: Hybrid quantum annealing heuristic method for solving job shop scheduling problem. In: Krzhizhanovskaya, V.V., et al. (eds.) ICCS 2020. LNCS, vol. 12142, pp. 502–515. Springer, Cham (2020). https://doi.org/10.1007/978-3-030-50433-5_39
9. Luo, S.: Dynamic scheduling for flexible job shop with new job insertions by deep reinforcement learning. Appl. Soft Comput. **91**, 106208 (2020)
10. Montazeri, M., Van Wassenhove, L.: Analysis of scheduling rules for an FMS. Int. J. Prod. Res. **28**(4), 785–802 (1990)
11. Priore, P., Gomez, A., Pino, R., Rosillo, R.: Dynamic scheduling of manufacturing systems using machine learning: an updated review. Ai Edam **28**(1), 83–97 (2014)
12. Samsonov, V., et al.: Manufacturing control in job shop environments with reinforcement learning. In: ICAART, vol. 2, pp. 589–597 (2021)

13. Santoro, G.E., Tosatti, E.: Optimization using quantum mechanics: quantum annealing through adiabatic evolution. J. Phys. A Math. Gen. **39**(36), R393 (2006)
14. Shiue, Y.R., Lee, K.C., Su, C.T.: Real-time scheduling for a smart factory using a reinforcement learning approach. Comput. Ind. Eng. **125**, 604–614 (2018)
15. Zhang, C., Song, W., Cao, Z., Zhang, J., Tan, P.S., Chi, X.: Learning to dispatch for job shop scheduling via deep reinforcement learning. In: Larochelle, H., Ranzato, M., Hadsell, R., Balcan, M.F., Lin, H. (eds.) Advances in Neural Information Processing Systems, vol. 33, pp. 1621–1632. Curran Associates, Inc. (2020). https://proceedings.neurips.cc/paper/2020/file/11958dfee29b6709f48a9ba0387a2431-Paper.pdf

Quantum Variational Multi-class Classifier for the Iris Data Set

Ilya Piatrenka and Marian Rusek(✉)

Institute of Information Technology, Warsaw University of Life Sciences—SGGW, ul. Nowoursynowska 159, 02-776 Warsaw, Poland
marian_rusek@sggw.edu.pl

Abstract. Recent advances in machine learning on quantum computers have been made possible mainly by two discoveries. Mapping the features into exponentially large Hilbert spaces makes them linearly separable—quantum circuits perform linear operations only. The parameter-shift rule allows for easy computation of objective function gradients on quantum hardware—a classical optimizer can then be used to find its minimum. This allows us to build a binary variational quantum classifier that shows some advantages over the classical one. In this paper we extend this idea to building a multi-class classifier and apply it to real data. A systematic study involving several feature maps and classical optimizers as well as different repetitions of the parametrized circuits is presented. The accuracy of the model is compared both on a simulated environment and on a real IBM quantum computer.

Keywords: Quantum computing · Hybrid classical-quantum algorithms · Variational quantum classifier · Artificial intelligence · Machine learning

1 Introduction

Classical machine learning techniques have made great strides in the past decade, enabled in large part by the availability of sufficiently powerful hardware. For example, the success of neural networks based on deep learning was possible in part because of powerful parallel hardware consisting of clusters of graphical processors [30]. Maybe the existence of quantum hardware might enable further advances in the field, making the development of new machine learning algorithms possible [8]. However, ideal fault-tolerant quantum computers [27] are still not available today. What we have are Noisy Intermediate Scale Quantum computers (NISQ) [5].

There are two major paradigms in the quantum computer world: quantum annealing and gate-based computation. Both allow for solving optimization problems by finding a minimum of certain objective functions. Thus, they can both potentially be used for machine learning tasks. D-Wave Systems is a major vendor of superconducting quantum annealing machines. Their recent machines

D. Groen et al. (Eds.): ICCS 2022, LNCS 13353, pp. 247–260, 2022.
https://doi.org/10.1007/978-3-031-08760-8_21

offer up to 5640 qubits [6,7]. A more general gate-based architecture potentially allows for execution of arbitrary algorithms not restricted to optimization problems. Recently IBM unveiled a 127 qubit gate-based machine operating on transmons[1]. Alternative physical realizations of gate-based quantum computers involve photonic systems and ion-traps [19]. Gate-based quantum circuits can be created using several open source software tools [10]. In this paper, a gate-based IBM 5 qubit transmon computer is used and programmed in Python with Qiskit library to solve a supervised machine learning problem.

To mitigate the influence of decoherence errors in NISQ computers, quantum circuits with limited depth are needed. This can be achieved when the work of a quantum computer is supplemented by a classical one. This yields hybrid quantum-classical computations. In a variational setting, the quantum circuit $U(\boldsymbol{\theta})$ is parametrized by a set of numbers $\boldsymbol{\theta}$. At each step of the algorithm, a cost function $C(\boldsymbol{\theta})$ is evaluated based on multiple measurements of the quantum circuit for a given input state $|\psi_0\rangle$. Then, an optimization algorithm running on a classical computer is used to update the parameters $\boldsymbol{\theta}$ to minimize the cost function $C(\boldsymbol{\theta})$. The final state $U(\boldsymbol{\theta}_{\text{opt}})|\psi_0\rangle = |\psi(\boldsymbol{\theta}_{\text{opt}})\rangle$ is a superposition that contains the solution to the problem with a high probability amplitude. Variational algorithms were introduced in 2014, with the variational eigensolver [21].

Another well known example of such a variational approach is the Quantum Approximate Optimization Algorithm (QAOA) [9]. It allows us to find approximate solutions to the Quadratic Unconstrained Binary Optimization (QUBO) problem. The final state $|\psi(\boldsymbol{\theta}_{\text{opt}})\rangle$ measured in the computational basis gives with high probability the bit string corresponding to the solution of the specific problem (e.g., MaxCut). The influence of noise on the performance of this algorithm on IBM quantum computers was studied in [2].

The interplay between quantum computing and machine learning has attracted a lot of attention in recent years. The use of an exponentially large feature space of dimension 2^n where n is the number of qubits that is only efficiently accessible on a quantum computer provides a possible path to quantum advantage. In [12] two algorithms have been proposed that process classical data and use the quantum state Hilbert space as the feature space. Both algorithms solve a problem of supervised learning: the construction of a binary classifier. One method, the quantum variational classifier, uses a variational quantum circuit with a classical optimizer to classify the data. The other method, a quantum kernel estimator, estimates the kernel function on the quantum computer and optimizes a classical Support Vector Machine (SVM). In this paper we extend the first method to encompass the multi-class classification case.

In [1] the effective dimension measure based on the Fisher information has been proposed. It can be used to assess the ability of a machine learning to train. It was found that a class of quantum neural networks is able to achieve a considerably higher capacity and faster training ability than comparable classical feedforward neural networks. A higher capacity is captured by a higher effective

[1] https://research.ibm.com/blog/127-qubit-quantum-processor-eagle.

dimension, whereas faster training implies that a model will reach a lower loss value than another comparable model for a fixed number of training iterations. This suggests an advantage for quantum machine learning, which was demonstrated both numerically and on real quantum hardware using the Iris Data Set [3]. In this paper we use this data set not only to check the model training speed but also its accuracy.

This paper consists of 5 main sections. In Sect. 2 the details of a variational hybrid quantum-classical model for multi-class classification are described. In Sect. 3, the results from training this model on the training subset of the Iris Data Set are presented and its accuracy is tested on test subset. Both a classical Qiskit simulator as well as a real IBM quantum computer are used. In Sect. 4 we discuss these results. In Sect. 5 we finish with some conclusions.

2 Model

2.1 Parametrized Quantum Circuits

Parameterized quantum circuits, where the gates are defined by tunable parameters, are fundamental building blocks of near-term quantum machine learning algorithms. They are used for two things:

- To encode the data, where the parameters are determined by the data being encoded.
- As a quantum model, where the parameters are determined by an optimization process.

As all quantum gates used in a quantum circuit are unitary, a parameterized circuit itself can be described as a unitary operation on qubits.

How do we choose parameterized quantum circuits that are good candidates for quantum machine learning applications? First, they need them to generalize well. This means that the circuit should be able to map to a significant subset of the states within the output Hilbert space. To avoid being easy to simulate on a classical computer, the circuit should also entangle qubits. In [28], the authors propose the measures of expressibility and entangling capability to discriminate between different parameterized quantum circuits. A strong correlation between classification accuracy and expressibility, and a weak correlation between classification accuracy and entangling capability has been found [13]. Note, that the use of parametrized quantum circuits with a high expressibility in supervised machine learning scenarios may lead to overfitting [24].

In the current era of near term quantum computing, there is limited error correction or mitigation and limited qubit connectivity. To accommodate these device constraints, in [14] a class of hardware efficient parameterized circuits was introduced. They are built from layers consisting of one two-qubit entangling gate and up to three single-qubit gates. Also, a particular qubit connection topology is used. These results have been extended in [12] where the authors introduced a general parameterized circuit, which includes layers of Hadamard gates interleaved with entangling blocks, and rotation gates. This unitary was chosen because it is classically difficult to compute, but tractable on near term quantum hardware. It can be used to encode data.

Data Encoding. Data representation is crucial for the success of machine learning models. For classical machine learning, the problem is how to represent the data numerically, so that it can be best processed by a classical machine learning algorithm. For quantum machine learning, this question is similar, but more fundamental: how to represent and efficiently input the data into a quantum system, so that it can be processed by a quantum machine learning algorithm. This is usually referred to as data encoding and is a critical part of quantum machine learning algorithms that directly affect their power.

One of the simplest data encoding methods is angle encoding. In this case, the number of features is equal to the number of qubits and the features are encoded into the rotation angles of qubits. This method only encodes one datapoint at a time. It does, however, only use a constant depth quantum circuit, making it amenable to current quantum hardware. For a general unitary operator from [12] this method corresponds to the parameters $k = 1, P_0 = Z$. In Qiskit the corresponding function is called `ZFeatureMap`. For $k = 2, P_0 = Z, P_1 = ZZ$ we get the Qiskit `ZZFeatureMap` circuit, which contains layers of Hadamard gates interleaved with qubit rotations and entangling blocks. These data encoding circuits were used in [1]. These authors call them "easy quantum model" and "quantum neural network" respectively.

In this paper, in addition to the feature map circuits described above, a slightly more sophisticated circuit is used. It is depicted in Fig. 1 and differs from the `ZZFeatureMap` by adding qubit rotations along Y axis. Using notation from [12] it corresponds to the parameters $k = 3, P_0 = Z, P_1 = Y, P_2 = ZZ$ and is generated using the Qiskit `PauliFeatureMap` function. It only encodes a datapoint $\boldsymbol{x}$ of 3 features, despite having 9 parameterized gates.

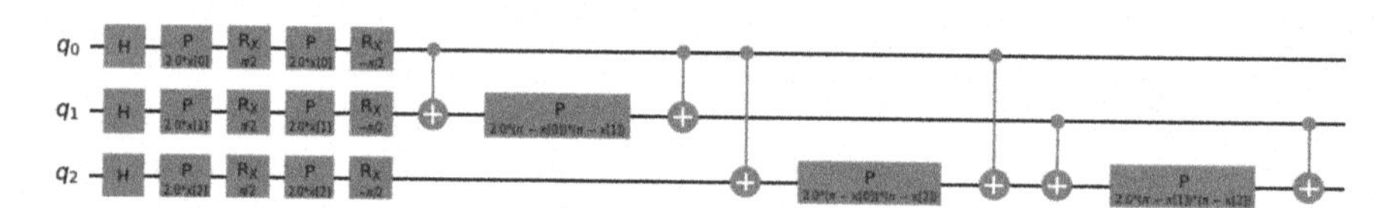

Fig. 1. `PauliFeatureMap` data encoding circuit for 3 qubits. H, P, and R_X are Hadamard, phase shift, and X rotation gates; x is the feature vector.

Quantum Model. The Qiskit `RealAmplitudes` circuit is a heuristic trial wave function used as Ansatz in chemistry applications or quantum model classification circuits in machine learning. The circuit consists of alternating layers of R_Y rotations and $CNOT$ entanglements. We use it in our paper with different numbers of repetitions. The number of variational parameters θ_j is equal to $(r+1)n$, where r is the number of repetitions of this circuit, and n is the number of qubits. This circuit can be regarded as a special case of the Qiskit `TwoLocal` circuit. The prepared trial quantum states will only have real amplitudes.

2.2 Quantum Variational Classification

Quantum supervised machine learning is the task of learning the parameters $\boldsymbol{\theta}$ of a variational circuit that maps an input feature vector $\boldsymbol{x}$ to an output $\hat{y}$ based on example training input-output pairs $(\boldsymbol{x}, y)$. The accuracy of the model can then be checked using a set of testing examples. In the case of a classification problem that asks us to assign data into specific categories, the outputs $\hat{y}$ are category numbers.

Label Assignment. Such a variational quantum classifier algorithm was introduced by multiple groups in 2018 [26]. For a binary classification problem, with binary output labels, the measured expectation value of $Z^{\otimes n}$ can be interpreted as the output of a classifier. The average is calculated over several shots of the algorithm. It can be shown, that this is equivalent to measuring the average value of Z acting on one qubit only.

Let us now extend this procedure to a multi-class classification. For a given feature vector $\boldsymbol{x}$ and variational parameters θ the output of the parmetrized circuit is measured in the computational basis. Thus the output of each shot of the algorithm is a bitstring $\boldsymbol{b}$, where $b_i \in \{0, 1\}$. The predicted category number can then be calculated as:

$$\hat{y} = \sum_{i=0}^{n-1} b_i \pmod K \tag{1}$$

where K is the number of categories. Note, that for two categories $K = 2$, Eq. (1) is just a parity function. Thus, in this case, the classification result is equivalent to the previously described one.

After measuring the same circuit again, a different set of bits $\boldsymbol{b}$ is obtained. Thus, the output from multiple shots of the experiment are the probabilities p_k of belonging the feature vector $\boldsymbol{x}$ to different categories $k = 0, \ldots K - 1$.

Cost Function. To compare the different parameters $\boldsymbol{\theta}$, we need to score them according to some criteria. We call the function that scores our parameters the "cost" or "loss" function, as bad results are more costly.

The combination of softmax with cross-entropy is a standard choice for a cost function to train neural network classifiers [22]. It measures the cross-entropy between the given true label y and the output of the neural network $\hat{y}$. The network's parameters are then adjusted to reduce the cross-entropy via back-propagation.

Cross-entropy measures the difference between two probability distributions q_k and p_k:

$$-\sum_{k=0}^{K-1} q_k \log_2 p_k \tag{2}$$

In our case, the target distribution $q_k = \delta_{ky}$ is just a delta function. Therefore, Eq. (2) becomes:

$$-\log_2 p_y \tag{3}$$

where p_k is the probability of measuring label k. In order to obtain the cost function, we sum expressions from Eq. (3) for all points from the training set. This calculation is done by Qiskit `OneHotObjectiveFunction`.

Similarly, the authors [1] also use a cross-entropy cost function for a binary classification problem.

2.3 Training Parameterized Quantum Circuits

Like classical models, we can train parameterized quantum circuit models to perform data classification tasks. The task of supervised learning can be mathematically expressed as the minimization of the objective function, with respect to the parameter vector $\boldsymbol{\theta}$. In the training phase, the quantum computer calculates the predicted labels. The classical computer compares them to the provided labels and estimates the success of our predictions using the objective function. Based on this cost, the classical computer chooses another value for θ using a classical optimization algorithm. This new value is then used to run a new quantum circuit, and the process is repeated until the objective function stabilizes at a minimum. Note, that finding a global minimum is not an easy task, as the loss landscape can be quite complicated [18].

There are many different types of algorithms that we can use to optimise the parameters of a variational circuit: gradient-based, evolutionary, and gradient-free methods. In this paper, we will be using gradient-based methods. For circuit-based gradients, there's a very nice theoretical result—the parameter shift rule [25]. It gives a very easy formula for calculating gradients on the quantum circuit itself. It is very similar to the equation for finite difference gradients except for the difference is not infinitesimally small.

In vanilla gradients, the Euclidean distance between the points is used, which doesn't take the loss landscape into account. With Quantum Natural Gradients [29], a distance that depends on the model based on Quantum Fisher Information is used instead. It allows to transform the steepest descent in the Euclidean parameter space to the steepest descent in the model space and thus approaches the target faster than vanilla gradient descent. However, this comes at the cost of needing to evaluate many more quantum circuits.

Simultaneous Perturbation Stochastic Approximation (SPSA) [11] is an optimization technique where to reduce the number of evaluations we randomly sample from the gradient. Since we don't care about the exact values but only about convergence, an unbiased sampling works on average quite well. In practise, while the exact gradient follows a smooth path to the minimum, SPSA will jump around due to the random sampling, but eventually it will converge, given the same boundary conditions as the gradient.

In this paper, in addition to SPSA, Constrained Optimization by Linear Approximation (COBYLA) [23], and Sequential Least SQuares Programming (SLSQP) [15] optimization algorithms are used.

3 Results

In this section we use the data from the Iris Data Set [3]. It contains 3 classes: *Iris-setosa*, *Iris-versicolor*, and *Iris-virginica*; 150 samples (50 for each class), and 4 features: *petal width*, *petal length*, *sepal width*, and *sepal length*. This data set was split into the training and test sets using `shuffle` and `train_test_split` functions from the `scikit` library. The test set constitutes 30% of the total data set, i.e., it contains 45 samples.

To obtain the results presented below, Python version 3.9.9 was used together with libraries listed in Table 1.

Table 1. Libraries used and their versions.

Library	Version
Qiskit	0.31.0
Qiskit-Machine-Learning	0.2.1
Qiskit-Terra	0.18.3
Qiskit-Aer	0.9.1
NumPy	1.21.1

3.1 Qiskit Simulation

As described in Sect. 2.1 we have considered circuits with 3 types of data encoding layers: `ZFeatureMap`, `ZZFeatureMap`, and `PauliFeatureMap`. One parametrized circuit `RealAmplitudes` was used. Both the data encoding and parametrized circuits were repeated 1, 2, and 4 times. This gives the total number of 27 variational classifier circuits. Each circuit was run and measured 1024 times.

Training Phase. In the training phase 3 classical optimizers from Sect. 2.3: SPSA, SLSQP, and COBYLA were used to tune the variational parameters on the Qiskit simulator. They were limited to 200 learning iterations each, but SLSQP always finished its work before 100 iterations. Sometimes, COBYLA or SPSA also finish their work earlier.

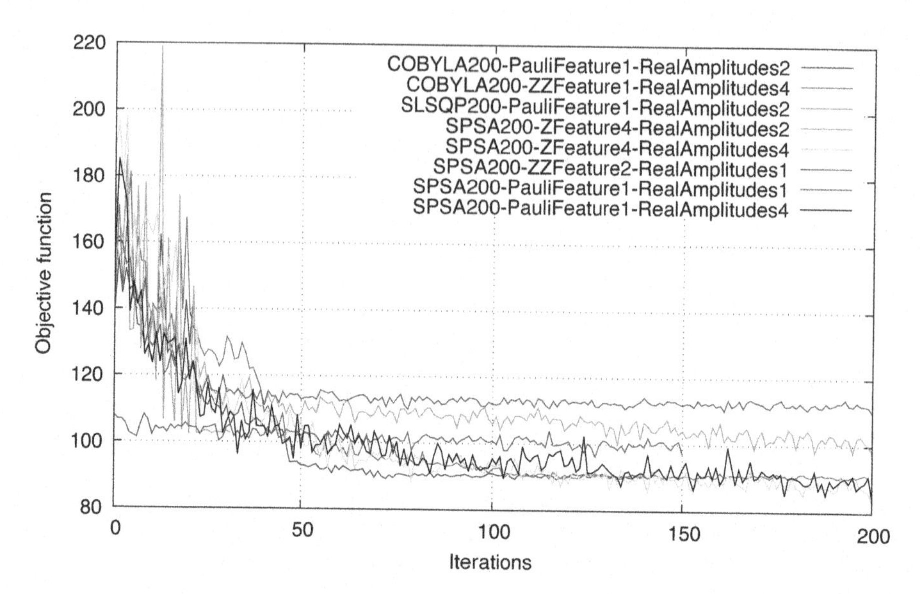

Fig. 2. Objective function value versus number of iterations as calculated by the Qiskit simulator. The cases ploted are marked in bold in Table 2. (Color figure online)

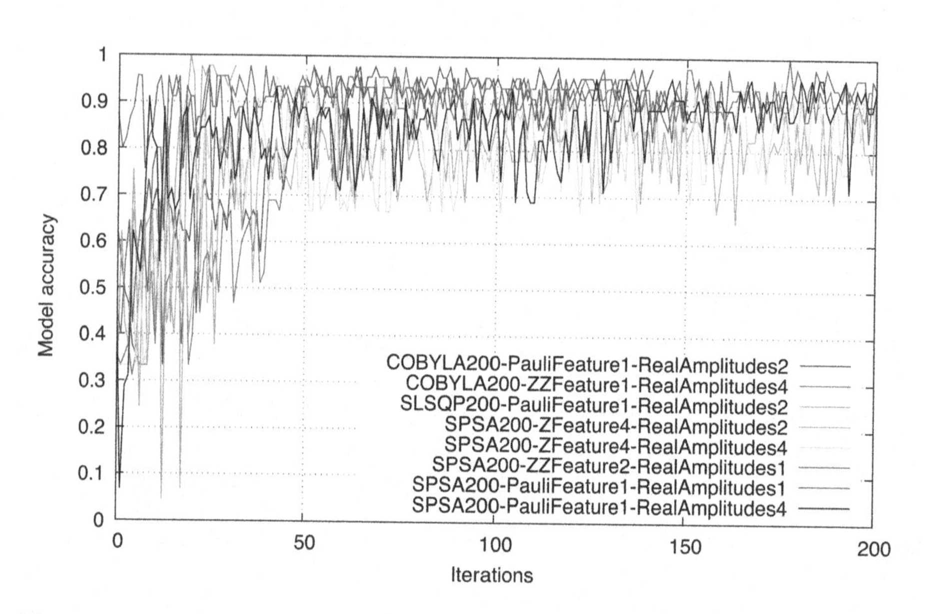

Fig. 3. Model accuracy versus number of iterations as calculated by the Qiskit simulator. The cases ploted are marked in bold in Table 2. (Color figure online)

Example results for selected cases (that also performed well on a real quantum computer) are shown in Fig. 2. The fastest case to stabilize is *SLSQP200-PauliFeature1-RealAmplitudes2* (blue line), which reached its target value of the objective function already after 31 iterations. Next come *COBYLA200-PauliFeature1-RealAmplitudes2* (violet line) and *SPSA200-ZZFeature2-RealAmplitudes1* (dark blue line), where 141 and 151 iterations were needed respectively. The result of *COBYLA200-PauliFeature1-RealAmplitudes2* (violet line) is among one of the lowest, was reached very fast, and is very stable with the number of iterations.

In general, if the loss landscape is fairly flat, it can be difficult for the optimization method to determine which direction to search. This situation is called a barren plateau [20]. For all 81 cases considered here, this phenomenon was not observed. The optimization algorithms were always able to stabilize at some minimum of the objective loss function.

Accuracy Check. Model accuracy is defined as the ratio of the correctly predicted classes to the number of points in the test set. The predicted class k is the one observed with the highest probability p_k in 1024 repetitions of the experiment. The results of its calculation on the test set are shown in Fig. 3. We see that the accuracy of the models optimized by SPSA oscillates a lot with the algorithm iteration number. The results of SLSPQ and COBYLA are by far more stable. The cases *COBYLA200-PauliFeature1-RealAmplitudes2* (violet line) and *SLSQP200-PauliFeature1-RealAmplitudes2* (blue line) seem to provide the highest and most stable accuracy. All the models shown here are able to score above 90%, therefore no overfitting is observed.

3.2 IBM Quito Quantum Computer

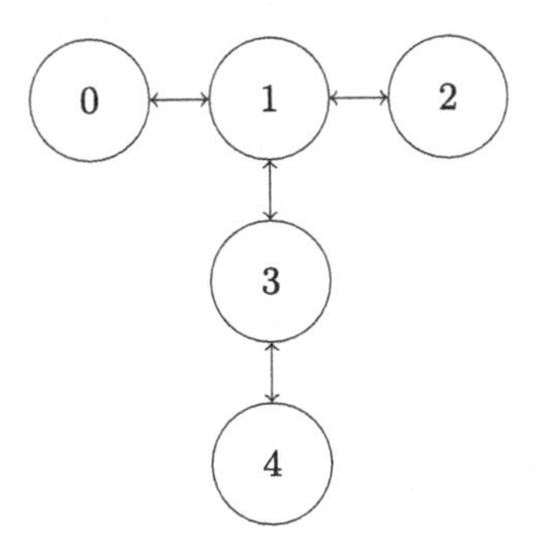

Fig. 4. Qubit connection topology of the IBM Quito quantum computer.

In this section we repeat the experiments on a real 5 qubit IBM Quito quantum computer. Its qubit connection topology is presented in Fig. 4. We need only

4 of them. Note that both the data encoding circuits and the parametrized circuit from Sect. 2.1 use an all-to-all qubit entanglement scheme (except for the `ZFeatureMap` where no entanglement is used). Therefore, physical qubits 0–3 are the optimal choice to map the 4 logical qubits of our circuits. By choosing this subset, each pair of qubits will be separated by at most one additional qubit. On the other hand, if qubits 1–4 were chosen, then some pairs of qubits would be separated by two additional qubits.

We have sorted the table of 81 results from Sect. 3.1 by descending accuracy. The first 32 cases have been taken and launched on qubits 0–3 of the IBM Quito quantum computer. For variational parameters the values computed during Qiskt simulation were substituted. No training on the quantum hardware was performed due to limited computer availability. The results of this experiment are presented in Table 2. Only 8 models scored accuracy above 60%. They are marked in bold (these cases were used in Figs. 2 and 3). For some models, the accuracy dropped below 33.3%, i.e., below random class assignment. Models with the lowest numbers of *CNOT* gates generally achieve the best results: *SPSA200-ZFeature4-RealAmplitudes4*, 24 gates, accuracy 73,3%; *SPSA200-PauliFeature1-RealAmplitudes1*, 18 gates, accuracy 71,1%; *SPSA200-ZFeature4-RealAmplitudes2*, 12 gates; accuracy 68,9%. This will be discussed in the next section.

Note, that the choice of an classical optimizer is very important for the quantum algorithm performance. The accuracy of *SPSA200-ZFeature4-RealAmplitudes2* dropped to 33.3% but *SPSA200-ZFeature4-RealAmplitudes4* performed much better—in this case the accuracy dropped only to 68.9%.

4 Discussion

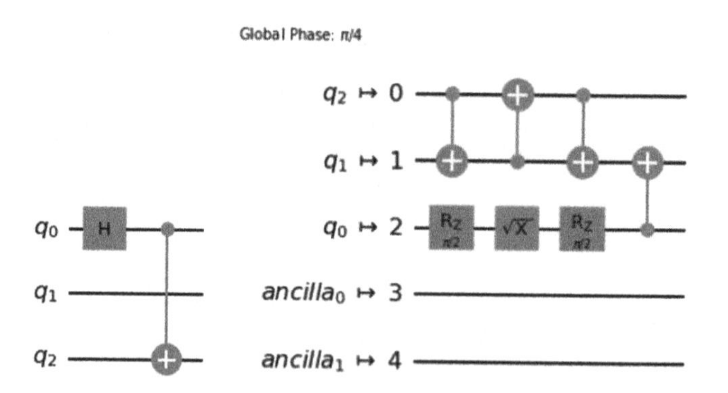

Fig. 5. Simple quantum circuit for Bell state production before (left) and after (right) transpilation.

On a real quantum computer, not every pair of qubits is directly connected. Therefore, executing a two-qubit gate on them leads to the inclusion of additional

Table 2. Model accuracy comparison between Qiskit simulation and execution on a real quantum computer.

Variational quantum circuit and optimization algorithm used for its training	Model accuracy on a Qiskit simulator	Model accuracy on a real quantum computer	Number of *CNOT* gates before transpilation	Number of *CNOT* gates after transpilation	Number of single qubit gates
COBYLA200-PauliFeature1-RealAmplitudes2	97.7%	**66.6%**	24	60	170
SLSQP200-ZZFeature4-RealAmplitudes1	97.7%	31.1%	54	135	184
SLSQP200-PauliFeature1-RealAmplitudes2	97.7%	**66.6%**	24	60	170
SLSQP200-PauliFeature1-RealAmplitudes4	97.7%	48.9%	36	88	226
SLSQP200-PauliFeature2-RealAmplitudes1	97.7%	57.8%	30	71	196
COBYLA200-ZZFeature2-RealAmplitudes4	95.5%	48.9%	48	129	220
COBYLA200-PauliFeature4-RealAmplitudes2	95.5%	33.3%	60	144	368
SPSA200-ZZFeature1-RealAmplitudes4	95.5%	55.6%	36	91	190
SPSA200-ZZFeature2-RealAmplitudes4	95.5%	37.8%	48	129	220
SPSA200-ZZFeature4-RealAmplitudes4	95.5%	35.6%	72	180	280
SLSQP200-ZZFeature2-RealAmplitudes2	95.5%	48.9%	36	101	164
SLSQP200-ZZFeature4-RealAmplitudes2	95.5%	33.3%	60	152	224
SLSQP200-ZZFeature2-RealAmplitudes4	95.5%	33.3%	48	129	220
SLSQP200-ZZFeature4-RealAmplitudes4	95.5%	35.6%	72	180	280
SLSQP200-PauliFeature4-RealAmplitudes2	95.5%	33.3%	60	144	368
SLSQP200-PauliFeature4-RealAmplitudes4	95.5%	33.3%	72	172	424
COBYLA200-ZFeature4-RealAmplitudes4	93.3%	33.3%	24	63	256
COBYLA200-ZZFeature1-RealAmplitudes4	93.3%	**66.7%**	36	91	190
COBYLA200-PauliFeature2-RealAmplitudes4	93.3%	40.0%	48	116	292
COBYLA200-PauliFeature4-RealAmplitudes4	93.3%	33.3%	72	172	424
SPSA200-ZFeature4-RealAmplitudes2	93.3%	**68.9%**	12	35	200
SPSA200-ZFeature4-RealAmplitudes4	93.3%	**73.3%**	24	63	256
SPSA200-ZZFeature2-RealAmplitudes1	93.3%	**62.2%**	30	84	124
SPSA200-ZZFeature4-RealAmplitudes1	93.3%	31.1%	54	135	184
SPSA200-PauliFeature1-RealAmplitudes1	93.3%	**71.1%**	18	43	130
SPSA200-PauliFeature4-RealAmplitudes1	93.3%	42.2%	54	127	328
SPSA200-PauliFeature4-RealAmplitudes2	93.3%	42.2%	60	144	368
SPSA200-PauliFeature1-RealAmplitudes4	93.3%	**68.9%**	36	88	226
SPSA200-PauliFeature4-RealAmplitudes4	93.3%	28.9%	72	172	424
SLSQP200-ZZFeature1-RealAmplitudes4	93.3%	33.3%	36	91	190
SLSQP200-PauliFeature2-RealAmplitudes2	93.3%	26.7%	36	88	236
SLSQP200-PauliFeature2-RealAmplitudes4	93.3%	33.3%	48	116	292
SLSQP200-PauliFeature4-RealAmplitudes1	93.3%	35.5%	54	127	328

helper gates to the circuit during the transpilation phase. For example, consider qubits 0 and 2 from Fig. 4. Let us launch on them a simple circuit for Bell state production that consists of one Hadamard gate H and one $CNOT$ gate. It is shown on the left of Fig. 5. During the transpilation phase, this single $CNOT$ gate is changed to four $CNOT$ gates. The result is presented on the right side of Fig. 5. The Hadamard gate H has been changed to 3 single qubit gates: $R_Z(\pi/2)$, $\sqrt{X}$, and $R_Z(\pi/2)$ because the IBM Quito quantum computer has no direct implementation for it. Notice the change of the qubit on which the single qubit gate operates—this is due to the phase-kickback effect [17].

At the moment the experiment from Sect. 3.2 was performed, the average error of the $CNOT$ gate of the IBM Quito computer was $1.052e{-}2$—a detailed error map for this this machine is given on the IBM Quantum Computing web page[2]. The circuits that performed best in Table 2 all have about 60 or less $CNOT$ gates after transpilation.

5 Conclusions

The subject of this study was to implement and study a variational quantum program for supervised machine learning. Multi-class classification of the Iris Data Set was performed without resorting to multiple binary classifications and One-vs-Rest or One-vs-One strategies. Model accuracy of different data encoding circuits and repetitions of the parametrized circuit was analyzed both on a simulator and on a real quantum computer. It seems that one repetition of `PauliFeatureMap` with Z, Y, and ZZ gates; and two repetitions of the `RealAmplitudes` is the candidate that is stable in training and gives good classification accuracy. This circuit consists of 60 $CNOT$ gates after transpilation on the IBM Quito computer and has 170 single qubit gates. 4 features of the data set are mapped onto a 16 dimensional Hilbert space. Encoding circuit parameters depend not only on the features themselves but also on their products. The parametrized circuit (ansatz) has the total number of 12 variational parameters that are trained using a classical optimizer. The best results seem to come from COBYLA and SLSQP. In future research we plan to implement classical artificial intelligence that would be responsible for an optimal circuit choice. There are already some first steps in this direction [16].

In [4] the authors proposed another hybrid classical-quantum variational approach for multi-class classification. Their quantum multi-class classifier is designed with multiple layers of entangled rotational gates on data qubits and ancilla qubits with adjustable parameters. The class number is predicted by measuring the ancilla qubits. Similarly to the `ZFeatureMap` data encoding uses only single-qubit rotations. Their circuit is created using PennyLane and run on the classical simulator. By training it on the Iris Data Set these authors were able to reach the accuracy of 92.10%. This is lower than the accuracy of all cases from Table 2 where the top accuracy od 97.7% was reached without using any ancilla qubits.

[2] https://quantum-computing.ibm.com/services?services=systems.

A large discrepancy between the accuracy of the model created on the Qiskit simulator and a real IBM Quito quantum computer has been found. It is due to gate errors and peculiar qubit connection topology. This accuracy drop persists even when the training is done on the quantum computer itself. We were quite surprised by these results because all IBM quantum computers with 7 or more qubits available today have a connection topology consisting of T-like structures similar to that from Fig. 4 and have similar error gates for the $CNOT$ gates. They also utilize the same gate set presented in Sect. 4. In [1] there was absolutely no problem in training a variational binary classifier on the Iris Data Set. The IBM Montreal 27 qubit quantum computer reached lower value of the cost function faster that a classical simulator (model accuracy was not calculated). The possible explanation is that for more qubits available the transpiler was able to choose physical qubits with lower $CNOT$ gate errors. Unfortunately we did't have access to that machine to check this. Another explanation is that these authors used linear qubit connectivity instead of all-to-all. We use the latter one similarly to [4]. In the future, fault tolerant quantum computers or computers with all-to-all qubit connection topology should eliminate these problems.

The code used in this paper as well as some additional results are available from a publicly accessible GitHub repository[3].

References

1. Abbas, A., Sutter, D., Zoufal, C., Lucchi, A., Figalli, A., Woerner, S.: The power of quantum neural networks. Nat. Comput. Sci. **1**(6), 403–409 (2021). https://doi.org/10.1038/s43588-021-00084-1
2. Alam, M., Ash-Saki, A., Ghosh, S.: Analysis of quantum approximate optimization algorithm under realistic noise in superconducting qubits. arXiv preprint arXiv:1907.09631 (2019)
3. Anderson, E.: The species problem in iris. Ann. Mo. Bot. Gard. **23**(3), 457–509 (1936)
4. Chalumuri, A., Kune, R., Manoj, B.S.: A hybrid classical-quantum approach for multi-class classification. Quantum Inf. Process. **20**(3), 1–19 (2021). https://doi.org/10.1007/s11128-021-03029-9
5. Córcoles, A.D., et al.: Challenges and opportunities of near-term quantum computing systems. arXiv preprint arXiv:1910.02894 (2019)
6. Dattani, N., Chancellor, N.: Embedding quadratization gadgets on Chimera and Pegasus graphs. arXiv preprint arXiv:1901.07676 (2019)
7. Dattani, N., Szalay, S., Chancellor, N.: Pegasus: the second connectivity graph for large-scale quantum annealing hardware. arXiv preprint arXiv:1901.07636 (2019)
8. Dunjko, V., Taylor, J.M., Briegel, H.J.: Quantum-enhanced machine learning. Phys. Rev. Lett. **117**(13), 130501 (2016)
9. Farhi, E., Goldstone, J., Gutmann, S.: A quantum approximate optimization algorithm. arXiv preprint arXiv:1411.4028 (2014)
10. Fingerhuth, M., Babej, T., Wittek, P.: Open source software in quantum computing. PLoS ONE **13**(12), e0208561 (2018)

[3] https://github.com/odisei369/quantum_learning_iris.

11. Gacon, J., Zoufal, C., Carleo, G., Woerner, S.: Simultaneous perturbation stochastic approximation of the quantum fisher information. Quantum **5**, 567 (2021)
12. Havlíček, V., et al.: Supervised learning with quantum-enhanced feature spaces. Nature **567**(7747), 209–212 (2019)
13. Hubregtsen, T., Pichlmeier, J., Bertels, K.: Evaluation of parameterized quantum circuits: on the design, and the relation between classification accuracy, expressibility and entangling capability. arXiv preprint arXiv:2003.09887 (2020)
14. Kandala, A., et al.: Hardware-efficient variational quantum eigensolver for small molecules and quantum magnets. Nature **549**(7671), 242–246 (2017)
15. Kraft, D., et al.: A software package for sequential quadratic programming (1988)
16. Kuo, E.J., Fang, Y.L.L., Chen, S.Y.C.: Quantum architecture search via deep reinforcement learning. arXiv preprint arXiv:2104.07715 (2021)
17. Lee, C.M., Selby, J.H.: Generalised phase kick-back: the structure of computational algorithms from physical principles. New J. Phys. **18**(3), 033023 (2016)
18. Li, H., Xu, Z., Taylor, G., Studer, C., Goldstein, T.: Visualizing the loss landscape of neural nets. In: Advances in Neural Information Processing Systems 31 (2018)
19. Linke, N.M., et al.: Experimental comparison of two quantum computing architectures. Proc. Natl. Acad. Sci. **114**(13), 3305–3310 (2017)
20. McClean, J.R., Boixo, S., Smelyanskiy, V.N., Babbush, R., Neven, H.: Barren plateaus in quantum neural network training landscapes. Nat. Commun. **9**(1), 1–6 (2018)
21. Peruzzo, A., et al.: A variational eigenvalue solver on a photonic quantum processor. Nat. Commun. **5**(1), 1–7 (2014)
22. Qin, Z., Kim, D., Gedeon, T.: Rethinking softmax with cross-entropy: neural network classifier as mutual information estimator. arXiv preprint arXiv:1911.10688 (2019)
23. Rios, L.M., Sahinidis, N.V.: Derivative-free optimization: a review of algorithms and comparison of software implementations. J. Glob. Optim. **56**(3), 1247–1293 (2013)
24. Salman, S., Liu, X.: Overfitting mechanism and avoidance in deep neural networks. arXiv preprint arXiv:1901.06566 (2019)
25. Schuld, M., Bergholm, V., Gogolin, C., Izaac, J., Killoran, N.: Evaluating analytic gradients on quantum hardware. Phys. Rev. A **99**(3), 032331 (2019)
26. Schuld, M., Petruccione, F.: Supervised Learning with Quantum Computers. QST, Springer, Cham (2018). https://doi.org/10.1007/978-3-319-96424-9
27. Shor, P.W.: Fault-tolerant quantum computation. In: Proceedings of 37th Conference on Foundations of Computer Science, pp. 56–65. IEEE (1996)
28. Sim, S., Johnson, P.D., Aspuru-Guzik, A.: Expressibility and entangling capability of parameterized quantum circuits for hybrid quantum-classical algorithms. Adv. Quant. Technol. **2**(12), 1900070 (2019)
29. Stokes, J., Izaac, J., Killoran, N., Carleo, G.: Quantum natural gradient. Quantum **4**, 269 (2020)
30. Wang, Y.E., Wei, G.Y., Brooks, D.: Benchmarking TPU, GPU, and CPU platforms for deep learning. arXiv preprint arXiv:1907.10701 (2019)

Post-error Correction for Quantum Annealing Processor Using Reinforcement Learning

Tomasz Śmierzchalski[1,2](✉), Łukasz Pawela[2], Zbigniew Puchała[2,3], Tomasz Trzciński[1,4,5], and Bartłomiej Gardas[2]

[1] Faculty of Electronics and Information Technology, Warsaw University of Technology, Nowowiejska 15/19, 00-665 Warsaw, Poland

[2] Institute of Theoretical and Applied Informatics, Polish Academy of Sciences, Bałtycka 5, 44-100 Gliwice, Poland

tsmierzchalski@iitis.pl

[3] Faculty of Physics, Astronomy and Applied Computer Science, Jagiellonian University, Łojasiewicza 11, 30-348 Kraków, Poland

[4] Faculty of Mathematics and Computer Science, Jagiellonian University, Łojasiewicza 6, 30-348 Kraków, Poland

[5] Tooploox Sp. z o.o., Tęczowa 7, 53-601 Wrocław, Poland

Abstract. Finding the ground state of the Ising model is an important problem in condensed matter physics. Its applications spread far beyond physic due to its deep relation to various combinatorial optimization problems, such as travelling salesman or protein folding. Sophisticated new methods for solving Ising instances rely on quantum annealing, which is a paradigm of quantum computation. However, commercially available quantum annealers are still prone to errors, and their ability to find low energetic states is limited. This naturally calls for a post-processing procedure to correct errors. As a proof-of-concept, this work combines the recent ideas revolving around the DIRAC architecture with the Chimera topology and applies them in a real-world setting as an error-correcting scheme for D-Wave quantum annealers. Our preliminary results show how to correct states output by quantum annealers using reinforcement learning. Our approach exhibits excellent scalability, as it can be trained on small instances. However, its performance on the Chimera graphs is still inferior to Monte Carlo methods.

Keywords: Quantum error correction · Quantum annealing · Deep reinforcement learning · Graph neural networks

1 Introduction

Many significant optimization problems can be mapped onto the problem of finding the ground state of the Ising model. Among them are Karp's 21 NP-complete problems [11], the travelling salesman [10], the protein folding problems [1] and

D. Groen et al. (Eds.): ICCS 2022, LNCS 13353, pp. 261–268, 2022.
https://doi.org/10.1007/978-3-031-08760-8_22

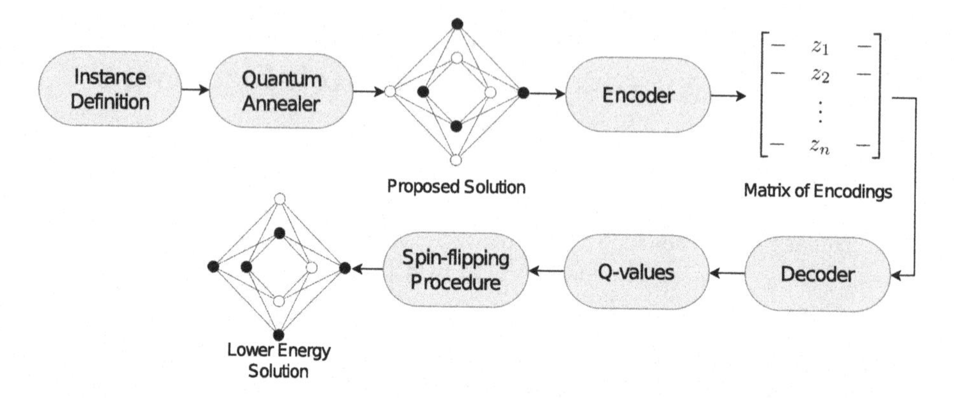

Fig. 1. Overview of our method. Arrows represent subsequent steps. First, we define the Ising instance. Then we obtain the proposed solution from a quantum annealer. White (black) nodes represent spin $\sigma_i = 1$ (-1). Edges represent couplings. In the next step, we encode such an instance using a graph neural network into the matrix of encodings, where each row z_i corresponds to the embedding of a vertex. This matrix is passed through a decoder to obtain Q-values of actions associated with each vertex. The spin flipping procedure involves "flipping" spins one by one according to Q-Values and recording the energy state after each step.

financial portfolio management [16] to name just a few. Promising new methods for solving Ising instances rely on quantum annealing.

The latter is a form of quantum computing, well-tailored for discrete optimization [7,17]. It is closely related to adiabatic quantum computation [12], a paradigm of universal quantum computation which relies on the adiabatic theorem [8] to perform calculations. It is equivalent, up to polynomial overhead, to the better-known gate model of quantum computation [12]. Nevertheless, commercially available quantum annealers are prone to various errors, and their ability to find low energetic states is limited.

Inspired by the recently proposed deep Reinforcement Learning (RL) method for finding spin glass ground states [3], here, we propose a new post-processing error correction schema for quantum annealers called *Simulated Annealing with Reinforcement* (SAwR). This procedure combines deep reinforcement learning with simulated annealing (SA). We employ a graph neural network to encode the Ising instance into an ensemble of low-dimensional vectors used for RL. The agent learns a strategy for improving solutions outputted by the D-Wave annealer. The process of finding the lower energy state involves "flipping" spins one by one according to the learned strategy and recording the energy state after each step. The solution is defined as the lowest energy state found during this procedure. Figure 1 presents an overview of this method. In SAwR, we start with SA, and at low temperature, we replace the Metropolis-Hasting criterion with a single pass of the spin flipping procedure.

2 Results

Ising Problem and Quantum Annealing. The Ising problem is defined on a graph $\mathcal{G} = (\mathcal{V}, \mathcal{E})$. Its Hamiltonian is given by [15]:

$$H_{\text{Ising}} = \sum_{\langle i,j \rangle \in \mathcal{E}} J_{ij}\sigma_i\sigma_j + \sum_{i \in \mathcal{V}} h_i\sigma_i, \tag{1}$$

where $\sigma_i \pm 1$ is the i-th spin, $\langle i, j \rangle$ denotes neighbours in $\mathcal{G}$, J_{ij} strength of interaction between i-th and j-th spin, h_i is an external magnetic field. Our goal is to find the ground state, which corresponds to the minimal energy of H_{Ising}, which is NP-hard. Quantum annealing is one of the methods for finding the ground state of (1). This can be achieved by the adiabatic evolution from the initial Hamiltonian H_X of the transverse field Ising model to the final Hamiltonian H_{Ising}. The Hamiltonian for such process is described by $\mathcal{H}(t) = A(t)H_X + B(t)H_{\text{Ising}}$, where $H_X = \sum_i \hat{\sigma_i}^x$ and $\hat{\sigma_i}^x$ is the standard Pauli X matrix acting on the i-th qubit. The function $A(t)$ decreases monotonically to zero, while $B(t)$ increases monotonically from zero, with $t \in [0, t_f]$, where t_f denotes the annealing time [6,17].

Reinforcement Learning Formulation. We consider a standard reinforcement learning setting defined as a Markov Decision Process [18] where an agent interacts with an environment over a number of discrete time steps $t = 0, 1, \ldots, T$. At each time step t, the agent receives a state s_t and selects an action a_t from some set of possible actions $\mathcal{A}$ according to its policy π, where π is a mapping from set of states $\mathcal{S}$ to set of actions $\mathcal{A}$. In return, the agent receives a scalar reward r_t and moves to the next state s_{t+1}. The process continues until the agent reaches a terminal state, s_T, after which the process restarts. We call one pass of such a process an episode. The return at time step t, denoted, $R_t = \sum_{k=0}^{T-t} \gamma^k r_{t+k}$ is defined as a sum of rewards that the agent will receive for the rest of the episode discounted by the discount factor $\gamma \in (0, 1]$. The goal is to maximize the expected return from each state s_t. The basic components, in the context of the Ising spin-glass model, are:

- **State**: s represents the observed spin glass instance, including the spin configuration σ, the couplings $\{J_{ij}\}$ and values of the magnetic field $\{h_i\}$.
- **Action**: $a^{(i)}$ refers to the flip of spin i, i.e., changing is value to the opposite. For example, after the agent performs an action, $a^{(i)}$, $\sigma_i = 1$ becomes -1.
- **Reward**: $r(s_t; a_t^{(i)}; s_{t+1})$ is the energy change after flipping spin i from state s_t to a new state s_{t+1}.

Starting at $t = 0$, an agent flips one spin during each time step, which moves him to the next state (different spin configuration). The terminal state s_T is met when the agent has flipped each spin. The solution is defined as the spin configuration σ corresponding to the lowest energy state found during this procedure. An action-value function $Q^\pi(s, a) = \mathbb{E}(R_t \mid s_t = s,\ a_t = a)$ is the expected return for selecting action a in state s and following policy π. The value $Q^\pi(s, a)$ is often called Q-value of action a in state s. The optimal action-value function, $Q^*(s, a) = \max_\pi Q^\pi(s, a)$ which gives the maximum action value for state

s and action a achievable by any policy. As learning of the optimal action-value function is in practice infeasible, We seek to learn the function approximator $Q(s, a; \Theta) \approx Q^*(s, a)$ where Θ is a set of learnable model parameters. We denote policy used in such approximation as π_Θ.

Model Architecture. Our model architecture is inspired by DIRAC (**D**eep reinforcement learning for sp**I**n-glass g**R**ound-st**A**te **C**alculation), an Encoder-Decoder architecture introduced in [3]. It exploits the fact that the Ising spin-glass instance is wholly described by the underlying graph. In this view, couplings J_{ij} become edge weights, external magnetic field h_i and spin σ_i become node weights. Employing DIRAC is a two-step process. At first, it encodes the whole spin-glass instance such that every node is embedded into a low-dimensional vector, and then the decoder leverages those embeddings to calculate the Q-value of every possible action. Then, the agent chooses the action with the highest Q-value. In the next sections, We will describe those steps in detail.

Encoding. As described above, the Ising spin-glass instance can be described in the language of graph theory. It allows us to employ graph neural networks [4,5], which are neural networks designed to take graphs as inputs. We use modified SGNN (**S**pin **G**lass **N**eural **N**etwork) [3] to obtain node embedding. To capture the coupling strengths and external field strengths (J_{ij} and h_i), which are crucial to determining the spin glass ground states, SGNN performs two updates at each layer: the edge-centric update and the node-centric update, respectively.

Let's $z_{(i,j)}$ denote embedding of edge (i, j) and $z_{(i)}$ embedding of node i. The edge-centric update aggregates embedding vectors from its adjacent nodes (i.e. for edge (i, j) this update aggregate embeddings $z_{(i)}$ and $z_{(j)}$), and then concatenates it with self-embedding $z_{(i,j)}$. The vector obtained in this way is then subject to non-linear transformation (ex. $\mathrm{ReLU}(x) = \max(0, x)$). Mathematically, it can be described by the following equation

$$z^{k+1}_{(i,j)} = \mathrm{ReLU}(\gamma_\theta(z^k_{(i,j)}) \oplus \phi_\theta(z^k_{(i)} + z^k_{(j)})), \tag{2}$$

where $z^k_{(i,j)}$ denotes encoding of edge (i, j) obtained after k layers. Similarly, $z^k_{(i)}$ denotes encoding of node i obtained after k layers, γ_θ and ϕ_θ are some differentiable functions of θ. Symbol $\oplus$ is used to denote concatenation operation.

The node-centric update is defined in a similar fashion. It aggregates embedding of adjacent edges and then concatenates it with self-embedding $z_{(i)}$. Later, we transform this concatenated vector to obtain the final embedding. Using notation from equation (2), the final result is the following:

$$z^{k+1}_{(i)} = \mathrm{ReLU}(\phi_\theta(z^k_{(i)}) \oplus \gamma_\theta(\mathrm{E}^k_i)), \quad \mathrm{E}^k_i = \sum_j z^k_{(i,j)}. \tag{3}$$

Edge features are initialized as edge weights $\{J_{ij}\}$. It is not trivial to find adequate node features, as node weights $\{h_i\}$ and spins σ_i are not enough.

We also included pooling layers not presented in the original design. We reasoned that after concatenation, vectors start becoming quite big, so we employ pooling layers to not only reduce the model size but also preserve the most essential parts of every vector. As every node is a potential candidate for action, we call the final encoding of node i its *action embedding* and denote it as Z_i. To represent the whole Chimera, we use *state embedding*, denoted as Z_s, which is the sum over all node embedding vectors, which is a straightforward but empirically effective way for graph-level encoding [9].

Decoding. Once all action embeddings Z_i and state embedding Z_s are computed in the encoding stage, the decoder will leverage these representations to compute an approximated state-action value function $Q(s, a; \Theta)$ which predicts the expected future rewards of taking action a in state s, and following the policy π_Θ till the end. Specifically, we concatenate the embeddings of state and action and use it as decoder input. In principle, any decoder architecture may be used. Here, we use a standard feed-forward neural network. Formally, the decoding process can be written as $Q(s, a^{(i)}; \Theta) = \psi_\Theta(Z_s \oplus Z_i)$, where ψ_Θ is a dense feed-forward neural network.

Training. We train our model on randomly generated Chimera instances. We found that the minimal viable size of the training instance is C_3. Smaller instances lack couplings between clusters, crucial in full Chimera, which leads to poor performance. We generate $\{J_{ij}\}$ and $\{h_i\}$ from a normal distribution $\mathcal{N}(0, 1)$ and starting spin configuration σ from a uniform distribution. To introduce low-energy instances, we employed the following pre-processing procedure. For each generated instance, with probability $p = 10\%$, we perform simulated annealing before passing the instance through SGNN.

Our goal is to learn approximation of optimal action-value function $Q(a, s; \Theta)$, so as the reinforcement learning algorithm we used standard n-step deep Q learning [13,14] with memory replay buffer. During the episode, we collect sequence of states action and rewards $\tau = (s_0, a_0, r_0, \ldots, s_{T-1}, a_{T-1}, r_{T-1}, s_T)$ with terminal state as final element. From those we construct n-step transitions $\tau_t^n = (s_t, a_t, r_{t,t+n}, s_{t+n})$ which we collect in memory replay buffer $\mathcal{B}$. Here $r_{t,t+n} = \sum_{k=0}^{k=n} \gamma^k r_{t+k}$ is return after n-steps.

Simulated Annealing with Reinforcement. Simulated annealing with reinforcement (SAwR) combines machine learning and classical optimization algorithm. Simulated annealing (SA) takes its name from a process in metallurgy involving heating a material and then slowly lowering the temperature to decrease defects, thus minimizing the system energy. In SA, we start in some state s and in each step, we move to a randomly chosen neighbouring state s'. If a move lowers energy $E(s)$ of the system, we accept it. If it doesn't, we use the so-called the Metropolis-Hasting criterion: $\mathbb{P}(\text{accept}\, s' \mid s) = \min(1, e^{-\beta \Delta E})$, where $\Delta E = E(s') - E(s)$ and β denotes inverse temperature. In our case, the move is defined as a single-spin flip. Simulated annealing tends to accept all possible moves at high temperatures. However, it likely accepts only those moves that lower the energy at low temperatures.

Our idea is to reinforce random sampling with the trained model. As a result, instead of using the Metropolis-Hasting criterion, we perform a single pass of the DIRAC episode at low temperatures, as described in the caption of Fig. 1.

3 Experiments

We collected data from the D-Wave 2000Q device using default parameters by generating 500 random instances of sizes 128, 512, 1152, and 2048 spins. Parameters $\{J_{ij}\}$ and $\{h_i\}$ were generated from a normal distribution $\mathcal{N}(0,1)$ and initial spin configuration, σ, from a uniform distribution. We used identical distributions for training instances. The low energy states of generated instances were obtained using quantum annealing. We have used three methods - simulated annealing, SAwR, and D-wave steepest descend post-processing [2], cf., Fig. 2.

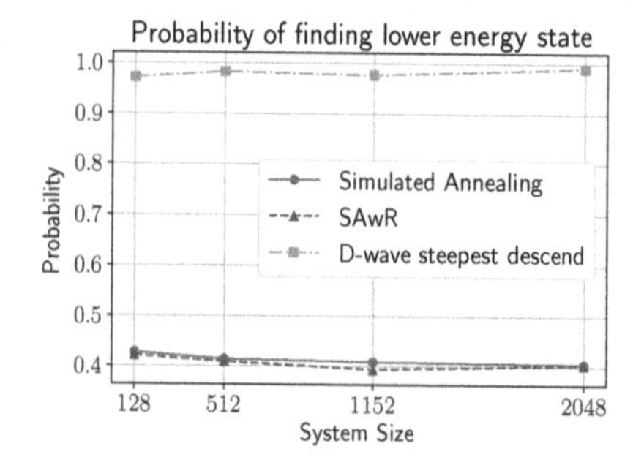

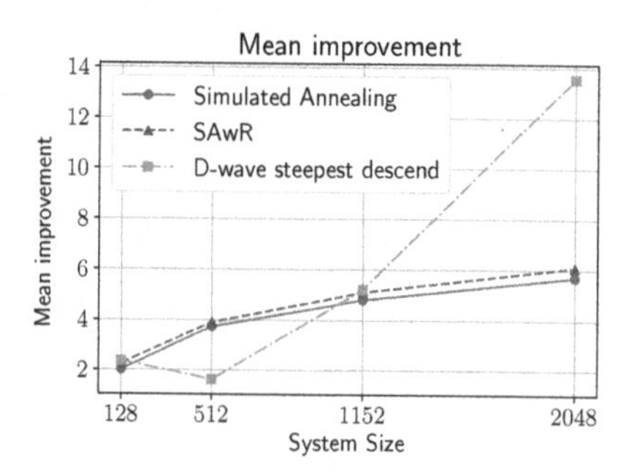

Fig. 2. Results of the experiments. SAwR is the simulated annealing with reinforcement. The probability of finding a lower energy state was computed over 500 random instances for each Chimera size. The value of the improvement was defined as the difference between the initial energy and the lowest energy found.

Two metrics were tested: the probability of finding lower energy states and the mean value of an improvement over the starting energy state. To compute the probability for each Chimera size, we started with proposed solutions obtained from quantum annealer and tried to lower them using different tested methods. Then we counted those instances for which a lower energy state was found. We define the value of the improvement as the difference between the initial energy and the lowest energy found.

SAwR achieved lower probabilities of finding a lower energy state. Although the difference between SAwR and traditional simulated annealing is slight, its consistency across all sizes suggests that it is systemic rather than random noise. It is interesting that, on average, SAwR was able to find a better low energy state than simulated annealing, but still, the difference is not significant.

Acknowledgments. TS acknowledges support from the National Science Centre (NCN), Poland, under SONATA BIS 10 project number 2020/38/E/ST3/ 00269.

This research was funded by National Science Centre, Poland (grant no 2020/39/B/ST6/01511 and 2018/31/N/ST6/02374) - TT, and Foundation for Polish Science (grant no POIR.04.04.00-00-14DE/ 18-00 carried out within the Team-Net program co-financed by the European Union under the European Regional Development Fund) - (LP, ZP, and BG).

References

1. Bryngelson, J.D., Wolynes, P.G.: Spin glasses and the statistical mechanics of protein folding. Proc. Natl. Acad. Sci. U.S.A. **84**(21), 7524–7528 (1987). https://doi.org/10.1073/pnas.84.21.7524
2. D-Wave Systems. Postprocessing. https://docs.dwavesys.com/docs/latest/c_qpu_pp.html. Accessed 8 Apr 2022
3. Fan, C., et al.: Finding spin glass ground states through deep reinforcement learning. arXiv preprint arXiv:2109.14411 (2021)
4. Gilmer, J., Schoenholz, S.S., Riley, P.F., Vinyals, O., Dahl, G.E.: Neural message passing for quantum chemistry. In: Precup, D., Teh, Y.W. (eds.) Proceedings of the 34th International Conference on Machine Learning, Proceedings of Machine Learning Research, vol. 70, pp. 1263–1272. PMLR (2017)
5. Hamilton, W.L.: Graph representation learning. Synthesis lectures on artificial intelligence and machine learning **14**(3), 1–159 (2020). https://doi.org/10.2200/S01045ED1V01Y202009AIM046
6. Jansen, S., Ruskai, M.B., Seiler, R.: Bounds for the adiabatic approximation with applications to quantum computation. J. Math. Phys. **48**(10), 102111 (2007). https://doi.org/10.1063/1.2798382
7. Kadowaki, T., Nishimori, H.: Quantum annealing in the transverse ising model. Phys. Rev. E **58**, 5355–5363 (1998). https://doi.org/10.1103/PhysRevE.58.5355
8. Kato, T.: On the adiabatic theorem of quantum mechanics. J. Phys. Soc. Jpn **5**(6), 435–439 (1950). https://doi.org/10.1143/JPSJ.5.435
9. Khalil, E., Dai, H., Zhang, Y., Dilkina, B., Song, L.: Combinatorial optimization algorithms over graphs. In: Guyon, I., et al. (eds.) Advances in Neural Information Processing Systems, vol. 30. Curran Associates, Inc. (2017). https://proceedings.neurips.cc/paper/2017/file/d9896106ca98d3d05b8cbdf4fd8b13a1-Paper.pdfLearning
10. Kirkpatrick, S., Toulouse, G.: Configuration space analysis of travelling salesman problems. J. Phys. **46**(8), 1277–1292 (1985). https://doi.org/10.1051/jphys:019850046080127700
11. Lucas, A.: Ising formulations of many NP problems. Front. Phys. **2**, 5 (2014). https://doi.org/10.3389/fphy.2014.00005
12. McGeoch, C.C.: Adiabatic quantum computation and quantum annealing: theory and practice. Synth. Lect. Quant. Comput. **5**(2), 1–93 (2014)
13. Mnih, V., et al.: Playing Atari with deep reinforcement learning. arXiv preprint arXiv:1312.5602 (2013)
14. Peng, J., Williams, R.J.: Incremental multi-step q-learning. In: Machine Learning Proceedings, pp. 226–232. Elsevier (1994). https://doi.org/10.1016/B978-1-55860-335-6.50035-0
15. Rams, M.M., Mohseni, M., Eppens, D., Jałowiecki, K., Gardas, B.: Approximate optimization, sampling, and spin-glass droplet discovery with tensor networks. Phys. Rev. E **104**, 025308 (2021). https://doi.org/10.1103/PhysRevE.104.025308

16. Rosenberg, G., Haghnegahdar, P., Goddard, P., Carr, P., Wu, K., De Prado, M.L.: Solving the optimal trading trajectory problem using a quantum annealer. IEEE J. Select. Top. Signal Process. **10**(6), 1053–1060 (2016). https://doi.org/10.1109/JSTSP.2016.2574703
17. Santoro, G.E., Martoňák, R., Tosatti, E., Car, R.: Theory of quantum annealing of an Ising spin glass. Science **295**(5564), 2427–2430 (2002). https://doi.org/10.1126/science.1068774
18. Sutton, R.S., Barto, A.G.: Reinforcement Learning: An Introduction. MIT Press, New York (2018). https://doi.org/10.1007/978-1-4615-3618-5

Simulations of Flow and Transport: Modeling, Algorithms and Computation

Adaptive Deep Learning Approximation for Allen-Cahn Equation

Huiying Xu, Jie Chen(✉), and Fei Ma

Department of Applied Mathematics, School of Science, Xi'an Jiaotong-Liverpool University, Suzhou 215123, China
jie.chen01@xjtlu.edu.cn
https://www.xjtlu.edu.cn/en/study/departments/academic-departments/applied-mathematics/

Abstract. Solving general non-linear partial differential equations (PDE) precisely and efficiently has been a long-lasting challenge in the field of scientific computing. Based on the deep learning framework for solving non-linear PDEs *physics-informed neural networks* (PINN), we introduce an adaptive collocation strategy into the PINN method to improve the effectiveness and robustness of this algorithm when selecting the initial data to be trained. Instead of merely training the neural network once, multi-step discrete time models are considered when predicting the long time behaviour of solutions of the Allen-Cahn equation. Numerical results concerning solutions of the Allen-Cahn equation are presented, which demonstrate that this approach can improve the robustness of original neural networks approximation.

Keywords: Deep learning · Adaptive collocation · Discrete time models · Physics informed neural networks · Allen-Cahn equation

1 Introduction

Non-linear partial differential equations (PDE) play an important role in numerous research areas including engineering, physics and finance. However, solving general non-linear PDEs precisely and efficiently has been a long-lasting challenge in the area of scientific computing [4,6,11,13,15]. Numerical methods such as finite difference and finite element methods (FEM) are popular approaches in solving PDEs, due to their capability in solving non-linear problems and great freedom they can provide in the choice of dispersion. These methods discretise continuous PDEs and create simple basis functions on the discretised domain Ω. Through solving the system of basis coefficients, they approximate the true solutions of the targeted problems. Notwithstanding significant advance of these methods achieved in the past few decades and their capability to deal with fairly

J. Chen—This work is partially supported by Key Program Special Fund in XJTLU (KSF-E-50, KSF-E-21) and XJTLU Research Development Funding (RDF-19-01-15).

D. Groen et al. (Eds.): ICCS 2022, LNCS 13353, pp. 271–283, 2022.
https://doi.org/10.1007/978-3-031-08760-8_23

complicated and oscillating problems, they consume a lot of time and computing resources, especially in the analysis and computation of complex problems.

Confronted with the shortcoming of traditional numerical methods in solving PDEs, researchers turned to deep learning techniques to simulate true results. Deep learning, a subset of machine learning containing powerful techniques that enable computers to learn from data, has attracted enormous interest among people in various fields. Deep learning is a great breakthrough of traditional machine learning algorithms on account of complicated non-linear combinations of input and output features using multiple hidden layers. During training process, the back-propagation algorithm increases weights of the combinations useful to obtain final output features, while gradually deprecate useless internal relations, making the approximation outcomes more accurate step by step. Great efforts have been made to obtain solutions of PDEs based on deep learning approaches. Neural networks are used to improve the precision of results obtained from finite difference method [8]. In [14], the authors apply regression analysis technique to develop a discretisation scheme based on neural networks in the solution of non-linear differential equations. To handle general non-linear PDEs, physics-informed neural networks (PINN) are trained to infer solutions, and the obtained surrogate models are differentiable in regard to all parameters and input data [11]. Deep Galerkin Method (DGM), a similar approach to PINN, is proposed to solve PDEs of high dimension through training a deep neural network to approach the differential operation, boundary conditions and initial condition [13]. Exploring the variational form of PDEs is another attractive approach to researchers [6]. Deep Ritz method (DRM) is one typical example of this approach, which casts governing equations in an energy-minimisation framework to solve PDEs [6].

The rest of this paper is organized as follows. In Sect. 2, we first briefly introduce artificial neural networks, and then provide a detailed explanation to PINNs. The idea of adaptive collocation is also introduced in this section. Using the examples of Allen-Cahn equation, we discuss how to simulate the results with the discrete time models of PINNs in Sect. 3, covering the structure of neural networks employed and the working mechanism. Adaptive collocation strategy is applied in selecting training points at initial and later stages when training PINNs. Several approximation outcomes are demonstrated, which are later compared to the results obtained from the algorithm without adaptive strategy. A conclusion and remark is presented in Sect. 4.

2 Methodology

2.1 Artificial Neural Networks

Neural networks are composed of three kinds of layers – input layer, hidden layers and output layer. The model with more than one hidden layer is referred to as deep neural network (DNN). Figure 1 displays the structure of a DNN with n hidden layers. The realisation of neural networks relies on two algorithms – feed-forward propagation and back-propagation. The former builds the framework for neural networks, while the latter optimises the constructed model to achieve desired results.

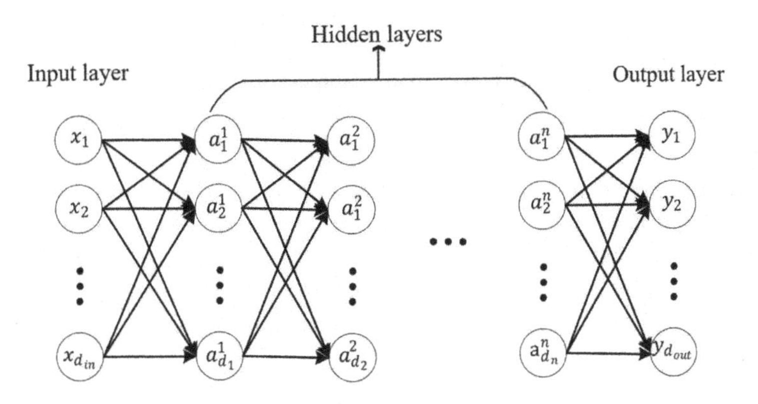

Fig. 1. Representation of a DNN as a graph. The number of nodes in the input layer and the output layer depends on the problem setting, so should be unchangeable, whereas the number of layers and nodes in each hidden layer could be modified to optimise the structure.

Feed-Forward Propagation Algorithm. In a *feed-forward neural network* where the input $\mathbf{x}$ contains d_{in} nodes, the output $\mathbf{y}$ contains d_{out} nodes, and the ith hidden layer $\mathbf{a}_i$ consists of d_i nodes with $i = 1, ..., n$:

$$\begin{aligned} \mathbf{a}_1 &= f(\mathbf{x}; \mathbf{w}_1, b_1) = \mathbf{x}^\top \mathbf{w}_1 + b_1, \\ \mathbf{a}_i &= f(\mathbf{a}_{i-1}; \mathbf{w}_i, b_i) = \mathbf{a}_{i-1}^\top \mathbf{w}_i + b_i, \\ \mathbf{y} &= f(\mathbf{a}_n; \mathbf{w}_{n+1}, b_{n+1}) = \mathbf{a}_n^\top \mathbf{w}_{n+1} + b_{n+1}, \end{aligned} \tag{1}$$

where $i = 2, 3, ..., n$, $\mathbf{w}_1, \mathbf{w}_2, ..., \mathbf{w}_{n+1}$ are the weights and $b_1, b_2, ..., b_{n+1}$ are the biases. Thus, combining the equations in Eq. (1) we get the the prediction for output value $\mathbf{y}$:

$$\mathbf{y} = f(f(\cdots f(\mathbf{x})\cdots)). \tag{2}$$

However, if the neural network is designed in this way, combining all these hidden layers will work the same as merely one hidden layer. Based on this concern, the linear function f here is transformed to a non-linear one, denoted by h known as an activation function. Sigmoid units, tanh units and rectified linear units (ReLU) are the three most popular ones, while ReLU as the default activation function is conventionally used in hidden layers [9]. ReLU is defined as follows:

$$h(x) = \max(0, x). \tag{3}$$

Since ReLU are quite close to linear functions, they are able to preserve more properties, facilitating the optimisation. Hence, in the renewed feed-forward neural network, the output value $\mathbf{y}$ can be obtained by the following:

$$\begin{aligned} \mathbf{a}_1 &= h(f(\mathbf{x}; \mathbf{w}_1, b_1)) = h(\mathbf{x}^\top \mathbf{w}_1 + b_1), \\ \mathbf{a}_i &= h(f(\mathbf{a}_{i-1}; \mathbf{w}_i, b_i)) = h(\mathbf{a}_{i-1}^\top \mathbf{w}_i + b_i), \\ \mathbf{y} &= h(f(\mathbf{a}_n; \mathbf{w}_{n+1}, b_{n+1})) = h(\mathbf{a}_n^\top \mathbf{w}_{n+1} + b_{n+1}). \end{aligned} \tag{4}$$

Back-Propagation Algorithm. After the framework for feed-forward neural network has been constructed, it comes to the training process to optimise the prediction of the output value $\mathbf{y}$, known as *back-propagation.* The most fundamental algorithm in back-propagation is *Gradient Descent.* In this part, the weights $\mathbf{w}$ and the biases b are trained through minimising the *cost function* J:

$$J(\mathbf{w}, b) = \frac{1}{d_{out}} \sum_{j=1}^{d_{out}} L(\hat{y}^{(j)}, y^{(j)}), \tag{5}$$

where y is the true value whereas $\hat{y}$ denotes the prediction value acquired from Eq. (4). The choice of function L depends on the problem to be solved. For example, cross-entropy loss function or exponential loss function is more suitable to be applied in binary classification problem. In regression problems, it is more appropriate to use L_2 loss defined as follows:

$$J(\mathbf{w}, b) = \frac{1}{d_{out}} \sum_{j=1}^{d_{out}} (\hat{y}^{(j)} - y^{(j)})^2. \tag{6}$$

Now, the problem is simplified to find:

$$(\mathbf{w}, b) = \arg\min_{\mathbf{w}, b} J(\mathbf{w}, b). \tag{7}$$

In Gradient Descent algorithm, the weights $\mathbf{w}$ and the biases b are trained following the formulas:

$$\mathbf{w} := w - \alpha \frac{\partial J(\mathbf{w}, b)}{\partial \mathbf{w}}, \qquad b := b - \alpha \frac{\partial J(\mathbf{w}, b)}{\partial b} \tag{8}$$

where α is the *learning rate.* It is a crucial hyperparameter deciding the rate of convergence. Apart from gradient descent algorithm, a few advanced optimisers can be used to optimise the training process, including Momentum, Adagrad, RMSprop and Adam.

2.2 Physics-Informed Neural Networks

In this work, we consider parametrised non-linear PDEs of the following general form:

$$u_t + \mathcal{N}[u] = 0, x \in \Omega, t \in [0, T], \tag{9}$$

where $u(t;x)$ is the latent solution, $\mathcal{N}[\cdot]$ represents a non-linear operator, and Ω is the domain in $\mathbb{R}^D$. Without the requirement to consider linearisation, prior assumptions, or division of local time interval, we can directly handle the non-linear problem in this setup. Additionally, a large variety of problems in mathematical physics are encapsulated in Eq. (9), including kinetic equations, diffusion processes, conservation laws, and so on [11].

Generally, we use the discrete time models of PINNs, in which q-stage Runge-Kutta methods [10] are applied to Eq. (9):

$$\begin{aligned} u(t_0+c_j\Delta t,x) &= u(t_0,x) - \Delta t\sum_{j=1}^{q} a_{ij}\mathcal{N}[u(t_0+c_j\Delta t,x)],\ i=1,\cdots,q, \\ u(t_0+\Delta t,x) &= u(t_0,x) - \Delta t\sum_{j=1}^{q} b_j\mathcal{N}[u(t_0+\Delta t,x)]. \end{aligned} \tag{10}$$

Here, Δt is a predetermined value and $[t_0, t_0+\Delta t]$ refers to the short time interval we are supposed to study on. The above equations depend on the parameters $\{a_{ij}, b_j, c_j\}$ which are fully determined by Runge-Kutta methods. To simplify the formulas, express Eq. (10) as

$$\begin{aligned} u_i^0(x) &= u(t_0+c_j\Delta t,x) + \Delta t\sum_{j=1}^{q} a_{ij}\mathcal{N}[u(t_0+c_j\Delta t,x)],\ i=1,\cdots,q, \\ u_{q+1}^0(x) &= u(t_0+\Delta t,x) + \Delta t\sum_{j=1}^{q} b_j\mathcal{N}[u(t_0+\Delta t,x)]. \end{aligned} \tag{11}$$

Thus, $u_i^0(x) = u_{q+1}^0(x) = u(t_0,x)$ for $i=1,\cdots,q$. Taking x as the input feature and the following

$$[u_1^0(x), u_2^0(x), \cdots, u_q^0(x), u_{q+1}^0(x)] \tag{12}$$

as output features, together with the setting of several hidden layers, we have constructed the basic structure of PINNs. Each node of the output features $u_i^0(x)$ for $i=1,\cdots,q$ and $u_{q+1}^0(x)$ are all equivalent to $u(t_0,x)$, i.e., the true value of u at the initial stage. Substituting $u(t_0,x)$ by what is obtained from the neural network in Eq. (10), we can predict the values along the short time interval, i.e., $u(t_0+c_j\Delta t,x)$ for $i=1,\cdots,q$ and $u(t_0+\Delta t,x)$.

After determining the input and output features of the neural network, we enable the algorithm to approach to the true solution through minimising the cost function J defined as follows:

$$J(\mathbf{w},b) = J_u + J_b, \tag{13}$$

where J_u refers to the difference between the approximation results $u_1^0(x)$, $\cdots$, $u_q^0(x)$, $u_{q+1}^0(x)$ and the true solution $u(t_0,x)$ of the PDE at initial stage, and J_b is the deviation of the prediction from the true value at boundary points at $t=t_0+c_1\Delta t,\cdots,t_0+c_q\Delta t, t_0+\Delta t$. Both J_u and J_b use mean squared error to reflect the variation.

2.3 Adaptive Collocation Strategy

Various methods of adaptive collocation can be employed to select training data of the neural networks. For instance, the authors use uniformly distributed training points at the initial stages when solving PDEs based on deep learning techniques, while add more points according to the residual values at later stages [2,3,7]. In [11], when training the discrete time algorithms of PINNs, certain number of random training points are generated from the domain Ω at the initial stage. In the case of short-time prediction, instead of choosing training points randomly, we propose to apply one approach of adaptive collocation strategy, and select points based on the real solution of the PDE. However, for the prediction of long-term solution, single-step of the discrete time algorithms may be inapplicable. In this case, we feed last estimation results of the model into the neural network as the sample data, and select training points from it adaptively.

Figure 2 illustrates the idea of our adaptive collocation strategy. The curve is assumed to be the true solution of the PDE at the initial stage, and the blue dots represent the selected training data. In the given simple example, both point-selection strategies choose 16 points from the domain, but the left picture sub-samples training data randomly, while the right one adaptively locates the points according to the value of u. Points have larger possibility to be selected around $u = 0$.

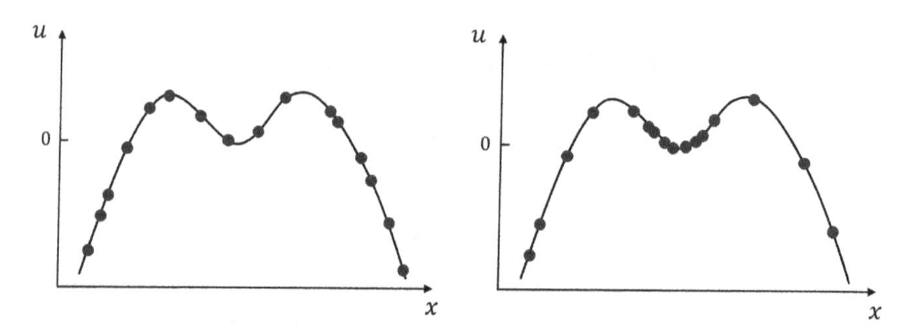

Fig. 2. Simple examples of sub-sampling training data. The left picture opts data randomly, while the right one uses adaptive collocation.

3 Numerial Experiments

3.1 Allen-Cahn Equation

Aiming to understate the capability of proposed discrete time models when handling non-linear PDEs, we take the time-fractional Allen-Cahn equation with period boundary conditions (14) as an example, which is a classical phase-field model.

$$\begin{aligned} &u_t - 10^{-4}u_{xx} + 5u^3 - 5u = 0,\ x\in[-1,1],\ t\in[0,2], \\ &u(0,x) = x^2\cos(\pi x), \\ &u(t,-1) = u(t,1), \\ &u_x(t,-1) = u_x(t,1). \end{aligned} \tag{14}$$

Originally, the Allen-Cahn equation was introduced to demonstrate the motion of a curved anti-phase boundary on multi-component alloy systems [1]. Through a phase-field approach, the Allen-Cahn equation has found its application in various research areas including fluid dynamics and complex moving interface problems [12]. For the Allen-Cahn equation (14), the non-linear operator in Eq. (9) is given by:

$$\mathcal{N}[u] = -10^{-4}u_{xx} + 5u^3 - 5u, \tag{15}$$

and $\mathcal{N}[u]$ is expressed as follows in the discrete form:

$$\begin{aligned} \mathcal{N}[u(t_0+c_j\Delta t,x)] &= -10^{-4}u_{xx}(t_0+c_j\Delta t,x) + 5[u(t_0+c_j\Delta t,x)]^3 \\ &\quad - 5u(t_0+c_j\Delta t,x),\ j=1,\cdots,q, \\ \mathcal{N}[u(t_0+\Delta t,x)] &= -10^{-4}u_{xx}(t_0+\Delta t,x) + 5[u(t_0+\Delta t,x)]^3 \\ &\quad - 5u(t_0+\Delta t,x). \end{aligned} \tag{16}$$

In the neural network that calculate solutions for the Allen-Cahn equation (14), the cost function J is defined as:

$$J(\mathbf{w},b) = J_u + J_b,$$

where

$$J_u = \sum_{i=1}^{q+1} |u_i^0(x) - u(t_0,x)|^2, \tag{17}$$

and

$$\begin{aligned} J_b = &\sum_{j=1}^{q} |u(t_0+c_j\Delta t,-1) - |u(t_0+c_j\Delta t,1)|^2 \\ &+ |u(t_0+\Delta t,-1) - |u(t_0+\Delta t,1)|^2 \\ &+ \sum_{j=1}^{q} |u_x(t_0+c_j\Delta t,-1) - |u_x(t_0+c_j\Delta t,1)|^2 \\ &+ |u_x(t_0+\Delta t,-1) - |u_x(t_0+\Delta t,1)|^2. \end{aligned} \tag{18}$$

Sample data are generated through simulating the Allen-Cahn equation (14) using the Chebfun package [5], which employs conventional spectral methods. An explicit Runge-Kutta integrator with step $\Delta t = 10^{-5}$ is used. Along the phase we study on, solutions of 2048 evenly distributed points in the domain $\Omega = [-1,1]$ are computed, forming the training and test data set.

3.2 Short-Time Prediction

In this subsection, we merely look on the prediction of solutions of the Allen-Cahn equation (14) from $t_0 = 0.1$ to $t_1 = 0.9$, i.e., $\Delta t = 0.8$. Figure 3 shows the simulation outcomes using the Chebfun package. Firstly, 200 training data are sub-sampled randomly from the 2048 points at the initial stage of $t_0 = 0.1$. To predict the solution of the equation at $t_1 = 0.9$, the discrete time models of PINNs are used with 4 hidden layers consisting of 200 nodes each. The output layer contains 101 neurons of $u(t_0 + c_j \Delta t, x)$ for $j = 1, \ldots, 100$ and $u(t_0 + \Delta t, x)$ corresponding to 100 stages of Runge-Kutta method. In Fig. 4, the location of sub-sampled initial training data and the prediction results at the final stage are displayed. The small relative error reflects effectiveness of current model. However, we wonder whether selecting points adaptively corresponding to the value of u at $t = 0.1$ could further improve the accuracy. To this end, among the sample input data $\{x_1, \cdots, x_{2048}\}$, x_k has bigger possibility to be selected when $u(t_0, x_k)$ is closer to 0 for $k = 1, \cdots, 2048$. The following formula shows the method we use to calculate the possibility of x_k to be selected. Denote this possibility by $\mathbb{P}(x_k)$.

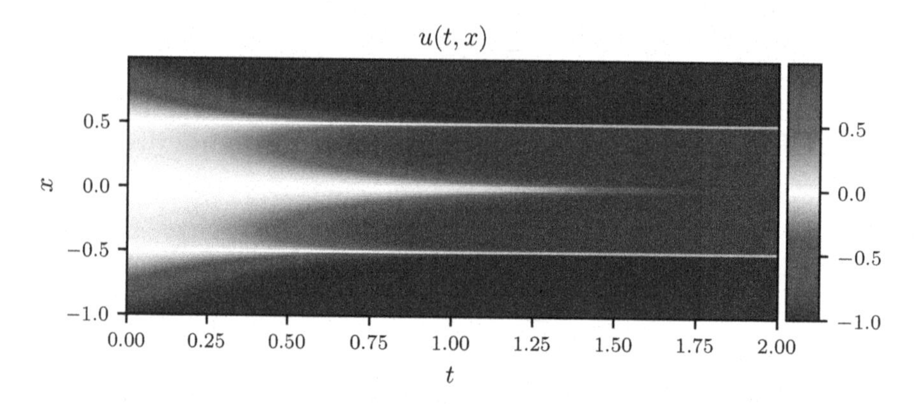

Fig. 3. Simulation results using conventional spectral models for $t \in [0, 2]$.

$$
\begin{aligned}
& f(x_k) = g(1 - |u(t_0, x_k)|) + \varepsilon, \; k = 1, \cdots, 2048, \\
& where \; g(z) = \begin{cases} z, & z \geq 0 \\ 0, & z < 0 \end{cases}, \\
& \mathbb{P}(x_k) = \frac{f(x_k)}{\sum_{m=1}^{2048} x_m}.
\end{aligned}
\tag{19}
$$

Similarly, the location of selected initial training data and the prediction results at the final stage are shown in Fig. 5. Indeed, accuracy of the estimated solution at $t = 0.9$ greatly improves compared to random data-selection approach.

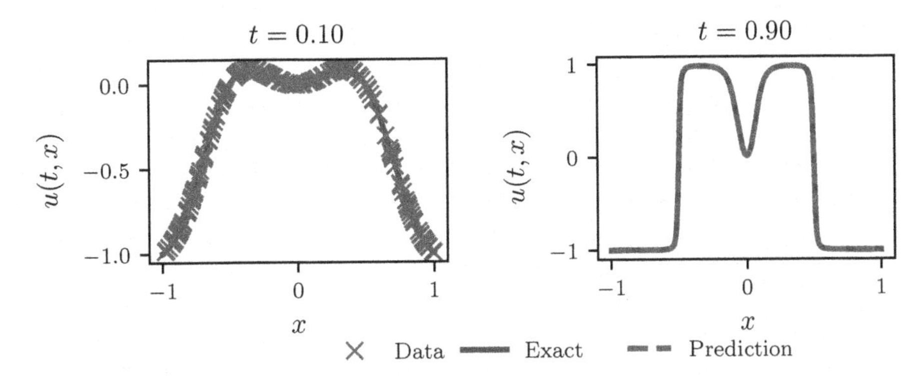

Fig. 4. The left picture depicts the 200 initial training points randomly sub-sampled from the data set of 2048 points, while the right one shows final prediction at $t_1 = 0.9$. The mean relative error on the whole data set is $1.013{\cdot}10^{-2}$.

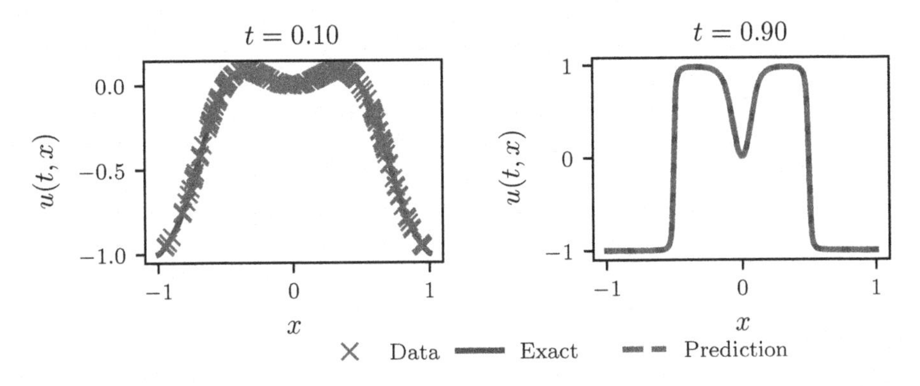

Fig. 5. The left picture depicts the 200 initial training points adaptively selected from the data set of 2048 points based on the value of $u(t_0, x_k)$ for $k = 1, \cdots, 2048$, while the right one shows final prediction at $t_1 = 0.9$. The mean relative error on the whole data set is $1.415{\cdot}10^{-3}$.

3.3 Long-Time Prediction

In Sect. 3.2, single-step discrete time model of PINNs is employed to predict solutions of the Allen-Cahn equation (14) from $t_0 = 0.1$ to $t_1 = 0.9$. In this part, we consider to expand the interval to $[0.1, 1.5]$. Figure 6 displays the location of randomly selected initial training data and the prediction of u at $t_2 = 1.5$ using the same model as in Sect. 3.2. Applying adaptive collocation strategy described in Eq. (19), approximated solution at t_2 is shown in Fig. 7. Easy to find from the figures that one-step discrete time models fed with randomly or adaptively generated training data does not work satisfactorily. In the case of long-time prediction, adaptive collocation strategy loses its effectiveness.

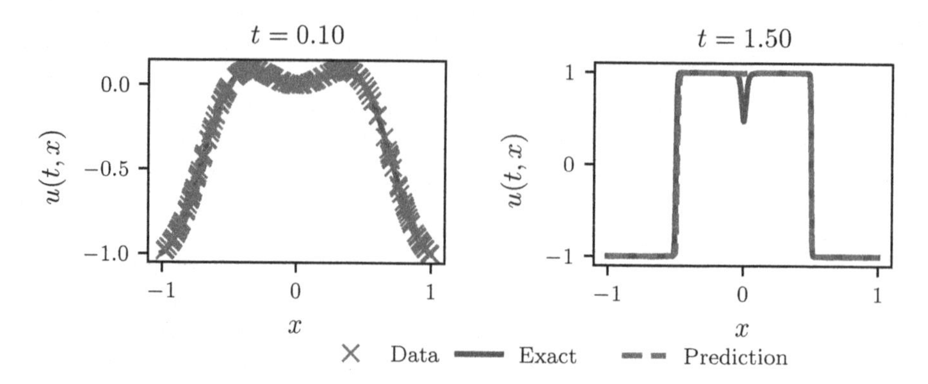

Fig. 6. The left picture depicts the 200 initial training points randomly sub-sampled from the data set of 2048 points, while the right one shows final prediction at $t_2 = 1.5$. The mean relative error on the whole data set is 0.1123.

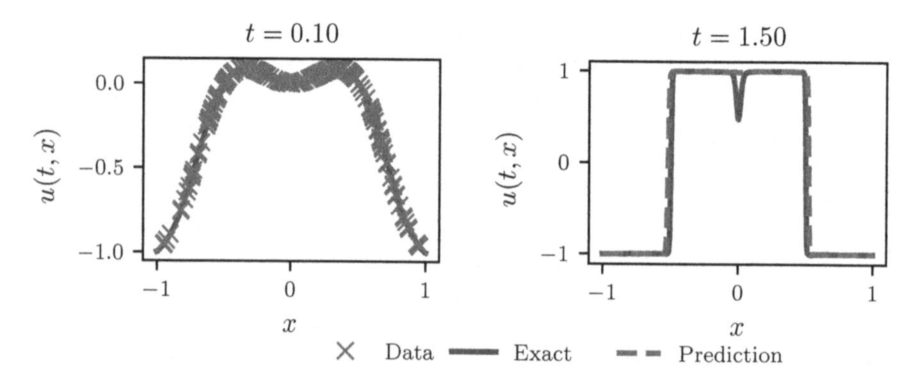

Fig. 7. The left picture depicts the 200 initial training points adaptively selected from the data set of 2048 points based on the value of $u(t_0, x_k)$ for $k = 1, \cdots, 2048$, while the right one shows final prediction at $t_2 = 1.5$. The mean relative error on the whole data set is 0.2447.

Given the undesirable approximation results, we employ multi-step discrete time models of PINNs to see whether they can be improved. Since the prediction results from $t_0 = 0.1$ to $t_1 = 0.9$ have been attained as shown in Fig. 4, we take the approximated solution $\bar{u}(t_1, x_k)$ for $k = 1, \cdots, 2048$ as the real value. Then we feed it into another neural network with the same structure as before in order to predict the solutions at $t_2 = 1.5$. In this trial, all training data are still sub-sampled randomly from the discretised domain. Figure 8 illustrates the results at the final stage. Then, based on the estimated solution at t_1 displayed in Fig. 5 using adaptive collocation, 200 training points at t_1 are selected adaptively according to the value of $\bar{u}_a(t_1, x_k)$ for $k = 1, \cdots, 2048$. Prediction of u at the final stage is shown in Fig. 9. Though the multi-step discrete time model still performs badly if sub-sample training data randomly at each step, the model yields desirable outcomes when selecting training points adaptively each time.

Hence, adaptive collocation strategy plays a role in improving the accuracy of estimated solutions of the Allen-Cahn equation in longer term.

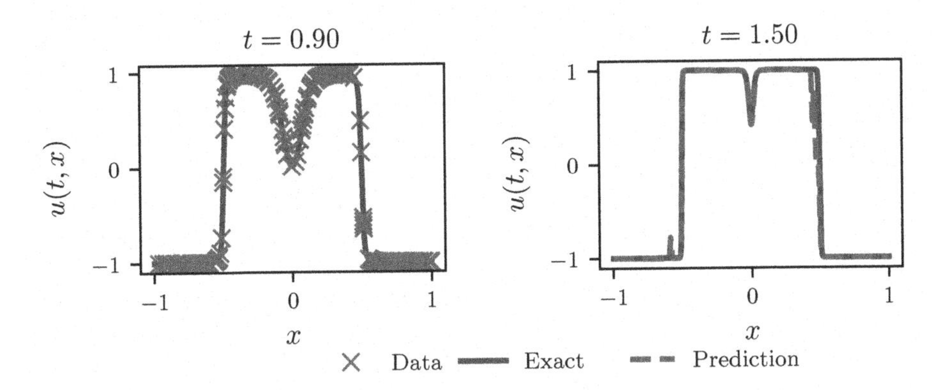

Fig. 8. The left picture depicts the 200 initial training points randomly sub-sampled from the data set of 2048 points at $t_1 = 0.9$, while the right one shows final prediction at $t_2 = 1.5$. The mean relative error on the whole data set is 0.1305.

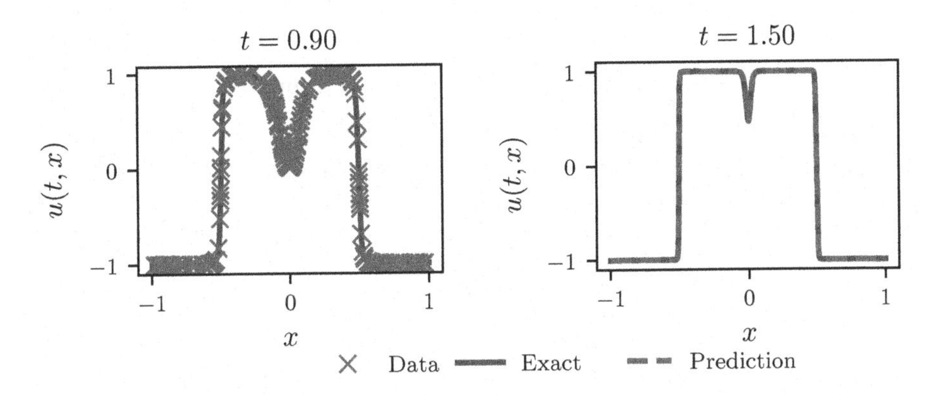

Fig. 9. The left picture depicts the 200 initial training points adaptively selected from the data set of 2048 points based on the value of $\bar{u}(t_1, x_k)$ for $k = 1, \cdots, 2048$, while the right one shows final prediction at $t_2 = 1.5$. The mean relative error on the whole data set is $1.940 \cdot 10^{-3}$.

4 Conclusion

In this study, adaptive collocation strategy is applied in the discrete time models of physics-informed neural networks to estimate solutions of the Allen-Cahn equation. A sample data set of 2048 evenly distributed points in the discretised domain $\Omega = [-1, 1]$ at certain time in discretised time domain $t \in [0, 2]$ is generated using Chebfun package which applies explicit Runge-Kutta integrator.

The accuracy of prediction results from $t_0 = 0.1$ to $t_1 = 0.9$ obtained via single-step discrete time model of PINNs shows the capability of the model to find solutions in short time period. Moreover, the model taking advantage of adaptive collocation approach even generates better simulation outcomes. In long-time prediction, estimation results demonstrate that single-step discrete time models cannot perform well. Through training the model twice with adaptive collocation strategy used to select training points, we provide quite precise approximation results at $t = 1.5$, reflecting the robustness of multi-step discrete time models of PINNs with adaptive collocation in long-time prediction.

When training the models, neural networks with varied hyper-parameters could be employed to see further improvement in the accuracy of prediction, and study of convergence properties of the neural networks remains open as possible research topics. Furthermore, the applicability of multi-step discrete time models of PINNs in other PDEs with larger time domain can be investigated in future work.

References

1. Allen, S.M., Cahn, J.W.: A microscopic theory for antiphase boundary motion and its application to antiphase domain coarsening. Acta Metallurgica **27**(6), 1085–1095 (1979)
2. Anitescu, C., Atroshchenko, E., Alajlan, N., Rabczuk, T.: Artificial neural network methods for the solution of second order boundary value problems. Comput. Mater. Continua **59**(1), 345–359 (2019). https://doi.org/10.32604/cmc.2019.06641
3. Bartels, S., Mller, R., Ortner, C.: Robust a priori and a posteriori error analysis for the approximation of Allen-Cahn and Ginzburg-Landau equations past topological changes. SIAM J. Numer. Anal. **49**(1), 110–134 (2011)
4. Beck, C.E.W., Jentzen, A.: Machine learning approximation algorithms for high-dimensional fully nonlinear partial differential equations and second-order backward stochastic differential equations. J. Nonlinear Sci. **29**(4), 1563–1619 (2019). https://doi.org/10.1007/s00332-018-9525-3
5. Driscoll, T.A., Hale, N., Trefethen, L.N.: Chebfun Guide. Oxford (2014)
6. Yu, B.: The deep Ritz method: A deep learning-based numerical algorithm for solving variational problems. CoRR abs/1710.00211 (2017). http://arxiv.org/abs/1710.00211
7. Feng, X., Wu, H.J.: A posteriori error estimates and an adaptive finite element method for the Allenccahn equation and the mean curvature flow. J. Sci. Comput. **24**(2), 121–146 (2005)
8. Gobovic, D., Zaghloul, M.E.: Analog cellular neural network with application to partial differential equations with variable mesh-size. In: Proceedings of IEEE International Symposium on Circuits and Systems - ISCAS 1994. vol. 6, pp. 359–362 (1994). https://doi.org/10.1109/ISCAS.1994.409600
9. Goodfellow, I., Bengio, Y., Courville, A.: Deep Learning. MIT Press, Cambridge (2016). http://www.deeplearningbook.org
10. Iserles, A.: A First Course in the Numerical Analysis of Differential Equations. Cambridge Texts in Applied Mathematics, 2nd edn. Cambridge University Press, Cambridge (2008). https://doi.org/10.1017/CBO9780511995569

11. Raissi, M., Perdikaris, P., Karniadakis, G.E.: Physics informed deep learning (part I): data-driven solutions of nonlinear partial differential equations. CoRR abs/1711.10561 (2017). http://arxiv.org/abs/1711.10561
12. Shen, J., Yang, X.: Numerical approximations of Allen-Cahn and Cahn-Hilliard equations. Discrete Continuous Dyn. Syst. **28**, 1669–1691 (2010). https://doi.org/10.3934/dcds.2010.28.1669
13. Sirignano, J., Spiliopoulos, K.: DGM: a deep learning algorithm for solving partial differential equations. J. Comput. Phys. **375**, 1339–1364 (2018)
14. Suzuki, Y.: Neural network-based discretization of nonlinear differential equations. Neural Comput. Appl. **31**(7), 3023–3038 (2017). https://doi.org/10.1007/s00521-017-3249-4
15. Zang, Y., Bao, G., Ye, X., Zhou, H.: Weak adversarial networks for high-dimensional partial differential equations. J. Comput. Phys. **411**, 109409 (2020)

DNS of Mass Transfer in Bi-dispersed Bubble Swarms

Néstor Balcázar-Arciniega(✉), Joaquim Rigola, and Assensi Oliva

Heat and Mass Transfer Technological Centre, Universitat Politècnica de Catalunya-BarcelonaTech, Colom 11, 08222 Terrassa (Barcelona), Spain
nestorbalcazar@yahoo.es, nestor.balcazar@upc.edu

Abstract. This work presents Direct Numerical Simulation of mass transfer in a bi-dispersed bubble swarm at high Reynolds number, by using a multiple marker level-set method. Transport equations are discretized by the finite-volume method on 3D collocated unstructured meshes. Interface capturing is performed by the unstructured conservative level-set method, whereas the multiple marker approach avoids the so-called numerical coalescence of bubbles. Pressure-velocity coupling is solved by the classical fractional-step projection method. Diffusive terms are discretized by a central difference scheme. Convective term of momentum equation, level-set equations, and mass transfer equation, are discretized by unstructured flux-limiters schemes. This approach improves the numerical stability of the unstructured multiphase solver in bubbly flows with high Reynolds number and high-density ratio. Finally, this numerical model is applied to research the effect of bubble-bubble interactions on the mass transfer in a bi-dispersed bubble swarm.

Keywords: Mass transfer · Bubbly flow · Unstructured flux-limiters · Unstructured meshes · Level-set method · Finite volume method · High-performance computing

1 Introduction

Mas transfer in poly-dispersed bubble swarms is frequent in nature and industry. For instance, bubbly flows are employed in chemical reactors to produce chemical products, as well as to improve mass transfer rates in the so-called unit operations of chemical engineering. Although empirical correlations have been reported to estimate mass transfer rates in bubbles [18], the interplay between fluid mechanics and mass transfer in turbulent bi-dispersed bubble swarms is not well understood yet. Indeed, beyond the

Néstor Balcázar, as a Professor Serra-Húnter (UPC-LE8027), acknowledges the Catalan Government for the financial support through this programme. The authors acknowledges the financial support of the *Ministerio de Economía y Competitividad, Secretaría de Estado de Investigación, Desarrollo e Innovación* (MINECO), Spain (PID2020-115837RB-100). Simulations were executed using computing time granted by the RES (IM-2021-3-0013, IM-2021-2-0020, IM-2021-1-0013, IM-2020-2-0002, IM-2019-3-0015) on the supercomputer MareNostrum IV based in Barcelona, Spain.

D. Groen et al. (Eds.): ICCS 2022, LNCS 13353, pp. 284–296, 2022.
https://doi.org/10.1007/978-3-031-08760-8_24

scientific motivation, understanding this phenomenon has practical importance in the design, optimization, and operation of industrial multiphase systems.

The development of supercomputers has promoted High-Performance computing (HPC) and Direct Numerical Simulation (DNS) of Navier-Stokes equations, as a pragmatic method to perform non-invasive numerical experiments of bubbly flows. In this sense, multiple numerical methods have been reported for DNS of two-phase flows, for instance: volume-of-fluid (VOF) methods [27], level-set (LS) methods [34,37], conservative level-set (CLS) methods [3,33], front tracking (FT) methods [41], and hybrid VOF/LS methods [6,38,39]. Some of these methods have been extended for interfacial heat transfer and mass transfer in gas-liquid multiphase flows, as reported in [2,15,16,21,22]. On the other hand, few works have been performed on DNS of mass transfer in bubble swarms [1,10,12,14,30,36]. Nevertheless, no previous studies of mass transfer in bi-dispersed bubble swarms have been reported in the context of the unstructured CLS method [10]. Therefore, this work aims to fill this lack in technical literature.

As advantages of present methodology, the unstructured CLS method [3,10] has been implemented on 3D collocated unstructured meshes, whereas the accumulation of mass conservation error inherent to standard level-set methods is circumvented. Furthermore, unstructured flux-limiters schemes as proposed in [3,7,10], are used to discretize the convective term of transport equations, avoiding numerical oscillations around discontinuities, and minimizing the so-called numerical diffusion [3,7,10]. Altogether, this numerical methods improve the numerical stability of the unstructured multiphase solver [3–7,10,12–14] in DNS of bubbly flows with high Reynolds number and high density ratio.

This paper is organized as follows: The mathematical model and numerical methods are presented in Sect. 2. Numerical experiments are presented in Sect. 3. Concluding remarks and future work are discussed in Sect. 4.

2 Mathematical Model and Numerical Methods

2.1 Incompressible Two-Phase Flow

The one-fluid formulation [41], is employed to introduce surface tension force as a singular terms in Navier-Stokes equations:

$$\frac{\partial}{\partial t}(\rho \mathbf{v}) + \nabla \cdot (\rho \mathbf{v}\mathbf{v}) = -\nabla p + \nabla \cdot \mu (\nabla \mathbf{v}) + \nabla \cdot \mu (\nabla \mathbf{v})^T + (\rho - \rho_0)\mathbf{g} + \mathbf{f}_\sigma, \quad (1)$$

$$\nabla \cdot \mathbf{v} = 0, \quad (2)$$

where $\mathbf{v}$ is the fluid velocity, p denotes the pressure field, ρ is the fluid density, μ is the dynamic viscosity, $\mathbf{g}$ is the gravitational acceleration, $\mathbf{f}_\sigma$ is the surface tension force per unit volume concentrated at the interface, subscripts d and c denote the dispersed phase (bubbles or droplets) and continuous phase respectively. Density and viscosity are constant at each fluid-phase, whereas a jump discontinuity is present at the interface Γ:

$$\rho = \rho_d H_d + \rho_c H_c, \ \mu = \mu_d H_d + \mu_c H_c. \quad (3)$$

Here H_c is the Heaviside step function that is one at fluid c (Ω_c) and zero elsewhere, whereas $H_d = 1 - H_c$. At discretized level physical properties are regularized in order to avoid numerical instabilities around the interface. On the other hand, bi-dispersed bubble swarms are simulated in a full-periodic cubic domain ($y - axis$ aligned to $\mathbf{g}$), therefore a force $-\rho_0 \mathbf{g}$ is included in momentum transport equation, Eq. (1), with $\rho_0 = V_\Omega^{-1} \int_\Omega (\rho_d H_d + \rho_c H_c)\, dV$, to avoid the acceleration of the entire flow field in the downward vertical direction [4,8,10,23],

2.2 Multiple Marker Unstructured CLS Method and Surface Tension

The unstructured conservative level-set method (UCLS) [3,10] developed for interface capturing on unstructured meshes is employed in this research. Furthermore, to avoid the numerical coalescence of bubbles, each fluid particle (bubble or droplet) is represented by a level-set function, as proposed in [4,7,8,10]. Therefore, the interface of the ith fluid particle is defined as the 0.5 iso-surface of the level-set function ϕ_i, where $i = 1, 2, ..., n_d$ and n_d is the total number of fluid particles in Ω_d. Since incompressible flow is assumed (Eq. 2), the ith interface transport equation can be written in conservative form as follows:

$$\frac{\partial \phi_i}{\partial t} + \nabla \cdot \phi_i \mathbf{v} = 0, \; i = 1, .., n_d. \tag{4}$$

Furthermore, a re-initialization equation is introduced to keep a sharp and constant level-set profile on the interface:

$$\frac{\partial \phi_i}{\partial \tau} + \nabla \cdot \phi_i (1 - \phi_i) \mathbf{n}_i^0 = \nabla \cdot \varepsilon \nabla \phi_i, \; i = 1, .., n_d. \tag{5}$$

where $\mathbf{n}_i^0$ denotes $\mathbf{n}_i$ evaluated at $\tau = 0$. Equation (5) is advanced in pseudo-time τ up to achieve the steady state. The compressive term of Eq. (5), $\phi_i(1 - \phi_i)\mathbf{n}_i^0$, forces the level-set function to be compressed onto the diffuse interface, along $\mathbf{n}_i$. The diffusive term, $\nabla \cdot \varepsilon \nabla \phi_i$, keeps the level-set profiles with characteristic thickness $\varepsilon = 0.5 h^{0.9}$, where h is the local grid size [3,7,10]. Geometrical properties of the interface, such as normal vectors $\mathbf{n}_i$ and curvatures κ_i, are computed as follows:

$$\mathbf{n}_i(\phi_i) = \frac{\nabla \phi_i}{||\nabla \phi_i||}, \; \kappa_i(\phi_i) = -\nabla \cdot \mathbf{n}_i, \; i = 1, .., n_d. \tag{6}$$

Surface tension forces are approximated by the Continuous Surface Force model [17], which has been extended to the multiple marker level-set method in [4,7,8,10]:

$$\mathbf{f}_\sigma = \sum_{i=1}^{n_d} \sigma \kappa_i(\phi_i) \mathbf{n}_i \delta_i^s = \sum_{i=1}^{n_d} \sigma \kappa_i(\phi_i) \nabla \phi_i. \tag{7}$$

where the regularized Dirac delta function is defined as $\delta_i = ||\nabla \phi||$ [3,4,7,8,10]. Finally, in order to avoid numerical instabilities at the interface, fluid properties in Eq. (3) are regularized by using a global level-set function ϕ [4,7], defined as follows:

$$\phi = min\{\phi_1, ..., \phi_{n_d}\}. \tag{8}$$

Thus, Heaviside functions presented in Eq. (3) are regularized as $H_d = 1 - \phi$ and $H_c = \phi$. In this work $0 < \phi \leq 0.5$ for Ω_d, and $0.5 < \phi \leq 1$ for Ω_c. Alternatively, if $0 < \phi \leq 0.5$ for Ω_c, and $0.5 < \phi \leq 1$ for Ω_d, then $H_d = \phi$ and $H_c = 1 - \phi$, whereas $\phi = max\{\phi_1, ..., \phi_{n_d}\}$ [10].

2.3 Mass Transfer

This work is focused in external mass transfer in bi-dispersed bubble swarms. Therefore, the concentration of chemical species in the continuous phase is computed by a convection-diffusion-reaction equation, as follows [10]:

$$\frac{\partial C}{\partial t} + \nabla \cdot (\mathbf{v}C) = \nabla \cdot (\mathcal{D}\nabla C) + \dot{r}(C), \tag{9}$$

where C denotes the chemical species concentration field, $\mathcal{D}$ denotes the diffusion coefficient or diffusivity which is equal to $\mathcal{D}_c$ in Ω_c and $\mathcal{D}_d$ elsewhere, $\dot{r}(C) = -k_1 C$ is the overall chemical reaction rate, k_1 is the first-order reaction rate constant. Furthermore, the concentration inside the bubbles is kept constant [1,10,21,36], whereas convection, diffusion and reaction of the mass dissolved from Ω_d exists only in Ω_c.

As proposed in [10], linear interpolation is applied to compute the concentration (C_P) at the interface cells, taking information from Ω_c (excluding interface cells), and imposing a Dirichlet boundary condition for the concentration at the interface ($\phi = 0.5$). Further details are reported in [10].

2.4 Numerical Methods

Transport equations are discretized by the finite-volume method on 3D collocated unstructured meshes, as introduced in [3,7,10]. For the sake of completeness, some points are remarked in what follows.

The convective term of momentum equation (Eq. (1)), level-set advection equation (Eq. (4)), and transport equation for concentration of chemical species (Eq. (9)), is explicitly computed, by approximating the fluxes at cell faces with unstructured flux-limiter schemes, as first proposed in [3,10]. As a consequence, approximation of convective term is written in the current cell Ω_P as follows: $(\nabla_h \cdot \beta\psi\mathbf{v})_P = \frac{1}{V_P}\sum_f \beta_f \psi_f \mathbf{v}_f \cdot \mathbf{A}_f$, where V_P is the volume of the current cell Ω_P, subindex f denotes the cell-faces, $\mathbf{A}_f = ||\mathbf{A}_f||\mathbf{e}_f$ is the area vector, $\mathbf{e}_f$ is a unit-vector perpendicular to the face f pointing outside the cell Ω_P [3,10]. Here, $\beta_f = \{\rho_f, 1, 1\}$, consistently with Eqs. (1, 4, 9), is approximated by linear interpolation. An especial interpolation is applied to $(\mathbf{v}_f \cdot \mathbf{A}_f)$ [6], to avoid the pressure-velocity decoupling on collocated meshes. Finally, $\psi_f = \{\mathbf{v}_f, C_f, \phi_f\}$ is computed as the sum of a diffusive upwind part (ψ_{C_p}) plus an anti-diffusive term [3,7,10]:

$$\psi_f = \psi_{C_p} + \frac{1}{2}L(\theta_f)(\psi_{D_p} - \psi_{C_p}). \tag{10}$$

where $L(\theta_f)$ is the flux limiter, $\theta_f = (\psi_{C_p} - \psi_{U_p})/(\psi_{D_p} - \psi_{C_p})$, C_p is the upwind point, U_p is the far-upwind point, and D_p is the downwind point [10]. Some of the flux-limiters ($L(\theta)$)) implemented in the unstructured multiphase solver [3–8,10], have the form [24,40]:

$$L(\theta_f) \equiv \begin{cases} \max\{0, \min\{2\theta_f, 1\}, \min\{2, \theta_f\}\} & \text{superbee,} \\ (\theta_f + |\theta_f|)/(1 + |\theta_f|) & \text{van Leer,} \\ \max\{0, \min\{4\theta_f, 0.75 + 0.25\theta_f, 2\}\} & \text{smart,} \\ 1 & \text{CD,} \\ 0 & \text{Upwind.} \end{cases} \quad (11)$$

An assessment of flux-limiters to discretize the convective term of transport equations on unstructured meshes is presented in [10]. In this research, Superbee flux-limiter is employed unless otherwise stated. From the flux-limiters remarked in Eq.(11), SUPERBEE, VAN-LEER, SMART and UPWIND preserve the numerical stability of the multiphase solver, as these schemes avoid numerical oscillations around discontinuities. Concerning the called numerical diffusion, the SUPERBEE flux-limiter is the less diffusive scheme, whereas the UPWIND scheme maximizes the numerical diffusion. Thus, the selection of the SUPERBEE flux-limiter scheme is crucial for bubbly flows with high Reynolds number and high-density ratio, preserving numerical stability and minimizing the numerical diffusion.

Compressive term of the re-initialization equation (Eq. (5)), is discretized at the cell Ω_P as follows [10]: $(\nabla \cdot \phi_i(1-\phi_i)\mathbf{n}_i^0)_P = \frac{1}{V_P}\sum_f (\phi_i(1-\phi_i))_f \mathbf{n}_{i,f}^0 \cdot \mathbf{A}_f$, where $\mathbf{n}_{i,f}^0$ and $(\phi_i(1-\phi_i))_f$ are linearly interpolated. The diffusive term of transport equations are centrally differenced [10]. Linear interpolation is used to find the cell-face values of physical properties and interface normals unless otherwise stated. Gradients are computed at cell centroids through the least-squares method using the information of the neighbor cells around the vertexes of the current cell (see Fig. 2 of [3]). For instance at the cell Ω_P, the gradient of the variable $\psi = \{v_j, C, \phi, ...\}$ is calculated as follows:

$$(\nabla\psi)_P = (\mathbf{M}_P^T \mathbf{W}_P \mathbf{M}_P)^{-1} \mathbf{M}_P^T \mathbf{W}_P \mathbf{Y}_P, \quad (12)$$

$\mathbf{M}_P$ and $\mathbf{Y}_P$ are defined as introduced in [3], $\mathbf{W}_P = \text{diag}(w_{P\to 1}, .., w_{P\to n})$ is the weighting matrix [29,32], defined as the diagonal matrix with elements $w_{P\to k} = \{1, ||\mathbf{x}_P - \mathbf{x}_k||^{-1}\}$, $k = \{1, .., n\}$, and subindex n is the number of neighbor cells. The impact of the selected weighting coefficient ($w_{P\to k}$) on the simulations is reported in our previous work [14].

The fractional-step projection method [19] is used to compute the pressure-velocity coupling. First, a predictor velocity ($\mathbf{v}_P^*$) is calculated at cell-centroids:

$$\frac{\rho_P \mathbf{v}_P^* - \rho_P^0 \mathbf{v}_P^0}{\Delta t} = \mathbf{C}_{\mathbf{v},P}^0 + \mathbf{D}_{\mathbf{v},P}^0 + (\rho_P - \rho_0)\mathbf{g} + \mathbf{f}_{\sigma,P}, \quad (13)$$

where the super-index 0 denotes the previous time-step, subindex P denotes the control volume Ω_P, $\mathbf{D_v} = \nabla \cdot \mu \nabla \mathbf{v} + \nabla \cdot \mu (\nabla \mathbf{v})^T$, and $\mathbf{C_v} = -\nabla \cdot (\rho \mathbf{v}\mathbf{v})$. Imposing the incompressibility constraint, $(\nabla \cdot \mathbf{v})_P = 0$, to the corrector step, Eq. (15), leads to a Poisson equation for the pressure at cell-centroids:

$$\left(\nabla \cdot \left(\frac{\Delta t}{\rho} \nabla p\right)\right)_P = (\nabla \cdot \mathbf{v}^*)_P, \; \mathbf{e}_{\partial\Omega} \cdot \nabla p|_{\partial\Omega} = 0. \quad (14)$$

which is solved by means of a preconditioned conjugate gradient method. A Jacobi pre-conditioner is used in this research. Here $\partial\Omega$ denotes the boundary of Ω, excluding regions with periodic boundary condition, where information of the corresponding periodic nodes is used [4, 10]. In a further step the updated velocity ($\mathbf{v}_P$) is computed at cell-centroids:

$$\frac{\rho_P \mathbf{v}_P - \rho_P \mathbf{v}_P^*}{\Delta t} = -(\nabla p)_P. \tag{15}$$

Furthermore, face-cell velocity $\mathbf{v}_f$ is interpolated [7, 10] to fulfill the incompressibility constraint and to avoid pressure-velocity decoupling on collocated meshes [35]. Then, $\mathbf{v}_f$ or some equivalent variable (e.g., $\mathbf{v}_f \cdot \mathbf{A}_f$) is employed to advect $\beta_f \psi_f$ on the convective term of transport equations [7, 10]. This approach benefits the numerical stability of the multiphase solver [3–8, 10], specially for bubbly flows with high density ratio and high Reynolds numbers, as demonstrated in our previous works [4, 8].

Temporal discretization of advection equation (Eq. (4)) and re-initialization equation (Eq. (5)) is performed by a TVD Runge-Kutta method [25]. Reinitialization equation (Eq. (5)), is solved for the steady state, using two iterations per physical time step to maintain the profile of the CLS functions [3, 6, 10].

The reader is referred to [3–5, 7, 8, 10, 13, 14] for further technical details on the finite-volume discretization of transport equations on collocated unstructured grids. Numerical methods are implemented in the framework of the parallel C++/MPI code TermoFluids [10]. The parallel scalability of the multiple marker level-set solver is reported in [8, 10].

3 Numerical Experiments

Bubbles regimes can be characterized by the following dimensionless numbers [18]:

$$Mo = \frac{g\mu_c^4 \Delta\rho}{\rho_c^2 \sigma^3}, \quad Eo = \frac{gd^2 \Delta\rho}{\sigma}, \quad Re_i = \frac{\rho_c U_{Ti} d}{\mu_c},$$
$$\eta_\rho = \frac{\rho_c}{\rho_d}, \quad \eta_\mu = \frac{\mu_c}{\mu_d}, \quad \alpha = \frac{V_d}{V_\Omega}, \quad \eta_d = \frac{d_b}{d_{b^*}}, \tag{16}$$

where Mo is the Morton number, Eo is the Eötvös number, Re is the Reynolds number, η_ρ is the density ratio, η_μ is the viscosity ratio, $\Delta\rho = |\rho_c - \rho_d|$ is the density difference between the fluid phases, subscript d denotes the dispersed fluid phase, subscript c denotes the continuous fluid phase, Since a bi-dispersed bubble swarm will be simulated, η_d denotes the ratio of bubble diameters, d_b is the diameter of bigger bubbles, d_{b^*} is the diameter of smaller bubbles, $d = d_b$ will be taken as the characteristic bubble diameter employed to define $\{Mo, Eo, Re, Da, Pe, t^*\}$, α is the bubble volume fraction, V_d is the volume of bubbles (Ω_d), V_Ω is the volume of Ω, and $t^* = t\sqrt{g/d}$ is the dimensionless time.

Numerical results will be reported in terms of the so-called drift velocity [10, 23], $U_{Ti}(t) = (\mathbf{v}_i(t) - \mathbf{v}_\Omega(t)) \cdot \hat{\mathbf{e}}_y$, which can be interpreted as the bubble velocity with respect to a stationary container, $\mathbf{v}_i(t)$ is the velocity of the ith bubble, $\mathbf{v}_\Omega(t)$ is the spatial averaged velocity in Ω.

Mass transfer with chemical reaction, $\dot{r}(C) = -k_1 C$, is characterized by the Sherwood number (Sh), the Damköler (Da) number, and Schmidt number (Sc) or Peclet number (Pe), defined in Ω_c as follows:

$$Sh = \frac{k_c d}{\mathcal{D}_c}, \; Sc = \frac{\mu_c}{\rho_c \mathcal{D}_c}, \; Pe = \frac{U_T d}{\mathcal{D}_c} = ReSc, \; Da = \frac{k_1 d^2}{\mathcal{D}_c}. \tag{17}$$

where k_c is the mass transfer coefficient in Ω_c.

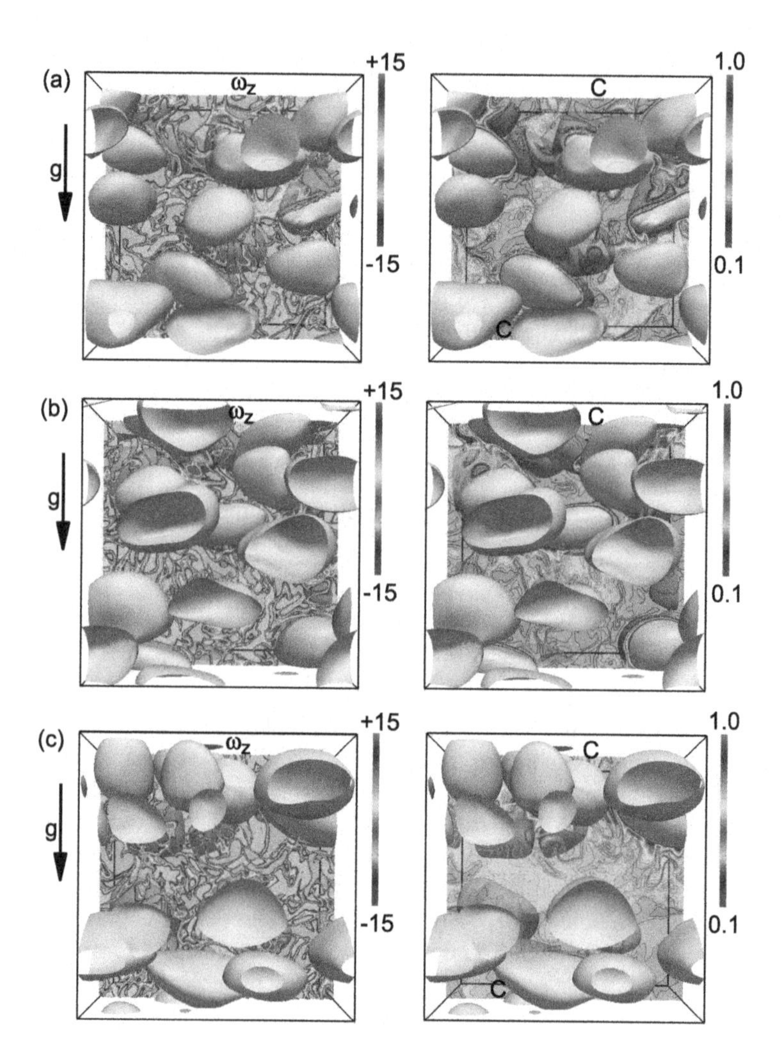

Fig. 1. Mass transfer in a bi-dispersed bubble swarm, $N_b = 16$ (total number of bubbles), in a full-periodic cube, $Eo = 4.0$, $d_b/d_{b^*} = 1.26$, $Mo = 5 \times 10^{-11}$, $\eta_\rho = \eta_\mu = 100$, $Sc = 1$, $Da = 170.7$, $\alpha = 19.6\%$, $\alpha_b = 13.1\%$, $\alpha_{b^*} = 6.54\%$. Vorticity ($\omega_z = \mathbf{e}_z \cdot \nabla \times \mathbf{v}$) and concentration ($C$) on the plane $x - y$ at (a) $t^* = tg^{1/2}d^{-1/2} = 7$, (b) $t^* = 14$, (c) $t^* = 21$.

3.1 Validations and Verifications

Multiple validations, verifications and extensions of the unstructured multiphase solver [3,10] are reported in our previous works, for instance: buoyancy-driven motion of single bubbles on unconfined domains [3,5,6], binary droplet collision with bouncing outcome [4], drop collision against a fluid interface without coalescence [4], bubbly flows in vertical channels [8,11], falling droplets [9], Taylor bubbles [26], thermocapillary-driven motion of deformable droplets [6], and liquid-vapor phase change [13]. A comparison of the unstructured CLS method [3] and coupled volume-of-fluid/level-set method [6] is reported in [9].

Concerning the mass transfer in single bubbles and mono-dispersed bubble swarms, on unconfined and confined domains, the reader is referred to our previous works [10,12,14], for validations and verifications of the multiple marker level-set solver employed in this research. Indeed, this work can be considered as a further step to perform Direct Numerical Simulation of mass transfer in bi-dispersed bubble swarms.

3.2 Mass Transfer in a Bi-dispersed Bubble Swarm

As a further step and with the confidence that the multiple marker level set solver has been validated [10,12–14], the DNS of mass transfer in a bi-dispersed bubble swarm is computed. The saturation of concentration of chemical species in Ω_c is avoided by the chemical reaction term in Eq. (9) [10,36]. Furthermore, a mass balance of the chemical species at steady state ($dC_c/dt = 0$) is employed to obtain the mass transfer coefficient (k_c) in Ω_c, as follows [10,12,14]:

$$k_c = \frac{V_c k_1 C_c}{(C_{\Gamma,c} - C_c) \sum_{i=1}^{n_d} A_i}. \tag{18}$$

Here $A_i = \int_\Omega \delta_i^s dV$ is the surface of the ith bubble, $C_c = V_c^{-1} \int_{\Omega_c} C dV$, and $\delta_i^s = ||\nabla \phi_i||$. Ω is a full-periodic cubic domain, with side length $L_\Omega = 3.18d$. Ω is discretized by 200^3 hexahedral control volumes, with grid size $h = L_\Omega/200$, distributed on 528 CPU-cores. As a consequence, bubbles are resolved with a grid size $h = d_b/63 = d_{b^*}/50$. In our previous works [10,12,14], it has been demonstrated that $h = d/35$ is enough to capture the hydrodynamics and mass transfer in gravity-driven bubbly flows [10]. Periodic boundary conditions are used on the $x-z$, $x-y$ and $y-z$ boundary planes. Bubbles are initially distributed in Ω following a random pattern, whereas fluids are quiescent. Since fluids are incompressible and bubble coalescence is not allowed, the void fraction ($\alpha = V_d/V_\Omega$) and number of bubbles are constant throughout the simulation.

Dimensionless parameters are $Eo = 4.0$, $d_b/d_{b^*} = 1.26$, $Mo = 5 \times 10^{-11}$, $\eta_\rho = \eta_\mu = 100$, $Sc = 1$, $Da = 170.7$, $\alpha = 19.6\%$, $\alpha_b = 13.1\%$, $\alpha_{b^*} = 6.54\%$, which corresponds to a bubbly flow with 16 bubbles distributed in Ω, 8 bubbles of diameter d_b and 8 bubbles of diameter d_b^*. Here α_b denotes the volume fraction of bigger bubbles, and α_{b^*} is the volume fraction of smaller bubbles. Figure 1 illustrates the mass transfer from a bi-dispersed swarm of 16 bubbles at $t^* = \{7, 14, 21\}$. Furthermore, concentration contours (C), and vorticity contours ($\omega_z = \hat{\mathbf{e}}_z \cdot \nabla \times \mathbf{v}$) are shown

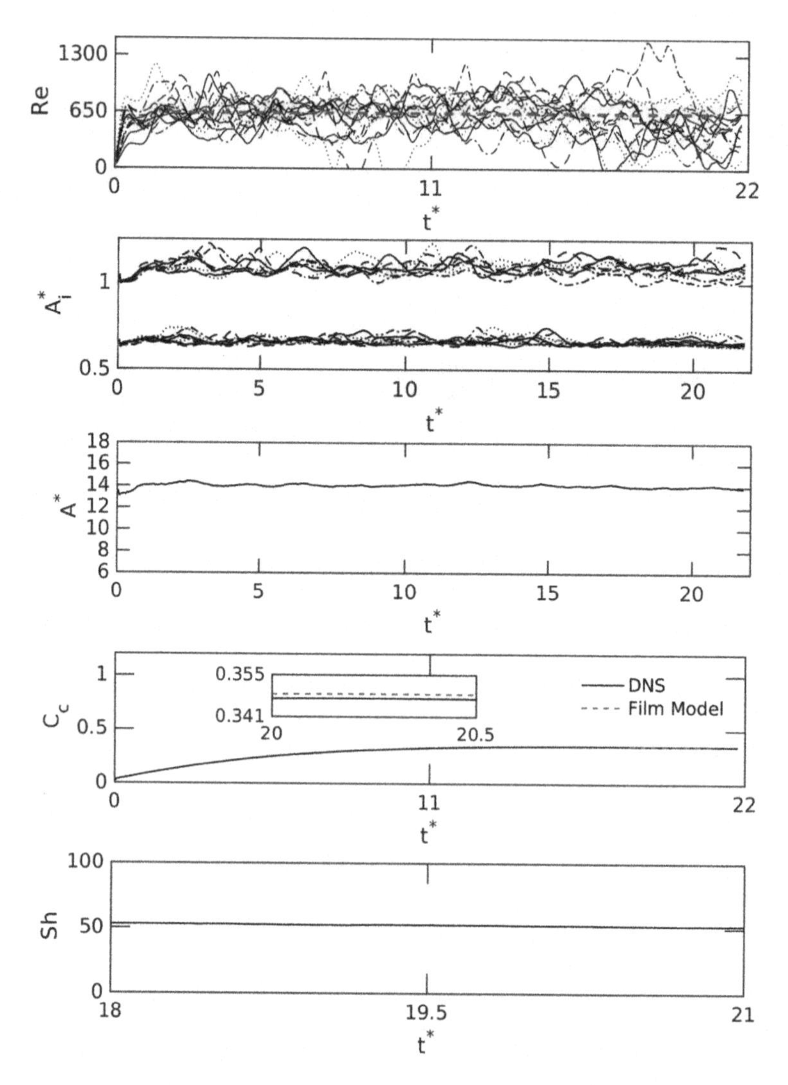

Fig. 2. Mass transfer from a bubble swarm, $N_b = 16$, in a full-periodic cube, $Eo = 4.0$, $d_b/d_{b^*} = 1.26$, $Mo = 5 \times 10^{-11}$, $\eta_\rho = \eta_\mu = 100$, $Sc = 1$, $Da = 170.7$, $\alpha = 19.6\%$, $\alpha_b = 13.1\%$, $\alpha_{b^*} = 6.54\%$. Time evolution of Reynolds number (Re) for each bubble (black lines), averaged Reynolds number for each bubble (continuous lines), time-averaged Reynolds number (red discontinuous line), normalized bubble surface $A_i^*(t)$, total interfacial surface of bubbles $A^*(t) = \sum_{i=1}^{n_d} A_i^*(t)$, spatial averaged concentration $C_c = V_c^{-1} \int_{\Omega_c} C dV$, and Sherwood number $Sh(t)$.

on the plane $x - y$. Figure 2 shows the time evolution of Reynolds number for each bubble and the time-averaged Reynolds number (discontinuous red line), normalized surface of each bubble $A_i^*(t) = A_i(t)/(4\pi d_b^2)$, total normalized surface of bubbles $A^*(t) = \sum_{i=1}^{n_d} A_i^*(t)$, space-averaged concentration of chemical species (C_c) in Ω_c,

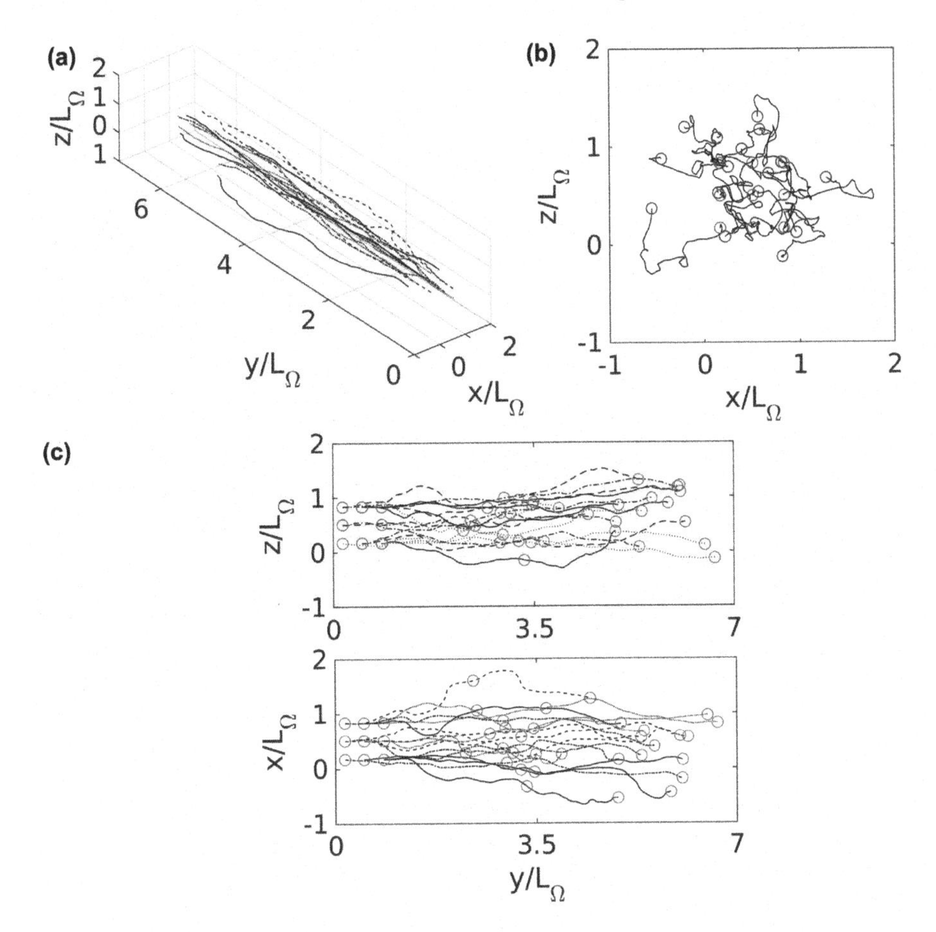

Fig. 3. Mass transfer from a bubble swarm, $N_b = 16$, in a full-periodic cube, $Eo = 4.0$, $d_b/d_{b^*} = 1.26$, $Mo = 5 \times 10^{-11}$, $\eta_\rho = \eta_\mu = 100$, $Sc = 1$, $Da = 170.7$, $\alpha = 19.6\%$, $\alpha_b = 13.1\%$, $\alpha_{b^*} = 6.54\%$. (a) 3D bubble trajectories. (b) Projection of bubble trajectories on the plane $x - z$. (c) Projection of bubble trajectories on the plane $x - y$ and $z - y$. Here L_Ω is the side-length of the periodic cubic domain.

and Sherwood number $Sh(t)$ at steady state ($dCc/dt = 0$). Strong deformation of bubble shapes ($A_i^*(t)$), bubble-bubble interactions, and path instabilities of bubbles at high Reynolds numbers, lead to fluctuations in $Re_i(t)$ as illustrated in Fig. 2. On the other hand, the Reynold number of the bi-dispersed bubble swarm, $\bar{Re} = n_d^{-1} \sum_{i=1}^{n_d} Re_i(t)$, tends to the steady-state. The spatial averaged concentration (C_c) achieves an steady-state value after initial transient effects, which demonstrates an equilibrium between mass transfer from the bubbles to Ω_c and the chemical reaction in Ω_c. As a consistency test, numerical results of the space-averaged concentration of chemical species in Ω_c compares very well with results obtained by the called film theory [10], as illustrated in Fig. 2. Furthermore, the mass transfer coefficient ($Sh \approx 50$) achieves the steady-state, once $dC_c/dt = 0$. Finally, Fig. 3 illustrates bubble trajectories, which indicate a repulsion effect between the bubbles.

4 Conclusions

DNS of mass transfer in a bi-dispersed bubble swarm has been performed using a parallel multiple-marker level-set method [4, 8, 10]. Numerical experiments demonstrate the reliability of this approach as an accurate tool for simulating bi-dispersed bubbly flows with mass transfer and chemical reaction in a full-periodic domain. The solver can reproduce the physics of bubble-bubble interactions in a long-time simulation of bubbly flows. This numerical approach avoids the numerical merging of bubbles, an artifact inherent to interface capturing methods, e.g., level-set, volume-of-fluid. Bubble-bubble interactions lead to a repulsion effect in horizontal alignment. On the other hand, when two bubbles are vertically aligned, their interactions follow the so-called drafting-kissing-tumbling mechanism observed in solid particles. This set of interactions induces a fluctuating velocity field known in the literature as bubble-induced turbulence. The time-averaged Reynolds number (Re) and Sherwood number (Sh) tend to the steady-state. Turbulence induced by the agitation of bubbles promotes the mixing of chemical species in the continuous phase. Furthermore, the spatially averaged concentration of chemical species tends to the steady-state, indicating a balance between chemical reaction in Ω_c and mass transfer from bubbles. Present results demonstrate that the multiple marker level-set method [10] is a predictive model to compute $Sh = Sh(Eo, Re, Da, \alpha_b \alpha_{b^*}, d_b/d_{b^*})$ in bi-dispersed bubbly flows. In future work, the model will be extended to simulate complex chemical reaction kinetics, as well as employed in parametric studies of $Sh = Sh(Eo, Re, Da, ...)$ to develop closure relations for models based on the averaged flow, e.g., two-fluid models [28].

References

1. Aboulhasanzadeh, B., Thomas, S., Taeibi-Rahni, M., Tryggvason, G.: Multiscale computations of mass transfer from buoyant bubbles. Chem. Eng. Sci. **75**, 456–467 (2012)
2. Alke, A., Bothe, D., Kroeger, M., Warnecke, H.J.: VOF-based simulation of conjugate mass transfer from freely moving fluid particles. In: Mammoli, A.A., Brebbia, C.A. (eds.) Computational Methods in Multiphase Flow V, WIT Transactions on Engineering Sciences, pp. 157–168 (2009)
3. Balcázar, N., Jofre, L., Lehmkhul, O., Castro, J., Rigola, J.: A finite-volume/level-set method for simulating two-phase flows on unstructured grids. Int. J. Multiphase Flow **64**, 55–72 (2014)
4. Balcázar, N., Lehmkhul, O., Rigola, J., Oliva, A.: A multiple marker level-set method for simulation of deformable fluid particles. Int. J. Multiphase Flow **74**, 125–142 (2015)
5. Balcázar, N., Lemhkuhl, O., Jofre, L., Oliva, A.: Level-set simulations of buoyancy-driven motion of single and multiple bubbles. Int. J. Heat Fluid Flow **56**, 91–107 (2015)
6. Balcázar, N., Lehmkhul, O., Jofre, L., Rigola, J., Oliva, A.: A coupled volume-of-fluid/level-set method for simulation of two-phase flows on unstructured meshes. Comput. Fluids **124**, 12–29 (2016)

7. Balcázar, N., Rigola, J., Castro, J., Oliva, A.: A level-set model for thermocapillary motion of deformable fluid particles. Int. J. Heat Fluid Flow Part B **62**, 324–343 (2016)
8. Balcázar, N., Castro, J., Rigola, J., Oliva, A.: DNS of the wall effect on the motion of bubble swarms. Procedia Comput. Sci. **108**, 2008–2017 (2017)
9. Balcázar, N., Castro, J., Chiva, J., Oliva, A.: DNS of falling droplets in a vertical channel. Int. J. Comput. Methods Exp. Meas. **6**(2), 398–410 (2018)
10. Balcázar, N., Antepara, O., Rigola, J., Oliva, A.: A level-set model for mass transfer in bubbly flows. Int. J. Heat Mass Transf. **138**, 335–356 (2019)
11. Balcázar, N., Lehmkuhl, O., Castro, J., Oliva, A.: DNS of the rising motion of a swarm of bubbles in a confined vertical channel. In: Grigoriadis, D.G.E., Geurts, B.J., Kuerten, H., Fröhlich, J., Armenio, V. (eds.) Direct and Large-Eddy Simulation X. ES, vol. 24, pp. 125–131. Springer, Cham (2018). https://doi.org/10.1007/978-3-319-63212-4_15
12. Balcázar, N., Antepara, O., Rigola, J., Oliva, A.: DNS of drag-force and reactive mass transfer in gravity-driven bubbly flows. In: García-Villalba, M., Kuerten, H., Salvetti, M.V. (eds.) DLES 2019. ES, vol. 27, pp. 119–125. Springer, Cham (2020). https://doi.org/10.1007/978-3-030-42822-8_16
13. Balcazar, N., Rigola, J., Oliva, A.: Unstructured level-set method for saturated liquid-vapor phase change. In: WCCM-ECCOMAS2020 (2021). https://www.scipedia.com/public/Balcazar_et_al_2021a, https://doi.org/10.23967/wccm-eccomas.2020.352
14. Balcázar-Arciniega, N., Rigola, J., Oliva, A.: DNS of mass transfer from bubbles rising in a vertical channel. In: Rodrigues, J.M.F., et al. (eds.) ICCS 2019. LNCS, vol. 11539, pp. 596–610. Springer, Cham (2019). https://doi.org/10.1007/978-3-030-22747-0_45
15. Bothe, D., Koebe, M., Wielage, K., Warnecke, H.J.: VOF simulations of mass transfer from single bubbles and bubble chains rising in the aqueous solutions. In: Proceedings of FEDSM03: Fourth ASME-JSME Joint Fluids Engineering Conference, Honolulu, 6–11 July 2003
16. Bothe, D., Fleckenstein, S.: Modeling and VOF-based numerical simulation of mass transfer processes at fluidic particles. Chem. Eng. Sci. **101**, 283–302 (2013)
17. Brackbill, J.U., Kothe, D.B., Zemach, C.: A continuum method for modeling surface tension. J. Comput. Phys. **100**, 335–354 (1992)
18. Clift, R., Grace, J.R., Weber, M.E.: Bubbles. Drops and Particles. Academin Press, New York (1978)
19. Chorin, A.J.: Numerical solution of the Navier-Stokes equations. Math. Comput. **22**, 745–762 (1968)
20. Coyajee, E., Boersma, B.J.: Numerical simulation of drop impact on a liquid-liquid interface with a multiple marker front-capturing method. J. Comput. Phys. **228**(12), 4444–4467 (2009)
21. Darmana, D., Deen, N.G., Kuipers, J.A.M.: Detailed 3D modeling of mass transfer processes in two-phase flows with dynamic interfaces. Chem. Eng. Technol. **29**(9), 1027–1033 (2006)
22. Davidson, M.R., Rudman, M.: Volume-of-fluid calculation of heat or mass transfer across deforming interfaces in two-fluid flow. Numer. Heat Transf. Part B Fundam. **41**, 291–308 (2002)
23. Esmaeeli, A., Tryggvason, G.: Direct numerical simulations of bubbly flows Part 2. Moderate Reynolds number arrays. J. Fluid Mech. **385**, 325–358 (1999)
24. Gaskell, P.H., Lau, A.K.C.: Curvature-compensated convective transport: SMART a new boundedness-preserving transport algorithm. Int. J. Numer. Methods **8**, 617–641 (1988)
25. Gottlieb, S., Shu, C.W.: Total Variation Dimishing Runge-Kutta Schemes. Math. Comput. **67**, 73–85 (1998)
26. Gutiérrez, E., Balcázar, N., Bartrons, E., Rigola, J.: Numerical study of Taylor bubbles rising in a stagnant liquid using a level-set/moving-mesh method. Chem. Eng. Sci. **164**, 102–117 (2017)

27. Hirt, C., Nichols, B.: Volume of fluid (VOF) method for the dynamics of free boundary. J. Comput. Phys. **39**, 201–225 (1981)
28. Ishii, M., Hibiki, T.: Thermo-Fluid Dynamics of Two-Phase Flow, 2nd edn. Springer, New York (2010). https://doi.org/10.1007/978-1-4419-7985-8
29. Jasak, H., Weller, H.G.: Application of the finite volume method and unstructured meshes to linear elasticity. Int. J. Numer. Meth. Eng. **48**, 267–287 (2000)
30. Koynov, A., Khinast, J.G., Tryggvason, G.: Mass transfer and chemical reactions in bubble swarms with dynamic interfaces. AIChE J. **51**(10), 2786–2800 (2005)
31. Lochiel, A., Calderbank, P.: Mass transfer in the continuous phase around axisymmetric bodies of revolution. Chem. Eng. Sci. **19**, 471–484 (1964)
32. Mavriplis, D.J.: Unstructured mesh discretizations and solvers for computational aerodynamics. In: 18th Computational Fluid Dynamics Conference, AIAA Paper, pp. 2007–3955, Miami (2007). https://doi.org/10.2514/6.2007-3955
33. Olsson, E., Kreiss, G.: A conservative level set method for two phase flow. J. Comput. Phys. **210**, 225–246 (2005)
34. Osher, S., Sethian, J.A.: Fronts propagating with curvature-dependent speed: algorithms based on Hamilton-Jacobi formulations. J. Comput. Phys. **79**, 175–210 (1988)
35. Rhie, C.M., Chow, W.L.: Numerical study of the turbulent flow past an airfoil with trailing edge separation. AIAA J. **21**, 1525–1532 (1983)
36. Roghair, I., Van Sint Annaland, M., Kuipers, J.A.M.: An improved front-tracking technique for the simulation of mass transfer in dense bubbly flows. Chem. Eng. Sci. **152**, 351–369 (2016)
37. Sussman, M., Smereka, P., Osher, S.: A level set approach for computing solutions to incompressible two-phase flow. J. Comput. Phys. **144**, 146–159 (1994)
38. Sussman, M., Puckett, E.G.: A coupled level set and volume-of-fluid method for computing 3D and axisymmetric incompressible two-phase flows. J. Comput. Phys. **162**, 301–337 (2000)
39. Sun, D.L., Tao, J.W.Q.: A coupled volume-of-fluid and level-set (VOSET) method for computing incompressible two-phase flows. Int. J. Heat Mass Transf. **53**, 645–655 (2010)
40. Sweby, P.K.: High resolution using flux limiters for hyperbolic conservation laws. SIAM J. Numer. Anal. **21**, 995–1011 (1984)
41. Tryggvason, G., et al.: A front-tracking method for the computations of multiphase flow. J. Comput. Phys. **169**, 708–759 (2001)
42. Winnikow, S.: Letter to the editors. Chem. Eng. Sci. **22**(3), 477 (1967)

Parallel Fluid-Structure Interaction Simulation

Meng-Huo Chen(✉)

Chung Cheng University, Chiayi City, Taiwan
kamil.wereszczynski@polsl.pl

Abstract. In this work we implement the parallelization of a method for solving fluid-structure interactions: one-field monolithic fictitious domain (MFD). In this algorithm the velocity field for solid domain is interpolated into fluid velocity field through an appropriate L^2 projection, then the resulting combined equations are solved simultaneously (rather than sequentially). We parallelize the finite element discretization of spatial variables for fluid governing equations and linear system solver to accelerate the computation. Our goal is to reduce the simulation time for high resolution fluid-structure interaction simulation, such as collision of multiple immersed solids in fluid.

Keywords: Fluid-structure interaction · Finite element · Parallel processing

1 Introduction

1.1 Fluid-Structure Interaction

Fluid-structure interaction (FSI) is a multiphysics problem that describes the interaction between a moving, sometimes deformable, structure and its surrounding incompressible fluid flow. In general, the solid materials deform largely and the deformations are strongly coupled to the flowing fluid. Numerical simulation of FSI is a computational challenge since the governing equations for solid and fluid regions are different and the algorithm needs to solve the locations of solid-fluid interfaces simultaneously with the dynamics in both regions where the kinematic (e.g., non-slipping) and dynamic (e.g., stress matched along the normal to solid-fluid interface) boundary conditions are imposed at the interface.

1.2 Numerical Scheme

In general, the numerical schemes for solving the FSI problems may be classified into two approaches: the monolithic/fully-coupled method [1–3] and the partitioned/segregated method [4,5]. In addition, each method can be categorized further depends on the way to handle the mesh: fitted (conforming) mesh methods and unfitted (nonconforming) mesh methods.

D. Groen et al. (Eds.): ICCS 2022, LNCS 13353, pp. 297–309, 2022.
https://doi.org/10.1007/978-3-031-08760-8_25

Fitted mesh methods require solid and fluid meshes match each other at the interface, and both fluid and the solid regions share the nodes on the interface. In this way both a fluid velocity and a solid velocity (or displacement) are defined on each interface node. Clearly the fluid and solid velocities should agree on the interface nodes. Partitioned/segregated approach using fitted mesh to solve the governing equations. The solid and fluid equations are sequentially solved and then the steps are iterated until the velocities become consistent at the interface. This approach is easier to implement but not robust or fail to converge for problems where the fluid and solid appear to have significant energy exchange. On the other hand, monolithic/fully-coupled scheme solve the fluid and solid equations simultaneously on fitted mesh and often use a Lagrange Multiplier to weakly enforce the continuity of velocity on the interface. This method provides accurate and stable solutions but is computational challenging since one needs to solve the large size of nonlinear algebraic systems arising from the implicit discretization of the fully-coupled solid and fluid equations.

For unfitted mesh methods the solid and fluid regions are represented by two separate meshes and normally these do not agree to each other on the interface. Since there is no clear boundary for the solid problem, one of the approach to address the issue is to use Fictitious Domain method (FDM). In FDM the region representing solid is treated as (fictitious) fluid whose velocity/displacement is constrained to be the same as that of the solid. This constraint is enforced using a distributed Lagrange multiplier (DLM). There appear to be two situations for using unfitted meshes approach: either avoid solving the solid equations (such as Immersed Finite Element Method), or solve them with additional variables (two velocity fields and Lagrange multiplier) in the solid domain.

In this article, we parallelize one-field Fictitious Domain method can be categorized as a monolithic approach using an unfitted mesh. The main idea of the one-field FDM is as follows: (1) One-field formulation: re-write the governing equations for solid in the form of fluid equations. (2) L^2 projection (isoparametric interpolation): combining the fluid and solid equations and discretize them in an augmented domain. Then the problem is solved on a single field. The existing one-field FDM algorithm used in sequential simulation provide reasonable running time for 2D and 3D problems with low to moderate resolution, while its performance on highly resolved meshes ($256 \times 256 \times 256$ grid points and above) is not desired. Our goal is to accelerate the simulation algorithm using parallel computing and hope to extend its applications on FSI problems requiring high resolution, such as the 3D models describing the collision of multiple immersed solids in fluid.

2 Governing Equations and FEM Discretization

In this section we use a 2D FSI model (Fig. 1) to describe the governing partial differential equations (PDEs) for FSI. Denote the regions representing solid and fluid by $\Omega_t^s \subset \mathbb{R}^d$ and $\Omega_t^f \subset \mathbb{R}^d$, respectively. The subscript t reveals that both regions are time dependent. $\Omega_t^s \cup \Omega_t^f$ is the fixed domain and the moving interface between solid and fluid is denoted by $\Gamma_t = \partial\Omega_t^s \cap \Omega_t^f$. Components of variables along spatial directions are indicated by subscripts i, j, k. In addition, the repeated indices are implicitly summed over. For instance, u_j^s and u_j^f represent the j-component (along the j direction) of the solid velocity and fluid velocity, respectively, σ_{ij}^s and σ_{ij}^f denote the ij-component for stress tensor of solid and fluid respectively, and $(u_i^s)^n$ is the i-component of solid velocity at time t^n. Quantities denoted by bold letters implies that variables are vectors or matrices.

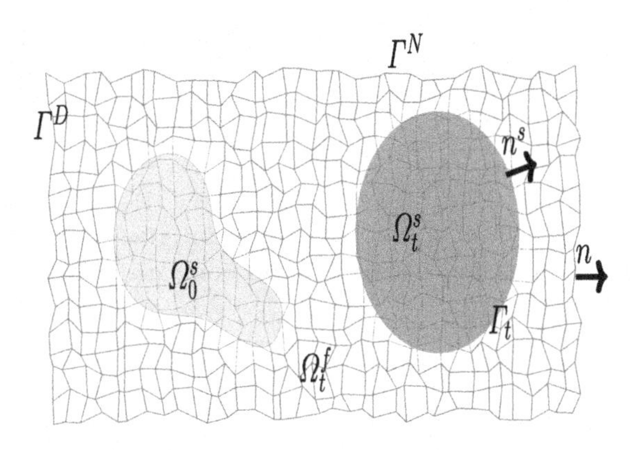

Fig. 1. 2D FSI problem and the boundary conditions.

Let ρ_s and ρ_f be the density of solid and fluid respectively, τ_{ij}^s and τ_{ij}^f be the deviatoric stress of the solid and fluid respectively, μ^f be the fluid viscosity, p^f be the fluid pressure, and g_i be the acceleration of gravity. Then the governing equations for incompressible fluid in Ω_t^f as shown in Fig. 1 are

$$\frac{\partial u_j^f}{\partial x_j} = 0, \tag{1}$$

$$\rho^f \frac{Du_i^f}{Dt} - \frac{\partial \sigma_{ij}^f}{\partial x_j} = \rho^f g_i, \tag{2}$$

$$\sigma_{ij}^f = \mu^f \left(\frac{\partial u_i^f}{\partial x_j} + \frac{\partial u_j^f}{\partial x_i} \right) - p^f \delta_{ij} = \tau_{ij}^f - p^f \delta_{ij}, \tag{3}$$

For the evolution of variables in solid domain Ω_t^s, we assume the solid is neo-Hookean incompressible solid and the governing equations are:

$$\frac{\partial u_j^s}{\partial x_j} = 0, \tag{4}$$

$$\rho^s \frac{Du_i^s}{Dt} - \frac{\partial \sigma_{ij}^s}{\partial x_j} = \rho^s g_i, \tag{5}$$

$$\sigma_{ij}^s = \mu^s \left(\frac{\partial x_i^s}{\partial X_k} \frac{\partial x_j^s}{\partial X_k} - \delta_{ij} \right) - p^s \delta_{ij} = \tau_{ij}^s - p^s \delta_{ij}. \tag{6}$$

In the governing Eqs. (4)–(6), μ^s and p^s are the solid shear modulus and the pressure of the solid, respectively, X_i represents the reference coordinates of the solid, and x_i refer to the current coordinates of the solid or fluid. Therefore the deformation tensor of the solid is denoted by $\mathbf{F} = [\partial x_i / \partial X_j]$. The incompressible neo-Hookean model described in Eqs. (4)–(6) can be used for predicting stress-strain behavior of materials undertaking large deformation [12]. Note that D/Dt in Eqs. (2) and (5) is the total time derivative and its form depends on what type of mesh (Eulerian or Lagrangian) used in the individual domain.

On the interface boundary Γ_t, the following conditions are imposed:

$$u_i^f = u_i^s, \tag{7}$$

$$\sigma_{ij}^f n_j^s = \sigma_{ij}^s n_j^s, \tag{8}$$

where n_j^s is the component of unit normal to the interface pointing outward, see Fig. 1. Dirichlet and Neumann boundary conditions are imposed on the boundaries for the fluid accordingly:

$$u_i^f = u_i^s, \qquad \text{on} \qquad \Gamma^D \tag{9}$$

$$\sigma_{ij}^f n_j = \bar{h}_i \qquad \text{on} \qquad \Gamma^N. \tag{10}$$

The initial conditions used in this work is:

$$u_i^f \mid_{t=0} = u_i^s \mid_{t=0} = 0 \tag{11}$$

in which we assume the system starts from the rest.

The finite element discretization of the governing equations starts with weak formulation of Eqs. (1), (2), (4) and (5). Let $(u, w)_\Omega = \int_\Omega uv \, \mathrm{d}\Omega$, and

$$u_i = \begin{cases} u_i^f & \text{in} \quad \Omega_t^f \\ u_i^s & \text{in} \quad \Omega_t^s \end{cases}$$

$$p = \begin{cases} p^f & \text{in} \quad \Omega_t^f \\ p^s & \text{in} \quad \Omega_t^s \end{cases}$$

Using constitution equations (3) and (6) with boundary condition in (10), integrating the stress terms by parts for the test functions $v_i \in H_0^1(\Omega)$ and $q \in L^2(\Omega)$, we get the weak formulation of the FSI system: finding $u_i \in H^1(\Omega)$ and $p \in L_0^2(\Omega)$ so that

$$\begin{aligned} \rho^f \left(\frac{Du_i}{Dt}, v_i \right)_\Omega + \left(\tau_{ij}^f, \frac{\partial v_i}{\partial x_j} \right)_\Omega - \left(p, \frac{\partial v_j}{\partial x_j} \right)_\Omega - \left(\frac{\partial u_i}{\partial x_j}, q \right)_\Omega \\ + (\rho^s - \rho^f) \left(\frac{Du_i}{Dt}, v_i \right)_{\Omega^s} + \left(\tau_{ij}^s - \tau_{ij}^f, \frac{\partial v_j}{\partial x_j} \right)_{\Omega^s} \\ = (\bar{h}, v_i)_{\Gamma^N} + \rho^f (g_i, v_i)_\Omega + (\rho^s - \rho^f)(g_i, v_i)_{\Omega^s}, \end{aligned} \tag{12}$$

$\forall v_i \in H_0^1(\Omega)$ and $\forall q \in H_0^1(\Omega)$.

The integrals in Eq. (12) are calculated in all domains as illustrated in Fig. 1. Note that the fluid region is approximated by an Eulerian mesh and solid region is represented by an updated Lagrangian mesh in order to track the solid deformation/motion, therefore in these two different domains the total time derivatives are:

$$\frac{Du_i}{Dt} = \frac{\partial u_i}{\partial t} + u_j \frac{\partial u_i}{\partial x_j} \quad \text{in} \quad \Omega \tag{13}$$

and

$$\frac{Du_i^s}{Dt} = \frac{\partial u_i^s}{\partial t} \quad \text{in} \quad \Omega^s. \tag{14}$$

Based on Eqs. (13) and (14), the time discretization of Eq. (12) becomes

$$\begin{aligned} \rho^f \left(\frac{u_i - u_i^n}{\Delta t} + u_j \frac{\partial u_i}{\partial x_j}, v_i \right)_\Omega + \left(\tau_{ij}^f, \frac{\partial v_i}{\partial x_j} \right)_\Omega - \left(p, \frac{\partial v_j}{\partial x_j} \right)_\Omega - \left(\frac{\partial u_j}{\partial x_j}, q \right)_\Omega \\ + (\rho^s - \rho^f) \left(\frac{u_i - u_i^n}{\Delta t}, v_i \right)_{\Omega_{n+1}^s} + \left(\tau_{ij}^s, \frac{\partial v_j}{\partial x_j} \right)_{\Omega_{n+1}^s} \\ = (\bar{h}_i, v_i)_{\Gamma^N} + \rho^f (g_i, v_i)_\Omega + (\rho^s - \rho^f)(g_i, v_i)_{\Omega_{n+1}^s}, \end{aligned} \tag{15}$$

where the superscript n of variable u_i represents the velocity at the nth time step. Note that we have replaced $\Omega_{t^{n+1}}^s$, the solid mesh at the $(n+1)$th time step, by Ω_{n+1}^s. By the spirit of splitting method introduced in [14], the above time evolution Eq. (15) can be viewed as the combination of two steps:

1. Convection step

$$\rho^f \left(\frac{u_i^\star - u_i^n}{\Delta t} + u_j^\star \frac{\partial u_i^\star}{\partial x_j}, v_i \right)_\Omega = 0 \tag{16}$$

2. Diffusion step

$$\begin{aligned} \rho^f \left(\frac{u_i - u_i^\star}{\Delta t}, v_i \right)_\Omega + \left(\tau_{ij}^f, \frac{\partial v_i}{\partial x_j} \right)_\Omega - \left(p, \frac{\partial v_j}{\partial x_j} \right)_\Omega - \left(\frac{\partial u_j}{\partial x_j}, q \right)_\Omega \\ + (\rho^s - \rho^f) \left(\frac{u_i - u_i^n}{\Delta t}, v_i \right)_{\Omega_{n+1}^s} + \left(\tau_{ij}^s - \tau_{ij}^f, \frac{\partial v_i}{\partial x_j} \right)_{\Omega_{n+1}^s} \\ = (\bar{h}_i, v_i)_{\Gamma^N} + \rho^f (g_i, v_i)_\Omega + (\rho^s - \rho^f)(g_i, v_i)_{\Omega_{n+1}^s}, \end{aligned} \tag{17}$$

where the intermediate field $u_i^\star$ obtained from the convection step is used in the diffusion step to solve the "correct" field u_i. To obtain the system of linear algebraic equations, it is necessary to linearize Eqs. (16) and (17). The details of the linearization of both equations are described in [13] and we list the final linearized form as follows:

1. Linearized weak form convection step using Talyor-Galerkin method:

$$(u_i^\star, v_i)_\Omega = \left(u_i^n - \Delta t u_j^n \frac{\partial u_i^n}{\partial x_j}, v_i \right)_\Omega - \frac{\Delta t^2}{2} \left(u_k^n \frac{\partial u_i^n}{\partial x_k}, u_j^n \frac{\partial v_i}{\partial x_j} \right)_\Omega \tag{18}$$

2. Linearized weak form of diffusion step

$$
\begin{aligned}
&\rho^f \left(\frac{u_i - u_i^\star}{\Delta t}, v_i \right)_\Omega + (\rho^s - \rho^f) \left(\frac{u_i^s - (u_i^s)^n}{\Delta t}, v_i \right)_{\Omega^s_{n+1}} \\
&+ \mu^f \left(\frac{\partial u_i}{\partial x_j} + \frac{\partial u_j}{\partial x_i}, \frac{\partial v_i}{\partial x_j} \right)_\Omega - \left(p, \frac{\partial v_j}{\partial x_j} \right)_\Omega - \left(\frac{\partial u_j}{\partial x_j}, q \right)_\Omega \\
&+ \mu^s \Delta t \left(\frac{\partial u_i}{\partial x_j} + \frac{\partial u_j}{\partial x_i} + \Delta t \frac{\partial u_i}{\partial x_k} \frac{\partial u_j^n}{\partial x_k} + \Delta t \frac{\partial u_i^n}{\partial x_k} \frac{\partial u_j}{\partial x_k}, \frac{\partial v_i}{\partial x_j} \right)_{\Omega^s_{n+1}} \\
&+ \Delta t^2 \left(\frac{\partial u_i}{\partial x_k} (\tau^s_{kl})^n \frac{\partial u_j^n}{\partial x_l} + \frac{\partial u_i^n}{\partial x_k} (\tau^s_{kl})^n \frac{\partial u_j}{\partial x_l}, \frac{\partial v_i}{\partial x_j} \right)_{\Omega^s_{n+1}} \\
&+ \Delta t \left(\frac{\partial u_i}{\partial x_k} (\tau^s_{kj})^n + (\tau^s_{il})^n \frac{\partial u_j}{\partial x_l}, \frac{\partial v_i}{\partial x_j} \right)_{\Omega^s_{n+1}} \\
&= (\bar{h}_i, v_i)_{\Gamma^N} + \rho^f (g_i, v_i)_\Omega + (\rho^s - \rho^f)(g_i, v_i)_{\Omega^s_{n+1}} \\
&+ \left(\mu^s \Delta t^2 \frac{\partial u_i^n}{\partial x_k} \frac{\partial u_j^n}{\partial x_k} + \Delta t^2 \frac{\partial u_i^n}{\partial x_k} (\tau^s_{kl})^n \frac{\partial u_j^n}{\partial x_l} - (\tau^s_{ij})^n, \frac{\partial v_i}{\partial x_j} \right)_{\Omega^s_{n+1}}
\end{aligned}
\tag{19}
$$

The spatial discretization in [13] uses a fixed Eulerian mesh for Ω and an updated Lagrangian mesh for Ω^s_{n+1} to discretize Eq. (19). The discretization Ω^h (for Ω) uses $\mathbf{P}_2$ (for velocities $\mathbf{u}$) $\mathbf{P}_1$ (for pressure p) elements (the Taylor-Hood element) with the corresponding finite element spaces

$$
\begin{aligned}
V^h(\Omega^h) &= \text{span}\{\psi_1, ..., \psi_{N^u}\} \in H^1(\Omega) \\
L^h(\Omega^h) &= \text{span}\{\phi_1, ..., \phi_{N^p}\} \in L^2(\Omega).
\end{aligned}
$$

The approximated solution $\mathbf{u}^h$ and p^h can be expressed in terms of these basis functions as

$$
\mathbf{u}^h(\mathbf{x}) = \sum_{i=1}^{N^\mathbf{u}} \mathbf{u}(\mathbf{x}_i)\psi_i(\mathbf{x}) \quad , \quad p^h(\mathbf{x}) = \sum_{i=1}^{N^p} p(\mathbf{x}_i)\phi_i(\mathbf{x}) \tag{20}
$$

The solid domain Ω^s_{n+1} at the $n+1$ time step is discretized as Ω^{sh}_{n+1} using linear triangular elements with the corresponding finite element spaces as:

$$
V^{sh}(\Omega^{sh}_{n+1}) = \text{span}\{\psi^s_1, ..., \psi^s_{N^s}\} \in H^1(\Omega^s_{n+1}), \tag{21}
$$

and approximate $\mathbf{u}^h(\mathbf{x}) \mid_{\mathbf{x} \in \Omega^{sh}_{n+1}}$ as:

$$
\mathbf{u}^{sh}(\mathbf{x}) = \sum_{i=1}^{N^s} \mathbf{u}^h(\mathbf{x}^s_i)\psi^s_i(\mathbf{x}) = \sum_{i=1}^{N^s} \sum_{j=1}^{N^u} \mathbf{u}(\mathbf{x}_j)\psi_j(\mathbf{x}^s_i)\psi^s_i(\mathbf{x}) \tag{22}
$$

where $\mathbf{x}^s_i$ is the nodal coordinate of the solid mesh.

Substituting (20), (22) and similar expressions for the test functions $\mathbf{v}^h$, q^h and $\mathbf{v}^{sh}$ into Eq. (18), we obtain the following matrix form:

$$\begin{bmatrix} \mathbf{A} & \mathbf{B} \\ \mathbf{B}^T & \mathbf{0} \end{bmatrix} \begin{bmatrix} \mathbf{u} \\ \mathbf{p} \end{bmatrix} = \begin{bmatrix} \mathbf{b} \\ \mathbf{0} \end{bmatrix}, \tag{23}$$

where

$$\mathbf{A} = \mathbf{M}/\Delta t + \mathbf{K} + \mathbf{D}^T(\mathbf{M}^s/\Delta t + \mathbf{K}^s)\mathbf{D} \tag{24}$$

and

$$\mathbf{b} = \mathbf{f} + \mathbf{D}^T\mathbf{f}^s + \mathbf{M}\mathbf{u}^\star/\Delta t + \mathbf{D}^T\mathbf{M}^s\mathbf{D}\mathbf{u}^n/\Delta t \tag{25}$$

In Eqs. (24) and (25), matrix $\mathbf{D}$ is the isoparametric interpolation matrix derived from Eq. (22) which can be expressed as

$$\mathbf{D} = \begin{bmatrix} \mathbf{P}^T & \mathbf{0} \\ \mathbf{0} & \mathbf{P}^T \end{bmatrix} \quad , \quad \mathbf{P}_{ij} = \psi_i(\mathbf{x}_j^s). \tag{26}$$

For the other matrices in (24) and (25), $\mathbf{M}$ and $\mathbf{K}$ are global mass matrix, global stiffness matrix from discretization of integrals in Ω^h. Similarly, $\mathbf{M}^s$ and $\mathbf{K}^s$ are mass matrix and stiffness matrix from discretization of integrals in Ω^{sh}. $\mathbf{u}$, $\mathbf{p}$, $\mathbf{f}$ are velocity, pressure and right hand side vectors, respectively. $\mathbf{B}$ and its transpose $\mathbf{B}^T$ represent the connections between pressure and velocities, which arise from the weak formulation in (19).

3 Parallelization of One-Field FDM Algorithm

3.1 Parallelism and Related Issues

The algorithm described in the previous section consists three parts: finite element discretization of the PDEs (calculation of element matrices, assembly of the global matrix), projecting matrix arising from solid equations into matrix arising from fluid equations (L^2 isoparametric projection) and solving the resulting system of linear algebraic equations. The computations of the first two parts are nearly embarrassingly parallel and require little or no communication, while the linear system solving using iterative scheme needs global communications (inner product) and communications from neighboring processes (matrix vector multiplications) for updating the residual and solution vectors.

3.2 Datatype and Computation Setup

The code implementation for solving the model described in the previous section is carried out in Campfire, where the structure grid generation is provided by PARAMESH. PARAMESH generates meshes as the union of blocks (arrays) of cells. Each block consists of internal cells and certain layer of guardcells (Fig. 2).

In finite element implementation each cell is treated as an (quadrilateral) element. For the 2D FSI simulation there are three variables to be solved: u (x-velocity), v (y-velocity) and p (pressure). For spatial variables we discretize the velocity field by quadratic quadrilateral ($P2$) elements and the pressure by linear quadrilateral ($P1$) elements. To accomplish storing the variable values in a 2D model, we employ four arrays from PARAMESH.

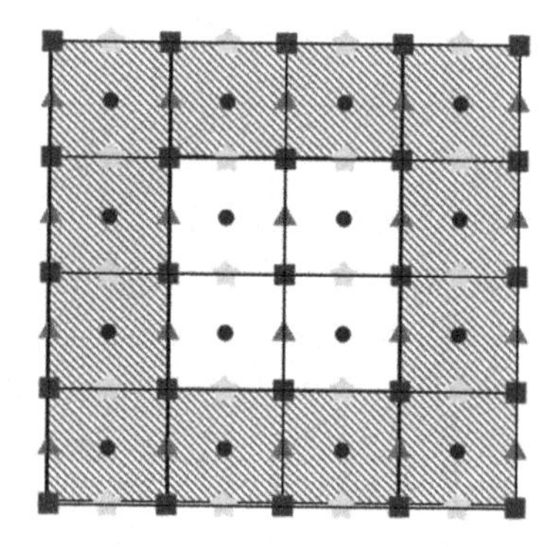

Fig. 2. A 2×2 block with one layer of guardcells

3.3 Linear System Solver

As known, the solution of a linear system of equations constitutes an important part of the algorithm for solving PDEs numerically. Thus the properties of the linear system solver is crucial to the performance of PDE simulations. Generally, iterative methods consist of three core computational steps, which are vector update, matrix-vector multiplications and inner product computation. Vector updates is inherently parallel and there is no communication required. Parallel algorithm for matrix-vector multiplication relies on the data structure of matrix (sparsity, organization etc.) and typically local data communications between neighboring processes are sufficient for correct computation. Inner product calculation requires global data communications but the complexity could be moderate ($O(\log_2 P)$, where P is the number of processes). The parallel implementation of these three computation kernels will provide significant performance enhancement on iterative methods. To accelerate the MINRES iteration we use incomplete LU as the preconditioner. The computations for applying preconditioners involve backward/forward substitutions, which is inherently sequential. In the parallel implementation the preconditioner becomes block Jacobi type and the matrix entries connecting the neighboring processes are ignored. This causes the convergence properties of parallel preconditioning deteriorate significantly from sequential preconditioning.

4 Numerical Experiments

In this work, we exam our parallel code on two 2D problems and compare the performance of parallel computation with the serial version in each case to access the parallel efficiencies. MINRES iterations stop when the 2-norm of the relative residual is less than 10^{-8} or the maximum number of iterations is reached (7500). For parallel simulations we use various number of processes up to 256 cores. All simulations were carried out on Taiwania, a supercomputer having a memory of 3.4 petabytes and delivering over 1.33 quadrillion flop/s of theoretical peak performance. The system has 630 compute nodes based on 40 core Intel Xeon Gold 6148 processors running at 2.4 GHz. Overall the system has a total of 25200 processor cores and 157 TB of aggregate memory. All the timing results are the average values from the running time of multiple simulations.

4.1 The Motion of a Disc in 2D Lid-Driven Cavity Flow

In the first case we consider the motion of a deformable disc in a lid-driven cavity flow in 2D domain. Zhao et al. [15] studied the simulation for validating their methods. The experiment parameters are shown in Table 1 and Fig. 3 is the graphic demonstration of the problem setup. Initially, a round stress-free disc of radius 0.2 is centered at (0.6, 0.5), then at $t = 0$ the top cavity wall starts moving horizontally at $u = 1$. For the material parameters, the density of the disc (ρ_s) and fluid (ρ_f) are both set to be 1. The elastic constant (μ_s) of the disc is 0.1 and the viscosity of the fluid (μ_f) is 0.01.

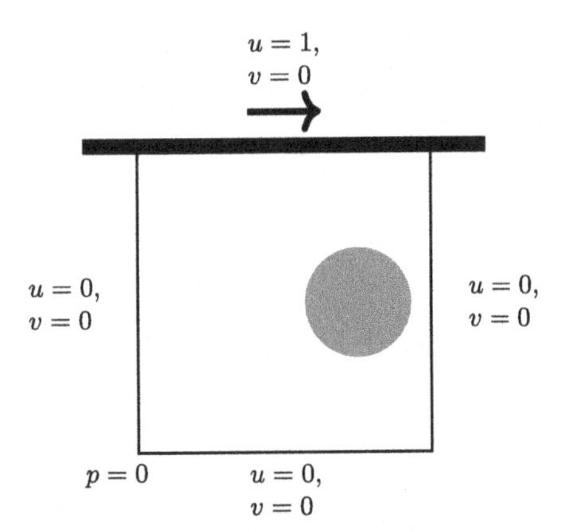

Fig. 3. A disc in lid-driven cavity flow with the boundary conditions.

Table 1. Parameters for 2D solid (disc) in a lid-driven cavity flow

Parameters	ρ_f	ρ_s	μ_f	μ_s
Values	1	1	0.01	0.1

The fluid mesh resolution of the numerical simulation is 128 × 128, solid mesh contains 31163 elements and 15794 nodes. $\Delta t = 0.01$, 800 time steps. The running time for the last 100 time steps and speedup are shown in Table 1. As seen from the table, our method gains a roughly 47 speedup for 256 cores parallel simulations. Comparing with serial simulation, we reduce the overall running time (need 800 time steps) from roughly 65 h to 1.5 h. In the table we also list the average number of preconditioned MINRES iterations for the first 100 time steps in each simulation to show the impact of the parallelized preconditioning on the convergence of MINRES iterations. Notice that the number of MINRES iterations reach the maximum threshold (10000) after $t = 0.3$, corresponding to the scenario when the disc is near the wall.

The solid deformation are visualized in Fig. 4. The motion and the deformation of the solid are nearly identical to the result from [13] and [15]. We see that the disc deformation is asymmetric about the disc's vertical center line, and lubrication forces prevent the disk from touching the lid. The solid body ends up in a fixed position near the center of the cavity, and the velocity field becomes steady.

For the parallel simulation we gained speed up factor of 44 for 256 processes, corresponding to parallel efficiency of 17%. The increased number of MINRES iteration explains the reduction of the parallel efficiency from 82% (4 processes) to 17% (256 processes) (Table 2).

Table 2. Parallel simulation results

Last 100 time steps	Serial	4	16	64	256
Time (min)	490	148	48	23	11
Speed up	1	3.3	10.2	21.3	44.5
Iterations (first 100 time steps)	2030	2420	2750	2980	3110

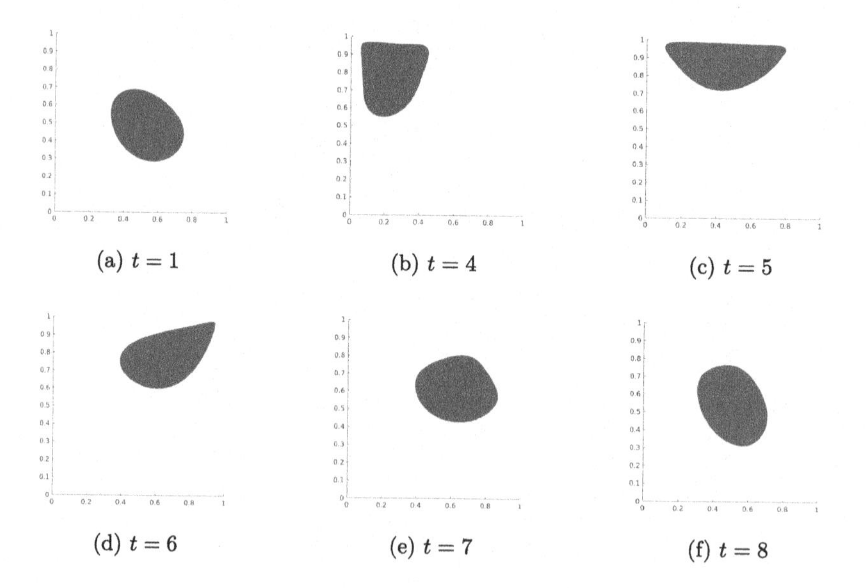

Fig. 4. The simulation results for a deformable solid motion in a lid-driven cavity flow.

4.2 Oscillation of a Flexible Leaflet Oriented Across the Flow Direction

The second example comes from [10–12], in which the problem was used for validation their methods. In this work we parallelize the one field monolithic FSI algorithm of this benchmark problem in [13]. The problem setup is described as follows. A leaflet is located at the center of the bottom horizontal side of a rectangular computational domain. The horizontal length and vertical length are 4 m and 1 m, respectively. The inflow velocity is in the x-direction and governed by $u_x = 15.0y(2-y)\sin(2\pi t)$. The computational domain and boundary conditions are illustrated in Fig. 5.

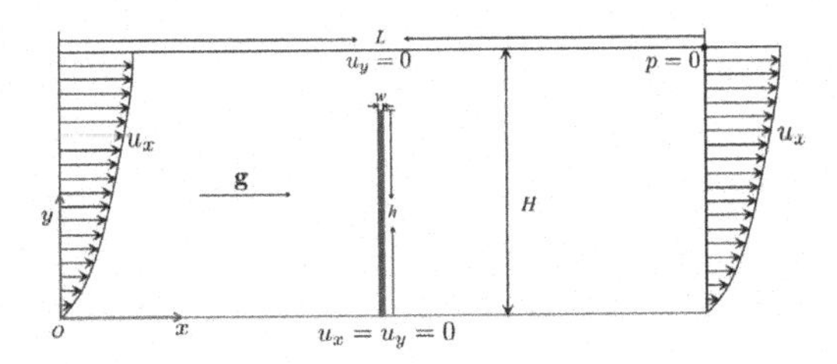

Fig. 5. Oscillation of a flexible leaflet across the flow direction. Source: [13]

The material parameters for the solid and fluid are listed in Table 3. The leaflet (solid region) is represented with 154 linear triangular elements with 116 nodes, and the corresponding fluid mesh is consist of 320 × 80 rectangular cells. The time step size for the evolution is $\Delta t = 5 \times 10^{-5}$ s. All leaflet simulations run from $t = 0$ to $t = 0.8$, which is 16000 time steps. Similar to the above example, we record simulation time for the middle 2000 time steps and the speedups of the parallel run are listed in Table 4. In this case the ratio of the number of solid mesh nodes to the number of fluid mesh nodes is much smaller than the example 1. For 256 processes simulation we gained a speed up factor of 28 corresponding to 11% parallel efficiency. Obviously the convergence of the preconditioned MINRES in this case is much worse than that in the example 1. This may attribute to the high elastic modulus (10^7 vs. 1) of the solid and the convergence of MINRES is more difficult (Fig. 6).

Table 3. Material parameters for oscillation of a flexible leaflet oriented across the flow direction.

Parameters	ρ_f	ρ_s	μ_f	μ_s
Values	100 kg/m^3	100 kg/m^3	10 N s/m^2	10^7 N/m^2

Table 4. Parallel simulation results: leaflet in 2D channel flow

Middle 2000 time steps	Serial	16	64	256
Time (min)	592	85	40	21
Speed up	1	11.2	14.8	28.2
Iterations (middle time steps)	2030	5010	>7500	>7500

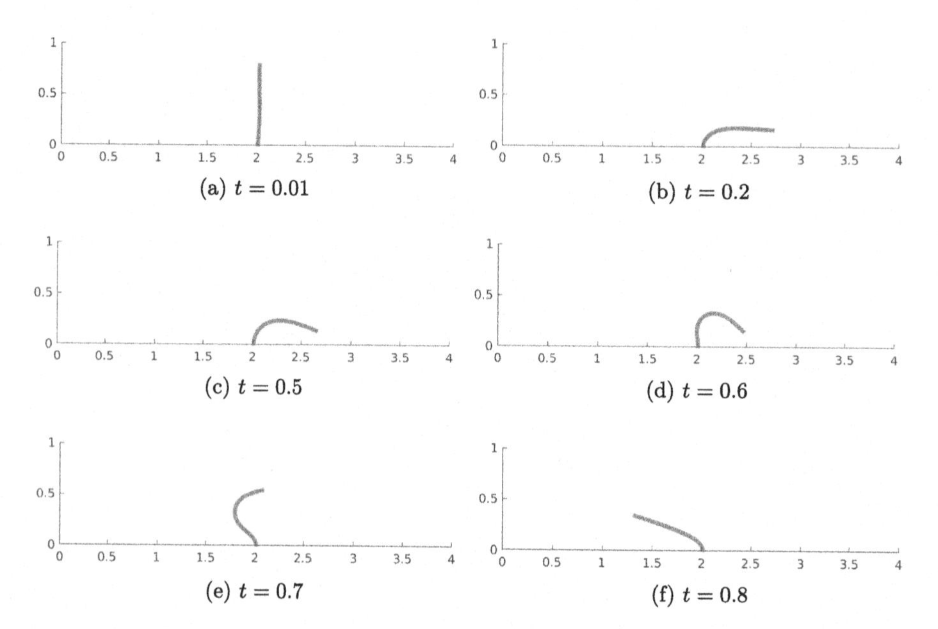

Fig. 6. The simulation results for oscillation of a flexible leaflet oriented across the flow direction.

5 Conclusion

In this research we have implemented the parallel computation on FSI 2D problem simulations. The convergence of preconditioned MINRES iteration is crucial to the parallel efficiency of the algorithm. For the first benchmark problem (2D disc in cavity-driven flow) the parallel simulation gain significant speed up although the parallel efficiency is not impressive. However, high elastic modulus of solid appeared in the second example (oscillating leaflet in channel flow) reduces the convergence performance of MINRES and hence the computation performance and parallel efficiency. One of the most important future work is to find preconditioners whose properties are minimally impacted by parallelization. With the modified preconditioner and algorithm we wish to extend our experiment to 3D FSI benchmark problems to verify our method.

References

1. Heil, M.: An efficient solver for the fully coupled solution of large displacement fluid-structure interaction problems. Comput. Meth. Appl. Mech. Eng. **193**(1–2), 1–23 (2004)
2. Heil, M., Hazel, A.L., Boyle, J.: Solvers for large-displacement fluid-structure interaction problems: segregated versus monolithic approaches. Comput. Mech. **43**(1), 91–101 (2008)
3. Muddle, R.L., Mihajlovic, M., Heil, M.: An efficient preconditioner for monolithically-coupled large-displacement fluid-structure interaction problems with pseudo-solid mesh updates. J. Comput. Phys. **231**(21), 7315–7334 (2012)
4. Kuttler, U., Wall, W.A.: Fixed-point fluid-structure interaction solvers with dynamic relaxation. Comput. Mech. **43**(1), 61–72 (2008)
5. Degroote, J., Bathe, K.J., Vierendeels, J.: Performance of a new partitioned procedure versus a monolithic procedure in fluid-structure interaction. Comput. Struct. **87**(11–12), 793–801 (2009)
6. Hou, G., Wang, J., Layton, A.: Numerical methods for fluid-structure interaction a review. Commun. Comput. Phys. **12**(2), 337–377 (2012)
7. Zhang, L., Gerstenberger, A., Wang, X., Liu, W.K.: Immersed finite element method. Comput. Meth. Appl. Mech. Eng. **193**(21), 2051–2067 (2004)
8. Zhang, L., Gay, M.: Immersed finite element method for fluid-structure interactions. J. Fluids Struct. **23**(6), 839–857 (2007)
9. Wang, X., Zhang, L.T.: Interpolation functions in the immersed boundary and finite element methods. Comput. Mech. **45**(4), 321–334 (2009)
10. Glowinski, R., Pan, T., Hesla, T., Joseph, D., Periaux, J.: A fictitious domain approach to the direct numerical simulation of incompressible viscous flow past moving rigid bodies: application to particulate flow. J. Comput. Phys. **169**(2), 363–426 (2001)
11. Yu, Z.: A DLM/FD method for fluid/flexible-body interactions. J. Comput. Phys. **207**(1), 1–27 (2005)
12. Baaijens, F.P.: A fictitious domain/mortar element method for fluid-structure interaction. Int. J. Numer. Meth. Fluids **35**(7), 743–761 (2001)
13. Wang, Y., Jimack, P., Walkley, M.: A one-field monolithic fictitious domain method for fluid-structure interactions. Comput. Meth. Appl. Mech. Eng. **317**, 1146–1168 (2017)
14. Zienkiewic, O.: The Finite Element Method for Fluid Dynamics, 6th edn. Elsevier, Amsterdam (2005)
15. Zhao, H., Freund, J.B., Moser, R.D.: A fixed-mesh method for incompressible flow-structure systems with finite solid deformations. J. Comput. Phys. **227**(6), 3114–3140 (2008)
16. Goodyer, C.E., Jimack, P.K., Mullis, A.M., Dong, H.B., Xie, Y.: On the fully implicit solution of a phase-field model for binary alloy solidification in three dimensions. Adv. Appl. Math. Mech. **4**, 665–684 (2012)
17. Bollada, P.C., Goodyer, C.E., Jimack, P.K., Mullis, A., Yang, F.W.: Three dimensional thermal-solute phase field simulation of binary alloy solidification. J. Comput. Phys. **287**, 130–150 (2015)
18. Wall, W.A.: Fluid-struktur-interaktion mit stabilisierten finiten elementen. Ph.D. thesis, Universitt Stuttgart (1999)

Characterization of Foam-Assisted Water-Gas Flow via Inverse Uncertainty Quantification Techniques

Gabriel Brandão de Miranda[1,2], Luisa Silva Ribeiro[1,2], Juliana Maria da Fonseca Façanha[3], Aurora Pérez-Gramatges[3,4], Bernardo Martins Rocha[1,2], Grigori Chapiro[1,2], and Rodrigo Weber dos Santos[1,2](✉)

[1] Graduate Program in Computational Modeling, Federal University of Juiz de Fora, Juiz de Fora, Brazil
rodrigo.weber@ufjf.edu.br

[2] Laboratory of Applied Mathematics (LAMAP), Federal University of Juiz de Fora, Juiz de Fora, Brazil

[3] Laboratory of Physical-Chemistry of Surfactants (LASURF), Pontifical Catholic University of Rio de Janeiro, Rio de Janeiro, Brazil

[4] Chemistry Department, Pontifical Catholic University of Rio de Janeiro, Rio de Janeiro, Brazil

Abstract. In enhanced oil recovery (EOR) processes, foam injection reduces gas mobility and increases apparent viscosity, thus increasing recovery efficiency. The quantification of uncertainty is essential in developing and evaluating mathematical models. In this work, we perform uncertainty quantification (UQ) of two-phase flow models for foam injection using the STARS model with data from a series of foam quality-scan experiments. We first performed the parameter estimation based on three datasets of foam quality-scans on Indiana limestone carbonate core samples. Then distributions of the parameters are inferred via the Markov Chain Monte Carlo method (MCMC). This approach allows propagating parametric uncertainty to the STARS apparent viscosity model. In particular, the framework for UQ allowed us to identify how the lack of experimental data affected the reliability of the calibrated models.

Keywords: Uncertainty quantification · Foam dynamics · Bayesian inference

1 Introduction

Oil extraction, the process by which usable oil is extracted and removed from underground, can be divided into three categories [5]. The first category is the primary oil recovery, which raises the reservoir pressure so that recovery occurs spontaneously. This process does not have a good yield since, on average, it recovers only 30% of the original volume of oil present in the reservoir. The

D. Groen et al. (Eds.): ICCS 2022, LNCS 13353, pp. 310–322, 2022.
https://doi.org/10.1007/978-3-031-08760-8_26

secondary oil recovery uses techniques such as injection water or gas into the reservoir through an injection well to push the oil out of the rock pores. The third category, also called enhanced oil recovery (EOR), uses more complex techniques such as thermal recovery (heating the oil to decrease its viscosity) and chemical recovery such as foam injection to reduce gas mobility and increase recovery.

EOR techniques have been increasingly used in the upstream oil industry, and in particular, one of the methods that stand out the most is foam injection. The alternating water and gas injection (WAG) process can be improved by using foams to reduce gas mobility, increase apparent viscosity, and improve recovery efficiency.

Several physical models of foam flow in porous media are available in literature [1,10,19]. Modeling of foam flow dynamics in porous media is very complex due to its non-Newtonian nature, its dependence on the foam texture, and the complex bubble generation/destruction process. In this work, the simplified version of the CMG-STARS model [6] is studied.

The process of estimating the model parameters is not straightforward, and several methods to this end have been proposed so far [3,11,12,18]. In [3] a manual process was used to adjust the foam flow parameters to apparent viscosity data. The proposed procedure works separately with data in the low and high-quality regimes and is based on the foam quality and apparent viscosity relation. The work of [11] used data weighting and constraints when employing nonlinear least-squares minimization methods for parameter estimation. The work of [12] used a combined approach with a graphical method and least-squares minimization techniques. In [18] the problem of fitting many parameters was replaced by a procedure based on linear regression and single-variable optimization, which avoids problems related to non-unique solutions and sensitivity issues of the initial estimates. It is important to remark that these methods did not perform any uncertainty quantification after estimating the parameters. To reduce the non-uniqueness and uncertainty of solutions, the work of [2] proposes an assisted/automated method to adjust the parameters of relative permeability measurements and provides a framework for a consistent uncertainty assessment of relative permeability measurements.

The present work used Bayesian inference techniques for parameter estimation, followed by uncertainty propagation to evaluate the uncertainties associated with the foam injection process numerically. The probability distributions of the parameters were estimated using the Markov Chain Monte Carlo (MCMC) method, which seeks to find the posterior distribution of the parameters given a dataset and a prior characterization of the parameters. In particular, we assessed the distributions of the parameters using data from a series of foam quality-scan experiments to characterize the parameters better.

The remaining of this manuscript is organized as follows: in Sect. 2 the experimental setup, the recorded data, and the methods used for parameter inference are reported; Sect. 3 presents the results obtained in terms of least-squares methods and Bayesian methods; and Sect. 4 ends this work with some conclusions and discussions.

2 Methods

This work uses Bayesian inference techniques for parameter estimation, also known as inverse uncertainty quantification (UQ). After this first step, uncertainty propagation or forward UQ is performed to evaluate the uncertainties associated with the numerical modeling of Enhanced Oil Recovery (EOR) based on the process of coinjection of foam. This section explains the experimental setup, the recorded data, and the methods used for parameter inference.

2.1 Experimental Setup

The experiments used in this work were described in [7], and part of the data was used in [16]. For clarity, the setup is briefly described here. Brine prepared by dissolving adequate amounts of salt in distilled water was used in the core-flooding. The concentrations are shown in Table 1. Before preparing the surfactant solution, the brine was degassed using a vacuum pump. The salts that were used to prepare the brine were purchased from Sigma-Aldrich Brasil and were reagent grade.

Table 1. Ionic composition of injection water (IW)

Ions	Na^+	K^+	Ca^{2+}	Mg^{2+}	SO_4^{2-}	Cl^-
Concentration (mg/L)	11008	393	132	152	41	17972

The surfactant chosen to perform the foam injection was sodium alpha-olefin sulfonate (Bioterge AS-40), which Stepan Brasil donated. It was used at a concentration of 0.1 wt%, with a critical micellar concentration (CMC) in IW at 20º and ambient pressure conditions 0.0017 wt%. Nitrogen (99.992% purity, Linde Brasil) was used for the gas phase.

A series of three foam quality-scan experiments were performed on a sample of Indiana limestone (Kocurek Industries, USA), which was the rock used in the experiments. The dimensions and petrophysical properties of the core used in this work are presented in Table 2.

The core was loaded onto the Hassler core support under confining pressure of 3.44 MPa (500 psi) vertically. It was aspirated for two hours and then saturated under vacuum with IW. Confinement pressure and pore pressure were increased simultaneously to 17.2 MPa (2500 psi) and 13.8 MPa (2000 psi), respectively. The core sample was left at this pressure for 24 h to saturate the core fully. Afterward, the pore pressure was decreased to 10 MPa (1500 psi), and then the brine permeability was measured. This procedure was done by injecting IW at different flow rates for pore volumes. After performing the permeability measurement, 0.1 wt%AOS surfactant solution was injected (all experiments used the same surfactant concentrations) and then through the core for at least 5 pore volumes (PV) to displace IW. The system temperature was raised to 60 °C.

Table 2. Dimensions and petrophysical properties of Indiana limestone, where L, D, PV, φ, and K are the length, diameter, pore volume, porosity, and permeability, respectively.

Properties	Experiment 1	Experiment 2	Experiment 3
L [m]	0.150	0.150	0.150
D [m]	0.0382	0.0382	0.0382
PV $[10^{-6}m^3]$	26.7	26.7	27.76
φ [-]	0.155	0.155	0.161
k $[m^2]$	2.70×10^{-13}	1.57×10^{-13}	2.91×10^{-13}
v[m/s]	1.45×10^{-5}	1.45×10^{-5}	2.40×10^{-5}

As the pressure and temperature were constant and there was no possibility of leakage, the nitrogen solution and surfactant were co-injected at constant superficial velocity (1.45×10^{-5} m/s) and injection flow rate (0.967 mL/min), but at different gas/liquid ratios. In Fig. 1 it is possible to see the schematic drawing of core-flood apparatus used for foam injection.

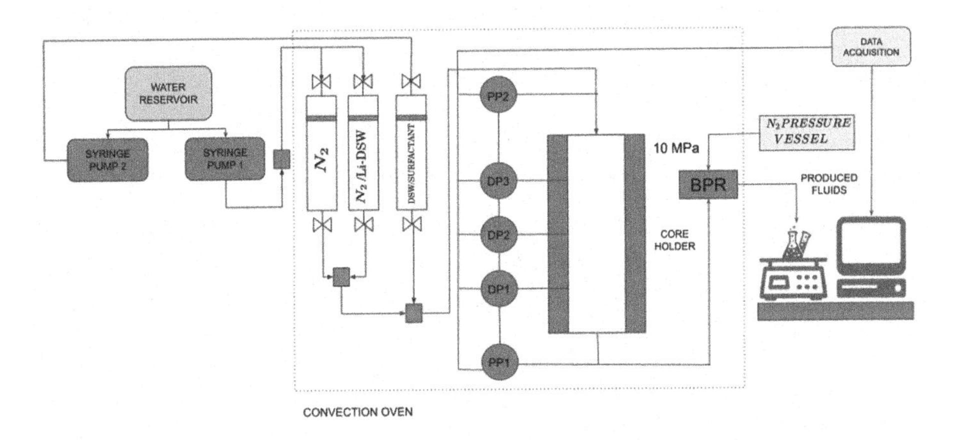

Fig. 1. Schematic drawing of core-flood apparatus.

2.2 Relative Permeabilities

Relative permeabilities were described by the Corey model for the two-phase flow of water and gas without surfactant, which are given by:

$$k_{rw} = k_{rw}^0 \left(\frac{S_w - S_{wc}}{1 - S_{wc} - S_{gr}} \right)^{n_w}, \quad \text{and} \quad k_{rg} = k_{rg}^0 \left(\frac{S_g - S_{gr}}{1 - S_{wc} - S_{gr}} \right)^{n_g}, \quad (1)$$

where n_w and n_g are the Corey exponents for water and gas, respectively, k_{rw}^0 and k_{rg}^0 are the end-point relative permeabilities for water and gas, respectively,

S_{wc} is the connate water saturation, and S_{gr} the residual gas saturation. Relative permeability data for high permeability Indiana Limestone found in the literature [13] were considered. The Corey parameters used in this work, which were fitted to the relative permeability data of [13] using the techniques described in [17], are given in Table 3.

Table 3. Relative permeability parameters for datasets.

Parameters	S_{wc}	S_{gr}	n_w	n_w	k_{rw}^0	k_{rg}^0
Values	0.4	0.293	2.98	0.96	0.302	0.04

2.3 STARS Model

To model the two-phase flow for foam flow, the CMG-STARS mathematical model [6] was used. In this model the foam effects are modeled considering a reduction factor that affects the mobility of the gas phase. The mobility reduction factor (MRF) term can describe the effects of surfactant concentration, water and oil saturations, shear-thinning, and other effects.

Let the mobility of gas and water phases be denoted by λ_g and λ_w, respectively. The total mobility λ_T is defined as $\lambda_T = \lambda_w + \lambda_g$. Thus, the apparent viscosity can be defined as the inverse of total relative mobility:

$$\mu_{app} = \lambda_T^{-1} = \left(\lambda_w + \frac{\lambda_g}{MRF}\right)^{-1}, \tag{2}$$

where the fact that mobility of the gas phase is influenced by the foam through the mobility reduction factor MRF is already taken into account.

The fractional gas flow is then redefined as follows including the MRF function:

$$f_g = \frac{\lambda_g}{MRF\left(\lambda_w + \frac{\lambda_g}{MRF}\right)} = \frac{\lambda_g}{MRF}\mu_{app}. \tag{3}$$

The gas mobility is given by:

$$\lambda_g = \frac{k_{rg}}{MRF\mu_g}, \qquad MRF = 1 + fmmobF_2, \tag{4}$$

where the F_2 term describes the effects of water saturation, and is given by:

$$F_2 = \frac{1}{2} + \frac{1}{\pi}arctan(sfbet(S_w - SF)), \tag{5}$$

where $fmmob$, $sfbet$, and SF are model parameters.

2.4 Procedures for Parameter Estimation

Two approaches were used in this work for parameter estimation: nonlinear least-squares minimization and Bayesian inference. To estimate parameters with nonlinear least-squares we used the Differential Evolution (DE) method implemented in the `lmfit` library [14] available in the Python programming language.

For Bayesian inference of the distributions of the parameters, the Markov Chain Monte Carlo (MCMC) [4] method was used. The prior distributions of the parameters of the model required for the MCMC were chosen considering the physical ranges of the parameters and knowledge available in the literature [8]. The `PyMC3` library [15] for Bayesian modeling was used for executing the MCMC method. For the inference process, four chains are built independently. The univariate slice sampler was adopted as the step function, 10^4 samples were drawn for each randomized parameter, and 10^3 samples were discarded from each of the final chains. The joint of these four chains describes a sample of the posterior distribution for each parameter.

Assuming θ as parameters of the STARS model and D as the data set, the MCMC attempts to estimate:

$$\mathbb{P}(D|\theta) = \frac{\mathbb{P}(D|\theta)\mathbb{P}(\theta)}{\mathbb{P}(D)} \tag{6}$$

where $\mathbb{P}(\theta)$ represents the prior knowledge of the input parameters θ, as a joint probability distribution; $\mathbb{P}(D|\theta)$ is the likelihood function; and $\mathbb{P}(D)$ is the evidence, a normalization factor for the posterior distribution.

To carry out the MCMC estimation, our prior knowledge about the parameters' distributions must be provided. Table 4 summarizes the priors adopted in this work, which was based on the choice used in previous works [16].

Table 4. Chosen prior distributions for the parameters of the CMG-STARS foam model used in the MCMC method.

fmmob	SF	sfbet
$\mathcal{U}(10, 1000)$	$\mathcal{U}(S_{wc}, 1 - S_{gr})$	$\mathcal{U}(10, 1000)$

2.5 Sensitivity Analysis

A variance based sensitivity analysis was used to assess how the input parameters $x_i \in \theta$ and their interactions contribute to the variations of any quantity of interest $\mathcal{Y}$. The main and total Sobol indices were used to this end. The first-order Sobol index, presented in the Eq. 7, expresses how any uncertain input x_i directly contributes to the variance of the output $\mathcal{Y}$.

$$S_i = \frac{\mathbb{V}[\mathbb{E}[\mathcal{Y}|x_i]]}{\mathbb{V}[\mathcal{Y}]} \tag{7}$$

To estimate the changes in $\mathbb{V}[\mathcal{Y}]$ considering the first and high order interactions of the i-th uncertain entry, the total Sobol index is evaluated. It is given by:

$$S_{T_i} = 1 - \frac{\mathbb{V}[\mathbb{E}[\mathcal{Y}|x_{-i}]]}{\mathbb{V}[\mathcal{Y}]} \tag{8}$$

where x_{-1} denotes the set of all input parameters except x_i. The sensitivity indices were computed with the `SAlib` library using the Saltelli method [9]. Bounds for parameters in the SA were defined as the bounds from the 90% confidence interval of the marginal posterior distributions.

3 Results

3.1 Least-Squares Estimates

First, we present some parameter estimates obtained with the nonlinear least-squares method to first characterize the foam flow in the core-flooding experiments. Figure 2 shows the results of the foam quality scan experiment in terms of the pressure drop versus time.

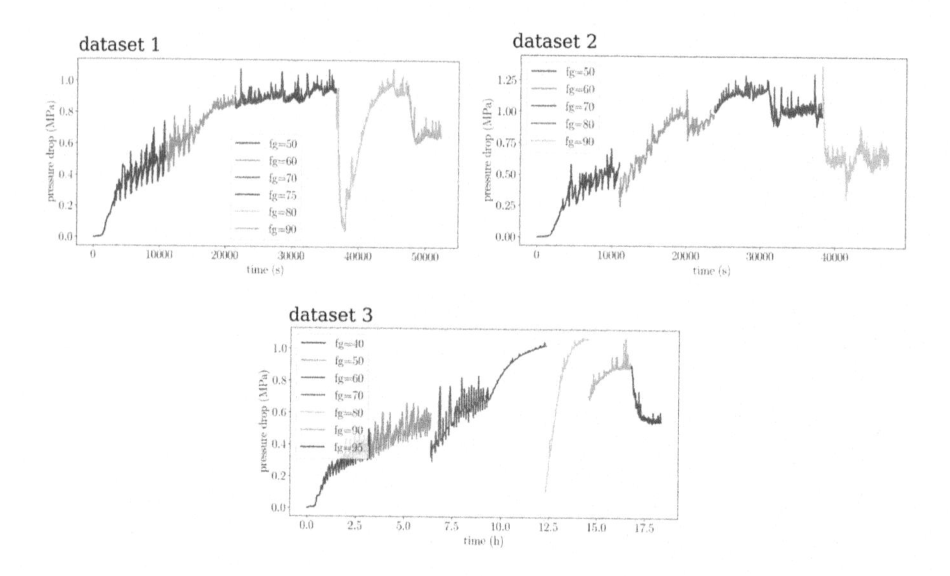

Fig. 2. Foam quality-scan experiment results

Figure 3 shows the steady-state experimental data for apparent viscosity as a function of foam quality for the three experiments previously described in Table 2. Fittings of the STARS model to the corresponding data for each dataset are also shown in Fig. 3. Model fittings and data for experiments #01, #02, and #03 are represented by blue, orange, and green, respectively, where dots represent the data and solid lines model evaluations.

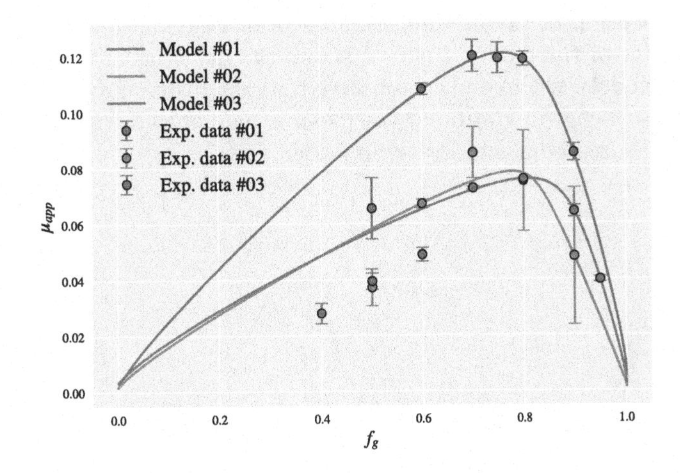

Fig. 3. Experimental data and STARS model evaluations for all the datasets.

Table 5 summarizes the parameter estimates obtained after applying the least-squares method in the three datasets. The parameters $fmmob$, SF, and $sfbet$ represent the reference mobility reduction factor, the water saturations around which weak foam collapses, and the sharpness from the transition between low- and high-quality foam regimes. The lower the $sfbet$ value, the smoother the transition from high to low quality, whereas larger values for $sfbet$ represent a sharp transition. The estimated values for $sfbet$ are in good agreement with the transition observed in the experimental data, where datasets #02 and #03 have a more sharp transition than dataset #01. The estimated values for SF for all datasets agree with two decimals places. The estimated values of $fmmob$ for all datasets are again in good agreement with the corresponding data, where for instance, the dataset #01 presents the highest apparent viscosity value among the datasets.

Table 5. STARS parameters estimated with nonlinear least-squares method.

Dataset/Parameter	$fmmob$	SF	$sfbet$
Dataset #01	292.71	0.44	367.88
Dataset #02	180.92	0.44	541.86
Dataset #03	173.10	0.44	419.77

3.2 Foam Model Parameters' Distributions

Next, to better characterize the parameters for each dataset, we performed a Bayesian inference using the MCMC method with the priors given in Table 4. After the execution of the method, the posterior distribution was obtained for each parameter of the model. Figure 4 shows the densities of the parameters of the STARS models, where one can observe that the distributions of $fmmob$ and SF are more concentrated around the mean value, whereas the distribution for $sfbet$ is more spread out and less symmetrical.

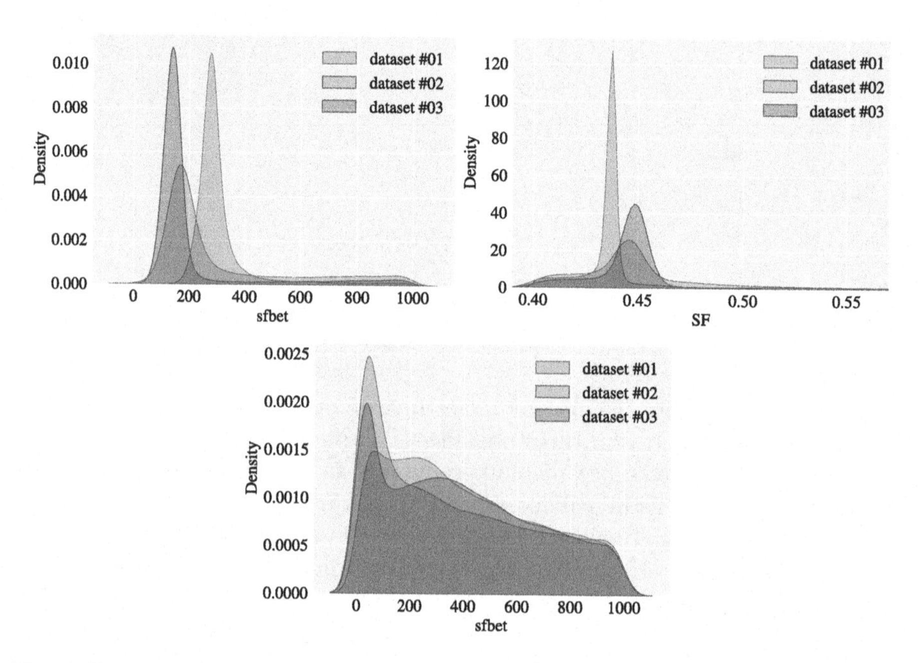

Fig. 4. Posterior distributions for $fmmob$, SF, and $sfbet$ parameters of the STARS foam model obtained by the MCMC method.

3.3 Forward Uncertainty Quantification Results

Figure 5 shows the propagation of uncertainty for the apparent viscosity for the STARS model. The shaded region represents the prediction interval, the solid lines represent the expected values, and the dots represent the experimental data. Experiment data from datasets #01, #02, and #03 are represented in blue, green, and red, respectively.

Analyzing the result of dataset #01 it is possible to observe that the expected value curves are close to the experimental data. For dataset #02 it is also possible to observe that the expected value curves are close to the experimental data, except when $f_g = 0.5$. The same is true for dataset #03, except for lower values of f_g. It is also possible to observe that the prediction interval observed in the low-quality regime is smaller than in the high-quality regime.

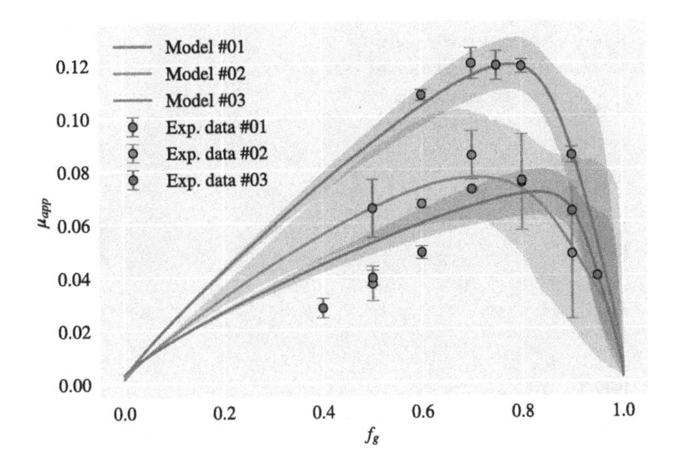

Fig. 5. Uncertainty quantification results of apparent viscosity

For dataset #02, the uncertainty range is large, jeopardizing the model's predictions. One hypothesis is that this dataset has a smaller number of data, i.e., the lack of data may have caused it. To confirm this hypothesis, we added two synthetic points generated with a significant random noise: thus, we define a normal distribution with a mean given by the point value obtained by the parameters estimated by the maximum a posteriori (MAP) estimator and a standard deviation of 25% of the MAP estimator around the mean found:

$$\mu_{app} \sim \mathcal{N}\left(\mu_{app}^{\text{MAP}}\left(SF^{\text{MAP}}, fmmob^{\text{MAP}}, sfbet^{\text{MAP}}\right), 0.25 \times \mu_{app}^{\text{MAP}}\right).$$

With this approach the dataset was augmented with the following data:

$$(f_g, \mu_{app}) \approx (0.402, 0.058), (0.970, 0.049).$$

With the modified dataset (experimental and synthetic), we performed inverse UQ using the MCMC method and then forward UQ using the STARS model. The results showed a significant reduction in the range of uncertainties, as presented in Fig. 6. Therefore, the hypothesis that the cause of significant uncertainty was due to the lack of data is probably correct.

3.4 Sensitivity Analysis

Observing the main and total Sobol indices with respect to apparent viscosity (μ_{app}) for the different datasets, as shown in Fig. 7, it is possible to notice that high order interactions between the parameters are negligible. It is also possible to observe that $fmmob$ dominates the output variance in the model for high water saturation values. Also, close to the expected value found for SF there is a significant change in its influence and the more uncertainty appears in its PDF (see Fig. 4), the larger is the range of S_w that the SF parameter dominates the sensitivity. For values of S_w below this region, the $sfbet$ parameter dominates the sensitivities.

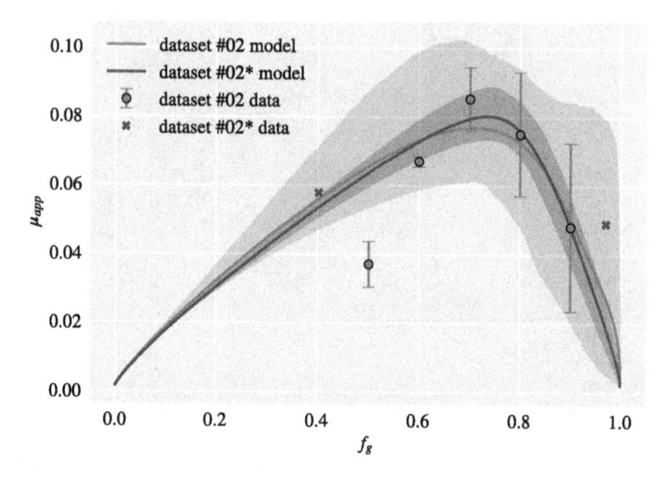

Fig. 6. Forward uncertainty quantification results of apparent viscosity for the augmented dataset.

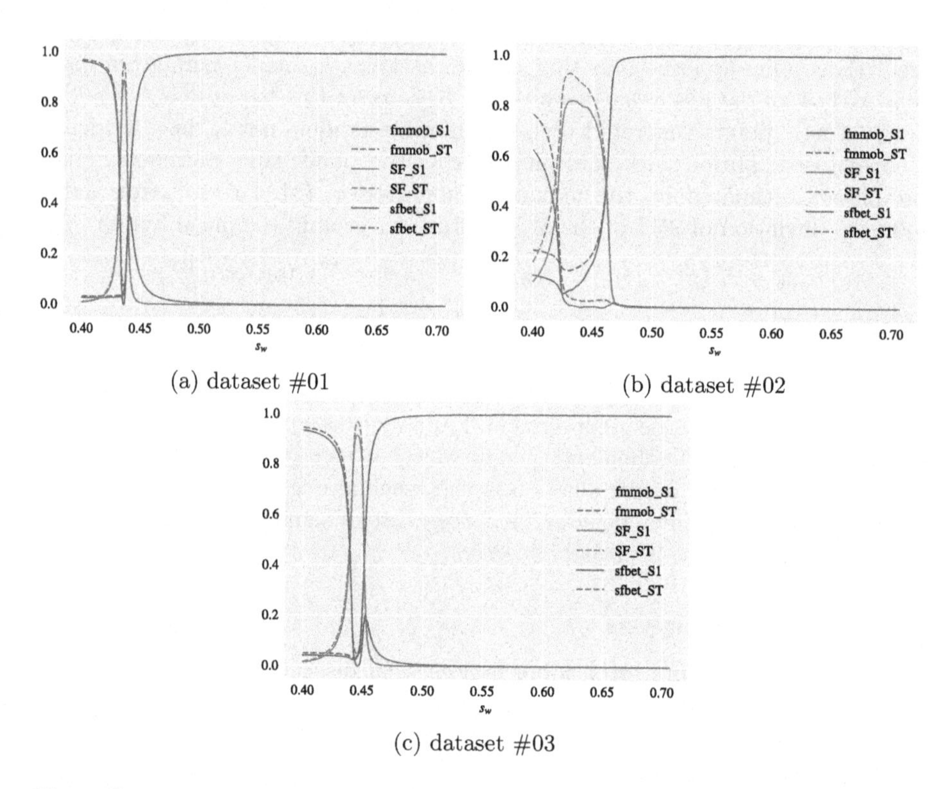

(a) dataset #01

(b) dataset #02

(c) dataset #03

Fig. 7. Sensitivity analysis using main (S_i) and total (S_{T_i}) Sobol indices for the apparent viscosity.

4 Conclusions

This work presented a framework for uncertainty quantification and sensitivity analysis of experimental data of core flooding and Bayesian model calibration in foam flow in porous media, referring to a series of three foam quality scans. The mathematical models of foam injection involve many parameters that control the complex physics of this process. The quantification of uncertainties is essential for the development of robust simulators. After performing the model calibration for the experimental data using an inverse Bayesian estimation, the direct UQ analysis for the apparent viscosity showed more significant uncertainties in dataset #02.

The addition of synthetic data made it possible to reduce model uncertainties significantly. In summary, it was possible to conclude through this work that the use of uncertainty quantification and sensitivity analysis contributes to understanding the phenomenon of foam flow in porous media. The use of these tools together can help to confront experiments and models to assess their quality and uncertainties and suggest new experiments to improve the model's reliability.

In the near future, the framework presented in this work will be used on other datasets to perform a more robust validation of the proposed methods and pipeline, which seeks to reduce model uncertainties. In addition, whereas the current work focuses on two-phase experiments, we expect it to be scalable in the sense that the presented pipeline can also be applied to databases of three-phase experiments.

Acknowledgements. The authors thank Dr. Ivan Landin for suggestions on improving the manuscript.

This research was carried out in association with the ongoing R&D projects ANP number 20358-8, "Desenvolvimento de formulações contendo surfactantes e nanopartículas para controle de mobilidade de gás usando espumas para recuperação avançada de petróleo" (PUC-Rio/Shell Brasil/ANP) and ANP number 201715-9, "Modelagem matemática e computacional de injeção de espuma usa em recuperação de petróleo" (UFJF/Shell Brazil/ANP) sponsored by Shell Brasil under the ANP R&D levy as "Compromisso de Investimentos com Pesquisa e Desenvolvimento", in partnership with Petrobras.

G.C. was supported in part by CNPq grant 303245/2019-0 and FAPEMIG grant APQ-00405-21. B. M. R. was supported in part by CNPq grant 310722/2021-7.

References

1. Ashoori, E., Marchesin, D., Rossen, W.: Roles of transient and local equilibrium foam behavior in porous media-traveling wave. In: ECMOR XII-12th European Conference on the Mathematics of Oil Recovery, p. 163. European Association of Geoscientists and Engineers (2010)
2. Berg, S., Unsal, E., Dijk, H.: Non-uniqueness and uncertainty quantification of relative permeability measurements by inverse modelling. Comput. Geotech. **132**, 103964 (2021)

3. Boeije, C., Rossen, W.: Fitting foam simulation model parameters to data. In: IOR 2013–17th European Symposium on Improved Oil Recovery, p. 342. European Association of Geoscientists and Engineers (2013)
4. Brooks, S.: Markov chain Monte Carlo method and its application. J. R. Stat. Soc. Ser. D (Stat.) **47**(1), 69–100 (1998)
5. Chen, Y., et al.: Switchable nonionic to cationic ethoxylated amine surfactants for CO_2 enhanced oil recovery in high-temperature, high-salinity carbonate reservoirs. SPE J. **19**(02), 249–259 (2014)
6. (CMG): Stars users manual; version 2019.10 (2019)
7. Facanha, J.M.F., Souza, A.V.O., Gramatges, A.P.: Comportamento de espumas em rochas carbonáticas análogas: comparação com curvas de traçador e efeito da permeabilidade. In: Rio Oil and Gas Expo and Conference. Brazilian Petroleum, Gas and Biofuels Institute - IBP (2020). https://doi.org/10.48072/2525-7579.rog.2020.039
8. Farajzadeh, R., Lotfollahi, M., Eftekhari, A.A., Rossen, W.R., Hirasaki, G.J.H.: Effect of permeability on implicit-texture foam model parameters and the limiting capillary pressure. Energy Fuels **29**(5), 3011–3018 (2015)
9. Herman, J., Usher, W.: SALib: an open-source python library for sensitivity analysis. J. Open Source Softw. **2**(9), 97 (2017). https://doi.org/10.21105/joss.00097
10. Kam, S.I.: Improved mechanistic foam simulation with foam catastrophe theory. Colloids Surf. A Physicochemical Eng. Aspects **318**(1–3), 62–77 (2008)
11. Lotfollahi, M., Farajzadeh, R., Delshad, M., Varavei, A., Rossen, W.R.: Comparison of implicit-texture and population-balance foam models. J. Nat. Gas Sci. Eng. **31**, 184–197 (2016)
12. Ma, K., Lopez-Salinas, J.L., Puerto, M.C., Miller, C.A., Biswal, S.L., Hirasaki, G.J.: Estimation of parameters for the simulation of foam flow through porous media. Part 1: the dry-out effect. Energy Fuels **27**(5), 2363–2375 (2013)
13. Mohamed, I., Nasr-El-Din, H., et al.: Formation damage due to CO_2 sequestration in deep saline carbonate aquifers. In: SPE International Symposium and Exhibition on Formation Damage Control. Society of Petroleum Engineers (2012)
14. Newville, M., Stensitzki, T., Allen, D.B., Rawlik, M., Ingargiola, A., Nelson, A.: LMFIT: non-linear least-square minimization and curve-fitting for python. In: Astrophysics Source Code Library, p. ascl-1606 (2016)
15. Salvatier, J., Wiecki, T.V., Fonnesbeck, C.: Probabilistic programming in python using PyMC3. PeerJ Comput. Sci. **2**, e55 (2016)
16. Valdez, A.R., et al.: Foam-assisted water-gas flow parameters: from core-flood experiment to uncertainty quantification and sensitivity analysis. Transp. Porous Media 1–21 (2021). https://doi.org/10.1007/s11242-021-01550-0
17. Valdez, A.R., Rocha, B.M., Chapiro, G., dos Santos, R.W.: Uncertainty quantification and sensitivity analysis for relative permeability models of two-phase flow in porous media. J. Pet. Sci. Eng. **192**, 107297 (2020)
18. Zeng, Y., et al.: Insights on foam transport from a texture-implicit local-equilibrium model with an improved parameter estimation algorithm. Ind. Eng. Chem. Res. **55**(28), 7819–7829 (2016)
19. Zitha, P., Du, D.: A new stochastic bubble population model for foam flow in porous media. Transp. Porous Media **83**(3), 603–621 (2010)

Numerical Simulation of an Infinite Array of Airfoils with a Finite Span

Takahiro Ikeda[1(✉)], Masashi Yamakawa[1], Shinichi Asao[2], and Seiichi Takeuchi[2]

[1] Kyoto Institute of Technology, Matsugasaki, Sakyo-Ku, Kyoto 606-8585, Japan
hirotori620@gmail.com

[2] College of Industrial Technology, 1-27-1 Amagasaki, Hyogo 661-0047, Japan

Abstract. Fish and birds can propel themselves efficiently by acting in groups and clarifying their hydrodynamic interactions will be very useful for engineering applications. In this study, concerning the work of Becker et al., we performed three-dimensional unsteady simulations of an infinite array of airfoils, which is one of the models of schooling, and we clarified the structure of the flow between them. The model velocities obtained from the simulations show a good correspondence with the experimental data of Becker et al. The vortex structure created by the airfoil is very complicated, and it is visually clear that the vortices generated from the left and right ends of the airfoil also contribute to the formation of the upward and downward flows.

Keywords: Computational fluid dynamics · Flapping airfoil · Self-propulsion

1 Introduction

Fish and birds can propel themselves by periodically moving or morphing their body. Many researchers and experts have conducted experiments and numerical simulations for oscillating airfoils to understand the propulsion mechanism hidden in their behavior [1]. In addition, it is known that migratory birds and small fishes such as sardines can travel with high energy efficiency by forming schools [2]. Inspired by them, Becker et al. [3] investigated the fluid-mediated interactions among the collective locomotion of self-propelled bodies through experiments and numerical simulations of array of flapping airfoils. In their study, the system was numerically calculated as the horizontal translational system of two-dimensional array of airfoils, since it is known that it shows a good comparison with the rotational system. Although it might be true, the flow structure of the three-dimensional array should be different from the two-dimensional one. The flow structure and the properties of the horizontal array of flapping airfoils with finite spans is still unclear.

Our research aims to clarify the flow structure of the three-dimensional horizontal array of flapping airfoils numerically. In this paper, the Moving-Grid Finite-Volume Method was adopted [4, 5]. This method can satisfy the physical conservation laws while moving or morphing the grids, so it is suitable for our study. To realize the self-propelled body in the simulation, we also adopt the concept of the Moving Computational Domain (MCD) method [6].

D. Groen et al. (Eds.): ICCS 2022, LNCS 13353, pp. 323–328, 2022.
https://doi.org/10.1007/978-3-031-08760-8_27

2 Simulation

2.1 Governing Equations

As the governing equations, the incompressible Navier Stokes (NS) equations are adopted. They are expressed as

$$\frac{\partial \mathbf{Q}}{\partial t} + \frac{\partial \mathbf{E_c}+\mathbf{E_v}}{\partial x} + \frac{\partial \mathbf{F_c}+\mathbf{F_v}}{\partial y} + \frac{\partial \mathbf{G_c}+\mathbf{G_v}}{\partial z} = 0, \tag{1}$$

where

$$Q = \begin{bmatrix} u \\ v \\ w \end{bmatrix}, \mathbf{E_c} = \begin{bmatrix} u^2+p \\ vu \\ wu \end{bmatrix}, \mathbf{F_c} = \begin{bmatrix} uv \\ v^2+p \\ wv \end{bmatrix}, \mathbf{G_c} = \begin{bmatrix} uw \\ vw \\ w^2+p \end{bmatrix}, \tag{2}$$

$$\mathbf{E}_\mathrm{v} = -\frac{1}{\mathrm{Re}}\begin{bmatrix} \partial u/\partial x \\ \partial v/\partial x \\ \partial w/\partial x \end{bmatrix}, \mathbf{F}_\mathrm{v} = -\frac{1}{\mathrm{Re}}\begin{bmatrix} \partial u/\partial y \\ \partial v/\partial y \\ \partial w/\partial y \end{bmatrix}, \mathbf{G}_\mathrm{v} = -\frac{1}{\mathrm{Re}}\begin{bmatrix} \partial u/\partial z \\ \partial v/\partial z \\ \partial w/\partial z \end{bmatrix}. \tag{3}$$

Here, the variables u, v, w are velocity and p is pressure.
Also, the mass conservation law is written as

$$\frac{\partial u}{\partial x} + \frac{\partial v}{\partial y} + \frac{\partial w}{\partial z} = 0. \tag{4}$$

The governing equations shown above are nondimensionalized using the characteristic length c, velocity U_∞, density ρ_∞ and viscosity μ_∞. The Reynolds number becomes

$$Re = \frac{\rho_\infty U_\infty c}{\mu_\infty}. \tag{5}$$

Here, $c = 0.06$ m is a cord length of the airfoil. $U_\infty = 0.1$ m/s, and the viscosity and the density are $\rho_\infty = 1000.0$ kg/m^3 and $\mu_\infty = 0.001$ Pa s, referring to the values in [3].

2.2 Numerical Approach

In this study, the moving-grid finite-volume method [4, 5] is used to solve the flow field with a moving object. In this method, the governing equations are integrated and discretized in a four-dimensional control volume of a unified space and time domain, so that the conservation law can be satisfied even when the grid is moving or deformed. In this study, we apply this method to the unstructured grid using cell-centered and collocated arrangement. In addition, the MCD method [6] is applied to analyze the free propulsion of an airfoil in this paper. To solve the incompressible NS equation discretized by the above method, the Fractional Step method [7] is adopted in this study. To solve the linear system of equations for both steps, the Lower-Upper Symmetric Gauss Seidel (LU-SGS) method [8] is used for the pseudo velocity and the Successive Over Relaxation (SOR) method is used for the pressure equations. All computations in this paper are conducted by in-house code.

2.3 Computational Model

The model simulated in this paper is an array of equally spaced airfoils in a row. In the numerical simulation of [3], the array is realized by a single flapping airfoil with periodic boundary conditions. In this paper, we also compute the similar one to the computation, but the difference is that the flow domain is three-dimensional, and the wing has a finite span length. Then, our computational model is shown in Fig. 1. The coordinate axes are x for the direction opposite to the airfoil's direction of motion, y for the direction of vibration, and z for the rest. The shape of the airfoil is NACA (National Advisory Committee for Aeronautics) 0017 which has a chord length of $c = 0.06$ m and a span length of 0.15 m. These dimensions are the same as in the experiment. The flow region has a length of $4c$ in span-wise direction, and the same lengths as the numerical simulation [3] in stream-wise and vertical direction.

The airfoil is given a prescribed vertical motion written in the dimensionless form as

$$y(t) = -\frac{A}{2c} sin(2\pi f t), \tag{6}$$

where the amplitude is $A = 0.09$ m and the f is a nondimensionalized frequency defined as

$$f = \frac{\bar{f} c}{U_\infty}. \tag{7}$$

The frequency varies in the range $0.1 \leq \bar{f} \leq 0.3$ Hz.

The propulsive force of the airfoil is calculated by integrating the viscous and pressure force acted on the surface. Using the force, propulsive velocity is determined by the Newton's second law described as

$$\rho V \frac{dv_{wing}}{dt} = F_{thrust}, \tag{8}$$

where $\rho = \rho_{wing}/\rho_\infty$. $\rho_{wing} = 10\rho_\infty$ is the density of the airfoil. V is the volume of the airfoil. The airfoil is fixed in the span-wise direction so that it can move only in the propulsive and vertical directions, therefore the dynamical motion of the airfoil is determined only by Eq. (6) and (8).

As for the boundary conditions, no-slip conditions are imposed on the blade surfaces, fixed velocity conditions ($\mathbf{u} = 0$) are upper and lower walls, and symmetric conditions are imposed on the sides. For inflow and outflow, periodic boundary conditions are used to reproduce a model with an infinite array of airfoils. All pressures are set to zero gradients, and the mean value of the pressures is fixed for the simulation.

Based on the above model, a spatial grid was created as follows. Figure 2 shows a cross section of the unstructured grid used in this study. MEGG3D [9] was used to create the unstructured grid for the computation. The number of elements is 2,879,012. Prismatic layers are created on the surface of the airfoil and upper and lower walls. The minimum grid width is wide enough to capture the viscous sub-layer, and the number of layers is 17 to reduce the size ratio to the tetra grid. All dimensions are nondimensionalized.

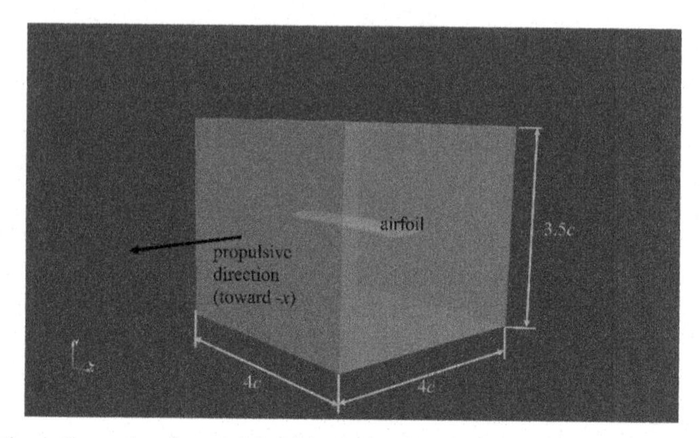

Fig. 1. The three-dimensional model for this study. the wing and the flow domain move toward the negative x direction with flapping in vertical (yellow axis in the figure) direction. (Color figure online)

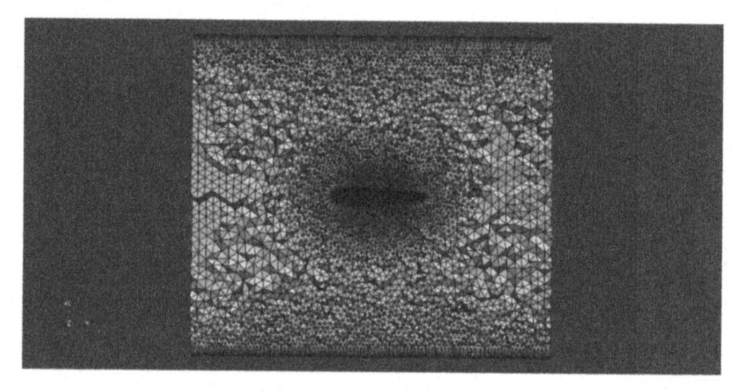

Fig. 2. Cross section of the unstructured grid for the computation of flapping airfoil.

3 Results

Figure 3 shows a comparison of the propulsive velocity calculated as described above with that of the experiment in [3], where the amplitude is $A = 0.1$ m. Since the original data in the experiment is expressed in terms of rotational frequency $F[1/\mathrm{s}]$, it is necessary to convert it to translational velocity for comparison. In this study, based on the dimensions of the water tank in the experiment in [3], the reference value of the propulsive velocity is calculated as follows,

$$v = 2\pi RF. \tag{9}$$

In Eq. (5), $R = 26$ cm is a distance from the axis of rotation to the mass center of each airfoil. From the figure, it can be seen that the speed of progression in the literature [3] and that in this study are in good agreement, although the amplitudes and the flow geometries are different. It is mentioned that the flows observed in rotational geometry compare well with those in translational one [3, 10], therefore it can be said that our result is also true.

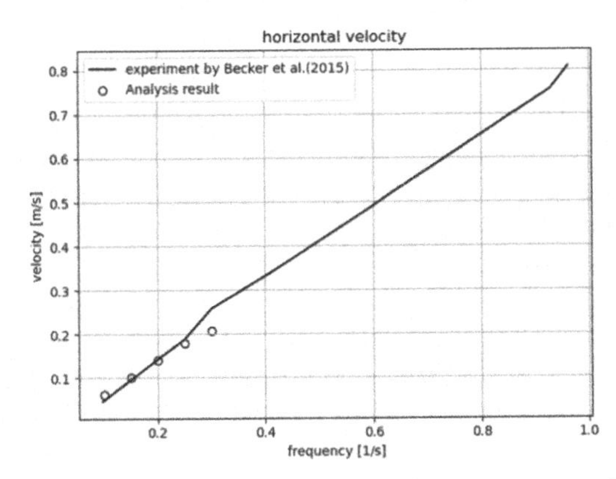

Fig. 3. The horizontal velocity of the airfoils. white circles are current results, and a solid line is the result by Becker et al.

Next, we describe the flow field. The isosurfaces of the Q-criterion are shown in Fig. 4 to clarify the vortex structure. The Q criterion is the second invariant of the velocity gradient tensor, and the region of $Q > 0$ is often used to identify vortices in the flow [11]. The color indicates the x component of the vorticity. In the figure, the airfoil is descending to the lower wall. Many vortices are generated in the spanwise direction of the airfoil surface, and these vortices form a large vortex at the trailing edge, but their shape is curved and not straight. The vortices generated by the airfoil are not only in the spanwise direction but also in the direction of motion. Since the airfoil is moving in the negative direction of x as it descends, the blue vortex on the left side of the airfoil has the same direction as the direction of motion, and the red vortex on the right side has the opposite direction. From the direction of the vortices on both sides of the airfoil, we can see that they also create a downward flow. We can also see that these vortices are colliding with the following airfoil in this visualization.

Fig. 4. Isosurfaces of Q criterion where the airfoils are descending at the frequency 0.2 Hz. In this figure, a single calculation result is displayed in a row to represent the infinite array of the airfoils. The figure is colored by the x component (stream-wise) of vorticity.

4 Conclusion

The model of Becker et al. to analyze bird and fish schools was reproduced to a limited extent in 3D unsteady simulations. It was found that the travel speeds of the model obtained from simulations using the airfoils corresponded to the experimental values. As for the flow field, it was found that the upward and downward flows generated by the airfoil affected the propulsion of the airfoil as in the experiment. In addition, the structure of the wake generated by the airfoil was clarified by visualizing the flow. Unlike the two-dimensional simulation in [3], it was found that the vortices facing the direction of the airfoil also contributed to the formation of the upward and downward flows.

For future works, we will conduct this unsteady three-dimensional simulation in a wide range of the flapping frequency and clarify the characteristics of the infinite array of the airfoils.

Acknowledgements. This publication was subsidized by JKA through its promotion funds from KEIRIN RACE and by JSPS KAKENHI Grant Number 21K03856.

References

1. Wu, X., Zhang, X., Tian, X., Li, X., Lu, W.: A review on fluid dynamics of flapping foils. Ocean Eng. **195**, 106712 (2020). https://doi.org/10.1016/j.oceaneng.2019.106712
2. Weihs, D.: Hydromechanics of fish schooling. Nature **241**, 290–291 (1973). https://doi.org/10.1038/241290a0
3. Becker, A.D., Masoud, H., Newbolt, J.W., Shelley, M., Ristroph, L.: Hydrodynamic schooling of flapping swimmers. Nat Commun **6**, 8514 (2015). https://doi.org/10.1038/ncomms9514
4. Mihara, K., Matsuno, K., Satofuka, N.: An Iterative finite-volume scheme on a moving grid: 1st report, the fundamental formulation and validation. Trans. Japan Society Mech. Eng. Series B **65**, 2945–2953 (1999). https://doi.org/10.1299/kikaib.65.2945
5. Inomoto, T., Matsuno, K., Yamakawa, M.: 303 Unstructured Moving-Grid Finite-Volume Method for Incompressible Flows. The Proceedings of The Computational Mechanics Conference **26**, 303–1-303–2 (2013) https://doi.org/10.1299/jsmecmd.2013.26._303–1
6. Watanabe, K., Matsuno, K.: Moving Computational domain method and its application to flow around a high-speed car passing through a hairpin curve. JCST **3**, 449–459 (2009). https://doi.org/10.1299/jcst.3.449
7. Kim, J., Moin, P.: Application of a fractional-step method to incompressible Navier-Stokes equations. J. Comput. Phys. **59**, 308–323 (1985). https://doi.org/10.1016/0021-9991(85)90148-2
8. Yoon, S., Jameson, A.: Lower-upper Symmetric-Gauss-Seidel method for the Euler and Navier-Stokes equations. AIAA J. **26**, 1025–1026 (1988). https://doi.org/10.2514/3.10007
9. Ito, Y.: Challenges in unstructured mesh generation for practical and efficient computational fluid dynamics simulations. Comput. Fluids **85**, 47–52 (2013)
10. Vandenberghe, N., Zhang, J., Childress, S.: Symmetry breaking leads to forward flapping flight. J Fluid Mech **506**, 147–155 (2004). https://doi.org/10.1017/S0022112004008468
11. Jeong, J., Hussain, F., Hussain, F.: On the identification of a vortex. JFM 285,69–94. J. Fluid Mech. **285**, 69–94 (1995) https://doi.org/10.1017/S0022112095000462

Simulation of Nearly Missing Helicopters Through the Computational Fluid Dynamics Approach

Momoha Nishimura(✉) and Masashi Yamakawa

Kyoto Institute of Technology, Matsugasaki 606-8585, Sakyo, Japan
d7821007@edu.kit.ac.jp

Abstract. This study achieves modelling two helicopters via computational fluid dynamics (CFD) and simulating the flow field that develops due to a near miss. The rotation of the main rotor and the translational movement of the helicopter are modelled in this study, and the long trajectory of the moving helicopter is realised in the simulation. Moreover, the interaction of flows around the two moving helicopters is also achieved by introducing the communication between multiple moving computational domains. Firstly, the validation test is conducted using a helicopter model with a rotating main rotor, where the results produced by our in-house code are compared with those computed by another CFD solver, *FaSTAR-Move*. This test verifies that the communication between the overlapping grids is reliably achieved in our simulation. In the simulation of nearly missing helicopters, two near-miss cases are computationally demonstrated, where the complex flow field which develops around the two helicopters is captured, and the disturbance in aerodynamic and moment coefficients exerted on the helicopters are observed. These results confirm the capability of this CFD approach for realising near-miss events on a computer.

Keywords: CFD · Moving grids · Overset methods · Helicopter

1 Introduction

Recently, the airspace in metropolitan areas has become extremely busy: there are commercial helicopters, police helicopters, and medevac flights, and the demand for air transport by aeroplanes has been increasing greatly. Moreover, considering the current expansion of drone industries and the launch of flying cars in the future, aircraft will need to share more congested airspace. While the danger of crashing need hardly be said, there is also the risk of near-miss flights, where aircraft are affected by the flow field that develops around other aircraft. In fact, the near collision of two helicopters has been reported [1]. Considering the increasing congestion of the future airspace, near-miss events will occur more frequently. Therefore, comprehending the aerodynamic effects exerted on aircraft and their behaviour in response to the effects in a near miss is essential to ensure the safety of aircraft as well as promoting legislation to introduce standards that increase the safety of aircraft in more congested airspace. However, since the flight

D. Groen et al. (Eds.): ICCS 2022, LNCS 13353, pp. 329–342, 2022.
https://doi.org/10.1007/978-3-031-08760-8_28

in this scenario involves high risk, flight tests by real aircraft are not easily manageable. On the other hand, computational fluid dynamics (CFD) allows an arbitrary scenario to be set and high-risk flight simulations to be conducted without risking human life. Thus, it is essential to further develop the CFD method to computationally examine the nearly missing aircraft.

The current CFD used for analysing the flow around aircraft is able to not only simulate fully fixed objects but also model their moving and deforming components and analyse the flow field in detail. For example, CFD simulations for helicopters analysed the flow generated by their rotors and its aerodynamic effects in combination with wind tunnel testing [2, 3]. The trajectory prediction of aircraft through CFD has also been reported [4]. The manoeuvring of aircraft has been simulated through the coupled simulation of CFD and flight dynamics that considers the configuration of the components, their function, and the controlling system. This approach is effective for analysing the flow field and the reaction of aircrafts to the flow field [5]. However, these simulations have primarily been used to model only one aircraft, and near-miss events involve multiple aircraft. This implies that the CFD method, which can model multiple moving objects and the interaction of the flow fields generated by them, is necessary for simulating near-miss events on a computer.

Therefore, in this paper, we simulate two helicopters in a near miss using the CFD method. Helicopters are chosen as models for the first application of the CFD approach for near-miss events based on an actual reported occurrence of a near miss involving helicopters [1]. The rotation of the main rotor and the translational movement of the helicopters are realised in the simulation, and the two helicopter models are crossed in a remarkably close context. The translation is simulated by a moving mesh method, in particular the moving-grid finite volume (MGFV) method [6]. This satisfies the geometric conservation law (GCL) condition by employing a unified space-time, four-dimensional control volume for the discretisation. The moving computational domain (MCD) method [7] is applied to modelling the helicopters that travel over long trajectories, which allows the complex flow field around moving helicopters to be realised instead of modelling fixed objects in the context of wind tunnel testing. The MCD method removes the spatial limitation for simulating objects with long trajectories that occurs due to computational cost. In this method, a large background mesh is not required, and thus objects can move freely in a three-dimensional space because the computational domains are moved in line with the motions of an object inside [8, 9]. Conventionally, it was difficult for the MCD method to gain information from outside the computational domain created for the enclosed object, and therefore the simulation was targeted at only one object. However, introducing the overset approach to the MCD method allowed communication between the domains created around each moving object [10]. In this study, the flow field variables are communicated between the main rotor grid and the fuselage grid of the helicopter, which model the rotation of the main rotor. There is also communication between the flow fields around the two moving helicopters to represent the flow interaction around them. Using these methods realised the near-miss flight simulation of helicopters and calculated the aerodynamic forces exerted on each helicopter during the near miss.

2 Numerical Methods

2.1 CFD Solver

The three-dimensional Euler equations for compressible flow are adopted as a governing equation:

$$\frac{\partial \boldsymbol{q}}{\partial t} + \frac{\partial \boldsymbol{E}}{\partial x} + \frac{\partial \boldsymbol{F}}{\partial y} + \frac{\partial \boldsymbol{G}}{\partial z} = \boldsymbol{0}, \tag{1}$$

where $\boldsymbol{q}$ represents a vector of conserved variables, and $\boldsymbol{E}$, $\boldsymbol{F}$, and $\boldsymbol{G}$ denote the inviscid flux vectors. t indicates time, and x, y, z are the coordinates. This system is closed by assuming the ideal gas law, where the ratio of specific heats $\gamma = 1.4$ is used.

Equation (1) is discretised by the cell-centred MGFV method, which uses a unified space-time, four-dimensional control volume for the discretisation [6, 10], yielding the following formulae with the variables in the current N step and the next $N + 1$ step:

$$\boldsymbol{q}^{N+1}(\tilde{n}_t)_6 + \boldsymbol{q}^{N}(\tilde{n}_t)_5 + \sum_{l=1}^{4}\left\{\boldsymbol{q}^{N+\frac{1}{2}}\tilde{n}_t + \boldsymbol{H}^{N+\frac{1}{2}}\right\}_l = 0$$

$$\boldsymbol{H} = \boldsymbol{E}\tilde{n}_x + \boldsymbol{F}\tilde{n}_y + \boldsymbol{G}\tilde{n}_z \tag{2}$$

$$\boldsymbol{q}^{N+\frac{1}{2}} = \frac{1}{2}\left(\boldsymbol{q}^{N} + \boldsymbol{q}^{N+1}\right), \boldsymbol{H}^{N+\frac{1}{2}} = \frac{1}{2}\left(\boldsymbol{H}^{N} + \boldsymbol{H}^{N+1}\right),$$

where $\widetilde{\boldsymbol{n}}= \left[\tilde{n}_t, \tilde{n}_x, \tilde{n}_y, \tilde{n}_z\right]$ represents the four-dimensional outward normal vector of the control volume. Roe's flux difference splitting (FDS) [11] is used to estimate the inviscid flux vector $\boldsymbol{H}_l$, and the MUSCL (monotonic upstream-centred scheme for conservation laws) scheme is applied to provide second-order accuracy. The primitive variables of $\boldsymbol{q}$ are reconstructed by the gradient, which is evaluated by the least-squares approach and Hishida's limiter [12]. $(\boldsymbol{q}\tilde{n}_t)_l$ in Eq. (2) is estimated by the following upwind scheme:

$$(\boldsymbol{q}\tilde{n}_t)_l = \frac{1}{2}\left[\boldsymbol{q}^{+}\tilde{n}_t + \boldsymbol{q}^{-}\tilde{n}_t - |\tilde{n}_t|\left(\boldsymbol{q}^{+} - \boldsymbol{q}^{-}\right)\right].$$

Unsteady flow is solved by a pseudo-time approach with the two-stage rational Runge-Kutta (RRK) scheme for the pseudo-time stepping.

2.2 MCD Method

The movement of objects with long trajectories is expressed by the MCD method [7]. Computational domains created around each object move in line with the motions of the object inside, and therefore this approach makes it possible for objects to move freely without any spatial limitations. However, because it is difficult for objects to obtain information from outside their computational domains, simulating the interaction of flow fields around multiple moving objects is challenging. Therefore, the overset approach described in the following section is introduced to allow communication between the computational domains (Fig. 1).

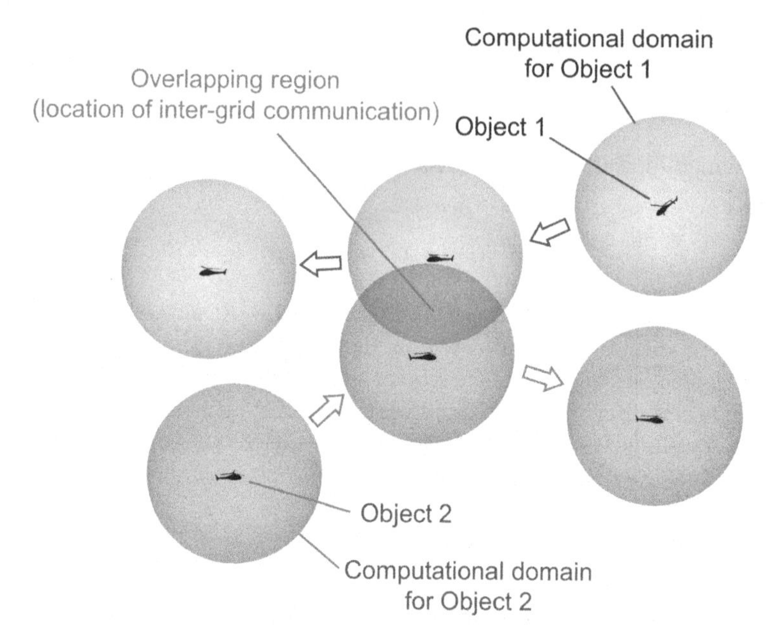

Fig. 1. The concept of the MCD method for multiple moving objects

2.3 Inter-grid Communication for Multiple Moving Computational Domains

The variables of each computational domain communicate in their overlapping region by applying the overset approach. The implicit hole cutting (IHC) of the overset approach can be described by the following procedure. First, the cell in the overlapping partner grid that includes the nodes of the target grid is determined. Here the KD-tree-based algorithm is used to search for the owner cells. Then, nodes and cells are classified using the method proposed by Nakahashi [13]. In this method, the node type is designated by comparing the distances between the node and the object in the target grid and between the same position in the overlapping partner grid and the object. The nodes which have shorter distances are designated as active cells and those with longer distances are designated as nonactive cells. A tetrahedron cell which consists of all nonactive nodes is a nonactive cell, for which variables are not computed. The cells which overlap the object of the partner grid are also designated as nonactive cells. Conversely, a cell whose nodes are all active is an active cell. The remaining cells are interpolation cells, where the flow field variables are interpolated from the overlapping partner grid, based on the values of so-called donor cells in the partner grid. The donor cells surround the interpolation cells. Inverse distance weighing is employed as the interpolation method.

3 Validation Test

The in-house code was validated by comparing the results computed by an unstructured overset grid CFD solver *FaSTAR-Move* [14, 15], which was developed by Japan Aerospace Exploration Agency (JAXA). *FaSTAR-Move* is not suitable for simulating objects which travel over long trajectories although it can model the deformation and short movement of objects by using the overset method. Therefore, in this test, a helicopter model is placed in uniform flow instead of using the MCD method and moving the whole model in a three-dimensional space. The computational grids created around the fuselage are fixed, while the other grids created around the main rotor rotate. The fuselage grid and the main rotor grid communicate with each other using the overset approach. This validation aims to confirm that the overset approach in the in-house code accurately interpolates the variables between the overlapping grids.

3.1 Helicopter Model

This study used a simplified helicopter model based on the AS-355 helicopter without the tail rotor. The origin of the body axes used for calculating the rolling, pitching, and yawing moment is the centre of mass, as illustrated in Fig. 2. The centre of mass was calculated from the polygons of the fuselage and the main rotor. The fuselage length of 11.2 m was normalised to 1 in the model.

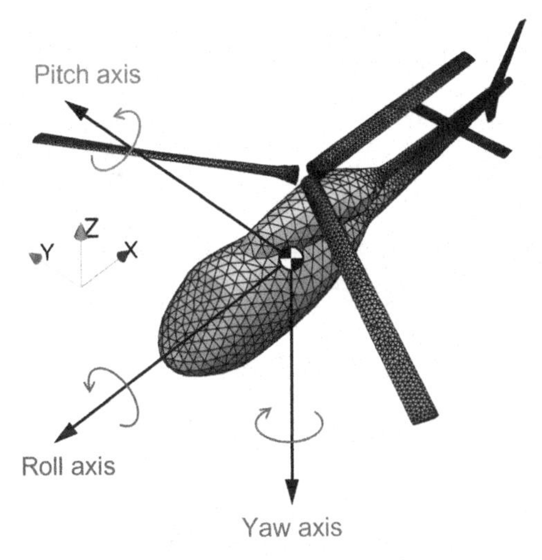

Fig. 2. Helicopter model

3.2 Computational Conditions

A cubic computational domain with a side of 30 L was created around the fuselage (Fig. 3a), where the fuselage length is L. Another computational domain was created around the main rotor (Fig. 3b). 407,079 cells for the fuselage and 151,927 cells for the main rotor were generated by an unstructured mesh generator *MEGG3D* [16].

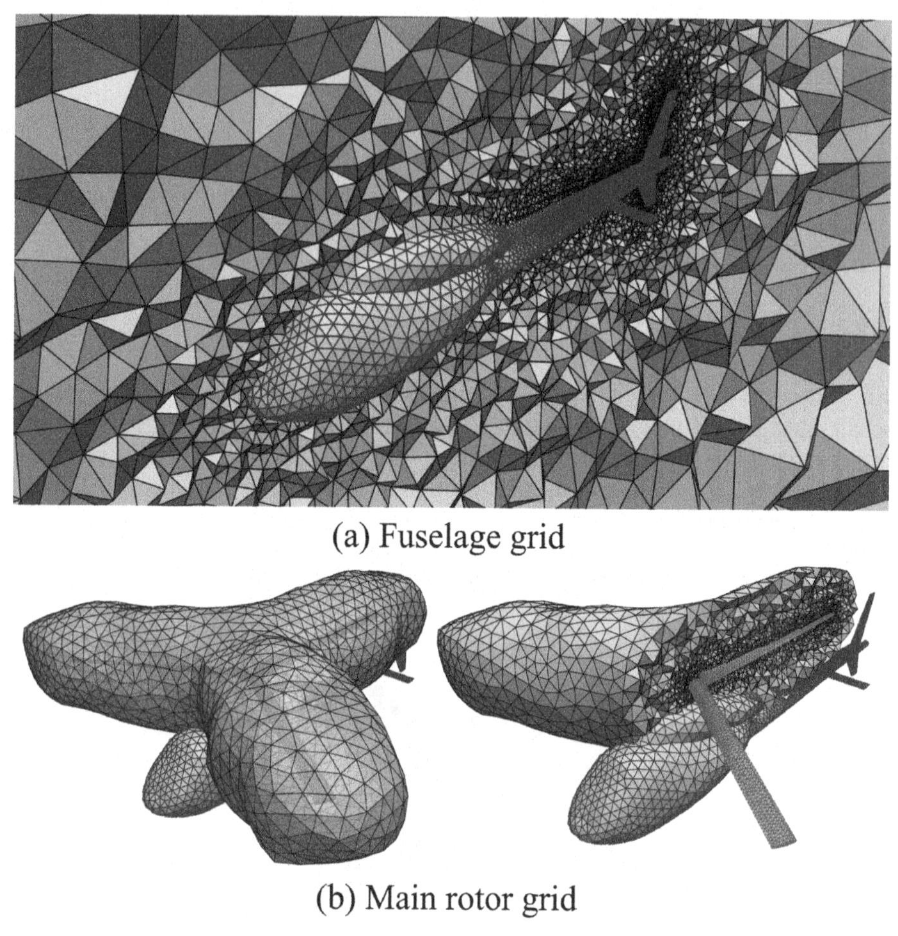

(a) Fuselage grid

(b) Main rotor grid

Fig. 3. Computational domains for (a) the fuselage and (b) main rotor

The velocity of the uniform flow is Mach 0.2 in the x direction, as the cruising speed of the helicopter is assumed to be approximately Mach 0.2. The main rotor and grids around it rotate counterclockwise at 395 rpm. In this study, the main rotor performs rotation only.

Table 1 shows the numerical method used in *FaSTAR-Move*. The numerical method of the in-house code is given in Table 1, and it is described in Sect. 2. The boundary conditions are the slip conditions at the surfaces of the fuselage and the main rotor, the uniform inflow conditions at the yz-plane of the outer boundary which the helicopter faces, and the Riemann invariant boundary conditions at the remaining outer boundaries of the fuselage grid. The variables are interpolated from the fuselage grid at the outer boundary of the main rotor grid unless the boundary cell is a nonactive cell.

Table 1. Numerical methods used for the validation test

	FaSTAR-Move	In-house code
Governing equation	3-D compressible Euler	3-D compressible Euler
Advection term	SLAU [17]	Roe's FDS
Reconstruction	Weighted Green-Gauss	Least square
Slope limiter	Hishida (van Leer)	Hishida (van Leer)
Time integration	LU-SGS (Lower-upper symmetric Gauss-Seidel)	RRK
Interpolation	Tri-linear interpolation	Inverse distance weighted interpolation

3.3 Results and Discussion

The graph in Fig. 4 depicts the time history of the pressure coefficient computed for the fuselage surface of the helicopter and compares it with the results provided by *FaSTAR-Move*. The position of the main rotor after it rotated for 1.0 s from the beginning of the calculation was set as 0°, and the figure shows the data while the main rotor rotates three complete rotations from the 0-deg position. Figure 4 indicates that the same trend can be seen in the pressure coefficient between the in-house code and *FaSTAR-Move*, although there is not exact agreement due to the differences of the numerical schemes and interpolation methods between the two software codes. Three peaks can be observed for one rotation, indicating that the pressure oscillation, which was raised by the flow generated by the three blades of the main rotor, was captured on the fuselage surface. The characteristics of the oscillation predicted in the present study, in particular the position

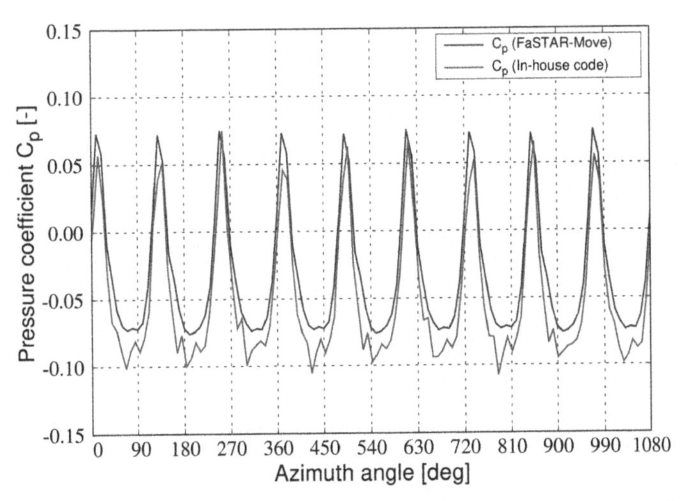

Fig. 4. Comparison of the pressure coefficients computed with *FaSTAR-Move* and the in-house code

and values of the coefficient peaks, are in line with the values predicted by *FaSTAR-Move*, instilling confidence that the flow field variables are accurately communicated between the overlapping grids in the in-house code.

4 Simulation of Nearly Missing Helicopters

4.1 Configurations of the Test Cases

Two cases were tested in which aerodynamic effects were assumed to be exerted on helicopters. Figure 5 shows the flight paths in the two cases. In Case 1, illustrated in Fig. 5a, the trajectories of the two helicopters intersect perpendicularly, where Helicopter 2 passes just behind the trajectory of Helicopter 1. Helicopter 2 flies slightly above Helicopter 1 to avoid colliding their blades. In Case 2, Helicopter 2 passes horizontally underneath Helicopter 1, where the trajectory of Helicopter 2 does not pass directly below that of Helicopter 1 but slightly shifted in the y direction, as depicted in Fig. 5b.

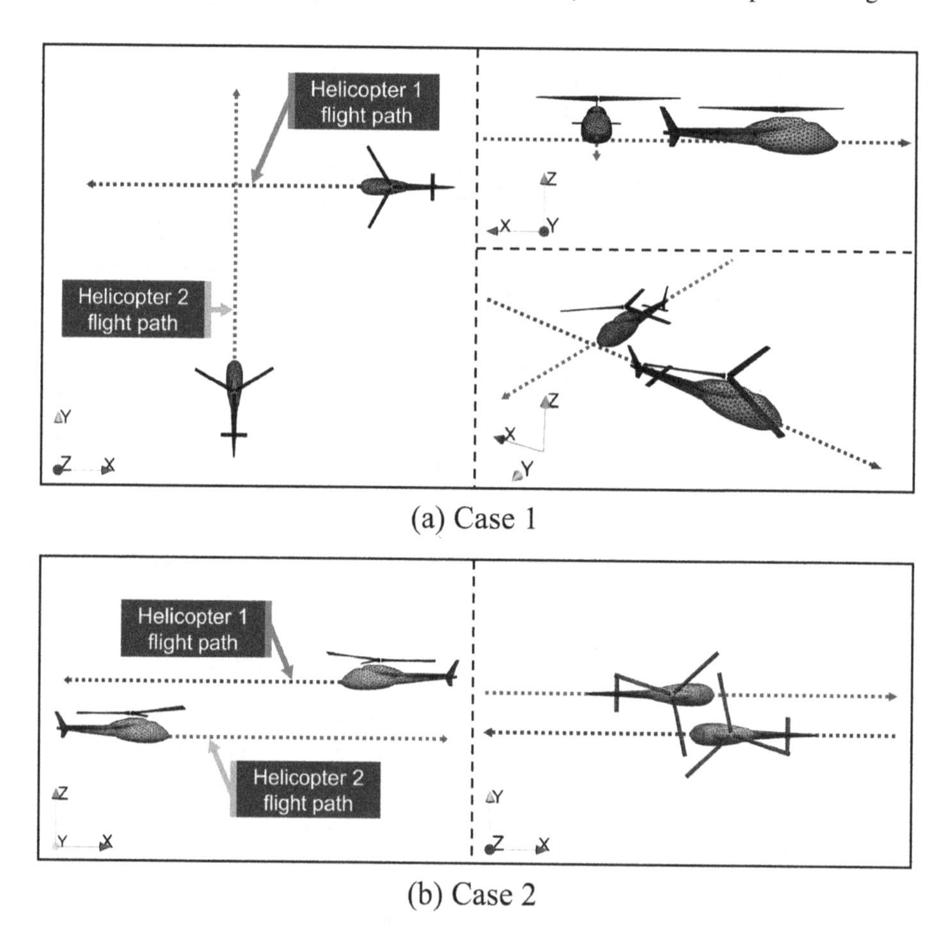

(a) Case 1

(b) Case 2

Fig. 5. Flight paths of the two test cases

4.2 Computational Conditions

The helicopter model described in Sect. 3.1 was used for the test cases of nearly missing helicopters. A spherical computational domain with a diameter of 20 L was created around the fuselage of each helicopter. Another computational domain was created around the main rotor, as in Fig. 3b. 350,777 cells for the fuselage and 135,059 cells for the main rotor were generated by *MEGG3D*. Two sets were prepared to model two helicopters, where a set includes the domain for the fuselage and another domain for the main rotor.

While the validation test involves two computational domains for one helicopter model, the simulation in this section involves four domains because there are two helicopter models. Here, the fuselage domain of Helicopter 1 is defined as Domain 1F, the main rotor domain of Helicopter 1 as Domain 1R, the fuselage domain of Helicopter 2 as Domain 2F, and the main rotor domain of Helicopter 2 as Domain 2R. The variables are communicated between Domains 1F and 1R and between Domains 2F and 2R to model each helicopter. There is also communication between Domains 1F and 2F to represent the interaction of the flow fields that develop around each helicopter. This definition is illustrated in Fig. 6, where the domains for the fuselages are shown smaller and not to scale for the sake of simplicity.

Table 2 shows the initial conditions, where no flow or turbulence is assumed in the atmosphere in which the helicopters fly. The boundary conditions are the slip conditions at the object surfaces and the Riemann invariant boundary conditions at the outer boundaries of the fuselage grids. The conditions of the outer boundaries of the main rotor grids are the same as in Sect. 3.2. The helicopters engage in translational motion at a cruising speed of Mach 0.2 in the direction shown in Fig. 5 as modelled by the MCD method. The main rotors and the grids around them rotate counterclockwise at 395 rpm. The main rotors rotate only. While the actual helicopters lean slightly in the direction of their travel when they move forward, the simulations in this study do not model this.

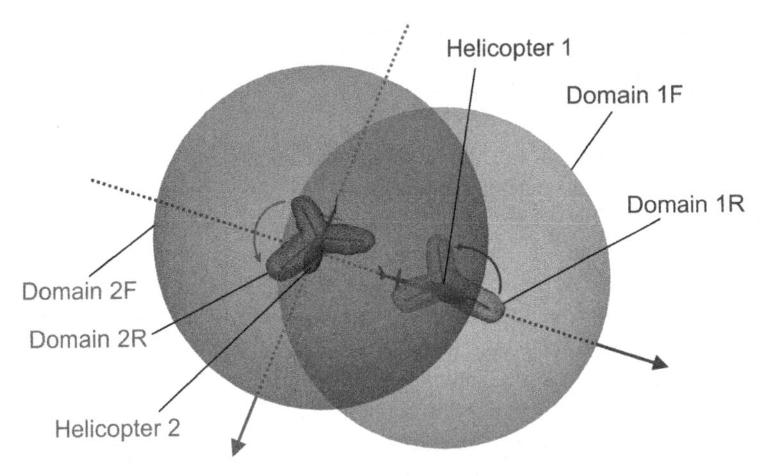

Fig. 6. Domain definitions used in the near-miss flight test cases

Table 2. Initial conditions for the simulation of nearly missing helicopters

Whole region	
Density (ρ)	1.0
Flow velocity x component (u)	0.0
Flow velocity y component (v)	0.0
Flow velocity z component (w)	0.0
Pressure (p)	ρ/γ

4.3 Results and Discussion

Case 1. Figures 7 and 8 illustrate the time histories of the aerodynamic coefficients and moment coefficients calculated for Helicopter 1, while Figs. 9 and 10 illustrate these calculated for Helicopter 2. The aerodynamic coefficients are computed with the following equations:

$$C_D = \frac{2F_D}{\rho V^2 S}, C_S = \frac{2F_S}{\rho V^2 S}, C_L = \frac{2F_L}{\rho V^2 S} \tag{3}$$

where C_D, C_S, C_L represent the drag coefficient, side-force coefficient, and lift coefficient, respectively. F_D, F_S, F_L indicate the drag, side force, and lift, respectively, which are the forces exerted on the helicopter in the x-, y-, and z-axes in Fig. 2, respectively. V denotes the characteristic speed, which in this study is the cruising speed of Mach 0.2. S is the characteristic area, which is represented by the main rotor disk area calculated using a main rotor radius of 5.35 m. The moment coefficients are computed with the following equations:

$$C_l = \frac{2M_l}{\rho V^2 SL}, C_m = \frac{2M_m}{\rho V^2 SL}, C_n = \frac{2M_n}{\rho V^2 SL} \tag{4}$$

where C_l, C_m, C_n represent the rolling moment coefficient, pitching moment coefficient, and yawing moment coefficient, respectively. M_l, M_m, M_n indicate the rolling moment, pitching moment, and yawing moment, respectively. L denotes the characteristic length, which is the fuselage length.

Figures 7 and 8 show that the aerodynamic coefficients and moment coefficients of Helicopter 1 do not change during the near miss. This suggests that Helicopter 1 is not affected by the flow that develops around Helicopter 2 because Helicopter 1 flies in front of Helicopter 2, where the flow field around Helicopter 2 does not develop.

On the other hand, Fig. 9 shows that the aerodynamic coefficients of Helicopter 2 are disturbed between 0.4 s and 0.8 s when it passes behind Helicopter 1. In particular, while the drag and side-force coefficients do not undergo a notable disturbance, the lift coefficient in Fig. 9 suggests that Helicopter 2 loses lift for approximately 0.3 s during the near miss. Figure 11 depicts the flow velocity z component distribution at 0.55 s in the xy-plane in the middle of the fuselage of the helicopters when the lift coefficient decreases noticeably. This indicates that Helicopter 2 flies into the area where the main rotor of Helicopter 1 generated a downward flow in the z direction. This flow field makes it difficult to create the ample pressure difference for obtaining the lift.

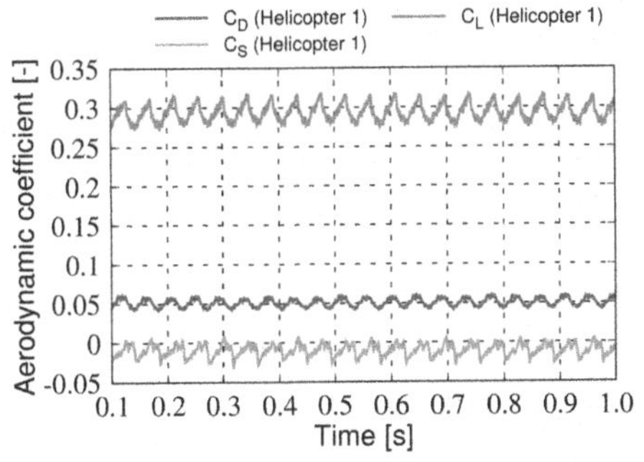

Fig. 7. Aerodynamic coefficient of Helicopter 1 (Case 1)

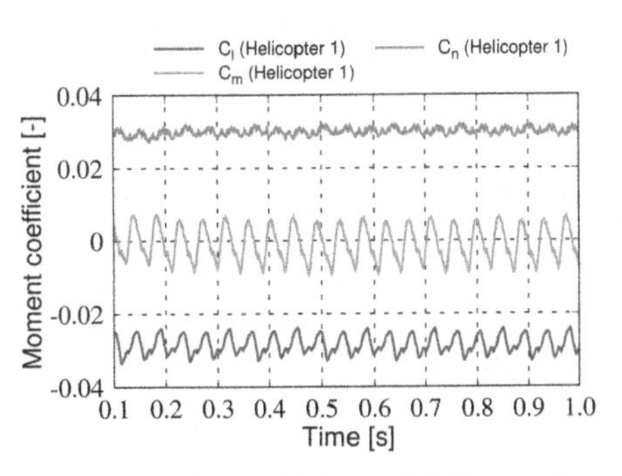

Fig. 8. Moment coefficient of Helicopter 1 (Case 1)

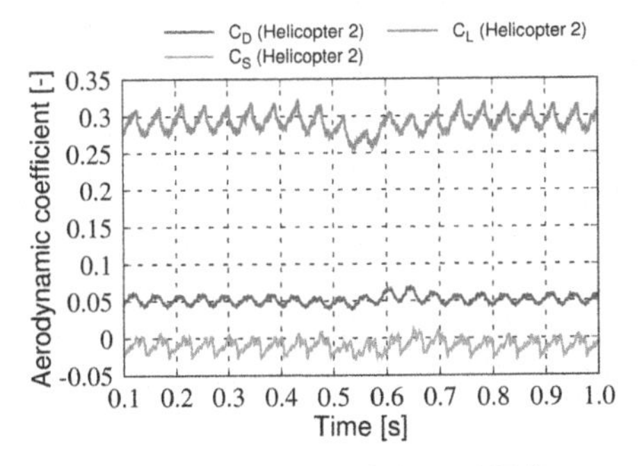

Fig. 9. Aerodynamic coefficient of Helicopter 2 (Case 1)

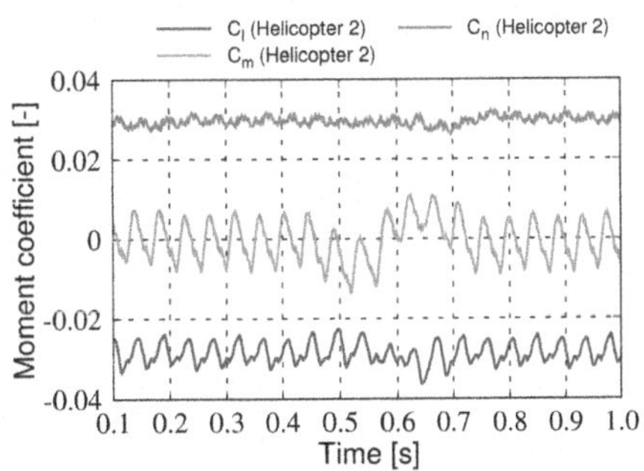

Fig. 10. Moment coefficient of Helicopter 2 (Case 1)

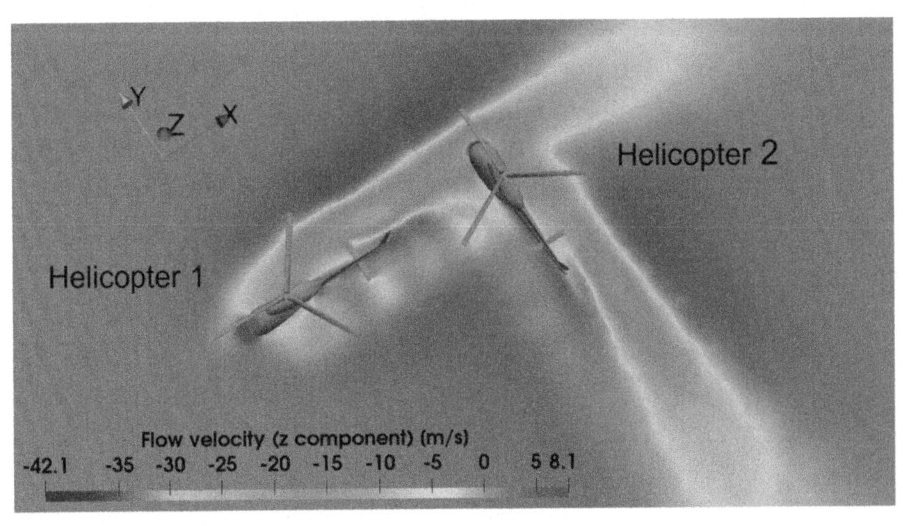

Fig. 11. Flow velocity z component distribution at 0.55 s (Case 1)

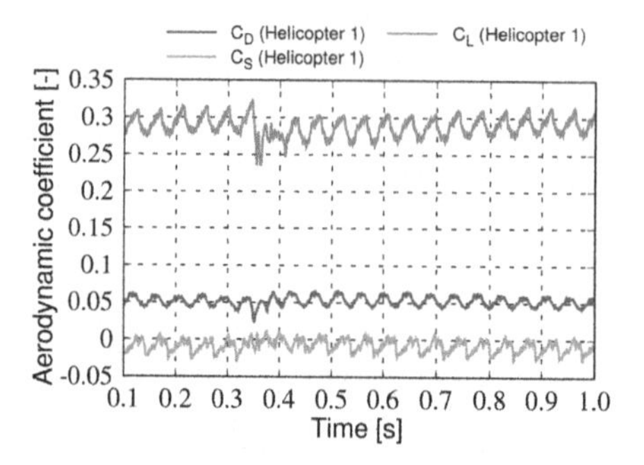

Fig. 12. Aerodynamic coefficient of Helicopter 1 (Case 2)

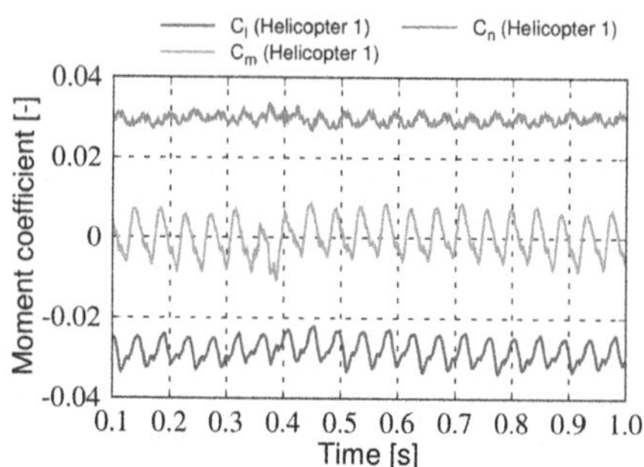

Fig. 13. Moment coefficient of Helicopter 1 (Case 2)

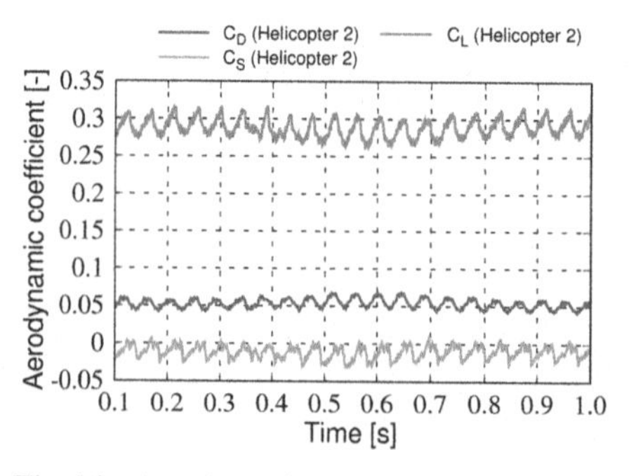

Fig. 14. Aerodynamic coefficient of Helicopter 2 (Case 2)

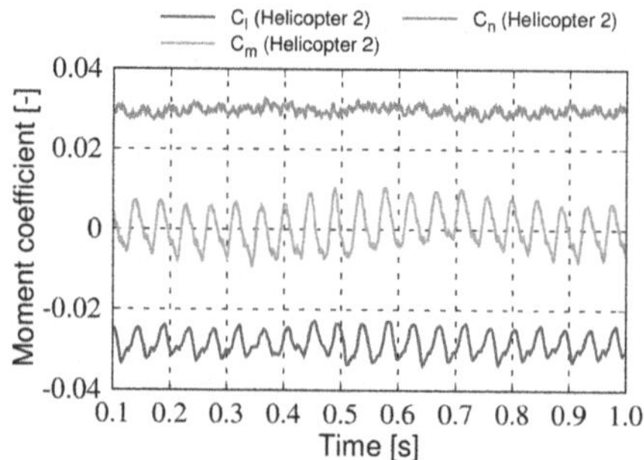

Fig. 15. Moment coefficient of Helicopter 2 (Case 2)

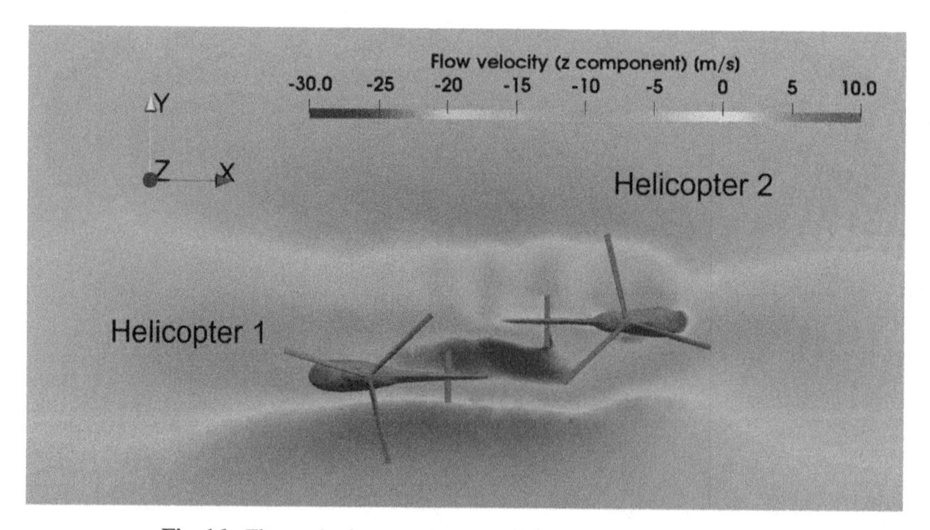

Fig. 16. Flow velocity z component distribution at 0.5 s (Case 2)

There is also a disturbance in the moment coefficients of Helicopter 2 in Fig. 10. The area through which Helicopter 2 passes during the near miss develops a complex flow field, with a spiral downward flow generated by the main rotor of Helicopter 1 as well as the flow that Helicopter 1's movement generates in the x direction. Helicopter 2 itself also creates its own flow with its movement and rotating main rotor. The interaction of their complex flow field affects the moment coefficients of Helicopter 2.

Case 2. Figures 12 and 13 illustrate the time histories of the aerodynamic coefficients and moment coefficients calculated for Helicopter 1 with Eq. (3), while Figures 14 and 15 illustrate these calculated for Helicopter 2 with Eq. (4).

Analysing the graphs, we can see a disturbance in the moment coefficients of the two helicopters, but they do not experience a significant disturbance. However, a key point revealed by the graphs is that both helicopters lose their lift during the near miss. Moreover, the lifts decrease for approximately 0.4 s, which is longer than that of Helicopter 2 in Case 1. Figure 16 depicts the flow velocity z component distribution at 0.5 s in the xy-plane in the middle of the fuselage of Helicopter 2. This figure indicates that both helicopters are affected by the flow generated by each other during the near miss, which results in the decrease in the lift coefficient of not only Helicopter 2 but also Helicopter 1. In addition, the helicopters just fly in the complex flow field due to their trajectories. Since Helicopter 2 in Case 1 only passes perpendicularly through the complex flow field generated by Helicopter 1, the lift recovers faster than in Case 2.

5 Conclusions

This study computationally demonstrates two helicopters and examines the aerodynamic effect exerted on them during a near miss. The MGFV method is applied to the moving mesh to represent the motion of the helicopters. The long trajectories of the moving helicopters are successfully modelled by the MCD method. The flows generated by the rotating main rotors, as well as the flow interaction around the helicopters, are achieved by applying the overset approach, which allows the communication of the moving computational domains. In the validation test, the results computed by the in-house code are in agreement with those of *FaSTAR-Move*, confirming that the flow field variables are accurately interpolated between the overlapping grids in this study. In the near-miss flight test case where two helicopters pass each other perpendicularly, the lift of Helicopter 2, which flies behind Helicopter 1, decreases. On the other hand, in the case where Helicopter 2 passes horizontally underneath Helicopter 1, the lift decrease occurs for both helicopters. The disturbance is observed in the moment coefficients in both cases. These results show that the complex flow field around nearly missing helicopters is realised although this study uses a simplified model, indicating that this CFD approach is capable of simulating near-miss events involving aircraft. Future work will introduce the function of each component of aircraft and study the reaction to the flow field in a near miss as well as the suggested manoeuvring.

Acknowledgements. This publication was subsidised by JKA through its promotion funds from KEIRIN RACE and by JSPS KAKENHI Grant Number 21K03856. A helicopter model from the VSP Hangar community (File ID# 359 [18]) was used in this study.

References

1. Australian Transport Safety Bureau. https://www.atsb.gov.au/publications/occurrence-briefs/2020/aviation/ab-2020-044/ Accessed 18 Feb 2022
2. Biava, M., et al.: CFD prediction of air flow past a full helicopter configuration. Aerosp. Sci. Technol. **19**(1), 3–18 (2012)
3. Steijl, R., et al.: Sliding mesh algorithm for CFD analysis of helicopter rotor–fuselage aerodynamics. Int. J. Numer. Meth. Fluids **58**(5), 527–549 (2008)
4. Crozon, C., et al.: Coupled flight dynamics and CFD—demonstration for helicopters in shipborne environment. The Aeronautical Journal **122**(1247), 42–82 (2018)
5. Takii, A., et al.: Turning flight simulation of tilt-rotor plane with fluid-rigid body interaction. J. Thermal Science Technol. **15**(2), JTST0021 (2020)
6. Matsuno, K.: Development and applications of a moving grid finite volume method. In: Topping, B.H.V., et al. (eds.) Developments and applications in engineering computational technology, Chapter 5, pp. 103–129. Saxe-Coburg Publications, Stirlingshire (2010)
7. Watanabe, K., et al.: Moving computational domain method and its application to flow around a high-speed car passing through a hairpin curve. J. Comput. Sci. Technol. **3**(2), 449–459 (2009)
8. Takii, A., et al.: Six degrees of freedom flight simulation of tilt-rotor aircraft with nacelle conversion. J. Comput. Sci. **44**, 101164 (2020)
9. Yamakawa, M., et al.: Optimization of knee joint maximum angle on dolphin kick. Phys. Fluids **32**(6), 067105 (2020)
10. Nishimura, M., Yamakawa, M., Asao, S., Takeuchi, S., Ghomizad, M.B.: Moving computational multi-domain method for modelling the flow interaction of multiple moving objects. Adv. Aerodynamics **4**(1), 1–18 (2022)
11. Roe, P.L.: Approximate Riemann solvers, parameter vectors, and difference schemes. J. Comput. Phys. **43**(2), 357–372 (1981)
12. Hishida, M., et al.: A new slope limiter for fast unstructured CFD solver FaSTAR (*in Japanese*). In: JAXA Special Publication: Proceedings of 42nd Fluid Dynamics Conference / Aerospace Numerical Simulation Symposium 2010, JAXA-SP-10–012 (2011)
13. Nakahashi, K., et al.: Intergrid-boundary definition method for overset unstructured grid approach. AIAA J. **38**(11), 2077–2084 (2000)
14. Hashimoto, A., et al.: Toward the Fastest Unstructured CFD Code "FaSTAR". In: 50th AIAA Aerospace Sciences Meeting including the New Horizons Forum and Aerospace Exposition 2012, AIAA 2012–1075 (2012)
15. Taniguchi, S., et al.: Numerical Analysis of Propeller Mounting Position Effects on Aerodynamic Propeller/Wing Interaction. AIAA SCITECH 2022 forum (2022)
16. Ito, Y.: Challenges in unstructured mesh generation for practical and efficient computational fluid dynamics simulations. Comput. Fluids **85**, 47–52 (2013)
17. Shima, E., et al.: Parameter-free simple low-dissipation AUSM-family scheme for all speeds. AIAA J. **49**(8), 1693–1709 (2011)
18. VSP Hangar. http://hangar.openvsp.org/vspfiles/359 Accessed 18 Feb 2022

Blood Flow Simulation of Left Ventricle with Twisted Motion

Masashi Yamakawa[1(✉)], Yuto Yoshimi[1], Shinichi Asao[2], and Seiichi Takeuchi[2]

[1] Faculty of Mechanical Engineering, Kyoto Institute of Technology, Kyoto, Japan
yamakawa@kit.ac.jp

[2] Department of Mechanical Engineering, College of Industrial Technology, Amagasaki, Japan

Abstract. To push out a blood flow to an aorta, a left ventricle repeats expansion and contraction motion. For more efficient pumping of the blood, it is known that the left ventricle also has twisted motion. In this paper, the influence of the twisted motion for a blood flow to an aorta was investigated. In particular, the relationship between the origin of cardiovascular disease and wall shear stress has been pointed out in the aorta region. Estimating the difference of the wall shear stress depend on the presence or absence of the twisted motion, the blood flow simulation was conducted. To express its complicated shape and the motion, the unstructured moving grid finite volume method was adopted. In this method, the control volume is defined for a space time unified domain. Not only a physical conservation law but also a geometric conservation law is satisfied in this approach. Then high accurate computation is conducted under the method. From the computation results, a remarkable difference of complicated vortex structures generated in the left ventricle was found as the influence of the left ventricular twisted motion. The vortex structures affected the blood flow leading into the aorta with the result that they generated a clear difference of the wall shear stress. The region where the difference occurred is aortic arch, then it corresponded with a favorite site of arteriosclerosis. Thus, the result showed the possibility that the simulation with the left ventricular twisted motion would be useful to specify causes of heart diseases.

Keywords: Computational fluid dynamics · Blood flow simulation · Left ventricle · Aorta · Twisted motion

1 Introduction

Serious disorders directly associated with the cause of death, for example arteriosclerosis or aneurism, are seen in a heart and vascular diseases. Then, the relation between the origin of the heart disease and blood flow has been pointed out. To specify causes of the diseases, flows in a heart or blood vessel have been studied through the method of experimentation or numerical simulation. Ku et al. [1] made a measurement the intimal thickening generated at the branching part of the human arteria carotis communis, as they were focused on the relations between the intimal thickening of an artery and the blood flow. Then, it was shown that the intimal thickening has a correlation with the time fluctuation of shear stress measured on a glass tube flow made from specimens

D. Groen et al. (Eds.): ICCS 2022, LNCS 13353, pp. 343–355, 2022.
https://doi.org/10.1007/978-3-031-08760-8_29

of blood vessel. Fukushima et al. [2] created a visualization of blood flow using the real blood vessel taken out from the body. The real blood vessel is made transparent by salicylic acid. Then, they determined whether vortex tube exist at the bifurcation of the blood vessel. While, Oshima et al. [3] have developed the simulation system the M-SPhyR to achieve multi scale and physics simulation. The system integrates image-based modeling, blood-borne material and interaction between blood flows and blood vessel walls. Using the system, they calculated the blood flows in the arterial circle of Willis as the cardinal vascular network of the brain. Then, they have succeeded to recreate the collateral flow which is an important function at the flow control depended on the arterial circle of Willis.

From the geometric view, the aorta connected to the left ventricle is comprised of three parts as the aorta ascendens expanding upward, the aortic arch taking a bend, and the aorta descendens expanding downward. Then, the three principal branched blood vessels expand from the aortic arch. While, the motion of the left ventricle are not only expansion and contraction but also twisted motion [4]. The twisted motion like wring a left ventricle can pump out blood efficiently, then the ejection fraction touches 70% despite the cardiac fiber's contraction factor of 20%. As the pumping mechanism of the left ventricle is gathering attention from engineering field, it is executing novel ideas regarding an industrial pump with twisted motion. Thus, the motion of the left ventricle wall is very interesting in terms of medical and engineering field [5].

It is easy to assume that a difference of left ventricular motion affects a blood flow not only in the left ventricle but also in the aorta. Then, the difference of the blood flow would affect the risk and favorite site of a cardiac disease. In this paper, the pulsatile flows at the left ventricle and the aorta are computed and estimated using the left ventricular movement with twisted motion. In particular, to satisfy a physical conservation law and a geometric conservation law, the unstructured moving grid finite volume method [6, 7] is adopted. In this method, a control volume is defined for a space time unified domain. The method made it possible to compute accurately for motion of the left ventricle and the aorta. Furthermore, the unstructured mesh approach was also able to express such the complicated shape. Then, the computation was carried out under the OpenMP parallel environment [8].

2 Numerical Approach

2.1 Governing Equations

As governing equations, the continuity equation and the Navier-Stokes equations for incompressible flows are adopted and written as follows:

$$\nabla \cdot \mathbf{q} = 0, \tag{1}$$

$$\frac{\partial \mathbf{q}}{\partial t}+\frac{\partial \mathbf{E}_a}{\partial x}+\frac{\partial \mathbf{F}_a}{\partial y}+\frac{\partial \mathbf{G}_a}{\partial z} = -\left(\frac{\partial \mathbf{E}_p}{\partial x}+\frac{\partial \mathbf{F}_p}{\partial y}+\frac{\partial \mathbf{G}_p}{\partial z}\right)+\frac{1}{\mathrm{Re}}\left(\frac{\partial \mathbf{E}_v}{\partial x}+\frac{\partial \mathbf{F}_v}{\partial y}+\frac{\partial \mathbf{G}_v}{\partial z}\right), \tag{2}$$

where $\boldsymbol{q}$ is the velocity vector, $\boldsymbol{E}_a$, $\boldsymbol{F}_a$, and $\boldsymbol{G}_a$ are advection flux vectors in the x, y, and z direction, respectively, $\boldsymbol{E}_v$, $\boldsymbol{F}_v$, and $\boldsymbol{G}_v$ are viscous-flux vectors, and $\boldsymbol{E}_p$, $\boldsymbol{F}_p$, and $\boldsymbol{G}_p$ are pressure terms. The elements of the velocity vector and flux vectors are

$$\mathbf{q}=\begin{bmatrix}u\\v\\w\end{bmatrix},\ \mathbf{E}_a=\begin{bmatrix}u^2\\uv\\uw\end{bmatrix},\ \mathbf{F}_a=\begin{bmatrix}uv\\v^2\\vw\end{bmatrix},\ \mathbf{G}_a=\begin{bmatrix}uw\\vw\\w^2\end{bmatrix},\ \mathbf{E}_p=\begin{bmatrix}p\\0\\0\end{bmatrix},$$
$$\mathbf{F}_p=\begin{bmatrix}0\\p\\0\end{bmatrix},\ \mathbf{G}_p=\begin{bmatrix}0\\0\\p\end{bmatrix},\ \mathbf{E}_v=\begin{bmatrix}\partial u/\partial x\\\partial v/\partial x\\\partial w/\partial x\end{bmatrix},\ \mathbf{F}_v=\begin{bmatrix}\partial u/\partial y\\\partial v/\partial y\\\partial w/\partial y\end{bmatrix},\ \mathbf{G}_v=\begin{bmatrix}\partial u/\partial z\\\partial v/\partial z\\\partial w/\partial z\end{bmatrix}, \tag{3}$$

where u, v, and w are the velocity components of the x, y, and z directions, respectively, and p is pressure. Re is the Reynolds number.

2.2 The Unstructured Moving-Grid Finite-Volume Method

In this simulation, expansion and contraction of the left ventricle and translation motion of the aorta are expressed using moving mesh approach. To assure a geometric conservation law in moving mesh, a control volume is defined in a space-time unified domain. For the discretization, Eq. (2) can be written in divergence form as

$$\tilde{\nabla}\cdot\tilde{\mathbf{F}}=\mathbf{0}, \tag{4}$$

where

$$\tilde{\nabla}=\begin{bmatrix}\frac{\partial}{\partial x}\\\frac{\partial}{\partial y}\\\frac{\partial}{\partial z}\\\frac{\partial}{\partial t}\end{bmatrix},\quad \tilde{\mathbf{F}}=\begin{bmatrix}\mathbf{E}_a+\mathbf{E}_p-\frac{1}{\mathrm{Re}}\mathbf{E}_v\\\mathbf{F}_a+\mathbf{F}_p-\frac{1}{\mathrm{Re}}\mathbf{F}_v\\\mathbf{G}_a+\mathbf{G}_p-\frac{1}{\mathrm{Re}}\mathbf{G}_v\\\mathbf{q}\end{bmatrix}. \tag{5}$$

The flow variables are defined at the center of the cell in the (x, y, z) space, as the approach is based on a cell-centered finite volume method. Thus, the control volume becomes a four-dimensional polyhedron in the (x, y, z, t)-domain. For the control volume, Eq. (4) is integrated using the Gauss theorem and written in surface integral form as:

$$\int_{\tilde{\Omega}}\tilde{\nabla}\cdot\tilde{\mathbf{F}}d\tilde{V}=\oint_{\partial\tilde{\Omega}}\tilde{\mathbf{F}}\cdot\tilde{\mathbf{n}}_u d\tilde{S}\approx\sum_{l=1}^{6}\left(\tilde{\mathbf{F}}\cdot\tilde{\mathbf{n}}\right)_l=\mathbf{0} \tag{6}$$

Here, $\tilde{\mathbf{n}}_u$ is an outward unit vector normal to the surface, $\partial\tilde{\Omega}$, of the polyhedron control volume $\tilde{\Omega}$, and $\tilde{\mathbf{n}}=(\tilde{n}_x,\tilde{n}_y,\tilde{n}_z,\tilde{n}_t)_l$, $(l = 1, 2,\ldots 6)$ denotes the surface normal vector of control volume, and its length is equal to the boundary surface area in four-dimensional (x, y, z, t) space. The upper and bottom boundary of the control volume ($l =$ 5 and 6) are perpendicular to the t-axis, and therefore they have only the $\tilde{n}_t$ component, and its length corresponds to the volume of the cell in the (x, y, z)-space at time t^n and t^{n+1}, respectively.

3 Computational Model and Conditions

3.1 Geometric Model of Left Ventricle and Aorta

A main function of the left ventricle is pumping blood to an aorta. The mitral valve and the aortic valve are put in the inlet and the outlet of the ventricle, respectively. The shape of the left ventricle is shown in Fig. 1. Bothe of the diameter of blood vessels at the mitral valve and the aortic valve are 3.0 cm. The length from the base of heart to the cardiac apex is 7.8 cm at lumen maximum volume. The cross-section shape of the left ventricle is ellipse. Then, the ratio of the major axis and minor axis on the ellipse is 5 to 4.

While, the aorta is comprised of three parts which are the ascending aorta expanding upward, the aortic arch taking a bend, and the descending aorta expanding downward. Furthermore, the three principal branched blood vessels which are called innominate artery, left common carotid artery and left subclavian artery expand from the aortic arch. Then the aortic arch itself curves three-dimensionally. In other words, the central axis of the aortic arch in not on a plane surface. Thus, the aorta is complicated shape with bending, bifurcation and three-dimensional torsion. In this paper, the shape of the aorta model is created, as shown Fig. 2.

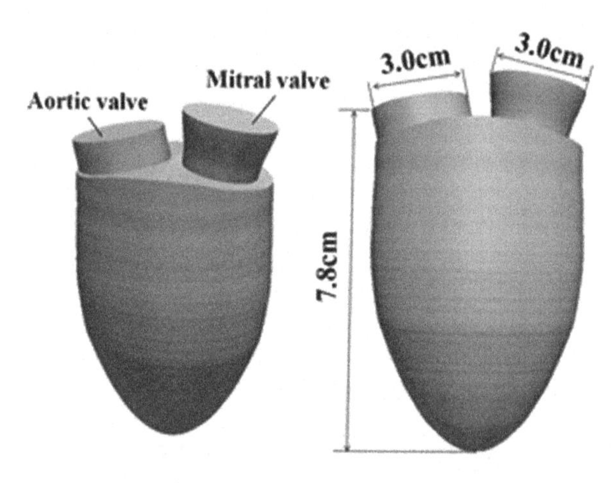

Fig. 1. Left ventricle model.

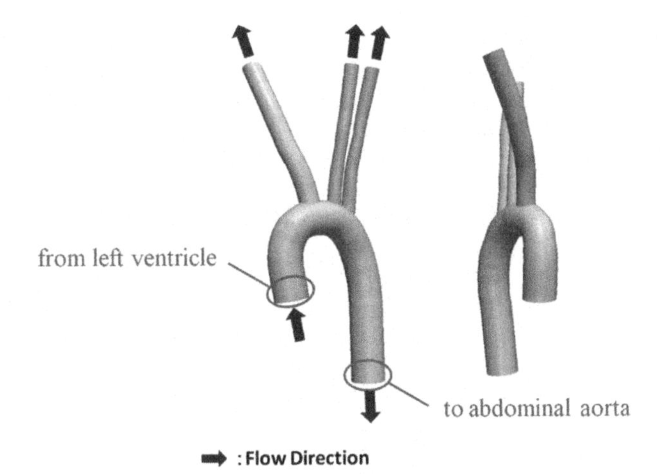

Fig. 2. Aorta model.

3.2 Motions of Left Ventricle and Aorta

The left ventricle is draining blood to the aorta by expansion and contraction. Then the heart rate is determined the systole and diastole of the heart. Then, a period from starting point to the next starting point of heart rate is called the cardiac cycle. If a pulse rate is 60 bpm, one cardiac cycle would be 1.0 s. Then, it is classified 0.49 s as the systole and 0.51 s as the diastole. The history of the left ventricle cavity volumetric change in one cardiac cycle is shown in Fig. 3. The expansion and contraction using moving mesh at the simulation are expressed under the history.

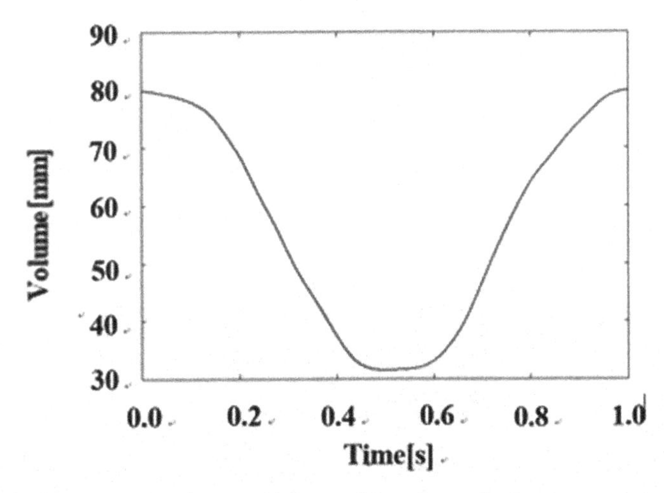

Fig. 3. History of left ventricle volumetric change.

In addition, the left ventricle has a twisted motion for effective pumping. In this paper, the influence of the twisted motion for a flow in the aorta is investigated with comparing with and without twisted motion. The case without twisted motion has just inward motion (contraction –expansion movement), as shown in Fig. 4. On the other hand, the case with twisted motion has torsion movement around its vertical axis in addition to inward motion, as shown in Fig. 5. In this case, the maximum torsion angle at base of heart is 3°. Then, the maximum torsion angle at apex of heart is 7.8°. These twist in the opposite direction respectively.

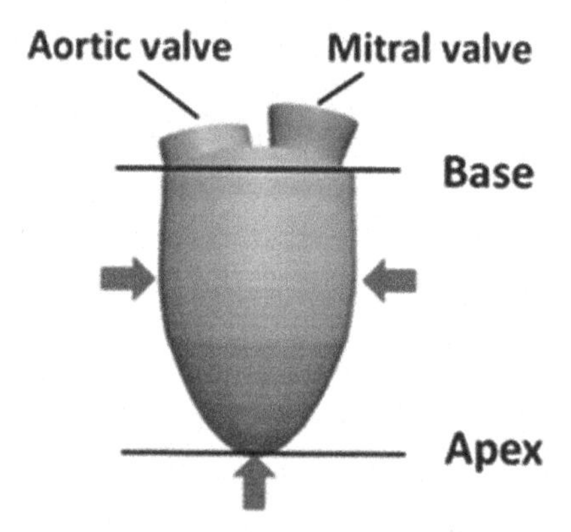

Fig. 4. Image of inward motion of the left ventricle (Case 1).

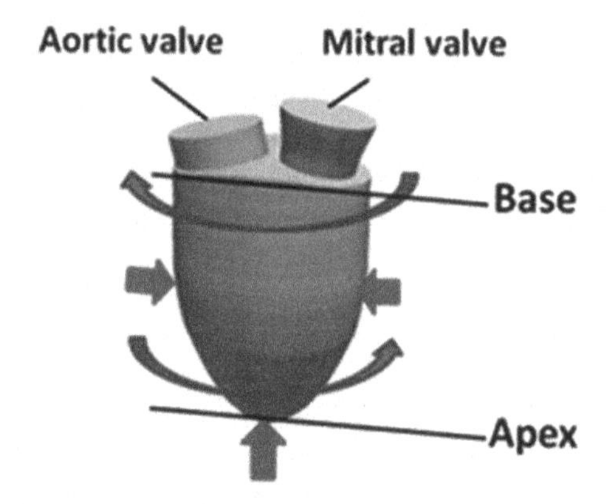

Fig. 5. Image of twisted motion in addition to inward motion (Case 2).

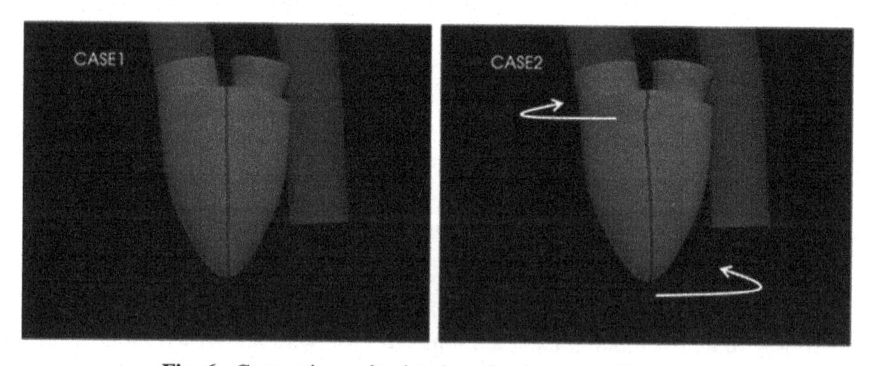

Fig. 6. Comparison of twisted motion by vertical blue line

The computation models with/without twisted motion at the minimum volume of the heart are compared in Fig. 6. Here, the twisted motion is confirmed by the vertical blue line in the figure. History of volume change in both cases are same. Difference of both cases are just shear stress for circumferential direction. Furthermore, translational motion [9, 10] for the left ventricle and the aorta is also adopted in this simulation.

3.3 Computational Conditions

The computational mesh is generated by MEGG3D [11] using tetrahedral and prism elements, as shown in Fig. 7. MEGG3D is a mesh generation software provided by Japan Aerospace Exploration Agency. The number of elements for the left ventricle model is 1,145,432 (tetrahedron: 755,042, prism: 390,390) and for the aorta is 1,631,657 (tetrahedron: 856,386, prism: 775,271). Then, the total number of the elements are 2,777,089.

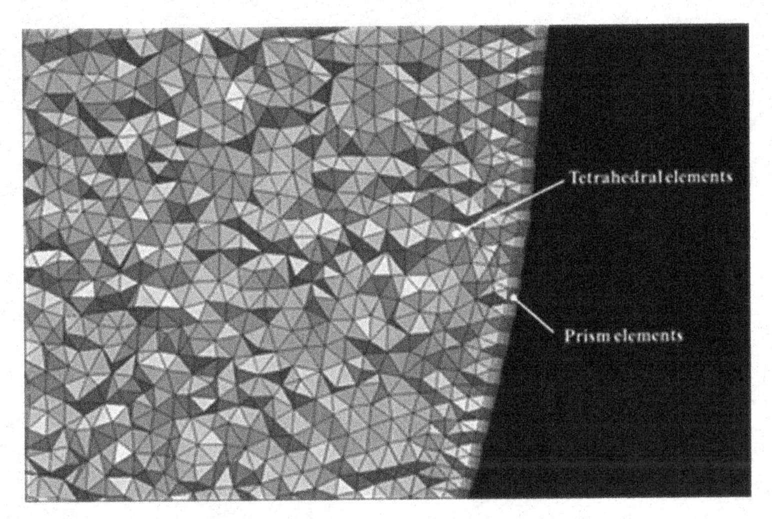

Fig. 7. Computational mesh near the wall

The heart rate is 60 bpm, and the Reynolds number is 2,030. As an initial condition, pressure $p = 0$ and velocity for x, y, z directions $u = v = w = 0$ are obtained for all elements. In the cardiac diastole, the mitral valve is open and the aortic valve is closed completely. Then, the velocity at the mitral valve is given as a linear extrapolation and pressure is fixed as $p = 0$. While, in the cardiac systole, the mitral valve is closed and the aortic valve is open completely. These open and closing motions are conducted instantly. On the four exit of blood vessels, velocity is determined as a linear extrapolation and pressure is zero. The velocity on all walls of the left ventricle and the aorta is given the moving velocity decided expansion, contraction, translational motion and twisted motion.

4 Computational Results

4.1 Comparison of Flows in the Left Ventricle

Figure 8 shows isosurface of Q-criterion in left ventricle at $t = 23.0$, 25.0 and 27.0. Here, the simulation starts at the beginning of diastolic phase. Then, one cycle (diastole and systole) is 10.0 as dimensionless time. Thus, the figure ($t = 23.0$, 25.0 and 27.0) is in third diastolic and systolic phase. There are some vortices generated by pulsation until second phase in the left ventricle. To the vortices area, ring vortex tube flow through the mitral valve is seen at $t = 23.0$. Then, the vortex tube becomes disrupted according to decline of blood inflow from the mitral valve, expanding to inside of the left ventricle with complicated eddy structure.

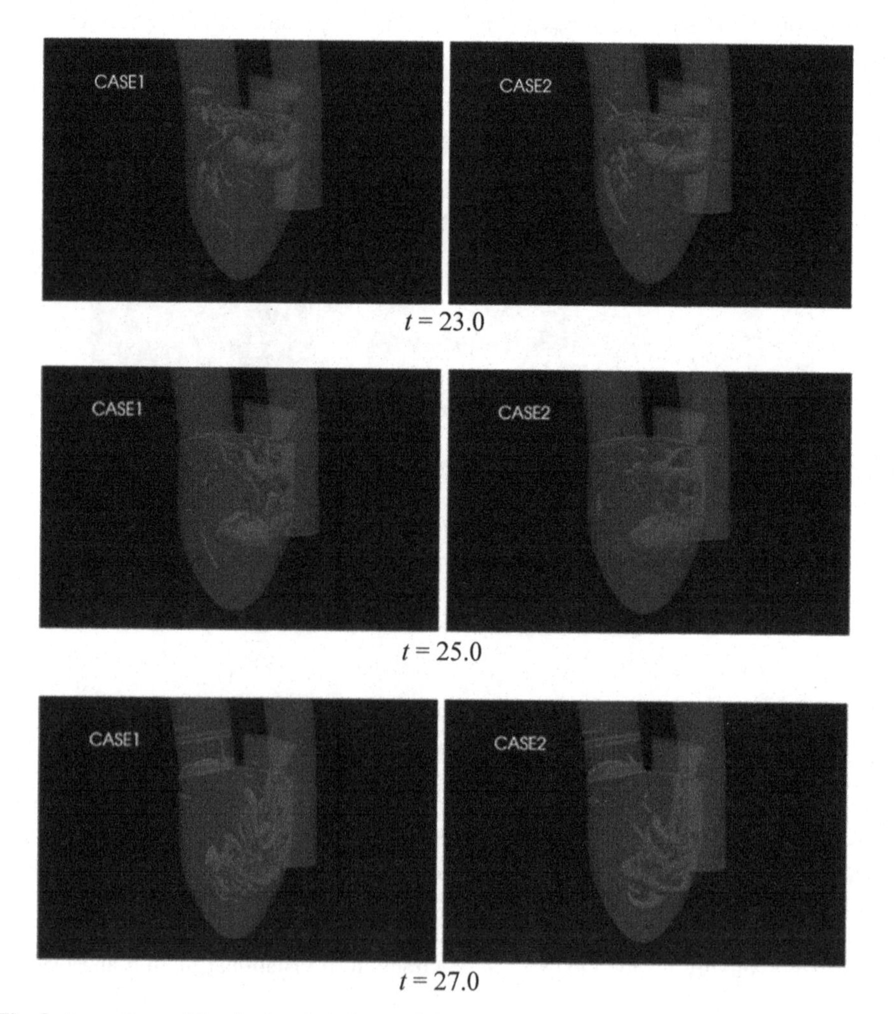

Fig. 8. Isosurface of Q-criterion in left ventricle ($t = 23.0$, 25.0, 27.0) (Case1: inward motion, Case2: inward and twisted motion)

In Fig. 8, the flow phenomenon between by inward motion (left: case1) and by inward + twisted motion (right: case2) should show the same trend. However, the detail eddy structures are markedly different. The difference is caused by addition of the twisted motion. Then, the difference should affect to flows in an aorta.

4.2 Shear Stress on the Wall of Aorta

To investigate the influence of flows caused by twisted motion of left ventricle, a flow in an aorta is simulated. Figure 9 shows shear stress on the wall of the aorta at $t = 25.0$, 27.0 and 29.0. They are in the systolic phase. The systole starts at $t = 25.0$. Thus, there is little wall shear stress at $t = 25.0$. At mid of systole ($t = 27.0$), a large change in a

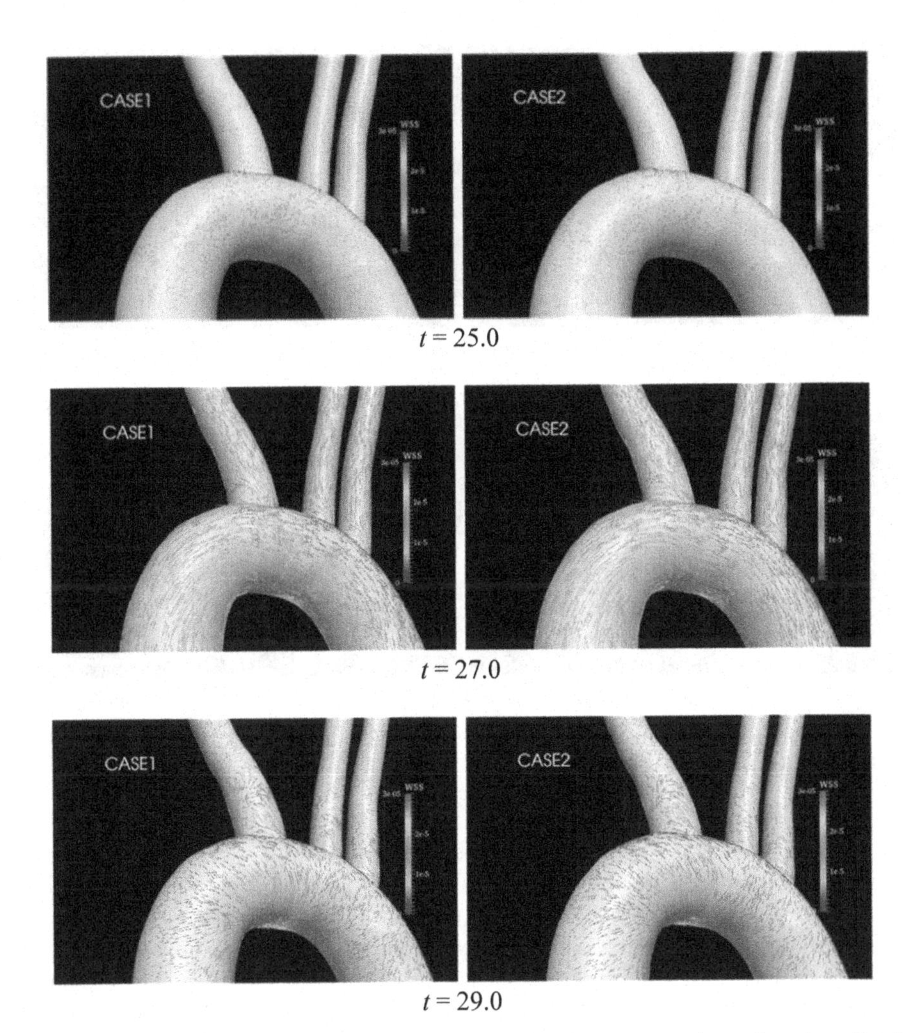

Fig. 9. Shear stress on the wall of aorta (t = 25.0, 27.0, 29.0) (Case1: inward motion, Case2: inward and twisted motion)

direction and a magnitude of wall shear stress is seen. It should be caused by complicated flow in the aorta.

From mid to end in systole at $t = 29.0$, we can see that the direction of wall shear stress reverses with mainstream at a part of the aortic arc. It is also confirmed from pressure contours of the left ventricle and aorta at $t = 27.0$ and 29.0, as shown in Fig. 10. It should be caused by decreasing of inner pressure of the left ventricle.

Here, Ku [1] reported that there is correlative relationship between rate of back stream of wall shear stress on the aorta for mainstream and the intimal thickening which is early involvement of atherosclerosis. On the other hand, it is known that an aortic arc is favorite site of arteriosclerosis. Thus, the simulation result which shows back stream of wall shear stress for mainstream is validity from a medical field.

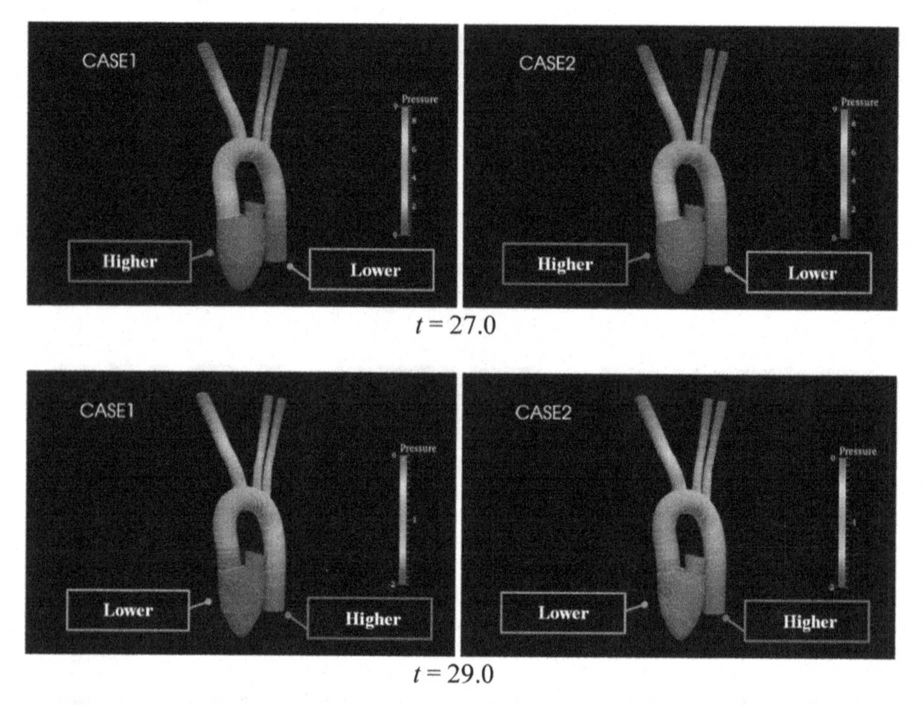

Fig. 10. Pressure contours on the left ventricle and aorta ($t = 27.0$ and 29.0) (Case1: inward motion, Case2: inward and twisted motion)

Although there is a slight difference between just inward motion and additional twisted motion on pressure contours at $t = 29.0$, there is almost no difference in both on other pressure contours and distribution of wall shear stress of aorta. To clarify the difference of influence on the aorta by flows from left ventricle, time averaged wall shear stress of the aorta in third systolic phase are shown in Fig. 11. Then, Fig. 12 shows difference between time averaged wall shear stress decided by inward motion and shear stress by additional twisted motion (difference between case1 and case2). As the figure, clear differences are seen at the bifurcation (Location 1) and at the bottom of aortic arc

(Location 2). It means that the difference of motions of left ventricle affects to the wall shear stress at the bifurcation and the bottom of aortic arc. The detail of the differences at the locations are shown in Table 1. We can see 17.9% difference of wall shear stress at the bifurcation (Location 1) and 6.2% difference of wall shear stress at the bottom of aortic arc (Location 2). These are great differences.

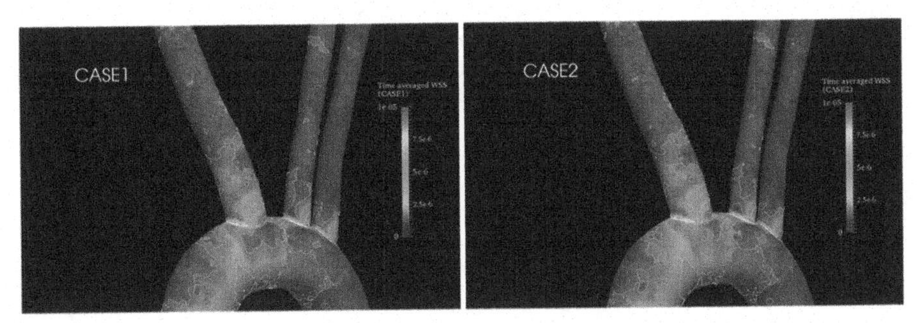

Fig. 11. Time averaged wall shear stress in third systolic phase (Case1: inward motion, Case2: inward and twisted motion)

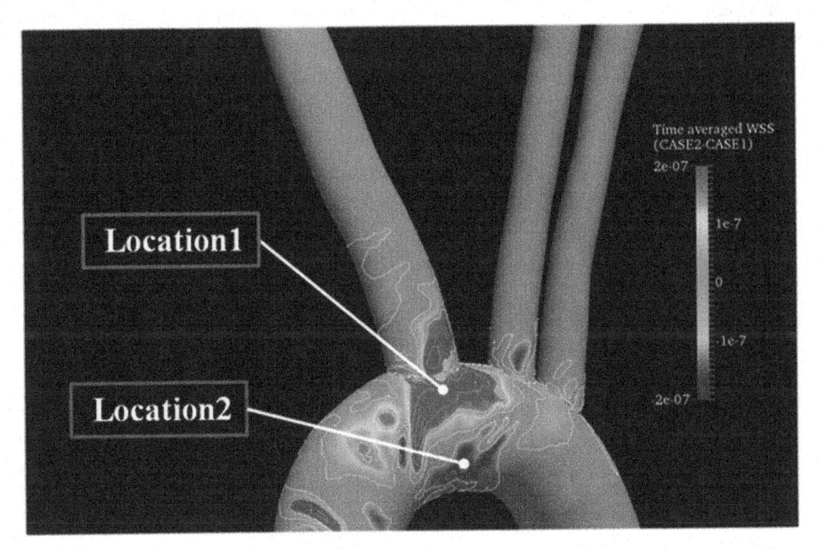

Fig. 12. Difference of time averaged wall shear stress for Case1 and Case2

As these results, we confirmed that motions of left ventricle affect to flows in the ventricle, and the flows also influence in the aorta. More than 10% difference of wall shear stress in the aorta caused by the twisted motion was also confirmed. In this paper, we adopted just one small motion. In fact, there are a lot of complicated motions of the left ventricle. Thus, difference of real motions of left ventricle should affect to risk and favorite site of vascular lesions. Therefore, when individual risk of vascular lesion for patients is referred, it is important to focus motions of left ventricle.

Table 1. Comparison between time averaged wall shear stress for Case1 and Case2

	A	B	C	D	E
Location1	1.46×10^{-6}	1.72×10^{-6}	2.62×10^{-7}	17.9	1.18
Location2	3.29×10^{-6}	3.09×10^{-6}	-2.04×10^{-7}	6.18	0.94

A: Time averaged wall shear stress for Case1 (dimensionless quantity)
B: Time averaged wall shear stress for Case2 (dimensionless quantity)
C: For Case2 – For Case1 (dimensionless quantity)
D: {|For Case2 – For Case1|/For Case1} × 100[%]
E: For Case2/For Case1

5 Conclusions

To express the influence to behavior of blood flow in a left ventricle and an aorta by twisted motion of the left ventricle, flows in the left ventricle and the aorta were computed using moving mesh method. First, it was shown that different motion between inward and additional twisted motion affect to the detail eddy structures in the left ventricle. From mid to end in systole, the reverse flow of wall shear stress for mainstream at a part of the aortic arc was seen. The correspondence of the result with the medical knowledge showed the validity of the simulation. Furthermore, more than 10% difference of wall shear stress in the aorta caused by the twisted motion was also confirmed. Although just twisted motion was adopted in this paper, there are a lot of complicated motions of the left ventricle in practical. Thus, difference of real motions of left ventricle should affect to risk and favorite site of vascular lesions. Therefore, when individual risk of vascular lesion for patients is referred, it is important to focus motions of left ventricle.

Acknowledgments. This publication was subsidized by JKA through its promotion funds from KEIRIN RACE.

References

1. Ku, D., et al.: Pulsatile flow and atherosclerosis in the human carotid bifurcation. Positive correlation between plaque location and low oscillating shear stress. Arteriosclerosis **5**, 293–302 (1985)
2. Fukushima, T., et al.: The horseshoe vortex: A secondary flow generated in arteries with stenosis, bifurcation, and branching. Biorheology **19**, 143–154 (1982)
3. Oshima, M., et al.: Multi-Scale & Multi-Physics Simulation of Cerebrovascular Disorders. J. Japan Society Fluid Mech. **26**(6), 369–374 (2007)
4. Omar, A.M., et al.: Left ventricular twist and torsion, Circulation: Cardio-vascular Imaging, **8**(8), e000009 (2015)
5. Liang, F., et al.: A multi-scale computational method applied to the quantitative evaluation of the left ventricular function. Comput. Biol. Med. **37**, 700–715 (2007)
6. Yamakawa, M., et al.: Numerical simulation for a flow around body ejection using an axisymmetric unstructured moving grid method. Comput. Therm. Sci. **4**(3), 217–223 (2012)

7. Yamakawa, M., et al.: Numerical simulation of rotation of intermeshing rotors using added and eliminated mesh method. Procedia Comput. Sci. **108C**, 1883–1892 (2017)
8. Yamakawa, M., et al.: Domain decomposition method for unstructured meshes in an OpenMP computing environment. Comput. Fluids **45**, 168–171 (2011)
9. Yamakawa, M., et al.: Blood flow simulation of left ventricle and aorta with translation motion, The Proc. of the International Conference on Computational Methods, vol. 6, pp. 7–17 (2019)
10. Fukui, T., et al.: Influence of geometric changes in the thoracic aorta due to arterial switch operations on the wall shear stress distribution. Open Biomed. Eng. J. **11**, 9–16 (2017)
11. Ito, Y.: Challenges in unstructured mesh generation for practical and efficient computational fluid dynamics simulations. Comput. Fluids **85**, 47–52 (2013)

Simulation of Virus-Laden Droplets Transmitted from Lung to Lung

Shohei Kishi(✉), Yuta Ida, Masashi Yamakawa, and Momoha Nishimura

Kyoto Institute of Technology, Matsugasaki, Sakyo-ku, Kyoto 606-8585, Japan
sk2.090412066@gmail.com

Abstract. In this study, we conducted a computational fluid dynamics analysis to estimate the trajectory of the virus-laden droplets. As numerical models, two human body models with airways were prepared. These models are represented by unstructured grids. Having calculated the unsteady airflow in the room, we simulated the trajectory of droplets emitted by the human speaking. In addition, inhaling the droplets into the lung of the conversation partner was simulated. The number of the droplets adhered to the respiratory lining of the partner was counted separately on the nasal cavity, oral cavity, trachea, bronchi, and bronchial inlet surface. The diameters of the droplets were also investigated in the same manner. It was noticeable that more than 80% of the droplets inhaled by the conversation partner adhered to the bronchial inlet surface. Also, the conversation partner did not inhale droplets larger than 35 μm in diameter. It was found that when the distance between two people was 0.75 m, more droplets adhered to the partner's torso.

Keywords: COVID-19 · Respiratory organ · Computational fluid dynamics

1 Introduction

The global spread of COVID-19 revealed a lack of conventional knowledge about airborne infections. In fact, it is only in recent years that research about the mechanisms of airborne infection has been a focus of attention. Therefore, the airborne transmission route of COVID-19 should be identified and knowledge about effective infection prevention must be supplemented. There are two main methods for engineering the airborne transmission pathways of COVID-19: experimental and numerical approaches. Experiments need to prepare two real people, one corona-infected and the other non-infected. However, this is impractical because of the high risks for the non-infected person.

On the other hand, computational simulations can be conducted safely and easily without preparing any real subjects. In particular, computational fluid dynamics (CFD) has been used to assess the risk of airborne infectious diseases. For example, Yamakawa et al. [1] simulated influenza infection using a respiratory and lung model represented by unstructured and moving grids. They concluded that infection caused by influenza viruses was more likely to occur with nasal breathing than with oral breathing. Ogura et al. [2] used the results of airflow simulations in a respiratory tract for modeling airflow

D. Groen et al. (Eds.): ICCS 2022, LNCS 13353, pp. 356–369, 2022.
https://doi.org/10.1007/978-3-031-08760-8_30

in a room. The coupled simulations of flow fields in the room and in the respiratory tract made it possible to predict the motion of virus-laden droplets affected by coughing and breathing. The result showed that large droplets flew downward, and small droplets flew upward in the room. Additionally, most of the droplets inhaled through the nose adhered to the nasal cavity in the respiratory tract. Srivastav et al. [3] used a three-branched airway model to simulate particle deposition. In brief, most of the droplets were deposited at the bifurcation due to inertial impaction. However, they simulated moving virus-laden droplets only in the respiratory tract system. Moreover, the particle size and number of droplets adhered to the respiratory lining are not presented. Detailed information on droplets adhered to the respiratory lining can accurately assess the infection risk.

Therefore, this study used a human model with airways to calculate the behavior of the virus. In addition, we clarified the place where the droplets emitted during talking adhered to the opponent's respiratory lining. The diameter of the adhered droplets and the number of droplets were also measured.

2 Numerical Approach

2.1 Flow Field and Heat Analysis

The flow and temperature fields were calculated by using the fluid simulation software *SCRYU/Tetra* [4]. The governing equations are the continuity equations, the incompressible Navier – Stokes equations, and the energy conservation equation. The continuity equation is defined by

$$\frac{\partial u_i}{\partial x_i} = 0. \tag{1}$$

The three-dimensional incompressible Navier – Stokes equations are given by

$$\frac{\partial(\rho u_i)}{\partial t} + \frac{\partial(u_j \rho u_i)}{\partial x_j} = -\frac{\partial p}{\partial x_i} + \frac{\partial}{\partial x_j}\mu\left(\frac{\partial u_i}{\partial x_j} + \frac{\partial u_j}{\partial x_i}\right) - \rho g_i \beta(T - T_0), \tag{2}$$

where $u_i(i, j = 1, 2, 3)$ represent the air velocity in the x, y, and z coordinates, and ρ indicates density, which is constant. t, p, and μ denote time, pressure, and viscosity coefficient, respectively. $g_i(i, j = 1, 2, 3)$ are the gravitational acceleration in the x, y, and z coordinates. β represent the body expansion coefficient. T and T_0 indicate temperature and base temperature, respectively. The energy conservation is defined by the following equation:

$$\frac{\partial(\rho C_p T)}{\partial t} + \frac{\partial(u_j \rho C_p T)}{\partial x_j} = \frac{\partial}{\partial x_j}\left(K\frac{\partial T}{\partial x_j}\right) + \dot{q}, \tag{3}$$

where C_p is the constant pressure specific heat. K represents the thermal conductivity, and $\dot{q}$ denotes the heat flux. The SIMPLEC algorithm [5] was used to solve Eqs. (1) and (2), handling with the coupling of velocity and pressure. The second order MUSCL method [6] was applied to the convective terms of Eqs. (2) and (3). The standard k–ε model is adopted as a RANS turbulence model since this study evaluates the main flow.

2.2 Movement of Virus-Laden Droplets

Droplet motions are modeled by following the method described in the study by Yamakawa [7]. First, we define the equation of transient motion of the droplets. After that, the change in radius of droplets due to evaporation and coalescence between droplets are considered. The reference [7] used random numbers every second to evaluate whether the virus was active or not, and then, droplets of inactive virus are removed accordingly. However, this study determined the lifespan of each virus-laden droplet by using a random number at the beginning of the calculation in order to reduce the computational cost. The virus-laden droplets are removed as soon as their lifetimes were over. In addition, this study omitted the position vector of the droplets in Eq. (15) of the reference [7] because two floating virus-laden droplets are in close proximity when they coalesced. Thus, the position vectors h_1 and h_2 of the two droplets before coalescence and the position vector h_3 of the coalesced droplet are approximated by

$$h_1 \approx h_2 \approx h_3. \tag{4}$$

3 Simulation of Droplets in the Room and the Respiratory Tract

3.1 Numerical Model

Figure 1 shows a model of two people sitting in a room and a magnified view of the computational grids for one person. The size of the room is 3.0 m × 3.0 m × 3.0 m and the two people sit face to face. The distance between the mouths of the two people is defined as x [m]. This study simulated the following cases: $x = 0.75$, 1.00, and 1.25. A 0.5 m × 0.5 m air vent is installed at the top of the room. Four different models are used in this simulation, as shown in Table 1.

Table 1. Definition of model type

Model type	Conditions for wearing a mask
(A)	Neither person wears a mask
(B)	Only the breathing person wears a mask
(C)	Only the talking person wears a mask
(D)	Both people wear masks

Figure 2 illustrates computational grids for the two human models, with (a) showing the sagittal plane, with (b) showing the cells around the face without a mask, and with (c) showing the cells around the mask. The grids were generated by *SCRYU/Tetra*. Figure 2(a) depicts the case where neither person wears a mask at $x = 0.75$. The surfaces of the airway and mask are covered with three prismatic layers. Computational grids are generated with approximately 1,000,000 cells for all the cases to reduce the computational cost. The smallest case involves 870,768 cells while the largest case includes 1,148,185 cells.

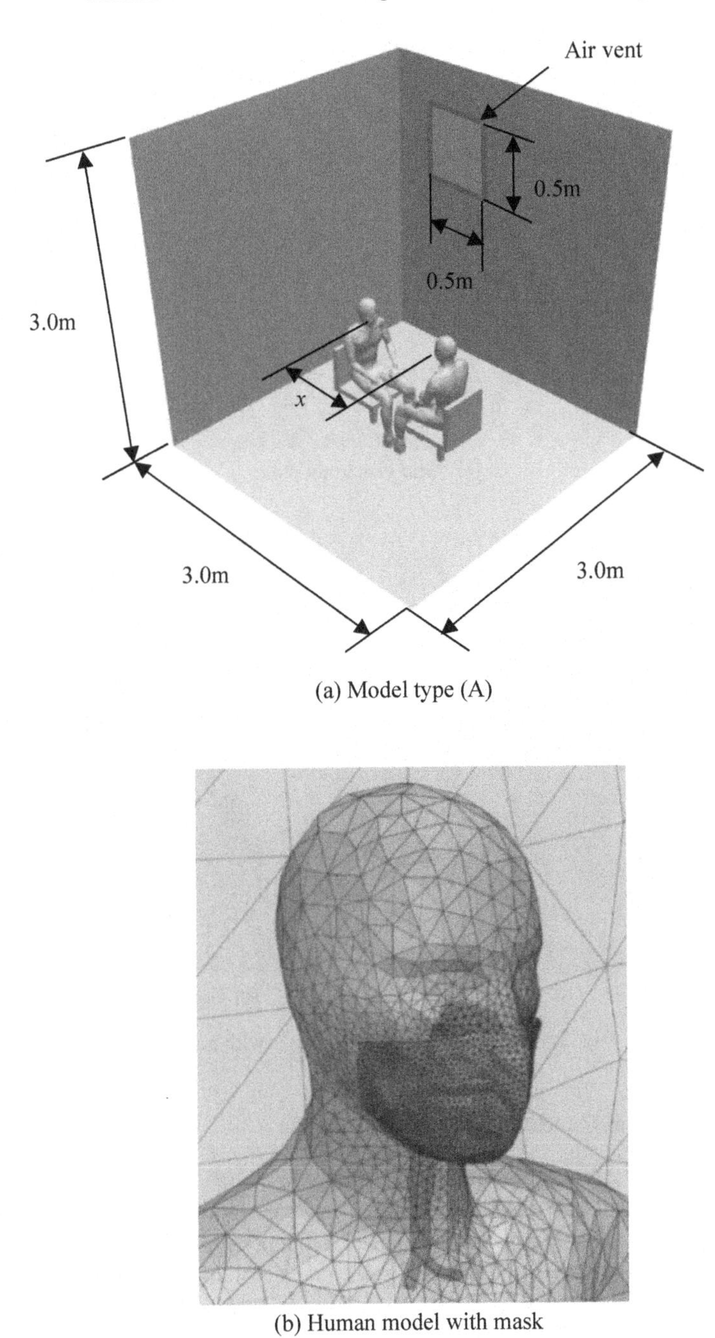

(a) Model type (A)

(b) Human model with mask

Fig. 1. Numerical model of the room and the human body with airway.

(a) Grid in sagittal plane

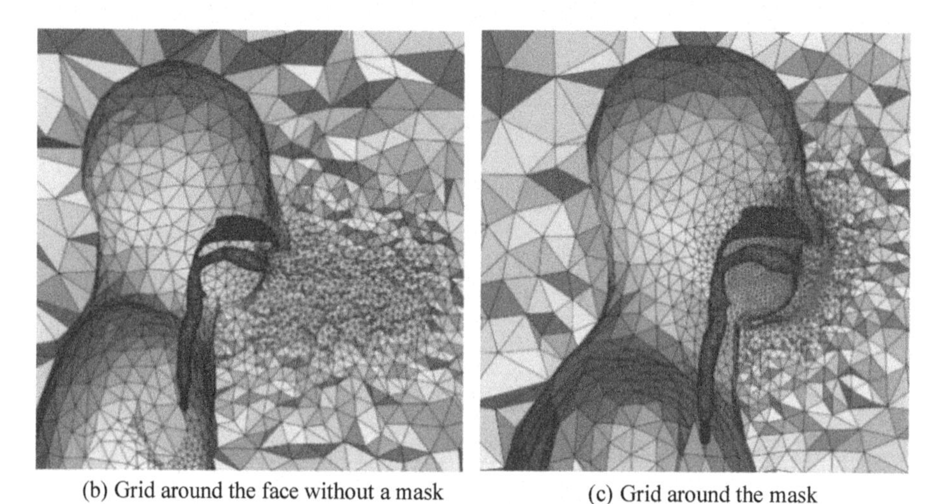

(b) Grid around the face without a mask (c) Grid around the mask

Fig. 2. Grid around the human body with airways.

3.2 Computational Conditions for Flow Analysis

Figure 3 shows the volume flow rate of breathing described in the Handbook of Physiology [8]. In this study, the positive volume flow rate represents intake air while the negative volume flow rate represents exhale air. This volume flow rate was given at the bronchial inlet surface of the breathing person as a boundary condition. Also, total pressure $p_T = 0$ was given at the air vent. The initial conditions are $p = 0.0$ and $u_1 = u_2 = u_3 = 0.0$.

In addition, Gupta et al. [9] defined the volume flow rate of talking, as shown in Fig. 4. This volume flow rate is given at the bronchial inlet surface of the talking person. The initial conditions are the same as the conditions for the breathing person. The talking person emits virus-laden droplets while the breathing person is exposed to the droplets. For other computational conditions, the room temperature is estimated to be $T_0 = 20\,°C$. The exhalation/inhalation temperature and human body surface temperature are assumed to be $T_1 = 30\,°C$.

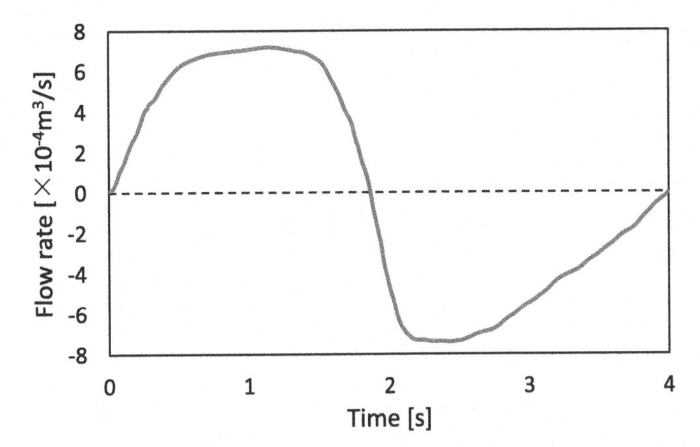

Fig. 3. Volume flow rate of breathing.

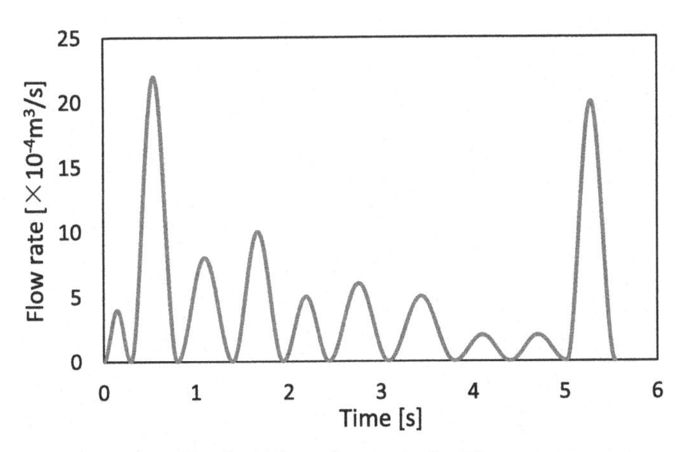

Fig. 4. Volume flow rate of talking.

3.3 Computational Conditions for Droplet Analysis

Figure 5 shows the number distribution of droplets by diameter. This was determined from the literature of Duguid et al. [10] and Bale et al. [11]. The literatures also showed that about 9600 droplets are released during a one-minute speech. Therefore, this study generated 16 droplets every 0.1 s from the bronchial entrance and simulated for 60 s. Figure 6 shows the initial generating position of droplets, with eight droplets generated on each side.

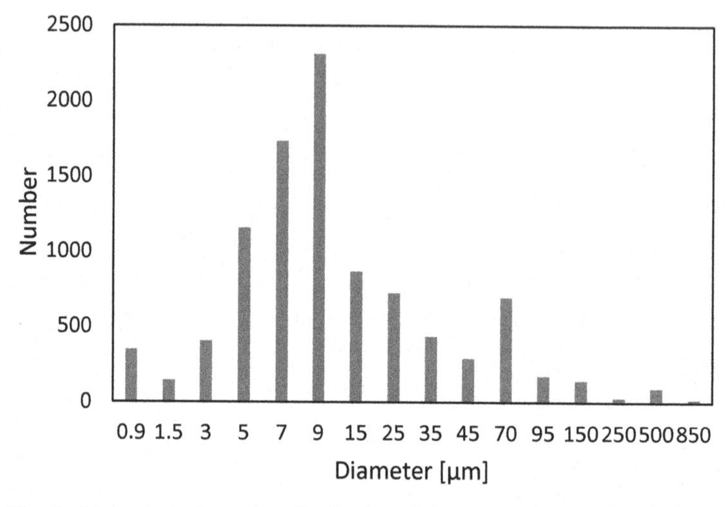

Fig. 5. Diameter and number distribution of droplets emitted during talking.

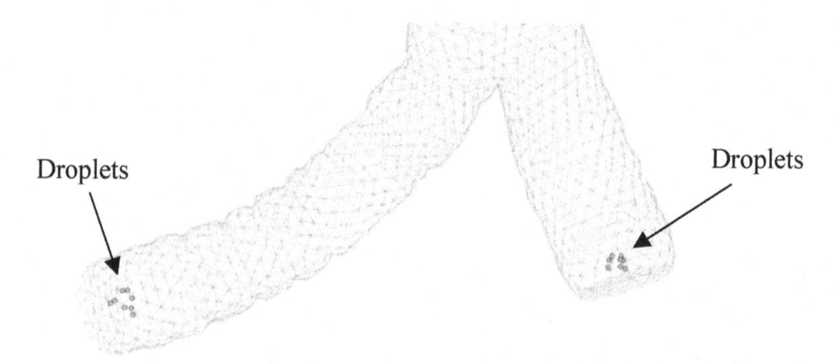

Fig. 6. Initial position for generating droplets.

The number and diameter of droplets adhered to the respiratory lining were examined as well. As shown in Fig. 7, the respiratory tract was divided into four regions: noses, mouth, trachea, and bronchi.The number and diameter of droplets adhered within each region, as well as to the bronchial inlet surface, were examined.

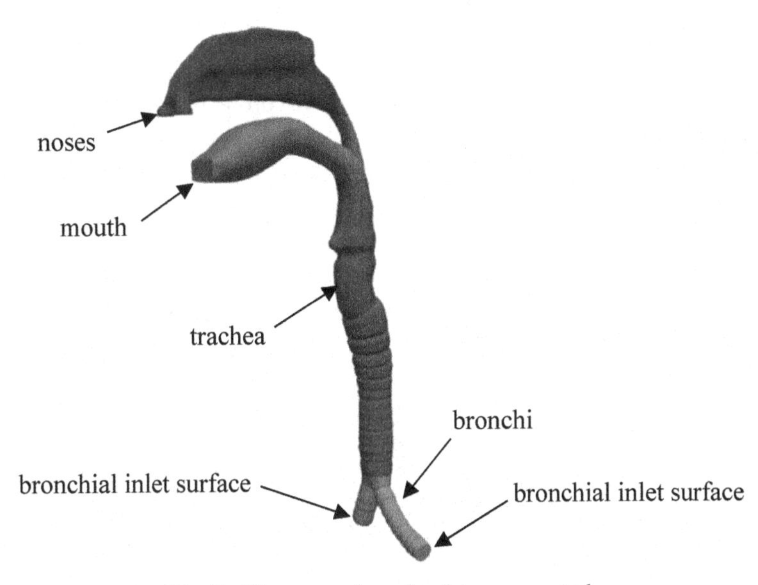

Fig. 7. The areas where droplets are counted.

3.4 Results of Droplet Analysis and Discussion

Table 2 shows the number of droplets which the breathing human aspirated, with the percentage of droplets adhered to the bronchial inlet surface in the total aspirated droplets. No droplets inhaled by the breathing person was observed in model type (C) and (D). The highest number of aspirated droplets was obtained at $x = 1.25$ in model (A). It is notable that the number of aspirated droplets at $x = 1.00$ in model (B) was higher than that in model (A). In addition, more than 80% of the droplets adhered to the bronchial inlet surface. In summary, the effective way to reduce the number of droplets inhaled by the breathing person is wearing a mask by the talking person. On the other hand, wearing a mask by only the breathing person could not completely block droplets emitted by the talking person.

Table 2. The number of aspirated droplets in the airway of the breathing human.

Model type	x[m]	Number of aspirated droplets	Number of droplets adhered to the bronchial inlet surface	Proportion [%]
(A)	0.75	70	63	90.0
	1.00	61	53	86.9
	1.25	89	72	80.9
(B)	0.75	64	59	92.2
	1.00	91	78	85.7
	1.25	33	31	93.9

Figure 8 shows the streamlines at $x = 1.00$ and $t = 1.1$ in model (B), when the volume flow rate of inhalation peaked. It can be seen that the flow from the outside to the inside of the mask slightly facilitates the suction of droplets, which leads to the entry of droplets into the airway at $x = 1.00$ in model (B). However, this phenomenon was not observed in other cases, thus the sufficient number of samples should be necessary to analyze it.

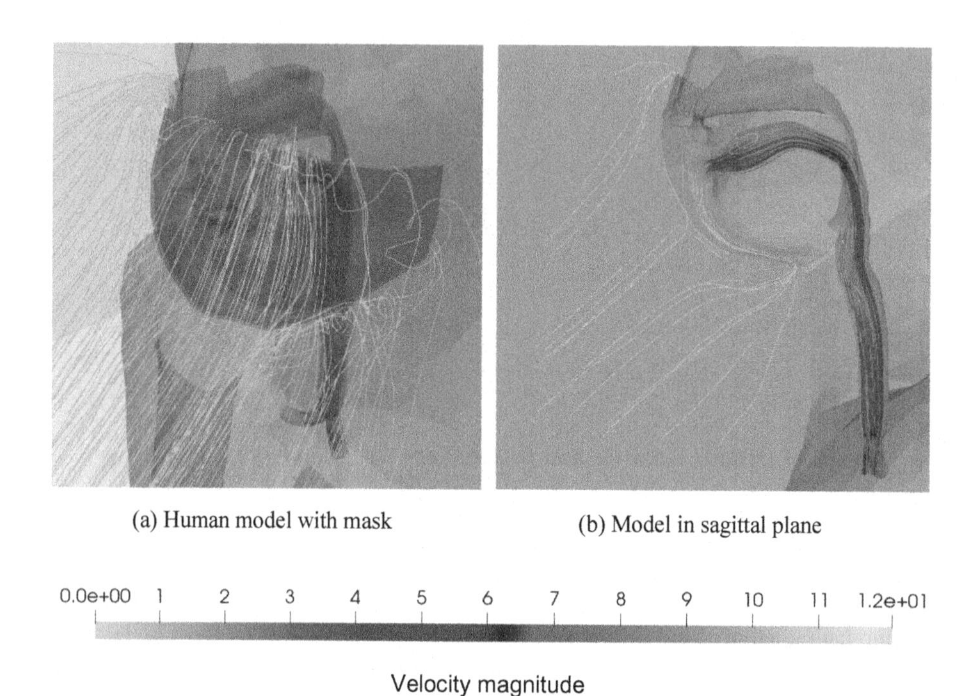

Fig. 8. Streamline at $x = 1.00$, $t = 1.1$ in model (B).

Figure 9 shows the distribution of droplets in model (A). The red particles represent floating droplets while the green particles represent adhered droplets. The droplets are released slightly downward due to the shape of the mouth. In addition, there is an updraft around the people due to the body heat, making the droplets prone to rise. However, many droplets adhere to the conversation partner's torso at $x = 0.75$ because the distance between the two people was too short for droplets to rise. On the other hand, many droplets are carried to the partner's mouth at $x = 1.25$.

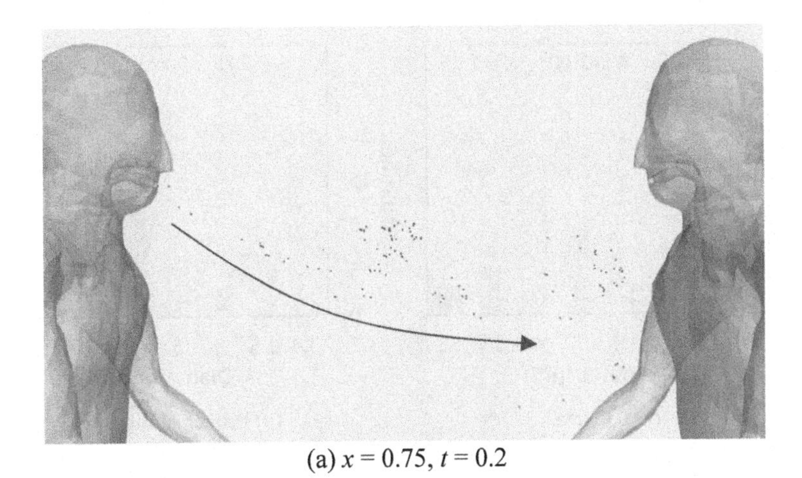

(a) $x = 0.75$, $t = 0.2$

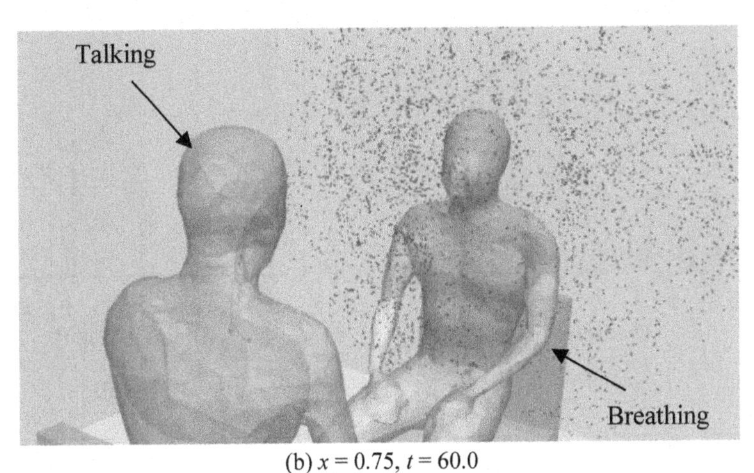

(b) $x = 0.75$, $t = 60.0$

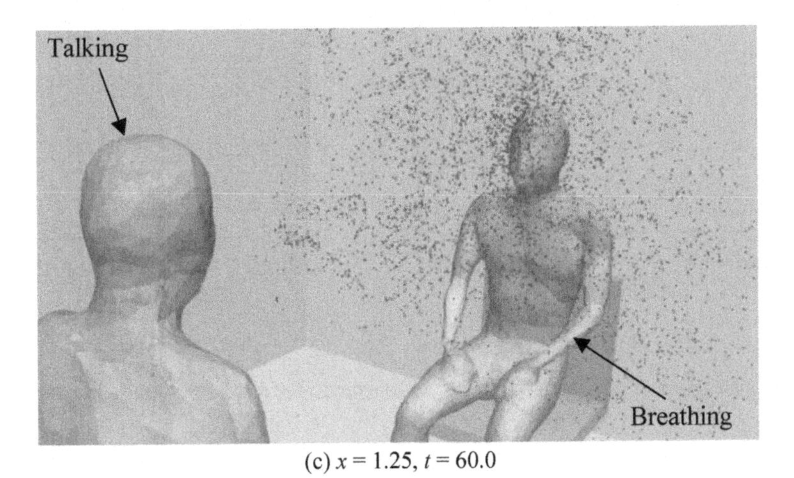

(c) $x = 1.25$, $t = 60.0$

Fig. 9. Droplets distribution in model (A).

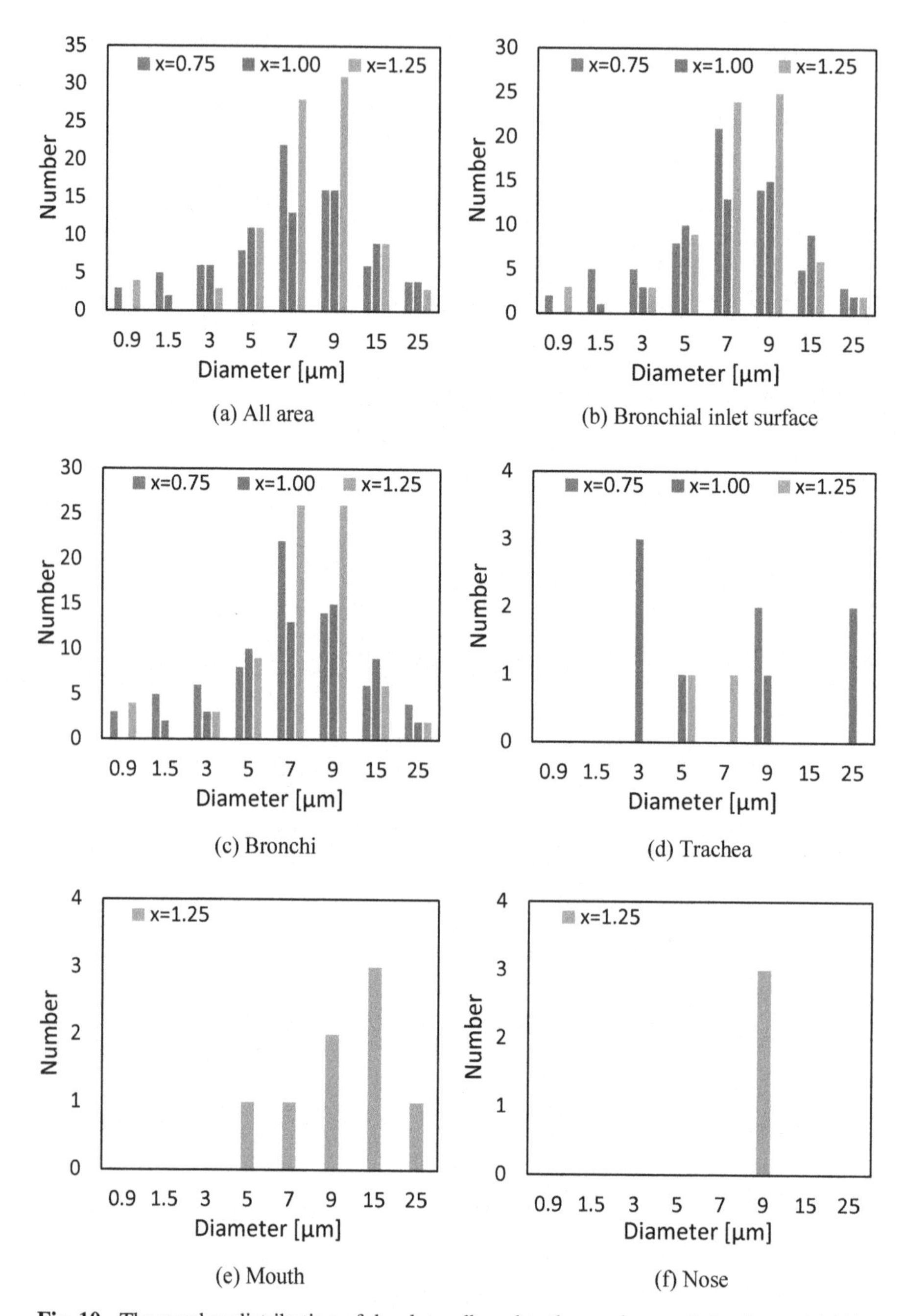

Fig. 10. The number distribution of droplets adhered to the respiratory lining in model (A).

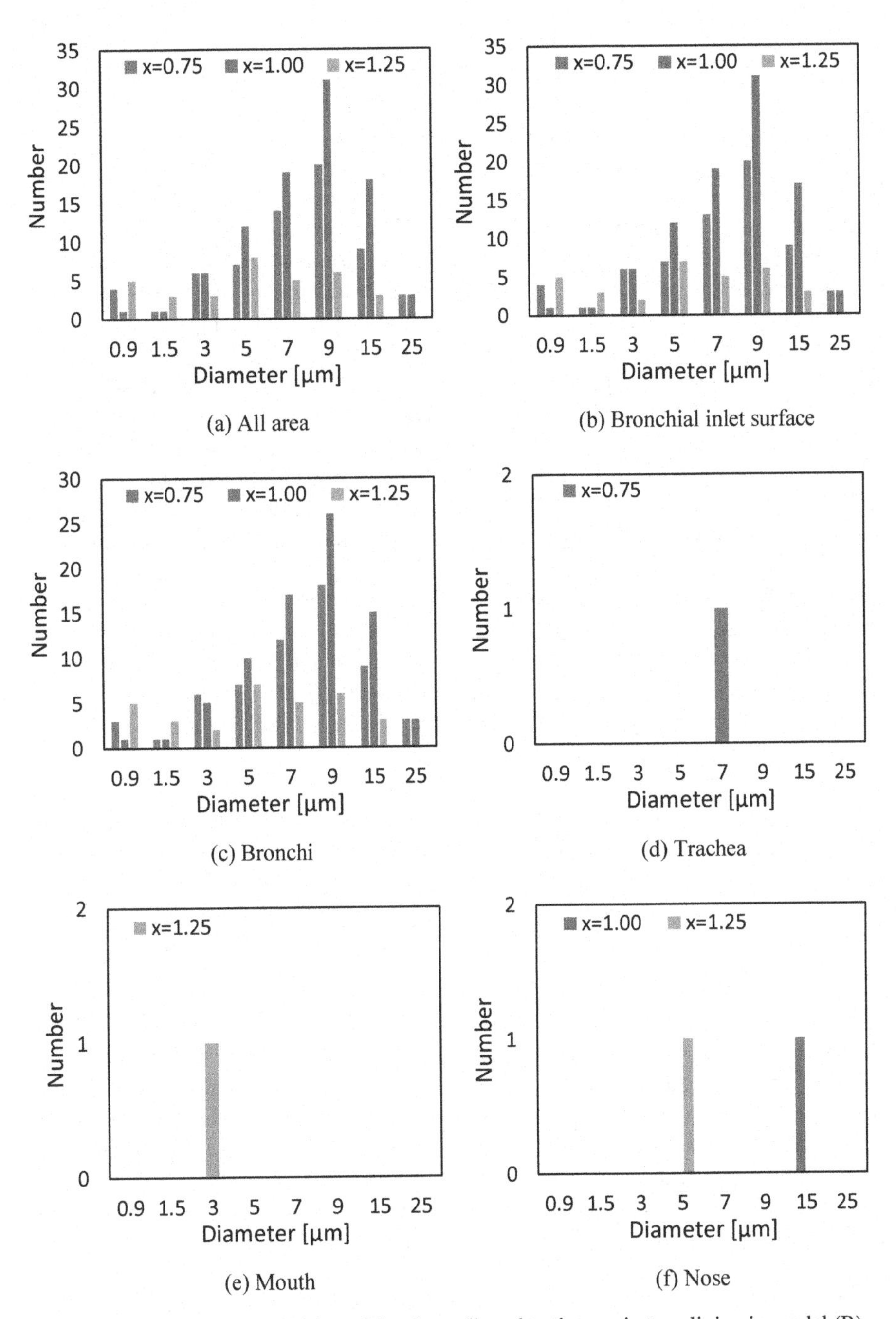

Fig. 11. The number distribution of droplets adhered to the respiratory lining in model (B).

Figures 10 and 11 show the numbers of droplets adhered to the respiratory lining in model (A) and in model (B), respectively. Figures 10(a) and 11(a) indicate that droplets larger than 35 μm adhered to nowhere. In addition, the adhered droplets increase at the diameters of 7 μm and 9 μm. The number distribution in Figs. 10(a) and 11(a) can be said to be reasonable, considering that the number of particles emitted during talking becomes high at the diameter of 7 μm and 9 μm, as illustrated in Fig. 5.

Figures 10(b) and 11(b) show the number of droplets adhered to the bronchial inlet surface. It can be seen that most of the aspirated droplets adhered to the bronchial inlet surface, compared with Figs. 10(a) and 11(a). The number of droplets adhered to the bronchial inlet surface was considerably high at $x = 1.00$ in Fig. 11(b). It can be considered that the flow from the outside into the mask accelerates the speed of droplets, and thus, they gained a high enough speed to reach the bronchial inlet surface.

Figures 10(c) and 11(c) indicate that a substantial number of droplets adhered to the bronchi compared to the other areas, which means that few droplets adhered to the areas other than the bronchi. The highest number of droplets adhered to the trachea was observed at $x = 1.00$, as shown in Fig. 10(d), while only one droplet adhered at $x = 0.75$ in Fig. 11(d). The number did not tend to increase at the diameters of 7 μm and 9 μm at trachea.

At the mouth in Fig. 10(e), the number of adhered droplets with the diameter of 15 μm peaked at $x = 1.25$. In contrast, in Fig. 11(e), only one droplet with the diameter of 3 μm adhered to the mouth at $x = 1.25$. Droplets adhered to the nose only when $x = 1.25$ in Fig. 10(f), while the adhesion was observed when $x = 1.00$ as well in Fig. 11(f). Not as many droplets adhered to the mouth and nose in model (B) as model (A). This suggests that the mask prevented droplet aspiration.

Droplets adhered only to the area deeper than the mouth at $x = 0.75$ and $x = 1.00$ in Figs. 10(c) and 10(d) while when $x = 1.25$ can observe droplet adhesion to the mouth and nose in Figs. 10(e) and 10(f). This implies that at $x = 1.25$, the droplets emitted during talking decelerate because the distance between the two people was too far, and thus, they lost power to reach the partner's area deeper than mouth.

4 Conclusions

This study simulated virus-laden droplets that are transmitted from lung to lung. Fluid flow and droplet behavior were analyzed for 12 models. In summary, the following three conclusions were obtained: First, most of the droplets aspirated by the breathing person adhered to the bronchial inlet surface. Second, droplets with a large diameter did not reach the inner wall of the airway. Third, the droplets adhered to the conversation partner's torso because of the insufficiently short distance for them to rise. This study will apply to various cases by changing computational conditions and model configurations.

Acknowledgments. This work was supported by JST CREST Grant Number JPMJCR20H7, by JKA through its promotion funds from KEIRIN RACE, and by JSPS KAKENHI Grant Number 21K03856.

References

1. Yamakawa, M., et al.: Influenza viral infection simulation in human respiratory Tract. In: The Proceedings of 29th International Symposium on Transport Phenomena, Paper ID:13, Honolulu Hawaii (2018)
2. Ogura, K., et al.: Coupled simulation of Influenza virus between inside and outside the body. In: The Proceedings of the International Conference on Computational Methods, Vol.7, pp. 71–82 (2020)
3. Srivastav, V., et al.: CFD Simulation of airflow and particle deposition in third to sixth generation human respiratory tract. In: Proceedings of 39th National Conference on Fluid Mechanics & Fluid Power, SVNIT Surat, Gujarat, India, Vol.227, (2012)
4. Watanabe, N., et al.: An 1D-3D Integrating Numerical Simulation for Engine Cooling Problem. SAE Technical Paper 2006–01–1603, (2006)
5. Doormal, V.J.P., Raithby, G.D.: Enhancements of the simple method for predicting incompressible fluid flows. Numerical Heat Transfer **7**(2), 147–163 (1984)
6. van Leer, B.: Towards the Ultimate conservation difference scheme 4: a new approach to numerical convection. J. Comp. Phys. **23**, 276–299 (1977)
7. Yamakawa, M., et al.: Computational investigation of prolonged airborne dispersion of novel coronavirus-laden droplets. J. Aerosol Sci. **155**, 105769 (2021)
8. Fenn, W.O., Rahn, H.: Handbook of Physiology, Section3: Respiration, American Physiological Society Washington DC (1965)
9. Gupta, J.K., et al.: Characterizing Exhaled Airflow from Breathing and Talking. Indoor Air 20.1 (2010)
10. Duguid, J.P., et al.: The size and the duration of air-carriage of respiratory droplets and droplet-nuclei. Epidemiol. Infect. **44**(6), 471–479 (1946)
11. Bale, R., et al.: Simulation of droplet dispersion in COVID-19 type pandemics on Fugaku. PASC'21, Article No.4, 1–11 (2021)

Smart Systems: Bringing Together Computer Vision, Sensor Networks and Machine Learning

A Personalized Federated Learning Algorithm for One-Class Support Vector Machine: An Application in Anomaly Detection

Ali Anaissi[✉], Basem Suleiman, and Widad Alyassine

School of Computer Science, The University of Sydney, Camperdown, Australia
ali.anaissi@sydney.edu.au

Abstract. Federated Learning (FL) has recently emerged as a promising method that employs a distributed learning model structure to overcome data privacy and transmission issues paused by central machine learning models. In FL, datasets collected from different devices or sensors are used to train local models (clients) each of which shares its learning with a centralized model (server). However, this distributed learning approach presents unique learning challenges as the data used at local clients can be non-IID (Independent and Identically Distributed) and statistically diverse which decrease learning accuracy in the central model. In this paper, we overcome this problem by proposing a novel personalized federated learning method based One-Class Support Vector Machine (FedP-OCSVM) to personalize the resulting support vectors at each client. Our experimental validation showed that our FedP-OCSVM precisely constructed generalized clients' models and thus achieved higher accuracy compared to other state-of-the-art methods.

1 Introduction

The emerging Federated Learning (FL) concept was initially proposed by Google for improving security and preventing data leakages in distributed environments [9]. FL allows the central machine learning model to build its learning from a broad range of data sets located at different locations. This innovative machine learning approach can train a centralized model on data generated and located on multiple clients without compromising the privacy and security of the collected data. Also, it does not require transmitting large amount of data which can be a major performance challenge especially for real-time applications. A good application of FL is in the civil infrastructures domain specifically in Structural Health Monitoring (SHM) applications where smart sensors are utilized to continuously monitor the health status of complex structures such as bridges to generate actionable insights such as damage detection.

This motivates for developing a more intelligent model that utilizes the centralized learning model but without the need to transmit the frequently-measured

D. Groen et al. (Eds.): ICCS 2022, LNCS 13353, pp. 373–379, 2022.
https://doi.org/10.1007/978-3-031-08760-8_31

data to one central model for processing unit. In this sense, we propose a federated learning approach for anomaly detection using One-class support vector machine (OCSVM) [12] which has been widely applied to anomaly detection and become more popular in recent years [2–4]. OCSVM has been successfully applied in many application domains such as civil engineer, biomedical, and networking [1,8], and produced promising results. Although our approach results in reducing data transmission and improving data security, it also raises significant challenges in how to deal with non-IID (Independent and Identically Distributed) data distribution and statistical diversity. Therefore, to address the non-IID challenge in our proposed FL approach, we developed a novel method to personalize the resulting support vectors from the FL process. The rationale idea of personalizing support vectors is to leverage the central model in optimizing the clients' models not only by using FL, but also by personalizing it w.r.t its local data distribution. The contribution of the work in this study is twofold.

- A novel method of learning OCVSM model in FL settings.
- A novel method to personalize the resulting support vectors to addresses the problem of non-IID distribution of data in FL.

2 Related Work

Federated Learning (FL) has gained a lot of interest in recent years and as a result, it has attracted AI researchers as a new and promising machine learning approaches. This FL approach attracts several well-suited practical problems and application areas due to its intrinsic settings where data needs to be decentralized and privacy to be preserved. For instance, McMahan *et al.* [11] proposed the first FL-based algorithm named *FedAvg*. It uses the local Stochastic Gradient Descent (SGD) updates to build a global model by taking average model coefficients from a subset of clients with non-IID data. This algorithm is controlled by three key parameters: C, the proportion of clients that are selected to perform computation on each round; E, the number of training passes each client makes over its local dataset on each round; and B, the local mini-batch size used for the client updates. Selected clients perform SGD locally for E epochs with mini-batch size B. Any clients which, at the start of the update round, have not completed E epochs (stragglers), will simply not be considered during aggregation. Subsequently, Li *et al.* [10] introduced the *FedProx* algorithm, which is similar to FedAvg. However, FedProx makes two simple yet critical modifications that demonstrated performance improvements. FedProx would still consider stragglers (clients which have not completed E epochs at aggregation time) and it adds a *proximal term* to the objective function to address the issue of statistical heterogeneity. Similarly, Manoj *et al.* [6] addressed the effects of statistical heterogeneity problem using a *personalization-based approach (FedPer)*. In their approach, a model is viewed as base besides personalization layers. The base layers will be aggregated as in the standard FL approach with any aggregation function, whereas the personalized layers will not be aggregated.

3 Personalized Federated Learning for OCSVM: FedP-OCSVM

3.1 OCSVM-FedAvg

In FL setting, learning is modeled as a set of C clients and a central server S, where each client learns based on its local data, and is connected to S for solving the following problem:

$$\min_{w \in \mathbb{R}^d} f(w) := \frac{1}{C} \sum_{c=1}^{C} f_c(w_c) \tag{1}$$

where f_c is the loss function corresponding to a client c that is defined as follows:

$$f_c(w_c) := \mathbb{E}[\mathcal{L}_c(w_c; x_{c,i})] \tag{2}$$

where $\mathcal{L}_c(w_c; x_{c,i}$ measures the error of the model w_c (e.g. OCSVM) given the input x_i. The Sequential Minimal Optimization (SMO) is often used in the support vector machine. However, in the case of the nonlinear kernel model as in OCSVM, SMO does not suit the FL settings well. Therefore, we propose a new method for solving the OCSVM problem in FL setting using the SGD algorithm.

The SGD method solves the above problem defined in Eq. 2 by repeatedly updating w to minimize $\mathcal{L}(w; x_i)$. It starts with some initial value of $w^{(t)}$ and then repeatedly performs the update as follows:

$$w^{(t+1)} := w^{(t)} + \eta \frac{\partial \mathcal{L}}{\partial w}(x_i^{(t)}, w^{(t)}) \tag{3}$$

In fact, the SGD algorithm in OCSVM focuses on optimizing the Lagrange multiplier $\alpha = [\alpha_1, \alpha_2, ..., \alpha_n]$ for all patterns x_i where $x_i : i \in [n], \alpha_i > 0$ are called support vectors. Thus, exchanging gradient updates in FL for averaging purposes is not applicable. Consequently, we modified the training process of SGD to share the coefficients of the features in the kernel space under the constraints of sharing an equal number of samples across each client C. In this sense, our SGD training process computes the kernel matrix $K = \phi(x_i, x_j)_{i,j=1,...,n}$ before looping through the samples. Then it computes the coefficients w after performing a number of epochs as follows:

$$w^{(t+1)} = \alpha K; \tag{4}$$

$$s.t \qquad \alpha = \alpha + \eta(1 - \sum_{i=1}^{n} w)$$

Each client performs a number of E epochs at each round to compute the gradient of the loss over its local data and to send the model parameters w^{t+1} to the central server S along with their local loss. The server then aggregates the gradients of the clients and applies the global model parameters update by computing the average value of all the selected clients model's parameters as follows:

$$w^{(t+1)} := \frac{1}{C} \sum_{i=1}^{C} w_C^{(t+1)}; \tag{5}$$

where C is the number of selected clients.

The server then share the $w^{(t+1)}$ to all selected clients in which each one performs another iteration to update $w^{(t+1)}$ but with setting $w_i^{(t)} = w^{(t+1)}$ as defined in the traditional FedAvg method.

3.2 Personalized Support Vectors

Our proposed approach may work well when clients have similar IID data. However, it is unrealistic to assume that since data may come from different environments or contexts in FL settings, thus it can have non-IID. Therefore, it is essential to decouple our model optimization from the global model learning in a bi-level problem depicted for personalized FL so the global model optimization is embedded within the local (personalized) models. Geometrically, the global model can be considered as a "*central point*", where all clients agree to meet, and the personalized models are the points in different directions that clients follow according to their heterogeneous data distributions. In this context, once the learning process by the central model is converged and the support vectors are identified for each client, we perform a personalized step to optimize the support vectors on each client. Intuitively, to generate a personalized client model, its support vectors must reside on the boundaries of the local training data (i.e. edged support vector). Thus, we propose a new algorithm to inspect the spatial locations of the selected support vector samples in the context of the FL settings explained above. It is intuitive that an edge support vector x_e will have all or most of its neighbors located at one side of a hyper-plane passing through x_e. Therefore, our edge pattern selection method constructs a tangent plane for each selected support vector $x_i : i \in [n], \alpha_i > 0$ with its k-nearest neighbors data points. The method initially selects the k-nearest data points to each support vector x_s, and then centralizes it around x_s by computing the norm vector v_i^n of the tangent plane at x_s . If all or most of the vectors are located at one side of the tangent plane), we consider x_s as an edge support vector denoted by x_e, otherwise, it is considered as an interior support vector and it is excluded from the selected original set of support vectors.

4 Experimental Results and Discussions

We validate our FedP-OCSVM method based on a real dataset collected from a Cable-Stayed Bridge in Australia[1] to detect potential damage. In all exper-

[1] The two bridges are operational and the companies which monitor them requested to keep the bridge name and the collected data about its health confidential.

iments, we used the default value of the Gaussian kernel parameter σ and $\nu = 0.05$.

We instrumented the Cable-Stayed Bridge with 24 uni-axial accelerometers and 28 strain gauges. We used accelerations data collected from sensors Ai with $i \in \{1, 2, \ldots, 24\}$. Figure 1 shows the locations of these 24 sensors on the bridge deck. Each set of sensors on the bridge along with one line (e.g. A1: A4) is connected to one client node and fused in a tensor node $\mathcal{T}$ to represent one client in our FL network, which results in six tensor nodes $\mathcal{T}$ (clients).

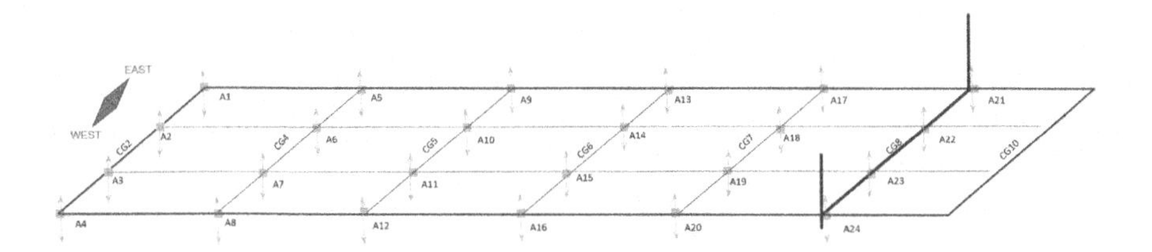

Fig. 1. The locations on the bridge's deck of the 24 Ai accelerometers used in this study. The cross girder j of the bridge is displayed as CGj [5].

This experiment generates 262 samples (a.k.a events) each of which consists of acceleration data for 2 s at a sampling rate 600 Hz. We separated the 262 data instances into two groups, 125 samples related to the healthy state and 137 samples for the damage state.

For each reading of the uni-axial accelerometer, we normalized its magnitude to have a zero mean and one standard deviation. The fast Fourier transform (FFT) is then used to represent the generated data in the frequency domain. Each event now has a feature vector of 600 attributes representing its frequencies. The resultant data at each sensor node $\mathcal{T}$ has a structure of 4 sensors $\times$ 600 features $\times$ 262 events.

We randomly selected 80% of the healthy events (100 samples) from each tensor node $\mathcal{T}$ for training multi-way of $\mathcal{X} \in \mathbb{R}^{4 \times 600 \times 100}$ (i.e. *training* set). The 137 examples related to the two damage cases were added to the remaining 20% of the healthy data to form a *testing* set, which was later used for the model evaluation.

We initially study the effect of the number of local training epochs E on the performance of the four experimented federated learning methods as suggested in previous works [7,11]. The candidate local epochs we consider are $E \in \{5, 10, 20, 30, 40, 50\}$. For each of the candidate E, we run all the methods for 40 rounds and report the final *F1-score* accuracy generated by each method. The result is shown in Fig. 2(b). We observe that conducting longer epochs on the clients improves the performance of FedP-OCSVM and FedPer, but it slightly deteriorates the performance of FedProx and FedAvg. The second experiment was to compare our method to FedAvg, FedPer and FedProx in terms of accuracy and the number of communication rounds needed for the global model to achieve

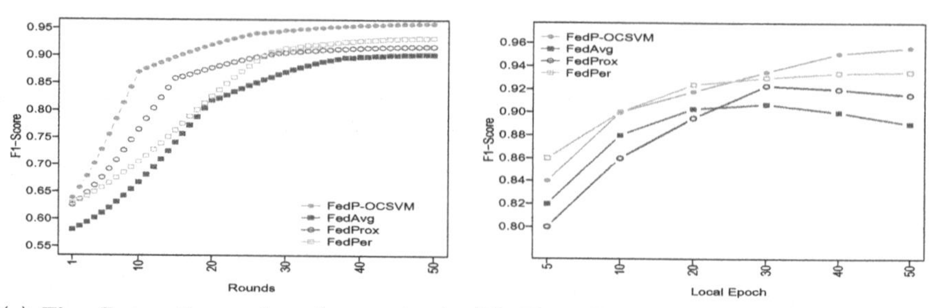

(a) The effect on the number of commuication rounds.

(b) The effect of number of local training epochs.

Fig. 2. Convergence rates of various methods in federated learning applied on Cable-Stayed Bridge with $T = 6$ clients.

Table 1. *F1-score* of various methods.

	FedP-OCSVM	FedProx	FedPer	FedAvg
Cable-Stayed Bridge	0.96 ± 0.02	0.92 ± 0.01	0.93 ± 0.03	0.90 ± 0.04

good performance on the test data. We set the total number of epochs E for FedP-OCSVM and FedPer to 50, and 30 for FedProx and FedAvg as determined by the first experimental study related to the local training epochs E. The results showed that FedP-OCSVM outperforms FedAvg, FedProx and FedPer in terms of local training models and performance accuracy. Table 1 shows the accuracy results of all experiments using *F1-score*. Although no data from the damaged state has been employed to construct the central model, each personalized local client model was able to identify the damage events related to "Car-Damage" and "Bus-Damage" with an average *F1-score* accuracy of 0.96 ± 0.02.

5 Conclusions

In this paper, we present a novel machine learning approach for an effective and efficient anomaly detection model in such applications like SHM systems that require information derived from many spatially-distributed locations throughout large infrastructure covering various points in the monitored structure. Our method employs a Federated Learning (FL) approach to OCSVM as an anomaly detection model augmented with a method to personalize the resulting support vectors from the FL process. Our experimental evaluation on a real bridge structure dataset showed promising damage detection accuracy by considering different damage scenarios. In the "Cable-Stayed Bridge" dataset, our FedP-OCSVM method achieved an accuracy of 96%. The experimental results of this case study demonstrated the capability of our FL-based damage detection approach with the personalization algorithm to improve the damage detection accuracy.

References

1. Anaissi, A., Goyal, M., Catchpoole, D.R., Braytee, A., Kennedy, P.J.: Ensemble feature learning of genomic data using support vector machine. PloS One **11**(6), e0157330 (2016)
2. Anaissi, A., et al.: Adaptive one-class support vector machine for damage detection in structural health monitoring. In: Kim, J., Shim, K., Cao, L., Lee, J.-G., Lin, X., Moon, Y.-S. (eds.) PAKDD 2017. LNCS (LNAI), vol. 10234, pp. 42–57. Springer, Cham (2017). https://doi.org/10.1007/978-3-319-57454-7_4
3. Anaissi, A., Khoa, N.L.D., Rakotoarivelo, T., Alamdari, M.M., Wang, Y.: Self-advised incremental one-class support vector machines: an application in structural health monitoring. In: International Conference on Neural Information Processing. pp. 484–496. Springer, Cham (2017). https://doi.org/10.1007/978-3-319-70087-8_51
4. Anaissi, A., Khoa, N.L.D., Rakotoarivelo, T., Alamdari, M.M., Wang, Y.: Adaptive online one-class support vector machines with applications in structural health monitoring. ACM Trans. Intell. Syst. Technol. **9**(6), 1–20 (2018)
5. Anaissi, A., Makki Alamdari, M., Rakotoarivelo, T., Khoa, N.: A tensor-based structural damage identification and severity assessment. Sensors **18**(1), 111 (2018)
6. Arivazhagan, M.G., Aggarwal, V., Singh, A.K., Choudhary, S.: Federated learning with personalization layers. arXiv preprint arXiv:1912.00818 (2019)
7. Chen, F., Luo, M., Dong, Z., Li, Z., He, X.: Federated meta-learning with fast convergence and efficient communication. arXiv preprint arXiv:1802.07876 (2018)
8. Khoa, N.L.D., Anaissi, A., Wang, Y.: Smart infrastructure maintenance using incremental tensor analysis. In: Proceedings of the 2017 ACM on Conference on Information and Knowledge Management, pp. 959–967. ACM (2017)
9. Konečný, J., McMahan, H.B., Yu, F.X., Richtárik, P., Suresh, A.T., Bacon, D.: Federated learning: Strategies for improving communication efficiency. arXiv preprint arXiv:1610.05492 (2016)
10. Li, T., Sahu, A.K., Zaheer, M., Sanjabi, M., Talwalkar, A., Smith, V.: Federated optimization in heterogeneous networks. arXiv preprint arXiv:1812.06127 (2018)
11. McMahan, B., Moore, E., Ramage, D., Hampson, S., Arcas, B.A.: Communication-efficient learning of deep networks from decentralized data. In: Artificial Intelligence and Statistics, pp. 1273–1282. PMLR (2017)
12. Schölkopf, B., Platt, J.C., Shawe-Taylor, J., Smola, A.J., Williamson, R.C.: Estimating the support of a high-dimensional distribution. Neural Comput. **13**(7), 1443–1471 (2001)

GBLNet: Detecting Intrusion Traffic with Multi-granularity BiLSTM

Wenhao Li[1,2] and Xiao-Yu Zhang[1,2](✉)

[1] Institute of Information Engineering, Chinese Academy of Sciences, Beijing 100093, China
{liwenhao,zhangxiaoyu}@iie.ac.cn

[2] School of Cyber Security, University of Chinese Academy of Sciences, Beijing 100093, China

Abstract. Detecting and intercepting malicious requests are some of the most widely used ways against attacks in the network security, especially in the severe COVID-19 environment. Most existing detecting approaches, including matching blacklist characters and machine learning algorithms have all shown to be vulnerable to sophisticated attacks. To address the above issues, a more general and rigorous detection method is required. In this paper, we formulate the problem of detecting malicious requests as a temporal sequence classification problem, and propose a novel deep learning model namely GBLNet, girdling bidirectional LSTM with multi-granularity CNNs. By connecting the shadow and deep feature maps of the convolutional layers, the malicious feature extracting ability is improved on more detailed functionality. Experimental results on HTTP dataset CSIC 2010 demonstrate that GBLNet can efficiently detect intrusion traffic with superior accuracy and evaluating speed, compared with the state-of-the-arts.

Keywords: Intrusion detection · Network security · Model optimization

1 Introduction

It has been a seesaw battle between intrusion traffic utilization and detection for decades. With the rapid development of network technology, many commercial applications are now transiting to a lightweight browser/server model (B/S). With such model, information is transported from a directory service via Hyper Text Transport Protocol (HTTP) or HTTP via HTTP over TLS (HTTPS), where TLS is an encryption protocol on transport layer. Although encryption technology has been popularized in recent years, it is argued that HTTP, with plaintext transmission, still dominates the intrusion traffic [9]. Most attackers who launch attacks on web applications pass the HTTP request method. As announced in [2], 80% of the Open Web Application Security Project (OWASP) top 10 network attacks are based on the HTTP, which lead to the vulnerability of servers and the leakage of user privacy data. It is more efficient to deploy a

D. Groen et al. (Eds.): ICCS 2022, LNCS 13353, pp. 380–386, 2022.
https://doi.org/10.1007/978-3-031-08760-8_32

HTTP based intrusion detection system than to repair a large number of web application vulnerabilities.

Motivated by the sensitive advantages of Bidirectional Long Short-Term Memory (BiLSTM) in temporal text processing and Convolutional Neural Networks in feature extracting, we take the temporal sequence classification problem into account to propose a novel deep learning model by girdling BiLSTM with multi-granularity convolutional features to detect malicious requests. It is worth mentioning that the model has greatly improved the convergence speed and convergence speed, which promotes the deployment of real-time updating dynamic intrusion detection systems. The main contributions of our work are as follows:

- A novel intrusion detector, namly GBLNet, is proposed to detect malicious HTTP request. GBLNet considers multi-granularity features extracted by Convolutional Nets to enhance the power of bidirectional LSTM.
- Encouraged by the enhanced CNNs-girdled BiLSTM, GBLNet shows superior efficiency when detecting malicious network traffic, compared with the existing detectors.

2 Detection with Convolution-Girdled BiLSTM

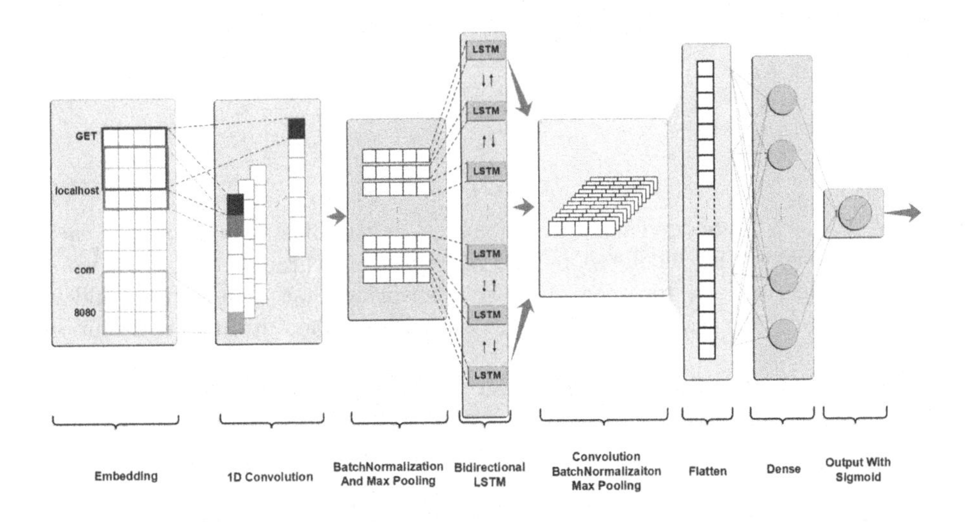

Fig. 1. Structure of GBLNet.

2.1 Detail of Model

We apply the Embedding layer as the first layer of our model. The Embedding layer can be divided into two parts. In Eq. 1, the first part projects each word in the sentence to a real-valued vector, and construct a model as follow:

$$f(w_t, ..., w_{t-n+1}) = \hat{p}(w_t|w_1^{t-1}) \tag{1}$$

where $f(w_t, ..., w_{t-n+1})$ is the trained model that represents the probability $\hat{p}(w_t|w_1^{t-1})$. The second part uses the word vector to construct a probability function instead of the previous function. The raw data input of the model is a vector processed by each word vector, as shown in Eq. 2:

$$f(w_{t-1}, ..., w_{t-n+1}) = g(C(w_{t-1}), ..., C(w_{t-n+1})) \tag{2}$$

where function C maps sequence of feature vectors to a conditional probability distribution function g. Each word vector X_w computed by the Embedding layer can be expressed as:

$$X_w = W_e{}^{d\times|V|}v^n \tag{3}$$

$$X_{1:L} = [x_1, x_2, x_3, ...x_L] \tag{4}$$

where v is the original input word and W_e is the trained embedding vectors. Containing all Xw, $X_{1:L}$ is the output of the Embedding layer.

One-dimensional convolutional layers are connected behind the Embedding layer. The input to the BiLSTM-prefixed CNN layer is an array of word vectors after Embedding. In the convolutional layer, the filter we used is $v \in \mathbb{R}^{3\times 100}$. The filter performs convolution on three word vectors of length 100. We apply 128 filters in convolutional layer with kernel size of 3:

$$f_j^{\iota} = h(\sum_{i\in M_j} X_{i:i+2}^{\iota-1} v_{i:i+2}^{\iota} + b_j^{\iota}) \tag{5}$$

$$F = [f_1, f_2, f_3, ...f_{n-2}] \tag{6}$$

where $X_{i:i+2}$ is embedded word vector and b_j^{ι} is the bias. The output of each filter is f_i, which is calculated by the filter moving through the set of word vectors. The step size for each move is 1, ensuring that each vector window $\{X_{1:3}, X_{2:4}, ...X_{n-2:n}\}$ can be scanned. F refers to the output of convolution layer.

We perform the BatchNormalization (BN) layer after the 1D convolution layer. BN layer fixes the size structure of F, and solves the gradient problem in the backward propagation process (gradient disappears and explosions) by normalizing activation to a uniform mean and variance, meanwhile, it maintains that different scale parameters should be more consistent in the overall update pace.

F_i is the linear transformation result of the normalize result. The values of γ and β are obtained by the BackPropagation (BP) algorithm.

The Max Pooling layer is connected behind the BN layer. The array after BN goes through a layer of neurons with ReLU activation function.

The output $\tilde{F}$ is a 349×128 two-dimensional array, which is performed by MaxPooling operation.

$$\tilde{F} = MaxPooling\{ReLU(F_1)\} \tag{7}$$

The BiLSTM layer is connected behind the CNN layer. The return sequences parameter is set to True, indicating that the output of each BiLSTM unit is valid and the output will be used as the input to the post-CNN. The BiLSTM layer has a internal structure can be expressed as:

$$c_k^t = i_k^t \circ z_k^t + f_k^t \circ c_k^{t-1}, k \in \{f, b\} \tag{8}$$

where the state of memory cell c_k^t can be affected by the previous state c_k^{t-1} and the input gate i_k^t. o_k^t is the output gate, computed by the input vector x^t and y_k^{t-1}, the output of the previous time step:

$$o_k^t = tanh(W_o^k x^t + R_o^k y_k^{t-1} + b_o^k), k \in \{f, b\} \tag{9}$$

where W_o^k and R_o^k are the weight vectors. y_k^t is the output of BiLSTM layer, of which calculated by o_k^t and the activation function (tanh):

$$y_k^t = o_k^t \circ tanh(c_l^t), k \in \{f, b\} \tag{10}$$

At the same time, in order to prevent over-fitting, dropout rate of 0.3 and recurrent dropout rate of 0.3 are added. The output of the BiLSTM layer is a 349×128 two-dimensional array.

The CNN that connected after BiLSTM is similar to the previous CNN layer structure. The number of filters in the convolutional layer is set to 128, the kernel size is 3, and the ReLU activation function is also used. We apply a BN layer before the pooling layer prevents gradient dispersion. The input of CNN is a two-dimensional array of 349×128 and the output is a two-dimensional array of 86×128.

Before accessing the output layer, we set up a Flatten layer to expand the two-dimensional array into a one-dimensional array and a hidden layer containing 64 neurons. An one-dimensional array obtained by Flatten is connected to this layer in a fully connected manner.

The output layer contains only one neuron activited by Sigmoid. Since detecting a malicious request is a binary problem, we chose Binary Crossentropy as the loss function of the model.

The output is a value between 0 and 1. The closer the output value is to 1, the greater the probability that the model will judge the input equest as a malicious attack. Conversely, the closer the value of the output is to 0, the greater the probability that the model will judge the input request as a normal request.

3 Experiments and Results

3.1 Dataset and Training

We evaluate GBLNet using the HTTP data set CSIC 2010. We randomly pick 80% (82416) of the whole dataset as the training dataset, including 57600 normal request and 24816 exception requests, and 20% (20604, 14400 normal request and 6204 exception requests) as the testing dataset. Each request contains up to 1400 words. For requests with less than 1400 words, we fill it to 1400.

In our experiment, four GTX 1080Ti graphics cards are used for training under the Ubuntu 16.04 operating system. The batch size during training is $64 \times N$ (N is 4, the number of GPU). Meanwhile, we used Keras API to build models based on TensorFlow and train the models for 5 epochs.

3.2 Results and Discussion

We evaluate with 4 commonly used metrics, including Accuracy (Acc), Precision (Pre), Recall (Re) and F1-score (F1).

Table 1. Accuracy, F1-score, Precision and Recall of different models include proposed deep learning methods and improved machine learning methods.

Model	Acc	F1	Pre	Re
RNN-IDS [7]	0.6967	0.8210	0.6967	1.0000
HAST-I [11]	0.9886	0.9919	0.9880	0.9958
HAST-II [11]	0.8177	0.8753	0.8301	0.9263
BiLSTM [6]	0.8314	0.8924	0.8083	0.9959
SAE [10]	0.8832	0.8412	0.8029	0.8834
PL-RNN [5]	0.9613	0.9607	0.9441	0.9779
BL-IDS [3]	0.9835	0.9858	0.9900	0.9817
DBN-ALF [1]	0.9657	0.9400	0.9648	0.93200
SVM [8]	0.9512	0.9371	0.9447	0.9296
LR [8]	0.9768	0.9414	0.9236	0.9598
SOM [4]	0.9281	0.7997	0.6977	0.9367
GBLNet	**0.9954**	**0.9967**	**0.9958**	**0.9977**

As shown in Table 1, first, we compare with the deep learning models and the optimized machine learning methods. The accuracy of our proposed model has achieved state of the art (99.54%), which is 29.87% higher than RNN-IDS (69.67%) and 17.77% higher than HAST-II (81.77%). It is also 0.68% higher than that of HAST-I (98.86%). Compared with the optimized machine learning methods, our model performs much better. The accuracy of our method is 6.73% higher than that of SOM, as well as slightly higher than that of SVM (0.95%) and LR (0.97%).

Secondly, we compare the performance among traditional machine learning approaches, including KNN, decision tree, naive bayes and random forest, demonstrated as Table 2. Although most traditional machine learning can achieve high accuracy, around 95%, our model is superior to them in all indicators.

Table 2. Comparison of proposed model and original machine learning methods.

Model	Acc	Pre	Re	F1
KNN	0.9317	0.9305	0.9760	0.9527
DecisionTree	0.9393	0.9579	0.9559	0.9569
NaiveBayes	0.7432	0.7787	0.8882	0.8298
RandomForest	0.9506	0.9627	0.9673	0.9650
GBLNet	**0.9954**	**0.9967**	**0.9958**	**0.9977**

Moreover, we also evaluate the models with convergence speed and training speed. Since the dynamic intrusion detection system, as an application type of firewall, needs to defense the malicious attack in real time, and the detection model should be continuously trained and updated, which emphasizes the cost on convergence speed and training speed should be smaller, the better.

Table 3. Consumed time compared with deep models

Model	Time consumption
RNN-IDS [7]	14 m 22 s
HAST-I [11]	37 m 78 s
HAST-II [11]	7 m 9 s
BiLSTM [6]	2 h 15 m
SAE [10]	2 h 47 m
PL-RNN [5]	28 m 17 s
BL-RNN [3]	31 m 74 s
DBN-ALF [1]	1 h 27 m
GBLNet	**30 m 30 s**

Table 3 presents the training time of different models mainly among the RNN-based and LSTM-Based models. GBLNet costs the least training time among LSTM-based models. It can be seen that the BiLSTM requires more than 2.25 h to train 5 rounds, while the GBLNet model uses 30 m 30 s. RNN-IDS, HAST-II reach shorter training time compared with GBLNet, however, RNN-IDS and HAST-II are far worse than our model in terms of accuracy. The results show the advantages of connecting the shadow and deep features maps of the convolutional layers, which plays an important role in speeding up the training by non-linear feature extractors.

4 Conclusion

This paper presents a novel strategy to detect malicious requests, and proposes a deep learning model named GBLNet, which girdles the bidirectional LSTM with multi-granularity convolutional features to fully consider the non-linear features of the malicious requests. Applying CNNs before BiLSTM to extract query features successfully maximizes the malicious features of the request queries, leading to much more accurate features representation than that of using BiLSTM to process the queries simply. By connecting the shadow and deep features map of the convolutional layers, GBLNet can guarantee better feature representations than other temporal models. Evaluations on real-world scenario prove that GBLNet can achieve superior detection rate and faster convergence speed, which promotes the application in the actual dynamic intrusion detection system.

Acknowledgment. This work was supported by the National Natural Science Foundation of China (Grant U2003111, 61871378).

References

1. Alrawashdeh, K., Purdy, C.: Fast activation function approach for deep learning based online anomaly intrusion detection. In: 2018 IEEE 4th International Conference on Big Data Security on Cloud (BigDataSecurity), IEEE International Conference on High Performance and Smart Computing, (HPSC) and IEEE International Conference on Intelligent Data and Security (IDS) (2018)
2. Fredj, O.B., Cheikhrouhou, O., Krichen, M., Hamam, H., Derhab, A.: An OWASP top ten driven survey on web application protection methods. In: International Conference on Risks and Security of Internet and Systems (2020)
3. Hao, S., Long, J., Yang, Y.: BL-IDS: detecting web attacks using BI-LSTM model based on deep learning. In: International Conference on Security and Privacy in New Computing Environments (2019)
4. Le, D.C., Zincir-Heywood, A.N., Heywood, M.I.: Unsupervised monitoring of network and service behaviour using self organizing maps. J. Cyber Secur. Mob. **8**(1), 15–52 (2019)
5. Liu, H., Lang, B., Liu, M., Yan, H.: CNN and RNN based payload classification methods for attack detection. Knowl.-Based Syst. **163**, 332–341 (2019)
6. Schuster, M., Paliwal, K.K.: Bidirectional recurrent neural networks. IEEE Trans. Signal Process. **45**, 2673–2681 (1997)
7. Shone, N., Ngoc, T.N., Phai, V.D., Shi, Q.: A deep learning approach to network intrusion detection. IEEE Trans. Emerg. Top. Comput. Intell. **2**, 41–50 (2018)
8. Smitha, R., Hareesha, K., Kundapur, P.P.: A machine learning approach for web intrusion detection: Mamls perspective. In: Soft Computing and Signal Processing (2019)
9. Tang, Z., Wang, Q., Li, W., Bao, H., Liu, F., Wang, W.: HSLF: HTTP header sequence based LSH fingerprints for application traffic classification. In: International Conference on Computational Science (2021)
10. Vartouni, A.M., Kashi, S.S., Teshnehlab, M.: An anomaly detection method to detect web attacks using stacked auto-encoder. In: 2018 6th Iranian Joint Congress on Fuzzy and Intelligent Systems (CFIS) (2018)
11. Wang, W., et al.: HAST-IDS: learning hierarchical spatial-temporal features using deep neural networks to improve intrusion detection. IEEE Access (2018)

AMDetector: Detecting Large-Scale and Novel Android Malware Traffic with Meta-learning

Wenhao Li[1,2], Huaifeng Bao[1,2], Xiao-Yu Zhang[1,2(✉)], and Lin Li[1,2]

[1] Institute of Information Engineering, Chinese Academy of Sciences, Beijing 100093, China
{liwenhao,baohuaifeng,zhangxiaoyu,lilin}@iie.ac.cn

[2] School of Cyber Security, University of Chinese Academy of Sciences, Beijing 100093, China

Abstract. In the severe COVID-19 environment, encrypted mobile malware is increasingly threatening personal privacy, especially those targeting on Android platform. Existing methods mainly focus on extracting features from Android Malware (DroidMal) by reversing the binary samples, which is sensitive to the deduction of the available samples. Thus, they fail to tackle the insufficiency of the novel DoridMal. Therefore, it is necessary to investigate an effective solution to classify large-scale DroidMal, as well as to detect the novel one. We consider few-shot DroidMal detection as DoridMal encrypted network traffic classification and propose an image-based method with meta-learning, namely AMDetector, to address the issues. By capturing network traffic produced by DroidMal, samples are augmented and thus cater to the learning algorithms. Firstly, DroidMal encrypted traffic is converted to session images. Then, session images are embedded into a high dimension metric space, in which traffic samples can be linearly separated by computing the distance with the corresponding prototype. Large-scale and novel DroidMal traffic is classified by applying different meta-learning strategies. Experimental results on public datasets have demonstrated the capability of our method to classify large-scale known DroidMal traffic as well as to detect the novel one. It is encouraging to see that, our model achieves superior performance on known and novel DroidMal traffic classification among the state-of-the-arts. Moreover, AMDetector is able to classify the unseen cross-platform malware.

Keywords: Malware detection · Network security · Meta-learning · COVID-19 · Privacy security

1 Introduction

With the prosperity of mobile applications, it becomes increasingly intractable to protect the security and privacy of mobile users [3]. The explosion of Android malware (DroidMal) is challenging the effectiveness and capability of detection

D. Groen et al. (Eds.): ICCS 2022, LNCS 13353, pp. 387–401, 2022.
https://doi.org/10.1007/978-3-031-08760-8_33

systems. Previous works managed to detect DroidMal by reversing engineering and focused on the extraction of the static features in binary [30]. By analyzing the binary of DroidMal, static features are considered as the representation of DroidMal, including CFG [15] (Call Flow Graph), static API calls [6], permissions, etc. Furthermore, the development of dynamic analysis offers an auxiliary way to detect DroidMal [25,26]. Dynamic features such as API call chains and system operations [16] can be included, by dynamically executing and hooking DroidMals in sandbox [19]. Overall, DroidMal detection methods based on source code analysis have achieved eye-catching performance after continuous optimization. However, it is generally not an easy task to obtain DroidMal samples, especially those from the emerging families. Moreover, advanced obfuscation techniques are applied to reinforce DroidMal and tremendously weaken the performance of DroidMal detectors [27].

Detecting the network traffic that generated by DroidMal is proved to be a more straightforward and efficient approach to tackle DroidMal [2]. DroidMal analysis problems are converted to the identification of malware network traffic, which skillfully tackles the insufficiency of samples and the diversity of obfuscation. Conventional methods with static IP/Port/Payload matching cannot satisfy the rapidly growing malicious traffic encryption, especially under the COVID-19 environment [24]. Therefore, many works equip machine learning algorithms, including Random Forest, Decision Tree, SVM, etc., to detect DroidMal network traffic [20]. Although such methods dramatically develop the detection systems, they suffer from the overwhelming demands on feature engineering and prior expert knowledge. Thus, they are not suitable to identify large-scale DroidMal traffic. The explosion of deep learning offers another solution to classify DroidMal traffic. Combining feature engineering and classification algorithm, the end-to-end models based on deep learning achieve superior performance compared with machine-learning-based ones [10]. In general, payloads in network traffic are automatically extracted by Deep Neural Networks (DNN) and fed to a Multi-layer perceptron (MLP) classifier [28]. Although such methods perform outstandingly on plaintext, they fail to detect encrypted malware traffic thus reach their limitations.

In general, three challenges remain unsolved in DroidMal traffic classification. **1)** It is difficult to attain emerging DroidMal and the corresponding network traffic, which leads to few-shot scenario in DroidMal Detection [4]. **2)** Conventional detectors mostly rely on pattern matching with plaintext, which are countered by the encryption techniques. **3)** Conventional techniques reach their limitations when tackling novel DroidMal. To address the issues, we propose a session-image-based model based on meta-learning, which can classify large-scale DroidMal encrypted traffic and detect unknown DroidMal traffic categories. First, we convert encrypted DroidMal network traffic to session images. Then, images are embedded into high dimension metric space based on meta-learning. Therefore, samples can be linearly classified by computing the distance between prototypes on high dimension separatable space.

Our contributions can be briefly summarized as follows:

1. We consider DroidMal detection as encrypted traffic classification to tackle the lack of DroidMal samples and propose AMDetector, a DroidMal-image model based on meta-learning to classify large-scale and novel DroidMal.
2. AMDetector is capable to classify large-scale DroidMal encrypted traffic while detecting unknown DroidMal categories and achieves superior performance compared with the state-of-the-arts.
3. It is encouraging to see that AMDetector attains the ability to perform cross-platform detection when classifying malware traffic generated by Windows.

The rest of the paper is organized as follows. In Sect. 2, we define the problems we focus on. In Sect. 3, AMDetector is described in detail. In Sect. 4, compare AMDetector with the state-of-the-art baselines and analyze the experimental results. Finally, we make a conclusion of our work and present the future enhancement.

2 Preliminaries

In this section, we first give the definition of DroidMal encrypted traffic classification. Then, we briefly introduce the meta-learning strategy on DroidMal encrypted traffic classification, which consists of the detection on known and novel DroidMal.

2.1 Problem Definition

We consider DroidMal detection as DroidMal encrypted traffic classification based on meta-learning. By running each DroidMal sample in sandbox, network traffic generated by the DroidMal can be captured with the help of network sniffer. Initially, DroidMal traffic dataset is separated into training set, validation set and testing set. First, we train our proposed model with training set with meta-learning strategy. Then, after each training epoch, model is evaluated on validation set to validate the effectiveness. Finally, a well-trained model is tested with testing set in order to verify the robustness and generalization. Overall, we aim to attain a well-trained model that gains the ability to classify large-scale known and novel DroidMal encrypted traffic precisely through a few testing samples.

We propose a traffic-based model with meta-learning to classify large-scale known and novel DroidMal encrypted traffic. Specifically, raw traffic of DroidMal is represented as image in the first part of our model. Then, images are embedded into high dimension metric space through convolutional mapping model. By computing the prototypes for each category, distance between samples and the corresponding prototype is obtained and contributes to the classification of known and novel DroidMal traffic.

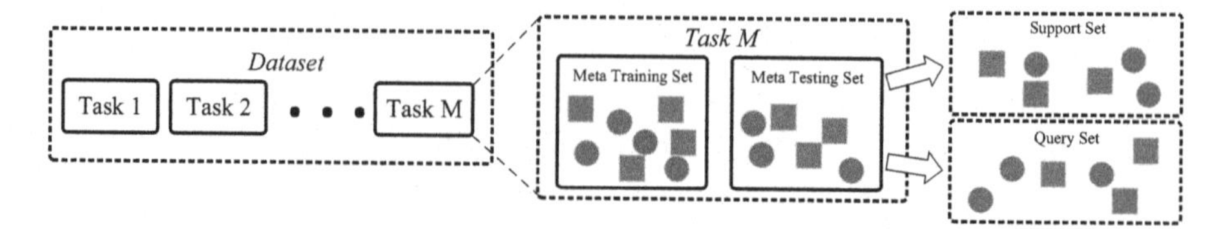

Fig. 1. Framework of meta-learning. The whole dataset is divided into M meta tasks, which contain meta training set and meta testing set. A support set and a query set are included in a meta set.

2.2 Traffic-Based Android Malware Classification with Meta-learning

In conventional machine learning (ML) and deep learning (DL) methods, models usually focus on a specific task. In the field of DoridMal encrypted traffic, a basic task is to train a classifier that can predict the labels of testing samples. Specifically, we train a classify with a training set $Set_D = \{(x_1, y_1), ..., (x_n, y_n)\}$ where x_n and y_n refer to traffic sample and the corresponding label, respectively. After optimizing the parameters, a well-trained classifier can determine the labels of the input samples. Generally, a conventional ML or DL classifier demands a great amount of available samples to converge, where $|Set_D|$ can be up to tens of thousands. Especially, when $|Set_D|$ is small, we can consider the task as a few-shot scenario.

Due to the scarce emerging DroidMal, capturing sufficient traffic samples can be a difficult task. Motivated by the previous work on few-shot learning [23], we employ meta-learning strategy to tackle the DoridMal traffic classification. In meta-learning, classifier learns to accomplish several meta tasks rather than focus on a specific one. For example, in a N-way-K-shot strategy, training set is split into meta tasks that consist of N categories with K samples. Demonstrated as Fig. 1, M meta tasks are separated. Each task consists of a meta training set and a meta testing set. In a single epoch, a meta batch is randomly picked up and considered as the meta tasks. Specifically, meta training set is used to optimize the model while the testing set is for validation. Theoretically, meta-learning classifier can be transferred to an unseen dataset with the help of its learning-to-learn strategy, which is capable to detect novel DroidMal traffic.

In the field of DroidMal encrypted traffic detection, we consider meta-tasks as multi-class classification. In general, an N-Way-K-Shot meta task consists of $N \times K$ samples. As shown in Fig. 1, a meta set is separated into a support set and a query set, which in our proposed model, the support set is for prototype computation while the query set is to optimize our model with a distance loss function.

We evaluate on a DoridMal encrypted dataset with 42 classes of DroidMal to classify large-scare known DroidMal traffic. Therefore, N classes with K samples are randomly selected from 42 classes when constructing a N-Way-K-Shot meta task in training process. During testing, the model can classify 42 classes of

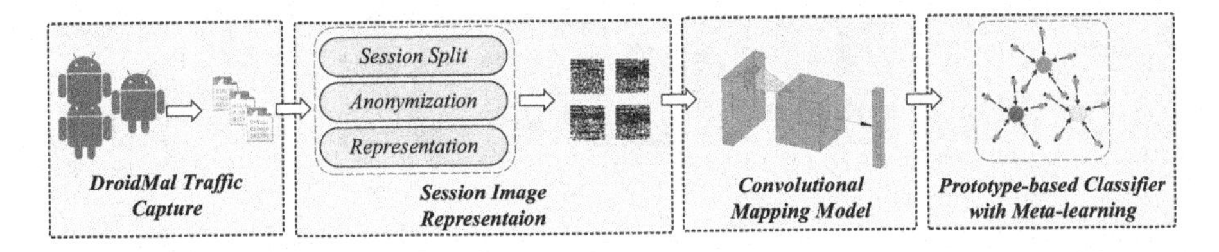

Fig. 2. Main framework of AMDetector.

DroidMal where $N = 42$ (AndMal2017 we evaluated on contains 42 classes of DroidMal).

In the task of novel DroidMal encrypted traffic detection, O of 42 classes are randomly chosen to be the novel categories, where O is the number of novel classes. During training, N classes are picked up from $42 - O$ classes to construct a meta task. Then, the model can classify known classes while detecting the novel ones where $N = 42$.

3 DroidMal Detection Based on Traffic Image with Meta-learning

The main structure of our proposed DroidMal detecting model (AMDetector) is shown in Fig. 2. Our model consists of Session Image Translation, a Convolutional Mapping Model and a Prototype-based Classifier. In the following sections, we will exhaustively introduce each part of our model.

3.1 Session Image Translation

In Session Image Translation, we aim to translate binary traffic data produced by DroidMal to session-based images to enhance the representation of DoridMal. In general, raw DroidMal traffic data are converted to 28×28 gray-scale images during this section.

Session Split: A unique traffic flow is constructed by abstracting a five-tuple, which formed as {Source IP, Destination IP, Source Port, Destination Port, Protocol}. During a time period, a session is the combination of two opposites flows that generated between two hosts. In real-world network, traffic is captured by sniffers and stored in traffic files (PCAP format), which are composed of diverse sessions. In this section, unique sessions are extracted separately from raw PCAP files and stored in discrete PCAPs, which contain a single session.

Anonymization: The irrelevant fields of network packets are masked during anonymization. Data captured from real-world network adheres to Open System Interconnection (OSI) reference model. Therefore, transmission information such as MAC address, IP address and TCP/UDP port, are included in every single

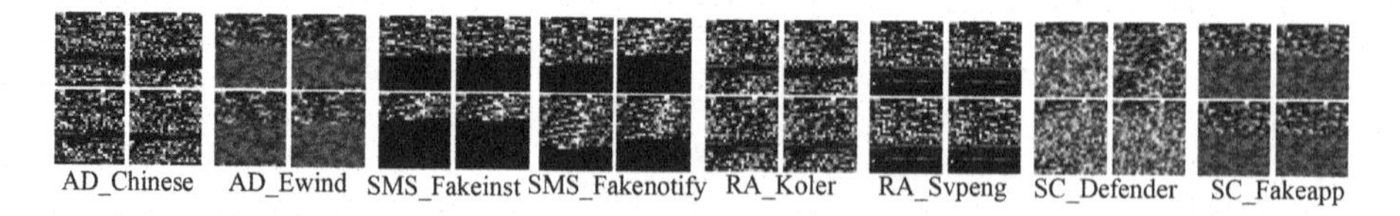

Fig. 3. Visualization of session images. Eight classes of DroidMal are illustrated, in form of {Family}_{subfamily}, e.g. `AD` stands for the family of `Advertisement`

packet. However, it is not appropriate to learn from these strong-relevant fields that can specify the labels in certain dataset but may change in different network environment, which leads to overfitting and limits the generalization of our model. Thus, such learning-irrelevant fields are masked by padding with 0x00.

Session Image Representation: In Session Image Representation, PCAPs that store a single session are converted to gray-scale image. Specifically, the first 784 bytes of a PCAP file are extracted and construct a 28×28 feature image. If a PCAP is smaller than 784 bytes, 0x00 will be padded to 784 bytes. Intuitively, binary with 0x00 and 0xFF are represented as a black pixel and a white pixel in session image, respectively.

Theoretically, the front part of a session contains the majority of connection features, thus reflects the intrinsic characteristics of a session to the most extent [5]. Meanwhile, translation of session image caters the input of CNN, as well as enhances the ability of representation. Shown as Fig. 3, we demonstrate part of the session images from different categories. We argue that different types of session images present discrimination obviously while the ones from the same class show high consistency, which proves the rationality of the session images.

3.2 Convolutional Mapping Model with Meta-learning

Convolutional mapping model embeds session images into high dimension metric space. In general, the mapping model promises to embed the original less-separated images into a high dimensional linearly separable spatial. Convolutional mapping model with meta-learning is intrinsically distinct from the conventional classifiers. Rather than classifies directly, convolutional mapping model gains the power to learn how to learn by accomplishing meta classification tasks. Theoretically, it is not essential to construct a complicated mapping model when employing meta-learning strategy, which effectively prevents overfitting when tackling few-shot problem and enhances the generalization and transferability.

With N-Way-K-Shot strategy, $N \times K$ session images with the size of 28×28 are fed into our model and embedded into 64-dimensional vectors. A concatenation of 4 convolutional sequences is concluded in our mapping model. Specifically, each sequence consists of a convolutional layer with a kernel of 3×3 windows, a batch normalization, and an element-wise rectified-linear non-linearity (ReLU) activation function. In order to improve the generalization of our model, max-pooling is applied before outputs. Finally, session images of 28×28 from meta tasks are embedded into 64-dimensional metric spatial.

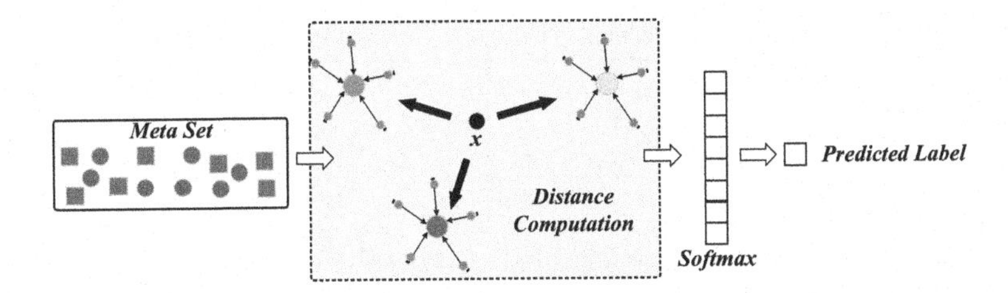

Fig. 4. Meta set is separated into support set and query set. Prototypes are attained by aggregating support set while query is used to compute the distance. Then, distance between each prototype is fed into Softmax and the predicted label is obtained.

3.3 Prototype-Based Classifier

Conventional DoridMal traffic classifiers attain the probability of each class by applying $Softmax$ directly. Then, the sample is classified into the category with the highest probability. However, conventional classification schema with $Softmax$ is not suitable for our N-Way-K-Shot meta learning, in which N is indefinite during training and testing. Inspired by Prototypical Nets [18], we propose a prototype-based classifier to adapt flexible N and K dynamically. Support set is used to compute prototypes of each class while query set is to optimize the model by computing the distance between prototypes.

Prototype Computation with Support Set. An N-Way-K-Shot meta task is separated into a support set and query set, which share the same classes of DroidMal. Prototypes of each class are obtained according to **Step 3** in **Algorithm 1**. For example, in a 5-Way-10-Shot meta task, a support set and a query set are randomly separated. Both of them contain 5 classes of DoridMal that embedded by convolutional mapping model. Then, 5 prototypes are attained by averaging the embedded vectors of each class in support set.

Model Optimization with Aggregation of Query Set. By computing the distance between vectors in query set and the corresponding prototypes, loss is obtained and is used to optimize convolutional mapping model, which endows the mapping model with the capability to aggregate the DoridMal of same class in high dimension metric spatial. Specifically, Euclidean distance is computed between vectors in query set and the corresponding prototypes. Loss is obtained by **Step 5** in **Algorithm 1**.

In summary, the training and testing process of our proposed Prototype-based Classifier is described as **Algorithm 1**.

3.4 Detecting Known and Novel Android Malware Traffic.

Employed with different strategies of meta-learning, AMDetector can classify large-scale known DoridMal encrypted traffic and the novel ones. Figure 4 illustrates the process of classification.

Algorithm 1. Training and Testing Framework of Prototype-based Classifier

Input:
The meta set D.
The mapping model $f(x)$
Unlabel sample x
Output:
Training: Mean of Loss L, Testing: Predicted Label $\hat{y}$ of x
1: Embedding samples in D, obtaining $f(D)$.
2: Randomly dividing $f(D)$ into Support Set D_S and Query Set D_Q .
3: Compute prototypes $\mathbf{P}$ of each class from Support set D_S with n classes, where $\mathbf{P} = \{P_N | P_N = \frac{1}{|D_N|} \sum_{x_i \in D_N} x_i, N = 1, 2, ..., n\}$ and $D_N \subseteq D_S$, where D_N refers to the samples with same label.
Training:
4: Computing the distance $Dst(D_Q, \mathbf{P})$ between samples from D_Q and the corresponding Prototype $P_i \in \mathbf{P}$.
5: Updating Loss by computing

$$L \leftarrow L + \frac{1}{|D_Q|}\left[Dst(f(x), \mathbf{P}) + \log \sum \exp(-Dst(f(x), \mathbf{P}))\right]$$

where $x \in D_Q$.
Testing:
6: Computing distance $Dst_x(x, \mathbf{P})$ between x and prototypes $\mathbf{P}$, where $Dst_x(x, \mathbf{P}) = \{Dst(x, P_N) | P_N \in \mathbf{P}, N = 1, 2, ..., n\}$.
7: Obtaining the predicted label $\hat{y}$ of x, where $\hat{y} = argmin(Dst_x)$.
8: **return** Training: L, Testing: $\hat{y}$

Known Android Malware Traffic Classification. In the field of known DoridMal encrypted traffic classification, label set in testing is the subset of that in training, where $Labels(Testing) \subseteq Labels(Training)$. Assuming that the number of classes in training set is $|Labels(Training)|$, the number in testing set $|Labels(Testing)| \leq |Labels(Training)|$. Figure 4 illustrates the process of classification. With N-Way-K-Shot detection strategy, N is set to $|Labels(Testing)|$ and $K = |\text{Testing Set}|$. Then, by calculating the distance between the query set and the prototype of the meta task in the testing set, the sample is classified into the category with the smallest distance between the corresponding prototype.

Novel Android Malware Traffic Detection. AMDetector is capable to classify large-scale DoridMal encrypted traffic precisely, as well as to detect the novel DoridMal traffic, even if the number of the available samples is small. Assuming that the available dataset is Set_U and the known set is Set_K, novel set $Set_N = Set_U - Set_K$, where $Set_K \cap Set_N = \varnothing$. Figure 4 demonstrates the framework of novelty detection. Model is trained with Set_K, applied with N-Way-K-Shot training strategy. During testing, N is set to $|Labels(Set_N)|$. The classification process is similar to that when classifying the known.

4 Experiment and Result

In this section, we evaluate our proposed model on two public datasets to verify the rationality and effectiveness of AMDetector. First, we briefly introduce the two datasets we use. Then, we describe the experimental setting and the state-of-the-arts, which are considered as baseline. Finally, we discuss the experimental results on known and novel DroidMal encrypted traffic classification in detail.

4.1 Experiment Preparation

Dataset Organization. We evaluated our model on two datasets, including AndMal2017 [12] and USTC2016 [22]. Specifically, AndMal2017 consists of 42 classes in 4 families of encrypted traffic generated by DoridMal. USTC2016 contains traffic produced by 10 classes of malware and 10 benign software.

AndMal2017 contains 42 classes of DroidMal from 4 major Android malware families, including Adware, Ransomware, Scareware and SMS Malware, of each contains around 10 classes of DroidMal. Encrypted traffic data is captured by running DroidMal in sandbox, which is used to represent the corresponding binary sample. After session image translation, around 2,000 images can be obtained from each class, which extends the insufficient binary samples of DroidMal.

USTC2016 is the second dataset included in our experiment, which consists of traffic data generated by 10 classes of malware and 10 classes of benign software from Windows. To verify the generalization of AMDetector, USTC-MW is regenerated by regrouping 10 classes of Windows malware traffic from the original dataset. USTC-MW contains 10 classes of malware traffic, including Cridex, Geodo, Htbot, Miuref, etc. After image translation, 1,000 samples from each class contribute to USTC-MW.

Evaluation Metrics. We compare AMDetector against the state-of-the-arts with four standard evaluation metrics that is commonly used in the task of classification, including Overall Accuracy (OA), Precision (Pr), Recall (Re) and F1-score (F1).

4.2 Experiment Design

Baseline Evaluation. AMDetector is compared against the state-of-the-art baselines for DoridMal traffic classification, including XGBoost-based method [17], RandomForest-based method [7], FlowPrint [8], conversation-level Features-based method [1], CNN-based method [21], GAN-based method [14], SiameseNet [11], FS-Net [13], TripletNet [9] and RBRN [29]. Note that part of the baselines cannot detect novel DroidMal, so they will be absent from evaluations of novelty detection (Sects. 4.4 and 4.5).

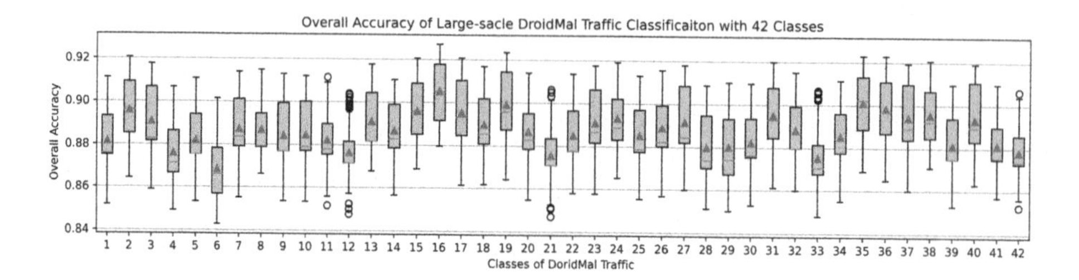

Fig. 5. Results on large-scale DoridMal traffic classification with 42 classes.

Table 1. Comparison results on large-scale DroidMal traffic classification with 42 known classes.

Methods	OA	Pr	Re	F1
XGBoost [17]	23.50	25.98	22.78	22.31
Random Forest [7]	27.43	27.22	26.39	26.46
FlowPrint [8]	20.26	21.77	19.84	19.61
Conversation Features [1]	66.71	39.71	41.09	40.38
CNN [21]	75.36	75.36	73.97	74.65
GAN [14]	66.52	67.60	66.44	67.02
FS-Net [13]	56.41	58.64	57.74	58.19
RBRN [29]	45.71	45.19	44.68	44.93
SiameseNet [11]	82.77	85.06	82.88	83.96
TripletNet [9]	67.33	67.76	69.93	68.83
AMDetector	**88.34**	**88.62**	**90.55**	**89.58**

Experimental Setting. We first evaluate on AndMal2017 to classify large-scale encrypted traffic from known DroidMal. Then, novelty detection on unknown DoridMal encrypted traffic is performed on the same dataset. Finally, cross-platform classification is evaluated on USTC-MW. Specifically, we employed Adam as the optimizer of our model and trained for 100 epochs with learning rate of 0.01. The experiments were performed using the following hardware and software platforms: Intel i7-9750 @2.6 GHz, 16 GB RAM, NVIDIA GeForce RTX2060; Windows 10, CUDA 10.1, and PyTorch 1.0.1.

4.3 Classification on Large-Scale Android Malware Encrypted Traffic

We evaluate the following well-designed experiments to validate the rationality and advancement of AMDetector. First, we evaluate on AndMal2017 to classify large-scale DroidMal encrypt traffic (42 classes detection). Specifically, 5-Way-10-Shot strategy is employed with a 42-classses training set when optimizing the model. During testing, N is set to 42 to evaluate large-scale classification.

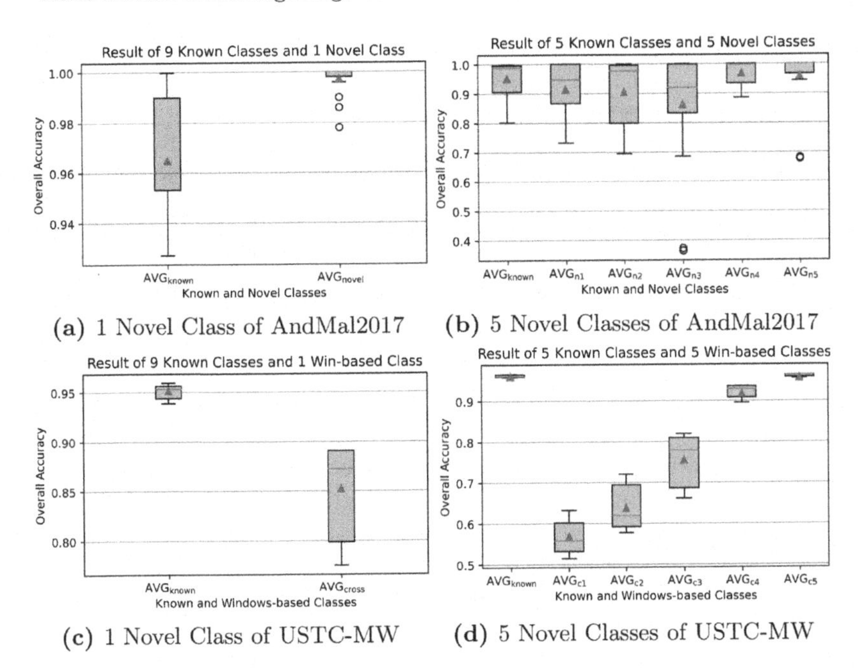

(a) 1 Novel Class of AndMal2017 (b) 5 Novel Classes of AndMal2017

(c) 1 Novel Class of USTC-MW (d) 5 Novel Classes of USTC-MW

Fig. 6. (a) and (b): Results of novelty detection on AndMal2017. (c) and (d): Results of cross-platform detection on USTC-MW.

Note that N is set to the number of classes included in testing set to predict the testing data integrally. Table 1 illustrates the classification results on 42 classes of DoridMal traffic, of which our model achieves superior performance compared against the state-of-the-arts. Also, Fig. 5 shows the results of 50 independent evaluations. It is not an easy task for conventional Softmax-based classifiers to classify large-scale of DroidMal traffic directly. However, AMDetector can maintain eye-catching performance, supported by the superiority of meta-learning strategy.

4.4 Novel Android Malware Encrypted Traffic Detection

In order to validate the ability to classify novel classes, we evaluate AMDetector on the regenerated DroidMal2017. Note that only the baselines that support novelty detection will be evaluated in this part. Two experiments are performed in this part, including 1-way and 5-way detection. In 1-way novelty detection, we randomly splintered 1 class out of the 42 classes of AndMal2017. Specifically, the number of classes in training set N_{known} is set to 41 and the number of classes in novel set N_{novel} is 1. During detection phase, 9 classes of known classes are randomly picked up to reconstruct a testing set, which consists of 9 known classes and 1 novel class. Similarly, in 5-way novelty detection, N_{konwn} and N_{novel} are set to 37 and 5, respectively. During detection, testing set that contains 5 known classes and 5

Table 2. Novel DroidMal traffic classification, where the $N_{novel} = 1, 5$ respectively.

Mode	$N_{novel} = 1, N_{known} = 9$				$N_{novel} = 5, N_{known} = 5$			
	OA	Pr	Re	F1	OA	Pr	Re	F1
CNN [21]	78.15	78.14	68.81	73.18	48.75	45.89	38.73	42.01
GAN [14]	72.07	72.48	45.32	55.77	42.09	43.19	25.87	32.36
SiameseNet [11]	92.76	93.33	92.76	93.04	89.26	90.00	89.26	89.63
Flowprint [8]	87.63	88.93	87.66	88.29	84.52	85.61	86.01	85.81
FS-Net [13]	92.11	93.41	93.69	93.55	91.87	90.99	91.42	91.20
TripletNet [9]	96.03	96.06	96.03	96.04	93.86	94.00	93.86	93.93
AMDetector	**99.77**	**99.80**	**99.83**	**99.81**	**95.06**	**95.32**	**95.95**	**95.63**

Table 3. Cross-platform malware traffic classificaiton, where the $N_{cross} = 1, 5$ respectively.

Mode	$N_{cross} = 1, N_{known} = 9$				$N_{cross} = 5, N_{known} = 5$			
	OA	Pr	Re	F1	OA	Pr	Re	F1
CNN [21]	57.32	81.09	31.93	45.82	10.92	34.73	10.72	16.38
GAN [14]	41.59	58.31	27.99	37.82	5.10	17.93	6.54	9.58
SiameseNet [11]	91.15	92.07	91.16	91.61	64.57	61.63	64.57	63.07
Flowprint [8]	82.71	83.06	83.75	83.40	71.85	72.82	72.09	72.45
FS-Net [13]	85.13	84.62	84.97	84.79	75.14	76.47	76.25	76.36
TripletNet [9]	89.89	90.47	89.87	90.18	69.08	68.25	69.08	68.66
AMDetector	**93.59**	**93.24**	**95.17**	**94.19**	**81.59**	**82.08**	**83.23**	**82.65**

novel classes is obtained. Figure 6a, 6b and Table 2 show the results after repetitive evaluation, where $N_novel = 1$ and 5, respectively. AVG_{known} refers to the average OA of the known classes while AVG_{c_n} refers to the average OA of the $n-th$ novel class. It is encouraged to see that AMDetector performs superiorly when detecting diverse novel classes. Meanwhile, AMDetector can achieve stable performance on known and novel classes classification synchronously.

4.5 Cross-Platform Malware Detection

In order to verify the generalization of AMDetector, AMDetector is trained with AndMal2017 and tested with USTC-MW. Note that only the baselines supporting novelty detection will be evaluated in this part. Following the same principle in Sect. 4.4, we randomly select 1 or 5 classes from USTC-MW and 9 or 5 classes from AndMal2017 to construct testing set, respectively. Figure 6c, 6d and Table 3 illustrate the results, where N_{novel} is set to 1 and 5, respectively. The results prove that our well-trained model can transfer to classify the unseen malware traffic, even if the traffic is generated from malware on Windows.

5 Conclusion

In this paper, we convert the few-shot Android malware detection to the problem of encrypted network traffic classification and proposed AMDetector, a session image-based model with meta-learning to tackle the above issues. First, DoridMal encrypted traffic is split with session and transferred to gray-scale images. Then, after embedding the images into a high dimension metric spatial, a prototype-based classifier is applied to separate the samples linearly. Well-designed experiments are evaluated on 2 public datasets and verify the efficiency and generality of our model when classifying large-scale DoridMal traffic, which is superior to the state-of-the-arts. Moreover, our model is capable to detect and classify the encrypted traffic that is unseen in training set, even if it is generated by the malware from different Operation Systems.

Acknowledgment. This work was supported by the National Natural Science Foundation of China (Grant U2003111, 61871378).

References

1. Abuthawabeh, M., Mahmoud, K.: Enhanced android malware detection and family classification, using conversation-level network traffic features. Int. Arab J. Inf. Technol. **17**(4A), 607–614 (2020)
2. Arora, A., Garg, S., Peddoju, S.K.: Malware detection using network traffic analysis in android based mobile devices. In: 2014 Eighth International Conference on Next Generation Mobile Apps, Services and Technologies (2014)
3. Arshad, S., Shah, M.A., Khan, A., Ahmed, M.: Android malware detection & protection: a survey. Int. J. Adv. Comput. Sci. Appl. **7**(2), 463–475 (2016)
4. Bai, Y., et al.: Unsuccessful story about few shot malware family classification and Siamese network to the rescue. In: Proceedings of ICSE (2020)
5. Celik, Z.B., Walls, R.J., McDaniel, P., Swami, A.: Malware traffic detection using tamper resistant features. In: MILCOM 2015–2015 IEEE Military Communications Conference (2015)
6. Chan, P.P.K., Song, W.-K.: Static detection of android malware by using permissions and API calls. In: Proceedings of ICML (2014)
7. Chen, R., Li, Y., Fang, W.: Android malware identification based on traffic analysis. In: Sun, X., Pan, Z., Bertino, E. (eds.) ICAIS 2019. LNCS, vol. 11632, pp. 293–303. Springer, Cham (2019). https://doi.org/10.1007/978-3-030-24274-9_26
8. van Ede, T., et al.: Flowprint: semi-supervised mobile-app fingerprinting on encrypted network traffic. In: Proceedings of NDSS (2020)
9. Hoffer, E., Ailon, N.: Deep metric learning using triplet network (2014)
10. Hou, S., Saas, A., Chen, L., Ye, Y.: Deep4MalDroid: a deep learning framework for android malware detection based on Linux Kernel system call graphs. In: 2016 IEEE/WIC/ACM International Conference on Web Intelligence Workshops (WIW) (2016)

11. Jmila, H., Khedher, M.I., Blanc, G., El Yacoubi, M.A.: Siamese network based feature learning for improved intrusion detection. In: Proceedings of ICONIP (2019)
12. Lashkari, A.H., Kadir, A.F.A., Taheri, L., Ghorbani, A.A.: Toward developing a systematic approach to generate benchmark android malware datasets and classification. In: 2018 International Carnahan Conference on Security Technology (ICCST) (2018)
13. Liu, C., He, L., Xiong, G., Cao, Z., Li, Z.: FS-Net: a flow sequence network for encrypted traffic classification. In: IEEE INFOCOM 2019-IEEE Conference on Computer Communications (2019)
14. Liu, Z., Li, S., Zhang, Y., Yun, X., Cheng, Z.: Efficient malware originated traffic classification by using generative adversarial networks. In: 2020 IEEE Symposium on Computers and Communications (ISCC) (2020)
15. Onwuzurike, L., Mariconti, E., Andriotis, P., De Cristofaro, E., Ross, G., Stringhini, G.: MaMaDroid: detecting android malware by building Markov chains of behavioral models (extended version). TOPS (2019)
16. Peiravian, N., Zhu, X.: Machine learning for android malware detection using permission and API calls. In: Proceedings of ICTAI (2013)
17. Sharan, A., Radhika, K.: Machine learning based solution for detecting malware android applications. Machine Learning (2020)
18. Snell, J., Swersky, K., Zemel, R.: Prototypical networks for few-shot learning. In: Proceedings of NeurIPS (2017)
19. Spreitzenbarth, M., Freiling, F., Echtler, F., Schreck, T.: Mobile-sandbox: having a deeper look into android applications. In: Proceedings of the 28th Annual ACM Symposium on Applied Computing (2013)
20. Tang, Z., Wang, Q., Li, W., Bao, H., Liu, F., Wang, W.: HSLF: HTTP header sequence based LSH fingerprints for application traffic classification. In: Paszynski, M., Kranzlmüller, D., Krzhizhanovskaya, V.V., Dongarra, J.J., Sloot, P.M.A. (eds.) ICCS 2021. LNCS, vol. 12742, pp. 41–54. Springer, Cham (2021). https://doi.org/10.1007/978-3-030-77961-0_5
21. Wang, W., Zhu, M., Zeng, X., Ye, X., Sheng, Y.: Malware traffic classification using convolutional neural network for representation learning. In: 2017 International Conference on Information Networking (ICOIN) (2017)
22. Wang, W., Zhu, M.: End-to-end encrypted traffic classification with one-dimensional convolution neural networks. In: 2017 IEEE International Conference on Intelligence and Security Informatics, ISI 2017, Beijing, China, 22–24 July 2017 (2017)
23. Wang, Y., Yao, Q., Kwok, J.T., Ni, L.M.: Generalizing from a few examples: a survey on few-shot learning. ACM Comput. Surv. **53**, 1–34 (2020)
24. Wang, Z., Fok, K.W., Thing, V.L.: Machine learning for encrypted malicious traffic detection: approaches, datasets and comparative study. Comput. Secur. **113**, 102542 (2022)
25. Wong, M.Y., Lie, D.: IntelliDroid: a targeted input generator for the dynamic analysis of android malware. In: NDSS (2016)
26. Yan, L.K., Yin, H.: DroidScope: seamlessly reconstructing the {OS} and Dalvik semantic views for dynamic android malware analysis. In: USENIX 2012 (2012)
27. Yang, W., Kong, D., Xie, T., Gunter, C.A.: Malware detection in adversarial settings: exploiting feature evolutions and confusions in android apps. In: Proceedings of ACSA (2017)
28. Yuan, Z., Lu, Y., Xue, Y.: DroidDetector: android malware characterization and detection using deep learning. Tsinghua Sci. Technol. **21**, 114–123 (2016)

29. Zheng, W., Gou, C., Yan, L., Mo, S.: Learning to classify: a flow-based relation network for encrypted traffic classification. In: Proceedings of WWW (2020)
30. Zhu, H.J., You, Z.-H.: DroidDet: effective and robust detection of android malware using static analysis along with rotation forest model. Neurocomputing **272**, 638–646 (2018)

Accessing the Spanish Digital Network of Museum Collections Through an Interactive Web-Based Map

Cristina Portalés[1(✉)], Pablo Casanova-Salas[1], Jorge Sebastián[2], Mar Gaitán[2], Javier Sevilla[1], Arabella León[2], Ester Alba[2], Rebeca C. Recio Martín[3], and Marta Tudela Sánchez[3]

[1] Institute of Robotics and Information and Communication Technologies, Universitat de València, 46010 València, Spain
cristina.portales@uv.es

[2] History of Art Department, Universitat de València, 46010 València, Spain

[3] Collections Area, Sub-Directorate General of State Museums, Ministry of Culture and Sport, 28004 Madrid, Spain

Abstract. Within the scope of the SeMap project, we have developed a web-based tool that aims to offer innovative dissemination of movable assets held in museums, linking them semantically and through interactive maps. SeMap is focused on depicting the objects that are catalogued in CER.ES, the Spanish Digital Network of Museum Collections, which offers a catalogue of 300,000+ objects. To properly represent such objects in the SeMap tool, which considers their semantic relations, we needed to preprocess the data embedded in catalogues, and to design a knowledge graph based on CIDOC-CRM. To that end, the collaboration among academia, heritage curators and public authorities was of high relevance. This paper describes the steps taken to represent the CER.ES objects in the SeMap tool, focusing on the interdisciplinary collaboration. We also bring the results of a usability test, that proves the developed map is usable.

Keywords: Interactive map · Cultural heritage · Interdisciplinarity

1 Introduction

Cultural objects have a history, but also a geography. For most art history researchers, geography poses great challenges [1]. They are not trained to use spatial analysis, not even on a basic level. They usually focus on extraordinary pieces and creators, the leading examples, leaving the rest to a discreet second (or third, fourth...) place. Their discourse often revolves solely around historical centers, omitting the peripheries in as much as they remain unrelated to those centers established by previous historiography. Nonetheless, the largest part of cultural objects are not those leading examples, but the huge amount of secondary or even trivial productions, including crafts and popular arts. Peripheries did have a life of their own, and many did not consider themselves to be peripheries at all.

D. Groen et al. (Eds.): ICCS 2022, LNCS 13353, pp. 402–408, 2022.
https://doi.org/10.1007/978-3-031-08760-8_34

Visualizing the geographical aspect of museum collections can, therefore, bring a new approach to art historical knowledge. The combination of Geographical Information Systems and collections' databases should provide innovative and powerful ways to visualize the spatial connections between their objects. Trade routes, production workshops, artists' travels, centers of distribution and consumption, formal or iconographic influence, patterns of modern collecting… many kinds of historical art or anthropological analysis can benefit greatly from digital systems that allow us to see heritage through geographical tools.

In this paper we present the results of SeMap [2], a research project that aims to offer innovative dissemination of movable cultural assets, linking them semantically to spatiotemporal maps. It is built on the objects that are catalogued in a network of Spanish museums set up by the Ministry of Culture and Education, which are currently accessible through a web portal [3]. SeMap brings a new way of accessing such information: on the one hand, because it provides a spatiotemporal map that expands the possibility of exploring data; on the other hand, because the data is embedded in a knowledge graph, allowing to retrieve knowledge from it, e.g., looking for similar objects. Additionally, SeMap combines different visualization and filtering strategies to offer an intuitive and innovative way to navigate through the map, combining GIS tools, graphical representations and knowledge-assisted visualization.

The paper is structured as follows. Section 2 introduces the SeMap project and the CER.ES collection, and explains how they have been integrated. Section 3 brings a usability evaluation of the tool. Finally, Sect. 4 offers some conclusions and outlines future work on this area.

2 Accessing Cultural Heritage Through an Interactive Map

2.1 The SeMap Project

SeMap is a research project that aims to offer an innovative dissemination of movable cultural assets kept in medium-size and small museums, linking them semantically and through a web-based interactive map [4], so users can access such data more intuitively. There are other works focused on visualizing cultural heritage assets on maps. To represent spatial features, most solutions use interactive icons and clustering, and also message boxes depicting text and images [5, 6]. Choropleth maps are also quite common, especially to show historical moments [7, 8]. Some of the maps show links to other pages or sidebar windows with detailed information of objects [9, 10]. SeMap use some of these strategies, such as representing the objects with interactive icons and clustering, providing a filtering menu, and allowing to inspect individual objects. Additionally, the data embedded in SeMap are semantically related. Few works can be cited in this regard in the field of mapping cultural heritage [11]. SeMap is focused on depicting the objects that are catalogued in CER.ES, the Spanish Digital Network of Museum Collections, which offers a catalogue of 300,000+ objects. Currently, CER.ES allows searching objects of its catalogue by typing a text or applying filters, and results are provided in the form of traditional lists or image thumbnails. The interactive map developed in SeMap offers an extended approach, where users can see georeferenced results of their queries and filter them further.

2.2 The CER.ES Collection

In Spain, the DOMUS [12] system was created in 1993 when the Documentary Standardization Commission was set up with the aim of establishing standardized and common management protocols for National Spanish museums, as well as developing an integrated automated system for museum documentation and management. The DOMUS system provides terminological control tools [13] that would serve for the correct identification, classification and description of the Cultural Heritage housed in Spanish museums [14], the first one being the Dictionary of Drawing and Printmaking, published by the National Chalcography in 1996. These controlled vocabularies can be classified into two groups [15]: specialized dictionaries, which bring together terminology specific to their corresponding subject area (ceramics, numismatics and furniture) and generic thesauri, which are applicable to the cataloguing of all types of movable and immovable cultural assets.

This system served as the basis for the creation of the cultural heritage thesauri and CER.ES, which includes 118 Spanish museum collections. It brings together information and images from the museums that make up the Digital Network of Spanish Museum Collections (Red Digital de Colecciones de Museos de España), its contents are also available on the HISPANIA network and EUROPEANA [16]. CER.ES can be consulted at [3] and belongs to the Ministry of Culture and Sport. It is a clear commitment to facilitate universal access to culture and to provide citizens with a legal offer of cultural content on the Internet. It is therefore a unified access to the cultural assets of Spanish museums, regardless of their administrative dependence or specialty. Through its website users can perform general and advanced searches in all the museums or make an advanced search, consult online catalogues, or specify objects by museum types, geographical location, or ownership. Currently, CER.ES offers more than 329,000 cultural assets and 580,000 images.

2.3 Integrating the CER.ES Collection in SeMap

The integration of the CER.ES collection in the interactive map developed under the SeMap project, has been possible thanks to the close collaboration between researchers at the Universitat de València and museum technicians from the Collections Area, Sub-Directorate General of State Museums (Subdirección General de Museos Estatales, SGME), from the Spanish Ministry of Culture and Sport.

Information Preprocessing. The total amount of data provided by the SGME team was 239,836 objects of various typologies and 78,074 elements of the document typology, which includes documents, photographs, films, etc.

Relationship to CER.ES dictionaries. The intended purpose of the data is that they can be consulted in a usable form, through the interface of a web application. However, this is very complex to reconcile with the number of terms in the different thesauri. For example, the Diccionario de Denominaciones de Bienes Culturales [13], used to determine the typology of an object, has 8,727 different terms. As the aim of the project is to disseminate cultural heritage to arrive to as many people as possible, a much smaller classification has been made. The thesauri used and the simplification applied are:

- Typology of the object, references to the "Bienes Materiales" CER.ES thesaurus. 8,276 terms were reduced to classification with 16 items.
- Material of the object, references to the "Materias" CER.ES thesaurus. 1,841 terms were simplified to a set of 21 elements.
- Techniques employed, references to the "Técnicas" CER.ES thesaurus. 1,355 terms were reduced to 20 items.

Georeferentiation. In SeMap, the provenance of the object and its current location, which is a museum, have been processed. Therefore, obtaining the current location has not been a problem. The problem was with its provenance, since each object has a place of provenance (country, toponym, etc.) and a name of the specific place or site from which it comes (monastery, site, etc.). The geolocation of provenance is a complex issue for several reasons. On the one hand, there were data without provenance information, on the other hand, there were data with low granularity (only country, or large administrative regions) and others with generic place names, with several possible references. To obtain the geographic location, an application was developed that introduced the information about the origin of the objects to the Geonames and Google Maps APIs.

Heterogeneity. In the information there are several data that have a high level of heterogeneity, because they have been introduced by humans with different criteria and format. The most affected fields have been the dimension and the dating of the object. With different units of measurement, separators, etc. This heterogeneity is easy to detect and to correct automatically.

Knowledge Graph Design. Since the supported information is related to cultural heritage, it was decided to use the model proposed by CIDOC-CRM as a starting point. This model is the most used to represent this type of data through a knowledge graph. As the model proposed by CIDOC-CRM is theoretical, SeMap decided to use an OWL (Web Ontology Language) implementation of this model [17]. The ontology schema used in the knowledge graph of SeMap can be found in [18].

2.4 Example of Use

In this section, we show an example of how the map can be used to access a specific object, after applying a filter and navigating through the visual results. This is exemplified in Fig. 1. Figure 1a shows a map view centred on Europe. We can see all the objects that are preserved in the database whose origin correspond to Europe and that meet the condition of "funerary". The colour of the circular markers (orange, yellow and green) indicates the number of objects that are grouped within a marker. Figure 1b depicts those objects that were already filtered. Green markers correspond to clusters with less than ten objects, yellow to more than ten and orange to more than one hundred objects. Figure 1c shows an additional zoom, representing a smaller area where there is an object. Additionally, the panel to the right shows the detailed information of such an object, after the user has clicked on its marker. Lastly, in Fig. 1d, the same object is depicted in the CER.ES website, which is automatically opened as a new window in the browser after a user clicks on "see in CER.ES" button (in Fig. 1c).

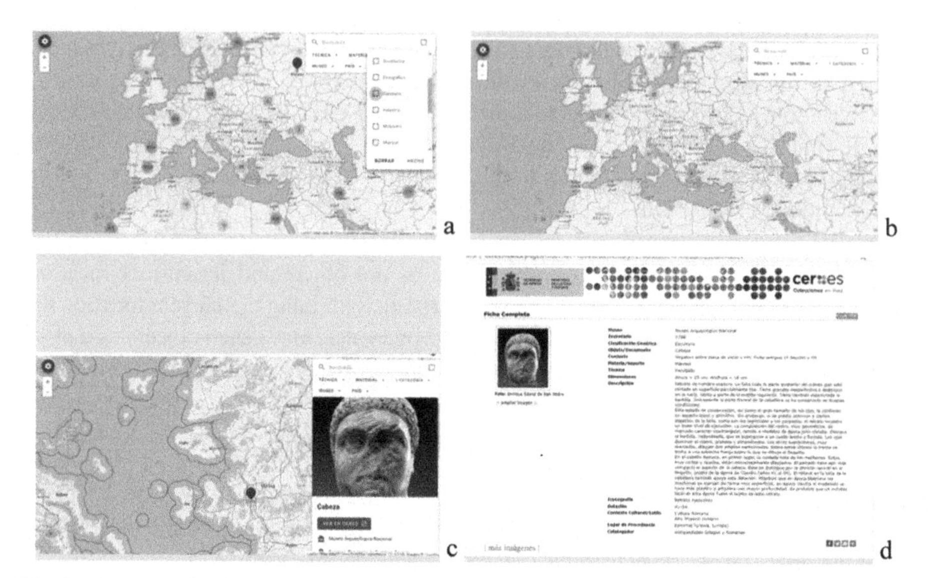

Fig. 1. An example of how to interact with the tool developed in SeMap, where: a) a view of a large area centred on Europe after filtering by category/funerary; b) depicts objects that were already filtered; c) a larger zoom representing a specific object; d) the same object of "c", as shown in CER.ES (Color figure online).

3 Usability Evaluation

In order to test the usability of our tool, we have performed an evaluation activity, based on the System Usability Scale (SUS) test [19]. People were recruited in written form, and an open call was provided at the project webpage and social networks. The evaluation was opened for three weeks. A total of 45 people participated in the evaluation.

We asked the participants to access the SeMap tool [4] and to do some specific tasks in order to test its functionalities (e.g., zooming, filtering, accessing specific objects, etc.). After that, they had to fulfill a questionnaire, which was anonymous. The questionnaire consisted of the SUS test [19], composed of ten questions (Q1 to Q10). For each question, participants must rate them with a score in the range 1–5, which means: 1: strongly disagree, 5: strongly agree. Odd questions (Q1, Q3, …, Q9) are positive, so the best possible answer would be 5 points; on the other hand, even questions (Q2, Q4, …, Q10) are formulated in a negative way, so the best possible answer would be 1 point. For this reason, the given scores by users are summarized in different graphs: in Fig. 2, the positive questions; in Fig. 3, the negative questions. In both cases, the best possible answers (5 and 1, respectively) are depicted in dark blue colour, so it is easier to analyse the results. As it can be seen, for the positive questions, the one receiving the maximum score (5 points) more times is Q3 (n = 11), that refers to the easiness of using the tool. Q7 and Q9 have similar satisfactory results, while Q5 brings the poorest result. For the five questions, at least 77.8% of the people (n = 35) rates them with a score of 3 (neutral value) or better (4 or 5 points). For the negative questions, the best rated is Q10, which refers to the easiness of learning how to use the tool. Q4 and Q8 also give very positive results, depicting that most users think that they would not need the support of a technical person to use the tool and that it is not cumbersome to use. For the five questions, at least 88.9% of the people (n = 40) rates them with a score of 3 (neutral value) or better (1 or 2 points). Comparing both graphs, it can be seen that overall, negative questions are better rated –in terms of the highest scores (in dark blue colour)– than positive questions. From the given scores (in the range 1–5), the SUS score is computed. This score ranges from 0 to 100, meaning 100 the best imaginable result. In our case, the SUS score reaches 70.7 points, which can be considered acceptable.

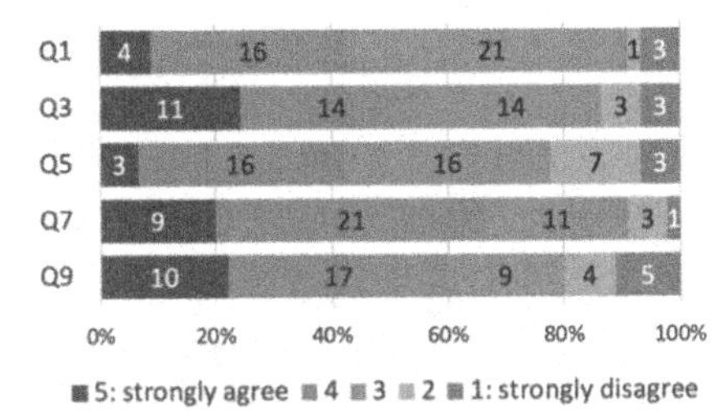

Fig. 2. Number of scores (from 1 to 5) for each positive SUS question.

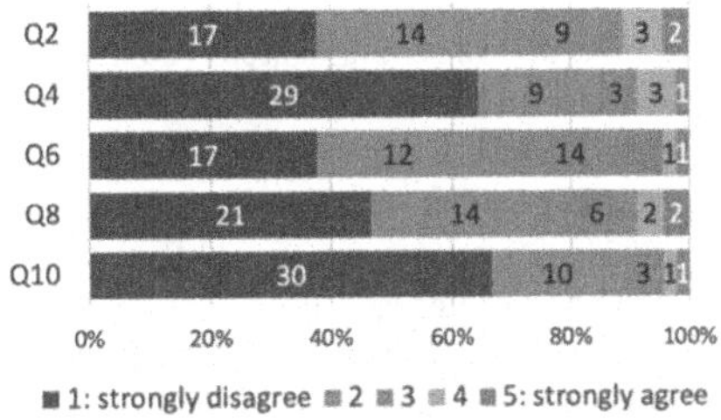

Fig. 3. Number of scores (from 1 to 5) for each negative SUS question.

4 Conclusions and Further Work

The work presented in this paper reports the collaborative work between disciplines and institutions that has led to the results of the SeMap project, an interactive web-based map that makes it possible to access the objects embedded in CER.ES (300,000 +), the Spanish digital network of museum collections. We have described the steps taken in order to visualize in a map the objects from CER.ES, namely, the information processing (relating dictionaries, georeferencing and heterogeneity) and the design of a knowledge graph. We have also shown the results of a usability test with 45 participants. The obtained satisfactory results prove the added value of stablishing such collaborations, among academia, heritage curators and public authorities.

Acknowledgements. The research leading to these results is in the frame of the project "SeMap: Acceso avanzado a los bienes culturales a través de la web Semántica y Mapas espacio-temporales", which has received funding from Fundación BBVA. Cristina Portalés is supported by the Spanish government postdoctoral grant Ramón y Cajal, under grant No. RYC2018–025009-I.

References

1. Sebastián Lozano, J.: Mapping art history in the digital era. The Art Bulletin. **103**, 6–16 (2021). https://doi.org/10.1080/00043079.2021.1882819
2. Portalés, C.: SeMap: La visualización espacio-temporal al servicio del patrimonio (2022). https://www.uv.es/semap/. Accessed 15 Feb 2022

3. Ministerio de Cultura y Deporte: Red Digital de Colecciones de Museos de España (2021). http://ceres.mcu.es/pages/SimpleSearch?index=true. Accessed 17 Nov 2021
4. Portalés, C.: SeMap Mapa (2022). https://www.uv.es/semap/mapa/. Accessed 15 Feb 2022
5. UNESCO World Heritage Centre: Interactive Map (2022). https://whc.unesco.org/en/interactive-map/. Accessed 12 Jan 2022
6. Pericles–Maritime Cultural Heritage (2022). https://mapyourheritage.eu/. Accessed 12 Jan 2022
7. Chronas: Enter History (2022). https://chronas.org. Accessed 12 Jan 2022
8. Historic Borders (2022). https://historyborders.app. Accessed 12 Jan 2022
9. English Heritage | A map of myth, legend & folklore (2022). https://mythsmap.english-heritage.org.uk. Accessed 12 Jan 2022
10. Sanborn Maps Navigator (2022). https://selenaqian.github.io/sanborn-maps-navigator/. Accessed 12 Jan 2022
11. Sevilla, J., Casanova-Salas, P., Casas-Yrurzum, S., Portalés, C.: Multi-purpose ontology-based visualization of spatio-temporal data: A case study on silk heritage. Appl. Sci. **11**, 1636 (2021). https://doi.org/10.3390/app11041636
12. El sistema integrado de documentación y gestión museográfica: DOMUS (2022). https://www.culturaydeporte.gob.es/cultura/mc/bellasartes/conoce-bellas-artes/exposicion-virtual-presentacion/exposicion-virtual-secciones/funciones-patrimonio/10domus.html. Accessed 20 Jan 2022
13. Tesauros - Diccionarios del patrimonio cultural de España – Portada (2022). http://tesauros.mecd.es/tesauros. Accessed 20 Jan 2022
14. Garrido, R.C.: Un modelo de normalización documental para los museos españoles: DOMUS y la red digital de colecciones de museos de España. Presented at the I Seminário de Investigação em Museologia dos Países de Língua Portuguesa e Es-panhola (2010)
15. Alquézar Yáñez, E.M.: Museos en internet: La experiencia de ceres. Boletín de la Anabad. **64**, 247–345 (2014)
16. Descubre el inspirador patrimonio cultural europeo (2022). https://www.europeana.eu/es. Accessed 20 Jan 2022
17. Schiemann, B., et al.: Erlangen CRM OWL (2022). http://erlangen-crm.org/. Accessed 20 Jan 2022
18. Sevilla, J.: javier-sevilla/semapOntology (2022). https://github.com/javier-sevilla/semapOntology. Accessed 20 Jan 2022
19. Brooke, J.: SUS-A quick and dirty usability scale. Usability Evaluation in Industry **189**, 194 (1996)

Federated Learning for Anomaly Detection in Industrial IoT-enabled Production Environment Supported by Autonomous Guided Vehicles

Bohdan Shubyn[1,2](✉), Dariusz Mrozek[1], Taras Maksymyuk[2], Vaidy Sunderam[3], Daniel Kostrzewa[1], Piotr Grzesik[1], and Paweł Benecki[1]

[1] Department of Applied Informatics, Silesian University of Technology, Gliwice, Poland
Bohdan.Shubyn@polsl.pl

[2] Department of Telecommunications, Lviv Polytechnic National University, Lviv, Ukraine

[3] Department of Computer Science, Emory University, Atlanta, GA 30322, USA
vss@emory.edu

Abstract. Intelligent production requires maximum downtime avoidance since downtimes lead to economic loss. Thus, Industry 4.0 (today's IoT-driven industrial revolution) is aimed at automated production with real-time decision-making and maximal uptime. To achieve this, new technologies such as Machine Learning (ML), Artificial Intelligence (AI), and Autonomous Guided Vehicles (AGVs) are integrated into production to optimize and automate many production processes. The increasing use of AGVs in production has far-reaching consequences for industrial communication systems. To make AGVs in production even more effective, we propose to use Federated Learning (FL) which provides a secure exchange of experience between intelligent manufacturing devices to improve prediction accuracy. We conducted research in which we exchanged experiences between the three virtual devices, and the results confirm the effectiveness of this approach in production environments.

Keywords: Federated learning · Predictive maintenance · Smart production · Artificial intelligence · Long-short term memory · Recurrent neural networks

1 Introduction

The main goal of Industry 4.0 is automated production with real-time decision-making. Modern manufacturing relies on a complex ecosystem that consists of many elements, various sensors, intelligent devices, people, and is a rich environment for data collection and analysis.

D. Groen et al. (Eds.): ICCS 2022, LNCS 13353, pp. 409–421, 2022.
https://doi.org/10.1007/978-3-031-08760-8_35

Leading technologies being rapidly adopted into production environments include Autonomous Guided Vehicles (AGVs). The use of AGVs in production systems has many advantages as it allows production lines to be automated and accelerates logistics. AI-driven analytics at the edge (i.e. edge computing) plays a significant role in coordinating a fleet of AGVs and enabling robust production cycles. These analytics cover the development and use of Machine Learning (ML) algorithms to analyze the behavior of AGVs on the edge IoT device to detect any anomalies, possible problems, or failures. However, AGVs operate as separate units, with own characteristics and sometimes in specific production environments. Thus, they gain experience during their operational cycles within different environments (e.g., different types of pavement on the floor, different temperature and air humidity in the room).

Real-time analysis of production data and advanced data exploration can provide remote condition monitoring and predictive maintenance tools to detect the first signs of failure in industrial environments long before the appearance of the early alarms that precede failures of AGVs in a short period. However, for the effective use of such approaches in real production, it is necessary to have large amounts of useful information, which is very difficult and expensive to obtain. Thus, to solve this problem and make AGVs in production more effective in detecting failures on a broader scale, we investigate the use of Federated Learning (FL), which allows the exchange of experience-data between AGVs.

The main idea of FL is that the same type of intelligent devices or AGVs in production has the opportunity to share experiences. As a result of sharing experience, it is possible to optimize production by increasing the amount of knowledge about various breakdowns of production, which allows better prediction and avoidance. To ensure security, and to prevent information from all these devices from being intercepted or stolen, it is transmitted in the form of neural network weights, which are suitable only for further processing at the highest level, without carrying directly helpful information.

Federated Learning originated from the fact that much of the data containing helpful information used to solve specific problems are challenging to obtain in quantities that would be sufficient to train a powerful model of deep learning. In addition to the helpful information needed to train the model, the data sets also contain other information that is not relevant to the problem. Moreover, Federated Learning benefits from the fact that IoT devices can store all the necessary information for training. Therefore, there is no need to store vast amounts of training data in the cloud, which improves decentralized, edge-based data processing.

In this paper, we show that FL improves the efficiency of failure prediction on edge IoT devices by building a global prediction model based on many local prediction models of particular AGVs. The rest of the paper is organized as follows. In Sect. 2, we review the related works. Section 3 describes a new approach to data exchange between devices, which allows intelligent devices to share experiences with one another to increase the accuracy of recurrent neural networks. In Sect. 5, we conduct a study that demonstrates the efficiency of the proposed approach in a smart production environment. And finally, Sect. 5 concludes the paper.

2 Related Works

Manufacturing companies use new technologies to monitor and better understand their operations, perform them in real-time, thus, turning *classical production* into *intelligent production.* Intelligent production is equipped with technology that ensures machine-machine (M2M) and machine-human (M2H) interaction in tandem with analytical and cognitive technologies so that decisions are made correctly and in a timely manner [2]. The most significant and influential technologies that facilitate conversion from classical production to smart production include Predictive Maintenance, Machine learning, Recurrent Neural Networks, and Federated Learning. Predictive Maintenance (PdM) monitors the state of production during its expected life cycle to provide advanced insights, which ensures the detection of anomalies that are not typical for the task. The purpose of predictive maintenance for manufacturing is to maximize their equipment parts' useful life, avoid unplanned downtime, and minimize planned downtime [9]. An excellent example of this technology is described in [5,7]. In [7], the authors rely on the Numenta Anomaly Benchmark (NAB) [1]. NAB was designed to fairly benchmark anomaly detection algorithms against one another. The approach proposed by the authors scored 64.71 points, while LSTM and GRU scored 49.38 and 61.06 points, respectively.

Predictive maintenance often applies Machine learning (ML) for anomaly detection. ML is a subset of artificial intelligence that is actively being used in industrial settings. The use of machine learning in production is described in detail in [6,10], showing that ML and Deep Learning (DL) can make current manufacturing systems more agile and energy-efficient and lead to optimization of many production processes.

The analysis of literature related to PdM shows that one of the most promising failure forecasting methods is Artificial Neural Networks (ANNs). In the case of manufacturing, it is even more appropriate to use Recurrent Neural Networks (RNNs). The most popular architectures of RNNs in the production environment are Gated recurrent unit (GRU) and Long short-term memory (LSTM). An example of using the GRU model for predictive analytics in intelligent manufacturing is presented in [12]. The authors proposed a hybrid prediction scheme accomplished by a newly developed deep heterogeneous GRU model, along with local feature extraction. Essien and Giannetti [4] proposed to use a novel deep ConvLSTM autoencoder architecture for machine speed prediction in an intelligent manufacturing process by restructuring the input sequence to a supervised learning framework using a sliding-window approach. Table 1 provides the summary of technologies and references to the literature related to intelligent production.

The most recent works for detecting anomalies in production environments rely on the idea of Distributed Learning and Federated Learning (FL) [8,13,14]. In [14], the authors introduced the architecture of the two-level FL named Real-Time Automatic Configuration Tuning (REACT) with local servers hosting the knowledge base for gathering the shared experience. This paper extends the idea by pushing the construction of the local models down to the edge IoT devices without exchanging the production and operational data. In this work, we show both alternative architectures and test the edge-based approach, in terms of the accuracy of performed prediction.

Table 1. A brief summary of technologies related to smart production idea.

Technology	Survey	Brief description
Predictive maintenance for manufacturing	[1,5,7,9]	The purpose of predictive maintenance (PdM) for manufacturing is to maximize the useful life of their equipment parts, avoiding unplanned downtime
Recurrent neural networks	[4,12]	Recurrent Neural Networks have effect of "memory", which will allow to create various patterns of errors in production, in order to understand what problems can await on it and solve them in real-time
AGVs in manufacturing	[3,11]	The use of AGV in production systems has many advantages as it allows production lines to be automated and accelerates logistics, moreover it can be introduced in almost all branches of industry and areas of production
Federated learning	[8,13,14]	The main idea of FL is that intellectual devices or AGVs in production has the opportunity to safely share experiences with each other

3 Federated Learning for Intelligent Manufacturing

Federated Learning allows companies to train machine learning models without moving data from devices where this data is generated, and therefore, it has the inherent characteristics of preserving the privacy of data and reducing the amount of transferred data. These characteristics are required for industrial IoT environments that need data processing solutions working in real-time. We have been introducing this idea in the production environments operating based on the fleet of AGVs that are manufactured by the AIUT company in Poland. Figure 1 shows the loaded Formica-1 AGV that we have been supplementing with edge-based AI/FL methods.

The complete process of exchanging data between AGVs is called a *round*. The round operates according to Algorithm 1 (also graphically visualized in Fig. 2). First, each AGV trains a local model on a specific data set locally (lines 1–4). In the second step, all AGVs send updated local models to the server (lines 5–7). Next, all local models are averaged on the server to create a global model that takes into account the experience of all AGVs (line 8). Finally, the server sends the updated global model back to the AGVs to update their local model with the new global model (lines 9–11).

Fig. 1. The Formica 1 AGV used in our tests.

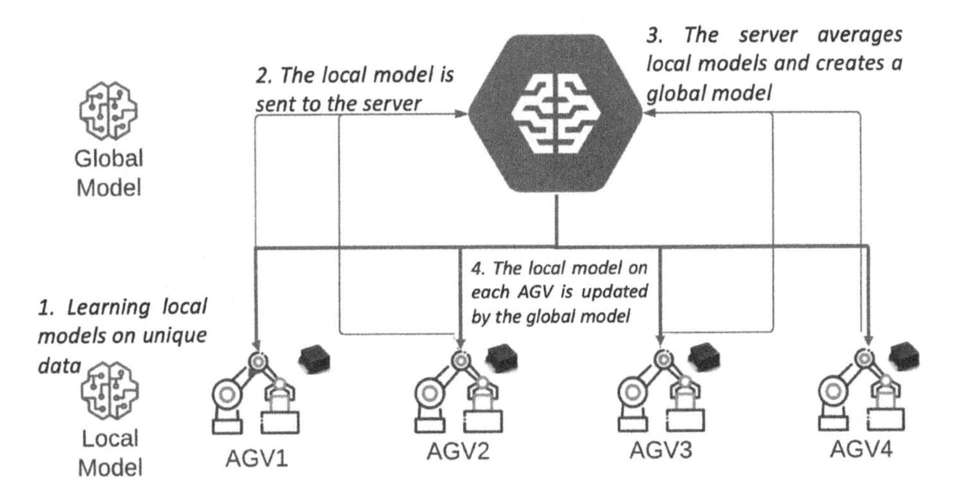

Fig. 2. The complete process of exchanging data between devices (Round)

Algorithm 1: Algorithm of the round

Data: lm (Local models on AGVs), gm (Global model), $AGVs$ (the fleet of AGVs), N (the number of AGVs), sgm (Server with a global model)
Result: $upAGVs$ (AGVs updated by global model)
1 **for** $i \leftarrow 1$ **to** $AGVs$ **do**
2 | Train the RNN of AGV_i locally on unique, AGV-specific data;
3 | $lm \leftarrow$ weights of the local RNN;
4 **end**
5 **foreach** $lm \in AGVs$ **do**
6 | Send lm to the sgm;
7 **end**
8 Build the gm by averaging lms on the sgm;
9 **foreach** $lm \in AGVs$ **do**
10 | Update lm by the sgm;
11 **end**

Given the technical characteristics of the operational, industrial environment for the AGVs, we have identified two main architectures for the integration of FL into production. They are suitable for the manufacturing ecosystem, and their choice depends on the specifics of the production.

3.1 AI on the Local Servers

In this case, each production line must have its own local database and computing resources, which will collect information and analyze all AGVs in this line (Fig. 3). Data from devices are sent to local servers (marked in blue in Fig. 3). These local servers create the global neural network model for this line, taking into account data from all devices in this line. The second step covers transferring the weights of neural networks from the local knowledge base to the data center (e.g., in the cloud), where a general global model is created. This global model takes the experience of all production lines (blue lines). And this global model of a neural network is sent back to all local knowledge bases (green lines). This approach is cheaper and suitable for improving prediction accuracy for static production lines.

3.2 AI on the On-Board IoT Devices

In this case, each AGV creates its own local neural network (Fig. 4). It learns from its unique data, thereby modifying the weights of neural networks. In the next step, the weights of the local neural networks are sent to the data center in the cloud (marked with blue lines), where one global model is created, taking into account the experience of all similar devices. Finally, the global model from the cloud is sent to all devices (marked with green lines). As a result, this process allows each AGV to gain experience from other AGVs, taking more anomalies and production-critical situations into account.

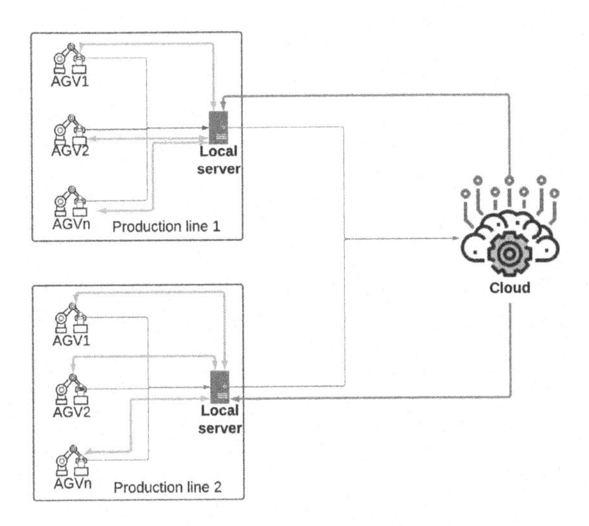

Fig. 3. Distributed architecture with AI/FL on the local servers.

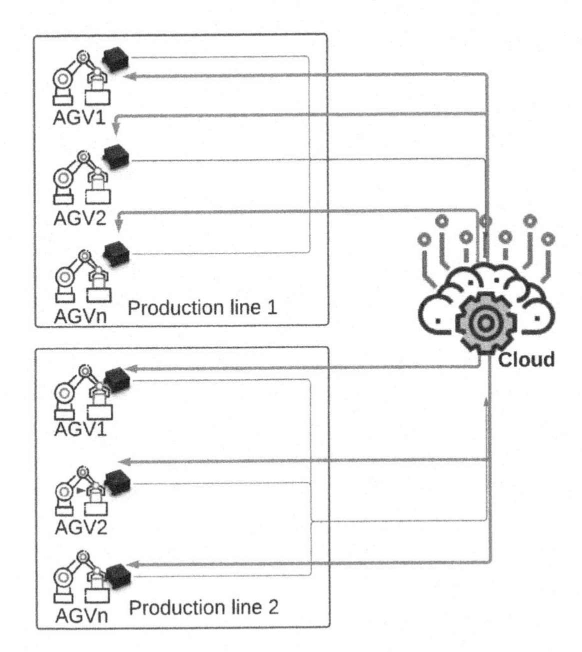

Fig. 4. Distributed architecture with AI on the on-board IoT devices.

Also, the data obtained in real-time are compared with those predicted by our FL-based neural network on the IoT devices themself, thus checking whether the AGV is working correctly. The advantages of this architecture are:

1. The capability to detect failures as quickly as possible;

2. Maximum data security, as data from AGVs are not transmitted to the cloud, and most of them will be processed locally. Only the weights of neural networks are transferred, from which it will not be possible to extract any information.

4 Testing Effectiveness of FL

We conducted several experiments with the real Formica-1 AGVs, obtaining various data from them, including momentary and cumulative power consumption, battery cell voltage, motor RPM, energy consumption and current consumption, cumulative distances, bearing temperatures, transportation pin actuator signals, and momentary frequencies. However, at present, this data set is not sufficient to train models that can be shared between other AGVs, since the work on embedding intelligence into the AGVs is still ongoing. Thus, we simulated the working environment with virtual AGVs (virtual clients). For this purpose, we used a Numenta Anomaly Benchmark (NAB) data set [1] that contains information from temperature sensors of an internal component of a large, industrial machine. The temperature is one of the essential monitored parameters for the proper operation of many production machines. For example, changes in the bearing temperature may suggest its failure and, consequently, the shutdown of the production machine or increased energy consumption and shorter operating time of the AGV. The data from the NAB data set were collected around the clock for 70 days at a sampling interval of 5 min. This data set allowed us to understand the problems of implementing FL in AGVs as well as average deviations of device temperature over time. For this study, we used FL architecture with the AI/FL implemented on the IoT device monitoring the AGV. This option does not require additional local servers for separate production lines. It also provides better security for industrial data, as all the data will be processed locally on the devices and won't be sent anywhere, reducing the communication needs (and the amount of transferred data). We divided the data set into four main parts. The first three parts of the data (each one of them was 30 percent of the data set) were used as training data for three different virtual clients. Then, we used the last part of this data set (10 percent) to test the efficiency of local models from the virtual clients and the global model to compare their effectiveness with each other.

4.1 Choosing Artificial Neural Network Model

Given the fact that we work with time series, we decided to use Recurrent Neural Networks (RNN). Therefore, we decided to use modified RNN architectures based on Long short-term memory (LSTM) cells.

A key component of the LSTM cell-based architecture is the state of the cell. It goes directly through the whole cell, interacting with several operations. The information can easily flow on it, without any changes. However, LSTM can remove information from the cell state using filters. Filters allow skipping information based on some conditions and consist of a sigmoid function layer and element-multiplication operation. LSTM is well-suited to predict time series given time lags of unknown duration. It trains the model by using back-propagation.

4.2 Comparison of Classical Machine Learning and Federated Learning

To verify the suitability of implementing FL in the AGV-based production environment, we decided to compare the effectiveness of three virtual clients trained with different parts of the training data set to the effectiveness of Federated Learning. For the FL-based approach, the model was obtained as a result of averaging the weights of neural networks of these clients.

The whole experiment was organized as follows:

1. We divided the whole data set into four parts. Three parts were used to conduct training on different virtual devices (using the LSTM). The fourth part of this data set was used to test the effectiveness of models.
2. On each IoT device (virtual client), we conducted the training and saved the trained model in the form of weights of neural networks.
3. The models of these three virtual devices were transferred to a separate device (which played the role of the general knowledge base), which averaged the models and created a global model based on the experience of all devices.
4. Local models on virtual devices were updated to a global model.
5. Based on the global model, we predicted the temperature of the device for a particular timestamp.

Our experiments also allowed us to compare the effectiveness of local models on virtual clients and global models obtained through Federated Learning. To verify the effectiveness of the built models, we used several metrics, including *Mean Squared Error* (MSE), *Mean absolute percentage error* (MAPE), and Root Mean Squared Error (RMSE). The effectiveness of the local models for virtual client 1, client 2, client 3 and the global model is shown in Fig. 5, Fig. 6, Fig. 7, and Fig. 8, respectively. We can observe that client 2 (Fig. 6) provides the best prediction results. The curves for the predicted and actual temperature are following a similar path. The MSE, MAPE, and RMSE are on the low level, which proves that this client predicts the temperature well. The results provided by clients 1 and 3 are not so good as for client 2. Especially above the timestamp 1,500, the temperature prediction is much worse, which is also visible in Figs. 5 and 7 and the values of the error metrics.

In order to see the difference in accuracy between all the models in more detail, we decided to show them all in one Fig. 9. The results show that the

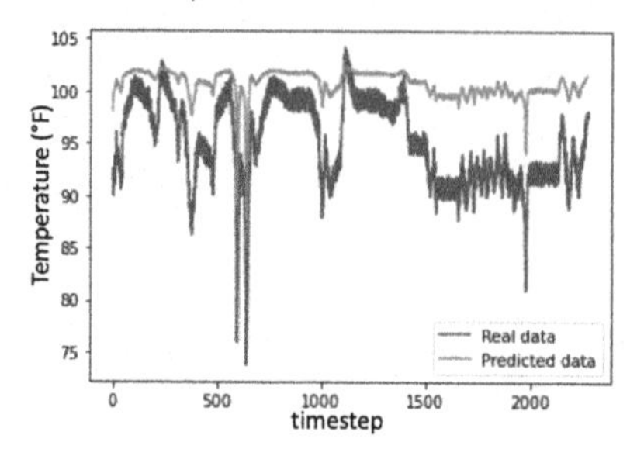

Fig. 5. Effectiveness of the local model for the virtual client 1.

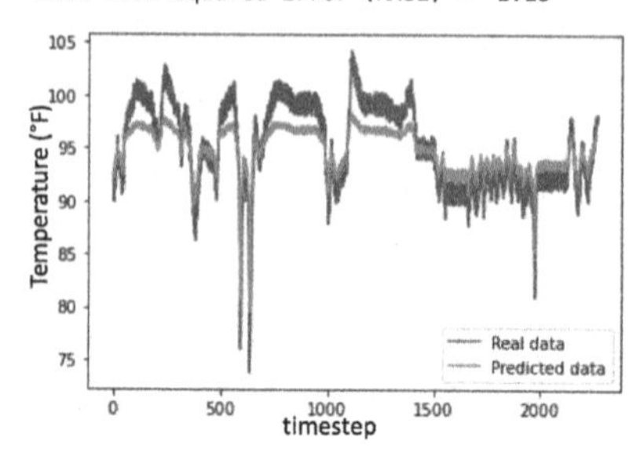

Fig. 6. Effectiveness of the local model for the virtual client 2.

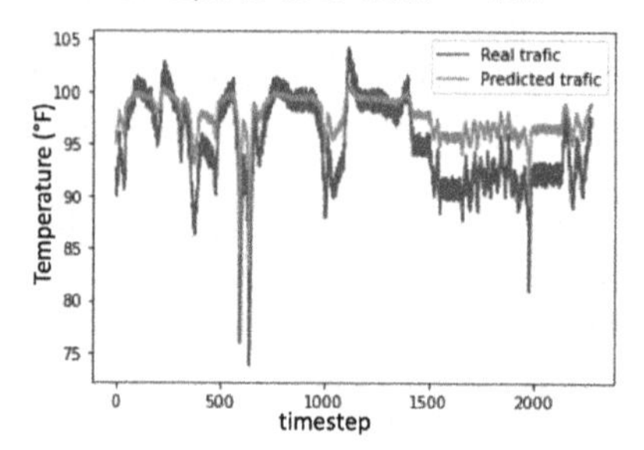

Fig. 7. Effectiveness of the local model for the virtual client 3.

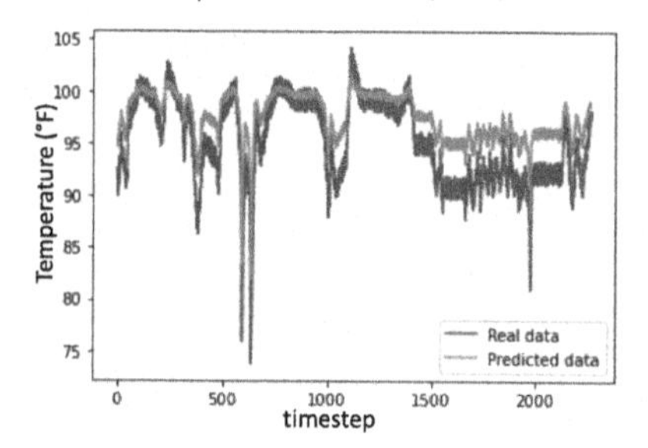

Fig. 8. Effectiveness of the global model.

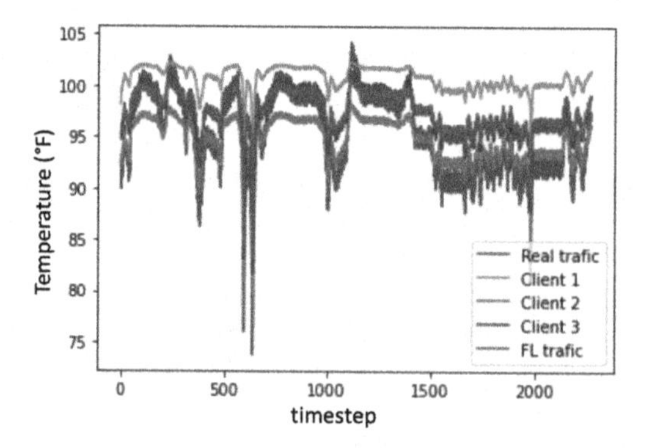

Fig. 9. Comparison of the prediction effectiveness of local models and the global model with real data.

accuracy of the prediction for the global model has increased compared to local models of virtual clients. In particular, we can see the most significant increase for the first client. Due to learning the neural network only on local data, the MSE was equal to 39.97. After averaging the models of three virtual clients and obtaining a global model, the MSE decreased to 9.43, which indicates better model performance. For the third client, the improvement of effectiveness is not so significant, but before the update of the prediction model, the value of MSE was equal to 11.81. For the second client, the value of MSE before updating the global model was 4.74. This result indicates that this client carried most of the valuable information at the time of testing. It could learn based on the most appropriate cases to predict the temperature level adequately. In contrast, clients 1 and 3 (for which the effectiveness is shown in Figs. 5 and 7) seem not to have many occasions to learn correctly, and thus, their prediction results are not perfect. As a result, the second client shared his experience with other clients, making the global model predictions for the test data set more accurate. However, we can also see that the aggregated experience includes also aggregated errors. This is visible in Fig. 9 for timestamps above 1,500, where we can observe the increased prediction error between the real temperature and the one that was predicted by the FL global model. The FL global model aggregates the wrong experience from clients 1 and 3, and this resulted in imperfect (but still better than for client 1 and 3) global model prediction accuracy in this period.

5 Discussion and Conclusions

Federated Learning is currently being actively integrated into Industry 4.0. In smart production, responding to changes in real-time is very important to ensure uninterrupted operation. Predictive Maintenance is used to predict anomalies and breakdowns in production. The more data is available, the more accurate the detection of anomalies and prediction of failures. However, managing such data requires the integration of efficient and secure mechanisms for data exchange between intelligent production devices. In order to provide a secure exchange of experience between smart devices both within one production and between different production environments, we proposed the use of Federated Learning. It provides the maximum safety for industrial data and allows increasing the effectiveness of predicting time-series parameters for big industrial machines or AGVs. In [14], the authors proposed to use a two-level FL architecture based on AI on the local servers, which is well suited for a static production line, demonstrating the benefits of such architecture for their case. However, in our case, we are dealing with AGVs that move independently in production. In this case, it is more appropriate to use the architecture of FL based on AI on the on-board IoT devices, which will provide a deeper understanding of the production environment for AGVs. Our results also show that despite aggregating some prediction errors the accuracy of predicting time-series parameters of the device increases after sharing experiences between AGVs. We have tested the proposed model on virtual clients and conducted experiments to evaluate the effectiveness of Federated Learning. The results show that the overall accuracy of prediction among all virtual clients is increased, which allows better detection of anomalies in autonomously controlled devices and leads us to the conclusion that this approach can be deployed on the AGVs, like the Formica-1 we smarticize.

Acknowledgements. The research was supported by the Norway Grants 2014–2021 operated by the National Centre for Research and Development under the project "Automated Guided Vehicles integrated with Collaborative Robots for Smart Industry Perspective" (Project Contract no.: NOR/POL-NOR/CoBotAGV /0027/2019-00), the Polish Ministry of Science and Higher Education as a part of the CyPhiS program at the Silesian University of Technology, Gliwice, Poland (Contract No.POWR.03.02.00-00-I007/17-00), and by Statutory Research funds of the Department of Applied Informatics, Silesian University of Technology, Gliwice, Poland (grants no. 02/100/BKM21/0015, 02/100/BKM22/0020 and 02/100/BK_22/0017).

References

1. Ahmad, S., Lavin, A., Purdy, S., Agha, Z.: Unsupervised real-time anomaly detection for streaming data. Neurocomputing **262**, 134–147 (2017) https://doi.org/10.1016/j.neucom.2017.04.070, https://www.sciencedirect.com/science/article/pii/S0925231217309864, online Real-Time Learning Strategies for Data Streams
2. Coleman, C., Damodaran, S., Deuel, E.: Predictive Maintenance and the Smart Factory. Deloitte University Press, Toronto (2017)
3. Cupek, R., et al.: Autonomous guided vehicles for smart industries – the state-of-the-art and research challenges. In: Krzhizhanovskaya, V.V., et al. (eds.) ICCS 2020. LNCS, vol. 12141, pp. 330–343. Springer, Cham (2020). https://doi.org/10.1007/978-3-030-50426-7_25
4. Essien, A., Giannetti, C.: A deep learning model for smart manufacturing using convolutional LSTM neural network autoencoders. IEEE Trans. Ind. Inf. **16**(9), 6069–6078 (2020). https://doi.org/10.1109/TII.2020.2967556
5. Klein, P., Bergmann, R.: Generation of complex data for AI-based predictive maintenance research with a physical factory model. In: Gusikhin, O., Madani, K., Zaytoon, J. (eds.) Proceedings of the 16th International Conference on Informatics in Control, Automation and Robotics, ICINCO 2019, Prague, Czech Republic, 29–31 July 2019, vol. 1, pp. 40–50. SciTePress (2019). https://doi.org/10.5220/0007830700400050
6. Kotsiopoulos, T., Sarigiannidis, P., Ioannidis, D., Tzovaras, D.: Machine learning and deep learning in smart manufacturing: the smart grid paradigm. Comput. Sci. Rev. **40**, 100341 (2021) https://doi.org/10.1016/j.cosrev.2020.100341, https://www.sciencedirect.com/science/article/pii/S157401372030441X
7. Malawade, A.V., Costa, N.D., Muthirayan, D., Khargonekar, P.P., Al Faruque, M.A.: Neuroscience-inspired algorithms for the predictive maintenance of manufacturing systems. IEEE Trans. Ind. Inf. **17**(12), 7980–7990 (2021) https://doi.org/10.1109/tii.2021.3062030
8. McMahan, B., Ramage, D.: Federated learning: collaborative machine learning without centralized training data. Google Res. Blog **3** (2017)
9. Pech, M., Vrchota, J., Bednář, J.: Predictive maintenance and intelligent sensors in smart factory: review. Sensors **21**(4) (2021). https://doi.org/10.3390/s21041470, https://www.mdpi.com/1424-8220/21/4/1470

10. Sharp, M., Ak, R., Hedberg, T.: A survey of the advancing use and development of machine learning in smart manufacturing. J. Manuf. Syst. **48**, 170–179 (2018) https://doi.org/10.1016/j.jmsy.2018.02.004, https://www.sciencedirect.com/science/article/pii/S0278612518300153, special Issue on Smart Manufacturing
11. Ullrich, G.: The history of automated guided vehicle systems (2015)
12. Wang, J., Yan, J., Li, C., Gao, R.X., Zhao, R.: Deep heterogeneous GRU model for predictive analytics in smart manufacturing: application to tool wear prediction. Comput. Ind. **111**, 1–14 (2019)
13. Yang, Q., Liu, Y., Chen, T., Tong, Y.: Federated machine learning: concept and applications (2019)
14. Zhang, Y., Li, X., Zhang, P.: Real-time automatic configuration tuning for smart manufacturing with federated deep learning. In: Kafeza, E., Benatallah, B., Martinelli, F., Hacid, H., Bouguettaya, A., Motahari, H. (eds.) ICSOC 2020. LNCS, vol. 12571, pp. 304–318. Springer, Cham (2020). https://doi.org/10.1007/978-3-030-65310-1_22

Acquisition, Storing, and Processing System for Interdisciplinary Research in Earth Sciences

Robert Brzoza-Woch(✉), Tomasz Pełech-Pilichowski, Agnieszka Rudnicka, Jacek Dajda, Ewa Adamiec, Elżbieta Jarosz-Krzemińska, and Marek Kisiel-Dorohinicki

AGH University of Science and Technology, 30-059 Kraków, Poland
rabw@agh.edu.pl

Abstract. The article presents the results of research carried out as part of the interdisciplinary cooperation of scientists in the field of geochemistry and computer science. Such a model of cooperation is justified and especially purposeful in resolving various environment protection tasks, including the issues of energy transformation and climate change. The research regards air quality monitoring case study conducted in Ochotnica, South Poland. The environmental data have been collected, stored, and processed using a set of sensor stations as well as a data storing and processing service. The stations and the service are very flexible, easily extensible, and they have been successfully designed, implemented, and tested by the authors of this paper in the mentioned air quality monitoring case study. The collaboration in the conducted research has been an opportunity to create and test in practice a comprehensive, versatile and configurable data acquisition and processing system which supports not only this use case, but also can be applied to a wide variety of general data acquisition and data analysis purposes. In this paper we discuss the analysis of acquired environmental data as well as a general approach to environmental data processing and visualization system.

Keywords: Environment protection · Data acquisition · Processing and visualization · Internet of Things · Data analysis

1 Introduction

Interdisciplinary cooperation between computer scientists and environmental geochemists provides a comprehensive and synergistic approach to solving scientific problems, i.e. regarding environmental protection or climate change.

1.1 Motivation

Air pollution is a global issue. It is extremely important and inextricably linked with premature deaths all over the world [9]. Poland is ranked as one of the most air-polluted countries in Europe, according to WHO [14]. As a result, the

D. Groen et al. (Eds.): ICCS 2022, LNCS 13353, pp. 422–435, 2022.
https://doi.org/10.1007/978-3-031-08760-8_36

inhabitants of the most polluted areas expect from the government more radical and effective measures, in particular regarding access to reliable and actual air quality data. This requires information acquired from densely placed air quality sensors. A larger number of sensors may generate more data [18], hence adequate storage, processing, analysis, and visualization capabilities are required to meet public expectations in assessing potential risk to human health and the environment. Moreover, all activities disseminating knowledge on environmental issues, through information and monitoring activities, indirectly help to build environmental awareness of the citizens and promote positive changes in the way of thinking, their decisions, as well as their everyday habits.

Considering the close relationship between air pollution, climate change and the issue of energy transformation, advanced data processing systems should be regarded as especially suitable for collecting and processing heterogeneous data thus obtaining information on, inter alia, potential sources of pollution. Consequently, this will allow for a quick and precise data analysis as well as the formulation of possible scenarios for planning and taking actions by municipalities. If the data processing system is advanced enough, those analyses can be to a large extent performed automatically. Moreover, the obtained air quality data can be further analyzed together with meteorological data, traffic intensity, and other parameters related to air emissions, in order to build local air quality forecasts models. Computer analysis of data allows to determine the trends in concentrations of pollutants in atmospheric air, to detect symptoms of possible air quality fluctuations, to show correlation between obtained data, or finally to perform any advanced analysis on individual components of the recorded data. Above mentioned features of air-monitoring systems should be suitable for the purpose of formulating remedial measures on the improvement of air quality, including promoting good practices, such as higher energy efficiency or replacement of home heating systems.

1.2 Available Solutions

According to the available literature, the problems regarding a comprehensive approach to air quality monitoring systems require a combination of data processing systems as well sensors which operate in the field. Such systems often utilize multiple concepts known from Internet of Things (IoT) and Wireless Sensor Networks (WSN) [5,7,25]. Multiple papers address different approaches to data acquisition sensor front-end issues. Descriptions of an air quality data collection system design and realization can be found for instance in [1,10,13,20,23]. There is a variety of solutions with interesting specific features, e.g. measuring vertical changes in the air quality [11] or the collection of air quality data from static and mobile sensors [16].

A vital aspect of a data acquisition system is the approach to its general architecture and methods of data aggregation and processing. The system complexity depends on planned overall system functionality and computing capabilities of data sources. In a basic approach just two or three layers can be sufficient (as e.g. in [2]): 1) a data source (e.g. a WSN), 2) a cloud infrastructure, and 3) an optional interface for end-users. When considering environment data acquisition systems which utilize resource-constrained and energy-efficient hardware platforms combined with short-range sensor communication, a hierarchical or layered architecture is often chosen, as described in for example [4,6,24].

The large-scale distributed data acquisition systems may produce significant amounts of data [18] which should be further processed in a central but not necessary centralized top layer part of the system. The architecture of the top layer, as well as its versatility, storage capabilities, and processing power are crucial for the implementation and application of advanced and useful algorithms, including prediction [3,22], techniques which make use of machine learning (e.g. [21]), advanced data analysis, visualization, and data fusion [8,15,19].

2 The Proposed Solution

The solutions briefly described in Sect. 1.2 mostly lack a genuinely comprehensive approach to the problem of complete systems for data acquisition, advanced processing, visualization, and a support for decision making. Instead, those topics are usually implemented, studied, and described as separate concepts. In contrast, the authors of this paper present a general, complete, and comprehensive approach to an advanced environment monitoring system. The general concept of the system has been adapted to the use case of air quality monitoring in close collaboration with specialists in geochemistry. The comprehensiveness of the presented solution was achieved thanks to the fact that as part of the cooperation in our team, all layers required for environmental monitoring and decision support have been developed and implemented: from the hardware layer of inexpensive sensor stations, through their embedded software, to the central system for high level results aggregation, processing, and visualization.

2.1 General Ideas of the Data Acquisition and Processing System

The important practical aspect of the described study was to collect air quality data obtained from the Ochotnica village, South Poland. Then the collected data could be visualized and further processed. The data acquisition has been done using air quality sensor stations designed and built by the authors of this article and additionally commercial sensor stations manufactured by Sensonar. Our sensors were initially calibrated and validated with the existing reference air monitoring stations in order to achieve reliable spatial and temporal resolution.

The data processing service is very flexible and allows for implementing multiple algorithms, visualization techniques, and automatic report generation. This, in turn, allows specialists and authorities to draw appropriate environmental conclusions necessary for applying further actions aimed at improving local air quality.

Figure 1 shows a simplified diagram of the general architecture of the data acquisition and processing system, which has been developed and then adapted for the air quality monitoring purposes.

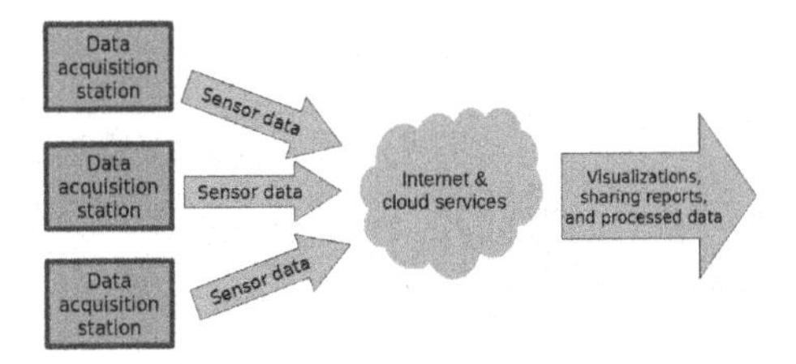

Fig. 1. A simplified diagram of a general architecture of the data acquisition and processing system.

The proposed solution uses versatile hardware and software platforms discussed further in this article. The solution can be used to facilitate individual decisions as well as to plan and implement a broader environmental policy, such as strategies for development and adaptation to climate change in cities, municipalities, and regions. The information flow for supporting the decision process, is presented in Fig. 2. The solution is based on a well-known approach: monitor-analyze-plan-execute with knowledge base (MAPE-K). We use the approach to implement a type of a feedback loop, which integrates environment and resource monitoring in order to facilitate the process of decision making by providing at least partially automatic data analysis. To provide input data, the proposed solution utilizes applied concepts of sensing and data transmission which are known from multiple IoT solutions.

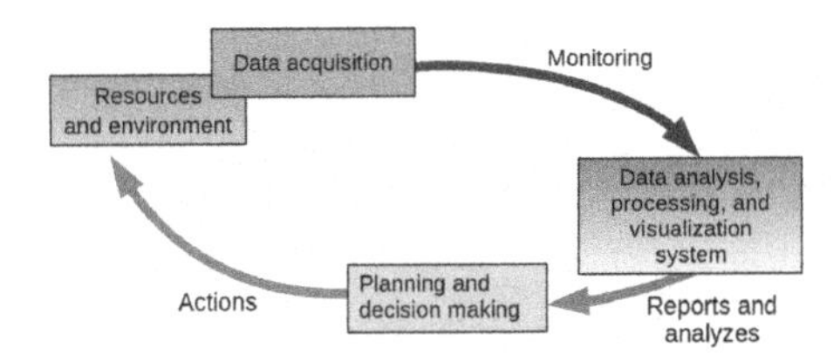

Fig. 2. The approach to the information flow in the proposed system.

The solution consists of the two initially mentioned subsystems which are to a large degree loosely-coupled and which may be used separately for different purposes as well. Further in this section we discuss important components of the system, mainly *DataHub*, which is responsible for general-purpose data storing, processing, and visualisation, and also *EnvGuard* which is the distributed sensor station system for environmental data acquisition.

2.2 DataHub

DataHub is a system primarily intended for *general-purpose* storage, processing, and visualization of data from various sources (also heterogeneous), processes and sets. It may perform advanced and specific analyses for different user groups.

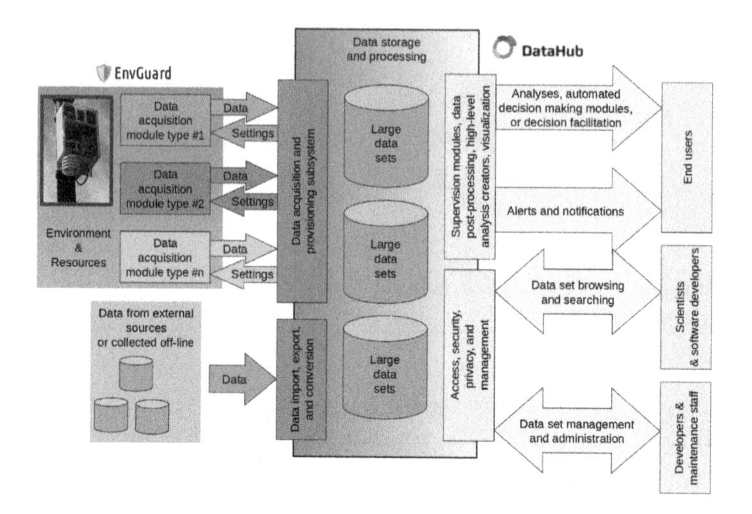

Fig. 3. The approach to the information flow in the proposed systems.

Its information flow block diagram and a sample usage context is presented in Fig. 3.

There are already available commercial solutions for similar purposes. However, their functionality and set of features may change, and they cannot be used to store very sensitive information without restrictive license agreements. That was the main reason to implement DataHub as our proprietary remote storage and processing system.

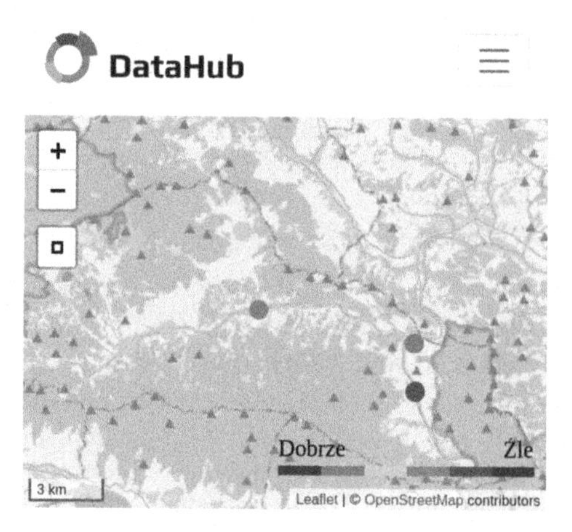

Fig. 4. View of the user-configurable map module for visualization of spatially distributed quantities (the GUI of the depicted map module is configured to display labels in Polish language).

The creators and developers of DataHub do not make any assumptions on data types to be supplied for the system. DataHub itself is also independent from any particular use cases and purposes. The system uses generic data format, particularly JSON, for reading and writing information to and from its internal datasets. The DataHub interface is also equipped with import and export modules for interaction with external data sources.

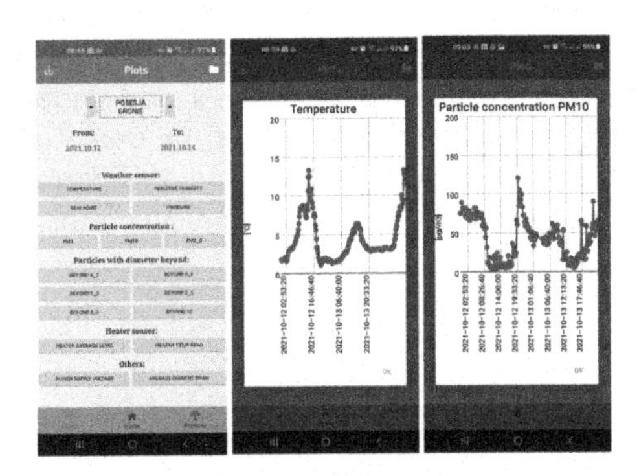

Fig. 5. Screenshots from the *AGH Stations* application for mobile devices.

DataHub functionality can be flexibly tailored to a wide variety of use cases, data types, and data processing methods, even in a large scale, depending on users' needs and proficiency. The customization of the DataHub data processing and visualization functionality can be done using several tools. General rule for the available tools is that the more technical knowledge and skills they require, the more use case-specific data processing and visualization can be achieved. This approach allowed us to create a system which can be operated by users with various specializations and roles. At the one side of this spectrum are regular end users who do not need any advanced technical knowledge and training. Those users may utilize generic built-in functionalities of DataHub to view data acquisition results or to set up basic visualizations. Modules for this group of users include built-in map visualization modules (Fig. 4).

Generic data interfaces of DataHub, such as a regular REST API and a basic Web interface, allow for accessing the acquired data from any device which supports communication over HTTP/REST. This simplifies implementation of custom applications and interfaces for data visualization using Web browsers and mobile devices. *AGH Stations* is one of such mobile applications. Sample screenshots of its user interface is shown in Fig. 5. The application is equipped with station configuration and deployment support functionality, so it is intended mainly for the technical and maintenance staff. However, its simplified basic version with limited features could be used by end users as well.

At the connection level, DataHub may require a secure connection to an internal institutional network. The connection can be established using a regular OpenVPN client. At the Web interface level DataHub also offers multiple levels of security. A user who can access the DataHub Web interface is by default also allowed to access data from endpoints configured as *public*. Data from such endpoints can be read without a personalized account and without further restrictions, using HTTP GET requests. If a user has an individual account in the DataHub service, then they may gain higher privilege levels. Currently the access to DataHub can be granted for educational purposes on individually negotiated terms and conditions. Each dataset may have assigned a group of users: at any privilege level users may have a read-only access to datasets as well as a right to read and execute optionally attached Jupyter Notebooks. A higher privilege level allows users to modify the dataset and endpoint parameters, as well to upload new Jupyter Notebook scripts to be executed. The highest configurable privilege level is intended for persons who perform administrative and maintenance tasks in a dataset and who may manage permissions for other users within the data set.

2.3 EnvGuard Air Quality Sensor Stations

The EnvGuard air sensor stations' role in the system is that they acquire data from the environment, check the validity of the measured quantities, pre-process the data, and send them to the DataHub service. As an additional feature, the sensor stations are able to communicate with each other and also to acquire data from DataHub.

Sensor stations can be perceived as versatile rugged industrial-grade IoT nodes. Instead of cutting edge hardware solutions, we decided to use proven technological solutions for the highest possible long-term reliability. The sensor stations have generic and well-defined internal interfaces between components and for external communication. That feature allows the stations to be set up very flexibly, efficiently, and even ad hoc.

To create a dense network of sensor stations and potentially allow community to participate in air pollution measurement and mitigation, the stations should be implemented as possibly least expensive solution, yet still offering a satisfactory level of accuracy. In our case, the latter problem has been to a large extent mitigated by carefully selecting sensors with a performance as close as possible to certified reference stations.

A general block diagram of the sensor station is shown in Fig. 6. Common basic components of each station include a particle concentration sensor and an environmental conditions sensor which measures at least temperature, relative humidity, and, optionally, atmospheric pressure. The utilized particle concentration sensor module is a very popular and easily available unit, model PMS7003. It measures the concentration of PM10, PM2.5 and PM1 particles. The station can also be equipped in extension sensor modules e.g. for NO_2 or SO_2 measurement capability and a compatible generic interface commonly used in embedded systems and IoT such as I^2C. However we try to rely mainly on the basic dust particle concentration and weather condition sensors to keep the sensor stations' hardware overall cost at possibly the lowest level.

The communication module used to exchange information with DataHub can be flexibly chosen. We have successfully implemented communication using not only on-board Wi-Fi modules available in the SBCs but also generic industrial mobile network communication modules and LoRa wireless connection. The sensor station power demands are in practice up to approx. 1 W. This encouraged

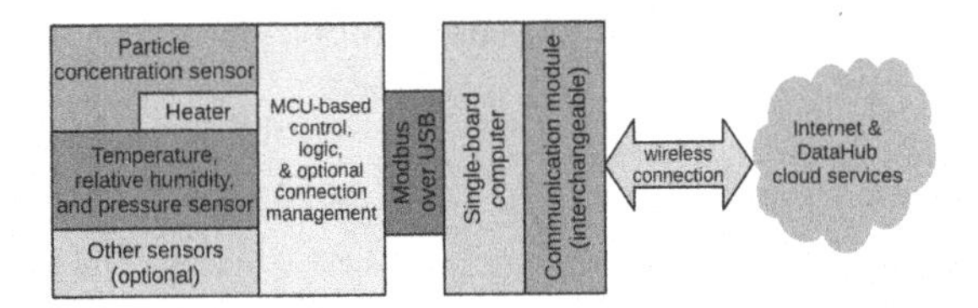

Fig. 6. A simplified general block diagram of the sensor station.

us to design a unique energy saving solution using a microcontroller (MCU). The MCU controls power supplies for all vital components of the sensor station while requiring very low power for itself, typically less than 50 µA in sleep mode. This way we achieved a very flexible power management subsystem providing either high computing power on-demand or extreme energy savings when needed.

All design-specific mechanical components of the designed sensor stations can be fabricated using a generic 3D printer. Moreover, the mechanical design of the station is optimized for low filament usage as well as quick and easy assembly in order to further decrease the stations' overall costs.

A sample sensor station intended and configured to operate in the EnvGuard system is depicted in Fig. 7. That particular station had been set-up in an inexpensive option with additional features which include UPS and mobile network communication.

Fig. 7. View of the sample sensor stations for the EnvGuard system.

Table 1. Air quality sensor stations location.

Location	GPS Lat	GPS Lon	GPS Alt
1. Ochotnica Górna, Primary School	49.512690	20.248250	604.0
2. Ochotnica Górna, Private property	49.518179	20.271481	566.0
3. Ochotnica Dolna, Municipal Office	49.525603	20.320180	489.0
4. Ochotnica Dolna Młynne, Primary School	49.552233	20.310992	592.0
5. Tylmanowa, Primary School, Sucharskiego	49.497837	20.403642	392.0
6. Tylmanowa, Primary School, Twardowskiego	49.514682	20.402704	388.0

2.4 The Conducted Air Pollution Measurements

Until now, the presented data acquisition and processing solution has been successfully used for actual measurements of air pollution in as many as 20 locations. This article, however, focuses on selected 6 particular locations, which might require an attention. The following parameters: PM10, PM2.5 and PM1 dust particle concentration were determined in: four primary schools in Ochotnica Górna, Ochotnica Dolna and Tylmanowa villages, also on the building of Municipal Office in Ochotnica Dolna, and on a private property in that area. Detailed locations of sensor stations are presented in Table 1

For the purposes of this paper we analyze data from the period of 1 year until October 2021. The acquisition process has been done using EnvGuard custom sensor stations and additionally with one *Sensonar AirSense Extended* station. The additional Sensonar station has been used as a reference station. The Sensonar station, in addition to PM 1, PM 2.5, PM 10 particle concentrations and ambient temperature, can also measure the following parameters: carbon monoxide CO, nitrogen dioxide NO_2, and sulfur dioxide SO_2. The Sensonar AirSense stations meet the requirements of the international ISO 37120: 2014E standard for pollution in cities. The stations are certified according to the BS EN ISO 9001: 2015 standard for the design, production, sale and distribution of gas, chemical and physical sensors for industry and administration.

3 Results and Discussion

Environmental data were collected from original air-monitoring sensors and then were processed and visualized with DataHub service. Sample graphs depicting research results, such as fractionation of PM dust, temperature and air humidity changes in selected measurement periods are presented in Fig. 8. Generating plots such as those is a relatively simple task when using DataHub, because DataHub supports direct execution of Python scripts from Jupyter Notebooks. In this particular case, a very popular *matplotlib* Python module has been employed to generate plots. Adequate software to execute Python code from Jupyter Notebooks is available directly on DataHub and it does not need to be installed on a user machine. The user is required to have only a generic Web browser and optionally a standard OpenVPN client.

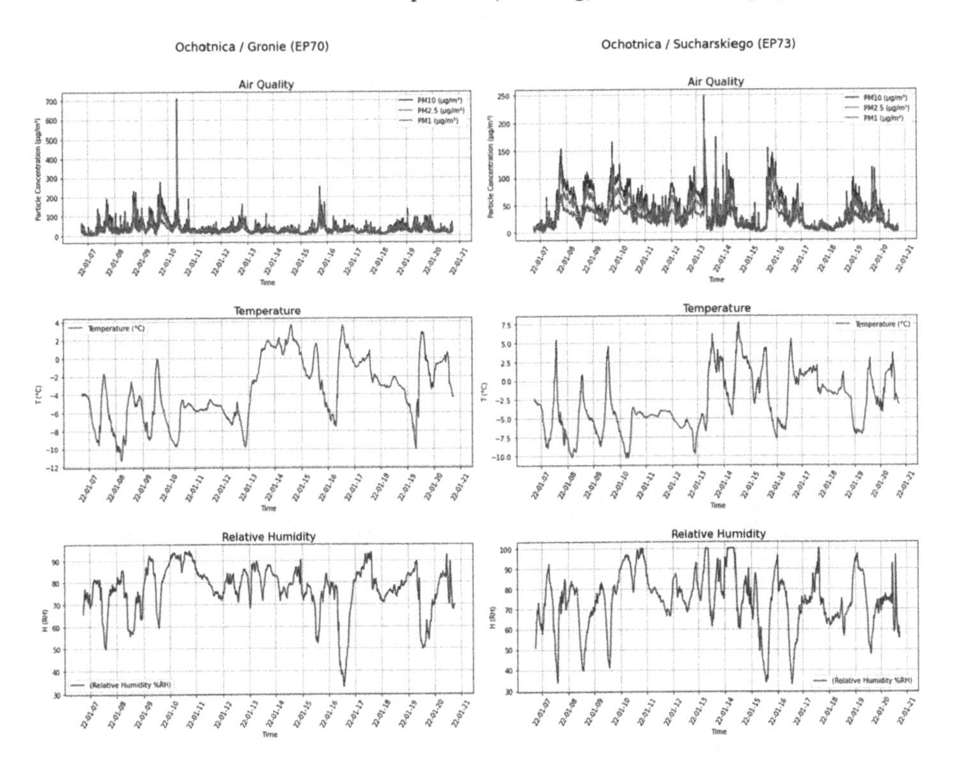

Fig. 8. Variability of physical parameters of ambient air over time.

Similarly, the users can implement various data processing algorithms relatively easily and effectively using multiple commonly available data processing modules for Python. Such additional modules can be easily included because the Jupyter Notebook execution environment in DataHub supports a flexible and on-demand installation process of additional modules imported in scripts. This feature makes it easy to generate also other means to visualize data such as correlation matrices using *pandas* and *numpy*. Selected matrices generated from Jupyter Notebook for DataHub are shown in Fig. 9.

The correlation matrices in this case provided researchers with useful information about which weather parameters are related in particular stations. This, in turn, allows for easier choices of environmental policies and even long-term urbanistic strategy planning by local decisive persons. Moreover, all the plots and visualizations such as those presented in this section can be made available for regular users of the DataHub service, even those who have basic technical knowledge, and who may be interested mostly in measurement results and automatic reports.

In practice, DataHub platform, which has been designed and implemented by the authors of this article, was found to be flexible and universal in terms of various physical outputs. Moreover, it does not have geographical limitations. Information about DataHub parameters are included in Table 2.

3.1 Results Analysis

This section contains an analysis of real-time environmental data using DataHub and formulation of recommendations for the residents and decision makers. One year real-time air quality data obtained from Ochotnica were processed via the DataHub service.

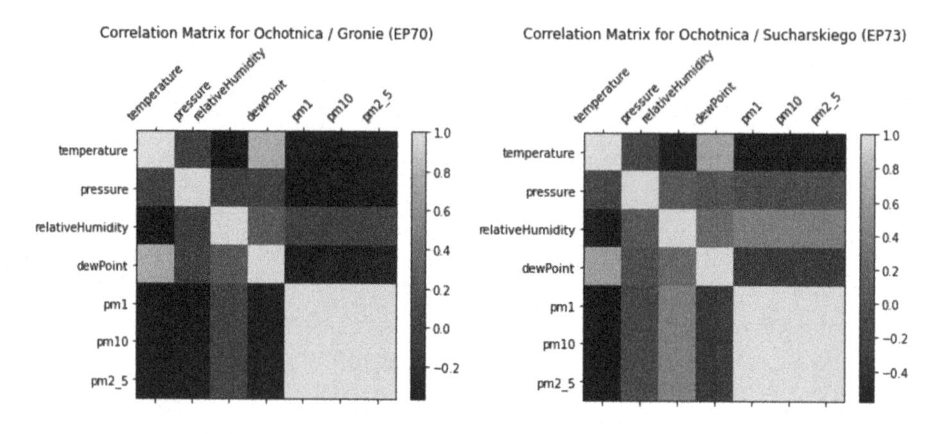

Fig. 9. Correlation matrices between certain physical ambient air parameters.

The acquisition, storing, and processing data from Ochotnica using DataHub service made it possible to determine the level of air pollution with such pollutants as: airborne dust, in particular PM 2.5, PM 10, PM 1, as well as CO, NO_2 and SO_2. The data for the described results have been acquired by both EnvGuard sensor stations as well as Sensonar reference stations, and they have been stored using DataHub service. The gathered data was a base for dedicated analysis of air pollution levels in the monitored area.

Detailed data analysis allowed us to identify the possible sources of low-emission pollution and to identify the places with the highest potential health exposure risk for its residents. Studies have shown that, according to standardized Common Air Quality Index (CAQI), the air quality in Ochotnica Dolna can be classified as good or moderate quality on the majority of days, with a significant number of days classified as of moderate to sufficient air quality, and to a lesser extent of bad or very bad air quality. Correlation was found between the temperature drop in the evening and night hours and elevated concentration of dust pollutants in the ambient air. This is probably due to the increased demand for heat and simply extensive use of various kinds of solid fuels for combustion in home furnaces. Moreover, the impact of traffic-related air pollution was also observed in the morning and afternoon/evening hours, which was directly related to increased frequency of work and home travels. In Ochotnica Dolna, an increase in NO_2 concentrations was found in the morning and afternoon-evening hours, which may be related to increased car traffic in the vicinity of the City Hall and the intersection towards Ochotnica Dolna - Młynne. An increased concentration

Table 2. DataHub settings and infrastructure parameters.

Parameter or a feature	Value	Explanation or additional information
Amount of data from API - currently being stored	38 GiB	As of January 20, 2022
Mass storage space for media & attachments	837 MiB	Includes readable information & attachments
Currently assigned maximum storage space for API data (PostgreSQL)	100 GiB	This parameter can and will be expanded according to the database storage space needs
Maximum request size	16 KiB	Request size with request body excluded
Maximum number of requests	5000	Reloads workers after this amount of managed requests
Initial number of workers in separate processes	4	Increased automatically up to 8 (built-in load balancing mechanisms); can be manually set to a higher value
Virtual machine for running the DataHub service	2 vCPUs, 4 GiB RAM, 4 GiB swap, 1 TiB of mass storage	These parameters can be flexibly configured because DataHub works on managed virtual machines

of CO was found in the morning and in the evening and night hours. However, no correlation was found between elevated concentration of CO and ambient air temperature dropping. High concentration of PM10, PM2.5 and PM1 was determined in the morning and evening hours, which may suggest, rather traffic-related source of its emission. In all analyzed locations, the permissible levels for PM10 and PM2.5 were exceeded in the evening and night hours (above 50 $\mu g/m^3$ for PM10 and above acceptable 25 $\mu g/m^3$ for PM2.5 [17] according to [12].

4 Conclusions and Future Work

The concept of the DataHub and EnvGuard systems presented in this paper emerged from the analysis of current literature, available solutions, as well as the expertise and the practical experience of geochemical scientists in cooperation with professional software developers, data analysts, as well as embedded and industrial IoT system engineers. We also closely cooperate with authorities interested in such systems in their respective regions and we have a very positive feedback on current DataHub and EnvGuard systems.

We have successfully developed concepts and practical implementation of the data acquisition, storage and processing solution. For users, developers, and scientists they facilitate data collection, storage, and processing with an emphasis on sharing and discussing the collected data sets and conclusions.

The novelty of our approach and research as well as our important contribution is the solution which addresses the important problem of air pollution in a very comprehensive way while in the reviewed available literature, authors usually concentrate on selected and specialized aspects of such systems. The resulting duo of the DataHub and EnvGuard systems has an ability not only to collect and store data acquired from the sensor stations, but also to facilitate the process of making decisions regarding environmental issues on the administrative level.

During our research and development collaboration as a team, we have successfully designed, built, and deployed the system which consists of DataHub service and includes more than 20 sensor stations. Most of the stations constantly acquire, process and transmit their results. The selected processing and visualization presented in this article are rather generic and simple. However, as future work related to DataHub, we are developing modules for automatic forecasting of air pollution using artificial intelligence solutions. We are also simultaneously working on (a) opening DataHub for an access for a larger group of users, and (b) a compact and even less expensive version of the sensor stations available for hobbyists and makers. We believe it may increase the popularity of the DataHub-EnvGuard systems and facilitate inclusion of community effort in the air quality measurements. That would contribute to the possibility of collecting air quality data on a larger scale and with a denser sensor network.

Acknowledgements. The research by Robert Brzoza-Woch, Tomasz Pełech-Pilichowski, Agnieszka Rudnicka, Jacek Dajda, Marek Kisiel-Dorohinicki was partially supported by the funds of Polish Ministry of Education and Science assigned to AGH University of Science and Technology. The research conducted by Ewa Adamiec and Elżbieta Jarosz-Krzemińska was supported by AGH University of Science and Technology, Grant No 16.16.140.315.

References

1. Ali, H., Soe, J., Weller, S.R.: A real-time ambient air quality monitoring wireless sensor network for schools in smart cities. In: 2015 IEEE First International Smart Cities Conference (ISC2), pp. 1–6. IEEE (2015)
2. Arroyo, P., Herrero, J.L., Suárez, J.I., Lozano, J.: Wireless sensor network combined with cloud computing for air quality monitoring. Sensors **19**(3), 691 (2019)
3. Ayele, T.W., Mehta, R.: Air pollution monitoring and prediction using IoT. In: 2018 Second International Conference on Inventive Communication and Computational Technologies (ICICCT), pp. 1741–1745. IEEE (2018)
4. Bai, X., Wang, Z., Sheng, L., Wang, Z.: Reliable data fusion of hierarchical wireless sensor networks with asynchronous measurement for greenhouse monitoring. IEEE Trans. Control Syst. Technol. **27**(3), 1036–1046 (2018)
5. Boppana, L., Lalasa, K., Vandana, S., Kodali, R.K.: Mongoose OS based air quality monitoring system. In: TENCON 2019–2019 IEEE Region 10 Conference (TENCON), pp. 1247–1252. IEEE (2019)
6. Brzoza-Woch, R., Konieczny, M., Nawrocki, P., Szydlo, T., Zielinski, K.: Embedded systems in the application of fog computing-levee monitoring use case. In: 2016 11th IEEE Symposium on Industrial Embedded Systems (SIES), pp. 1–6. IEEE (2016)

7. Cynthia, J., Saroja, M., Sultana, P., Senthil, J.: IoT-based real time air pollution monitoring system. Int. J. Grid High Perform. Comput. (IJGHPC) **11**(4), 28–41 (2019)
8. Deng, X., Jiang, Y., Yang, L.T., Lin, M., Yi, L., Wang, M.: Data fusion based coverage optimization in heterogeneous sensor networks: a survey. Inf. Fusion **52**, 90–105 (2019)
9. EEA: EEA report No. 28/2016, air quality in Europe - 2016 report. https://www.eea.europa.eu/publications/air-quality-in-europe-2016. Accessed 20 Jan 2020
10. Gryech, I., Ben-Aboud, Y., Guermah, B., Sbihi, N., Ghogho, M., Kobbane, A.: MoreAir: a low-cost urban air pollution monitoring system. Sensors **20**(4), 998 (2020)
11. Gu, Q., R Michanowicz, D., Jia, C.: Developing a modular unmanned aerial vehicle (UAV) platform for air pollution profiling. Sensors **18**(12), 4363 (2018)
12. Guerreiro, C., Colette, A., Leeuw, F.: Air quality in Europe: 2018 report. European Environment Agency
13. Idrees, Z., Zou, Z., Zheng, L.: Edge computing based IoT architecture for low cost air pollution monitoring systems: a comprehensive system analysis, design considerations & development. Sensors **18**(9), 3021 (2018)
14. IQAir: World's most polluted cities 2020. https://www.iqair.com/world-most-polluted-cities. Accessed 20 Jan 2020
15. Lau, B.P.L., et al.: A survey of data fusion in smart city applications. Inf. Fusion **52**, 357–374 (2019)
16. Mendez, D., Perez, A.J., Labrador, M.A., Marron, J.J.: P-sense: a participatory sensing system for air pollution monitoring and control. In: 2011 IEEE International Conference on Pervasive Computing and Communications Workshops (PERCOM Workshops), pp. 344–347. IEEE (2011)
17. Minister of the Environment: Regulation of the minister of the environment of 24 august 2012 on levels of some pollutants. Air J. Laws 2012, Item 1031 (2012)
18. Plageras, A.P., Psannis, K.E., Stergiou, C., Wang, H., Gupta, B.B.: Efficient IoT-based sensor big data collection-processing and analysis in smart buildings. Futur. Gener. Comput. Syst. **82**, 349–357 (2018)
19. Qi, L., et al.: Privacy-aware data fusion and prediction with spatial-temporal context for smart city industrial environment. IEEE Trans. Industr. Inform. **17**(6), 4159–4167 (2020)
20. Raipure, S., Mehetre, D.: Wireless sensor network based pollution monitoring system in metropolitan cities. In: 2015 International Conference on Communications and Signal Processing (ICCSP), pp. 1835–1838. IEEE (2015)
21. ul Samee, I., Jilani, M.T., Wahab, H.G.A.: An application of IoT and machine learning to air pollution monitoring in smart cities. In: 2019 4th International Conference on Emerging Trends in Engineering, Sciences and Technology (ICEEST), pp. 1–6. IEEE (2019)
22. Shaban, K.B., Kadri, A., Rezk, E.: Urban air pollution monitoring system with forecasting models. IEEE Sens. J. **16**(8), 2598–2606 (2016)
23. Yi, W.Y., Lo, K.M., Mak, T., Leung, K.S., Leung, Y., Meng, M.L.: A survey of wireless sensor network based air pollution monitoring systems. Sensors **15**(12), 31392–31427 (2015)
24. Yu, T., Wang, X., Shami, A.: Recursive principal component analysis-based data outlier detection and sensor data aggregation in iot systems. IEEE Internet Things J. **4**(6), 2207–2216 (2017)
25. Zhang, H.F., Kang, W.: Design of the data acquisition system based on STM32. Procedia Comput. Sci. **17**, 222–228 (2013)

A Study on the Prediction of Evapotranspiration Using Freely Available Meteorological Data

Pedro J. Vaz[1,4], Gabriela Schütz[1,2], Carlos Guerrero[1,3], and Pedro J. S. Cardoso[1,4](✉)

[1] Universidade do Algarve, Faro, Portugal
{pjmartins,gschutz,cguerre,pcardoso}@ualg.pt
[2] CEOT – Centre for Electronics, Optoelectronics and Telecommunications, Faro, Portugal
[3] MED – Mediterranean Institute for Agriculture, Environment and Development, Évora, Portugal
[4] LARSyS – Laboratory of Robotics and Engineering Systems, Lisbon, Portugal

Abstract. Due to climate change, the hydrological drought is assuming a structural character with a tendency to worsen in many countries. The frequency and intensity of droughts is predicted to increase, particularly in the Mediterranean region and in Southern Africa. Since a fraction of the fresh water that is consumed is used to irrigate urban fabric green spaces, which are typically made up of gardens, lanes and roundabouts, it is urgent to implement water waste prevention policies. Evapotranspiration (ETO) is a measurement that can be used to estimate the amount of water being taken up or used by plants, allowing a better management of the watering volumes but, the exact computation of the evapotranspiration volume is not possible without using complex and expensive sensor systems.

In this study, several machine learning models were developed to estimate reference evapotranspiration and solar radiation from a reduced-feature dataset, such has temperature, humidity, and wind. Two main approaches were taken: (i) directly estimate ETO, or (ii) previously estimate solar radiation and then inject it into a function or method that computes ETO. For the later case, two variants were implemented, namely the use of the estimated solar radiation as (ii.1) a feature of the machine learning regressors and (ii.2) the use of FAO-56PM method to compute ETO, which has solar radiation as one of the input parameters. Using experimental data collected from a weather station located in Vale do Lobo, south Portugal, the later approach achieved the best result with a coefficient of determination (R^2) of 0.975 over the test dataset. As a

This work was supported by the Portuguese Foundation for Science and Technology (FCT), projects UIDB/50009/2020 – LARSyS, UIDB/00631/2020 – CEOT BASE and UIDP/00631/2020 – CEOT PROGRAMÁTICO, UIDB/05183/2020 – MED and by project GSSIC – Green Spaces SMART Irrigation Control, grant number ALG-01-0247-FEDER-047030. Particular thanks to GSSIC project's companies Visualforma - Tecnologias de Informação, S.A. and Itelmatis, Lda.

D. Groen et al. (Eds.): ICCS 2022, LNCS 13353, pp. 436–450, 2022.
https://doi.org/10.1007/978-3-031-08760-8_37

final notice, the reduced-set features were carefully selected so that they are compatible with online freely available weather forecast services.

Keywords: Evapotranspiration · Machine learning · Public garden · Smart irrigation

1 Introduction

The hydrological drought is assuming a structural character with a tendency to worsen in regions such as Algarve, Portugal. The problem is not particular to the region occurring, e.g., in most countries of the Mediterranean basin. The climate report, "Climate Change and Land", from August 2019, by the Intergovernmental Panel on Climate Change (IPCC) [18], predicts that, due to climate change, the frequency and intensity of droughts will increase, particularly in the Mediterranean region and in Southern Africa.

A fraction of the fresh water that is consumed by humans is used to irrigate green spaces in the urban fabric, which are typically made up of gardens, lanes and roundabouts, as well as green spaces in hotel and resort chains. The irrigation methodology of these green spaces is commonly done using basic irrigation controllers that are configured according to the experience of those responsible for maintaining the green spaces, without the use of information regarding climate, plants, or soils, as well as data from sensors, nearby weather stations, and from a weather forecast application programming interface (API) that can provide real-time and predicted information. Furthermore, common irrigation controllers have no connectivity, are stand-alone solutions where irrigation schedules are pre-programmed, and only in more complete versions allow irrigation inhibition by means of a rain detection sensor. This is the typical profile that can be found in the overwhelming majority of green space irrigation control systems.

Evapotranspiration (ETO) is a measurement that can be used to estimate the amount of water being taken up or used by plants, allowing a better management of the watering volumes. However, its exact computation is not possible without using complex and expensive sensor systems being many times estimated by formulas or other methods. The use of one over the other depends many times on the available weather parameters.

This paper presents part of a framework to estimate ETO supported by the use of machine learning, acquired intelligence, meteorological data from weather stations on the field, as well as meteorological data and forecasts from APIs available on the internet. The framework will include the computation of crop water requirements derived from the ETO prediction methods, providing an optimal irrigation schedule in terms of start(s) and duration(s), in order to optimize water expenditure, energy expenditure, and the well-being of the crop. This allows the development of an intelligent irrigation solution, technologically differentiated from other platforms on the market, using innovative communications technology, hardware and software, aggregating devices such as probes, field controllers, meteorological stations, among others. The development of the full framework will be done in project GSSIC – Green Spaces SMART Irrigation Control which is developing an innovative intelligent irrigation solution, in terms of reducing water consumption, reducing reaction time in solving problems, increasing efficiency in detecting anomalies, and maintaining the quality of green spaces.

The main contribution of this paper is the study and proposal of a set of methods to estimate ETO and solar radiation using features commonly available in most open weather forecast APIs.

The paper is structured as follows. The next section presents the problem's background and the methodologies used by others to tackle the problem in study. Section 3 explores the dataset and explains the computational setup. The Sect. 4 presents the proposed methods and the associated performance analysis. The final section presents some conclusion and establishes some future work.

2 Problem's Background

Water requirements depend on the reference evapotranspiration (ETO) which is one of the fundamental parameters for irrigation scheduling as well as improving the management and use of water resources. Prediction of reference evapotranspiration for the following days plays a vital role in the design of intelligent irrigation scheduling, as it is proportional to the amount of water that will have to be restored during the irrigation period [5].

Some of the main characteristics that distinguish crop evapotranspiration (ET_c) from ETO are (i) the crop cover density and total leaf area, (ii) the resistance of foliage epidermis and soil surface to the flow of water vapor, (iii) the aerodynamic roughness of the crop canopy, and (iv) the reflectance of the crop and soil surface to short wave radiation [1]. In this context, known the crop coefficient (K_c), the crop evapotranspiration (ET_c) value for a specific time period can be estimated by

$$\mathrm{ET}_c = K_c\mathrm{ETO}. \tag{1}$$

The crop coefficient can be simple or have two components, one representing the basal crop coefficient (K_{cb}) and another representing the soil surface evaporation component (K_e), being computed by

$$K_c = K_s K_{cb} + K_e, \tag{2}$$

where $K_s \in [0,1]$ is used to introduce a K_c reduction in cases of environmental stresses such as lack of soil water or soil salinity [1].

It is thus clear that to make a prediction of a crop's water requirements (ET_c), it is necessary to accurately estimate the reference evapotranspiration (ETO), which is the evapotranspiration of a reference surface, defined as hypothetical grass with a uniform height of 0.12 m, a fixed surface resistance of 70 sm^{-1}, and an albedo (reflection coefficient) of 0.23 [2].

Historically several deterministic methods have been developed to estimate evapotranspiration using single or limited weather parameters and are generally categorized as: temperature, radiation or combination based. Temperature based methods include Thorntwait [20], Blaney-Criddle [3] and Hargreaves and Samani [7]; Radiation methods include Priestley-Taylor [15] and Makkink [11]; Combination methods include Penman [14], modified Penman [4] and FAO56-PM [2].

Shahidian et al. [17] give an overview of several methods and compare their performance under different climate conditions. For most of the methods, the authors concluded that when applied to climates different from those on which they were developed and tested they can yield a poor performance and may require the adjustment of empirical coefficients to accommodate local climate conditions, which is not ideal.

The *Food and Agriculture Organization of the United Nations* (FAO) recommends using the FAO-56 Penman-Monteith (FAO-56PM) formula as a reference method for estimating ETO [2]. To give a deeper idea of the involved parameters, the formula is given by

$$ETO = \frac{0.408\Delta(R_n - G) + \gamma\frac{900}{T+273}u_2(e_s - e_a)}{\Delta + \gamma(1 + 0.34u_2)}, \quad (3)$$

where R_n is the net radiation at crop surface $[MJm^{-2}day^{-1}]$, G is the soil heat flux density $[MJm^{-2}day^{-1}]$, T is the air temperature at 2 m height $[^oC]$, u_2 is the wind speed at 2 m height $[ms^{-1}]$, e_s is the saturation vapor pressure $[kPa]$, e_a is the actual vapor pressure $[kPa]$, $e_s - e_a$ is the saturation vapor pressure deficit $[kPa]$, Δ is the slope vapor pressure curve $[kPa^oC^{-1}]$, and γ is the psychrometric constant $[kPa^oC^{-1}]$. Being based on physical principles, the formula has become widely adopted as a standard for ETO computation since it performs well under different climate types. However, to compute ETO using FAO56-PM the following main meteorological parameters are required: temperature, solar radiation, relative humidity, and wind speed.

All parameters can be easily obtained from weather forecast APIs except for solar radiation. Solar radiation forecasting APIs are, at the moment, not common and present a high-cost penalty. So, apart from the water availability in the topsoil, being the evaporation from a cropped soil mainly determined by the fraction of the solar radiation reaching the soil surface [2], there is the need to (i) develop alternative methods for ETO estimation using limited meteorological parameters, that do not require solar radiation and are compatible with the weather parameters obtained by freely available weather forecast and historical weather data APIs, or (ii) to estimate the solar radiation itself and use it as an approximation on the solar radiation dependent methods. This is also important since in most situations a proper functioning, maintained, and calibrated weather station, with solar radiation measurement capability is not close to the area of interest.

Recently, as an alternative, several authors have used machine and deep learning to estimate ETO. For instance, Granata [6] compared three different evapotranspiration models, which differ in the input variables, using data collected in Central Florida, a humid subtropical climate. For each of this models four variants of machine learning algorithms were applied: M5P Regression Tree,

Bagging, Random Forest, and Support Vector Machine (SVM). However, all three models included as input variable the net solar radiation. Wu and Fan [22] evaluated eight machine learning algorithms divided in four classes: neuron based (MLP – Multilayer Perceptron, GRNN – General Regression Neural Network, and ANFIS – Adaptive Network-based Fuzzy Inference System), kernel-based (SVM, KNEA – Kernel-based Non Linear Extension of Arps decline model), tree-based (M5Tree – M5 model tree, Extreme Gradient Boosting – XGBoost), and curve based (MARS – Multivariate Adaptive Regression Spline). The methods were applied to data collected from 14 weather stations in various climatic regions of China and used only temperature or temperature and precipitation as input to the models. Daily ETO estimates were satisfactory, but can be possibly improved by including further weather parameters and using different machine learning algorithms. Ferreira *et al.* [5] used six alternative empirical reduced-set equations, such as Hargreaves and Samani [7], and compared the estimated values with the ones from an Artificial Neural Network (ANN) and SVM model. Data was collected from 203 weather stations and used for daily ETO estimation for the entirety of Brazil. Temperature or temperature and humidity was used as input features. They concluded that in general ANN was the best performing model when including, as input features, data from up to four previous days. Results were good considering that only temperature or temperature and humidity were used as inputs.

In this study, we explore and develop machine learning based ETO prediction models supported on data from a weather station placed in the Algarve region, in south Portugal, as it will be described in Sect. 3.

3 Dataset and Experimental Setup

Data from February 2019 up to September 2021 was collected from a weather station that uses sensors from Davis Instruments, located in Vale do Lobo, in south Portugal. The following weather parameters were periodically measured throughout the day and stored with a daily resolution: temperature (minimum, maximum, and average), dew point (minimum, maximum, and average), relative humidity (minimum, maximum, and average), solar radiation (maximum and average), wind speed (minimum, maximum, and average), wind direction, atmospheric pressure (minimum, maximum, and average), rain intensity, and precipitation.

The time series was split using a ratio of 75% for training and 25% for testing, naturally without shuffling. This results in test data starting from February 4, 2021 onward. Furthermore, train data was divided into 10 folders using time series cross-validation [9] and a grid search was used to tune the hyperparameters for each model that was used. As foreseeable, all presented model evaluation metric values are obtained using the test data that the models never saw while training. In this context, for model statistical evaluation and performance comparison the coefficient of determination (R^2), mean absolute error (MAE) and mean absolute percentage error (MAPE) were used. Considering y_t the actual value and $\hat{y}_t$ the estimated value at instants $t = 1, 2, ..., n$, and $\overline{y}$ the mean value of the actual samples, they are defined as $R^2 = 1 - (\sum_{t=1}^{n}(y_t - \hat{y}_t)^2)/(\sum_{t=1}^{n}(y_t - \overline{y})^2)$, $MAE = \frac{1}{n}\sum_{t=1}^{n}|y_t - \hat{y}_t|$, and $MAPE = \frac{1}{n}\sum_{t=1}^{n}\left|\frac{y_t - \hat{y}_t}{y_t}\right| \times 100\%$.

The FAO-56PM equation, see Eq. (3), was used to compute the target ETO from the data collected from the weather station. When using solar radiation as target, the average solar radiation from the weather station was used.

During the conduction of this study the following widely known machine learning regression models were used: Ordinary Least Squares (OLS), Ridge regression (Ridge), Lasso regression (Lasso), k-Nearest Neighbors (kNN), Support Vector Machine (SVM), Decision Tree (Tree), and Random Forest (Forest) [10,19]. Table 9 (Appendix A) summarizes the sets of hyperparameters used in the grid search procedure, being the final configurations presented in the corresponding sections.

Finally, to carry out the study, Python v3.8.2, Numpy v1.20.3 [8], Pandas v1.3.2 [12,16], Scikit-learn v0.24.2 [13], and PyET v1.0.1 [21] were used. Pandas library was used for data analysis and manipulation, PyET to compute the reference evapotranspiration using the FAO56-PM method, and Sklearn is a widely used python machine learning framework that includes regressors, data preprocessing, and model metrics evaluation tools.

4 Models for ETO Estimation: Tuning and Feature Selection

This section divides in the following way. Firstly, in Sect. 4.1, a baseline method for ETO estimation using the referred ML algorithms and having as input features all weather parameters that are provided by the weather station is presented. Then, and while still using the measured solar radiation, a first attempt is made at reducing the number of input features that are used, while maintaining similar model performance metrics. In Sect. 4.2 ML ETO estimation models that do not use solar radiation as a feature are explored. In general, solar radiation either as a measurement or as a forecast is not available, hence the need to develop models that do not use it as an input feature. Finally, since solar radiation seems to be one of the main ETO drivers, in Sec. 4.3 ML solar radiation estimation models that use a reduced feature set are explored. Then, two different approaches are taken: (i) inject the solar radiation estimation into another ML model to predict ETO or (ii) use FAO56-PM formula to compute ETO having as input the estimated solar radiation.

4.1 An ETO Baseline Using ML Methods and Measured Solar Radiation

To establish a baseline, the ML regression models were trained using all features available in the data collected from Vale do Lobo weather station, including the measured solar radiation. Furthermore, using the set of parameters described in

Table 9, the conducted grid search established the parameters values outlined in Table 10 (Appendix A) as the best configurations. Table 1 summarizes the attained metrics results. It can be seen that the best performing methods are OLS, Ridge, and Random Forest regressors with the best R^2 equal to 0.981, which corresponds to a MAE of 0.18 mm and a MAPE of 5.51%.

In a second phase, with the objective of reducing the feature set served as input to the algorithms (recall that, besides algorithms constraints, this reduction is important since many weather stations and weather APIs do not provide data for all relevant weather parameters), analysis of the Lasso coefficients and of the Random Forest feature importance was conducted, resulting in a new model with a reduced-set of features. As it can be seen on Table 2, it was observed that when using maximum and minimum temperature, average humidity, average wind, and average solar radiation as features, the models had similar performance to the previous results, and some even improved their metrics values. In this case, random forest gives the best R^2 score being closely followed by OLS. This is an important result since, except for solar radiation, these features are easily obtained through weather forecast APIs.

4.2 ETO Estimation Using ML Methods with Limited Set of Features (Excluding Solar Radiation)

In a first attempt to use ML algorithms to directly estimate ETO without using solar radiation as a feature, and using a tuning strategy similar to the

Table 1. Comparison of several regression methods for ETO estimation using all available features, including measured solar radiation, namely: $Month$, Day, $TempMax$, $TempAvg$, $TempMin$, $HumidityMax$, $HumididtyAvg$, $HumidityMin$, $DewpointMax$, $DewpointAvg$, $DewpointMin$, $PressureMax$, $PressureAvg$, $PressureMin$, $WindMax$, $WindAvg$, $WindGust$, $RainIntensity$, $Precipitation$, $SolarRadiationAVG$, and $SolarRadiationMax$.

	OLS	Ridge	Lasso	kNN	SVM	Tree	Forest
R^2	**0.981**	0.979	0.967	0.910	0.967	0.938	0.972
MAE (mm)	**0.18**	0.19	0.22	0.40	0.21	0.34	0.20
MAPE (%)	**5.51**	5.60	6.41	10.67	6.60	9.18	5.54

Table 2. Comparison of several regression methods for ETO estimation using limited features, but including measured solar radiation, namely: $TempMax$, $TempMin$, $HumididtyAvg$, $WindAvg$, and $SolarRadiationAVG$.

	OLS	Ridge	Lasso	kNN	SVM	Tree	Forest
R^2	0.969	0.962	0.967	0.934	0.967	0.933	**0.971**
MAE (mm)	**0.21**	0.23	0.22	0.31	0.22	0.32	**0.21**
MAPE (%)	6.60	6.52	6.70	7.70	6.72	8.91	**5.89**

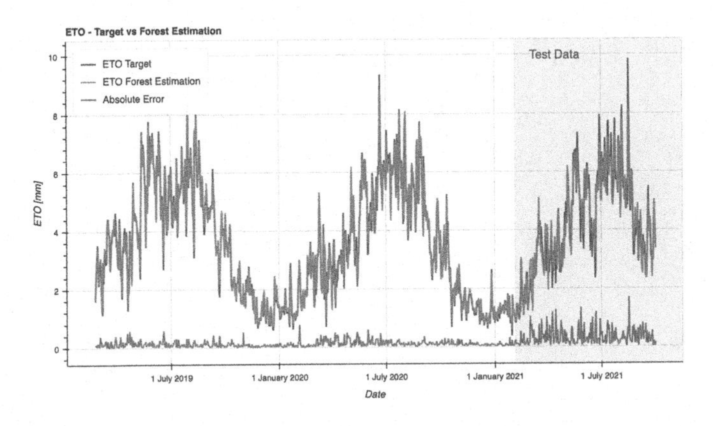

Fig. 1. Target ETO vs Random Forest estimation where solar radiation, actual or estimated, was not used as feature.

one described in Sect. 4.1 (to simplify our explanation and avoid an exhaustive description, due to space constraints, only final settings and conclusions are summarized), it was found that $Month \in \{1, 2, \ldots, 12\}$ was an important feature. I.e., when comparing with the feature-set used to obtain the results in Table 2, it can be seen that the used features are similar except for adding $Month$ and dropping the average solar radiation. Table 3 shows the results obtained with this set of features, and it can be clearly seen that Random Forest is the best performing model with an R^2 of 0.936, a MAE of 0.32 mm, and a MAPE of 9.11%. Figure 1 sketches the ETO target, the ETO estimated using the Random Forest model, and the absolute error. The plot includes the predictions for the full dataset but, the shadowed region corresponds to the test data, the one used to compute the metrics, being visible the increase of the absolute error for those dates.

Maintaining the hyperparameters tuning strategy, attempts were made to improve the models' performance by doing some feature engineering, namely with new features constructed by: (i) computing the inverse of the features values (justified by the fact that some features appear in the denominator of reference FAO56-PM equation, Eq. (3)), (ii) polynomial features, and (iii) adding time lags. However, the success was minor and not noticeable to be presented here but, the idea was not abandoned as will be seen in the next sections.

4.3 ETO Estimation Using Approximated Solar Radiation Values

In order to try to improve the limitation and results obtained in the previous sections, a different approach was tried. The idea was to use a reduced-set of features to previously estimate solar radiation and then either inject that solar radiation prediction into another ML regressor (with the same reduced-set features) or use FAO56-PM formula, Eq. (3), to simply approximate the ETO values. Both approaches are presented next.

Table 3. Comparison of several regression methods for ETO estimation using limited features, namely: $Month$, $TempMax$, $TempMin$, $HumididtyAvg$, $WindAvg$.

	OLS	Ridge	Lasso	kNN	SVM	Tree	Forest
R^2	0.856	0.855	0.859	0.893	0.855	0.814	**0.936**
MAE (mm)	0.53	0.53	0.53	0.43	0.53	0.56	**0.32**
MAPE (%)	15.15	15.20	14.78	11.93	15.16	16.24	**9.11**

Estimating Solar Radiation Using ML Methods. In this section the solar radiation measured in the weather station was used as the target, i.e., the value to be estimated. Following the same tuning procedures as before (namely, the analysis of Lasso coefficients and Random Forest feature importance), the conclusion was that the best configuration for solar radiation estimation was attained for the Random Forest method with the following features: month, day, maximum and minimum temperature, average humidity, average wind speed, and average dew point. More precisely, the results presented in Table 4 show that the Random Forest model is the one with more satisfactory performance, with an R^2 of 0.814, a MAE of 21.31 W/m^2, and a MAPE of 11.29%.

Again, further attempts were made to improve the models' performance by doing feature engineering such as polynomial features, inverse of features, and adding time lags. Of these, only polynomial features were helpful in improving models' performance. After individually analyzing the features' relevance for the models, it was found that by adding the following reduced-set polynomial feature $Month^2 \times Day$, the performance metrics were improved for all models, except Ridge and Lasso. The justification for such is not obvious and, as such, will not be discussed here. Detailed in Table 5, Random Forest is still the best performing model, now with an R^2 of 0.822, a MAE of 20.63 W/m^2, and a MAPE of 10.99%. Figure 2 plots the target solar radiation, the approximated solar radiation obtained with the Random Forest method, and absolute error curves (shadowed is the test set). This improved solar radiation estimation will be used next to predict the ETO values.

Table 4. Comparison of several regression methods for average solar radiation estimation using limited features, namely: $Month$, Day, $TempMax$, $TempMin$, $HumididtyAvg$, $WindAvg$, $DewpointAvg$.

	OLS	Ridge	Lasso	kNN	SVM	Tree	Forest
R^2	0.532	0.505	0.382	0.580	0.312	0.594	**0.814**
MAE (W/m^2)	39.05	40.48	45.59	36.30	48.09	30.67	**21.31**
MAPE (%)	19.61	20.55	22.32	19.33	23.26	16.34	**11.29**

Table 5. Comparison of several regression methods for average solar radiation estimation using polynomial features, namely: *Month*, *Day*, *TempMax*, *TempMin*, *HumididtyAvg*, *WindAvg*, *DewpointAvg*, $Month^2 \times Day$.

	OLS	Ridge	Lasso	kNN	SVM	Tree	Forest
R^2	0.553	0.313	0.375	0.590	0.400	0.605	**0.822**
MAE (W/m^2)	38.04	49.08	46.53	32.48	43.87	30.36	**20.63**
MAPE (%)	18.99	24.74	23.4	16.69	22.21	16.33	**10.99**

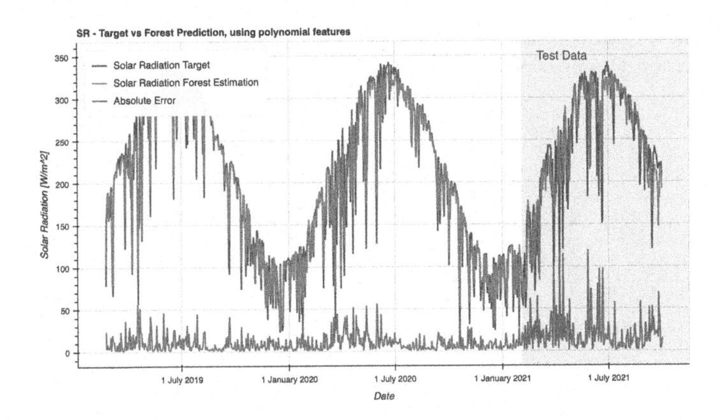

Fig. 2. Target solar radiation vs Random Forest estimation with polynomial features.

ETO Estimation Using ML and the Approximated Solar Radiation. The predicted values from the best performing solar radiation estimation model in the previous section, which was the Random Forest model with polynomial restricted features (see Table 5), were injected as a feature together with maximum temperature, average humidity and average wind into the early studied methods to estimate the ETO, being the obtained results summarized in Tab. 6. It can be seen that the Random Forest is the best performing model, with an R^2 of 0.951, a MAE of 0.26 mm and a MAPE of 7.44%. This result is close to the before presented ETO baseline that used ML and limited features, but included the measured solar radiation (see Table 2). As a reference, in that case, the MAPE was equal to 5.89%. As before, some feature engineering was tested but brought no further improvement. Figure 3 (top) plots the target ETO, the estimated ETO, and corresponding error curves for the train and test (shadowed) dataset.

ETO Estimation Using FAO56-PM Equation and the Approximated Solar Radiation. To finalize our study, a hybrid approach was tested. In this case, the predicted solar radiation is used with the FAO56-PM equation to estimate target ETO, being the results shown on Table 7. With an R^2 of 0.975, MAE of 0.18 mm and MAPE of 5.51% over the unseen test data, this result is better

Table 6. Comparison of several regression methods for ETO estimation using limited features and previously estimated solar radiation, namely: $TempMax$, $HumididtyAvg$, $WindAvg$, SolarRadAVG_prediction_forest

	OLS	Ridge	Lasso	kNN	SVM	Tree	Forest
R^2	0.944	0.944	0.893	0.934	0.937	0.921	**0.951**
MAE (mm)	0.30	0.30	0.38	0.32	0.31	0.37	**0.26**
MAPE (%)	9.36	8.99	9.64	8.58	9.60	10.54	**7.44**

Table 7. Result obtained when computing ETO using FAO56-PM equation and using as solar radiation the previously calculated prediction from best performing Random Forest model.

	FAO56-PM + SR_pred
R^2	0.975
MAE (mm)	0.18
MAPE (%)	5.51

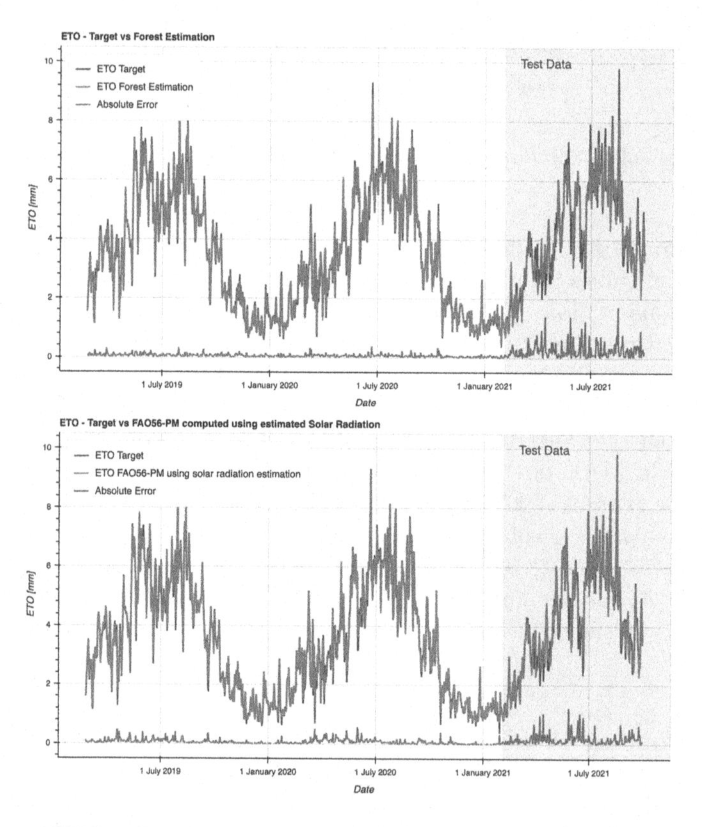

Fig. 3. Target ETO vs Random Forest estimation (top) and FAO56-PM equation (bottom) using estimated solar radiation.

than any of the previously obtained ones, even better than the ML reduced-set baseline (see Sect. 4.1) that used the weather station measured solar radiation as a feature. Figure 3 (bottom) plots ETO target, estimated ETO, and error curves, being evident the improvement in the error when compared with the top plot.

In short, Table 8 presents an overview of the best ETO estimators that where previously presented. Comparing the first two columns it can be concluded that when the measured solar radiation is available, the use of a reduced-set has low impact on model performance. Further, the use of previously estimated solar radiation (last two columns) improves results when solar radiation measurement is not available. The hybrid method (last column) gives similar results to those of the ML baseline when using all the weather parameters provided by the weather station, and gives better performance than the reduced-set ML baseline that used the actual measured solar radiation.

Table 8. Overview of the best ETO estimators for each method that was presented.

	Measured solar radiation		No solar radiation	Estimated solar radiation	
	Table 1	Table 2	Table 3	Table 6	Table 7
R^2	**0.981**	0.971	0.936	0.951	0.975
MAE (mm)	**0.18**	0.21	0.32	0.26	**0.18**
MAPE (%)	**5.51**	5.89	9.11	7.44	**5.51**

5 Conclusion and Future Work

In this study, several ML models and a hybrid approach for the ETO estimation were tested with different degrees of success. Since solar radiation is the main ETO driver, as stated by several authors and also concluded by us, models were also developed for estimating solar radiation using features usually available in the common weather forecast APIs. This allowed both the injection of the previously estimated solar radiation in ML regressors to estimate ETO, but also the possibility to use the hybrid approach where solar radiation is previously estimated and then FAO56-PM algorithm is used to finally compute ETO. The latter yielded the best results, with an R^2 of 0.975, a MAE of 0.18 mm and an MAPE of 5.51%, which when compared with other authors works, is a good result considering the limited weather parameter features that were used.

Future work will include the use of other prediction methods (such as, recurrent neural network models) and a more extensive dataset, by using the existing weather station infrastructure that is installed in the Algarve region, in south

Portugal. The objective will be to develop local and pooled models of ETO predictors for the Algarve region. Also, since all limited feature models here presented are compatible with freely available weather forecast APIs a study needs to be made to assess the impact of using such APIs as input data to the ML models here developed.

A Machine Learning Algorithms Parameters

Table 9. Sets of hyperparameters used in the grid search procedure (according to the available parameters in the Scikit-learn suite [13])

Model	Hyperparameters	Range explored
OLS	fit_intercept	{False; True}
	normalize	{False; True
Ridge	alpha	$\{10^{-4}, 10^{-3}, \ldots, 10^{2}\}$
	fit_intercept	{False; True}
	normalize	{False; True}
Lasso	alpha	$\{10^{-4}, 10^{-3}, \ldots, 10^{2}\}$
	fit_intercept	{False; True}
	normalize	{False; True}
KNN	n_neighbours	{1, 2, ..., 7}
	weights	{uniform; distance}
	leaf_size	{1, 3, 5, 10, 20, 30, 40}
SVM	C	{0.01, 0.1, 0.5, 1.0, 10.0, 100}
	max_iter	10000
	fit_intercept	{False; True}
DT	splitter	{best; random}
	criterion	{mse; friedman_mse; mae; poisson}
Forest	n_estimators	{10, 100, 250, 500, 750, 1000}
	min_samples_leaf	{1, 2, 3, 5, 10}
	max_depth	{3, 5, 10}
	criterion	{mse}
	max_features	{None; sqrt; log2}

Table 10. Tunned parameters

Model	Hyperparameter	Table 1	Table 2	Table 3	Table 4	Table 5	Table 6
OLS	fit_intercept	True	False	False	True	True	False
	normalize	True	False	False	False	True	False
Ridge	alpha	100	0.1	100	100	1	0.1
	fit_intercept	True	True	False	True	True	True
	normalize	False	True	False	False	True	True
Lasso	alpha	0.1	0.1	0.1	20	100	0.01
	fit_intercept	True	False	False	False	False	True
	normalize	False	False	False	False	False	True
KNN	n_neighbours	4	2	2	4	2	2
	weights	distance	distance	distance	distance	distance	distance
	leaf_size	1	1	1	1	1	1
SVM	C	0,01	0,01	0,01	0,1	0,1	0.01
	max_iter	10000	10000	10000	10000	10000	10000
	fit_intercept	True	False	False	False	True	False
DT	splitter	Random	best	Random	best	best	Random
	criterion	friedman_mse	mae	friedman_mse	friedman_mse	friedman_mse	friedman_mse
Forest	n_estimators	100	1000	500	100	100	1000
	min_samples_leaf	1	1	1	1	1	1
	max_depth	10	10	10	10	10	10
	criterion	mse	mse	mse	mse	mse	mse
	max_features	None	None	None	None	None	None

References

1. Allen, R.: Crop coefficients. In: Encyclopedia of Water Science (2003)
2. Allen, R.G., Pereira, L.S., Raes, D., Smith, M., et al.: Crop evapotranspiration-guidelines for computing crop water requirements-FAO irrigation and drainage paper 56. FAO, Rome **300**(9), D05109 (1998)
3. Blaney, H.F., Criddle, W.D.: Determining consumptive use and irrigation water requirements. No. 1275, US Department of Agriculture (1962)
4. Doorenbos, J.: Guidelines for predicting crop water requirements. FAO Irrig. Drainage Paper **24**, 1–179 (1977)
5. Ferreira, L.B., da Cunha, F.F., de Oliveira, R.A., Filho, E.I.F.: Estimation of reference evapotranspiration in Brazil with limited meteorological data using ANN and SVM: a new approach. J. Hydrol. **572**, 556–570 (2019). https://doi.org/10.1016/j.jhydrol.2019.03.028
6. Granata, F.: Evapotranspiration evaluation models based on machine learning algorithms. Agric. Water Manag. **217**, 303–315 (2019)
7. Hargreaves, G.H., Samani, Z.A.: Estimating potential evapotranspiration. J. Irrig. Drain. Div. **108**(3), 225–230 (1982)
8. Harris, C.R., et al.: Array programming with NumPy. Nature **585**(7825), 357–362 (2020). https://doi.org/10.1038/s41586-020-2649-2
9. Hyndman, R.J., Athanasopoulos, G.: Forecasting: Principles and Practice, 3rd edn. OTexts, Australia (2021)
10. Kishore Ayyadevara, V.: Pro Machine Learning Algorithms, 1st edn. APRESS, New York (2018)

11. Makkink, G.F.: Testing the Penman formula by means of lysimeters. J. Inst. Water Eng. **11**, 277–288 (1957)
12. McKinney, W.: Data structures for statistical computing in Python. In: van der Walt, S., Millman, J. (eds.) Proceedings of the 9th Python in Science Conference, pp. 56–61 (2010). https://doi.org/10.25080/Majora-92bf1922-00a
13. Pedregosa, F., et al.: Scikit-learn: machine learning in Python. J. Mach. Learn. Res. **12**, 2825–2830 (2011)
14. Penman, H.L.: Natural evaporation from open water, bare soil and grass. Proc. R. Soc. London Ser. A. Math. Phys. Sci. **193**(1032), 120–145 (1948)
15. Priestley, C.H.B., Taylor, R.J.: On the assessment of surface heat flux and evaporation using large-scale parameters. Monthly Weather Rev. **100**(2), 81–92 (1972)
16. Reback, J., Jbrockmendel, McKinney, W., Van Den Bossche, J., et al.: pandas-dev/pandas: Pandas 1.4.1 (2022). https://doi.org/10.5281/ZENODO.3509134
17. Shahidian, S., Serralheiro, R., Serrano, J., Teixeira, J., Haie, N., Santos, F.: Hargreaves and other reduced-set methods for calculating evapotranspiration. IntechOpen (2012)
18. Shukla, P., et al.: IPCC, 2019: climate change and land: an IPCC special report on climate change, desertification, land degradation, sustainable land management, food security, and greenhouse gas fluxes in terrestrial ecosystems. Intergovernmental Panel on Climate Change (2019)
19. Skiena, S.S.: The Data Science Design Manual. Springer, Cham (2017). https://doi.org/10.1007/978-3-319-55444-0
20. Thornthwaite, C.W.: An approach toward a rational classification of climate. Geogr. Rev. **38**(1), 55–94 (1948)
21. Vremec, M., Collenteur, R.: PyEt - open source python package for calculating reference and potential evaporation (v1.0.1). Zenodo (2021)
22. Wu, L., Fan, J.: Comparison of neuron-based, kernel-based, tree-based and curve-based machine learning models for predicting daily reference evapotranspiration. PLoS ONE **14**(5), e0217520 (2019)

Detecting SQL Injection Vulnerabilities Using Nature-inspired Algorithms

Kevin Baptista, Anabela Moreira Bernardino(✉), and Eugénia Moreira Bernardino

Computer Science and Communication Research Center (CIIC), School of Technology and Management, Polytechnic of Leiria, 2411-901 Leiria, Portugal
2190371@my.ipleiria.pt, {anabela.bernardino, eugenia.bernardino}@ipleiria.pt

Abstract. In the past years, the number of users of web applications has increased, and also the number of critical vulnerabilities in these web applications. Web application security implies building websites to function as expected, even when they are under attack. SQL Injection is a web vulnerability caused by mistakes made by programmers, that allows an attacker to interfere with the queries that an application makes to its database. In many cases, an attacker can see, modify or delete data without proper authorization. In this paper, we propose an approach to detect SQL injection vulnerabilities in the source code, using nature-based algorithms: Genetic Algorithms (GA), Artificial Bee Colony (ABC), and Ant Colony Optimization (ACO). To test this approach empirically, we used web applications purposefully vulnerable as Bricks, bWAPP, and Twitterlike. We also perform comparisons with other tools from the literature. The simulation results verify the effectiveness and robustness of the proposed approach.

Keywords: SQL Injection · Nature-inspired algorithms · Genetic algorithms · Swarm optimization algorithms

1 Introduction

The Open Web Application Security (OWASP) Top 10 is a standard awareness document for developers and web application security that lists the top 10 web application security risks for 2021 [1]. The focus of this paper is on SQL Injection.

OWASP defines SQL Injection as a type of injection attack that occurs when untrusted data is sent to an application as part of a query [1]. The main goal for the attacker is to trick the interpreter into executing unintended queries to execute unauthorized actions like obtaining unauthorized data.

In the last years, several combinatorial optimization problems have arisen in the communication networks field. In many cases, to solve these problems it is necessary the use of emergent optimization algorithms [2]. We developed an automated tool, based on the application of nature-based algorithms that tries to find the greatest number of vulnerabilities in the shortest time possible. We implemented Genetic Algorithm (GA),

D. Groen et al. (Eds.): ICCS 2022, LNCS 13353, pp. 451–457, 2022.
https://doi.org/10.1007/978-3-031-08760-8_38

Artificial Bee Colony (ABC), and Ant Colony Optimization (ACO). The main purpose of this tool is to be used in a white box scenario, having access to the code base. Thus, it could help developers to find potential vulnerabilities in their codebase. To empirically evaluate our approach, we used several open-source PHP projects that are known to have certain vulnerabilities, such as Bricks, bWAPP, and Twitterlike.

We compare the results obtained by our approach with manual analysis, the results obtained in previous works by the same authors [3, 4], and with the tools Web Application Protection (WAP) and SonarPHP that use a static analysis approach to detect vulnerabilities in PHP web applications.

The paper is organized as follows. Section 2 presents some related works. Section 3 descriebes the problem representation. Section 4 describes the proposed algorithms and, Sect. 5 discusses the experiments conducted and the results obtained. Section 6 presents the conclusions and some directions for future work.

2 Related Work

An intensive study was done before developing the approach presented in this document. Since this article has a limited number of pages, we only highlight the most important works in the area.

Mckinnel et al. [5] made a comparative study of several Artificial Intelligence algorithms in exploiting vulnerabilities. They compiled several works in the area to compare several algorithms, including unsupervised algorithms, reinforcement Learning, GA, among others. The authors conclude that GA performs better over time due to the evolutionary nature of generations. They suggest that its applicability needs to take into account a good definition of the fitness function in order to obtain better results.

Manual penetration tests, although effective, can hardly meet all security requirements that are constantly changing and evolving [6]. Furthermore, they require specialized knowledge which, in addition to presenting a high cost, is typically slower. The alternative is to use automated tools that, although faster, often do not adapt to the context and uniqueness of each application. In [6], the author developed a reinforcement learning strategy capable of compromising a system faster than a brute force and random approach. He concluded that it is possible to build an agent capable of learning and evolving over time so that it can penetrate a network. Its effectiveness was equal to human capacity. Finally, he concludes that although the initial objective was long, there are still several directions to be explored, from the use of different algorithms for both an offensive (red team) and a defensive (blue team) security perspective. It suggests the application of game theory concepts [7], especially treating a problem like a Stackelber Security Game. These techniques have been successfully applied in various security domains such as finding the optimal allocation for airport security given the attackers' knowledge.

Alenezi and Javed [8] analyzed several open-source projects in order to identify vulnerabilities. They concluded that most of these errors are due to negligence on the part of developers as well as the use of bad practices. The authors suggest the development of a framework that encourages programmers to follow good practices and detect possible flaws in the code.

In [9], the authors propose a solution to detect XSS (cross-site scripting) vulnerabilities based on the use of GAs as well as a proposal to remove the vulnerabilities found during the detection phase. The aim, therefore, was to find as many vulnerabilities as possible with as few tests as possible. The results obtained with GAs were compared with other static analysis tools.

In [3] and [4] the same authors of this paper proposed an approach to detect SQL injection vulnerabilities in the source code, using GA [3] and swarm-based algorithms [4]. In these works were used different representations of the individuals. We also compare our results with these results.

3 Problem Representation

In order to use a nature-inspired algorithm as an optimization algorithm to find SQL Injection vulnerabilities, the process was divided into two steps: (1) identification of all SQL queries; and (2) use of GA, ABC, or ACO to generate attack vectors to be injected in the queries.

In order to obtain all non-parametrized queries, first, it is necessary to perform a search to list all PHP files recursively in a given folder. Afterward, all variables in the code are indexed and their history is kept. This step is crucial to capture SQL queries that are parametrized, but still vulnerable because the vulnerabilities occurred before in the code. These queries and all non-parametrized queries are kept in a list to be used in the second phase by the algorithm.

The main goal of the second phase is to find a vector that could compromise one of the queries listed in the previous step. So, the algorithm starts by initializing all the needed parameters, then it either goes through the GA, ABC, or ACO. Each one has its specific steps and parameters, but the fitness calculation is common to all of them. More information about these steps can be found in [3, 4].

Here, each individual is made up of a set of five genes. Each gene is a String (values are derived from SQL injection database which was constructed based on resources [10] and [11]). In Fig. 1, there is a possible representation for the individual, where each gene is an attack vector.

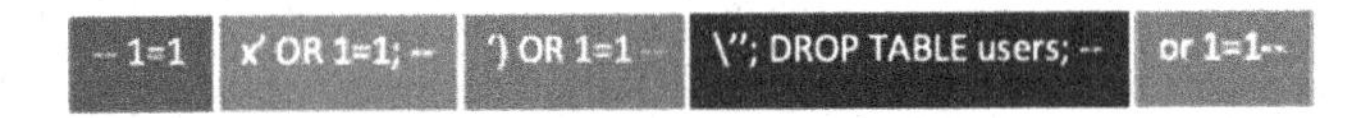

Fig. 1. Example of an individual.

Following this approach, an individual can generate up to *N* different attack vectors, where *N* is the number of genes in the individual. individuals (bees in ABC and ants in ACO). As represented in Fig. 1, each gene in the individual is going to be tested as an attack vector.

The queries executed for this scenario are illustrated in Table 1 (an individual with 5 genes executes 5 queries). In this example, only two are vulnerable.

Table 1. Executed queries.

Query	Vulnerable
SELECT * FROM users WHERE ua ='--1 = 1'	No
SELECT * FROM users WHERE ua ='x' or 1 = 1; --'	Yes
SELECT * FROM users WHERE ua =") OR 1 = 1--'	No
SELECT * FROM users WHERE ua ='\";DROP TABLE users; --'	Yes
SELECT * FROM users WHERE ua ='or 1 = 1--'	No

A fitness value is used to evaluate the performance of an individual. An individual with bigger fitness means that it is able to crack successfully more queries. The fitness function used for this problem is as follows:

$$fitness(i) = \frac{U_{vul} * 5 + G_{vul} * 2}{totalGenes} \quad (1)$$

where i is the individual being tested, *totalGenes* is the total number of genes in an individual, U_{vul} is the number of unique vulnerabilities detected by the individual, and G_{vul} is the number of genes that detected at least one vulnerability.

4 Algorithms

The nature-inspired algorithms are used to generate attack vectors to be injected into the queries. Evolutionary Algorithms (EAs) are randomized search heuristics, inspired by the natural evolution of species [2, 12]. The main idea is to simulate the evolution of candidate solutions for an optimization problem. GA is an example of an EA [2].

The basic concept of GA is designed to simulate processes in a natural system necessary for evolution, specifically those that follow the principles first laid down by Charles Darwin - the survival of the fittest [12]. GAs are EAs that use techniques inspired by evolutionary biology, such as inheritance, mutation, selection, and crossover. More information about the steps of this algorithm can be found in [2, 12].

The area of Swarm Intelligence (SI) relies on the collective intelligence of agents that interactively explore the search space. Some of the best-known areas of swarm intelligence are ACO, Particle Swarm Optimization, and bees-inspired algorithms [13, 14].

The ACO algorithm is essentially a system based on agents which simulate the natural behavior of ants, including mechanisms of cooperation and adaptation [16–18]. In real life, ants indirectly communicate among themselves by depositing pheromone trails on the ground, influencing the decision processes of other ants. This simple communication form among individual ants causes complex behaviors and capabilities in the colony. The real ant behavior turns into an algorithm establishing a mapping between (1) the real ant search and the set of feasible solutions to the problem; (2) the amount of food in a source and the fitness function; and (3) the pheromone trail and an adaptive memory [18]. The pheromone trails serve as distributed, numerical information which ants use to probabilistically build solutions to the problem to be solved and which they adapt during the execution of the algorithm to reflect their search experience. More information about the steps and the formulas used to initialize/update the pheromone trails of this algorithm can be found in [4, 13, 14, 17].

The minimal model of swarm-intelligent forage selection in a honey-bee colony, that ABC algorithm adopts, consists of three kinds of bees: employed bees, onlooker bees, and scout bees [19, 20]. In ABC each iteration of the search consists of four steps: (1) sending the employed bees onto their food sources and evaluating their nectar amounts; (2) after sharing the nectar information of food sources, selecting food source regions by the onlookers and evaluating the nectar amount of these food sources; (3) determining the scout bees and then sending them randomly onto possible new food sources; and (4) memorizing the best food source [19, 20]. These four steps are repeated through a predetermined number of iterations defined by the user. In a robust search process, the exploration and exploitation processes must be carried out together. In the ABC algorithm, while onlookers and employed bees carry out the exploitation process in the search space, the scouts control the exploration process. More information about the steps of this algorithm can be found in [2, 4, 19, 20].

5 Experimental Results

In order to implement the approach described, we develop a tool in Java. To conduct the experiments, we collected different open-source web applications purposefully vulnerable as Bricks, bWAPP, and Twitterlike. All experiments were performed on a Raspberry PI 4 Model B with 8GB of RAM and a quad-core 64-bits of 1.5 Ghz.

In order to obtain the best combination of parameters, several smoke tests were performed. Table 2 shows the best combination of parameters obtained for the algorithms GA, ACO, and ABC.

Table 2. Best parameters.

Alg	Parameters
GA	Max generations: 50; population size: 20; elitism: no; selection method: tournament (size 6); crossover method: one cut (probability: 0.5); mutation probability: 0.05
ACO	Max iterations (*mi*): 30; number of ants: 80; *mi* without improvement: 2; number of modifications: 5; Q: 100; q probability: 0.1; pheromone evaporation rate: 0.3; pheromone influence rate: 0.3
ABC	Max iterations: 100; number of employed bees: 75; number of onlooker bees: 100; number of scout bees: 10% of employed bees; number of modifications: 11

We have first identified SQL Injection vulnerabilities manually since there is not an official number of these vulnerabilities. The biggest problem with a manual approach is that it takes a long time to detect all the vulnerabilities.

The results produced by our approach were compared with manual analysis and with the tools WAP and SonarPHP that use a static analysis approach. We also compare our results with the results obtained in previous works of the same authors that also use GA [3], ACO [4], and ABC [4]. The solutions in these works are represented differently. These results are shown in Table 3.

Table 3. Best results.

Project	Manual	GA	ABC	ACO	GA [3]	ABC [4]	ACO [4]	WAP	SonarPHP
Bricks	12	**12**	**12**	**12**	11	11	11	11	**12**
bWAPP	56	**52**	33	33	30	47	51	15	14
Twitterlike	17	**13**	**13**	**13**	12	10	10	5	9

As we were able to see, when comparing with static analysis, our approach was able to identify more vulnerabilities. All the algorithms implemented present better results in comparison with static analysis. GA, using the representation of the individuals shown in this paper, proves to be more efficient in terms of the number of vulnerabilities detected.

As we can see most vulnerabilities are detected with GA, however, for the case of bWAPP and Twitterlike GA did not find all vulnerabilities. This is probably due to the fact that an individual has a fixed genome in terms of size, which sometimes leads to invalid queries. An approach with a dynamic genome size could potentially bring better results in terms of total SQL injection vulnerabilities found.

6 Conclusion

SQL injection can cause serious problems in web applications. In this paper we presented an approach to detect SQL Injection vulnerabilities in the code base, using a white-box approach. To optimize the search of potential vulnerabilities in the code we use nature-inspired algorithms: GA, ACO, and ABC. The optimization problem was formulated to find the best set of attack vectors. The proposed approach was divided into two steps, the first being a pre-analysis of queries in the source code that were used for the next phase, in which the various algorithms were used. This abstraction of concepts is quite convenient, allowing scalability of functionalities and the opportunity to implement several algorithms.

With more optimizations, better adaptations to the code of the algorithms, we think that it will be possible to improve the results obtained. The tool should be expanded to support multiple languages. At this phase, our approach only supports PHP applications. The scope of this article was constrained to SQL injection, but other vulnerabilities could potentially benefit from this approach.

Acknowledgements. This work was supported by national funds through the Portuguese Foundation for Science and Technology (FCT) under the project UIDB/04524/2020.

References

1. Stock, A., Glas, B., Smithline, N., Gigler, T.: OWASP Top Ten. OWASP Homepage. https://owasp.org/www-project-top-ten/ (2022). Accessed 18 Feb 2022
2. Yang, X.-S.: Nature-Inspired Optimization Algorithms, 1st edn. Elsevier (2014)
3. Batista, K., Bernardino, A.M., Bernardino, E.M.: Exploring SQL injection vulnerabilities using genetic algorithms. In: Proceedings of the XV International Research Conference, Lisboa (2021)
4. Batista, K., Bernardino, E.M., Bernardino, A.M.: Detecting SQL injection vulnerabilities using artificial bee colony and ant colony optimization. Lecture Notes in Networks and Systems. Springer (2022)
5. McKinnel, D.R., Dargahi, T., Dehghantanha, A., Choo, K.-K.R.: A systematic literature review and meta-analysis on artificial intelligence in penetration testing and vulnerability assessment. Comput. Electr. Eng. **75**, 175–188 (2019)
6. Niculae, S.: Applying Reinforcement Learning and Genetic Algorithms in Game-Theoretic Cyber-Security. Master Thesis (2018)
7. Nguyen, T.H., Kar, D., Brown, M., Sinha, A., Jiang, A.X., Tambe, M.: Towards a science of security games. In: Toni, B. (ed.) Mathematical Sciences with Multidisciplinary Applications. SPMS, vol. 157, pp. 347–381. Springer, Cham (2016). https://doi.org/10.1007/978-3-319-31323-8_16
8. Alenezi, M., Javed, Y.: Open source web application security: A static analysis approach. In: Proceedings of 2016 International Conference on Engineering and MIS (2016)
9. Tripathi, J., Gautam, B., Singh, S.: Detection and removal of XSS vulnerabilities with the help of genetic algorithm. Int. J. Appl. Eng. Res. **13**(11), 9835–9842 (2018)
10. Friedl, S.: SQL Injection Attacks by Example. http://www.unixwiz.net/techtips/sql-injection.html (2017). Accessed 18 Feb 2022
11. Mishra, D.: SQL Injection Bypassing WAF. https://www.owasp.org/index.php/SQL_Injection_Bypassing_WAF (2022). Accessed 18 Feb 2022
12. Eiben, A.E., Smith, J.E.: Introduction to Evolutionary Computing. Springer, Heidelberg (2015). https://doi.org/10.1007/978-3-662-44874-8
13. Kennedy, J., Eberhart, R.C., Shi, Y.: Swarm Intelligence. Morgan Kaufmann, San Francisco (2001)
14. Wahab, M.N.A., Nefti-Meziani, S., Atyabi, A.: A comprehensive review of swarm optimization algorithms. PLoS ONE **10**(5), e0122827 (2015)
15. Karaboga, D., Akay, B.: A survey: algorithms simulating bee swarm intelligence. Artif. Intell. Rev. **31**, 61 (2009)
16. Dorigo, M.: Ottimizzazione, apprendimento automatico, ed algoritmi basati su metafora naturale (Optimisation, learning and natural algorithms). Doctoral dissertation. Dipartimento di Elettronica e Informazione, Politecnico di Milano, Italy (1991)
17. Dorigo, M., Maniezzo, V., Colorni, A.: The ant system: Optimization by a colony of cooperating agents. IEEE Trans. Syst. Man Cybern. **26**, 29–41 (1996)
18. Gambardella, L.M., Taillard, E.D., Dorigo, M.: Ant colonies for the quadratic assignment problem. J. Operational Research Society **50**(2), 167–176 (1999)
19. Karaboga, D.: An idea based on honey bee swarm for numerical optimization, Technical report TR06. Erciyes University, Engineering Faculty, Computer Engineering Department (2005)
20. Karaboga, D., Akay, B.: A comparative study of artificial bee colony algorithm. Appl. Math. Comput. **214**, 108–132 (2009)

On-Edge Aggregation Strategies over Industrial Data Produced by Autonomous Guided Vehicles

Piotr Grzesik[1(✉)], Paweł Benecki[1], Daniel Kostrzewa[1], Bohdan Shubyn[1,2], and Dariusz Mrozek[1]

[1] Department of Applied Informatics, Silesian University of Technology, Gliwice, Poland
pj.grzesik@gmail.com

[2] Department of Telecommunications, Lviv Polytechnic National University, Lviv, Ukraine

Abstract. Industrial IoT systems, such as those based on Autonomous Guided Vehicles (AGV), often generate a massive volume of data that needs to be processed and sent over to the cloud or private data centers. The presented research proposes and evaluates the approaches to data aggregation that help reduce the volume of readings from AGVs, by taking advantage of the edge computing paradigm. For the purposes of this article, we developed the processing workflow that retrieves data from AGVs, persists it in the local edge database, aggregates it in predefined time windows, and sends it to the cloud for further processing. We proposed two aggregation methods used in the considered workflow. We evaluated the developed workflow with different data sets and ran the experiments that allowed us to highlight the data volume reduction for each tested scenario. The results of the experiments show that solutions based on edge devices such as Jetson Xavier NX and technologies such as TimescaleDB can be successfully used to reduce the volume of data in pipelines that process data from Autonomous Guided Vehicles. Additionally, the use of edge computing paradigms improves the resilience to data loss in cases of network failures in such industrial systems.

Keywords: Cloud computing · Edge computing · Automated guided vehicles · Data aggregations · Internet of things · TimescaleDB · Edge analytics

1 Introduction

In recent years, we've observed rapid growth of adoption of IoT systems for various use cases, such as manufacturing [11,18,23], environmental monitoring [7,13], smart cities applications [8,16], agriculture [12,15], health monitoring [17,19] among others. In most of the mentioned cases, a massive volume of data is generated. That data often needs to be sent over to the cloud or private

D. Groen et al. (Eds.): ICCS 2022, LNCS 13353, pp. 458–471, 2022.
https://doi.org/10.1007/978-3-031-08760-8_39

data centers, usually over the Internet, which can be challenging if the network connection is unstable or offers limited bandwidth. In order to address that problem, new computing paradigms such as fog [21] and edge [22] computing have been introduced. The main goal of these paradigms is to reduce the volume of data that needs to be sent over to the cloud by performing data processing directly on devices that are closer to the source of data. Additionally, it enables such systems to react faster to the changes in the system, even in situations where the Internet connection is slow, unreliable, or even not available most of the time. Such edge devices are often responsible for ingestion of the data, storage, aggregation of selected metrics, and sending the results to the cloud for further processing.

Focusing on manufacturing and production, we see the growing popularity of Autonomous Guided Vehicles (AGVs) [6] that allow to modernize production lines, improve internal factory logistics, ensure better safety of factory workers, which in turn enables more flexible and agile production systems. These vehicles need to process data coming from various sensors such as lidars, cameras, optical encoders in a time-efficient manner, which we believe can be improved with edge computing-based techniques.

This paper aims to evaluate two selected approaches to data aggregation performed directly on low-powered edge devices, such as Jetson Xavier NX, in the context of using them for enhancing the industrial data acquisition pipeline from AGVs. In the article, we focused on techniques that allow reducing the volume of data that will need to be sent over to the cloud for further processing. The paper is organized as follows. In Sect. 2, we review the related works. In Sect. 3, we describe the considered workflow, data models, and approaches to data aggregation. Section 4 contains a description of the testing environment. In Sect. 5, we present the performed experiments along with the results. Finally, Sect. 6 concludes the results of the paper.

2 Related Works

Scientific literature shows a variety of use cases for running data aggregations and processing directly on edge devices. In [9], Luca Greco et al. propose an edge-based analytical pipeline for real-time analysis of wearable sensor data. The authors of the research selected Raspberry Pi as an edge computing device, Apache Cassandra as a database, and Apache Kafka and Apache Flink for data processing software. They conclude that selected software, except Apache Flink, provided suitable performance and that the proposed processing pipeline architecture can be successfully used for anomaly detection in real-time.

Real-time processing with Raspberry Pi edge devices is also presented by Abdelilah Bouslama et al. [5] for medical, sensor data. The authors proposed a medical system dedicated to monitoring patients with reduced mobility or located in isolated areas. For implementing the proposed solution, the authors used Amazon Web Services cloud offering and edge devices running Node-Red applications.

One of the important problems while processing data streams is detecting data patterns directly on edge devices instead of doing all processing in the cloud. Eduard Renart et al. [20] proposed a processing framework dedicated to stream processing of data from smart city environments. The authors compare their solution to a single-cloud deployment of Apache Storm and Apache Kafka and conclude that for considered workflows, their solution can offer up to 78% latency reduction and 56% computation time reduction.

In another research, Fatos Xhafa et al. [25] mention the challenges related to processing IoT data streams in the context of edge computing. They focus on processing data coming from cars and experimentally evaluate an infrastructure based on a Raspberry Pi device using Node-Red. The authors conclude the paper with performance tests of the proposed infrastructure. They highlight that a single Raspberry Pi is not suitable for the selected use cases as it cannot process all incoming data in a time-efficient manner.

However, Raspberry Pi is frequently used as the edge device, which is visible in Hikmat Yar et al. [26]. The authors proposed a smart home automation system based on the Raspberry Pi device, which serves as a central controlling unit, analytical engine, and storage system for data generated by smart home devices. Thanks to processing sensor data directly at the edge, authors managed to reduce bandwidth, computation, and storage costs.

Edge computing is also used in other areas. M. Safdar Munir et al. [14] proposed an intelligent irrigation system based on the edge computing paradigm. The authors take advantage of edge servers to collect data from sensors, validate it, and preprocess before sending it to the cloud service for further evaluation against trained machine learning models. The authors highlight that the use of the edge computing paradigm allows reducing the volume of data sent over and improves the overall speed and efficiency of the whole system.

Zhong Wu and Chuan Zhou in [24] propose a system dedicated to detecting the riding posture of equestrian athletes to propose improvements to it in real-time, using the edge computing paradigm for data processing. After performing packet loss rate, driving, and riding tests, the authors conclude that the proposed solution can be successfully used in practical applications.

In another paper, Hong Zhang et al. [27] described an object tracking system dedicated to smart cities, based on IoT and edge computing paradigm. The major contribution is the proposal of a correlation filter algorithm for lightweight computation tracking that can successfully run on low-power consumption IoT devices such as Raspberry Pi or Xilinx SoC platforms. The proposed solution offers better tracking accuracy and robustness than comparable existing systems.

In [10], Grzesik et al. evaluate the possibility of running metagenomic analysis in real-time in the edge computing environment. The authors evaluate Jetson Xavier NX in multiple power modes, running basecalling and classification workloads. After performance experiments, the authors conclude that Jetson Xavier NX can serve as a portable, energy-efficient device capable of running metagenomics experiments.

The above examples show that there is a lot of interest in performing data processing and aggregations directly on edge devices, taking advantage of the edge computing paradigm. This paper aims to expand knowledge in this area in the context of processing industrial data from AGVs. We show two strategies for reducing the amount of data transferred to the data center for further analysis, relying on low-powered boards such as Jetson Xavier NX [1] and software such as TimescaleDB [3].

3 Analytical Workflow, Data Models, and Aggregation Methods

The analytical workflow consists of a few steps. Firstly, the AGV client periodically retrieves the data from each AGV and persists it in a local database. A separate process is responsible for running aggregations on data retrieved from the local storage. The results from these aggregations are also persisted separately in a local database. Lastly, the third process is responsible for retrieving the aggregated data, performing optional filtering, and sending the aggregated data to the cloud for further processing. The workflow diagram is presented in Fig. 1.

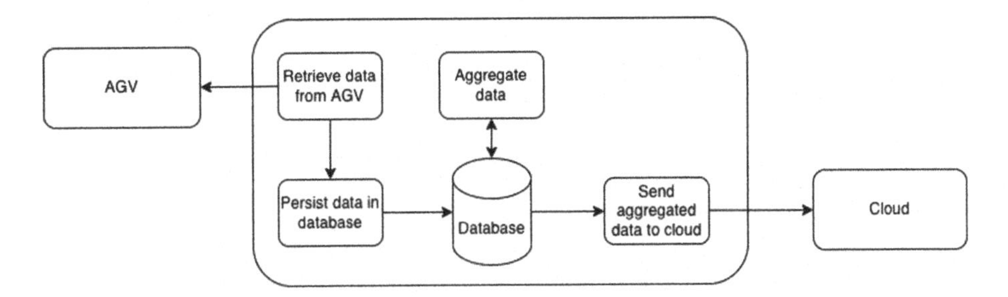

Fig. 1. Diagram of the aggregation workflow on the edge device.

3.1 Data Models and Aggregation Approaches

Each data point collected from the AGV consists of metrics such as battery cell voltage, momentary and cumulative power, energy, and current consumption, cumulative distances, and momentary frequencies. Each reading is additionally timestamped and tagged with the unique AGV identifier. Table 1 shows the structure of a single data point with corresponding data types, their sizes in bytes, and aggregation methods that will be applied to that field. The total size of a single data point from AGV is equal to 84 bytes. The data readings were obtained during real-world experiments with the Formica-1 AGV (Fig. 2) performing different types of workflows and it was collected with the intention of using it for performing predictive maintenance analysis in a cloud-based environment with the use of machine learning algorithms.

Table 1. Data model of raw readings from AGVs

Value	Type	Size in bytes	Aggregation methods
AGV ID	Integer	4	–
Timestamp	Timestamp	8	–
Momentary current consumption	Decimal	8	AVG, MAX, MIN
Battery cell voltage	Decimal	8	LAST
Momentary power consumption	Decimal	8	AVG, MAX, MIN
Momentary energy consumption	Decimal	8	AVG, MAX, MIN
Cumulative energy consumption	Decimal	8	LAST
Momentary frequency left	Decimal	8	AVG, MAX, MIN
Momentary frequency right	Decimal	8	AVG, MAX, MIN
Cumulative distance left	Decimal	8	LAST
Cumulative distance right	Decimal	8	LAST

Table 2. Data model of aggregated AGV readings

Value	Type	Size in bytes
AGV ID	Integer	4
Aggregation timestamp	Timestamp	8
Average momentary current	Decimal	8
Max momentary current	Decimal	8
Min momentary current	Decimal	8
Last battery voltage	Decimal	8
Average momentary power	Decimal	8
Min momentary power	Decimal	8
Max momentary power	Decimal	8
Min momentary energy	Decimal	8
Max momentary energy	Decimal	8
Average momentary energy	Decimal	8
Last cumulative energy	Decimal	8
Average momentary frequency left	Decimal	8
Max momentary frequency left	Decimal	8
Min momentary frequency left	Decimal	8
Average momentary frequency right	Decimal	8
Max momentary frequency right	Decimal	8
Min momentary frequency right	Decimal	8
Last cumulative distance left	Decimal	8
Last cumulative distance right	Decimal	8

3.1.1 Window Aggregation

The first considered method of reducing the volume of the data is called window aggregation and is presented in Algorithm 1. Firstly, the data points collected from AGVs $p \in P$ are filtered against *agvid* for each of N AGVs and the aggregation window start ws and end we, which is shown in line 6 of the algorithm. Afterward, the filtered readings $p' \in P'$ are sorted by the *timestamp* property. Next, each field of the filtered readings $p' \in P'$ is aggregated over time, which is shown in line 8 of the algorithm. For particular types of values, we have decided that we will only report the last recorded value in the aggregation period, while for others, we decided to report average, maximum, and minimal values in the aggregation period. Afterward, the resulting aggregation point is appended to the result array A. After processing data for all AGVs in a given time window, the aggregation window start ws and end we are incremented by time window size Δt (lines 14–15). The processing stops after the algorithm reaches the end aggregation timestamp te (line 3). The structure of aggregated data A, along with the selected aggregation operations for each data point property, is presented in Table 2. The total size of a single aggregation data point is equal to 164 bytes.

Algorithm 1: Data aggregation algorithm

Data: P (all raw readings from AGVs), Δt (aggregation window size), ts (start aggregation timestamp), te (end aggregation timestamp), N (the number of AGVs)

Result: A (all aggregated data points)

```
1  ws ← ts ;                                    /* aggregation window start */
2  we ← ts + Δt ;                               /* aggregation window end */
3  agvid ;                                      /* identifier of a single AGV */
4  while we < te do
5      for agvid ← 1 to N do
6          P' ← ∅ ;                             /* readings set for aggregation window */
7          foreach p ∈ P do
8              if p.agvid = agvid and p.timestamp > ws and
                p.timestamp <= we then
9                  P' ← P' ∪ p;
10             end
11         end
12         Sort set of filtered readings p' ∈ P' by timestamp property;
13         Aggregate each field of filtered readings p' ∈ P' over time;
14         Append aggregation points to result array A;
15     end
16     ws ← ws + Δt;
17     we ← te + Δt;
18 end
```

3.1.2 Window Aggregation with Delta Updates

Another method of reducing the volume of the data we considered is called *window aggregation with delta updates* and operates according to Algorithm 2. The aggregation step is the same as in Sect. 3.1.1. However, when sending the aggregation data, an additional step is introduced in which all aggregated readings from AGVs $a \in A$ are first filtered against *agvid* for each of N AGVs. The result of that operation is assigned to the AF property, which is shown in line 5 of the algorithm. Then, the readings $af \in AF$ are sorted by the timestamp property. In the next step, starting in line 9 of the algorithm, the aggregation results $AF[i]$ are compared with aggregation results for the previous time window $AF[i-1]$ when constructing the data point p. If the value for the field did not change compared to the previous time window, it is not included in the resulting data point p. The data point p is appended to the array of delta updates D with data points that will be sent to the cloud for further processing.

Algorithm 2: Delta updates optimization

```
   Data: A (all aggregated readings from AGVs), N (the number of AGVs)
   Result: D (delta updates data points)
 1 for agvid ← 1 to N do
 2 |   AF ← ∅;
 3 |   foreach a ∈ A do
 4 |   |   if a.agvid = agvid then
 5 |   |   |   AF ← AF ∪ a;
   |   |   |   ;                                  /* readings set for given AGV */
 6 |   |   end
 7 |   end
 8 |   Sort set of filtered readings af ∈ AF by timestamp property and assign
 |       result to AF;
 9 |   for i ← 0 to len(AF) do
10 |   |   if i = 0 then
11 |   |   |   D ← D ∪ AF[i];
12 |   |   else
13 |   |   |   Compare each property of AF[i] with AF[i − 1] and include only
   |   |   |     the fields that are different in the resulting point p;
14 |   |   |   D ← D ∪ p;
15 |   |   end
16 |   end
17 end
```

4 Testing Environment

The testing environment consists of a group of autonomous guided vehicles equipped with sensors and a central edge device that serves as a database, AGV client, and an analytical engine for the AGV data. Figure 3 presents the diagram of the described edge computing system. The AGV that was used for data collection was Formica 1, produced by AIUT Ltd. and is presented on Fig. 2.

Fig. 2. Autonomous Guided Vehicle Formica 1.

The edge computing device selected for this experiment was Jetson Xavier NX, with its full technical specification presented below [1]:

- CPU - 6-core NVIDIA Carmel ARM®v8.2 64-bit CPU 6 MB L2 + 4 MB L3
- GPU - NVIDIA Volta™ architecture with 384 NVIDIA® CUDA® cores and 48 Tensor cores
- Memory - 8 GB 128-bit LPDDR4x 51.2 GB/s
- OS Storage - SDHC card (32 GB, class 10)
- DB Storage - Solid State Drive, PNY 500 GB M.2 PCIe NVMe XLR8 CS3030
- OS - Ubuntu 18.04.5 LTS with kernel version 4.19.140-tegra

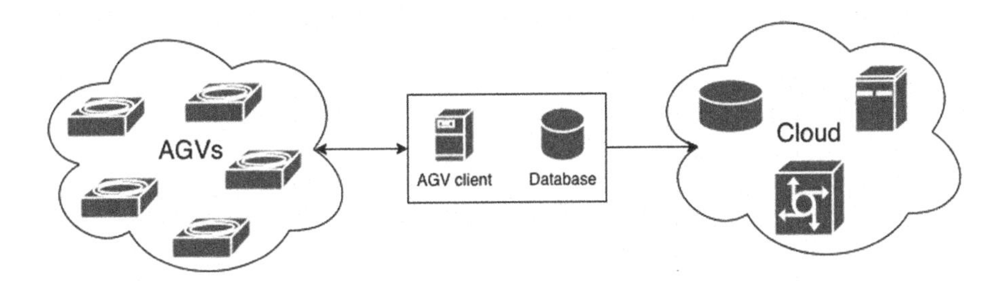

Fig. 3. Diagram of the edge computing system.

For persistence and analytical aggregations, we selected TimescaleDB, which is an open-source, written in C, time-series database, distributed as an extension to PostgreSQL [2]. It extends traditional tables to form data structures called hypertables, which are abstractions for single, continuous tables. Internally, hypertables are split into chunks, which are implemented using standard PostgreSQL tables [4] and represent data for specific time intervals. It offers support for all PostgreSQL client libraries and SQL operations. It can be used as a drop-in replacement of traditional relational databases that additionally provides significant improvements for storing and processing time-series data, namely, fast data ingestion and support for analytical queries over time windows [3].

5 Performance Experiments

To experimentally test the data reduction rates with the aggregation strategies presented in the paper, we simulated the production environment with 10 AGVs generating data based on the actual sensor readings obtained from operational cycles of the Formica 1 vehicle. From each AGV, we read 1,380 data points in regular time intervals. In the experiment, we used three such data sets, with readings generated every 200, 500, and 1,000 ms, which resulted in 41,400 data points used for testing. The generated data was based on actual readings recorded during real-world tests with the Formica-1 AGV.

For the purposes of running the experiments, we prepared several Python software packages developed to carry out each step of the workflow. Firstly, the generated data was persisted to the database to three separate tables, each for different data set based on sampling frequency (200, 500, and 1000 ms between reads). Then, the second package was responsible for generating data aggregates, taking advantage of native database aggregation functions, and the results were persisted to separate tables. Aggregations were run for each data set with different aggregation time windows - 2, 5, and 10 s. Lastly, for each of the aggregation approaches, we computed the total number of bytes that would have to be sent to the cloud for further processing and compared the results against the baseline of the number of bytes that would be sent if we forwarded all readings directly to the cloud.

Firstly, we analyzed results obtained for data readings that were collected every 200 ms. With aggregations being done every 2 s, the total size of data that needs to be sent to the cloud was reduced from the baseline of 1,159,200 bytes to 226,320 bytes with just aggregations and to 153,680 bytes with additional optimization of sending only data that changed between aggregation results (delta updates). For aggregations done every 5 s, we measured 91,840 bytes for the aggregation alone and 73,920 bytes with delta updates, where for the time window of 10 s, the results were 45,920 and 40,640 bytes, respectively. The results for that data set are presented in Fig. 4.

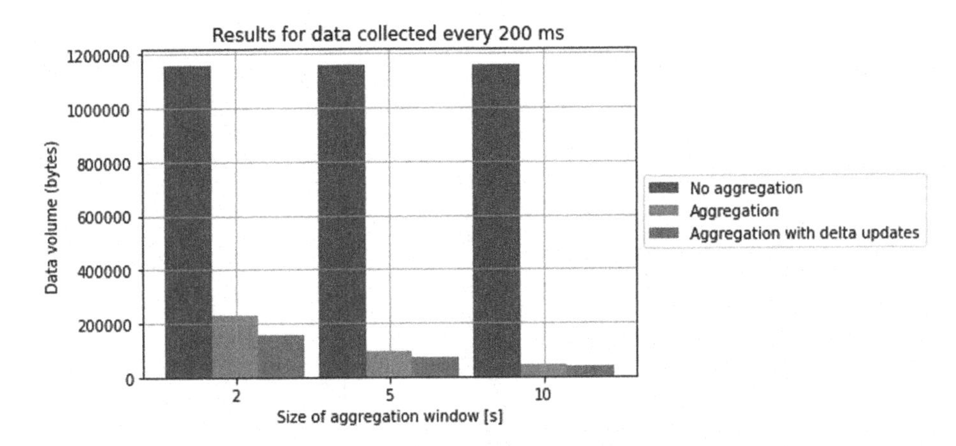

Fig. 4. Dependency between data volume for various data aggregation scenarios and different sizes of the aggregation window for data readings collected every 200 ms.

For data readings collected every 500 ms, we've obtained the following results. The baseline number of bytes was once again 1,159,200 bytes, and it was reduced to 565,800 and 323,960 bytes for a 2-seconds time window, to 226,320 and 153,680 bytes for 5-seconds time window, and finally to 113,160 and 86,040 bytes for 10-seconds aggregation window for aggregation and aggregation with delta updates. The results for that data set are presented in Fig. 5.

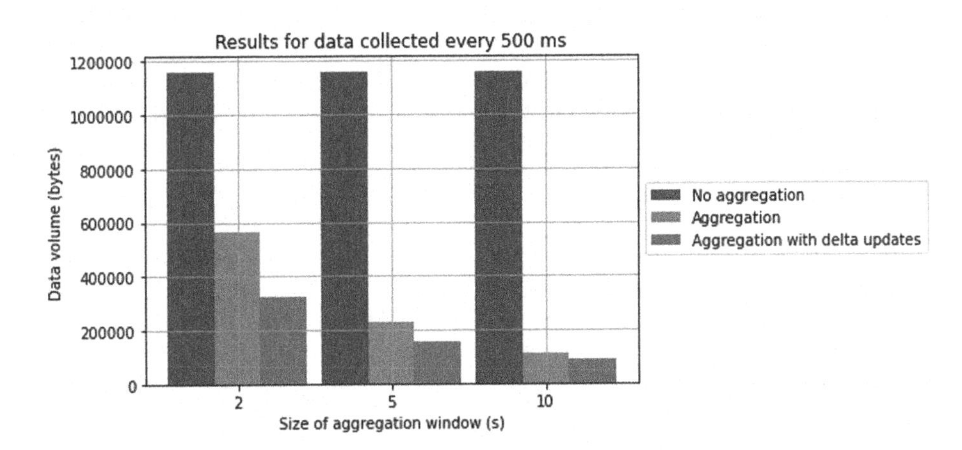

Fig. 5. Dependency between data volume for various data aggregation scenarios and different sizes of the aggregation window for data readings collected every 500 ms.

Lastly, we have considered the data set that contained readings collected every 1000 ms, which had the same baseline number of 1,159,200 bytes. For the 2-seconds aggregation window, we've achieved a reduction to 1,131,600 and 538,880 bytes, for the 5-seconds window to 452,640 and 275,680 bytes, and for the 10-seconds aggregation window, the results were 226,320 and 153,680 bytes, respectively for aggregation and aggregation with delta updates optimization. The results for that data set are presented in Fig. 6.

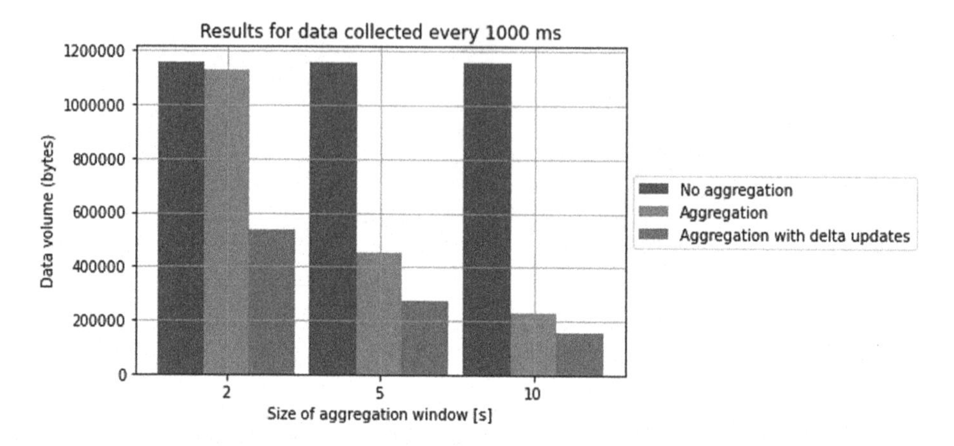

Fig. 6. Dependency between data volume for various data aggregation scenarios and different sizes of the aggregation window for data readings collected every 1000 ms.

Thanks to proposed aggregation techniques, we achieved data volume reduction from 2% for 1000 ms data set and window aggregation method done every 2 s to as much as 81.1% for 200 ms data set and window aggregation with delta updates method done every 10 s. The data size reductions for each data set and each aggregation method are summarized in Table 3. We also observe that the delta updates technique is more effective for smaller aggregation windows, which is best visible for a 1000 ms data set. Here, it helped to improve data volume reduction from 2% to 44.9% for a 2-s aggregation window.

Table 3. Data size reduction for considered aggregation methods

	Aggregation			Agg. with delta updates		
Aggregation time window [s]	2	5	10	2	5	10
Data size reduction for 200 ms data set [%]	67.6	77.3	80.7	72.8	78.6	81.1
Data size reduction for 500 ms data set [%]	43	67.6	75.8	60.5	72.9	77.8
Data size reduction for 1000 ms data set [%]	2	51.2	67.6	44.9	64	72.9

6 Concluding Remarks and Future Work

The decision to take advantage of the edge-based data processing pipeline can have multiple benefits for the overall industrial system. As shown in the experiments presented in the paper, both proposed aggregation strategies can greatly help with reducing the volume of readings data from AGVs that need to be send over to the cloud for further processing. In addition to reducing the use of the network bandwidth, using edge storage system based on TimescaleDB database improves resiliency to network outages, because the data is additionally persisted on the edge device, which prevents potential data loss in such scenarios. The aggregated data available at the edge of the network can also be used to make quicker decisions based on the state of the system and improve reaction time to changing environments. In our paper, we proved that the solution based on such edge devices like Jetson Xavier NX and using technologies such as TimescaleDB could successfully help reduce the volume of data in industrial data acquisition pipelines for Autonomous Guided Vehicles. In the future works, we plan to expand the proposed algorithms to evaluate if we can achieve better efficiency and data volume reduction, while preserving data quality, exploring techniques such as GPU acceleration.

Acknowledgments. The research was supported by the Polish Ministry of Science and Higher Education as a part of the CyPhiS program at the Silesian University of Technology, Gliwice, Poland (Contract No. POWR.03.02.00-00-I007/17-00), the Norway Grants 2014-2021 operated by the National Centre for Research and Development under the project "Automated Guided Vehicles integrated with Collaborative Robots for Smart Industry Perspective" (Project Contract no.: NOR/POL-NOR/CoBotAGV/0027/2019-00) and by Statutory Research funds of Department of Applied Informatics, Silesian University of Technology, Gliwice, Poland (grant No BK/RAu7/2022).

References

1. Jetson Xavier NX specification. https://developer.nvidia.com/embedded/jetson-xavier-nx-devkit. Accessed 5 Feb 2022
2. PostgreSQL documentation. https://www.postgresql.org/about/. Accessed 9 Feb 2022
3. TimescaleDB documentation. https://docs.timescale.com/latest/introduction. Accessed 9 Feb 2022
4. TimescaleDB: SQL made scalable for time-series data (2017). https://pdfs.semanticscholar.org/049a/af11fa98525b663da18f39d5dcc5d345eb9a.pdf
5. Bouslama, A., Laaziz, Y., Tali, A., Mohamed, E.: AWS and IoT for real-time remote medical monitoring. Int. J. Intell. Enterp. **6**, 293–310 (2019)
6. Cupek, R., et al.: Autonomous guided vehicles for smart industries – the state-of-the-art and research challenges. In: Krzhizhanovskaya, V.V., et al. (eds.) ICCS 2020. LNCS, vol. 12141, pp. 330–343. Springer, Cham (2020). https://doi.org/10.1007/978-3-030-50426-7_25
7. Fadhel, M., Sekerinski, E., Yao, S.: A comparison of time series databases for storing water quality data. In: Auer, M.E., Tsiatsos, T. (eds.) IMCL 2018. AISC, vol. 909, pp. 302–313. Springer, Cham (2019). https://doi.org/10.1007/978-3-030-11434-3_33
8. Gaur, A., Scotney, B., Parr, G., McClean, S.: Smart city architecture and its applications based on IoT. Procedia Comput. Sci. **52**, 1089–1094 (2015)

9. Greco, L., Ritrovato, P., Xhafa, F.: An edge-stream computing infrastructure for real-time analysis of wearable sensors data. Future Gener. Comput. Syst. **93**, 515–528 (2019). https://www.sciencedirect.com/science/article/pii/S0167739X18314031
10. Grzesik, P., Mrozek, D.: Metagenomic analysis at the edge with Jetson Xavier NX. In: Paszynski, M., Kranzlmüller, D., Krzhizhanovskaya, V.V., Dongarra, J.J., Sloot, P.M.A. (eds.) ICCS 2021. LNCS, vol. 12745, pp. 500–511. Springer, Cham (2021). https://doi.org/10.1007/978-3-030-77970-2_38
11. Hu, L., Miao, Y., Wu, G., Hassan, M.M., Humar, I.: iRobot-Factory: an intelligent robot factory based on cognitive manufacturing and edge computing. Future Gener. Comput. Syst. **90**, 569–577 (2019). https://www.sciencedirect.com/science/article/pii/S0167739X1831183X
12. Jaiganesh, S., Gunaseelan, K., Ellappan, V.: IOT agriculture to improve food and farming technology. In: 2017 Conference on Emerging Devices and Smart Systems (ICEDSS), pp. 260–266 (2017)
13. Liu, X., Nielsen, P.S.: Air quality monitoring system and benchmarking. In: Bellatreche, L., Chakravarthy, S. (eds.) DaWaK 2017. LNCS, vol. 10440, pp. 459–470. Springer, Cham (2017). https://doi.org/10.1007/978-3-319-64283-3_34
14. Munir, M.S., Bajwa, I.S., Ashraf, A., Anwar, W., Rashid, R.: Intelligent and smart irrigation system using edge computing and IoT. Complexity **2021**, 6691571 (2021). https://doi.org/10.1155/2021/6691571
15. Nandyala, C.S., Kim, H.K.: Green IoT agriculture and healthcare application (GAHA). Int. J. Smart Home **10**(4), 289–300 (2016)
16. Neelakandan, S., Berlin, M., Tripathi, S., Devi, V.B., Bhardwaj, I., Arulkumar, N.: IoT-based traffic prediction and traffic signal control system for smart city. Soft. Comput. **25**(18), 12241–12248 (2021)
17. Paul, A., Pinjari, H., Hong, W.H., Seo, H., Rho, S.: Fog computing-based IoT for health monitoring system. J. Sens. **2018**, 1–7 (2018)
18. Raileanu, S., Borangiu, T., Morariu, O., Iacob, I.: Edge computing in industrial IoT framework for cloud-based manufacturing control. In: 2018 22nd International Conference on System Theory, Control and Computing (ICSTCC), pp. 261–266 (2018)
19. Rajavel, R., Ravichandran, S.K., Harimoorthy, K., Nagappan, P., Gobichettipalayam, K.R.: IoT-based smart healthcare video surveillance system using edge computing. J. Ambient Intell. Humaniz. Comput. (2021). https://doi.org/10.1007/s12652-021-03157-1
20. Renart, E.G., Diaz-Montes, J., Parashar, M.: Data-driven stream processing at the edge. In: 2017 IEEE 1st International Conference on Fog and Edge Computing (ICFEC), pp. 31–40 (2017)
21. Sabireen, H., Neelanarayanan, V.: A review on fog computing: architecture, fog with IoT, algorithms and research challenges. ICT Express **7**(2), 162–176 (2021)
22. Singh, S.: Optimize cloud computations using edge computing. In: 2017 International Conference on Big Data, IoT and Data Science (BID), pp. 49–53, December 2017
23. Wang, X., Garg, S., Lin, H., Kaddoum, G., Hu, J., Alhamid, M.F.: An intelligent UAV based data aggregation algorithm for 5G-enabled internet of things. Comput. Netw. **185**, 107628 (2021). https://www.sciencedirect.com/science/article/pii/S138912862031255X
24. Wu, Z., Zhou, C.: Equestrian sports posture information detection and information service resource aggregation system based on mobile edge computing. Mob. Inf. Syst. **2021**, 4741912, July 2021. https://doi.org/10.1155/2021/4741912

25. Xhafa, F., Kilic, B., Krause, P.: Evaluation of IoT stream processing at edge computing layer for semantic data enrichment. Future Gener. Comput. Syst. **105**, 730–736 (2020). https://www.sciencedirect.com/science/article/pii/S0167739X19321296
26. Yar, H., Imran, A.S., Khan, Z.A., Sajjad, M., Kastrati, Z.: Towards smart home automation using IoT-enabled edge-computing paradigm. Sensors **21**(14) (2021). https://www.mdpi.com/1424-8220/21/14/4932
27. Zhang, H., Zhang, Z., Zhang, L., Yang, Y., Kang, Q., Sun, D.: Object tracking for a smart city using IoT and edge computing. Sensors **19**(9) (2019). https://www.mdpi.com/1424-8220/19/9/1987

Performance of Explainable AI Methods in Asset Failure Prediction

Jakub Jakubowski[1(✉)], Przemysław Stanisz[1], Szymon Bobek[2], and Grzegorz J. Nalepa[2]

[1] AGH University of Science and Technology, 30-059 Krakow, Poland
jjakubow@agh.edu.pl

[2] Jagiellonian Human-Centered Artificial Intelligence Laboratory (JAHCAI), Institute of Applied Computer Science, Jagiellonian University, 30-348 Krakow, Poland

Abstract. Extensive research on machine learning models, which in the majority are black-boxes, created a great need for the development of Explainable Artificial Intelligence (XAI) methods. Complex machine learning (ML) models usually require an external explanation method to understand their decisions. The interpretation of the model predictions are crucial in many fields, i.e., predictive maintenance, where it is not only required to evaluate the state of an asset, but also to determine the root causes of the potential failure. In this work, we present a comparison of state-of-the-art ML models and XAI methods, which we used for the prediction of the RUL of aircraft turbofan engines. We trained five different models on the C-MAPSS dataset and used SHAP and LIME to assign numerical importance to the features. We have compared the results of explanations using stability and consistency metrics and evaluated the explanations qualitatively by visual inspection. The obtained results indicate that SHAP method outperforms other methods in the fidelity of explanations. We observe that there exist substantial differences in the explanations depending on the selection of a model and XAI method, thus we find a need for further research in XAI field.

Keywords: Machine learning · Explainable artificial intelligence · Predictive maintenance

1 Introduction

In the last ten years, we observe a tremendous growth of products and research papers related to machine learning and data mining. The increasing number of real-life applications of these techniques is driven by multiple factors. From the technical perspective, cloud computing allows to train complex models on specially designed clusters. Progress in the field of big data enables researchers to store and process an enormous amount of data. Development of machine learning techniques, especially deep learning, affected in the improvement of the model accuracy with the significant growth in areas like image classification,

D. Groen et al. (Eds.): ICCS 2022, LNCS 13353, pp. 472–485, 2022.
https://doi.org/10.1007/978-3-031-08760-8_40

natural language processing, speech recognition, decision making, and time series analysis. Furthermore, artificial intelligence (AI) models are now much more accessible thanks to the development of open source frameworks in a variety of programming languages including Python, R, and C++.

For manufacturing companies, artificial intelligence is one of the driving factors in the transition to Industry 4.0 [4], which is the key challenge in the years to come. The applications of artificial intelligence models in the manufacturing industry include process control, production planning, network traffic monitoring, and predictive maintenance (PdM). The potential benefits for companies to use machine learning for monitoring the health of their assets are very high as nowadays relatively simple techniques are used. Most of the machine's equipment is now replaced either in a corrective or preventive manner. The first aims to replace the element after its failure, whereas the goal of the second method is to replace an element after a predefined period of time, before it fails. Both strategies have significant drawbacks, which justifies the need for new solutions. The corrective approach is not suitable for critical assets, which failure may cause safety issues or significant financial losses. On the other hand, the preventive approach may increase the total operating cost due to the increased frequency of asset replacements. Using machine learning for estimating the condition of the machine may be a promising alternative, which can give the industry substantial gains. The topic of RUL prediction with the use of artificial intelligence approaches was widely studied by researchers in the last years, especially in the field of deep learning [23].

One of the major issues with machine learning models is their black-box nature, which impedes the understanding of the model and the result. This lack of transparency may impact the trustworthiness of the model during the development phase as well as during its operation in the production environment (real-life applications), especially when there is a need to understand factors influencing the model decision. Except for that, there are also legal concerns, which may oblige companies and institutions to provide explanations for the model prediction whenever it affects user [6].

To address these issues, Explainable AI (XAI) methods, which try to explain the prediction of black-box models, gained popularity among researchers in recent years. Despite the rapid growth in the field of XAI, there are concerns about its efficiency in giving the right explanations, which lead to the conclusion that black-box models should not be used in any high-stake decisions [20]. Another alternative is to use glass-box models, which are models inherently interpretable, thus they do not require any additional mechanism to provide the explanations.

In this research, we evaluate the performance of black-box explainability methods and compare it with the results obtained by the interpretable machine learning model. We focus on two explainability metrics - stability and consistency. Those metrics can be used for quantitative assessment of the explanations produced by different models. They allow to compare the explanations within one model and between different models. This paper is a part of Explainable

Predictive Maintenance (XPM) project, which is devoted to the use of XAI in predictive maintenance solutions. In the project we focus on four different real-life cases, which namely are: steel manufacturing, city subway, wind farms and trucks maintenance. This paper constitutes a preliminary work in the area of evaluation of XAI methods for PdM. Hence, to assure reproducibility, we have based this study on a public data set, which describes a degradation of turbofan engine (CMAPSS) (provided by NASA). In particular, we aim at analysing which XAI methods and ML models are suitable to predict the RUL of the tubrofan engine and provide acceptable explanations of that decision.

The rest of the paper is organized as follows. In Sect. 2, we provide an overview of state-of-the-art machine learning, explainability methods, interpretable models, and metrics used for the evaluation of model explanations. In Sect. 3, we present the failure prediction case, which we use in the study – this is the dataset coming from the simulations of an aircraft turbofan engine. We also present our approach towards predicting asset failure with explanations and evaluate the quality of those explanations. In Sect. 4, we present the results obtained in this study and in Sect. 5 we summarize our research and point out directions for further investigation.

2 Explainable Artificial Intelligence

Machine learning models combine complex mathematical algorithms with the data coming from a certain process to build a general mathematical model, which is able to make a correct prediction on previously unseen data. In most cases, complex models are able to achieve very high accuracy scores in the certain problem, but generally they are significantly more difficult to explain [7].

2.1 Explaining Black-Box Models

Explainable AI algorithms are able to build an understanding of the black-box models by applying different methods, i.e., input perturbations, to find the driving factors of the prediction. The process of producing explanations depends on factors like the characteristics of the explained model, data structure, and prediction type, i.e., image data need different methods of explanation than tabular data as the values of each pixel are not understandable by humans straightaway – they need to be visualized.

The methods might be either model-specific or model-agnostic. In the first case, only a specific predefined type of models can be explained – examples of such methods are Grad-CAM [22], which is designed to give visual explanations of deep learning models, and RFEX, which focuses strictly on the explanation of Random Forest Classifier [17]. The model-agnostic approach is not based on selected AI methods, but aims to build framework for explanations of any model. Examples of such explanation methods are Local Interpretable Model-Agnostic Explanations (LIME) [18], SHapley Additive exPlanations (SHAP) [12], Anchors [19].

Another way of dividing explainability methods is into local and global explanations. Local explanations aim to understand why the model made a certain prediction in a selected case (one observation), while the global explanations try to give an overall understanding of the model as a whole.

The XAI methods may also differ based on the way of presenting the explanation to the end-user. SHAP gives a numerical value of feature importance, which tries to evaluate how the prediction of the model changes under the certain condition (value of the selected feature), while Anchors explain the prediction in the form of rules.

2.2 Glass-Box Models

The problem of explainability does not exist in the case of glass-box models, which are interpretable without further need of using explainability methods [20]. This may increase the reliability of the machine learning model, because the explanations do not rely on the external method (such as SHAP), which may also be not trustworthy for the end-user of the solution.

One of the simplest and most widely used interpretable models are linear models (i.e., linear regression, logistic regression). However, their performance is known to be relatively poor on more complex data sets. An extension of the linear models, which may increase their accuracy are, for example, Generalized Additive Models (GAMs) [8], which instead of using linear relationships between features, use many nonlinear equations, which are summed to give the prediction. Comparing them with linear models trade-off between accuracy and explainability is observed – at the cost of higher performance, GAMs are less interpretable. Moreover, their accuracy is generally not as high as state-of-the-art algorithms [13].

A promising algorithm, which tries to achieve high accuracy and interpretability is Explainable Boosting Machine (EBM) [15]. It is based on the idea of General Additive Models, but uses more advanced machine learning techniques like boosting and bagging to improve the accuracy of the model.

2.3 Explainability Metrics

In the previous sections, we have highlighted that even though explainability methods give more insight into the prediction of the model, they may not be trustworthy themselves. Thus, there is a need to derive metrics, which can be used to validate the performance of the models and XAI methods. The two base requirements, which must be fulfilled to perceive model explanations as trustworthy are that (1) similar observations should lead to similar explanations within a certain model and (2) the explanations for a given observation should be similar irrespective of the machine learning model and explanation method.

The first criterion assures that small changes in the observations will not lead to high changes in the explanations. This is referred as stability or robustness of the explainable model. Alvarez-Melis and Jaakkola have proposed to use a metric based on Lipschitz continuity to calculate the stability of the explanation at a given point [2]:

$$\tilde{L}_x(x_i) = \max_{x_j \in \mathcal{N}_\in (x_i) \leq \epsilon} \frac{\|\Phi_i - \Phi_j\|_2}{\|x_i - x_j\|_2} \tag{1}$$

where $\tilde{L}_x(x_i)$ is the stability of the point x_i, Φ_i and Φ_j are the feature importance vectors of points x_i and x_j (each feature of an observation has a feature importance assigned to it by XAI method), $\mathcal{N}_\in(x_i)$ is the neighborhood of x_i, which is defined as all points which have distance (defined as L2 norm) to x_i smaller than ϵ.

The general idea behind this metric is to find all points in the neighborhood of the point x_i and find the maximum dissimilarity, which is defined as the Euclidean distance between explanations divided by Euclidean distance between the points. The lower the value of stability, the better performance of the model at a given point. The major drawback of this metric is that its value is relative, therefore it is not possible to conclude on the stability of one model without having a comparison with other models.

The second criterion is used for the comparison of different models and validate their consistency with each other. It assumes that if we have two different models (or explanation methods) then we expect to have similar explanations for the same observation. If this is not the case, then either one model (at least) is not making a good prediction or the explanation method is not properly finding relevant features. The consistency metric has been proposed in [3]. For the comparison of two different explanations, equation takes the following form:

$$C(\Phi_{1,i}, \Phi_{2,i}) = \frac{1}{\|\Phi_{1,i} - \Phi_{2,i}\|_2 + 1} \tag{2}$$

where $C(\Phi_{1,i}, \Phi_{2,i})$ if the consistency of ith observation, $\Phi_{1,i}$ and $\Phi_{2,i}$ are the feature importance vectors of ith observation for the first and second model respectively.

In a perfect scenario, when all feature importances are equal, the consistency is equal to 1 and it drops as the distance between explanations increases. Theoretically, the lowest possible value of consistency defined in such a way is 0. Nevertheless, the value of consistency is also dependent on the magnitude of the feature importance vector, therefore it may be affected by feature engineering, i.e., scaling.

3 Asset Failure Prediction

3.1 C-MAPSS Data Set

Commercial Modular Aero-Propulsion System Simulation (C-MAPSS) [5] is a software developed by National Aeronautics and Space Administration (NASA) to simulate the behaviour of turbofan engines. Based on this tool, Saxena et al. [21] have prepared a dataset, which consists of run-to-failure simulations of hundreds of turbofan engine units.

The dataset consists of 21 features, i.e., temperature, pressure, fan speed, fuel and coolant flow, and 3 operational parameters (settings). Each observation is the average measurement from a simulated flight of a single unit. As the number of completed cycles (flights) of a certain unit increases, the gradual degradation is observed. The turbofan engines may operate with different external conditions (up to six) and exhibit one of two failures – high-pressure compressor (HPC) or fan degradation. The simulation dataset is divided into four subsets named FD001-FD004, which contain data of different complexity – from simpler (one external condition and one type of failure) to more complex (six external conditions and two types of failure) cases.

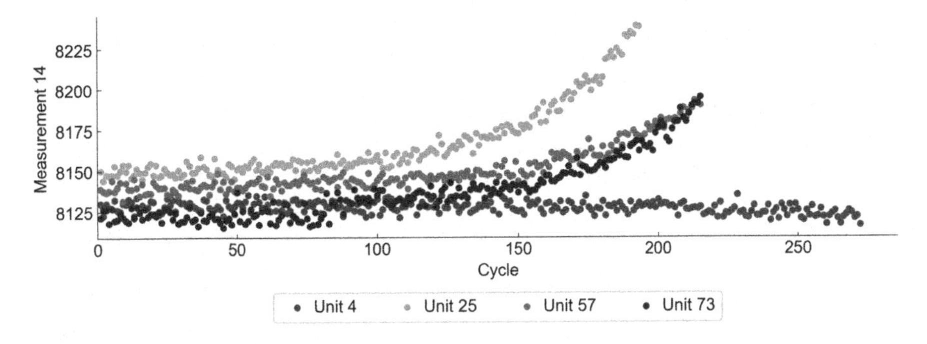

Fig. 1. Exemplary plot showing how the value of selected feature changes with the progress of engine degradation (FD003).

For each turbofan engine, the RUL at a given point may be determined based on the total number of completed cycles. This way, a prediction model, which determines the state of health of the unit may be developed. Figure 1 presents how the selected measurement deviates from normal working conditions as the number of cycles increases and the turbofan engine deteriorates. Figure 2 shows how the distributions of some measurements differ in normal working conditions and at the end of life. In most cases, a shift towards lower or higher values is observed as the unit undergoes failure.

3.2 Failure Prediction

The prediction of unit failure may be considered as a problem of finding the remaining useful life (RUL) estimation of the unit given the current working conditions. The RUL prediction on the C-MAPSS dataset has been widely studied and multiple prediction models were proposed. Most recent research is mainly focused on the development of different deep learning architectures [1,10,11,16], but more classical ML approaches were also studied [9,14].

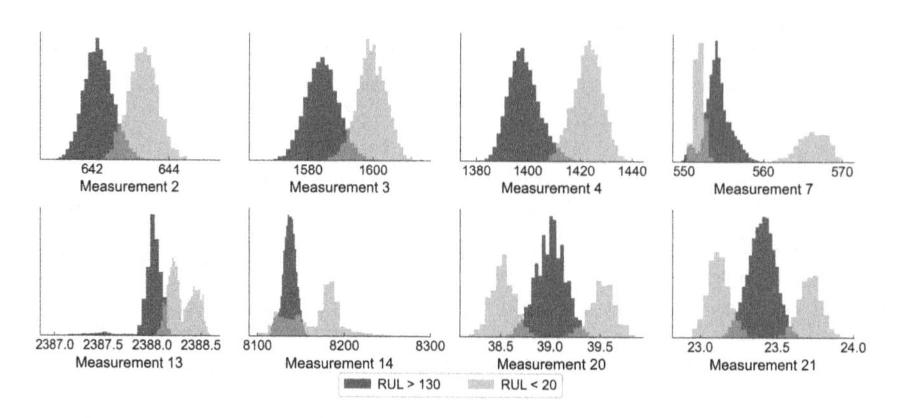

Fig. 2. Comparison of selected measurements distribution in normal (RUL $>$ 130) and degradation (RUL $<$ 20) conditions (FD003).

Our work is dedicated mostly to the topic of Explainable AI, rather than the development of new model architectures, thus we have used more commonly used state-of-the-art techniques, which are known for their performance on various types of data and prediction problems. The algorithms selected for the prediction task are XGBoost (XGB), Random Forest (RF), Support Vector Machine (SVM), and Multi-layer Perceptron (MLP). We also use Explainable Boosting Machine [15] as a glass-box alternative for the models listed above.

The dataset on which we have trained the models is the FD003, which contains a simulation of 100 turbofan engine units in a run-to-failure manner with over 24,000 observations in total. Each simulated flight is made under the same external conditions with two possible types of failure. For each unit, we calculate the RUL based on the known value of the cycle for each observation and known total number of cycles for a given unit. The feature scaling is performed in the following manner – for each unit we take the first 50 cycles (which are assumed to be always healthy working conditions) and scale all features linearly to the [0, 1] range. Then we apply the obtained scale to all observations of the given unit. With such an approach, we are able to find the relative change of the measurements in relation to the baseline, which are the first 50 cycles. The first 50 cycles are then removed from the training and test data sets, as in practice when we calculate RUL for a new unit, those points could not be predicted, as the baseline is not yet known. We also apply a rectification of RUL, which is a common practice in the case of such problems. Whenever the value of RUL exceeds 130, it is limited to 130. This is required, because when RUL is higher than 130 cycles, there are no signs of asset degradation and thus no model is able to precisely distinguish between, i.e., RUL $=$ 300 and RUL $=$ 130.

For every model, we conduct hyperparameter tuning using a grid search method to find a set of parameters, which assure the high accuracy of the model. The model accuracy is determined with the root mean square error ($RMSE$) as the model evaluation metric. The cross-validation technique with 5 folds is used to eliminate the possibility of model overfitting – data is divided into train and test sets in such a way that every unit can be only in one of these two sets at a time.

3.3 Failure Explanation

After training the models with the best found hyperparameters, we use SHAP and LIME methods to get explanations of every model for several randomly selected units.

Then, we compare the explanations in several manners. As the evaluation of the robustness, we determine the stability range for every model and compare them together. To evaluate the stability in a more qualitative manner, we plot how the feature importance for different models changes as RUL decreases in each selected unit. We expect that for the explanations to be trustworthy, the explanation for the RUL prediction should be similar throughout the whole degradation process. Otherwise, the end-user of the XAI model may be misguided and lose trust in the predictions.

We also calculate the mean consistency between every combination of the two models to check which models give the most consistent explanations.

Additional issue in the comparison of the explanation methods is that the feature importance values produced by different XAI methods cannot be directly compared, as the meaning of the feature importance magnitude might be different. To overcome this problem, we have scaled the feature importances for each method in a following manner:

$$a_E = P_{95}(\|x_i\| : \{x_i \in X_E\}) \tag{3}$$

$$\Phi^s_{E,m,i} = \frac{\Phi_{E,m,i}}{a_E} \tag{4}$$

where a_E is the scaling factor for an explanation method E, $P_n(\mathrm{X})$ denotes the nth percentile of multiset X, X_E is the concatenated multiset with all the feature importances of a given explanation method (for all models) and $\Phi_{E,m,i}$ is the feature importance of ith observation for a ML model m and explanation method E – superscitpt s denotes a scaled value.

We have decided on the 95th percentile to minimize the effect of the outliers on the final results. The scaling factors were determined for each method without distinguishing between the models to preserve the same scale within the explanation method, i.e., we assume SHAP values for all models are comparable without scaling. Although feature importance in all explanation methods may be positive or negative, in all cases $\Phi_i = 0$ means that a certain feature has no impact on the result. Thus, it is important to assure during scaling that this point does not shift, what could be achieved with, i.e., a simple min-max normalization.

4 Results and Discussion

In this section, we present the results of our study. We have trained five different models to predict the remaining useful life on the FD003 dataset. The models were trained and evaluated using Python programming language and sckit-learn library. Table 1 presents the root mean square error ($RMSE$) and coefficient of determination (R^2) calculated on the test dataset for all models used in this study.

Table 1. Metrics of machine learning models on FD003 dataset.

	XGB	RF	SVM	MLP	EBM
RMSE	**14.8**	15.0	15.1	15.1	15.4
R^2	0.878	0.867	0.884	**0.887**	0.862

All models achieved comparable performance on the test dataset and their accuracy is acceptable to use them for the prediction problem defined. From the test dataset, we have randomly selected 10 units and produced explanations for them with the use of SHAP and LIME methods – in contrast to Explainable Boosting Machine algorithm, which is interpretable. We have only explained the observations, which had an actual RUL below 130, as we are particularly interested in the explanations for the low RUL values – the explanations for the healthy turbofan unit have no practical significance.

In Fig. 3 we present how the feature importance values of a selected measurement changes as RUL decreases. SHAP values reflect the progressing degradation process, however for SVM, XGB, and RF the explanation scores converge to a certain value, while for MLP a decrease is still observed. Based on the behaviour of the measurement, MLP response seems to be more intuitive. In the case of LIME explanations, we observe three intervals: stability at high RUL values, fluctuations at intermediate RUL values, and stability at lower RUL values. Such a response does not seem to be of practical use, as in the intermediate RUL interval the state of the engine is not well explained. Changes in the explanation scores of EBM are similar to SHAP values, however higher fluctuations are observed.

The distribution of stability (as defined in Eq. 1 – with $\epsilon = 2.0$) is presented in the Fig. 4. The best stability was obtained for the models explained with the SHAP method. The lowest median was achieved by Multi-layer perceptron model, nevertheless the results for other models (explained with SHAP) are comparable.

In the Fig. 5 we presents the mean consistency between each model. The results are relatively far from 1.0, which may indicate the models are very far from being consistent, which raises the issue of their fidelity. The highest consistency score is observed between the three pairs of models: XGB and RF with SHAP, XGB and RF with LIME, SVM and MLP with SHAP. This leads to

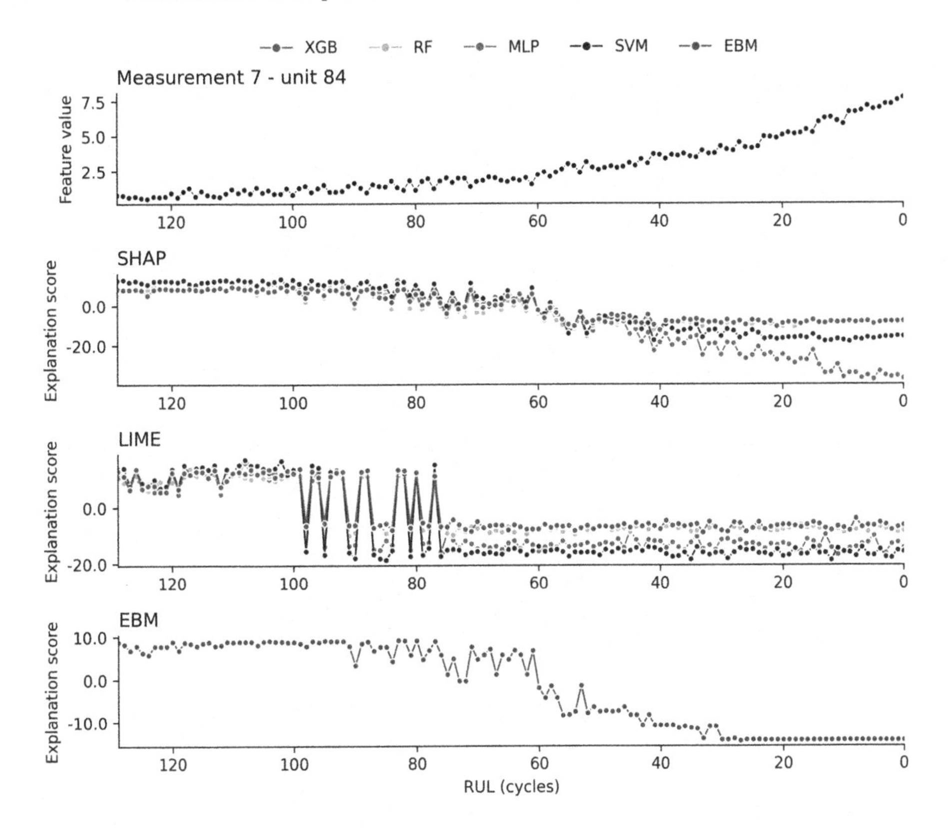

Fig. 3. Explanation scores of the Measurement 14 for a randomly selected unit.

observations that consistency is higher if we compare two models using the same XAI method and that kind of model (tree-based or not) also impacts the consistency. The explanations of the same model with different XAI techniques give lower consistency than in the case of the pairs mentioned above. It shows that the selection of important features is not only driven by the model training, but is also dependent on the choice of the XAI method.

In the analyzed dataset, each unit undergoes one of the two failures, which should also be visible in measurements and explanations – different measurements may impact the RUL depending on the type of failure that is occuring in the engine. Thus, we expect to have two clusters of failure data, each having observations from one failure. Those clusters should be both visible in the measurements as well as explanations. In the Fig. 6 we present the visualization of all features and explanations for the MLP model by reduction of the data dimensionality with the use of the Principal Component Analysis (PCA)

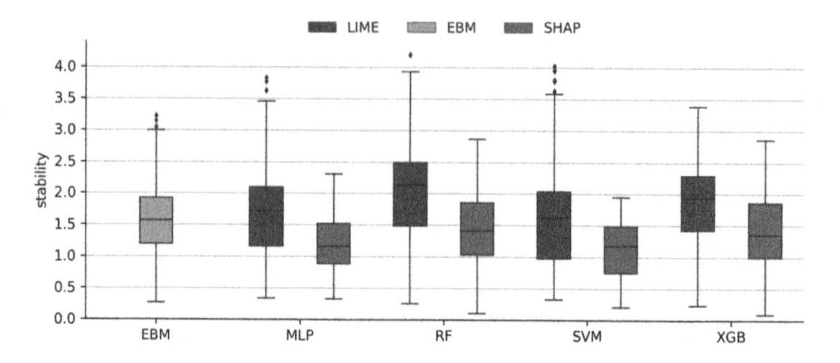

Fig. 4. Stability distribution of the investigated models for 10 randomly selected turbofan units. The lower is the median and the variance, the better.

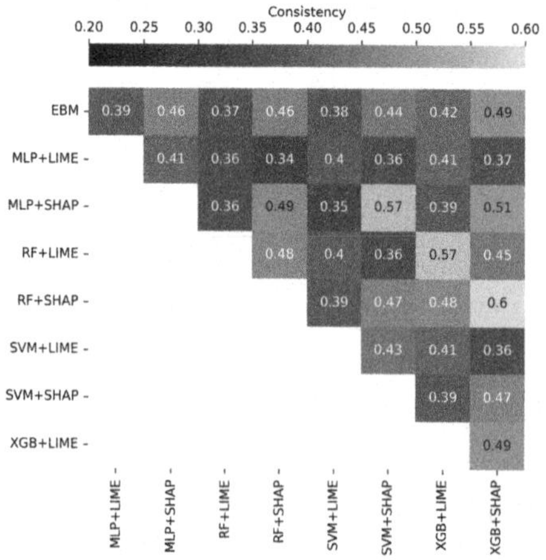

Consistency: 0.20 0.25 0.30 0.35 0.40 0.45 0.50 0.55 0.60

	MLP+LIME	MLP+SHAP	RF+LIME	RF+SHAP	SVM+LIME	SVM+SHAP	XGB+LIME	XGB+SHAP
EBM	0.39	0.46	0.37	0.46	0.38	0.44	0.42	0.49
MLP+LIME		0.41	0.36	0.34	0.4	0.36	0.41	0.37
MLP+SHAP			0.36	0.49	0.35	0.57	0.39	0.51
RF+LIME				0.48	0.4	0.36	0.57	0.45
RF+SHAP					0.39	0.47	0.48	0.6
SVM+LIME						0.43	0.41	0.36
SVM+SHAP							0.39	0.47
XGB+LIME								0.49

Fig. 5. Mean consistency between the models for 10 randomly selected turbofan units.

method. The visualization of the measurements implies that we are dealing with two distinct failures. There is a common starting point (normal condition) and as the RUL decreases, the measurements move in different directions of the plane. The behaviour of the SHAP values is similar to measurements, which shows that there is a consistency between them. On the other hand, in the case of LIME, four distinct regions are present, which cannot be simply explained by the distribution of the dataset. Nevertheless, it is still a noticeable shift between normal and failure points. This may be driven by the fact that LIME is known to be affected by small perturbations in the dataset [2].

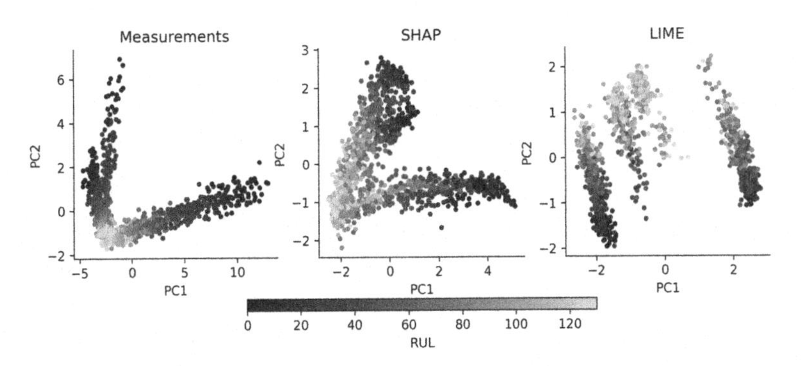

Fig. 6. PCA visualization of the measurements and explanations.

5 Conclusion and Future Works

In this paper, we have discussed the problem of Explainable AI in the predictive maintenance case. We have focused on the comparison of feature importances assigned by SHAP and LIME explainability methods for different types of black-box models. We also used Explainable Boosting Machine, which is a glass-box model and can provide explanations without the utilization of external explainability methods.

The results show that the SHAP method outperforms LIME and EBM in terms of stability and consistency of the explanations. The most promising results were obtained with MLP model, where the response of the XAI model was in our opinion the most reasonable, which was also confirmed by one of the best results in the stability metric. We have observed that the LIME method is not performing well in terms of stability and its results might be affected by small changes in the model, which is coherent with other studies. Explainable Boosting Machine has shown comparable performance in terms of prediction accuracy as the other techniques used (it was slightly less accurate than the rest of algorithms), but the stability of the explanations was worse than in the case of SHAP method. This implies that the glass-box models may not always perform better in terms of stability than a combination of black-box with XAI method.

The study has shown that there exist differences between the explanations that depend on the XAI method and ML algorithm. Not only different models result in different feature importances assigned by XAI methods, but there also exists a disagreement between XAI explanations for the same model. This indicates that the differences in the XAI methods are not coming only from the internal decisions of ML model, but also from the XAI methods themselves. Thus, there is a need for further research on the machine learning models and XAI methods, which will lead to the production of accurate and trustworthy algorithms for predictive maintenance tasks. The situation when unstable and inconsistent algorithms are used in the production environment may lead to the loss of trustworthiness of ML models by the end-users of those algorithms.

In future work, we plan to further investigate the topic of XAI methods in predictive maintenance applications with a special focus on remaining useful life estimation. Our next works will be devoted to the real-life use cases we plan to investigate in the Explainable Predictive Maintenance Project. We want to focus on the explainability problem in more complex deep learning architectures, which show promising potential in PdM use cases, i.e., convolutional, LSTM, Transformer Networks or ensembles of them. We also see a great need for further research on explainability metrics – the current metrics give some valuable information. However, we observe that they cannot be evaluated in a straightforward way, i.e., they depend on the magnitude of the feature importance vectors, and they do not provide information on the source of bias in explanations. We also plan to investigate more in-depth differences between the explanations of SHAP, LIME, EBM and others explanation methods.

Acknowledgements. This paper is funded from the XPM (Explainable Predictive Maintenance) project funded by the National Science Center, Poland under CHIST-ERA programme Grant Agreement No. 857925 (NCN UMO-2020/02/Y/ST6/00070).

References

1. Abid, K., Sayed-Mouchaweh, M., Cornez, L.: Deep ensemble approach for RUL estimation of aircraft engines. In: Hasic Telalovic, J., Kantardzic, M. (eds.) MeFDATA 2020. CCIS, vol. 1343, pp. 95–109. Springer, Cham (2021). https://doi.org/10.1007/978-3-030-72805-2_7
2. Alvarez-Melis, D., Jaakkola, T.S.: On the robustness of interpretability methods. CoRR abs/1806.08049 (2018). http://arxiv.org/abs/1806.08049
3. Bobek, S., Bałaga, P., Nalepa, G.J.: Towards model-agnostic ensemble explanations. In: Paszynski, M., Kranzlmüller, D., Krzhizhanovskaya, V.V., Dongarra, J.J., Sloot, P.M.A. (eds.) ICCS 2021. LNCS, vol. 12745, pp. 39–51. Springer, Cham (2021). https://doi.org/10.1007/978-3-030-77970-2_4
4. Frank, A.G., Dalenogare, L.S., Ayala, N.F.: Industry 4.0 technologies: implementation patterns in manufacturing companies. Int. J. Prod. Econ. **210**, 15–26 (2019). https://doi.org/10.1016/j.ijpe.2019.01.004
5. Frederick, D., DeCastro, J., Litt, J.: User's guide for the commercial modular aero-propulsion system simulation (C-MAPSS). NASA Technical Manuscript **2007–215026** (2007)
6. Goodman, B., Flaxman, S.: European union regulations on algorithmic decision-making and a "right to explanation". AI Mag. **38**(3), 50–57 (2017). https://doi.org/10.1609/aimag.v38i3.2741
7. Gunning, D., Aha, D.: DARPA's explainable artificial intelligence (XAI) program. AI Mag. **40**(2), 44–58 (2019). https://doi.org/10.1609/aimag.v40i2.2850
8. Hastie, T., Tibshirani, R.: Generalized additive models. Stat. Sci. **1**(3), 297–310 (1986). https://doi.org/10.1214/ss/1177013604
9. Khelif, R., Chebel-Morello, B., Malinowski, S., Laajili, E., Fnaiech, F., Zerhouni, N.: Direct remaining useful life estimation based on support vector regression. IEEE Trans. Industr. Electron. **64**(3), 2276–2285 (2017). https://doi.org/10.1109/TIE.2016.2623260

10. Li, X., Ding, Q., Sun, J.Q.: Remaining useful life estimation in prognostics using deep convolution neural networks. Reliab. Eng. Syst. Saf. **172**, 1–11 (2018). https://doi.org/10.1016/j.ress.2017.11.021
11. Listou Ellefsen, A., Bjørlykhaug, E., Æsøy, V., Ushakov, S., Zhang, H.: Remaining useful life predictions for turbofan engine degradation using semi-supervised deep architecture. Reliab. Eng. Syst. Saf. **183**, 240–251 (2019). https://doi.org/10.1016/j.ress.2018.11.027
12. Lundberg, S.M., Lee, S.I.: A unified approach to interpreting model predictions. In: Guyon, I., et al. (eds.) Advances in Neural Information Processing Systems, vol. 30, pp. 4765–4774. Curran Associates, Inc. (2017)
13. Molnar, C.: Interpretable Machine Learning, 2 edn. (2022). https://christophm.github.io/interpretable-ml-book
14. Mosallam, A., Medjaher, K., Zerhouni, N.: Data-driven prognostic method based on Bayesian approaches for direct remaining useful life prediction. J. Intell. Manuf. **27**(5), 1037–1048 (2014). https://doi.org/10.1007/s10845-014-0933-4
15. Nori, H., Jenkins, S., Koch, P., Caruana, R.: InterpretML: a unified framework for machine learning interpretability. CoRR abs/1909.09223 (2019). http://arxiv.org/abs/1909.09223
16. de Oliveira da Costa, P.R., Akçay, A., Zhang, Y., Kaymak, U.: Remaining useful lifetime prediction via deep domain adaptation. Reliab. Eng. Syst. Saf. **195**, 106682 (2020). https://doi.org/10.1016/j.ress.2019.106682
17. Petkovic, D., Altman, R., Wong, M., Vigil, A.: Improving the explainability of random forest classifier - user centered approach. In: Biocomputing 2018, pp. 204–215. WORLD SCIENTIFIC (2017). https://doi.org/10.1142/9789813235533_0019
18. Ribeiro, M.T., Singh, S., Guestrin, C.: "why should i trust you?": explaining the predictions of any classifier. In: Proceedings of the 22nd ACM SIGKDD International Conference on Knowledge Discovery and Data Mining, KDD 2016, pp. 1135–1144. Association for Computing Machinery, New York (2016). https://doi.org/10.1145/2939672.2939778
19. Ribeiro, M.T., Singh, S., Guestrin, C.: Anchors: high-precision model-agnostic explanations. In: Proceedings of the AAAI Conference on Artificial Intelligence, vol. 32, no. 1 (2018)
20. Rudin, C.: Stop explaining black box machine learning models for high stakes decisions and use interpretable models instead. Nat. Mach. Intell. **1**(5), 206–215 (2019). https://doi.org/10.1038/s42256-019-0048-x
21. Saxena, A., Goebel, K., Simon, D., Eklund, N.: Damage propagation modeling for aircraft engine run-to-failure simulation. In: 2008 International Conference on Prognostics and Health Management, pp. 1–9 (2008). https://doi.org/10.1109/PHM.2008.4711414
22. Selvaraju, R.R., Cogswell, M., Das, A., Vedantam, R., Parikh, D., Batra, D.: Grad-CAM: visual explanations from deep networks via gradient-based localization. Int. J. Comput. Vision **128**(2), 336–359 (2019). https://doi.org/10.1007/s11263-019-01228-7
23. Wang, Y., Zhao, Y., Addepalli, S.: Remaining useful life prediction using deep learning approaches: a review. Procedia Manuf. **49**, 81–88 (2020). https://doi.org/10.1016/j.promfg.2020.06.015. Proceedings of the 8th International Conference on Through-Life Engineering Services – TESConf 2019

Streaming Detection of Significant Delay Changes in Public Transport Systems

Przemysław Wrona, Maciej Grzenda(✉), and Marcin Luckner

Faculty of Mathematics and Information Science, Warsaw University of Technology, ul. Koszykowa 75, 00-662 Warszawa, Poland
{P.Wrona,M.Grzenda,M.Luckner}@mini.pw.edu.pl

Abstract. Public transport systems are expected to reduce pollution and contribute to sustainable development. However, disruptions in public transport such as delays may negatively affect mobility choices. To quantify delays, aggregated data from vehicle location systems are frequently used. However, delays observed at individual stops are caused inter alia by fluctuations in running times and the knock-on effects of delays occurring in other locations. Hence, in this work, we propose both a method detecting significant delays and a reference architecture, relying on the stream processing engines in which the method is implemented. The method can complement the calculation of delays defined as deviation from schedules. This provides both online rather than batch identification of significant and repetitive delays, and resilience to the limited quality of location data. The method we propose can be used with different change detectors, such as ADWIN, applied to a location data stream shuffled to individual edges of a transport graph. It can detect in an online manner at which edges statistically significant delays are observed and at which edges delays arise and are reduced. Such detections can be used to model mobility choices and quantify the impact of regular rather than random disruptions on feasible trips with multimodal trip modelling engines. The evaluation performed with the public transport data of over 2000 vehicles confirms the merits of the method and reveals that a limited-size subgraph of a transport system graph causes statistically significant delays.

Keywords: Stream processing · Drift detection · Public transport · GPS sensors

1 Introduction

Public transport (PT) is expected to contribute to sustainable development by reducing pollution and road congestion. However, disruptions may negatively affect nominal and perceived journey time [12]. Hence, disruptions such as delays or missed connections are quantified to measure the performance of a PT system. Importantly, frequent public transport disruptions may negatively affect mobility choices. Developments in automatic vehicle location (AVL) [10] systems have largely increased the availability of spatio-temporal datasets documenting both

D. Groen et al. (Eds.): ICCS 2022, LNCS 13353, pp. 486–499, 2022.
https://doi.org/10.1007/978-3-031-08760-8_41

the location of individual vehicles and real arrival and departure times. Such data is typically used for real time monitoring of public transport services, and improving operations [8]. Public transport schedules, such as schedules published in General Transit Feed Specification (GTFS) format can be compared against real departure times of PT vehicles. This provides for the aggregation of delays. As an example, in [8] delays at individual stop points were aggregated to provide features such as total delays per stop point, the number of times a bus was delayed at a stop point and the average delay. Next, histograms of delays and maps of locations with delays and significant delays were produced. Recently, the newly available large volumes of delay report records have been used for more in-depth analysis of delay data. In [10], a proposal to discretise delay changes into hour time bins and delay time bins was made. This was to consider and normalise the values associated with each bin separately. Importantly, the calculations were performed for each edge representing a sequence of two stops consecutively visited by a vehicle. Agglomerative clustering was used to identify clusters of PT system edges grouping edges similar in terms of delays observed at these edges.

Importantly, the majority of works on long term delay analysis rely on batch processing of data sets collected in the preceding periods. In [8], this included hierarchical clustering and non-linear regression for delay prediction. In [10], agglomerative clustering of edges was used, which allowed the identification of stop pairs between which minor or major delay changes were observed under different probabilities throughout the entire period under consideration.

Many studies on long term analysis were based on averaging delay data. However, delays in a PT system may occur due to various reasons such as traffic light conditions preventing a vehicle from passing a crossroads, accidents, road reconstruction, too demanding schedules or demand fluctuations affecting boarding times at individual stops. Furthermore, some of the delays may be reported due to limited precision of location data obtained from GPS receivers and wrongly suggesting that a vehicle has not yet (or has already) departed from the stop. Hence we propose a Streaming Delay Change Detection (SDCD) method. The method can be used with varied change detectors applied to delay data to identify how frequently statistically significant delays occur at individual edges of a PT system. The SDCD method we propose relies on stream processing, i.e. identifying changes in delay distribution in near-real-time rather than through batch processing of historical data, and can be used with high volume data streams.

The primary contributions of this work are as follows:

- We propose the SDCD method to monitor and detect changes in delay distribution, and propose two variants of the method to detect changes during entire days and individual time slots.
- We evaluate the method with real Warsaw public transport data and make the implementation of the method and data available to the research community[1].

[1] The source code of the SDCD method and the data used in this work are available at https://github.com/przemekwrona/comobility-sdcd.

2 Related Works

2.1 Quantifying Delays and Change Detectors

Delays in public transport systems are typically analysed based on the data sets aggregating delays observed for individual vehicles at stop points such as bus stops [5,8] or edges defined by two consecutive stop points [10]. Some studies go beyond calculating average delay values. As an example, Szymanski et al. proposed using bins of variable lengths for aggregating delay values e.g. grouping delays of $[-10.5\,\text{min},\ -5.5\,\text{min}]$ in a single bin [10].

Yap et al. in [12] note the difference between the change in PT system performance caused by stochastic demand or supply fluctuations i.e. the change referred to as disturbance, and disruption, which is the change caused by distinctive incidents or events. Both these changes are examples of perturbations. Importantly, disruptions can propagate in the PT system and their consequences can be observed even in distant locations. Due to complex demand-supply interactions, in the case of urban PT networks, simulations-based models are often necessary to predict the impact of disruptions [12].

The volume of vehicle location data collected from AVL systems is growing. It reached 12 mln records reported in a study for Stockholm [8], 16 mln for Wroclaw [10] or even 2.9 bln of records collected for Warsaw over approximately 30 months [5]. This inspired research into the use of big data frameworks for the storage and processing of location and delay records. A survey of related works and a proposal for a unified architecture serving storage and analytical needs of IoT data with emphasis on vehicle location data can be found in [5].

In parallel, developments in stream rather than batch processing of high volume and velocity data raised interest in change detection methods applicable to data streams. One of the popular detectors is **ADWIN** proposed in [1]. In ADWIN, the adaptive window approach is used for streaming data and applied to detect changes in the average value of a stream of bits or real-valued numbers. The size of sliding windows used for change detection is not constant and defined a priori, but depends on the rate of detections. Thus for stationary data, the window is growing, but in the case of detection, the window is narrowing to discard historical data. The only parameter of the detector is a confidence value $\delta \in (0, 1)$, controlling the sensitivity of the detection, i.e. influencing the ratio of false positives. A change is detected with a probability of at least $1 - \delta$.

Another recently proposed approach to concept drift detection relies on the Kolmogorov-Smirnov test applied to sliding windows populated with recent data instances from a data stream [7]. The parameters for the **KSWIN** detector are the probability α of the test statistic and the sizes of two sliding windows used for the detection of a difference between the distributions of data present in the two windows. Concept change is detected when the distance between the empirical cumulative distribution functions (eCDFs) of the two differently sized windows exceeds the α-dependant threshold.

Research into change detectors is largely inspired by the need to detect when an update of the learning model is needed to adapt the model to concept drift.

A family of methods monitoring the mean estimated from real values with an explicit focus on monitoring the values of performance measures of learning models was proposed in [2]. The methods rely on **HDDM** algorithms proposed in the study and use Hoeffding's inequality to report warnings and actual drifts based on two confidence levels – the parameters of the change detector.

The change detection methods applicable to data streams were not used until very recently for transport data. Among the first works of this kind, Moso et al. in [6] addressed the problem of collecting message exchanges between vehicles and analysing trajectories. Variations of trajectories from normal ones were detected to identify anomalies. This recent study is among the first studies exploiting the use of Page-Hinkley and ADWIN change detection methods to process Cooperative Awareness Messages produced by vehicles in order to perform road obstacle detection. Out of the two methods, ADWIN yielded promising results, which is unlike Page-Hinkley, which additionally required parameter tuning.

2.2 Analysing Multi-modal Connections and the Impact of Perturbations on Travel Times

Delays of individual PT vehicles not only have an impact on travel time, but also may cause lost transfers. As some trips require multiple connections and multimodal routes, to estimate travel times under static schedules and real conditions, simulation software is needed. A popular solution is to use OpenTripPlanner (OTP)[2] - an open-source and cross-platform multi-modal route planner. OTP gives the ability to analyse varied transport data. That includes modifications of schedules (also in real-time) and changes to the street network. Importantly, OTP can model the effects of road infrastructure changes and examine the consequences of temporary changes in schedules [13].

Several recent scientific works used OTP as an analytic tool. Lawson et al. examined a "blended data" approach, using an open-source web platform based on OTP to assist transit agencies in forecasting bus rider-ship [3]. Ryan et al. used OTP to examine the critical differences between the two representations of accessibility, calculating door-to-door travel times to supermarkets and healthcare centres [9]. To perform connection planning both static PT schedules made available in GTFS format [9,11] and real feed of vehicle arrival and departure times in the form of GTFS Realtime[3] [4] can be used. In particular, a comparison of travel times estimated by OTP under planned schedules and real departure times provided in GTFS Realtime can be made.

While the stream of real arrival and departure times, including possible delays, can be forwarded to a modelling environment such as OTP, this does not answer whether delays exemplify systematic problems at some edges of the PT graph or occasional fluctuations. Hence, in our study, we focus on detecting statistically significant perturbations in the performance of a public transport

[2] http://www.opentripplanner.org/.

[3] https://developers.google.com/transit/gtfs-realtime.

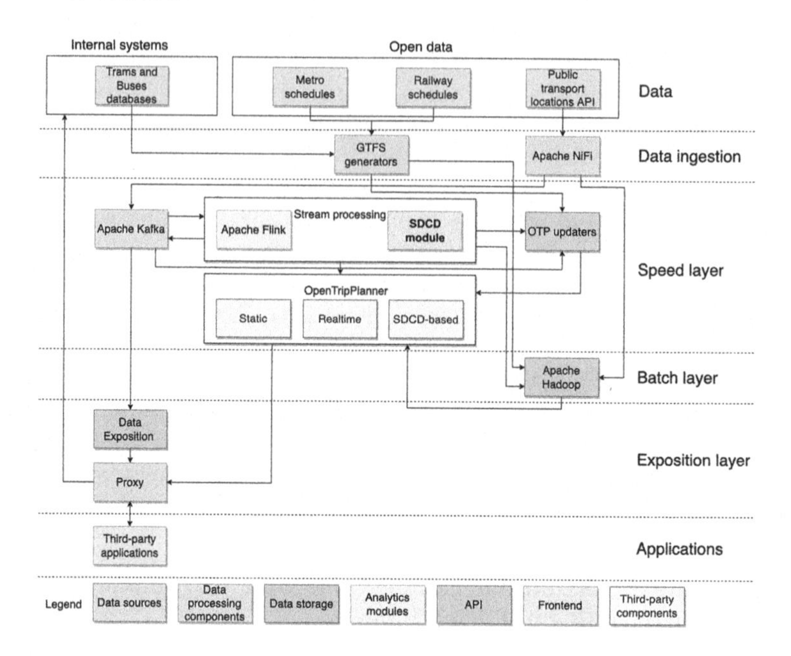

Fig. 1. The architecture of delay detection and impact modelling system

system. In this way, we aim to reduce the risk of reporting disruptions caused by stochastic fluctuations, unless these disruptions occur frequently. Hence, rather than averaging delays possibly observed occasionally and caused by limited precision of GPS readouts, variability in the number of passengers or traffic light conditions, we aim to identify these delays which occur frequently and over longer periods. To make this possible, we propose a method applying change detectors to data collected at individual edges of public transport graph and the architecture within which the method can be implemented.

Furthermore, let us note that such detections can provide basis for generating schedules reflecting regular statistically significant delays and using them in a simulation environment such as OTP.

3 Architecture of Delay Detection and Modelling System

To validate the approach proposed in this work, we implemented SDCD method as a part of IoT platform collecting and analysing sensor data, including data from AVL systems. The platform we used for the collection and processing of vehicle location and delay data is an update and extension of USE4IoT architecture [5]. Let us note that without the loss of generality, by delays we mean both arrivals before and after scheduled time. The USE4IoT is an Urban IoT platform designed as an extension of Lambda architecture. It fulfils the requirements of the Lambda pattern and adds extra layers matching the needs of smart cities.

Figure 1 presents the architecture of the part of the system related to the SDCD method. In the analysed case, input data comes from the open data portal of the city of Warsaw[4] and additional open sources. Some data is collected online from a public transport localisation stream. Other data, such as timetables, are downloaded periodically. The data ingestion layer is responsible for collecting the data from the various data sources. It requires a combination of components and frameworks. In the case of USE4IoT, Apache NiFi is used to poll the data sources for the most recent data and convert new data records into data streams.

Big data, including location data streams, are archived using the Hadoop Distributed File System to store tabular data, including data collected as online data streams and timetables downloaded daily. The vehicle location streams are redirected to stream processing engines through Apache Kafka to ensure high throughput and resilience to downstream performance. Next, the Apache Flink application is used to process and merge the location of vehicles with PT timetables. The architecture provides stream processing with a mean delay of less than 2.5 s [5].

In this work, we propose three OTP instances, each serving different needs. The OTP instances are updated from three types of sources. Static GTFS data is created based on static schedules and uploaded into the first OTP instance. The stream analytics modules detect delays and untypical events and supplies OTP with a real-time GTFS. Therefore, the real-time OTP instance can calculate multi-modal connections considering current vehicle location and delays. Finally, we propose the SDCD module to detect statistically important changes in public transport delays. This can provide the basis for GTFS files containing credible schedules, i.e. the schedules reflecting statistically significant delays that update the static departure times, to be used in the SDCD-based OTP instance. In this way, a comparison between travel options and times under a) static schedules, b) real-time situation and c) schedules reflecting significant delays possibly observed frequently over preceding days can be made.

Finally, the entire architecture was created to make it possible to forward the results through the data exposition layer to the application layer, possibly including third-party applications. However, the core part of the solution, which we focus on in this work, is the SDCD method providing the basis for online detection of statistically significant delay changes.

4 Streaming Delay Change Detection

Let L denote the set of PT lines, each defined by a sequence of bus or tram stops $\{s_{l,1}, \ldots, s_{l,j}\}$. Let us note that when describing PT system, we will rely on the notation similar to the notation proposed in [12]. In our case, we assume that $s_{l,j} = s_{l,1}$ i.e. a line is defined by a loop, while stops visited in one direction are not necessarily the stops visited by a vehicle travelling in the opposite direction.

Let PT network be a directed graph $G = (S, E)$, where S is the set of all stops in the network e.g. in the urban PT network and E is the set of edges. An edge $(s_i, s_j), i \neq j$ exists i.e. $(s_i, s_j) \in E$ if and only if at least one line l exists such that the two consecutive stops of the line are s_i and s_j.

[4] https://api.um.warszawa.pl.

Algorithm 1: Streaming delay change detection algorithm

```
input  : stream: a stream of vehicle locations S_1, S_2, ...
         GetDetectorId(V): a function returning the identifier of the change
         detector to be used
         Δ ∈ {TRUE, FALSE} - the parameter defining whether to process
         Δd() or d() delays
output: detections: A stream of detected changes in delay stream
1  ConceptChangeDetectors ← {}          # Initialize empty map of detectors
2  i = 1;
3  while stream has next element do
4      V ← S_i
5      K ← GetDetectorId(V)
6      D ← ConceptChangeDetectors.getDetector(K)
7      if D is NULL then
8          D ← newConceptChangeDetector()
9          ConceptChangeDetectors.putDetector(K, D)
10     if Δ then
11         delay ← Δd(V)
12     else
13         delay ← d(V)
14     D.addDelay(delay)
15     if D.detectedChange () then
16         detections.save(V, D.identifier)
17     i = i + 1;
```

Furthermore, let $\mathcal{S}_1, \mathcal{S}_2, \ldots$ be the stream of location records received over time from an AVL system. Without the loss of generality, we assume each $\mathcal{S}_i$ contains both current geocoordinates of a vehicle course v, the line l operated by the vehicle, and the identifiers of two most recently visited stops s_i, s_{i-1} by the vehicle. Real departure times and planned departure times as defined in static schedules are also available for both of these stops. These are denoted by $t_{\mathrm{R}}(s, v)$ and $t_{\mathrm{S}}(s, v)$, respectively. Hence, $d(\mathcal{S}_j) = d(s_i, v) = t_{\mathrm{R}}(s_i, v) - t_{\mathrm{S}}(s_i, v)$ denotes delay i.e. the difference between real and planned departure time for a vehicle course v observed at stop s_i i.e. the most recently left stop during course v. Let us note that if raw data from AVL include no line identifiers, they can be retrieved from schedule data. Furthermore, if needed stop identifiers can be identified based on past vehicle coordinates and stop coordinates of the line served by the vehicle. The data of vehicles not in service are skipped.

Our approach to detect changes in a stream of delays uses change detectors such as detectors relying on the ADWIN algorithm [1]. The SDCD method is defined in Algorithm 1. As an input stream, we use the location stream of PT vehicles $\mathcal{S}_1, \mathcal{S}_2, \ldots$ described above. Furthermore, the location stream can be shuffled into substreams linked to individual edges of PT graph or bins linked to a combination of an edge and an hour $h = 0, \ldots, 23$ of the day. In the first case, which we call edge-based, all vehicle location records describing vehicles visiting the sequence of two stops defining an edge will be gradually processed by one change detector. In the bin-based approach, all records related to an edge and time of the day defined by a one-hour time slot will be processed together. Hence, the intuition behind the edge-based approach is to identify delays and delay reductions as they appear over time. In this case, the detector is recognised by pair of stops. Thus, for each pair of stops, one detector that collects data all the time is created.

We propose bin-based approach to identify possible changes in delays at the same time of the day, e.g. between 8:00 and 8:59 over consecutive days and occurring at one edge of PT graph. In this case, the detector identification is extended by the hour that comes from the current vehicle timestamp. Hence, at most 24 detectors are created for each pair of stops visited in a row. The two approaches of defining detector keys are formally defined in Algorithm 2.

Moreover, we propose to calculate delay change between stops, defined as $\Delta d(\mathcal{S}_j) = d(s_i, v) - d(s_{i-1}, v)$. Let us note that $d(\mathcal{S}_j) > 0$ may be accompanied by $\Delta d(\mathcal{S}_j) = 0$ or even $\Delta d(\mathcal{S}_j) < 0$. As an example, it is possible that a delayed vehicle ($d(\mathcal{S}_j) > 0$) has reduced its delay when travelling between stops s_{i-1} and s_i i.e. $\Delta d(\mathcal{S}_j) < 0$. Hence, the third parameter of Algorithm 1 is whether to detect changes in $d()$ or $\Delta d()$ streams of values.

Algorithm 2: The functions calculating detector identifiers.

```
1 Function getDetectorIdForEdge(V):
2     return V.getCurrStopId().join(V.getPrevStopId())
3 Function getDetectorIdForEdgeAndTime(V):
4     return V.getCurrStopId().join(V.getPrevStopId()).join(V.getHour())
```

During the algorithm initialisation, we create an empty map of detectors (Line 1). Every time data for a new detector key, i.e. new edge or new bin, is encountered in the stream for the first time, we create a new change detector object (Lines 7–9). Next, we add the value of delay expressed in seconds to the detector and check if the detector detected a change in the stream. Detected change is saved together with detector key, i.e. edge identifier in the case of edge-based, and edge and time slot in the bin-based approach.

Finally, once significant perturbations in the performance of a public transport system defined by repetitive detections of delays are identified, we propose to develop SDCD-based schedules, i.e. the public transport schedules reflecting significant perturbations observed at individual edges. Next, by comparing the behaviour of a public transport system under static schedules and SDCD-based schedules, the impact of significant perturbations can be assessed. For example, bus delays may cause missing a scheduled connection at a transfer stop and largely increase overall travel time.

5 Results

5.1 Reference Data

The data used to validate the SDCD method comes from the Warsaw Public Transport public API, which provides the current position of vehicles every 10 s, yielding 2.0–2.5 GB of data each day. The average daily number of records over the period selected to illustrate the results of this study exceeds 4 million (839 thousand for trams and 3.17 million for buses), out of which approx. 1.1 million are departure records. The ratio of records linked with a static schedule reached 92%. The remaining records represent inter alia vehicles not in service.

The public transport vehicles travel an average of 14.6 thousand edges E daily, defined by the two following stops. An average edge of the public transport network is visited by 54 vehicles (the median) per day. However, the actual range is from a single vehicle course per edge to over ten thousand on some city centre edges. The median delay at an edge reaches 104 s, which is considered acceptable according to criteria adopted by the local public transport authority.

5.2 Change Detections

In the first experiments, Algorithm 1 was used with three change detectors – ADWIN [1], KSWIN [7], and HDDM [2] – used to perform change detection. We selected the two latter methods to enable the comparison of different detectors including the ADWIN change detector, i.e. the detector already used for a related problem of road obstacle detection in [6]. In the case of HDDM methods, HDDM_A was selected for the experiments. This relies on a lower number of parameters than HDDM_W, which additionally requires the weight given to recent data to be set. Setting such a weight would require additional hyperparameter tuning. Hence, for the sake of simplicity, HDDM and HDDM_A will be used interchangeably in the remainder of this study. In all experiments, the implementation of detectors from the `scikit-multiflow` library was used, and default settings of the ADWIN detector were applied. In the case of KSWIN and HDDM, the same confidence setting as for ADWIN was applied.

Figure 2 presents the edges whose detectors reported at least one delay change during the reference period selected to visualise the results of this study, i.e. 18th December 2021 (Saturday) to 21st December 2021 (Tuesday). The detectors are organised into two types. The first type of detector analysed delay $d()$ at the destination stop s_i of edge $e = (s_{i-1}, s_i)$, hereafter referred to as *delay*. The second type analysed changes of delay $\Delta d()$ observed at an edge e, referred to as Δ*delay* in the remainder of this work. In this experiment, one detector analysed the data from entire days to find delay changes, i.e. an edge-based approach was used. For the sake of clarity the edges at which detections occurred are depicted by points placed in the destination stop s_i of edge $e = (s_{i-1}, s_i)$.

The ADWIN algorithm (Fig. 2a) detected the smallest number of delay changes in comparison to KSWIN (Fig. 2b) and HDDM (Fig. 2c). Interestingly, all algorithms detected more accelerations ($d() < 0$) than delays ($d() > 0$). It

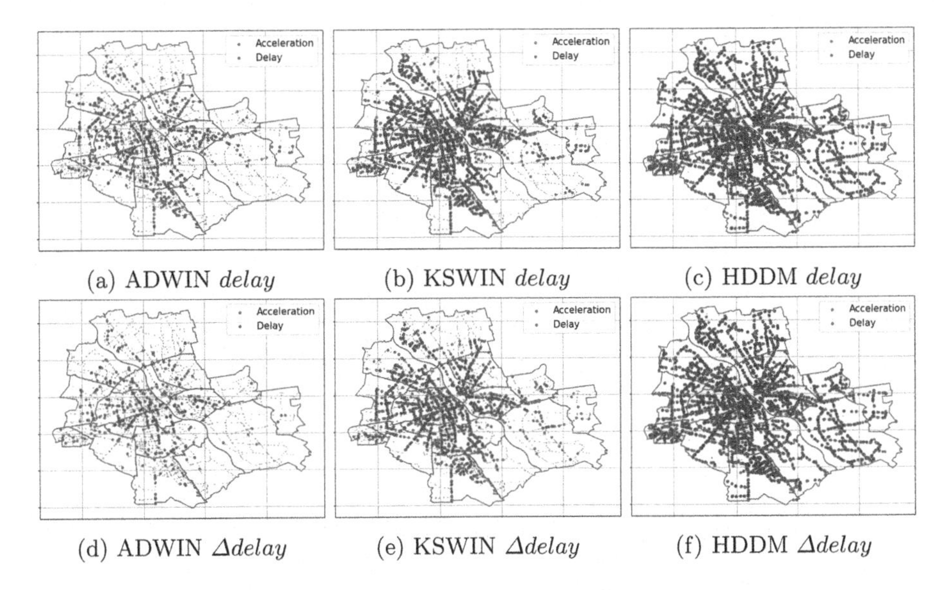

(a) ADWIN *delay* (b) KSWIN *delay* (c) HDDM *delay*

(d) ADWIN $\Delta delay$ (e) KSWIN $\Delta delay$ (f) HDDM $\Delta delay$

Fig. 2. The locations at which changes in delays were detected in the location stream between 18th December 2021 and 21st December 2021 (best viewed in color). Edge-based approach. The City of Warsaw area. (Color figure online)

may look positive. Still, accelerations are rare compared to the number of all edges. A possible explanation of the larger number of acceleration events than of delay events is that major delays are easy to attain at even short distances, but reducing them inevitably takes more time i.e. longer distances over which delay reduction has to be attained by the drivers, which is reflected by a larger number of acceleration edges.

The detection of delay changes $\Delta d()$ is more diverse. While the ADWIN (Fig. 2d) and HDDM (Fig. 2f) detect both directions of changes, the KSWIN (Fig. 2e) detects mostly accelerations. It may be caused by the fact that the KSWIN is comparing eCDF functions while the other algorithms compare the mean values. Once again, the ADWIN algorithm detected the smallest number of changes.

In the second experiment – instead of a single detector working throughout the entire period – the bin-based approach divided the records related to an edge into one-hour time slots. The rest of the conditions stay the same as in the previous experiment. The results are presented in Fig. 3. For the ADWIN (Fig. 3a and Fig. 3d) and KSWIN (Fig. 3b and Fig. 3e) algorithms the number of detected changes drops rapidly compared to edge-based approach. This effect shows that most detected delay changes are statistically important compared to other periods of the day but are rather typical for the specific hour, which sounds reasonable because of dynamic traffic changes during the day. However, the drop out effect is not observed for the HDDM algorithm (Fig. 3c and Fig. 3f), which may suggests that the HDDM algorithm detects again too many events.

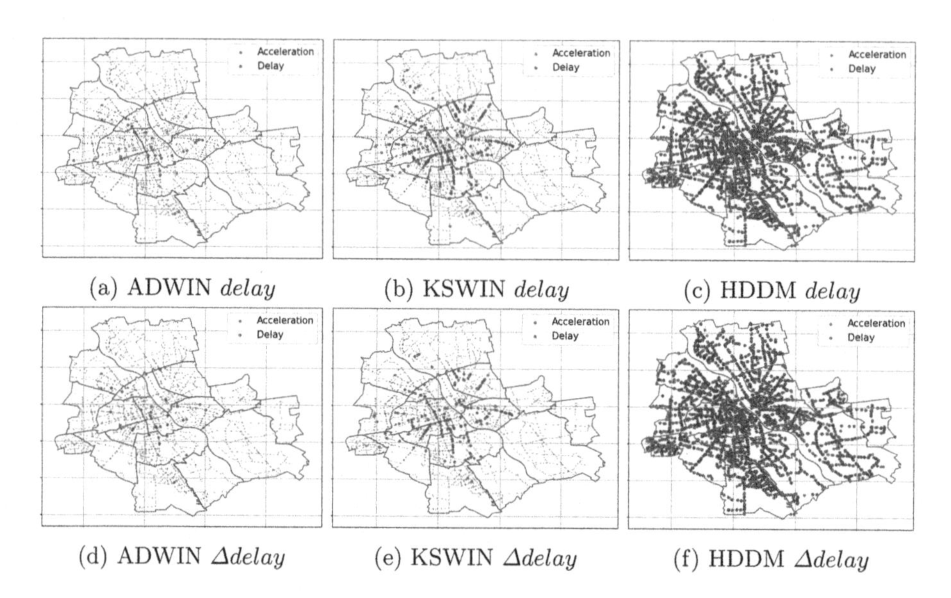

(a) ADWIN *delay* (b) KSWIN *delay* (c) HDDM *delay*

(d) ADWIN $\Delta delay$ (e) KSWIN $\Delta delay$ (f) HDDM $\Delta delay$

Fig. 3. The locations at which changes in delays were detected in the location stream between 18th December 2021 and 21st December 2021 (best viewed in color). Bin-based approach. The City of Warsaw area. (Color figure online)

The ADWIN algorithm is the least demanding in the context of parametrisation. Moreover, the results of other algorithms are counterintuitive when KSWIN detects only accelerations or the bin-based approach does not reduce the number of HDDM detections. Therefore, the ADWIN was selected for further analysis.

Table 1 presents statistics for all ADWIN delay $d()$ and delay change $\Delta d()$ detections in the edge-based approach. The number of detections is relatively small compared to the daily throughput and the number of analysed edges. When one compares the medians, the detected delays $d() > 0$ are comparable to the median delays of 104 s (see Sect. 5.1). Their standard deviation is relatively high and similar to the median. Therefore, the detected changes have a local character in the sense of a detection value (which would not necessarily be an exception in another location), but are globally shifted to reductions, which are taken as exceptions in contrast to the global delay level.

The proportion between reductions and increases is more balanced for Δdelay change detections. A very small median and several times higher standard deviation reveals that many detections concern minor delay change only, which additionally helps focus on these edges at which major delay change occurs.

To sum up, the statistical results show that statistically significant delay changes are rare for thousands of analysed connections. In practice, it is recommended to use both types of detections ($d()$ and $\Delta d()$) with an additional cut off of the small absolute values to focus on delays which are both statistically significant and high.

Table 1. Delay changes detected with SDCD algorithm. ADWIN detector.

Delay type	Date	Departure records	Increases	Reductions	Median[s]	STD[s]
$d()$	2021-12-18	1181271	5	1059	131.0	125.0
	2021-12-19	1242939	10	666	110.0	84.0
	2021-12-20	1256871	7	862	107.0	91.0
	2021-12-21	1049178	6	680	131.0	101.0
$\Delta d()$	2021-12-18	1181271	249	365	6.0	32.0
	2021-12-19	1242939	199	299	4.0	29.0
	2021-12-20	1256871	219	336	5.0	55.0
	2021-12-21	1049178	202	310	5.0	45.0

5.3 Peak Hours Analysis

To show how delay change detections can provide for more locally focused analysis, let us analyse detections observed during two separate periods containing the morning and evening rush hours. Figure 4 compares detections between 6 am and 10 am (Fig. 4a and Fig. 4b), and 4 pm and 8 pm (Fig. 4c and Fig. 4d).

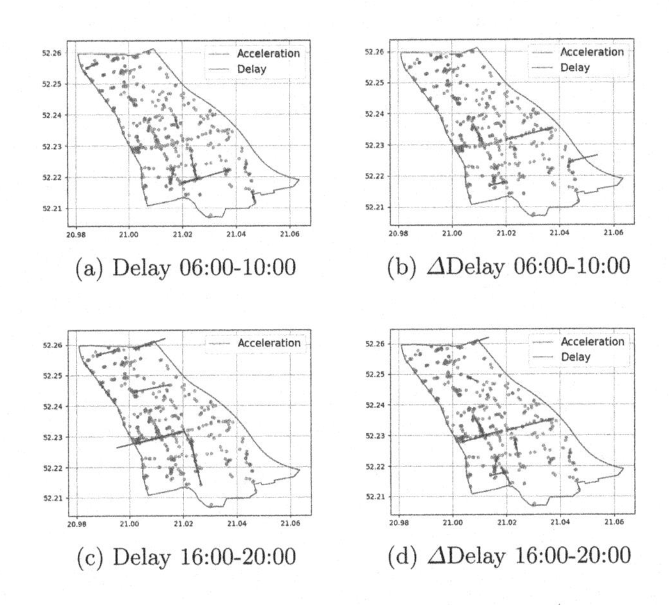

(a) Delay 06:00-10:00 (b) ΔDelay 06:00-10:00

(c) Delay 16:00-20:00 (d) ΔDelay 16:00-20:00

Fig. 4. Changes in the stream of delay and Δdelay values on 21st December 2021 (best viewed in color). Edge-based approach. The axis of the plots are longitude and latitude in degrees. (Color figure online)

Comparison of morning and evening delay detections (Fig. 4a and Fig. 4c) shows that some segments have acceleration or delay detected both in the morning and evening. That shows some segments of the traffic infrastructure with issues regardless of the time of the day. Comparison of delay changes detections (Fig. 4b and Fig. 4d) shows that some segments changed the direction of detected delay as a direction of traffic jams changes between the morning and evening rush hours. Finally, a comparison of both types of detectors shows segments with a delay detected by both types of detectors. In fact, there is a segment (in a rectangle defined by 21–21.02 Longitude and 52.21–52.22 Latitude), which illustrates the edges at which updates to static schedules could possibly be made.

6 Conclusions and Future Works

Delays in public transport may have a significant impact on mobility choices and discourage many citizens from using public transport services. However, reports of delays based on vehicle location data may be caused both by inevitable temporal fluctuations and limited precision of GPS-based readouts. Furthermore, delays may occur due to short-term events such as a street temporarily partly blocked due to maintenance work.

To identify and focus on statistically significant delays, in this work rather than aggregating delays we propose the SDCD method. This makes it possible to detect delays possibly arising or reduced in another part of the PT system and propagated to the location of interest, as well as delays arising or reduced at an edge of interest. Furthermore, we evaluate change detectors in terms of their usability to identify ADWIN as the most promising solution. The methods we propose rely on scalable online processing of location records rather than batch processing of historical data. Hence, they can also provide a basis for the detection of temporary problems irrespective of their causes, such as traffic jams or broken PT vehicles.

The SDCD method is a part of the system integrating big data frameworks, which ensure the scalability of the solution, and including OpenTripPlanner instances. In the future, we will focus on how delay detections performed with the SDCD method can be aggregated to identify long-term trends. One of the objectives of such aggregation is to identify those city areas in which statistically significant delays may have a substantial impact on the perception of public transport quality. Hence, the development of features quantifying the frequency and impact of delays both propagated to and arising at different edges of a public transport graph on feasible connections is planned.

Acknowledgements. This research has been supported by the CoMobility project. The CoMobility benefits from a 2.05 million€ grant from Iceland, Liechtenstein and Norway through the EEA Grants. The aim of the project is to provide a package of tools and methods for the co-creation of sustainable mobility in urban spaces.

References

1. Bifet, A., Gavaldà, R.: Learning from time-changing data with adaptive windowing. In: Proceedings of the 2007 SIAM International Conference on Data Mining, pp. 443–448. Society for Industrial and Applied Mathematics (2007). https://doi.org/10.1137/1.9781611972771.42
2. Frias-Blanco, I., del Campo-Avila, J., Ramos-Jimenez, G., Morales-Bueno, R., Ortiz-Diaz, A., Caballero-Mota, Y.: Online and non-parametric drift detection methods based on Hoeffding's bounds. IEEE Trans. Knowl. Data Eng. **7**, 810–823 (3 2015). https://doi.org/10.1109/TKDE.2014.2345382
3. Lawson, C.T., Muro, A., Krans, E.: Forecasting bus ridership using a "Blended Approach". Transportation **48**(2), 617–641 (2021). https://doi.org/10.1007/s11116-019-10073-z
4. Liebig, T., Piatkowski, N., Bockermann, C., Morik, K.: Predictive trip planning-smart routing in smart cities. In: CEUR Workshop Proceedings, vol. 1133, pp. 331–338 (2014)
5. Luckner, M., Grzenda, M., Kunicki, R., Legierski, J.: IoT architecture for urban data-centric services and applications. ACM Trans. Internet Technol. **20**(3) (2020). https://doi.org/10.1145/3396850
6. Moso, J.C., et al.: Anomaly detection on roads using C-ITS messages. In: Krief, F., Aniss, H., Mendiboure, L., Chaume-tte, S., Berbineau, M. (eds.) Communication Technologies for Vehicles - 15th International Workshop, Nets4Cars/Nets4Trains/Nets4Aircraft 2020, Bordeaux, France, 16–17 November 2020, Proceedings. LNCS, vol. 12574, pp. 25–38. Springer, Cham (2020). https://doi.org/10.1007/978-3-030-66030-7_3
7. Raab, C., Heusinger, M., Schleif, F.M.: Reactive soft prototype computing for concept drift streams. Neurocomputing **416**, 340–351 (2020). https://doi.org/10.1016/j.neucom.2019.11.111
8. Raghothama, J., Shreenath, V.M., Meijer, S.: Analytics on public transport delays with spatial big data. In: Proceedings of the 5th ACM SIGSPATIAL International Workshop on Analytics for Big Geospatial Data - BigSpatial 2016, pp. 28–33. ACM Press (2016). https://doi.org/10.1145/3006386.3006387
9. Ryan, J., Pereira, R.H.: What are we missing when we measure accessibility? Comparing calculated and self-reported accounts among older people. J. Transp. Geogr. **93**(March), 103086 (2021). https://doi.org/10.1016/j.jtrangeo.2021.103086
10. Szymanski, P., Zolnieruk, M., Oleszczyk, P., Gisterek, I., Kajdanowicz, T.: Spatio-temporal profiling of public transport delays based on large-scale vehicle positioning data from GPS in Wrocław. IEEE Trans. Intell. Transp. Syst. **19**, 3652–3661 (2018). https://doi.org/10.1109/TITS.2018.2852845
11. Waldeck, L., Holloway, J., van Heerden, Q.: Integrated land use and transportation modelling and planning: a South African journey. J. Transp. Land Use **13**(1), 227–254 (2020). https://doi.org/10.5198/jtlu.2020.1635
12. Yap, M., Cats, O., Törnquist Krasemann, J., van Oort, N., Hoogendoorn, S.: Quantification and control of disruption propagation in multi-level public transport networks. Int. J. Transp. Sci. Technol. (2021). https://doi.org/10.1016/j.ijtst.2021.02.002
13. Young, M.: OpenTripPlanner - creating and querying your own multi-modal route planner (2021). https://github.com/marcusyoung/otp-tutorial

Software Engineering for Computational Science

Learning I/O Variables from Scientific Software's User Manuals

Zedong Peng[1], Xuanyi Lin[2], Sreelekhaa Nagamalli Santhoshkumar[1], Nan Niu[1](✉), and Upulee Kanewala[3]

[1] University of Cincinnati, Cincinnati, OH 45221, USA
{pengzd,nagamasa}@mail.uc.edu, nan.niu@uc.edu
[2] Oracle America, Inc., Redwood Shores, CA 94065, USA
linx7@mail.uc.edu
[3] University of North Florida, Jacksonville, FL 32224, USA
upulee.kanewala@unf.edu

Abstract. Scientific software often involves many input and output variables. Identifying these variables is important for such software engineering tasks as metamorphic testing. To reduce the manual work, we report in this paper our investigation of machine learning algorithms in classifying variables from software's user manuals. We identify thirteen natural-language features, and use them to develop a multi-layer solution where the first layer distinguishes variables from non-variables and the second layer classifies the variables into input and output types. Our experimental results on three scientific software systems show that random forest and feedforward neural network can be used to best implement the first layer and second layer respectively.

Keywords: Scientific software · User manual · Software documentation · Classification · Machine learning

1 Introduction

The behavior of scientific software, such as a neutron transport simulation [9] and a seismic wave propagation [16], is typically a function of a large input space with hundreds of variables. Similarly, the output space is often large with many variables to be computed. Rather than requiring stimuli from the users in an interactive mode, scientific software executes once the input values are entered as a batch [46].

Recognizing the input/output (I/O) variables is a prerequisite for software engineering tasks like metamorphic testing [14]. In metamorphic testing, a change in some input variable(s) is anticipated to lead to a predictable effect on certain output variable(s). For instance, a metamorphic test case for the Storm Water Management Model (SWMM) [43] is: the surface water runoff is expected to decrease when bioretention cell is added [20], where "bioretention cell" is an input variable and "runoff" is an output variable.

D. Groen et al. (Eds.): ICCS 2022, LNCS 13353, pp. 503–516, 2022.
https://doi.org/10.1007/978-3-031-08760-8_42

In the previous study, we identified I/O variables manually from the user manual of SWMM [31]; however, this manual work was tedious and labor-intensive. The total cost of consolidating 807 input and 164 output variables was approximately 40 human-hours; yet we found that the I/O classification was not exclusive, e.g., 53 variables introduced in the user manual of SWMM [34] were both input and output variables.

To automate the I/O variable identification from a scientific software system's user manual, we investigate the use of machine learning (ML) in this paper. Specifically, we build on the experience of our manual work to codify the natural-language features that are indicative of the variable types (I, O, both I and O). We then develop a two-layer ML approach by first distinguishing variables from non-variables, followed by the classification of the variable types. We report the experimental results of applying our ML solution to the user manuals of three different scientific software systems, and further reveal the most influential ML features for classifying I/O variables.

It is worth noting that user manuals represent only one source of I/O variable identification and yet a complementary source to source code. Not only are user manuals amenable to natural language processing techniques, but they tend to be relatively stable in the face of frequent code changes. For instance, since the user manual of SWMM v 5.1 was written in September 2015 [34], the code release has been updated five times from v 5.1.011 to v 5.1.015 [48]. While there exists an inherent tradeoff between code's transitory nature and user manual's stable status, our objective is to automatically process documentation in support of scientific software engineering tasks such as metamorphic testing.

The main contribution of this paper work is the two-layer ML approach along with the natural-language features employed in these layers. The ML solution could reduce the cost of identifying I/O variables from scientific software's user manuals and other types of documentation, e.g., user manuals [18] and release notes [19]. In what follows, we provide background information in Sect. 2. Section 3 presents our two-layer ML solution together with the exploited features, Sect. 4 describes the experimental results, and finally, Sect. 5 draws some concluding remarks and outlines future work.

2 Background

Documentation provides valuable sources for software developers. Aghajani *et al.* [2] showed that user manual was found helpful for most of the 15 surveyed software engineering tasks (especially the operations and maintenance tasks) by at least one fifth of the 68 industrial practitioners. In scientific software development, a user manual describes the scope, purpose, and how-to of the software. The survey by Nguyen-Hoan *et al.* [23] showed that 70% of scientific software developers commonly produced user manuals, and Pawlik *et al.* [27] confirmed that it is most likely that user manuals will be prepared when scientific software is expected to be used outside a particular, typically limited, group.

Using software documentation to support metamorphic testing was first proposed by Chen *et al.* [8] in order to form I/O relations by systematically enumerating a pair of distinct complete test frames of the input domain. Zhou *et al.* [50] relied on the online specifications of the search engines to construct five I/O relations for metamorphic testing, whereas Lin *et al.* [17,18] exploited user forums to find such relations. No matter which documentation source is used, the identification of I/O variables themselves remains manual in these contemporary approaches.

Despite the lack of automated support for variable classification in the context of metamorphic testing, researchers have shown promise in building ML solutions to analyze software requirements written in natural languages: determining which statements actually represent requirements [1], separating functional and nonfunctional requirements [10], recognizing the temporal requirements that express the time-related system behaviors and properties [7], tracing requirements [47], just to name a few. Motivated by these ML solutions, we next present an automated approach to classifying variables from scientific software's user manuals.

3 Classifying Variables via Machine Learning

We gained experience in manually identifying variables from SWMM's user manual [31]. SWMM, created and maintained by the U.S. Environmental Protection Agency (EPA), is a dynamic rainfall-runoff simulation model that computes runoff quantity and quality from primarily urban areas. Figure 1-a shows SWMM's integrated environment for defining study area input data, running hydrologic, hydraulic, and water quality simulations, and viewing the results. Although graphical user interfaces like Fig. 1-a help visualize some I/O information, user manual more completely introduces the I/O variables of the scientific software.

Figure 1-b shows an excerpt of SWMM user manual [34]. This excerpt is also annotated with the results from our variable classifier. We notice that variables are nouns or noun phrases (NPs); however, not every noun or NP is a variable. Indeed, non-variable greatly outnumbers variable, and this generally holds in scientific software's user manual. For this reason, we build a multi-layer variable classifier, the architecture of which is shown in Fig. 2. Compared to a single, holistic classifier, the multi-layer architecture enables different ML algorithms to tackle problems at different granularity levels, thereby better handling imbalanced data and becoming more scalable for hierarchical training and classification [5].

As shown in Fig. 2, the preprocessing involves a few steps. If the user manual is a PDF file, we use the ExtractPDF tool [37] to convert it into a plain text file. We then apply NLTK in Python [26] to implement the tokenizer that breaks the text file's content into tokens (e.g., words, numbers, and symbols). NLTK is further used to split the tokens into sentences based on conventional delimiters (e.g., period and line break), and to assign the part-of-speech (PoS) tags (e.g., noun and verb) to each token. Our final preprocessing step uses the TextBlob Python library [39] to extract NPs.

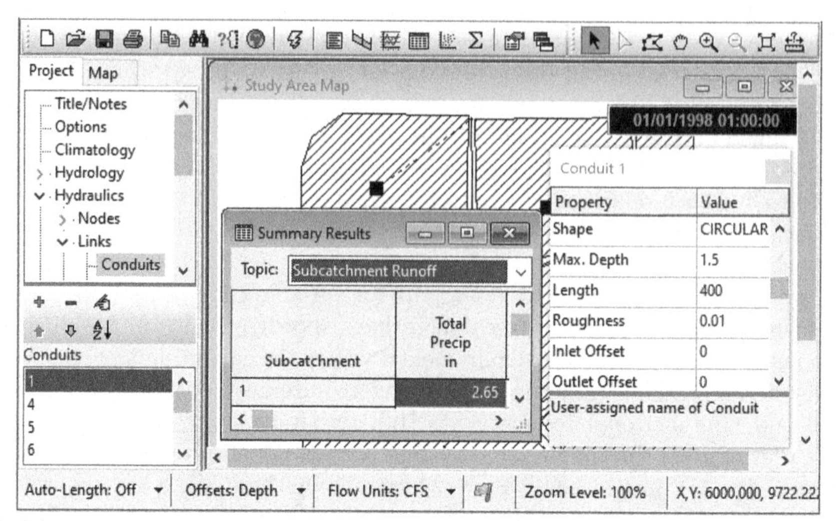

(a) Storm Water Management Model (SWMM) [43] performs single event or long-term runoff simulations

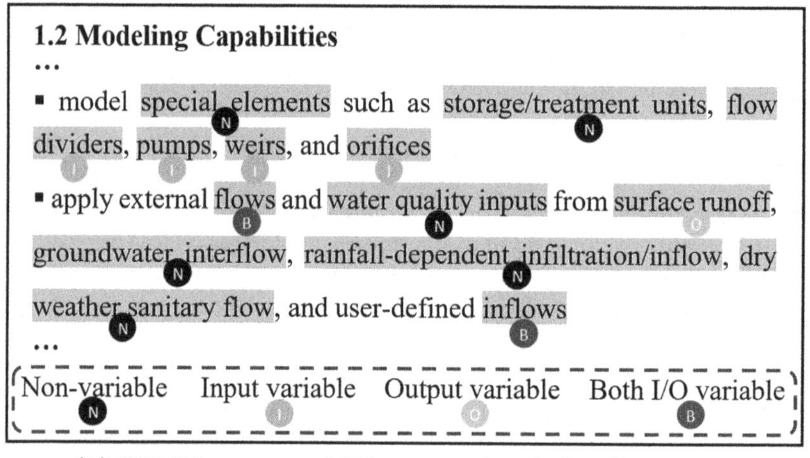

(b) SWMM user manual [34] annotated with the I/O variables

Fig. 1. SWMM screenshot and its user manual excerpt.

The nouns and NPs are candidates for variable classification, and our classifier shown in Fig. 2 works at the *sentence* level. Supervised learning is used at both layers: the first layer distinguishes which nouns or NPs are variables and which ones are not, and the second layer further predicts if a variable is the scientific software's input, output, or both input and output. The both category is of particular interest because the same variable can be one simulation's input and another's output. In SWMM, for example, "pollutant washoff" may be a key input variable for a wastewater treatment engineer, but at the same time can be an important output variable that a city manager researching low impact development (LID) pays attention to.

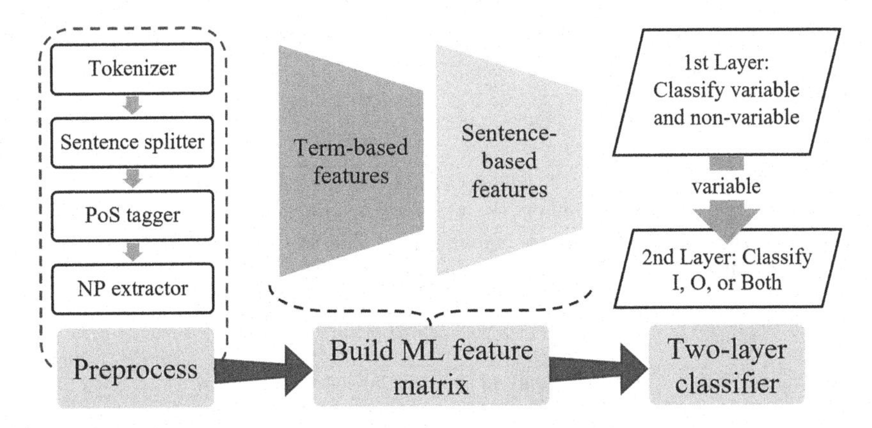

Fig. 2. Two-layer variable classifier: first layer distinguishes variables from non-variables, and second layer decides if a variable is input, output, or both.

As noun/NP and sentence represent the unit to classify and the context of classification respectively, our feature design is driven by the terms and the sentences. We group the features into term-based and sentence-based in Tables 1 and 2 respectively. These tables also provide our intuition behind each feature, leading up to the feature's inclusion and exclusion of the first layer (variable or non-variable) and the second layer (I, O, or both). We orient our features

Table 1. Term-based features for variable learning (• means the feature is used in a specific layer)

ID	Name	Description	First layer	Second layer
f1	hasNumerical-StatisticalTerm	A noun or NP containing a statistical term or symbol (e.g., "percent", "%", "ratio", "max", etc.) could imply an I/O variable	•	
f2	hasInitTerm	A noun or NP containing "initial" or "init" could indicate an I variable	•	•
f3	hasSummary-Term	A noun or NP containing statistical terms (e.g., "final", "average", and "total") could indicate an O variable	•	•
f4	hasInputIn-Heading	If the heading of chapter, section, etc. to which the noun or NP belongs contains "input", the noun or NP is likely to be an I variable	•	•
f5	hasOutputIn-Heading	If the heading of chapter, section, etc. to which the noun or NP belongs contains "output", "result", or "summary", the noun or NP is likely to be an O variable	•	•
f6	hasHighFre-quency	If a noun or NP is part of the top 1% of most frequently occurred terms in the user manual, it is likely a domain concept of the software	•	

Table 2. Sentence-based features for variable learning (• means the feature is used in a specific layer)

ID	Name	Description	First layer	Second layer
f7	singleNoun	If a non-heading sentence contains only one noun or NP, it is likely a variable	•	
f8	nounWith-NearUnit	If a sentence contains a noun or NP and a unit (e.g., inch or inches) within the 5-term neighborhood, the sentence likely introduces the noun or NP variable with unit	•	
f9	beginsWith-Noun	If a non-heading sentence begins with a noun or NP, then the sentence likely provides details about the noun or NP details about the noun/NP variable	•	
f10	hasNoun-Subject	If a non-heading sentence has a noun or NP subject, the sentence likely explains the noun or NP variable or describes the variable's operations	•	
f11	hasBEVerb-WithNoun	If a sentence has the linking verb (e.g., "is") followed by a noun or NP, the sentence likely provides definitional information about the noun or NP variable	•	
f12	nounWith-NearNum	If a sentence has a noun or NP and a value within the 5-term neighborhood, the sentence likely introduces the default value of the noun or NP input variable	•	•
f13	nounSubject-WithNum	If a non-heading sentence has a noun or NP subject followed by a second sentence containing a numerical value, then it is likely the first sentence introduces the noun or NP input variable and the second sentence gives the default value		•

around syntactic and semantic properties that are meaningful irrespectively of the exact characteristics of individual user manual. Of particular note are f8 and f12 that associate a noun/NP with a measuring unit or a numerical value within the 5-term neighborhood. It is worth bearing in mind that such a distance of five terms is not a parameter but a property of the English language [21].

Classification is carried out according to the feature matrix at either layer, and the output is the identification of scientific software's I/O variables. Examining which ML algorithms best implement our multi-layer variable classifier is addressed next in the evaluation section.

4 Experimental Evaluation

4.1 Research Questions

We set out to answer three research questions.

RQ_1: Which ML algorithms yield the most accurate I/O variable classifications?

As ML classification algorithms can be broadly grouped into mathematical, hierarchical, and layered categories [38], we compare the classification accuracy measured by recall, precision, and F-measure of five ML algorithms: logistic regression and support vector machine (mathematical category), decision tree and random forest (hierarchical category), and feedforward neural network (layered category).

RQ_2: Is the two-layer approach better than the single-layer counterpart for classifying I/O variables?

We compare multi-layer with single-layer in I/O variable classifications. We implement the ML algorithms with scikit-learn in Python [36] and tune these algorithms to maximize classification accuracy.

RQ_3: What are the most influential ML features for classifying I/O variables?

We assess the importance of the features in Tables 1 and 2 via information gain [49], which measures how efficient a given feature is in our multi-layer variable classifier. The higher the information gain, the more discriminative power a feature has. Our analysis here also offers insights into the performance of a baseline classifier relying on the user manual headings (i.e., f4 and f5 of Table 1). These two features make use of terms like "input" and "output" directly, and hence provide straightforward classification cues.

4.2 Subject Systems

Our experiments are conducted on three subject systems. We choose these scientific software systems due to their sustained developments, and open accesses to their user manuals. Another reason was our familiarity to the water domain [6,12,22] and the EPA SWMM system [30–32]. Table 3 lists the characteristics of the three subject systems.

Table 3. Subject system characteristics

Subject system	Basic information			User manual		Answer set size			
	Years of existence	Written in	# of LoC	source	# of pages	Non-variable	Input (I) variable	Output (O) variable	Both I and O
SWMM [43]	49	C	46,291	[34]	353	13,665	807	164	53
SWAT [40]	30	Fortran	84,296	[3]	650	24,894	1,006	454	78
MODFLOW [44]	36	Fortran	61,945	[41]	188	12,602	601	172	47

Besides SWMM simulating water runoff quantity and quality in primarily urban areas [43], the Soil & Water Assessment Tool (SWAT) is a small watershed to river basin-scale model widely used in regional management (e.g., simulating soil erosion prevention and non-point source pollution control) [40]. In contrast, the original scope of the Modular Hydrologic Model (MODFLOW) is solely limited to groundwater-flow simulation, though the simulation capabilities have since been expanded to include groundwater/surface-water coupling, solute transport, land subsidence, and other nonfunctional aspects [25,44].

Although written in different programming languages and developed by different teams [4], the size of all three systems can be considered to be medium (between 1,000 and 100,000 LoC) according to Sanders and Kelly's study of scientific software [35]. Due to the thousands of users worldwide, each software maintains authoritative user manual where I/O variables are comprehensively documented [3,34,41].

As our multi-layer variable classifier (cf. Fig. 2) uses supervised learning, labeled data are required for training. Three researchers therefore spent about 100 human-hours in total constructing the subject systems' answer sets manually. They first individually and independently labeled a randomly chosen chapter from each user manual, resulting in a 0.87 Fleiss' κ and hence achieving a strong inter-rater agreement [11]. The discrepancies were resolved in a joint meeting, and a protocol was agreed upon. The researchers then applied the protocol to label the remaining user manuals separately. From Table 3, we can see that each system contains hundreds of variables, confirming scientific software's large input and output spaces [46].

4.3 Results and Analysis

To answer $\mathbf{RQ}_1$, we use the manually labeled data to train the ML algorithms in a ten-fold cross validation procedure. The average recall, precision, and F-measure across the ten-fold validation are reported in Table 4. As F-measure is the harmonic mean of recall and precision, Table 4 highlights in **bold** the best accuracy. Among the five ML algorithms considered, random forest best implements the first layer of our variable classifier where a binary decision (variable or non-variable) is made. This result is in line with Ibarguren *et al.*'s experience that random forest almost always has lower classification error in handling uneven data sets [13]. Feedforward neural network, as shown in Table 4-b, is advantageous to implementing our second layer where variables are further classified into I, O, or both. Logistic regression can also be an option since it achieves the highest F-measure in SWAT. Table 4 demonstrates the modularity of our multi-layer variable classifier where different ML algorithms can be deployed for different classification tasks.

To answer $\mathbf{RQ}_2$, we compare our solution to a single-layer alternative. In particular, we train the five ML algorithms with all the 13 features of Tables 1 and 2 to classify four variable types at once. Table 5 shows that our multi-layer classifier outperforms the best single-layer solution where the average measure across ten-fold validation are presented. The multi-layer solution achieves better performances especially in classifying I and O variables of the scientific software. We therefore suggest random forest and feedforward neural network to be the first layer and second layer ML algorithm respectively for classifying variables from scientific software's user manual.

Table 4. ML algorithm selection results (**RQ**$_1$)

(a) First layer distinguishing variables from non-variables

Accuracy of ML		SWMM	SWAT	MODFLOW
Logistic regression	Recall	0.755	0.767	0.693
	Precision	0.778	0.769	0.717
	F-measure	0.765	0.768	0.704
Support vector machine	Recall	0.762	0.820	0.678
	Precision	0.827	0.820	0.802
	F-measure	0.784	0.820	0.713
Decision tree	Recall	0.763	0.821	0.680
	Precision	0.828	0.821	0.804
	F-measure	0.785	0.821	0.715
Random forest	Recall	0.812	0.847	0.750
	Precision	0.828	0.852	0.804
	F-measure	**0.815**	**0.849**	**0.768**
Feedforward neural network	Recall	0.766	0.783	0.640
	Precision	0.778	0.804	0.816
	F-measure	0.771	0.793	0.673

(b) Second layer classifying I, O, or both-I-and-O variables

Accuracy of ML		SWMM	SWAT	MODFLOW
Logistic regression	Recall	0.582	0.826	0.643
	Precision	0.819	0.729	0.776
	F-measure	0.620	**0.763**	0.659
Support vector machine	Recall	0.573	0.815	0.643
	Precision	0.814	0.727	0.777
	F-measure	0.614	0.757	0.660
Decision tree	Recall	0.582	0.810	0.643
	Precision	0.819	0.724	0.776
	F-measure	0.620	0.753	0.659
Random forest	Recall	0.620	0.815	0.643
	Precision	0.786	0.727	0.776
	F-measure	0.661	0.757	0.659
Feedforward neural network	Recall	0.705	0.815	0.696
	Precision	0.819	0.723	0.780
	F-measure	**0.745**	0.756	**0.702**

Table 5. Comparing F-measures ($\mathbf{RQ}_2$)

Accuracy of ML (F-measure)		Non-variable	Input (I) variable	Output (O) variable	Both I and O
SWMM	Multi-layer	0.895	0.811	0.753	0.672
	Single-layer (random forest)	0.900	0.396	0.499	0.652
SWAT	Multi-layer	0.940	0.916	0.741	0.632
	Single-layer (decision tree)	0.940	0.570	0.488	0.586
MOD-FLOW	Multi-layer	0.917	0.713	0.731	0.661
	single-layer (random forest)	0.916	0.132	0.190	0.628

Table 6. Feature importance ($\mathbf{RQ}_3$)

(a) Top-3 features ranked by the information gain scores

First layer			Second layer		
SWMM	SWAT	MODFLOW	SWMM	SWAT	MODFLOW
f7	f7	f7	f4	f5	f5
f12	f4	f4	f13	f12	f4
f8	f11	f5	f12	f4	f12

(b) Information gain ranking of baseline features: f4 and f5

	First layer			Second layer		
	SWMM	SWAT	MODFLOW	SWMM	SWAT	MODFLOW
f4	6th	2nd	2nd	1st	1st	2nd
f5	7th	4th	3rd	4th	3rd	1st

Our answers to $\mathbf{RQ}_3$ are listed in Table 6. The singleNoun feature (f7) of Table 2 has the highest information gain and hence exhibits the most discriminative power in distinguishing variables from non-variables. Other important features in the first layer include keywords in headings (f4 and f5) and neighboring terms or verbs in a sentence (f8, f11, and f12). Surprisingly, f1—used only in the first layer—is not among the most discriminative features, and neither is f13 in the second layer for SWAT and MODFLOW. As shown in Table 6-b, keywords-in-the-heading as baseline features work well in the second layer, but other features play complementary, and often more dominant, roles in ML-based variable classification. From Table 6, we acknowledge the effectiveness of the simple baseline features, and also emphasize the important role played by sentence-based features such as f7 and f12.

4.4 Threats to Validity

We discuss some of the most important factors that must be considered when interpreting our experimental results. A threat to internal validity concerns the quality of the scientific software's user manual. Like other approaches based on software documentation [2], our ML-based I/O classification could be hindered if mistakes exist in the user manual. For this reason, we share our manual-labeling results in an institution's digital preservation repository [29] to facilitate reproducibility.

A factor affecting our study's external validity is that our results may not generalize to other scientific software systems from SWMM, SWAT, and MODFLOW. In fact, the three systems that we studied are within the water domain and developed by government agencies. Producing user-oriented documentation has become a requirement for scientists-developers mandated by organizations like the U.S. EPA [42] and the U.S. Geological Survey (USGS) [45]. Therefore, it is interesting to extend the variable classification via ML to other scientific software systems.

5 Conclusions

In this paper, we have presented an automatic approach that classifies I/O variables from the user manual. Our evaluations on SWMM, SWAT, and MODFLOW show the accuracy of ML-based I/O classification, and support the hierarchy and modularity of a multi-layer ML solution. Specifically, we recommend random forest for distinguishing variables from non-variables, and feedforward neural network for further classifying the variables into input and output types.

Our future work includes expanding the experimentation to other scientific software systems, building more efficient ML solutions by using a subset of the most important features, releasing the solutions as cloud-based tools [33], integrating the classified variables into metamorphic testing [28], and performing theoretical replications [15,24]. Our goal is to better support scientists in improving the effectiveness and efficiency of their software development and maintenance activities.

Acknowledgments. *We thank the EPA SWMM team, especially Michelle Simon, for the research collaborations. We also thank the anonymous reviewers for their constructive comments.*

References

1. Abualhaija, S., Arora, C., Sabetzadeh, M., Briand, L.C., Vaz, E.: A machine learning-based approach for demarcating requirements in textual specifications. In: International Requirements Engineering Conference, pp. 51–62 (2019)
2. Aghajani, E., et al.: Software documentation: the practitioners' perspective. In: International Conference on Software Engineering, pp. 590–601 (2020)

3. Arnold, J.G., Kiniry, J.R., Srinivasan, R., Williams, J.R., Haney, E.B., Neitsch, S.L.: Soil & Water Assessment Tool (SWAT) Input/Output Documentation (Version 2012). https://swat.tamu.edu/media/69296/swat-io-documentation-2012.pdf. Accessed 06 Mar 2022
4. Bhowmik, T., Niu, N., Wang, W., Cheng, J.-R.C., Li, L., Cao, X.: Optimal group size for software change tasks: a social information foraging perspective. IEEE Trans. Cybern. **46**(8), 1784–1795 (2016)
5. Burungale, A.A., Zende, D.A.: Survey of large-scale hierarchical classification. Int. J. Eng. Res. Gen. Sci. **2**(6), 917–921 (2014)
6. Challa, H., Niu, N., Johnson, R.: Faulty requirements made valuable: on the role of data quality in deep learning. In: International Workshop on Artificial Intelligence and Requirements Engineering, pp. 61–69 (2020)
7. Chattopadhyay, A., Niu, N., Peng, Z., Zhang, J.: Semantic frames for classifying temporal requirements: an exploratory study. In: Workshop on Natural Language Processing for Requirements Engineering (2021)
8. Chen, T.Y., Poon, P.-L., Xie, X.: METamorphic relation identification based on the category-choice framework (METRIC). J. Syst. Softw. **116**, 177–190 (2016)
9. Clarno, K., de Almeida, V., d'Azevedo, E., de Oliveira, C., Hamilton, S.: GNES-R: global nuclear energy simulator for research task 1: high-fidelity neutron transport. In: American Nuclear Society Topical Meeting on Reactor Physics: Advances in Nuclear Analysis and Simulation (2006)
10. Dalpiaz, F., Dell'Anna, D., Aydemir, F.B., Çevikol, S.: Requirements classification with interpretable machine learning and dependency parsing. In: International Requirements Engineering Conference, pp. 142–152 (2019)
11. Fleiss, J.L., Cohen, J.: The equivalence of weighted kappa and the intraclass correlation coefficient as measures of reliability. Educ. Psychol. Measur. **33**(3), 613–619 (1973)
12. Gudaparthi, H., Johnson, R., Challa, H., Niu, N.: Deep learning for smart sewer systems: assessing nonfunctional requirements. In: International Conference on Software Engineering: Software Engineering in Society, pp. 35–38 (2020)
13. Ibarguren, I., Pérez, J.M., Muguerza, J., Gurrutxaga, I., Arbelaitz, O.: Coverage-based resampling: building robust consolidated decision trees. Knowl. Based Syst. **79**, 51–67 (2015)
14. Kanewala, U., Chen, T.Y.: Metamorphic testing: a simple yet effective approach for testing scientific software. Comput. Sci. Eng. **21**(1), 66–72 (2019)
15. Khatwani, C., Jin, X., Niu, N., Koshoffer, A., Newman, L., Savolainen, J.: Advancing viewpoint merging in requirements engineering: a theoretical replication and explanatory study. Requir. Eng. **22**(3), 317–338 (2017). https://doi.org/10.1007/s00766-017-0271-0
16. Li, Y., Guzman, E., Tsiamoura, K., Schneider, F., Bruegge, B.: Automated requirements extraction for scientific software. In: International Conference on Computational Science, pp. 582–591 (2015)
17. Lin, X., Peng, Z., Niu, N., Wang, W., Liu, H.: Finding metamorphic relations for scientific software. In: International Conference on Software Engineering (Companion Volume), pp. 254–255 (2021)
18. Lin, X., Simon, M., Peng, Z., Niu, N.: Discovering metamorphic relations for scientific software from user forums. Comput. Sci. Eng. **23**(2), 65–72 (2021)
19. Lin, X., Simon, M., Niu, N.: Releasing scientific software in GitHub: a case study on SWMM2PEST. In: International Workshop on Software Engineering for Science, pp. 47–50 (2019)

20. Lin, X., Simon, M., Niu, N.: Scientific software testing goes serverless: creating and invoking metamorphic functions. IEEE Softw. **38**(1), 61–67 (2021)
21. Maarek, Y.S., Berry, D.M., Kaiser, G.E.: An information retrieval approach for automatically constructing software libraries. IEEE Trans. Softw. Eng. **17**(8), 800–813 (1991)
22. Maltbie, N., Niu, N., Van Doren, M., Johnson, R.: XAI tools in the public sector: a case study on predicting combined sewer overflows. In: ACM Joint European Software Engineering Conference and Symposium on the Foundations of Software Engineering, pp. 1032–1044 (2021)
23. Nguyen-Hoan, L., Flint, S., Sankaranarayana, R.: A survey of scientific software development. In: International Symposium on Empirical Software Engineering and Measurement, pp. 1–10 (2010)
24. Niu, N., Koshoffer, A., Newman, L., Khatwani, C., Samarasinghe, C., Savolainen, J.: Advancing repeated research in requirements engineering: a theoretical replication of viewpoint merging. In: International Requirements Engineering Conference, pp. 186–195 (2016)
25. Niu, N., Yu, Y., González-Baixauli, B., Ernst, N., Leite, J., Mylopoulos, J.: Aspects across software life cycle: a goal-driven approach. Trans. Aspect-Orient. Softw. Develop. **V1**, 83–110 (2009)
26. NLTK. Natural Language Toolkit. https://www.nltk.org. Accessed 06 Mar 2022
27. Pawlik, A., Segal, J., Petre, M.: Documentation practices in scientific software development. In: International Workshop on Cooperative and Human Aspects of Software Engineering, pp. 113–119 (2012)
28. Peng, Z., Kanewala, U., Niu, N.: Contextual understanding and improvement of metamorphic testing in scientific software development. In: Int. Symp. Emp. Softw. Eng. Measur. pp. 28:1–28:6 (2021)
29. Peng, Z., Lin, X., Niu, N.: Data of Classifying I/O Variables via Machine Learning. https://doi.org/10.7945/85j1-qf68. Accessed 06 Mar 2022
30. Peng, Z., Lin, X., Niu, N.: Unit tests of scientific software: a study on SWMM. In: International Conference on Computational Science, pp. 413–427 (2020)
31. Peng, Z., Lin, X., Niu, N., Abdul-Aziz, O.I.: I/O associations in scientific software: a study of SWMM. In: International Conference on Computational Science, pp. 375–389 (2021)
32. Peng, Z., Lin, X., Simon, M., Niu, N.: Unit and regression tests of scientific software: a study on SWMM. J. Comput. Sci. **53**, 101347:1–101347:13 (2021)
33. Peng, Z., Niu, N.: Co-AI: a Colab-based tool for abstraction identification. In: International Requirements Engineering Conference, pp. 420–421 (2021)
34. Rossman, L.A.: Storm Water Management Model User's Manual Version 5.1. https://www.epa.gov/water-research/storm-water-management-model-swmm-version-51-users-manual. Accessed 06 Mar 2022
35. Sanders, R., Kelly, D.: Dealing with risk in scientific software development. IEEE Softw. **25**(4), 21–28 (2008)
36. Scikit-learn. Machine Learning in Python. https://scikit-learn.org/stable/ Accessed 06 Mar 2022
37. Spikerog SAS. ExtractPDF. https://www.extractpdf.com. Accessed 06 Mar 2022
38. Suthaharan, S.: Machine Learning Models and Algorithms for Big Data Classification. ISIS, vol. 36. Springer, Boston (2016). https://doi.org/10.1007/978-1-4899-7641-3
39. TextBlob. Simplified Text Processing. https://textblob.readthedocs.io. Accessed 06 Mar 2022

40. United States Department of Agriculture. Soil & Water Assessment Tool (SWAT). https://data.nal.usda.gov/dataset/swat-soil-and-water-assessment-tool. Accessed 06 Mar 2022
41. United States Department of the Interior & United States Geological Survey. Modular Hydrologic Model (MODFLOW) Description of Input and Output (Version 6.0.0). https://water.usgs.gov/ogw/modflow/mf6io.pdf. Accessed 06 Mar 2022
42. United States Environmental Protection Agency. Agency-wide Quality System Documents. https://www.epa.gov/quality/agency-wide-quality-system-documents. Accessed 06 Mar 2022
43. United States Environmental Protection Agency. Storm Water Management Model (SWMM). https://www.epa.gov/water-research/storm-water-management-model-swmm. Accessed 06 Mar 2022
44. United States Geological Survey. Modular Hydrologic Model (MODFLOW). https://www.usgs.gov/software/software-modflow. Accessed 06 Mar 2022
45. United States Geological Survey. Review and Approval of Scientific Software for Release (IM OSQI 2019–01). https://www.usgs.gov/about/organization/science-support/survey-manual/im-osqi-2019-01-review-and-approval-scientific. Accessed 06 Mar 2022
46. Vilkomir, S.A., Swain, W.T., Poore, J.H., Clarno, K.T.: Modeling input space for testing scientific computational software: a case study. In: International Conference on Computational Science, pp. 291–300 (2008)
47. Wang, W., Niu, N., Liu, H., Niu, Z.: Enhancing automated requirements traceability by resolving polysemy. In: International Requirements Engineering Conference, pp. 40–51 (2018)
48. Wikipedia. Storm Water Management Model. https://en.wikipedia.org/wiki/Storm_Water_Management_Model. Accessed 06 Mar 2022
49. Witten, I.H., Frank, E., Hall, M.A.: Data Mining: Practical Machine Learning Tools and Techniques. Morgan Kaufmann (2016)
50. Zhou, Z., Xiang, S., Chen, T.Y.: Metamorphic testing for software quality assessment: a study of search engines. IEEE Trans. Softw. Eng. **42**(3), 264–284 (2016)

Distributed Architecture for Highly Scalable Urban Traffic Simulation

Michał Zych, Mateusz Najdek(✉), Mateusz Paciorek, and Wojciech Turek

AGH University of Science and Technology, Krakow, Poland
najdek@agh.edu.pl

Abstract. Parallel computing is currently the only possible method for providing sufficient performance of large scale urban traffic simulations. The need for representing large areas with detailed, continuous space and motion models exceeds the capabilities of a single computer in terms of performance and memory capacity. Efficient distribution of such computation, which is considered in this paper, poses a challenge due to the need of repetitive synchronization of the common simulation model. We propose an architecture for efficient memory and communication management, which allows executing simulations of huge urban areas and efficient utilization of hundreds of computing nodes. In addition to analyzing performance tests, we also provide general guidelines for designing large-scale distributed simulations.

Keywords: Urban traffic simulation · Simulation scalability · HPC

1 Introduction

The demand for reliable simulations of complex social situations draws attention of scientists towards the problems of simulation algorithms efficiency. Providing valuable results requires detailed models of society members and their environment, which, together with the need for simulating large scenarios fast, exceeds the capabilities of sequential algorithms. However, the problem of parallel execution of such simulations is not straightforward. The considered computational task is focused on repetitive modifications of one large data structure, which, when performed in parallel, must be properly synchronized. There are a few successful examples [10,14] of parallel spacial simulation algorithms, however, efficient utilization of HPC-grade hardware for simulating real-life scenarios with continuous space and motion models remains an open problem.

In this paper we present the experiences with scaling the SMARTS simulation system [11], which is presumably the first distributed simulator for continuous urban traffic model. This work is the continuation of the preliminary research presented in [8], where basic scalability issues were identified and corrected, making it possible to execute a complex simulation task on 2400 computing cores of a supercomputer. Here we propose a redesigned architecture of the simulation system, which aims at overcoming limitations of the original approach.

D. Groen et al. (Eds.): ICCS 2022, LNCS 13353, pp. 517–530, 2022.
https://doi.org/10.1007/978-3-031-08760-8_43

The proposed architecture, together with other improvements introduced to the SMARTS simulation system, made it possible to execute a simulation of 100,000 square kilometers of intensively urbanised area with almost 5 million cars. The system efficiently utilized 9600 computing cores, which is presumably the largest hardware setup, executing a real-life scenario of urban traffic simulation, reported so far.

The experiences with discovering the reasons for scalability limitations of the original implementation may be also valuable in other applications. Therefore, we provide a summary of guidelines for implementing super-scalable spacial simulations. The guidelines concern the architecture of a distributed simulation system, communication protocols and data handling. We believe that the summary may be valuable for researchers and engineers willing to use HPC-grade hardware for large-scale simulations.

2 Scalable Traffic Simulations

The attempts towards implementing improvements of parallel urban traffic simulation have been present in the literature since the end of the 20th century – popular microscopic model of traffic, the Nagel-Schreckenberg model, was the first to be executed in parallel by its creator in 1994 [7]. Further research by the same team [12] suggested that the global synchronization in parallel traffic simulation algorithm might not at all be necessary. Despite that, several centralized approaches to the problem were tried later on. Methods described in [6] and [9] have shown efficiency improvements only up to a few computing nodes. Identified problem of scalability limitations caused by synchronization was addressed in [2] by increased time between global synchronization. It allowed better efficiency, but also influenced the simulation results, which cannot be accepted. The work presented in [13] proposed a method for solving this issue by introducing a corrections protocol. This, however, again resulted in poor scalability.

Achieving significant scalability requires dividing the computational tasks into parts, which are scale-invariant [1]. This refers to the number of computations but also to the volume of communication, which cannot grow with the number of workers. Centralized synchronization of parallel computation guarantees just the opposite.

Presumably, the first architecture of distributed traffic simulation system, which follows these guidelines was described in [16]. The proposed Asynchronous Synchronization Strategy assumes that each computing worker communicates with a limited and constant number of other workers – those responsible for adjacent fragments of the environment. The paper reports linear scalability up to 32 parallel workers.

The proposed distribution architecture was used and extended in [14], where traffic simulation scaled linearly up to 19200 parallel workers executed on 800 computing nodes. The simulation of over 11 million cars run at 160 steps per second, which is presumably the largest scenario tested so far in the domain of traffic simulations. It opened new areas of application for such simulations, like the real-time planning presented in [15]. While proving the possibility of creating efficient HPC-grade traffic simulation, the implementation used discrete model and did not support real-life scenarios. In order to provide these experiences in a useful tool, we decided to investigate the possibility of redesigning the architecture of an existing, highly functional traffic simulation tool. We selected the SMARTS simulator [11], which supports importing of real cities models and simulated continuous traffic models. The details of the original design and the introduced improvements will be presented in the following sections.

3 The SMARTS System

SMARTS *(Scalable Microscopic Adaptive Road Traffic Simulator)* [11] is a traffic simulator designed with an intent to support a continuous spatial model of the environment with a microscopic level of details. The simulation is executed in time steps representing a short duration of real-world time. During the simulation, each driver makes its own decisions based on one of the implemented decision models and the state of the environment. Two decision models are used: the IDM (*Intelligent Driver Model*) [4] predicts appropriate acceleration based on the state of the closest preceding vehicle while the MOBIL (*Minimizing Overall Braking Induced by Lane Changes*) [3] triggers lane changes. SMARTS is implemented in Java. The most important features of SMARTS are:

- Loading data from OSM (OpenStreetMap[1]).
- Static or dynamic traffic lights control.
- Commonly-used right-of-way rules.
- Different driver profiles that can affect the car-following behavior.
- Public transport modelling.
- Graphical user interface for easy configuration and visualization of results.
- Various types of output data, such as the trajectory of vehicles and the route plan of vehicles.

SMARTS provides a distributed computing architecture, which can accelerate simulations using multiple processes running at the same time. The initial architecture of SMARTS (shown in Fig. 1) includes a server process and one or more worker processes, which are responsible for performing the actual computation. Worker processes communicate with each other through TCP connections using sockets. The architecture had a significant drawback, as it forced to run each worker in its own separate virtual machine within shared cluster node, which required environment map to be loaded from the drive and parsed by each worker individually, causing increased heap memory usage.

SMARTS also uses two mechanisms to synchronize the simulation. The first mechanism is a decentralized simulation mode. The very first version of SMARTS ran in a centralized mode, where the server needed to wait for confirmation messages from all the workers and to instruct all the workers to proceed at each

[1] https://www.openstreetmap.org/.

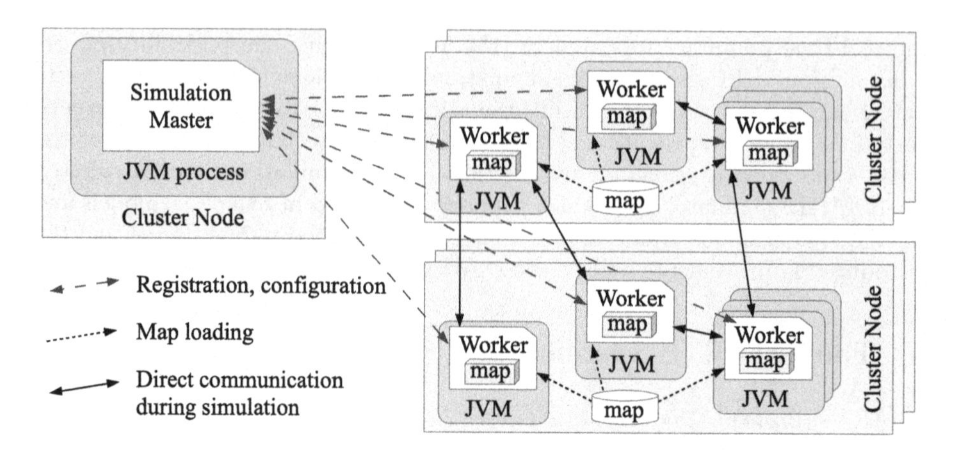

Fig. 1. Initial architecture of the simulation system.

time step. Later, the decentralized mode was added. In this mode, the workers do not need to communicate with the server during the simulation stage.

The second mechanism is called Priority Synchronous Parallel (PSP) model, where a worker uses a two-phase approach to reduce the impact of the slowest worker [5]. During the first phase, the worker simulates the vehicles in a high priority zone at the boundary of its simulation area. This zone covers all the vehicles that may cross the boundary of the area at the current time step. During the second phase, the worker sends the information about the border-crossing vehicles to its neighbors while simulating the rest of the vehicles, i.e., the vehicles that are within the boundary but are not in the high priority zone.

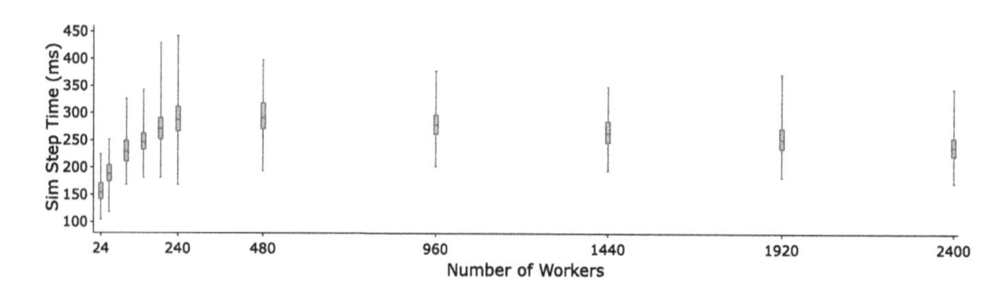

Fig. 2. Simulation step times for problem size increasing in proportion to the number of workers.

The original version of SMARTS [11] was only able to utilize up to 30 workers, depending on simulated environment size, which also determined the limits of the original scalability tests. After our initial improvements, presented in [8], we were able to successfully run simulation that involved the number of vehicles growing in proportion to the increasing computing power (1000 vehicles per core), with execution times shown in Fig. 2. The experiment was based on the road network in 59 km × 55 km area in Beijing, China. The real-world map model was prepared from OSM map. The road network graph contained approximately 220 thousand nodes and 400 thousand directional edges, which represent roads in this model. The whole simulation process consisted of 1500 steps with a step duration equal to 200 ms. It is equivalent to 5 min of real-time traffic. The execution times of the last 1000 steps were measured and the average time of one in every 50 steps was recorded. The software architecture of this solution was not yet optimized for High Performance Computing Systems, although it showed promising results. All average execution times of simulation steps were within similar range and the median stabilized on the level of about 250 ms.

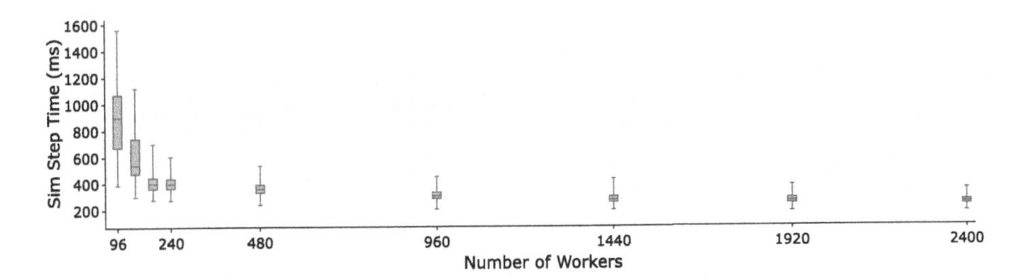

Fig. 3. Simulation step times with fixed problem size for increasing number of workers.

We also prepared an experiment to analyze the strong scalability for the original architecture. It involved a predefined number of 2.4 million vehicles in total, distributed across whole environment. Figure 3 shows that in each case the increasing number of workers, and hence the computing power, resulted in reduction of average simulation time of a single step, therefore improving overall simulation performance.

The original architecture exposed significant limitations in these experiments. The scalability was not satisfactory and the system was unable to run scenarios larger that the one used in the tests.

4 New Communication Architecture

Despite the improvements proposed in the previous paper [8], SMARTS still have limitations that prevent efficient executions of large scenarios. The main problems are: the inability to run simulations above 2400 cores and memory leaks when simulating huge areas.

These memory problems are mainly caused by the requirements posed by the Worker-based architecture, to store the entire map in each Worker. A computing node with multiple cores to exceed all available memory when simulating a large map. It is important to note that only the area simulated by a Worker is changed, thus the rest of the map is used for read-only operations (e.g. path planning).

The other crucial problem is the impossibility to start the simulation correctly if more than 2400 Workers are used. The server trying to handle all server-worker connections, needs to create too many threads, which consequently causes an error at the initialization stage.

This paper proposes a new architecture for the simulator, which is shown in Fig. 4. A proxy instance, the WorkerManager, was created to handle the communication with the server and to manage the lifecycle of the Workers that are under its control. By using this type of architecture, the number of JVMs created on a node was reduced (1 JVM per node instead of 1 per core). By appropriately assigning tasks to the proxy instance, the main problems of the simulator can be eliminated.

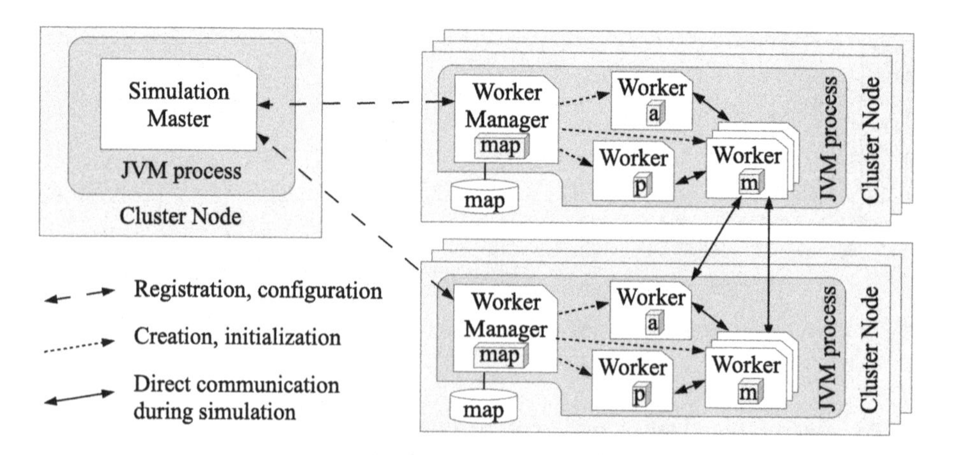

Fig. 4. Modified architecture of the simulation system.

By using a WorkerManager instance, all read-only objects can be stored only once in node memory. While the case is obvious for any kind of constants in the simulation, it is not possible to store the whole map only once, because each Worker uses part of it to execute the simulation. In this case, as shown in Fig. 4, the following solution was selected: WorkerManager stores the whole map area, which is used in read-only mode to generate vehicle routes during the simulation, while at the initialization stage individual Workers copy only the part of the map they are responsible for and a small fragment on the boundary of the map division between Workers.

The use of separate simulation zones for individual Workers has also improved simulation performance. Each core now searches for vehicles only in the part of the map for which it is responsible. As a result, the work required to search for vehicles in each step of the simulation has been reduced, which has a positive impact on its efficiency.

Another significant change due to the new architecture is the possibility of aggregating server-worker communication. As shown in Fig. 1, in the previous implementation the communication between the server and the Worker was direct. As can be seen in Fig. 4, the way of communication was changed by creating a dedicated proxy instance. The server currently communicates only with WorkerManagers. This results in a reduction of threads created by the server which are responsible for handling communication. In addition, the server, thanks to the message from the WorkerManager, possesses information which Worker is controlled by which WorkerManager. Thanks to this, it aggregates messages that should be sent to each of the Workers and sends them as a single message to the proxy instance. In addition, for optimization purposes, the common parts of messages were separated so as not to duplicate the information sent to the WorkerManager.

What is also important, the communication between individual Workers during the simulation has not been changed in any way. It still takes place directly in an asynchronous way. The server sends information about all Workers, so they can establish direct communication.

Through the use of proxy instances and communication aggregation, a new algorithm for the initialization and finalization of the simulation was created. The new method of initialization can be described as follows:

- Creating WorkerManagers and registering them to the server.
- The server creates a map using the OSM data and divides it into a desired number of fragments. Then it sends all simulation settings, such as path to the file containing the simulation map, number of Workers and other simulation properties.
- WorkerManagers create a map of the simulation using the file and validate the map (the server sends a hash of the map it created from the file). After this stage is completed they send back a message to the server about the successful creation of the map.
- Server orders the creation of Workers.
- WorkerManagers create a desired number of Workers and send their data to the server.
- Server assigns map fragments and initial number of vehicles for each fragment to individual Workers. Then this information is sent to WorkerManagers.
- WorkerManagers pass the message to individual Workers. Each Worker copies the corresponding fragment of the map and generates vehicle routes. When every Worker finishes this step, the WorkerManager sends a message about being ready to start the simulation.
- After receiving all messages about readiness, server orders to start the simulation stage.

Using the new architecture allowed the simulation to run correctly with up to 9600 computing cores. In addition, the server sends far fewer messages at the initialization stage than before, which has a positive impact on the performance of this stage of the entire process. Moreover, by storing common parts such as a simulation map only once, it is now possible to run much larger test scenarios.

Another improvement is the new distribution of the initial number of cars to individual Workers. Previously, when calculating the number of cars, the value was always rounded down and the last Worker was assigned the rest of the cars.

In the case of the largest scenario tested in this study, such a division resulted in the last Worker receiving about 9000 cars, while the rest received only 499. This caused an uneven imbalance from the very beginning of the simulation. The presented example shows how important the seemingly non-obvious details are during the design of systems for such a large scale.

The extended versions of the SMARTS system, used for conducting the presented experiments, is available online[2].

5 Evaluation

In order to measure the scalability of the simulation using the proposed architecture, two experiments were prepared. Both experiments used the same simulation scenario: cars navigating the road network from a real-world area. The map represented the area of size $101,000\ \text{km}^2$ in north-eastern China, containing Beijing and neighboring cities. The map was created with use of OSM data.

In both experiments each worker process was assigned to one computing core, thus the names "worker" and "core" are used interchangeably when the quantities are discussed. The experiments are explained in detail in their respective sections below.

The evaluation was executed with the use of the Prometheus supercomputer located in the AGH Cyfronet computing center in Krakow, Poland. Prometheus is the 440th fastest supercomputer (TOP500[3] list for November 2021). It is a peta-scale (2.35 PFlops) cluster using HP Apollo 8000 nodes connected via InfiniBand FDR network. Each node contains two Xeon E5-2680v3 12-core 2.5 GHz CPUs. The total number of available cores is 55,728, accessible as HPC infrastructure with a task scheduling system.

5.1 Constant Number of Cars

The first experiment followed the usual rules of strong scalability evaluation: the size of the problem was constant and the number of computing power was increased to analyze the scalability of the solution. The number of cars was set to 4.8M. The numbers of cores used for the execution of the experiment were: 240, 480, 1200, 2400, 4800, 7200, and 9600, allocated in groups of 24 per node due to the architecture of used computing environment. The size of the experiment precluded the execution on number of cores lesser than 240, one of the reasons being the portion of a problem assigned to single node exceeding memory limits.

[2] https://github.com/zychmichal/SMARTS-extension.
[3] https://www.top500.org/system/178534/#ranking.

Figure 5 shows the measured execution times, sampled from all workers in various iterations. As can be observed in Fig. 5a, the execution on 240 cores resulted in several values that do not match the general distribution of measured times, achieving more than four times the median value. These values are caused by Java Garbage Collector invocations, which were required due to the size of tasks assigned to single nodes. After increasing the number of nodes such situations did not occur.

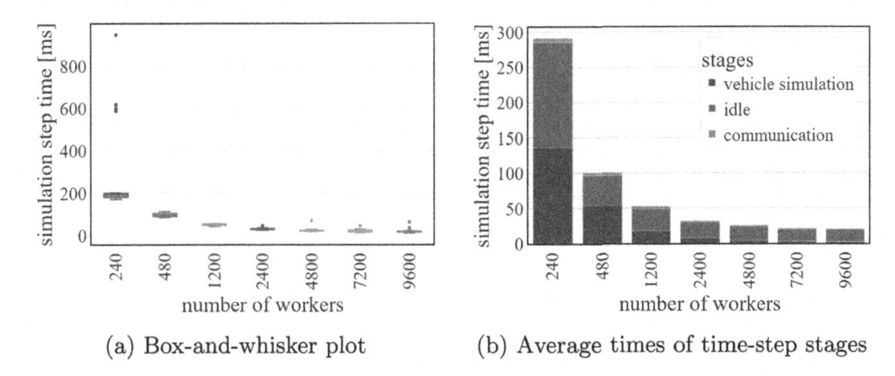

(a) Box-and-whisker plot (b) Average times of time-step stages

Fig. 5. Simulation time-steps with constant number of cars for increasing number of workers.

The measured times were analyzed using the *speedup* and *efficiency* metrics. The speedup should be calculated with the reference to single-core execution of the program. As the execution of the experiment on a single worker is not feasible, a different way of determining reference value was used. Assuming constant amount of work to be performed in each experiment, the average time of vehicle simulation (i.e. the time not spent in communication or waiting for other workers) in 240 cores run, multiplied by 240, was chosen as the approximation of single-core execution time.

Figure 6 shows the described metric values derived from the results presented in Fig. 5. As can be observed in Fig. 6a, the speedup does not follow the ideal line that could be observed for embarrassingly parallel problems. It is expected, as both communication and waiting for other workers contribute to the longer execution times, which impacts the speedup. Nevertheless, by eliminating the bottleneck in form of single point of synchronization, the speedup grows when new resources are added, achieving ca. 2000 for 9600 cores.

The decrease in the benefits from adding new cores is also seen in Fig. 6b, which shows the efficiency based on the measured times. The outlying values for 240 cores impacting the average cause the efficiency for this execution to be significantly worse than expected, which can be additionally observed in the large standard deviation in efficiency for this run.

The most significant cause for this loss of efficiency is the uneven distribution of workload. The time of idle waiting for the other workers (Fig. 5b) reduces as

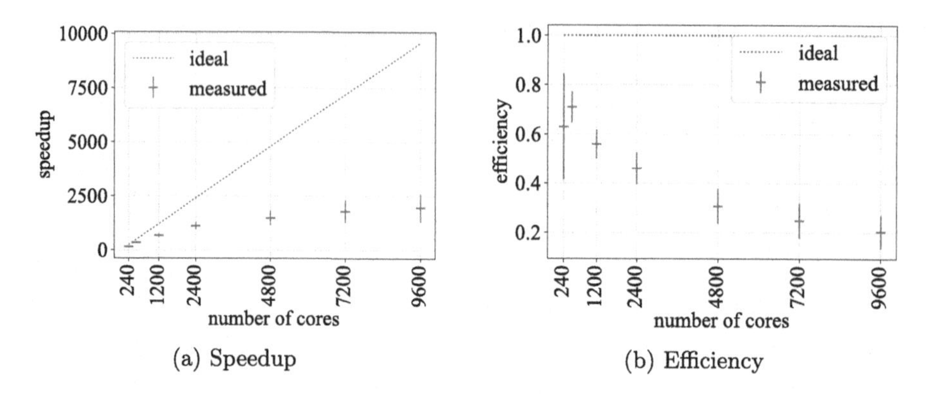

(a) Speedup (b) Efficiency

Fig. 6. Speedup and efficiency in experiment with constant problem size. Blue markers show mean value and standard deviation, red dotted lines show ideal values for linear scalability. (Color figure online)

the cores are added, but its proportion to the total time of simulation increases. This limits the benefits that can be obtained by adding the resources. The uneven load might result from the fluctuations in the car density—if an area simulated by any given worker contains less cars than the average, then some other workers have to simulate more cars than the average, and all other workers will have to wait until the most loaded one finishes its work.

The results presented in Fig. 3 were obtained in an experiment similar to this one, although with half the number of cars and smaller road network. The median execution time of a single time step for the largest number of cores (2400) was ca. 250 ms. Therefore, a good approximation of expected average execution time of a single step for the same scenario as the experiment presented above would be ca. 500 ms. However, the same number of cores yielded average execution time of ca. 60 ms. The new architecture is clearly outperforming the old one, achieving the execution time 5 times better for large numbers of computing resources.

5.2 Growing Number of Cars

The second experiment was inspired by the weak scalability evaluation. However, due to the predefined and irregular simulation environment, it was impossible to scale the environment to match the computing power. Therefore, the only parameter that was changed was the number of cars. As the size of the environment does influence the complexity and the time of the simulation, the resulting experiment does not conform strictly to the rules of weak scalability.

The numbers of cores used in consecutive executions were the following: 24, 48, 120, 240, 480, 1200, 2400, 4800, 7200, and 9600. The number of cars was kept at level of 500 per core.

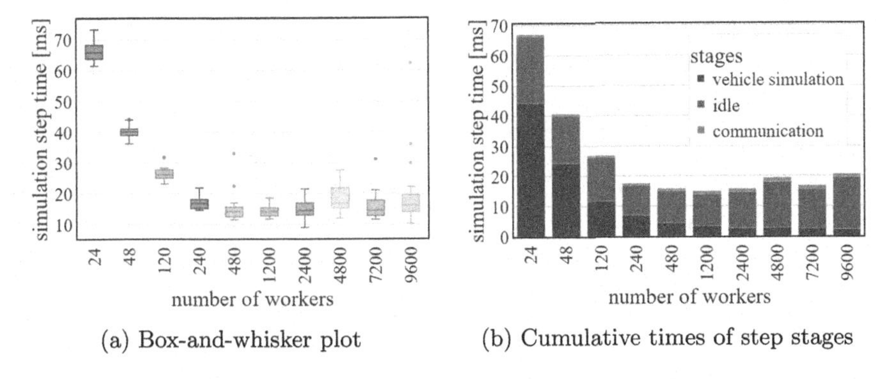

(a) Box-and-whisker plot (b) Cumulative times of step stages

Fig. 7. Simulation time-steps with number of cars growing with the number of workers.

Figure 7 shows the measured execution times, obtained in the same way as in Fig. 5. As the Fig. 5a shows, for small numbers of workers the size of the environment to process influences the time significantly. However, for 240 workers and above, the execution time levels out, which is a desired behaviour in this test. Adding new tasks and new resources allows preserving the execution time constant.

Once again it is possible to compare to the results from previous architecture, presented in Fig. 2. In this case, the number of cars per core was twice the value in the described experiment. The median execution time of a single time step for the largest number of workers (2400) was ca. 250 ms. The expected average execution time for the number of cars in this experiment calculated from this result would be ca. 125 ms. Once again, the results obtained using the new architecture are far better, around 8 times faster execution is observed.

6 Design Guidelines for Scalable Spacial Simulations

In this section we summarize the conclusions from our previous [8] work and the architecture patterns of the current work, which tends towards more general design guidelines for scalable spacial simulation. As authors of the work [1] observed, to be able to achieve super-scalability in the parallelization of algorithms, they should be scale invariant by design. This applies to all aspects of the task size, including the amount of calculations, but also the volume of possessed and process data, the number and the volume of messages exchanged with other tasks and the number of connections. Besides these requirements concerning the computation itself, serious problems can occur during initialization and results collecting phases.

Although the initial size of tasks can be even in spacial simulations, preserving the evenness during simulation is non-trivial. Simulated entities are transferred between workers, altering the balance and causing the need of waiting for most loaded stragglers. Dynamic load balancing is a complex challenge in this area.

Memory-related issues exhibit their significance only when large scenarios are considered. They are often neglected at the beginning of parallel program development, which magnifies their impact in large scale. Therefore, at the design stage, all data structures should be divided into three categories: task-independent, static task-dependent and dynamically growing task-dependent. All task-dependent structures have to be split between workers in order to support starting large simulations and preserve the scale-invariant assumption. Dynamically growing structures (e.g. simulation results) have to be explicitly managed and serialized when necessary. Failing to address this issue causes run-time problems, which are far more complex to identify.

Probably the most important element of a salable design is the communication architecture and synchronization schema. Centralized communication cannot be used during the simulation process – there are no exceptions here. If it is necessary to distribute messages across all the workers then proper communication topology should be introduced and messages should be aggregated wherever possible. In our case worker managers on each cluster node are responsible for aggregation of incoming and outgoing communication with the server. It is very important to ensure only a limited number of connections per worker, but also per all other elements of the system – the managers tend to become bottlenecks also. When proper simulation occurs, it is a key to ensure direct communication between workers. The algorithm has to ensure limited number of this communication type by proper division of the task.

Another important condition of scalable spacial simulation is the proper synchronization approach. In every case, central synchronization should be removed and implicit synchronization strategy should be used instead. After computing a simulation step, each worker sends messages to its neighbours and expects them to do the same. After that, the computation continues.

An additional issue, which currently prevents us from following all the presented guidelines, is the feature of the simulation algorithm, which requires access to the whole model. In the considered traffic simulation new cars need a path, which is calculated using the whole map. In general, such situations violate the scale-invariant condition and should be disallowed. In our case we could require require all the paths in advance. Otherwise we propose to extract the operations, which require access to the whole model into a single entity located at each computing node to reduce the memory demand.

The described guidelines concern the execution of the simulation task itself, which is not the only source of scalability problems. The phases of initialization and finalization can also burden the efficiency or prevent the simulation from running at all. A few typical challenges to point out here are: model division, distribution and loading and results collecting. Distributing the model data and collecting the results can become bottlenecks, when implemented as one-to-many communication. Our experience show, that far better results are achieved by using a shared file system for both these tasks, while the communication is used only for configuring them.

The model division algorithm has to split the model into fragments for specified number of nodes and cores. Existing algorithms are typically sequential, which is a huge waste of resources, waiting idle for the simulation task. Research on parallel splitting methods continues, however, a simpler approach can be adopted in the meantime. By using a faster and less accurate stochastic algorithm, running in many instances on all available nodes, we can often compute a better solution in shorter time. One should also keep in mind that event a perfect initial division does not solve the imbalance problem.

7 Conclusions and Further Work

The presented distributed architecture, together with proper data handling mechanisms allowed simulating great-scale real-life scenario with continuous space and motion model. The use of HPC-grade hardware provides significant performance improvements, but is also necessary due to memory required by the simulation model. The conclusions from the presented work formulate a set of guidelines for achieving super-scalability in spacial simulations.

The presented results clearly show, that there is still space for further improvements in the area of load balancing. Uneven distribution of work between computing nodes results in significant waste of computational power. This problem poses significant challenge for future research because of high dynamics of load changes in this particular problem.

Acknowledgments. The research presented in this paper was funded by the National Science Centre, Poland, under the grant no. 2019/35/O/ST6/01806. The research was supported by PL-Grid Infrastructure.

References

1. Engelmann, C., Geist, A.: Super-scalable algorithms for computing on 100,000 processors. In: Sunderam, V.S., van Albada, G.D., Sloot, P.M.A., Dongarra, J.J. (eds.) ICCS 2005. LNCS, vol. 3514, pp. 313–321. Springer, Heidelberg (2005). https://doi.org/10.1007/11428831_39
2. Kanezashi, H., Suzumura, T.: Performance optimization for agent-based traffic simulation by dynamic agent assignment. In: Proceedings of the 2015 Winter Simulation Conference (WSC 2015), pp. 757–766. IEEE Press, Piscataway (2015)
3. Kesting, A., Treiber, M., Helbing, D.: General lane-changing model MOBIL for car-following models. J. Transp. Res. Board **1999**(1), 86–94 (2007)
4. Kesting, A., Treiber, M., Helbing, D.: Enhanced intelligent driver model to access the impact of driving strategies on traffic capacity. Trans. Roy. Soc. Lond. A **368**(1928), 4585–4605 (2010)
5. Khunayn, E.B., Karunasekera, S., Xie, H., Ramamohanarao, K.: Straggler mitigation for distributed behavioral simulation. In: 2017 IEEE 37th International Conference on Distributed Computing Systems (ICDCS), pp. 2638–2641. IEEE (2017)

6. Klefstad, R., Zhang, Y., Lai, M., Jayakrishnan, R., Lavanya, R.: A distributed, scalable, and synchronized framework for large-scale microscopic traffic simulation. In: Proceedings of the 2005 IEEE Intelligent Transportation Systems, 2005, pp. 813–818 (2005)
7. Nagel, K., Schleicher, A.: Microscopic traffic modeling on parallel high performance computers. Parallel Comput. **20**(1), 125–146 (1994)
8. Najdek, M., Xie, H., Turek, W.: Scaling simulation of continuous urban traffic model for high performance computing system. In: Paszynski, M., Kranzlmüller, D., Krzhizhanovskaya, V.V., Dongarra, J.J., Sloot, P.M.A. (eds.) ICCS 2021. LNCS, vol. 12742, pp. 256–263. Springer, Cham (2021). https://doi.org/10.1007/978-3-030-77961-0_22
9. O'Cearbhaill, E.A., O'Mahony, M.: Parallel implementation of a transportation network model. J. Parallel Distrib. Comput. **65**(1), 1–14 (2005)
10. Paciorek, M., Turek, W.: Agent-based modeling of social phenomena for high performance distributed simulations. In: International Conference on Computational Science, pp. 412–425. Springer, Cham (2021). https://doi.org/10.1007/978-3-319-70087-8_51
11. Ramamohanarao, K., et al.: SMARTS: Scalable microscopic adaptive road traffic simulator. ACM Trans. Intell. Syst. Technol. **8**(2), 1–22 (2016)
12. Rickert, M., Nagel, K.: Dynamic traffic assignment on parallel computers in transims. Future Gen. Comput. Syst. **17**(5), 637–648 (2001)
13. Toscano, L., D'Angelo, G., Marzolla, M.: Parallel discrete event simulation with erlang. In: Proceedings of the 1st ACM SIGPLAN Workshop on Functional High-performance Computing (FHPC 2012), pp. 83–92. ACM, New York (2012)
14. Turek, W.: Erlang-based desynchronized urban traffic simulation for high-performance computing systems. Future Gen. Comput. Syst. **79**, 645–652 (2018)
15. Turek, W., Siwik, L., Byrski, A.: Leveraging rapid simulation and analysis of large urban road systems on HPC. Transp. Res. Part C: Emerg. Technol. **87**, 46–57 (2018)
16. Xu, Y., Cai, W., Aydt, H., Lees, M., Zehe, D.: An asynchronous synchronization strategy for parallel large-scale agent-based traffic simulations. In: Proceedings of the 3rd ACM SIGSIM Conference on Principles of Advanced Discrete Simulation (SIGSIM PADS 2015), pp. 259–269. ACM, New York (2015)

Automated and Manual Testing in the Development of the Research Software RCE

Robert Mischke, Kathrin Schaffert, Dominik Schneider, and Alexander Weinert(✉)

Institute for Software Technology, German Aerospace Center (DLR), 51147 Cologne, Germany
{robert.mischke,kathrin.schaffert,dominik.schneider, alexander.weinert}@dlr.de

Abstract. Research software is often developed by individual researchers or small teams in parallel to their research work. The more people and research projects rely on the software in question, the more important it is that software updates implement new features correctly and do not introduce regressions. Thus, developers of research software must balance their limited resources between implementing new features and thoroughly testing any code changes.

We present the processes we use for developing the distributed integration framework RCE at DLR. These processes aim to strike a balance between automation and manual testing, reducing the testing overhead while addressing issues as early as possible. We furthermore briefly describe how these testing processes integrate with the surrounding processes for development and releasing.

Keywords: Research Software Engineering · Software Testing · RCE

1 Introduction

More and more research is supported by software [8], which ranges wildly in scope and maturity. Software may be developed and supported by major companies, it may result from internal research projects, or it may be a proof-of-concept script developed by individual researchers. Particularly software resulting from internal research projects may be used by numerous research projects, while still being maintained by a handful of original developers.

To be able to implement new features required by new research projects, while simultaneously maintaining existing features, developers require effective and efficient software testing. Such testing enables developers to detect issues early in development, thus simplifying their resolution. Such testing, however, requires infrastructure and non-trivial resource investment to effectively spot issues [25]. Developers should, e.g., not test features they implemented themselves, since they are likely to subconsciously avoid addressing edge cases [20].

A. Weinert—Authors are listed in alphabetical order.

D. Groen et al. (Eds.): ICCS 2022, LNCS 13353, pp. 531–544, 2022.
https://doi.org/10.1007/978-3-031-08760-8_44

Thus, effective testing requires additional resources, which are unavailable in typical research projects. Hence, these projects have to divide their existing resources between development of new features, maintenance of existing ones, and rigorous testing. Each of these activities has a non-negligible overhead.

One such research software project is RCE [6], a distributed scientific integration framework which we develop at the Institute for Software Technology at the German Aerospace Center (DLR). We describe the testing processes we employ to validate changes to RCE and to recognize regressions as soon as possible. These tests comprise automated tests as well as manual ones that address hard-to-test areas of RCE's functionality. We show how we balance the need to thoroughly test changes to RCE with the overhead incurred by such tests.

Related Work. Software testing is a core topic discussed in virtually all education on software development [2] as well as an active field of research [21]. Current research directions range from automated test case generation [3] over automated exploration of edge cases [9,16] to automated formal verification of software [5]. Most work in this area focuses on technical aspects of testing and verification, but does not consider embedding testing into software development processes.

The role of software engineers developing research software has been the topic of investigation in recent years [8]. To the best of our knowledge, there does not exist literature on testing such software effectively and efficiently.

Afzal, Alone, Glocksien, et al. have identified Software Test Process Improvement (STPI) approaches [1] and studied selected STPI approaches at a large Swedish car manufacturer. However, the STPI approaches that are effective and efficient for a large commercial company are likely very different from ones that are effective and efficient for research software engineering teams.

Structure of this Work. First, in Sect. 2, we give some background on RCE and its userbase. Subsequently, in Sect. 3 we briefly describe the development process of RCE and explain where testing takes place. The focus of this work lies on Sect. 4, where we present the technologies and processes we use for testing RCE. Afterwards, in Sect. 5, we briefly describe the process of releasing the tested changes to users before closing in Sect. 6 with a conclusion and an outlook on possible avenues for future work.

2 Background

RCE is a general tool for engineers and scientists from a wide range of disciplines to create and execute distributed workflows as well as evaluate the obtained results. [6] It is available free of charge as an open source software at https://rcenvironment.de/, its source code is available at https://github.com/rcenvironment. RCE supports scientists and engineers to simulate, analyze and evaluate complex multidisciplinary systems like airplanes or satellites. RCE achieves this goal by enabling them to create automated and distributed workflows containing their own specific tools. Thus, RCE mainly serves to 1) integrate tools and compose them into workflows, 2) share the integrated tools within a network, and to 3) execute the developed workflows and manage the data flow across different network topologies.

A comprehensive introduction of RCE and its features is out of scope of this work. Instead, we refer to Flink, Mischke, Schaffert, et al. [11] for a user's description of RCE and to Boden, Flink, Först, et al. [6] for technical details.

To implement the features described above in various IT environments, RCE can be used *interactively* via Graphical User Interface (GUI) or on servers as a *headless* terminal-only application. The interactive mode is typically used to design and execute workflows and to inspect their results. The GUI is split up into *views*, each supporting a different task requested by the user (cf. Fig. 1). The headless mode is typically used to provide tools in form of a tool server, or to forward communication between RCE instances. RCE runs on Microsoft Windows and Linux operating systems in both modes. Supporting multiple operating systems and window managers increases development and testing efforts.

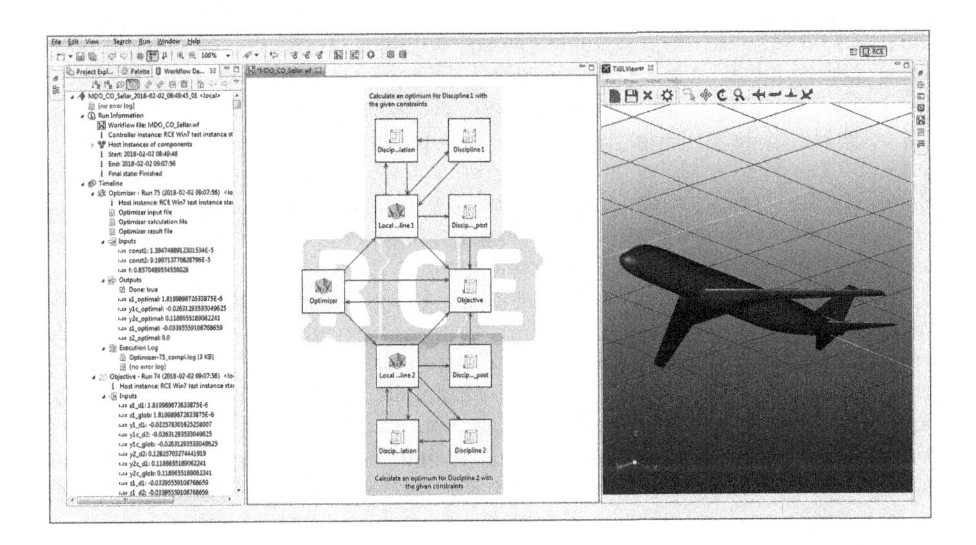

Fig. 1. A screenshot of RCE's GUI.

Users usually work with an RCE network consisting of multiple RCE instances. They integrate discipline-specific software tools into RCE by defining their inputs and outputs and compose them into a workflow by matching types of inputs and outputs. Thus, the outputs of one tool are used as inputs for another tool. Tools can be shared in the network, whose topology is freely configurable.

One complication in testing RCE arises from its diverse userbase, in terms of the activities and tasks [19] that users perform with RCE as well as in terms of their prior knowledge. A comprehensive classification of users with respect to these properties is out of the scope of this work. We instead refer to Boden, Mischke, Weinert, et al. [7] for a more comprehensive classification. Instead of focusing on the tasks and activities of users, we give a brief overview over the three most relevant types of users. Furthermore, Boden, Flink, Först, et al. give an overview over research projects RCE is involved in [6].

Firstly, a part of RCE's userbase consists of individual researchers from numerous fields who use RCE on their local working machine for their respective research. These users have diverse operating systems and system environments and typically interact with the GUI of RCE. Secondly, there exist groups of users comprised of researchers at different institutions that collaborate on a research project [23]. These groups have built expansive RCE-based server infrastructure connecting their respective organizational networks. Finally, there exist research projects that use RCE merely as an execution backend. Researchers interact with RCE via custom-made domain-specific interfaces [24].

3 Development Process

Currently, the core development team of RCE consists of four applied researchers, equivalent to about three full-time positions related to RCE. These developers are not only involved in the software development of RCE, but are also embedded in research projects. Here, they support and discuss current applications and use cases with users. They also collect feedback and input for roadmap planning. Most members are also involved in non-development related tasks, e.g., they give software development training or supervise students. Each of these activities has a non-negligible time commitment.

To avoid additional process overhead, we do not follow a prescribed development method, e.g., waterfall method, V-Model, or Scrum. Instead, we put large emphasis on mutual communication, following the core ideas of agile software development [4]. We coordinate collaboration in regular group and point-to-point meetings. Within these meetings we continuously adapt the development roadmap to new or changing requirements using the *Mantis* bug tracker [17] for documentation and communication. Further discussions are held as needed, e.g., for pair programming, knowledge transfer, or architectural decisions.

From these meetings, the schedule for releasing new versions is generated and continuously adjusted. There are dedicated meetings to decide about the scope of each release regarding new features, bugfixes, and/or other improvements.

Following semantic versioning [22], releases fall into one of three categories: *Maintenance* releases only address issues with existing functionalities or contain internal changes. *Minor* releases may contain new features or changes to the user experience provided they do not break backwards compatibility. Finally, *major* releases may contain changes that potentially break backwards compatibility.

Before a release, we review all code changes made since the last release. Additionally, we identify features that may have inadvertently been affected by these changes. All new and modified parts of the code have to be tested carefully. This can be done with automated and manual tests. We describe the whole RCE testing process in more detail in the following section.

4 Testing

Maintaining the quality of a complex software project like RCE would be infeasible without automated tests. Specifying and setting up these tests requires significant up-front effort. However, once implemented, all automated tests can be applied at any time with minimal effort. This addresses two important goals:

Firstly, any behavior of the software that is tested automatically does not need to be tested manually anymore. Over the lifetime of a project, this saves large amounts of manual testing time. This quickly offsets the initial setup effort.

Secondly, every existing test explicitly defines expectations about the system behavior. An automated test setup allows these to be frequently validated. This greatly reduces the risk of unintended changes in the system's behavior, e.g., due to code or environment changes. These automated tests thus allow developers to change code without fear of breaking existing functionality. Especially in a complex software with many dependencies like RCE, this is crucial for project maintainability. We describe our setup for automated tests in Sect. 4.1.

One form of automated testing that we do not currently employ is automated GUI testing. This is because on one hand, such tests must be sensitive enough to actually detect functional regressions. On the other hand, this sensitivity greatly increases the maintenance effort: Even minor non-functional changes to the user interface often require adaptations to the tests. Moreover, the GUI inherently depends on its system environment and thus, automated GUI-tests require dedicated infrastructure. In our estimation, the potential benefits of automated GUI tests do not outweigh these downsides.

Automated tests alone do not suffice to test RCE and its GUI comprehensively. The more user-facing a feature is, the more its expected behavior comprises simplicity, intuitiveness, and efficiency of its graphical representation. These properties are subjective and thus hard to define in automated tests. Finally, automated tests do not account for blind spots of the developer, e.g., the interaction of distinct GUI features or the overall consistency of the GUI. To validate these requirements effectively and efficiently, we use manual test cases. We write these in plain English and prompt the tester to explore existing behavior for regressions. We describe our manual testing process in Sect. 4.2.

4.1 Automated Tests

Automated tests can differ vastly regarding their implementation, test scope, and execution environment. The automated tests used in RCE team can be roughly divided into *unit* tests, *integration* tests, and *behavior-driven* tests.

Unit tests validate expectations about isolated, fine-granular parts of the software's code. Integration tests work on the combination of several code parts and usually test more complex interactions between them. While this distinction seems clear in theory, it is often less clear-cut in practice. Thus, we combine both types in our code repository with no explicit technical distinction between them. So far, we have not encountered any downsides from this approach.

More relevant, in contrast, is the distinction between fast-running and long-running tests. Combining both would slow down the execution of the fast-running tests, and lead to them being executed less often. To avoid this, we manually mark long-running test cases on the code level. Our testing infrastructure is configured to exclude these from the frequent standard test runs, which are executed multiple times daily whenever there are code changes. The slow tests, instead, are only executed at longer intervals, e.g., about weekly.

Both types of tests use the *JUnit* framework [15], with integration tests additionally using *EasyMock* [10] for setting up more complex test scenarios.

Unit and integration tests are used by many software projects, and can be considered a standard practice in today's software development. Even integration tests, however, only work on relatively small subsets of the overall application. For this reason, we complement them with behavior-driven tests to validate high-level, functional aspects of RCE. Such test cases include the verification of

- the execution and outcome of test workflows, including both success and intentional failure cases,
- the proper startup process of the application itself, e.g., for detecting stability or concurrency issues,
- the effect of certain command line parameters on the application,
- the resulting state after connecting RCE instances via their network features,
- the effect of authorization operations, e.g., changing access control settings of workflow components,
- and the presence or absence of specific log entries and/or warnings.

Listing 1.1 shows a very simple example of such a behavior-driven test case. It is written using the *Gherkin* syntax [27], which is a standardized text format for specifying human-readable test cases. A core aspect of this syntax is the standardization of test flows by breaking them down into "Given/When/Then" clauses. These clauses define the test environment or preconditions, test activities to be performed, and expected outcomes, respectively. "And" clauses are a syntactical addition for better readability, and repeat the type of the previous clause. Lines starting with "@" are tags that can be used for test organization, e.g., for identifying individual tests or defining sets of tests to execute together.

For executing tests written in this syntax, we use the Java version of the *Cucumber* framework [26]. As the activity and result phrases (e.g., "starting all instances" and "should consist of") are highly application-specific, these can not be provided by the generic testing framework. Instead, it provides tools to define them, and to map them to specific execution code, which has to be written manually. The advantage of this is that the relatively complex execution code is only created once. Once this is done, the semantic behavior to be tested can be described and maintained with much lower effort. Additionally, these semantic test descriptions are much easier to read, understand, and validate than equivalent tests defined in program code. To a certain degree, this even allows test cases to be written and maintained by non-developers autonomously.

To ensure that the actual behavior of the complete application is tested, all behavior-driven tests are run on standalone product builds of the application

```
@Network02
@NoGUITestSuite
Scenario: Basic networking between three instances (auto-start
        ↪ connections, no relay flag)

Given instances "NodeA, NodeB, NodeC" using the default build
And   configured network connections "NodeA->NodeC [autoStart],
        ↪ NodeB->NodeC [autoStart]"

When  starting all instances
Then  all auto-start network connections should be ready within 20 seconds
And   the visible network of "NodeA" should consist of "NodeA, NodeC"
And   the visible network of "NodeB" should consist of "NodeB, NodeC"
```

Listing 1.1. A basic behavior-driven test case for RCE. The symbol ↪ does not occur in the actual code but indicates an added linebreak for the sake of presentation.

code. This required building a fairly complex infrastructure to download, configure, and start/stop application instances automatically. Just like implementing the semantic phrases described above, however, this was a one-time effort. While we cannot quantify this exactly, we are certain from observation that this was quickly offset by reduced test creation time and lower manual testing effort.

4.2 Manual Tests

In this section we describe our manual testing process that we have established over the last years. This process has proven successful for our distributed and small developer team. Moreover, working from home became more popular during the COVID-19 pandemic, which further increased the distribution of development teams. First, we will describe the roles we have identified in our project, then we will give a brief overview of our general testing process. After that, we will have a deeper look into the process of a test session. Finally, we will present the main principles for such sessions that we found for the RCE project.

Roles in Manual Testing. In this section we introduce the roles involved in our manual testing process, namely the *developers*, the *testers* and the *test manager*.

In our experience it is unusual for research software projects to have a dedicated testing team. The typically small team size along with the motivation to produce high-quality research results means that teams concentrate on conducting their research tasks. Often, software testing—particularly manual testing—is not given the required attention. Nevertheless, manual tests are inevitable for developing high-quality software due to the reasons described above.

Below we describe our approach to manage manual test phases in our small team with high workload. When we talk about small teams, we mean teams with fewer than ten team members. In previous years, the RCE team was about the size of four to ten team members. Due to our team size, we do not usually split into developers and testers, i.e., every developer is also a tester and vice versa. At least during the release phases when manual testing becomes more important, these roles overlap. The dedicated role of the test manager coordinates the manual testing process. In RCE, the test manager is also part of the development team and thus even takes on three roles during testing.

The developers are responsible for providing test cases for their work. This has two aspects. On the one hand, they write automated tests as described in Sect. 4.1. On the other hand, they identify software parts for manual testing, write manual test cases, and inform the test manager about them.

The test manager reviews the tests and provides feedback. In addition, the test manager schedules and prepares the test sessions and guides the team through them. They also track the test progress and keep an eye on any unfinished tasks for both developers and testers. To manage the overall testing process, we use the test management software *Testrail* [12]. Testrail is a web-based management solution that lets users manage, track and organize test cases and test plans. During test sessions, the testers are responsible for completing the test cases assigned to them by the test manager. Furthermore, they give feedback on the test results to the team, especially to the responsible developer. Since every tester is also a developer, the team is doubly challenged during such phases. Everyone needs to test the application, while development activities—such as bug fixes or further improvements—need to continue in parallel. It is a special challenge to organize the small team and the increased workload well during this time. Especially when the release has to meet a certain deadline, it is critical to avoid personal bottlenecks as much as possible, which can be hard as not all tasks can be easily shifted between team members.

Manual Testing Process. To minimize the time for manual testing, we test in several phases: The *Pretesting*, the *Release Testing* and the *Final Testing*.

Each new feature must pass a Pretesting phase. Depending on the feature size this pretest takes a few hours to one or two days. One or two testers are recruited from the development team to test the new feature on a maximum of two operating systems. These pretests are used to detect and correct as many errors as possible before the actual testing phase. Only when these tests are passed the new feature is discussed for inclusion in the next release. This procedure reduces the test effort during the Release Testing to a minimum. Minor developments, e.g., bug fixes or smaller improvements, are only pretested by the developer.

The main testing phase is the Release Testing, where all new developments come together and are tested in their entirety. The new developments comprise not only new features, but also improvements to existing features and bug fixes. Each Release Testing lasts several weeks and ends when the team decides that the software under test is ready for release. We describe how the team comes to this decision in the next section. Release Testing does typically not exceed four weeks. In contrast to Pretestings, the entire team usually participates in these tests. During these weeks, testing and development work alternate. Our Test Session Process (cf. Sect. 4.2) supports the team in balancing and prioritizing their tasks.

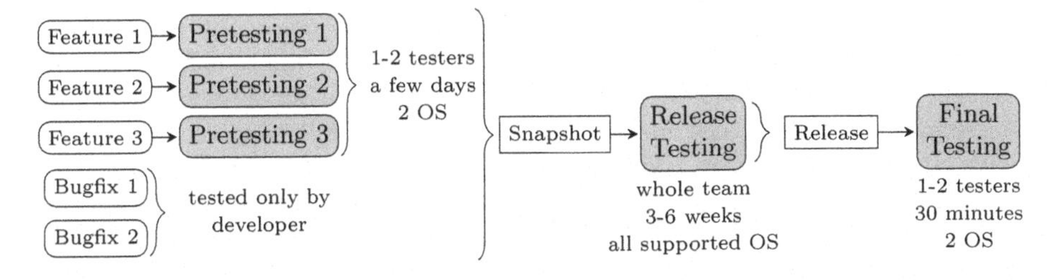

Fig. 2. The manual testing process. Nodes with gray background denote testing activities, while rounded nodes with white background denote code changes. Square nodes with white background denote software artifacts.

We test up to ten configurations, each comprising an operating system, a desktop environment, a certain Java Runtime (JRE) version, and the choice whether to test RCE with its GUI or in headless mode.

At the end of the Release Testing the final product is built. Before it is officially released, we apply the Final Testing to guard against the possibility of otherwise undetected errors having crept in. There could, e.g., have been upgrades of build tools or system libraries, IT problems like running out of disk space causing build steps to finish incompletely, or simply code changes causing unexpected side effects which have not been caught by test cases. To have a chance of detecting these, two team members take one last look at the product. This takes about half an hour and focuses on the basic functionalities of RCE.

We now take a closer look at the Release Testing phases.

Test Session Process. Before Release Testing there are some organizational tasks for the test manager. First, they check whether manual test cases exist for all new features and minor enhancements included in the release. If so, the test plan can be created. Otherwise, the test manager reminds the responsible developers to add their test cases. In addition, the test manager discusses the assignment of testers to platforms so that the testers can prepare their testing environment. As we test for different operating systems, this typically involves setting up one or more virtual machines to simulate these platforms.

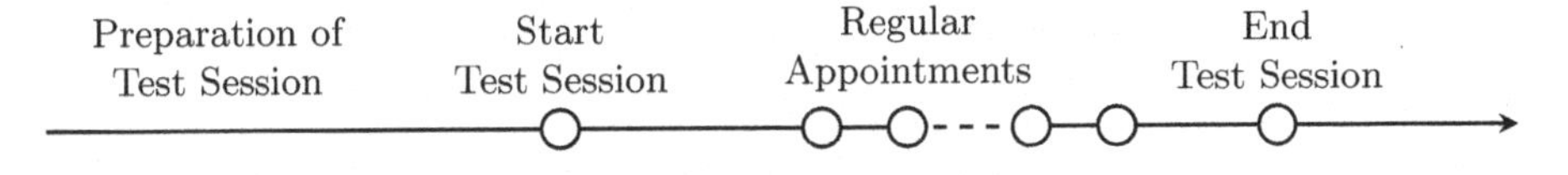

Fig. 3. Test session process

The testing then starts with a kick-off meeting. In this meeting, the testers discuss all necessary details for the testing as a team and clarify any questions, e.g., final organization issues or comprehension questions for specific test cases.

Developers share knowledge about their new features, point out what aspect testers should focus on, or give a brief overview of a new functionality. The test manager moderates the meetings and reminds the team of our common principles for testing. We discuss these principles in detail in the following section.

During the test session the whole team meets twice a week for about half an hour each. These meetings serve to focus the team on the most important tasks for the days ahead. The team takes a look at the test progress together. The test manager presents intermediate test results and checks for "blind spots" that have not yet been tested sufficiently. Testers draw the team's attention to critical bugs they have detected, especially if those bugs prevent further testing. Developers report on bug fixes and point out test cases to be retested.

Beyond that, we use these meetings to prevent employee overload. If someone in the team has so much development work that the test scope can no longer be met, the test manager coordinates support. The test manager adjusts the test plan accordingly and reassigns the test cases.

As the test phase progresses, fewer and fewer tasks are outstanding. This phase has no pre-determined deadline, but rather comprises iteratively resolving and retesting critical bugs. Towards the end, everyone ensures no important tasks have been overlooked. Once Release Testing has ended, the team cleans up, at most two team members perform Final Testing, and we begin review.

During this review, the entire team gathers feedback, suggestions and ideas on how we can optimize our process in the future. The test manager then evaluates this collection and develops proposals that are presented to the team. This review process continuously contributes to the improvement of our testing process.

Principles for Testing. Within the RCE project we have agreed on some testing principles. These are to 1) ensure that assigned test cases reach a *final state*, to 2) stick to the *Test Case Life Cycle*, and to 3) look outside the box.

Testers have to take care that all their assigned test cases have been executed at the end of the test session (Principle 1. During the test session a test case can reach different states. We list all possible states and their meanings in Table 1.

Table 1. States of test cases.

Passed	Test case was executed and no errors were observed
Passed with Remarks	Similar to "Passed", but the tester discovered some improvements for a future release
Not applicable	Test case was not executed because it is not applicable on the current test configuration
Won't test	Test case was not executed because it has been tested sufficiently on other configurations
Failed	Test case failed
Failed & Blocked	Test case failed and blocked on all other configurations
Retest	Test case ready to retest
Waiting for new build	Test case ready to retest in the next build
Blocked	Test case blocked by the developer due to ongoing development work
Failed & Postponed	Test case failed and fix is postponed to a future release

States are final states if they cannot reach any other state afterwards. Final states are for example "passed", "won't test" or "failed & postponed". When a test session starts, all test cases are "untested". Tester cycles them through different states before they reach a final state. In doing so they follow the Test Case Life Cycle (Principle 2. Some states lead to an action that must be performed by the tester or developer while others are intermediate or final states. Figure 4 shows all states with their actions. The states on the left-hand side can be assigned by the testers. The states on the right-hand side are to be set exclusively by the responsible developers or by the test manager.

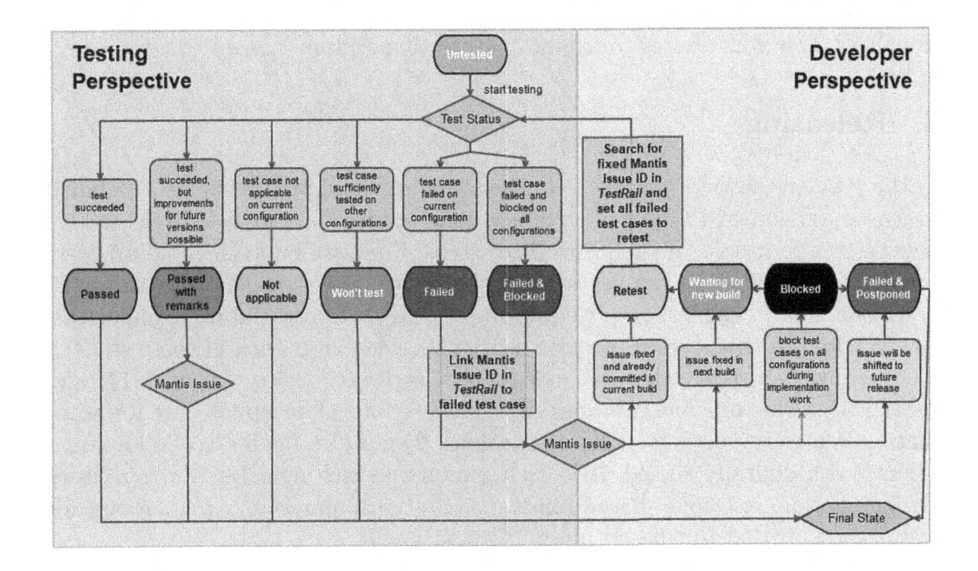

Fig. 4. The test case life cycle

We now give an example of a possible life cycle of a test case. A tester picks an "untested" test case and executes it. During testing the tester finds a bug in the tested feature, sets the test case to "failed" and creates a Mantis issue. The tester references the Mantis issue ID in the failed test case. This issue describes the erroneous behavior in detail and is assigned to the responsible developer. During the regular appointments, the tester reminds the team about the found bug if it has not been noticed yet. Here the team also discusses the priority of found bugs and whether they need to be fixed for the current release. If this is not the case, the test case is set to "failed & postponed", a final state. Otherwise the responsible developer fixes the bug and sets the status to "waiting for new build". The associated test case can be easily found via the issue reference. With the next build, the developer or the test manager change the state of the test case to "retest". This status, in turn, triggers the tester to perform the test once again. This cycle can be repeated several times until the test is finally "passed".

The states presented here were established iteratively via our review process. Using of these states has proven successful in the development of RCE.

Finally, we have the commitment in the team that test cases should not be understood as a step-by-step instruction. Rather, they indicate an area or functionality of the software to be tested and provide suggestions on how to test. Often, of course, there are some instructions to follow, e.g., to create the required test setup. Apart from that, every tester should always think outside the box (Principle 3 and also test functionalities "close" to the described one. They should keep in mind the following questions: 1) Is there anything I could test that is not described in the test case? 2) Are there interactions with other functionalities not considered in the tests? 3) How can I try to abuse the functionality? This contributes to exploratory testing and increases test coverage.

5 Releasing

In the previous section we have described how we validate the changes included in a new version of RCE. After having done so, we then move to release this new version to users. This requires three steps: First, we build the final artifacts that are distributed to the users. Afterwards, we deploy these artifacts to the download site. In the last step we inform existing and potential users about this new version. In this section, we give a brief overview over each of these steps.

First, we construct the artifacts to be distributed using *Jenkins* [14] and *Maven* [18]. This pipeline compiles the codebase into 1) an update site for automatic updates to existing RCE installations, 2) zip files for both Windows and Linux, both digitally signed, 3) a .deb package as well as a Debian repository for installation on Debian-based Linux distributions, and 4) an .rpm package for installation on Red Hat-based Linux distributions.

We publish these artifacts to our website at https://rcenvironment.de. We moreover mirror the source code of the newly released artifact to Github at https://github.com/rcenvironment for accessibility and create a Github release containing this artifact.

Finally, we inform users about the newly released version. To do so, we first collect a list of changes with respect to the previous version. For this, we consult both the Mantis changelog as well as a manually compiled list of new features created by developers in the wiki. We inform users by compiling and publishing 1) a technical changelog on Github, 2) a nontechnical changelog on rcenvironment.de, 3) an email newsletter pointing users to the nontechnical changelog, and 4) a tweet at https://twitter.com/RCEnvironment pointing users to the nontechnical changelog. We publish these items in order, where each item contains a reference to one or more items published before.

6 Conclusion and Future Work

In this work we have presented our processes for ensuring that changes made to RCE implement the desired feature, integrate well with "neighboring" features, and do not cause regressions in already existing features. We have shown how we test changes, bugfixes, and new features affecting various aspects of RCE, from its technical foundations to its GUI. We introduced our manual testing process, with the focus on small research development teams where each team member performs multiple roles. We have continuously developed and optimized this process in recent years and have successfully used it in numerous release phases.

In future work, we are looking to expand the scope of automated behavior-driven testing of RCE. In particular, we aim to improve the scalability of our distributed test scenarios. To this end, our test orchestration setup shall be adapted to make use of cloud infrastructure. This will allow larger test scenarios to be defined and executed, and create more options for intense load testing.

Moving to this cloud-based infrastructure also opens up possibilities for new testing approaches. First steps have already been taken towards employing aspects of *Chaos Testing*, e.g., actively introducing network and/or instance failures into test networks to verify RCE's robustness against them. [13] Thus, we will be able to replicate real-world circumstances more closely, which will allow even more thorough testing, leading to increased robustness of the final product.

We are also looking to expand the scope of our automated testing towards RCE's GUI functionality. Automated low-level GUI testing, however, is notoriously difficult to get right, as discussed in the introduction to Sect. 4. In light of these known problems, we have started to experiment with approaches that mix direct code access with GUI element testing. These involve combining simulated UI events with dedicated test-supporting code embedded in the application. We plan to continue these experiments to further reduce manual testing while keeping the maintenance effort below that of low-level GUI testing.

References

1. Afzal, W., Alone, S., Glocksien, K., Torkar, R.: Software test process improvement approaches: a systematic literature review and an industrial case study. J. Syst. Softw. **111**, 1–33 (2016). https://doi.org/10.1016/j.jss.2015.08.048
2. Ammann, P., Offutt, J.: Introduction to Software Testing. Cambridge University Press (2016). https://doi.org/10.1017/9781316771273
3. Anand, S., et al.: An orchestrated survey of methodologies for automated software test case generation. J. Sys. Softw. **86**(8), 1978–2001 (2013). https://doi.org/10.1016/j.jss.2013.02.061
4. Beck, K., et al.: Manifesto for agile software development (2001)
5. Beyer, D.: Software verification: 10th comparative evaluation (SV-COMP 2021). In: TACAS 2021, pp. 401–422 (2021). https://doi.org/10.1007/978-3-030-72013-1_24
6. Boden, B., Flink, J., Först, N., Mischke, R., Schaffert, K., Weinert, A., Wohlan, A., Schreiber, A.: RCE: An integration environment for engineering and science. SoftwareX **15**, 100759 (2021). https://doi.org/10.1016/j.softx.2021.100759
7. Boden, B., Mischke, R., Weinert, A., Schreiber, A.: Supporting the composition of domain-specific software via task-specific roles. In: ICW 2020. ACM (2020). https://doi.org/10.1145/3397537.3399576

8. Brett, A., et al.: Research Software Engineers: State Of The Nation Report 2017 (2017). https://doi.org/10.5281/ZENODO.495360
9. Böhme, M., Pham, V.T., Nguyen, M.D., Roychoudhury, A.: Directed greybox fuzzing. In: CCS 2017. ACM (2017). https://doi.org/10.1145/3133956.3134020
10. EasyMock Developers: EasyMock. https://easymock.org/, repository at https://github.com/easymock/easymock
11. Flink, J., Mischke, R., Schaffert, K., Schneider, D., Weinert, A.: Orchestrating tool chains for model-based systems engineering with RCE. In: Aerospace Conference 2022. IEEE (2022)
12. Gurock Software GmbH. TestRail https://www.gurock.com/testrail/
13. Hampel, B.: Automatisierte Anwendung von Chaos Engineering Methoden zur Untersuchung der Robustheit eines verteilten Softwaresystems. Bachelor Thesis, Leipzig University of Applied Sciences (2021)
14. Jenkins Developers. Jenkins. https://www.jenkins.io/, repository at https://github.com/jenkinsci/jenkins
15. JUnit Developers. JUnit https://junit.org/, repository at https://github.com/junit-team/
16. Li, J., Zhao, B., Zhang, C.: Fuzzing: a survey. Cybersecurity **1**(1), 1–13 (2018). https://doi.org/10.1186/s42400-018-0002-y
17. MantisBT Developers. MantisBT https://mantisbt.org/, repository at https://github.com/mantisbt/mantisbt
18. Maven Developers. Apache Maven https://maven.apache.org/, repository at https://gitbox.apache.org/repos/asf/maven-sources.git
19. Norman, D.A.: Human-centered design considered harmful. Interactions **12**(4), 14–19 (2005). https://doi.org/10.1145/1070960.1070976
20. Oliveira, D., Rosenthal, M., Morin, N., Yeh, K.C., Cappos, J., Zhuang, Y.: It's the psychology stupid: how heuristics explain software vulnerabilities and how priming can illuminate developer's blind spots. In: ACSAC 2014. ACM (2014). https://doi.org/10.1145/2664243.2664254
21. Orso, A., Rothermel, G.: Software testing: a research travelogue (2000–2014). In: FOSE 2014. ACM (2014). https://doi.org/10.1145/2593882.2593885
22. Preston-Werner, T.: Semantic versioning 2.0. https://semver.org/
23. Risse, K.: Virtual Product House-Integration and Test Centre for Virtual Certification, https://www.dlr.de/content/en/articles/aeronautics/aeronautics-research/virtual-product-house.html. Accessed 4 Feb 2022
24. Risueño, A.P., Bussemaker, J., Ciampa, P.D., Nagel, B.: MDAx: Agile generation of collaborative MDAO workflows for complex systems. In: AIAA AVIATION 2020 FORUM. American Institute of Aeronautics and Astronautics (2020). https://doi.org/10.2514/6.2020-3133
25. Health, R.T.I.: Social, and Economics Research: The Economic Impacts of Inadequate Infrastructure for Software Testing. Tech. rep, NIST (2002)
26. SmartBear Software Inc. Cucumber. https://cucumber.io/, repository at https://github.com/cucumber/
27. SmartBear Software Inc. Gherkin https://cucumber.io/docs/gherkin/

Digging Deeper into the State of the Practice for Domain Specific Research Software

Spencer Smith(✉) and Peter Michalski

McMaster University, 1280 Main Street West, Hamilton, ON L8S 4K1, Canada
smiths@mcmaster.ca
http://www.cas.mcmaster.ca/~smiths/

Abstract. To improve the productivity of research software developers we need to first understand their development practices. Previous studies on this topic have collected data by surveying as many developers as possible, across a broad range of application domains. We propose to dig deeper into the state of the practice by instead looking at what developers in specific domains create, as evidenced by the contents of their repositories. Our methodology prescribes the following steps: i) Identify the domain; ii) Identify a list of candidate software; iii) Filter the list to a length of about 30 packages; iv) Collect repository related data on each package, like number of stars, number of open issues, number of lines of code; v) Fill in the measurement template (the template consists of 108 questions to assess 9 qualities (including the qualities of installability, usability and visibility)); vi) Rank the software using the Analytic Hierarchy Process (AHP); vii) Interview developers (the interview consists of 20 questions and takes about an hour); and, viii) Conduct a domain analysis. The collected data is analyzed by: i) comparing the ranking by best practices against the ranking by popularity; ii) comparing artifacts, tools and processes to current research software development guidelines; and, iii) exploring pain points. We estimate the time to complete an assessment at 173 person hours. The method is illustrated via the example of Lattice Boltzmann Solvers, where we find that the top packages engaged in most of recommended best practices, but still show room for improvement with respect to providing API documentation, a roadmap, a code of conduct, programming style guide and continuous integration.

Keywords: Research software · Software quality · Empirical measures · Software artifacts · Developer pain points · Lattice Boltzmann Method

1 Introduction

Research software is critical for tackling problems in areas as diverse as manufacturing, financial planning, environmental policy and medical diagnosis and treatment. However, developing reliable, reproducible, sustainable and fast research

D. Groen et al. (Eds.): ICCS 2022, LNCS 13353, pp. 545–559, 2022.
https://doi.org/10.1007/978-3-031-08760-8_45

software is challenging because of the complexity of physical models and the nuances of floating point and parallel computation. The critical importance of research software, and the challenges with its development, have prompted multiple researchers to investigate the state of development practice. Understanding the creation of research software is critical for devising future methods and tools for reducing development time and improving software quality.

Previous studies on the state of the practice for research software have often focused on surveying developers [7,13,16]. Although surveys provide valuable information, they are limited by their reliance on what developers say they do, rather than directly measuring what they actually do. Therefore, we propose a state of the practice assessment methodology that investigates the work products developers create by digging into the software repositories they create.

Although other studies [6,27] mine research software repositories to estimate productivity, code quality and project popularity, the other studies focus on automation and on code related artifacts. To gain deeper insight, at the expense of taking more time, we will relax the automation requirement and use manual investigation where necessary. We will also expand our assessment beyond code to include other artifacts, where artifacts are the documents, scripts and code that we find in a project's public repository. Example artifacts include requirements, specifications, user manuals, unit test cases, system tests, usability tests, build scripts, API (Application Programming Interface) documentation, READMEs, license documents, and process documents.

The surveys used in previous studies have tended to recruit participants from all domains of research software. This is what we will call *research software in general*, as opposed to specific domain software. The surveys may distinguish participants by programming language (for instance, R developers [16]), or by the role of the developers (for instance postdoctoral researchers [12]), but the usual goal is to cast as wide a net as possible. Case studies [2,19], on the other hand, go more in depth by focusing on a few specific examples at a time. For our new methodology, we propose a scope between these two extremes. Rather than focus on assessing the state of the practice for research software in general, or just a few examples, we will focus on one scientific domain at a time. The practical reason for this scope is that digging deep takes time, making a broad scope infeasible. We have imposed a practical constraint of one person month of effort per domain.[1] Focusing on one domain at a time has more than just practical advantages. By restricting ourselves to a single domain we can bring domain knowledge and domain experts into the mix. The domain customized insight provided by the assessment has the potential to help a specific domain as they adopt and develop new practices. Moreover, measuring multiple different domains facilitates comparing and contrasting domain specific practices.

Our methodology is built around 8 research questions. Assuming that the domain has been identified (Sect. 2.1), the first question is:

[1] A person month is considered to be 20 working days (4 weeks in a month, with 5 days of work per week) at 8 person hours per day, or $20 \cdot 8 = 160$ person hours.

RQ1: What software projects exist in the domain, with the constraint that the source code must be available for all identified projects? (Sects. 2.2, 2.3)

We next wish to assess the representative software to determine how well they apply current software development best practices. By *best practices* we mean methods, techniques and tools that are generally believed to improve software development, like testing and documentation. As we will discuss in Sect. 2.4, we will categorize our best practices around software qualities. Following best practices does not guarantee popularity, so we will also compare our ranking to how the user community itself ranks the identified projects.

RQ2: Which of the projects identified in RQ1 follow current best practices, based on evidence found by experimenting with the software and searching the artifacts available in each project's repository? (Sects. 2.4)
RQ3: How similar is the list of top projects identified in RQ2 to the most popular projects, as viewed by the scientific community? (Sect. 4)

To understand the state of the practice we wish to learn the frequency with which different artifacts appear, the types of development tools used and the methodologies used for software development. With this data, we can ask questions about how the domain software compares to other research software.

RQ4: How do domain projects compare to research software in general with respect to the artifacts present in their repositories? (Sect. 5)
RQ5: How do domain projects compare to research software in general with respect to the use of tools? (Sect. 6)
RQ6: How do domain projects compare to research software in general with respect to the processes used? (Sect. 7)

Only so much information can be gleaned by digging into software repos. To gain additional insight, we need to interview developers (Sect. 2.5), as done in other state of the practice assessments [10], to learn:

RQ7: What are the pain points for developers? (Sect. 8)
RQ8: How do the pain points of domain developers compare to the pain points for research software in general? (Sect. 8)

Our methodology answers the research question through inspecting repositories, using the Analytic Hierarch Process (AHP) for ranking software, interviewing developers and interacting with at least one domain expert. We leave the measurement of the performance, for instance using benchmarks, to other projects [9]. The current methodology updates the approach used in prior assessments of domains like Geographic Information Systems [24], Mesh Generators [23], Seismology software [26], and statistical software for psychology [25]. Initial tests of the new methodology have been done for medical image analysis software [3] and for Lattice Boltzmann Method (LBM) software [11]. The LBM example will be used to illustrate the steps in the methodology.

2 Methodology

The assessment is conducted via the following steps, which depend on interaction with a Domain Expert partner, as discussed in Sect. 2.6.

1. Identify the domain of interest. (Sect. 2.1)
2. List candidate software packages for the domain. (Sect. 2.2)
3. Filter the software package list. (Sect. 2.3)
4. Gather the source code and documentation for each software package.
5. Measure using the measurement template. (Sect. 2.4)
6. Use AHP to rank the software packages. (Sect. 2.4)
7. Interview the developers. (Sect. 2.5)
8. Domain analysis. (Sect. 2.7)
9. Analyze the results and answer the research questions. (Sects. 3–8)

We estimate 173 h to complete the assessment of a given domain [22], which is close our goal of 160 person hours.

2.1 How to Identify the Domain?

To be applicable the chosen domain must have the following properties:

1. The domain must have well-defined and stable theoretical underpinnings.
2. There must be a community of people studying the domain.
3. The software packages must have open source options.
4. The domain expert says there will be at least 30 candidate packages.

2.2 How to Identify Candidate Software from the Domain?

The candidate software to answer RQ1 should be found through search engine queries, GitHub, swMATH and scholarly articles. The Domain Expert (Sect. 2.6) should also be engaged in selecting the candidate software. The following properties are considered when creating the list:

1. The software functionality must fall within the identified domain.
2. The source code must be viewable.
3. The repository based measures should be available.
4. The software must be complete.

2.3 How to Filter the Software List?

If the list of software is too long (over around 30 packages), then filters are applied in the priority order listed. Copies of both lists, along with the rationale for shortening the list, should be kept for traceability purposes.

1. Scope: Software is removed by narrowing what functionality is considered to be within the scope of the domain.

2. Usage: Software packages are eliminated if their installation procedure is missing or not clear and easy to follow.
3. Age: Older software packages (age being measured by the last date when a change was made) are eliminated, except in the cases where an older software package appears to be highly recommended and currently in use. (The Domain Expert should be consulted on this question as necessary.)

2.4 Quantitative Measures

We rank the projects by how well they follow best practices (RQ2) via a measurement template [22]. For each software package (each column in the template), we fill-in the rows. This process takes about 2 h per package, with a cap of 4 h. An excerpt of the template is shown in Fig. 1. In keeping with scientific transparency, all data should be made publicly available.

Summary Information								
Software name?	DL_MESO	SunlightLB	MP-LABS	LIMBES	LB3D-Prime	LB2D-Prime	laboetie	Musubi
Number of developers	unclear	2	1	unclear	1	1	2	unknown
License?	terms of use	GNU GPL	GNU GPL	GNU GPL	unclear	unclear	GNU GPL	BSD
Platforms?	Windows, OS X, Linux	Linux	Linux	Unix	Windows, Linux	Windows, Linux	Linux	Windows, OS X, Linux
Software Category?	private	public	public	public	public	public	public	public
Development model?	freeware	open source	freeware	freeware	freeware	freeware	unclear	freeware
Programming language(s)?	FORTRAN, C++, Java	C, Perl, Python	FORTRAN, Markdown	FORTRAN	C	C, Shell	FORTRAN, Wolfram Markdown	Fortran
...	...	...	...	...	...	...	...	...
Installability								
Installation instructions?	yes	yes	yes	yes	yes	yes	yes	yes
Instructions in one place?	yes	yes	yes	yes	yes	yes	yes	yes
Linear instructions?	yes	yes	yes	yes	yes	yes	yes	yes
Installation automated?	yes, makefile	yes, makefile	yes, makefile	yes, makefile	yes, makefile	yes, makefile	yes, makefile	yes
Descriptive error messages?	yes	yes	no	n/a	n/a	no	n/a	n/a
Number of steps to install?	8	6	6	4	2	4	4	10
Numbe extra packages?	4	4	3	1	2	2	2	5
Package versions listed?	yes	no	no	no	no	no	no	no
Problems with uninstall?	unavail	unavail	unavail	unavail	unavail	unavail	unavail	unavail
...	...	...	...	...	...	...	...	...
Overall impression (1..10)?	9	7	6	8	7	5	7	8
...	...	...	...	...	...	...	...	...
Surface Reliability								
...	...	...	...	...	...	...	...	...

Fig. 1. Excerpt of the top sections of the measurement template

The full template [22] consists of 108 questions categorized under 9 qualities: (i) installability; (ii) correctness and verifiability; (iii) surface reliability; (iv) surface robustness; (v) surface usability; (vi) maintainability; (vii) reusability; (viii) surface understandability; and, (ix) visibility/transparency.

The questions were designed to be unambiguous, quantifiable and measurable with limited time and domain knowledge. The measures are grouped under headings for each quality, and one for summary information (Fig. 1). The summary section provides general information, such as the software name, number of developers, etc. Several of the qualities use the word "surface". This is to highlight that, for these qualities, the best that we can do is a shallow measure. For instance, we do not conduct experiments to measure usability. Instead, we are looking for an indication that usability was considered by looking for cues in the documentation, such as getting started instructions, a user manual or documentation of expected user characteristics.

Tools are used to find some of the measurements, such as the number of files, number of lines of code (LOC), percentage of issues that are closed, etc. The tool GitStats is used to measure each software package's GitHub repository for the number of binary files, the number of added and deleted lines, and the number of commits over varying time intervals. The tool Sloc Cloc and Code (scc) is used to measure the number of text based files as well as the number of total, code, comment, and blank lines in each GitHub repository. At this time our focus is on simple metrics, so we do no need the static analysis capabilities of tools like Sonargraph or SonarCloud.

Virtual machines (VMs) are used to provide an optimal testing environments for each package [23] because with a fresh VM there are no worries about conflicts with existing libraries. Moreover, when the tests are complete the VM can be deleted, without any impact on the host operating system. The most significant advantage of using VMs is that every software install starts from a clean slate, which removes "works-on-my-computer" errors.

Once we have measured each package, we still need to rank them to answer RQ2. To do this, we used the Analytical Hierarchy Process (AHP), a decision-making technique that uses pair-wise comparisons to compare multiple options by multiple criteria [17]. In our work AHP performs a pairwise analysis between each of the 9 quality options for each of the (approximately) 30 software packages. This results in a matrix, which is used to generate an overall score for each software package for the given criteria [23].

2.5 Interview Developers

Several of the research question (RQ5, RQ6 and RQ7) require going beyond the quantitative data from the measurement template. To gain the required insight, we interview developers using a list of 20 questions [22]. The questions cover the background of the development teams, the interviewees, and the software itself. We ask the developers how they organize their projects and about their understanding of the users. Some questions focus on the current and past difficulties, and the solutions the team has found, or will try. We also discuss the importance of, and the current situation for, documentation. A few questions are about specific software qualities, such as maintainability, usability, and reproducibility. The interviews are semi-structured based on the question list. Each interview should take about 1 h.

The interviewees should follow standard ethics guidelines for consent, recording, and including participant details in the report. The interview process presented here was approved by the McMaster University Research Ethics Board under the application number MREB#: 5219. For LBM we were able to recruit 4 developers to participate in our study.

2.6 Interaction with Domain Expert

Our methodology relies on engaging a Domain Expert to vet the list of projects (RQ1) and the AHP ranking (RQ2). The Domain Expert is an important member of the assessment team. Pitfalls exist if non-experts attempt to acquire an authoritative list of software, or try to definitively rank software. Non-experts have the problem that they can only rely on information available on-line, which has the following drawbacks: i) the on-line resources could have false or inaccurate information; and, ii) the on-line resources could leave out relevant information that is so in-grained with experts that nobody thinks to explicitly record it. Domain experts may be recruited from academia or industry. The only requirements are knowledge of the domain and a willingness to be involved.

2.7 Domain Analysis

For each domain a table should be constructed that distinguishes the programs by their variabilities. In research software the variabilities are often assumptions.

3 Measuring and Ranking the LBM Domain

For the LBM example the initial list had 46 packages [11]. To answer RQ1, this list was filtered by scope, usage, and age to decrease the length to 24 packages, as shown in Table 1. This table also shows the domain analysis, in terms of variabilities, for LBM software [11].

To answer RQ2 the 24 LBM packages were ranked for the 9 qualities [11]. For space reasons we will only show the overall ranking in Fig. 2. In the absence of information on priorities, the overall ranking was calculated with an equal weight between all qualities. The LBM data is available on Mendeley.

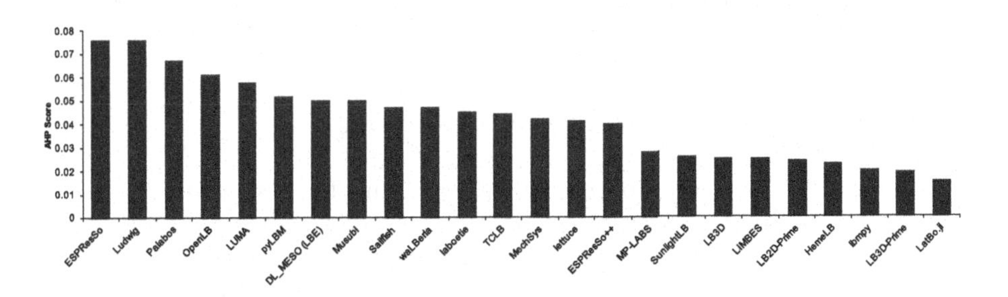

Fig. 2. AHP overall score

Table 1. Features of Software Packages (Dim for Dimension (1, 2, 3), Pll for Parallel (CUDA, MPI, OpenMP (OMP)), Com for Compressible (Yes or No), Rflx for Reflexive Boundary Condition (Yes or No), MFl for Multi-fluid (Yes or No), Turb for Turbulent (Yes or No), CGE for Complex Geometries (Yes or No), OS for Operating System (Windows (W), macOS (M), Linux (L)))

Name	Dim	Pll	Com	Rflx	MFl	Turb	CGE	OS
DL_MESO (LBE)	2, 3	MPI/OMP	Y	Y	Y	Y	Y	W, M, L
ESPResSo	1, 2, 3	CUDA/MPI	Y	Y	Y	Y	Y	M, L
ESPResSo++	1, 2, 3	MPI	Y	Y	Y	Y	Y	L
HemeLB	3	MPI	Y	Y	Y	Y	Y	L
laboetie	2, 3	MPI	Y	Y	Y	Y	Y	L
LatBo.jl	2, 3	–	Y	Y	Y	N	Y	L
LB2D-Prime	2	MPI	Y	Y	Y	Y	Y	W, L
LB3D	3	MPI	N	Y	Y	Y	Y	L
LB3D-Prime	3	MPI	Y	Y	Y	Y	Y	W, L
lbmpy	2, 3	CUDA	Y	Y	Y	Y	Y	L
lettuce	2, 3	CUDA	Y	Y	Y	Y	Y	W, M, L
LIMBES	2	OMP	Y	Y	N	N	Y	L
Ludwig	2, 3	MPI	Y	Y	Y	Y	Y	L
LUMA	2, 3	MPI	Y	Y	Y	Y	Y	W, M, L
MechSys	2, 3	–	Y	Y	Y	Y	Y	L
MP-LABS	2, 3	MPI/OMP	N	Y	Y	N	N	L
Musubi	2, 3	MPI	Y	Y	Y	Y	Y	W, L
OpenLB	1, 2, 3	MPI	Y	Y	Y	Y	Y	W, M, L
Palabos	2, 3	MPI	Y	Y	Y	Y	Y	W, L
pyLBM	1, 2, 3	MPI	Y	Y	N	Y	Y	W, M, L
Sailfish	2, 3	CUDA	Y	Y	Y	Y	Y	M, L
SunlightLB	3	–	Y	Y	N	N	Y	L
TCLB	2, 3	CUDA/MPI	Y	Y	Y	Y	Y	L
waLBerla	2, 3	MPI	Y	Y	Y	Y	Y	L

4 Comparison to Community Ranking

To address RQ3 we need to compare the ranking by best practices to the community's ranking. Our best practices ranking comes from the AHP ranking (Sect. 2.4). We estimate the community's ranking by repository stars and watches. The comparison will provide insight on whether best practices are rewarded by popularity. However, inconsistencies between the AHP ranking and the community's ranking are inevitable for the following reasons: i) the overall quality ranking via AHP makes the unrealistic assumption of equal weighting between quality factors; ii) stars are not a particularly good measure of popularity, because of how people use stars and because young projects have less time to accumulate them [28]; iii) and, as for consumer products, there are more factors influencing popularity than just quality.

Table 2 compares the AHP ranking of the LBM package to their popularity in the research community. Nine packages do not use GitHub, so they do not have a star measure. Looking at the stars of the other 15 packages, we can observe a pattern where packages that have been highly ranked by our assessment tend to have more stars than lower ranked packages. The best ranked package by AHP (ESPResSo) has the second most stars, while the ninth ranked package (Sailfish) has the highest number of stars. Although the AHP ranking and the community popularity estimate are not perfect measures, they do suggest a correlation between best practices and popularity.

Table 2. Excerpt from repository ranking metrics [11]

Name	Our ranking	Repository stars	Repository star rank	Repository watches	Repository watch rank
ESPResSo	1	145	2	19	2
Ludwig	2	27	8	6	7
Palabos	3	34	6	GitLab	GitLab
OpenLB	4	N/A	N/A	N/A	N/A
LUMA	5	33	7	12	4
pyLBM	6	95	3	10	5
DL_MESO (LBE)	7	N/A	N/A	N/A	N/A
Musubi	8	N/A	N/A	N/A	N/A
Sailfish	9	186	1	41	1
...	...	...	...	...	...
LatBo.jl	24	17	10	8	6

5 Comparison of Artifacts to Other Research Software

We answer RQ4 by comparing the artifacts that we observe to those observed and recommended for research software in general. While filling in the measurement template (Sect. 2.4), the domain software is examined for the presence of artifacts, which are then categorized by frequency as: common (more than 2/3 of projects), uncommon (between 1/3 and 2/3), and rare (less than 1/3). The observed frequency of artifacts should then be compared to the artifacts recommended by research software guidelines, as summarized in Table 3.

As Table 3 shows, the majority of LBM generated artifacts correspond to general recommendations from research software developers. Areas where LBM developers could improve include providing: API documentation, a product roadmap, a code of conduct, code style guide, and uninstall instructions.

Table 3. Comparison of recommended artifacts in software development guidelines to artifacts in LBM projects (C for Common, U for Uncommon and R for Rare)

	[30]	[18]	[1]	[32]	[20]	[8]	[29]	[5]	[14]	LBM
LICENSE	✓	✓	✓	✓	✓		✓	✓	✓	C
README		✓	✓	✓	✓		✓	✓	✓	C
CONTRIBUTING		✓	✓	✓	✓		✓	✓	✓	U
CITATION				✓				✓	✓	U
CHANGELOG		✓		✓	✓		✓			U
INSTALL					✓		✓	✓	✓	C
Uninstall								✓		R
Dependency list			✓		✓			✓		C
Authors							✓	✓	✓	C
Code of conduct							✓			
Acknowledgements							✓	✓	✓	R
Code style guide		✓					✓	✓	✓	R
Release Info.		✓				✓	✓			U
Prod. Roadmap						✓	✓	✓		R
Getting started					✓		✓	✓	✓	C
User manual			✓				✓			U
Tutorials							✓			C
FAQ							✓	✓	✓	R
Issue track		✓	✓		✓	✓	✓		✓	C
Version control		✓	✓	✓	✓	✓	✓	✓	✓	C
Build scripts		✓		✓	✓	✓	✓		✓	C
Requirements		✓				✓			✓	
Design Doc.		✓	✓		✓		✓	✓	✓	U
API Doc.					✓		✓	✓	✓	R
Test plan		✓				✓				U
Test cases	✓	✓	✓		✓	✓	✓	✓	✓	U

6 Comparison of Tools to Other Research Software

Software tools are used to support the development, verification, maintenance, and evolution of software, software processes, and artifacts [4, p. 501]. To answer RQ5 we summarize tools that are visible in the repositories and that are mentioned during the developer interviews. The tools for LBM are as follows:

Development Tools: Continuous Integration (CI), Code Editors, Development Environments, Runtime Environments, Compilers, Unit Testing Tools, Correctness Verification Tools

Dependencies: Build Automation Tools, Technical Libraries, Domain Specific Libraries
Project Management Tools: Collaboration Tools, Email, Change Tracking Tools, Version Control Tools, Document Generation Tools

Once the data on tools is collected, the use of two specific tools should be compared to research software norms: version control and CI. A little over 10 years ago version control was estimated to be used in only 50% of research software projects [13]. More recently, version control usage rates for active mesh generation, geographic information system and statistical software packages were close to 100% [21]. Almost every software guide cited in Table 3 includes the advice to use version control. For LBM packages 67% use version control. The high usage in LBM software matches the trend in research software in general. CI is rarely used in LBM (3 of 24 packages or 12.5%). This contrasts with the frequency with which CI is recommended in guidelines [1,5,29].

7 Comparison of Processes to Other Research Software

The interview data on development processes is used to answer RQ6. This data should be contrasted with the development process used by research software in general. The literature suggests that scientific developers naturally use an agile philosophy [2,19], or an amethodical process [10]. Another point of comparison is on the use of the frequently recommended peer review approach [8,14,30].

The LBM example confirms an informal, agile-like, process. The development process is not explicitly indicated in the artifacts. However, during interviews one developer (ESPResSo) told us their non-rigorous development model is like a combination of agile and waterfall. A loosely defined process makes sense for LBM software, given the small self-contained teams. One of the developers (ESPResSo) also noted that they use an ad hoc peer review process.

8 Developer Pain Points

To answer RQ7 and RQ8, we ask developers about their pain points and compare their responses to the literature [16,31]. Pain points to watch for include: cross-platform compatibility, scope bloat, lack of user feedback, dependency management, data handling concerns, reproducibility, and software scope.

An example pain point noted for LBM is a lack of development time. Small development teams are common for LBM software packages (Fig. 1). Lack of time is also highlighted by other research software developers [15,16,31].

9 Threats to Validity

The measures in the measurement template [22] may not be broad enough to accurately capture some qualities. For example, there are only two measures of surface robustness. Similarly, reusability is assessed by the number of code files and LOC per file, assuming that a large number of relatively small files implies modularity. Furthermore, the measurement of understandability relies on 10 random source code files. It is possible that the 10 files that were chosen to represent a software package may not be representative.

Another risk is missing or incorrect data. Some software package data may have been missed due to technology issues like broken links. Some pertinent data may not have been specified in public artifacts, or may be obscure within an artifact or web-page. For the LBM example, the use of unit testing and CI was mentioned in the artifacts of only three (ESPResSo, Ludwig, Musubi) packages. However, interviews suggested a more frequent use of both unit testing and CI in the development processes. Given that we could not interview a representative for every project, to keep the measures consistent we did not modify our initial repository inspection based CI count.

10 Concluding Remarks

To improve the development of research software, both in terms of productivity, and the resulting software quality, we need to understand the current state of the practice. An exciting strategy to approach this goal is to assess one domain at a time, collecting data from developer interviews and digging deeply into their code repositories. By providing feedback specific to their domain, the developer community can be drawn into a dialogue on how to best make improvements going forward. Moreover, they can be encouraged to share their best practices between one another, and with other research software domains.

We have outlined a methodology for assessing the state of the practice for any given research software domain based on a list of about 30 representative projects. In addition to interviews, we use manual and automated inspection of the artifacts in each project's repositories. The repository data is collected by filling in a 108 question measurement template, which requires installing the software on a VM, running simple tests (like completing the getting started instructions (if present)), and searching the code, documentation and test files. Using AHP the projects are ranked for 9 individual qualities (installability, correctness and verifiability, reliability, robustness, usability, maintainability, reusability, surface understandability, visibility/transparency) and for overall quality. Perspective and insight is shared with the user community via the following: i) comparing the ranking by best practices against an estimate of the community's ranking of popularity; ii) comparing artifacts, tools and processes to current research software development guidelines; and, iii) exploring pain points via developer interviews. Using our methodology, spreadsheet templates and AHP tool, we estimate (based on our experience with using the process) the time to complete an assessment for a given domain at 173 person hours.

For the running example of LBM we found that the top packages engaged in most of recommended best practices, including examples of practising peer review. However, we did find room for improvement with respect to providing API documentation, a roadmap, a code of conduct, programming style guide, uninstall instructions and CI.

In the future, we would like reduce measurement time down by increased automation. We would also like to conduct a meta-analysis, where we look at how different domains compare to answer new research questions like: What lessons from one domain could be applied in other domains? What differences exist in the pain points between domains? Are there differences in the tools, processes, and documentation between domains?

References

1. Brett, A., et al.: Scottish COVID-19 response consortium, August 2021. https://github.com/ScottishCovidResponse/modelling-software-checklist/blob/main/software-checklist.md
2. Carver, J.C., Kendall, R.P., Squires, S.E., Post, D.E.: Software development environments for scientific and engineering software: a series of case studies. In: ICSE 2007: Proceedings of the 29th International Conference on Software Engineering, pp. 550–559. IEEE Computer Society, Washington, DC (2007). https://doi.org/10.1109/ICSE.2007.77
3. Dong, A.: Assessing the state of the practice for medical imaging software. Master's thesis, McMaster University, Hamilton, ON, Canada, September 2021
4. Ghezzi, C., Jazayeri, M., Mandrioli, D.: Fundamentals of Software Engineering, 2nd edn. Prentice Hall, Upper Saddle River (2003)
5. van Gompel, M., Noordzij, J., de Valk, R., Scharnhorst, A.: Guidelines for software quality, CLARIAH task force 54.100, September 2016. https://github.com/CLARIAH/software-quality-guidelines/blob/master/softwareguidelines.pdf
6. Grannan, A., Sood, K., Norris, B., Dubey, A.: Understanding the landscape of scientific software used on high-performance computing platforms. Int. J. High Perform. Comput. Appl. **34**(4), 465–477 (2020). https://doi.org/10.1177/1094342019899451
7. Hannay, J.E., MacLeod, C., Singer, J., Langtangen, H.P., Pfahl, D., Wilson, G.: How do scientists develop and use scientific software? In: Proceedings of the 2009 ICSE Workshop on Software Engineering for Computational Science and Engineering, SECSE 2009, pp. 1–8. IEEE Computer Society, Washington, DC (2009). https://doi.org/10.1109/SECSE.2009.5069155
8. Heroux, M.A., Bieman, J.M., Heaphy, R.T.: Trilinos developers guide part II: ASC software quality engineering practices version 2.0, April 2008. https://faculty.csbsju.edu/mheroux/fall2012_csci330/TrilinosDevGuide2.pdf
9. Kågström, B., Ling, P., Van Loan, C.: GEMM-based level 3 BLAS: high-performance model implementations and performance evaluation benchmark. ACM Trans. Math. Softw. (TOMS) **24**(3), 268–302 (1998)
10. Kelly, D.: Industrial scientific software: a set of interviews on software development. In: Proceedings of the 2013 Conference of the Center for Advanced Studies on Collaborative Research, CASCON 2013, pp. 299–310. IBM Corp., Riverton (2013). http://dl.acm.org/citation.cfm?id=2555523.2555555

11. Michalski, P.: State of the practice for lattice Boltzmann method software. Master's thesis, McMaster University, Hamilton, Ontario, Canada, September 2021
12. Nangia, U., Katz, D.S.: Track 1 paper: surveying the U.S. National Postdoctoral Association regarding software use and training in research, pp. 1–6. Zenodo, June 2017. https://doi.org/10.5281/zenodo.814220
13. Nguyen-Hoan, L., Flint, S., Sankaranarayana, R.: A survey of scientific software development. In: Proceedings of the 2010 ACM-IEEE International Symposium on Empirical Software Engineering and Measurement, ESEM 2010, pp. 12:1–12:10. ACM, New York (2010). https://doi.org/10.1145/1852786.1852802
14. Orviz, P., García, Á.L., Duma, D.C., Donvito, G., David, M., Gomes, J.: A set of common software quality assurance baseline criteria for research projects (2017). https://doi.org/10.20350/digitalCSIC/12543
15. Pinto, G., Steinmacher, I., Gerosa, M.A.: More common than you think: an in-depth study of casual contributors. In: 2016 IEEE 23rd International Conference on Software Analysis, Evolution, and Reengineering (SANER), vol. 1, pp. 112–123 (2016). https://doi.org/10.1109/SANER.2016.68
16. Pinto, G., Wiese, I., Dias, L.F.: How do scientists develop and use scientific software? An external replication. In: Proceedings of 25th IEEE International Conference on Software Analysis, Evolution and Reengineering, pp. 582–591, February 2018. https://doi.org/10.1109/SANER.2018.8330263
17. Saaty, T.L.: The Analytic Hierarchy Process: Planning, Priority Setting, Resource Allocation. McGraw-Hill Publishing Company, New York (1980)
18. Schlauch, T., Meinel, M., Haupt, C.: DLR software engineering guidelines (2018). https://doi.org/10.5281/zenodo.1344612
19. Segal, J.: When software engineers met research scientists: a case study. Empir. Softw. Eng. **10**(4), 517–536 (2005). https://doi.org/10.1007/s10664-005-3865-y
20. Smith, B., Bartlett, R., Developers, x.: xSDK community package policies, December 2018. https://doi.org/10.6084/m9.figshare.4495136.v6
21. Smith, W.S.: Beyond software carpentry. In: 2018 International Workshop on Software Engineering for Science (held in conjunction with ICSE 2018), pp. 1–8 (2018)
22. Smith, W.S., Carette, J., Michalski, P., Dong, A., Owojaiye, O.: Methodology for assessing the state of the practice for domain X. arXiv:2110.11575, October 2021
23. Smith, W.S., Lazzarato, A., Carette, J.: State of practice for mesh generation software. Adv. Eng. Softw. **100**, 53–71 (2016)
24. Smith, W.S., Lazzarato, A., Carette, J.: State of the practice for GIS software. arXiv:1802.03422, February 2018
25. Smith, W.S., Sun, Y., Carette, J.: Statistical software for psychology: comparing development practices between CRAN and other communities. arXiv:1802.07362 33 p. (2018)
26. Smith, W.S., Zeng, Z., Carette, J.: Seismology software: state of the practice. J. Seismolog. **22**(3), 755–788 (2018)
27. Sood, K., Dubey, A., Norris, B., McInnes, L.C.: Repository-analysis of open-source and scientific software development projects (2019). https://doi.org/10.6084/m9.figshare.7772894.v2
28. Szulik, K.: Don't judge a project by its GitHub stars alone, December 2017. https://blog.tidelift.com/dont-judge-a-project-by-its-github-stars-alone
29. Thiel, C.: EURISE network technical reference (2020). https://technical-reference.readthedocs.io/en/latest/
30. USGS: USGS software planning checklist, December 2019. https://www.usgs.gov/media/files/usgs-software-planning-checklist

31. Wiese, I.S., Polato, I., Pinto, G.: Naming the pain in developing scientific software. IEEE Softw. **37**, 75–82 (2019). https://doi.org/10.1109/MS.2019.2899838
32. Wilson, G., Bryan, J., Cranston, K., Kitzes, J., Nederbragt, L., Teal, T.K.: Good enough practices in scientific computing. CoRR arXiv:1609.00037 (2016)

A Survey on Sustainable Software Ecosystems to Support Experimental and Observational Science at Oak Ridge National Laboratory

David E. Bernholdt, Mathieu Doucet, William F. Godoy(✉), Addi Malviya-Thakur, and Gregory R. Watson

Oak Ridge National Laboratory, Oak Ridge, TN 37830, USA
godoywf@ornl.gov
https://www.ornl.gov

Abstract. In the search for a sustainable approach for software ecosystems that supports experimental and observational science (EOS) across Oak Ridge National Laboratory (ORNL), we conducted a survey to understand the current and future landscape of EOS software and data. This paper describes the survey design we used to identify significant areas of interest, gaps, and potential opportunities, followed by a discussion on the obtained responses. The survey formulates questions about project demographics, technical approach, and skills required for the present and the next five years. The study was conducted among 38 ORNL participants between June and July of 2021 and followed the required guidelines for human subjects training. We plan to use the collected information to help guide a vision for sustainable, community-based, and reusable scientific software ecosystems that need to adapt effectively to: i) the evolving landscape of heterogeneous hardware in the next generation of instruments and computing (*e.g.* edge, distributed, accelerators), and ii) data management requirements for data-driven science using artificial intelligence.

Keywords: Scientific software ecosystem · Experimental and observational science EOS · Sustainability · Survey

1 Introduction

Computational science and engineering (CSE) is well-established as a compute, data, and software-intensive approach to scientific research with decades of history behind it. Traditionally, the software and data aspects of CSE have been

This manuscript has been authored by UT-Battelle, LLC, under contract DE-AC05-00OR22725 with the US Department of Energy (DOE). The publisher acknowledges the US government license to provide public access under the DOE Public Access Plan (https://energy.gov/downloads/doe-public-access-plan).
G. R. Watson—Contributed equally to this work.

D. Groen et al. (Eds.): ICCS 2022, LNCS 13353, pp. 560–574, 2022.
https://doi.org/10.1007/978-3-031-08760-8_46

marginalized, as incentive systems have emphasized novel scientific results over considerations related to the software and data that underpin them or indeed the skills and expertise of the people who develop the software. However, in recent years, there has been a trend to improve this through an increased understanding of the importance of the software and data to achieve high quality, trustworthy, and reasonably reproducible scientific results.

Experiment and observation have a far longer history in the conduct of scientific research than computationally-based approaches. As computing has become more capable and accessible, experimental and observational science (EOS) has also progressively expanded its use of software and computing for instrument control, data collection, reduction, analysis; data management; and other activities. It would not be a stretch to say that modern EOS relies on software as much as computational science, but with differences in complexity and scale. Therefore, as new approaches in instrumentation, and enhanced computational capabilities make feasible novel and more complex experiments, the reliance on software and computing in many areas of EOS is growing rapidly [2,3,13,18]. In addition, EOS researchers are increasingly seeking to harness high-performance computing (HPC) that had historically been the province of computational scientists to deal with rapidly increasing data volumes. Modeling and simulation techniques originated in the CSE community are widely used to help understand and interpret EOS data [9,18]. As a result, the EOS community is transforming towards an increasing software- and compute-intensity level.

This paper represents the results of a small survey targeting EOS-focused staff within one organization that is focused heavily on EOS: Oak Ridge National Laboratory (ORNL). ORNL is the largest multi-program laboratory under the U.S. Department of Energy (DOE) Office of Science. The survey attempts to identify challenges facing EOS software stakeholders and how they anticipate the situation evolving over a five-year time frame. The structure of the paper is as follows: Sect. 2 presents background information on the need for software ecosystems in related fields, while a brief description of the current landscape of software ecosystems in scientific computing is also presented in anticipation of the survey responses. Next, an overview of the survey structure, methodology and questions is described in Sect. 3. Next, results from the survey and a discussion is presented in Sect. 4. This section is perhaps the most important in the report as we attempt to craft a narrative and interpretation by correlating the overall answers. Lastly, our conclusions from the survey results and analysis are presented in Sect. 5 outlining the most important takeaways from the collected data.

2 Background

Software ecosystems are an increasingly important component of scientific endeavors, particularly as computational resources have become more available to include simulations and data analysis in research [17]. Nevertheless, there is still debate of what constitutes a "software ecosystem". Dhungana *et al.* [7]

points out the similarities between software and natural ecosystems. Monteith *et al.* [19] indicates that scientific software ecosystems incorporate a large environment that includes not only software developers, but also scientists who both use and extend the software for their research endeavors. Therefore, it is important to acknowledge the different needs and goals of these groups when considering the broad breadth of scientific software. Hannay *et al.* [10] asserts that there is a great deal of variation in the level of understanding of software engineering concepts when scientists develop and use software. Kehrer and Penzenstadler [15] explore individual, social, economic, technical and environmental dimensions to provide a framework for sustainability in research software. As described by Dongarra *et al.* [8], the high-performance computing (HPC) community identified in the last decade the need for an integrated ecosystem in the exascale era. Efforts resulted in the DOE Exascale Computing Project (ECP) [1]. Within the scope of ECP, the Extreme-scale Scientific Software Stack (E4S) [12] aims to reduce barriers associated with software quality and accessibility in HPC.

Few studies have been carried out to assess EOS software ecosystems, so it is necessary to draw from experiences in other communities. A critical part of a software ecosystem are the developers themselves, since they play a crucial role that requires establishing sustainable collaborative relationships within the community. Sadi *et al.* [20] draws from test cases on mobile platforms to provide an in-depth analysis of developers' objectives and decision criteria for the design of sustainable collaborations in software ecosystems. The guidelines to pursue requirements for a software ecosystem is discussed by Kaiya [14] to allow for a decrease in effort, increase of gain, sustainability and increase in participation of developers in an engaging conversation. Empirical assessments of how modern software and data practices are used in science are provided in recent studies [11,16,22]. Software ecosystems can be vast and have been continuously evolving through the development of new processes, inventions, and governance [5]. Therefore it is important that scientific communities develop a tailored plan that relies on software reuse and development of an interdependent [6] and component-centric [21] software ecosystem that understands the goals and needs within the context of EOS.

3 Survey Overview

3.1 Survey Motivation

As a preeminent scientific research organization, ORNL uses and creates a great deal of the software used to undertake CSE. Software is central to modern EOS for data acquisition, reduction, analysis, distribution, and related modeling and simulation used for CSE. However, instruments and sensors are growing more capable and sophisticated and experiments more complex. At the same time, the computing landscape is also changing, with a transition towards "edge" systems that are situated closer to the experiments, and the increasing use of cloud and HPC resources. These factors drive complexity in software and consequently create greater challenges for developers. In order to better understand this changing landscape, we decided to conduct a *Software Ecosystem Survey*. The goal was to collect a range of information about immediate and future development needs, the environmental factors influencing these needs, and the skills and training necessary for developer teams to be able to effectively work in these changing environments. We hope to use the information collected from the survey to improve participation and collaboration in software ecosystems and aid in creating sustainable software as we enter an exciting new frontier of scientific discovery.

3.2 Survey Design

The survey is organized into 10 major sections for a total of 34 questions. Responses are completely anonymous and meet the requirements for conducting ethical human-subject research studies. The survey design was focused on multiple-choice questions and answers, allowing the taker to register quickly and efficiently. The survey is exhaustive and might take up to 45 min to complete.

The survey begins with the description and motivation so that participants understand what the survey is trying to achieve as well as what information is expected from them. The survey then asks a series of questions to provide background information and project demographics to understand the nature of the software project and its contributors. A series of questions relating to the technical approach used by the project are asked in order to understand current software and data needs and challenges. Next, there are questions to understand the current skill levels and how new skills are acquired among software project participants. Finally, questions about the future demographics, future technical approach, and preparations for the future are formulated to understand the leading technological disruptions and the significant challenges projects will be facing in the next five years. In addition, we formulated questions on confidence to address current and future challenges at a personal and team level. The survey was developed using Google Forms™ and the responses were recorded over the course of a few weeks.

3.3 Selection of Participants

We invited individuals at ORNL working across diverse scientific domains developing research software to participate in this survey. This included developers from nuclear energy, biostatistics, transportation, building technology, geospatial analytics, among several others. Almost all of the invited participants had experience developing scientific software and were part of the broader software development community. The participants' anonymity was maintained by excluding any personal or work related information.

4 Survey Results

We received a total of 38 responses to the survey (raw results available at [4]). None of the questions were required, so the number of responses to specific

Table 1. Responses to time percentage spent *using* or *developing* software.

Research software activity	Time percentage				
	0–20%	21–40%	41–60%	61–80%	81–100%
Using	34.2%	23.7%	18.4%	13.2%	10.5%
Developing	21.2%	13.2%	23.7%	13.2%	28.9%

questions may in some cases, be fewer, which we will indicate as necessary. For questions about the project demographics and technical approach, we asked both about the current situation and the situation expected five years from now.

4.1 Background Information

This survey section consists of three questions to identify the respondents' relationship with scientific software activities and roles. The first question attempts to classify the respondents' work in relation to software. Overall, 71% of respondents indicated that research software is a primary product of their work, while 29% indicated it is not. The other two questions asked respondents to characterize the portion of their work time spent *using* or *developing* research software. Table 1 shows that 67% of the respondents spend at least 41% of their time as developers. We take this as a good indication that the survey respondents are in our target audience.

4.2 Project Demographics

In the remainder of the survey, we asked respondents to focus on the single research software development project that they considered most significant in their work, as the Project Demographics section of the survey was intended to characterize aspects of their particular project. Overall, 82% of respondents reported that they were users of the software as well as developers, whereas 18% were exclusively developers. This is consistent with our informal observations of computational science and engineering, where the majority of developers are also users. Looking five years out, respondents had essentially the same expectations (81% and 19%).

Table 2 illustrates responses to questions about the size of the development team, in terms of the overall number of active developers on the project and the number of those developers the respondent interacts with on a regular (weekly) basis. The results show that roughly two-thirds (58%) of projects are comprised of no more than 3 developers, and even on larger projects, the majority (68%) regularly interact with no more than 3 team members. However, there are also a considerable minority of software project teams (18%) with more than 10 developers.

We also tried to characterize the organizational breadth of the project teams. Table 3 shows that the majority of projects are of multi-institutional nature. Of

Table 2. Number of developers in a project and how many of these interact weekly with the respondent.

Developers that are:	People					
	0	1	2–3	4–5	6–10	>10
Active including the respondent	n/a	10.5%	47.4%	18.4%	5.3%	18.4%
Interacting weekly with the respondent	7.9%	21.1%	39.5%	18.4%	10.5%	2.6%

Table 3. Organizational breadth of research software projects. Organizational units within ORNL are listed from smallest (group) to largest (directorate). The "ORNL" response denotes teams spanning multiple directorates.

Breadth all developers within	Typical size	Current percentage	Expected in five years
Group	8–10 staff	13%	8%
Section	3–4 groups	8%	0%
Division	3–4 sections	0%	3%
Directorate	2–4 divisions	8%	3%
Whole lab	8 science directorates	16%	3%
Multiple institutions		55%	84%

the project teams comprised exclusively of ORNL staff members, the largest number included staff from multiple directorates (the largest organizational level at ORNL), but the next largest number included staff from a single group (the smallest organizational level). Our informal observation is that in CSE at ORNL, the majority of projects are also multi-institutional. This is an indicator that while some of the software serves ORNL specific scientific purposes, a large portion exposes the team to outside organizations that could help leverage common software development activities via collaboration. It is interesting to observe that in five years, there are expectations for a strong shift towards more broadly based project groups, particularly multi-institutional teams.

The DOE Office of Science defines a user facility as "a federally sponsored research facility available for external use to advance scientific or technical knowledge."[1] The DOE stewards a significant number of the user facilities in the United States, nine of which are hosted at ORNL[2] and were targeted in our distribution of the survey. We asked respondents whether their research software was intended for use at a user facility (whether at ORNL, within the DOE

[1] https://science.osti.gov/User-Facilities/Policies-and-Processes/Definition.
[2] https://www.ornl.gov/content/user-facilities.

Table 4. Characteristics considered to be of moderate or higher importance to the software projects.

Characteristic	Current percentage	Expected in five years
Functionality	97%	97%
Usability	87%	89%
Maintainability or sustainability	87%	89%
Performance	89%	92%
Portability	53%	34%
Security	21%	34%

Table 5. Analysis of free-form responses asking for additional important characteristics not included in the original six. The authors consider most of the proposed characteristics could be included in one of the original six, while two could not.

Response	Occurrences	Original characteristic
Unique capabilities	1	Functionality
Documentation	2	Usability
Intuitive	1	Usability
Robustness	1	Usability
Extensibility	1	Maintainability or sustainability
Accuracy	1	*n/a*
Correctness	1	*n/a*

system, or elsewhere). Only 8% of respondents indicated that their software did *not* target a user facility. The remainder indicated that the software was used in their own work at a user facility (13%), or by multiple users (47%). A quarter of respondents (26%) indicated that their software was part of the software suite that the facility offers to its users.

Finally, we asked the respondents to rate the importance of various software characteristics as summarized in Table 4. The importance of most of these characteristics is expected to be about the same looking out five years, except that portability drops and security increases in importance. Given an opportunity to suggest additional "moderate importance or higher" characteristics, we received 7 responses (some multiple listing characteristics). We consider that some of the responses could be consolidated into the original list of characteristics, while others would be new additions. The free-form responses are characterized in Table 5.

4.3 Project Technical Approach

This section included six questions related to the technical aspects of the project and its environment.

First, we asked respondents to indicate the various technical categories applied to their software project. Multiple categories could be selected, and a free-response option was allowed so respondents could add categories not included in the original list. Table 6 summarizes the results. Not surprisingly, for a survey focused on EOS, data processing and analysis is by far the most common category used to describe the software. The prominence of data reduction and interaction, each considered applicable to 50% of projects. Interestingly, tools and infrastructure, modeling, and simulation were also similarly prominent (50% and 47%, respectively). "Modeling and simulation" is a term often used to characterize applications in CSE, and may indicate fairly routine use of these techniques in the analysis of experimental and observational data. The number of projects categorized as tools and infrastructure was unexpectedly large. We plan to explore both of these categories in greater depth in follow-up studies. It is interesting to note that looking out five years, nearly every category increases, suggesting an expectation that the projects will broaden in terms of the capabilities and thus more categories will apply in the future. The largest growth areas are expected to be in numerical libraries and data acquisition.

Table 6. Categories applicable to the focus software project. The last two were provided as free-form additions to the list.

Category	Current percentage	Expected in five years
Data acquisition	31%	47%
Data reduction	50%	55%
Data processing and analysis	74%	79%
Data interaction (e.g., Jupyter notebooks, graphical or web interfaces, etc.)	50%	55%
Data dissemination or sharing	21%	29%
Modeling and simulation	46%	60%
Numerical libraries	21%	45%
Tools and infrastructure	50%	55%
Deep learning, machine learning, text analysis	3%	3%
Optimization	3%	3%

We also asked several questions intended to elicit the importance of various technologies or approaches to the software projects. The first question asked for an assessment of the importance of eight explicitly named technologies (on a 4-point scale), while the second questions was request for a free-form response.

Table 7 shows the number and percentage of the 38 survey respondents who indicated that each technology was of moderate or higher importance (responses of 3 or 4). This question was intended to gauge the extent of technologies we thought might be "emergent" in this community. We were not surprised to see that continuous integration/deployment was essential to most of the projects (76%), followed by numerical libraries (66%) as many EOS efforts require fairly sophisticated numerical approaches. Data storage and interaction technologies were also important (55% and 63%, respectively). Cloud-based deployment is only relevant to roughly one-third of the respondents (32%) and only a small minority (5%) rated cloud application programming interface (API) services as important, despite growing trends in cloud computing technologies in the last decades. This can be interpreted due to either lack of expertise in available cloud technologies, or that the cost of migrating operations might not justify the added value to the funded science deliverables. This is something we plan to explore further in follow-up studies. Looking ahead five years, all of the listed technologies are expected to increase in their importance to software projects. The largest growth is expected in the importance of cloud API services, with numerical libraries, data dissemination or sharing, and cloud deployment and operation technologies also significantly increased in importance.

Table 7. Technologies considered to be of moderate or higher importance to the software projects.

Technology	Current percentage	Expected in five years
Data storage	55%	66%
Data interaction (e.g., Jupyter notebooks, graphical or web interfaces, etc.)	63%	68%
Data dissemination or sharing	39%	58%
Continuous integration/continuous deployment	76%	79%
Server-based deployment and operation	45%	53%
Cloud deployment and operation (using cloud resources to make the software available to users)	32%	42%
Numerical libraries (using library-based APIs as part of your software solution, for example BLAS, solvers, etc.)	66%	82%
Use of cloud API services (using cloud-based APIs as part of your software solution, for example, GCP Life Sciences API, Vision API, Trefle API, GBIF API, etc.)	5%	37%

The free-form version of the question was included with the expectation that a wide range of technologies would be considered useful to different projects. A total of 33 of the 38 respondents answered this question, listing a total of 69 distinct items, many appearing in multiple responses. For the sake of brevity, we have assigned the responses to categories, which are listed in Table 8. We note that there is no unique and unambiguous way to categorize the tools and

Table 8. Important technologies for the respondents' software projects, as categorized by the authors. There were a total of 69 unique technologies identified by the respondents.

Technology category	Responses	Percentage
AI tools	6	9%
Application frameworks	9	13%
Build and test tools	5	7%
Computational notebook tools	2	3%
Data tools	8	12%
Deployment tools	1	1%
Hardware	2	3%
Libraries	15	22%
Python modules	7	10%
Software development tools	14	20%

technologies named in the responses, but in this paper, our goal is to provide an overview; the specific responses are available in the survey dataset [4]. We note that application frameworks, libraries, and software development tools play significant roles in the software projects surveyed. Python is also prominent, as are data tools, and artificial intelligence (AI) tools.

We also asked respondents to indicate the programming languages used in their projects. Overall, 37 of the 38 survey respondents answered this free-form question, naming a total of 58 distinct languages, 22 unique. Python and C++ were the most prominent languages, cited by 78% and 54% of the responding projects, respectively. C and Javascript followed, listed in 16% and 14% of responses, respectively; Java and bash were the only other languages listed more than once (8% and 5%). Programming for GPUs was noticeable as well, but diverse approaches were named (CUDA, HIP, Kokkos, OpenACC, OpenMP).

Finally, in this section of the survey, we asked about the computational environments targeted by the project. Respondents could select from five predefined responses as well as being able to provide a free-form response (four were received). The results are summarized in Table 9. Here, we see that individual computers dominate, but in many cases, accelerators (*e.g.*, GPUs, or FPGAs) are used. However usage of larger shared resources is also high. About a third of project target cloud computing, which is consistent with the response to an earlier question in which a third of respondents also rated cloud deployment and operation technologies as important to their projects (Table 7). The use of GPU accelerators is consistent with the use of GPU programming languages as well. Looking out five years, we see that respondents are generally expecting decreased use of single computers in favor of most other types of hardware, most significantly large-scale cloud and HPC environments.

4.4 Skills and Training

This section gauges their assessment of their own knowledge and that of their project team as a whole in various, as well as the areas in which new knowledge would be important. We requested responses on a 4-point scale ranging from "not knowledgeable" (1) to "very knowledgeable" (4), or "not important" (1) to "very important" (4). Table 10 summarizes the responses of 3–4 for personal knowledge, team knowledge, and the importance of acquiring new knowledge. Respondents are fairly confident in both their own and their team's knowledge of most areas. The weakest area was computer hardware, which also scored the lowest in terms of the importance of improving knowledge. Respondents were somewhat more comfortable with their (and their teams') knowledge of the science, algorithms, and software tools for their work than with software development practices. However software development practices and algorithms rated slightly higher than the science and software tools in terms of areas needing more knowledge.

Table 9. Hardware environments targeted by the projects. Respondents could select any choices that applied, the last two entries were provided as free-form responses.

Hardware environment target	Current percentage	Expected in five years
Single computer (laptop, desktop, or server)	68%	60%
Single computer with computational accelerator (e.g., GPU, FPGA, etc.)	40%	50%
Shared organizational resource (e.g., multiprocessor server or cluster, "edge" systems, etc.)	63%	66%
Lab-level, national, or commercial cloud resources (e.g., CADES cloud, AWS, etc.)	34%	53%
Lab-level, or national HPC resources (e.g., CADES condos, OLCF, etc.)	47%	63%
Neuromorphic processors	3%	3%
Quantum processors	0%	3%
Dedicated computational infrastructure for online analysis of experimental data (cpu + gpu + fpga)	3%	0%
Resources/infrastructure frozen for the duration of the experiment	0%	3%

4.5 Preparing for the Future

The final section of the survey asked respondents about their concerns about their projects in the next five years, and their overall confidence level in being able to deal with the changes they anticipate.

Table 10. Self-assessed level of knowledge in various areas for the respondent and their team, and the perceived importance of gaining new knowledge in the area.

Area	Personal knowledge	Team knowledge	New knowledge
The scientific context of the software	84%	95%	69%
The algorithms and methods required to implement the software	84%	84%	74%
The software frameworks, libraries, and tools to support the implementation of the software	79%	82%	71%
Software development best practices	68%	79%	74%
The computer hardware to which the software is targeted (including cloud computing, if appropriate)	58%	66%	55%

Table 11. Levels of concern about various possible changes in their projects over the next five years. Responses were on a 4-point scale from "not concerned" (1) to "very concerned" (4). We summarize the responses of moderate or high concern (3–4).

Aspect	Responses	Percentage
Changes in the hardware environment	23	61%
Changes in the scientific context	12	32%
Changes in the required functionality of the software	23	61%
Changes in the required usability of the software	21	55%
Changes in the required maintainability or sustainability of the software	23	61%
Changes in the required performance of the software	25	66%
Changes in or increased importance of data storage technologies	18	47%
Changes in or increased importance of numerical libraries	16	42%
Changes in or increased importance of data interaction technologies (e.g., Jupyter notebooks, graphical or web interfaces, etc.)	19	50%
Changes in or increased importance of continuous integration/continuous deployment technologies	17	44%
Changes in or increased importance of cloud for making the software available to users	16	42%
Changes in or increased importance of the use of cloud API service technologies	13	34%

As we see in Table 11, the most significant area of concern has to do with the performance of the software, followed by changes in the hardware environment and the requirements for the maintainability or sustainability of the software. Concerns about changes in the hardware environment here are noteworthy, given the fact that in Table 10 respondents were both least confident in their personal and their team's knowledge of the hardware, *and* ranked it the least important area in which to gain more knowledge. The areas of least concern were the scientific context of the project (which was also an area of high confidence in Table 10) and in cloud service APIs. The low concern about cloud service APIs may be explained by their low importance in Table 7.

When asked how confident they were as individuals in their ability to deal with the changes anticipated over the next five years, a significant portion (63%) responded that they were moderately confident (3 on a 4-point scale), and one-fifth of the respondents (19%) were highly confident. When asked about their *team's* ability to deal with the coming changes, the majority (79%) are confident, with 42% being moderately confident and 37% highly confident.

5 Conclusions

Our survey attempts to provide empirical evidence to understand the landscape of the software supporting EOS activities at a major research laboratory. Results suggest that the field is still focused on classical computing approaches. More recent developments in computing that are the norm outside of science, like AI, edge computing, and cloud-based services, are still not being used extensively in existing projects. On the other hand, the field is clearly moving towards a broad collaborative approach. Projects tend to be multi-institutional and benefit a wider array of users. At the same time, most projects are comprised of a handful of developers, and most project interactions tend to be between a few developers. This highlights the niche nature of highly-specialized scientific software. We have also seen that data dissemination and sharing are becoming a focus across a range of scientific endeavors. In addition, projects feel confident that they will tackle challenges in the next five-year time frame. We understand these results as a reflection of the alignment to sponsors' expectations, for which its impact on ORNL EOS deliverables must justify investments in aspects of scientific software shown in this survey. In the future, we would like to use the insights gained from this survey to better support EOS developers and develop guidelines for a sustainable EOS software and data ecosystem.

References

1. Exascale computing project, September 2017. https://exascaleproject.org/
2. Abbott, R., et al.: Open data from the first and second observing runs of Advanced LIGO and Advanced Virgo. SoftwareX **13**, 100658 (2021). https://doi.org/10.1016/j.softx.2021.100658
3. Special issue on software that contributed to gravitational wave discovery. SoftwareX (2021). https://www.sciencedirect.com/journal/softwarex/special-issue/103XKC9DRLV
4. A survey on sustainable software ecosystems to support experimental and observational science at oak ridge national laboratory (2022). https://doi.org/10.6084/m9.figshare.19529995

5. Bartlett, R., et al.: xSDK foundations: toward an extreme-scale scientific software development kit. Supercomput. Front. Innov. **4**(1), 69–82 (2017). https://doi.org/10.14529/jsfi170104
6. Bavota, G., Canfora, G., Penta, M.D., Oliveto, R., Panichella, S.: The evolution of project inter-dependencies in a software ecosystem: the case of apache. In: 2013 IEEE International Conference on Software Maintenance, pp. 280–289 (2013). https://doi.org/10.1109/ICSM.2013.39
7. Dhungana, D., Groher, I., Schludermann, E., Biffl, S.: Software ecosystems vs. natural ecosystems: learning from the ingenious mind of nature. In: Proceedings of the Fourth European Conference on Software Architecture: Companion Volume, pp. 96–102 (2010)
8. Dongarra, J., et al.: The International Exascale Software Project roadmap. Int. J. High Perform. Comput. Appl. **25**(1), 3–60 (2011). https://doi.org/10.1177/1094342010391989
9. Enders, B., et al.: Cross-facility science with the Superfacility Project at LBNL. In: 2020 IEEE/ACM 2nd Annual Workshop on Extreme-scale Experiment-in-the-Loop Computing (XLOOP), pp. 1–7 (2020). https://doi.org/10.1109/XLOOP51963.2020.00006
10. Hannay, J.E., MacLeod, C., Singer, J., Langtangen, H.P., Pfahl, D., Wilson, G.: How do scientists develop and use scientific software? In: 2009 ICSE Workshop on Software Engineering for Computational Science and Engineering, pp. 1–8 (2009). https://doi.org/10.1109/SECSE.2009.5069155
11. Heaton, D., Carver, J.C.: Claims about the use of software engineering practices in science: a systematic literature review. Inf. Softw. Technol. **67**, 207–219 (2015). https://doi.org/10.1016/j.infsof.2015.07.011
12. Heroux, M.A.: The extreme-scale scientific software stack (e4s). Technical report, Sandia National Lab. (SNL-NM), Albuquerque, NM, United States (2019)
13. Ivezić, Ž., et al.: LSST: from science drivers to reference design and anticipated data products. Astrophys. J. **873**(2) (2019). https://doi.org/10.3847/1538-4357/ab042c
14. Kaiya, H.: Meta-requirements for information system requirements: lesson learned from software ecosystem researches. Procedia Comput. Sci. **126**, 1243–1252 (2018). https://doi.org/10.1016/j.procs.2018.08.066. Knowledge-Based and Intelligent Information & Engineering Systems: Proceedings of the 22nd International Conference, KES-2018, Belgrade, Serbia
15. Kehrer, T., Penzenstadler, B.: An exploration of sustainability thinking in research software engineering. In: Chitchyan, R., Penzenstadler, B., Venters, C.C. (eds.) Proceedings of the 7th International Workshop on Requirements Engineering for Sustainable Systems (RE4SuSy 2018) co-located with the 26th International Conference on Requirements Engineering (RE 2018), Banff, Alberta, Canada, 20 August, 2018. CEUR Workshop Proceedings, vol. 2223, pp. 34–43. CEUR-WS.org (2018). http://ceur-ws.org/Vol-2223/paper5.pdf
16. Lamprecht, A.L., et al.: Towards fair principles for research software. Data Sci. **3**(1), 37–59 (2020)
17. Manikas, K., Hansen, K.M.: Software ecosystems-a systematic literature review. J. Syst. Softw. **86**(5), 1294–1306 (2013)
18. Megino, F.B., et al.: Integration of titan supercomputer at OLCF with ATLAS production system. In: Journal of Physics: Conference Series **898**, 092002 (2017). https://doi.org/10.1088/1742-6596/898/9/092002

19. Monteith, J.Y., McGregor, J.D., Ingram, J.E.: Scientific research software ecosystems. In: Proceedings of the 2014 European Conference on Software Architecture Workshops, pp. 1–6 (2014)
20. Sadi, M.H., Dai, J., Yu, E.: Designing software ecosystems: how to develop sustainable collaborations? In: Persson, A., Stirna, J. (eds.) CAiSE 2015. LNBIP, vol. 215, pp. 161–173. Springer, Cham (2015). https://doi.org/10.1007/978-3-319-19243-7_17
21. dos Santos, R.P., Werner, C.M.L.: Revisiting the concept of components in software engineering from a software ecosystem perspective. In: Proceedings of the Fourth European Conference on Software Architecture: Companion Volume, ECSA 2010, pp. 135–142. Association for Computing Machinery, New York (2010). https://doi.org/10.1145/1842752.1842782
22. Storer, T.: Bridging the chasm: a survey of software engineering practice in scientific programming. ACM Comput. Surv. (CSUR) **50**(4), 1–32 (2017)

Solving Problems with Uncertainty

Derivation and Computation of Integro-Riccati Equation for Ergodic Control of Infinite-Dimensional SDE

Hidekazu Yoshioka[1] (✉) and Motoh Tsujimura[2]

[1] Graduate School of Natural Science and Technology, Shimane University, Nishikawatsu-cho 1060, Matsue 690-8504, Japan
yoshih@life.shimane-u.ac.jp

[2] Graduate School of Commerce, Doshisha University, Karasuma-Higashi-iru, Imadegawa-dori, Kyoto 602-8580, Japan
mtsujimu@mail.doshisha.ac.jp

Abstract. Optimal control of infinite-dimensional stochastic differential equations (SDEs) is a challenging topic. In this contribution, we consider a new control problem of an infinite-dimensional jump-driven SDE with long (sub-exponential) memory arising in river hydrology. We deal with the case where the dynamics follow a superposition of Ornstein–Uhlenbeck processes having distributed reversion speeds (called supOU process in short) as found in real problems. Our stochastic control problem is of an ergodic type to minimize a long-run linear-quadratic objective. We show that solving the control problem reduces to finding a solution to an integro-Riccati equation and that the optimal control is infinite-dimensional as well. The integro-Riccati equation is numerically computed by discretizing the phase space of the reversion speed. We use the supOU process with an actual data of river discharge in a mountainous river environment. Computational performance of the proposed numerical scheme is examined against different discretization parameters. The convergence of the scheme is then verified with a manufactured solution. Our paper thus serves as new modeling, computation, and application of an infinite-dimensional SDE.

Keywords: Infinite-dimensional stochastic differential equation · Stochastic control in infinite-dimension · Integro-riccati equation

1 Introduction

Optimal control of stochastic partial differential equations, namely infinite-dimensional stochastic differential equations (SDEs), has recently been a hot research topic from both theoretical and engineering sides because of their rich mathematical properties and importance in applied problems [1]. Such examples include but are not limited to shape optimization under uncertainty [2], evolution theory of age-dependent population dynamics [3], and portfolio management [4].

D. Groen et al. (Eds.): ICCS 2022, LNCS 13353, pp. 577–588, 2022.
https://doi.org/10.1007/978-3-031-08760-8_47

The main difficulty in handling a control problem of an infinite-differential SDE comes from the infinite-dimensional nature of the optimality equation. Indeed, in a conventional control problem of a finite-dimensional SDE, finding an optimal control reduces to solving an optimality equation given as a finite-dimensional parabolic partial differential equation. Its solution procedure can be constructed by a basic numerical method like a finite difference scheme [5]. By contrast, a control problem of an infinite-differential SDE involves a partial differential equation having an infinite dimension, which cannot be solved numerically in general. A tailored numerical scheme is necessary to handle the infinite-dimensional nature [6, 7]; such schemes have not always been applied to problems with actual system dynamics. This issue is a bottleneck in applications of infinite-differential SDEs in engineering problems. Hence, demonstrating a computable example of interest in an engineering problem can be useful for better understanding the control of infinite-differential SDEs.

The objectives of this paper are to present an infinite-dimensional SDE arising in hydrology and environmental management, and to formulate its ergodic linear-quadratic (LQ) control problem. The system governs temporal evolution of river discharge as a superposition of Ornstein–Uhlenbeck processes (supOU process) as recently identified in Yoshioka [8]. Markovian stochastic modeling of river discharge has long been a standard method for assessing streamflows [9]. However, some researchers including the first author recently found that the Markovian assumption is often inappropriate for discharge time series of actual perennial river environments due to the sub-exponential auto-correlation [8]. This sub-exponential auto-correlation is consistent with the supOU process as an infinite-dimensional SDE, which is why we are focusing on this specific stochastic process.

The goal of our control problem is to modulate the discharge considering a water demand with a least effort in long-run. This can be the simplest management problem of water resources in which maintaining the water depth or discharge near some prescribed level is preferred. The LQ nature allows us to reduce the infinite-dimensional optimality equation to a two-dimensional integro-Riccati equation which is computable by a collocation method in space [10] combined with a forward Euler method in time. This integro-Riccati equation itself has not been derived in the literature so far. Focusing on an actual parameter set, we provide computational examples of the optimal control along with their well-posedness and optimality. Our problem is simple but involves several nontrivial scientific issues to be tackled in future. We believe that this contribution would advance modeling problems with uncertainty from a viewpoint of infinite-dimensional SDEs.

2 Control Problem

2.1 Uncontrolled System

We consider a control problem of discharge at a point in a river, which is a continuous-time and continuous-state scalar variable denoted as X_t at time $t \geq 0$ with an initial condition $X_0 \geq 0$. Our formulation is based on the SDE representation of supOU processes suggested in Barndorff-Nielsen [11] and later justified in Barndorff-Nielsen and

Stelzer [12]. The assumptions made in our SDE are based on the physical consideration of river discharge as a jump-driven process [8].

The system dynamics without control follow the distributed SDE

$$\mathrm{d}X_t = \underline{X} + \int_0^{+\infty} Y_t(\lambda)\pi(\mathrm{d}\lambda),\ t > 0 \tag{1}$$

with a minimum discharge $\underline{X} \geq 0$ and $Y_t(\cdot)$ $(t \in \mathbb{R})$ governed by

$$\mathrm{d}Y_t(\lambda) = -\lambda Y_t(\lambda)\mathrm{d}t + \mathrm{d}L_t(\lambda),\ t > 0. \tag{2}$$

Here, π is a probability measure of a positive random variable absolutely continuous with respect to $\mathrm{d}\lambda$ on the half line $\lambda > 0$, such that

$$\int_0^{+\infty} \frac{\pi(\mathrm{d}\lambda)}{\lambda} < +\infty, \tag{3}$$

and $L_t(\cdot)$ $(t > 0)$ is a pure positive-jump space-time Lévy process corresponding to an ambit field whose background Lévy measure $v = v(\mathrm{d}z)$ is a finite-variation type:

$$\int_0^{+\infty} \min\{1, z\} v(\mathrm{d}z),\ \int_0^{+\infty} z^2 v(\mathrm{d}z) < +\infty. \tag{4}$$

The conditions (3) and (4) are imposed to well-define jumps of the SDE (2) and to guarantee boundedness of the statistical moments of discharge [8]. The expectation of $\mathrm{d}L_t(\lambda)\mathrm{d}L_t(\theta)$ $(t > 0,\ \lambda, \theta > 0)$ is formally given by

$$\mathbb{E}[\mathrm{d}L_s(\lambda)\mathrm{d}L_s(\theta)] = \begin{cases} M_1^2(\mathrm{d}s)^2 & (\theta \neq \lambda) \\ M_2\delta_{\{\theta=\lambda\}}\mathrm{d}\lambda/\pi(\mathrm{d}\lambda)\mathrm{d}s & (\theta = \lambda) \end{cases},\ M_k = \int_0^{+\infty} z^k v(\mathrm{d}z)\ (k = 1, 2) \tag{5}$$

with the Dirac delta δ. The noise process associated with the supOU process therefore is not of a trace class [13], suggesting that the system dynamics are highly irregular. This point will be discussed in the next section.

The SDE representation (1) implies that the river discharge is multi-scale in time because it is a superposition of infinitely many independent OU processes having different reversion speeds λ on the probability measure π (i.e., different values of the decay speed of flood pulses). More specifically, in the supOU process, each jump of X associates a corresponding λ generated from π [11], allowing for the existence of flood pulses decaying with different speeds. In principle, this kind of multi-scale nature cannot be reproduced by simply using a classical OU process because it has only one decay speed. The supOU processes are therefore expected to be a more versatile alternative to the classical OU ones. The parameters of the densities π and v were successfully identified in Yoshioka [8], which will be used later.

2.2 Controlled System

The controlled system is the SDE (1) with Y now governed by

$$\mathrm{d}Y_t(\lambda) = (-\lambda Y_t(\lambda) + u_t(\lambda))\mathrm{d}t + \mathrm{d}L_t(\lambda),\ t > 0. \tag{6}$$

Here, u is a control variable progressively measurable with respect to a natural filtration generated by $L_t(\cdot)$ ($t > 0$), and satisfies the square integrability conditions

$$\limsup_{T\to+\infty}\frac{1}{T}\mathbb{E}\left[\int_0^T\int_0^{+\infty}(u_s(\lambda))^2\pi(\mathrm{d}\lambda)\mathrm{d}s\right],\ \limsup_{T\to+\infty}\frac{1}{T}\mathbb{E}\left[\int_0^T X_s^2\mathrm{d}s\right] < +\infty. \tag{7}$$

Our objective functional is the following long-run LQ type:

$$J(u) = \limsup_{T\to+\infty}\frac{1}{T}\mathbb{E}\left[\frac{1}{2}\int_0^T\left\{\left(X_s-\overline{X}\right)^2 + w\int_0^{+\infty}(u_s(\lambda))^2\pi(\mathrm{d}\lambda)\right\}\mathrm{d}s\right] \tag{8}$$

with a target discharge $\overline{X} > 0$ representing a water demand and $w > 0$ is a weight balancing the two terms: the deviation from the target and control cost. The objective is to find the minimizer $u = u^*$ of $J(u)$: $H = \inf_u J(u) \geq 0$. Note that the jumps are not controlled as they represent uncontrollable inflow events from upstream.

2.3 Integro-Riccati Equation

By a dynamic programming argument [e.g., 14], we infer that the optimality equation of the control problem is the infinite-dimensional integro-partial differential equation

$$\begin{aligned}&-H+\inf_{u(\cdot)}\left\{\int_0^{+\infty}(-\lambda Y(\lambda)+u(\lambda))\nabla V(Y)\pi(\mathrm{d}\lambda)+\frac{w}{2}\int_0^{+\infty}(u(\lambda))^2\pi(\mathrm{d}\lambda)\right\}\\&+\int_0^{+\infty}\left(\int_0^{+\infty}V\left(Y(\cdot)+z\delta_{\{\omega=(\cdot)\}}\frac{\mathrm{d}(\cdot)}{\pi(\mathrm{d}(\cdot))}\right)\pi(\mathrm{d}\omega)-V(Y)\right)v(\mathrm{d}z) \qquad , Y\in L^2(\pi),\\&+\frac{1}{2}\int_0^{+\infty}\int_0^{+\infty}Y(\lambda)Y(\theta)\pi(\mathrm{d}\lambda)\pi(\mathrm{d}\theta)-\overline{X}\int_0^{+\infty}Y(\lambda)\pi(\mathrm{d}\lambda)+\frac{1}{2}\overline{X}^2=0\end{aligned} \tag{9}$$

where $L^2(\pi)$ is a collection of square integrable functions with respect to π, and $\nabla V(Y)$ is the Fréchet derivative identified as a mapping from $L^2(\pi)$ to $L^2(\pi)$. The infimum in (9) must be taken with respect to functions belonging to $L^2(\pi)$. By calculating the inf term, (9) is rewritten as

$$\begin{aligned}&-H-\int_0^{+\infty}\lambda\nabla V(Y)Y(\lambda)\pi(\mathrm{d}\lambda)-\frac{1}{2w}\int_0^{+\infty}(\nabla V(Y))^2\pi(\mathrm{d}\lambda)\\&+\int_0^{+\infty}\left(\int_0^{+\infty}V\left(Y(\cdot)+z\delta_{\{\omega=(\cdot)\}}\frac{\mathrm{d}(\cdot)}{\pi(\mathrm{d}(\cdot))}\right)\pi(\mathrm{d}\omega)-V(Y)\right)v(\mathrm{d}z) \qquad , Y\in L^2(\pi)\\&+\frac{1}{2}\int_0^{+\infty}\int_0^{+\infty}Y(\lambda)Y(\theta)\pi(\mathrm{d}\lambda)\pi(\mathrm{d}\theta)-\overline{X}\int_0^{+\infty}Y(\lambda)\pi(\mathrm{d}\lambda)+\frac{1}{2}\overline{X}^2=0\end{aligned} \tag{10}$$

with (a candidate of) the optimal control as a minimizer of the infimum in the first line of (9):

$$u^*(Y) = -\frac{1}{w}\nabla V(Y),\ Y\in L^2(\pi). \tag{11}$$

A formal solution to (9) is a couple (V, h) of smooth $V : L^2(\pi) \to \mathbb{R}$ and $h \in \mathbb{R}$.

Invoking the LQ nature of our problem suggests the ansatz: for any $Y \in L^2(\pi)$,

$$V(Y) = \frac{1}{2}\int_0^{+\infty}\int_0^{+\infty}\Gamma(\theta,\lambda)Y(\lambda)Y(\theta)\pi(\mathrm{d}\lambda)\pi(\mathrm{d}\theta) + \int_0^{+\infty}\gamma(\lambda)Y(\lambda)\pi(\mathrm{d}\lambda) \quad (12)$$

with symmetric $\pi\otimes\pi$-integrable Γ and π-integrable γ. Substituting (12) into (10) yields our integro-Riccati equation

$$0 = -(\theta+\lambda)\Gamma(\theta,\lambda) - \frac{1}{w}\int_0^{+\infty}\Gamma(\theta,\omega)\Gamma(\omega,\lambda)\pi(\mathrm{d}\omega) + 1,\ \lambda,\theta > 0, \quad (13)$$

$$0 = -\lambda\gamma(\lambda) - \frac{1}{w}\int_0^{+\infty}\Gamma(\lambda,\tau)\gamma(\tau)\pi(\mathrm{d}\tau) + \frac{1}{2}M_1\int_0^{+\infty}\Gamma(\lambda,\tau)\pi(\mathrm{d}\tau) - \overline{X},\ \lambda > 0, \quad (14)$$

$$h = -\frac{1}{w}\int_0^{+\infty}(\gamma(\tau))^2\pi(\mathrm{d}\tau) + \frac{1}{2}M_2\int_0^{+\infty}\Gamma(\theta,\theta)\pi(\mathrm{d}\theta) + M_1\int_0^{+\infty}\gamma(\lambda)\pi(\mathrm{d}\lambda) + \frac{1}{2}\overline{X}^2. \quad (15)$$

In summary, we could reduce an infinite-dimensional Eq. (9) to the system of finite-dimensional integral Eqs. (13)–(15). This integro-Riccati equation is not found in the literature to the best of our knowledge. The integro-Riccati equation is not solvable analytically, motivating us to employ a numerical method for approximating its solution, which is now the triplet $(\Gamma(\cdot,\cdot), \gamma(\cdot), H)$. Note that the three Eqs. (13)–(15) are effectively decoupled with respect to the three solution variables. With this finding, we can solve them in the order from (13), (14), to (15). This structure also applies to our numerical method.

2.4 Remarks on the Optimality

The optimality of the integro-Riccati Eq. (13)–(15) follows "formally" by the verification argument [10] based on an Itô's formula for infinite-dimensional SDEs, suggesting that the formula (11) gives an optimal control and $h = H$. To completely prove the optimality, one must deal with the irregularity of the driving noise process that is not of a classical trace class. In particular, possible solutions to the optimality Eq. (9) should be limited to a functional space such that the non-local term having the Dirac delta is well-defined. The linear-quadratic ansatz (12) meets this requirement, while it is non-trivial whether this holds true in more complicated control problems of supOU processes. One may replace this term by regularizing the correlation of the space-time noise to avoid the well-posedness issue; however, this method may not lead to a tractable mathematical model such that the statistical moments and auto-correlation are found explicitly, and hence critically degrades usability of the model in practice. These issues are beyond the scope of this paper because they need sophisticated space-time white noise analysis [e.g., 15].

3 Computation with Actual Data

3.1 Computational Conditions

We show computational examples with an actual discharge data set at an observation station in a perennial mountainous river, Tabusa River, with the mean 2.59 (m^3/s) and variance 61.4 (m^6/s^2). The supOU process was completely identified and statistically examined in Yoshioka [8]. The identified model uses a gamma distribution for π and a tempered stable distribution for ν, both of which were determined by a statistical analysis of moments and auto-correlation function. The model correctly fits the auto-correlation with the long-memory behaving as $l^{-0.75}$ for a large time lag l, generates the average, standard deviation, skewness, and kurtosis within the relative error $6.23 \cdot 10^{-3}$ to $8.28 \cdot 10^{-2}$, and furthermore captures the empirical histogram.

The Eq. (13)–(15) is discretized by the collocation method [10]. The measure π is replaced by the discrete one $\pi(\mathrm{d}\lambda) \to \pi_n(\mathrm{d}\lambda)$ as follows ($1 \le i \le n$):

$$\pi_n(\mathrm{d}\lambda) = \sum_{i=1}^{n+1} c_i \delta_{\{\lambda=\lambda_i\}},\ c_i = \int_{\eta_{n,i-1}}^{\eta_{n,i}} \pi(\mathrm{d}\lambda),\ \lambda_i = \frac{1}{c_i}\int_{\eta_{n,i-1}}^{\eta_{n,i}} \lambda\pi(\mathrm{d}\lambda),\ \eta_{n,i} = \overline{\eta}\frac{i}{n^\beta} \quad (16)$$

for a fixed resolution $n \in \mathbb{N}$ and parameters $\overline{\eta} > 0$ and $\beta \in (0, 1)$, where we define $c_{n+1} = \sum_{i=1}^{n} c_i$ and $\lambda_{n+1} = +\infty$. The parameter $\overline{\eta}$ in the last equation of (16) specifies the degree of domain truncation, while the parameter β modulates the degree of refinement of discretization as the resolution n increases; choosing a larger β means a slower refinement of the discretization.

Replacing π by π_n in (13)–(15) at each node $(\lambda, \theta) = (\lambda_i, \lambda_j)$ $(1 \le i, j \le n)$ leads to a system of nonlinear system governing $\Gamma(\lambda_i, \lambda_j)(1 \le i, j \le n)$, $\gamma(\lambda_i)(1 \le i \le n)$, and $h(= H)$. Instead of directly inverting this system, we add a temporal partial differentials $\frac{\partial \Gamma}{\partial \tau}$ and $\frac{\partial \gamma}{\partial \tau}$ to the left-sided of (13) and (14), respectively with a pseudo-time parameter τ.The temporal discretization is based on a forward Euler scheme with the increment of pseudo-time $1/(24n)$ (day). Stability of numerical solution is maintained with this increment. The system is discretized from initial conditions $\Gamma \equiv 0$ and $\gamma \equiv 0$ until it becomes sufficiently close to a steady state with the increment smaller than 10^{-10} between each successive pseudo-time steps. The quantity H in (15) is then evaluated using the resulting Γ and γ.

3.2 Computational Results

Because the system (13)–(15) does not admit an explicit solution, we examine convergence of numerical solutions against the following manufactured solution that is obtained by adding appropriate source terms to the right-sides of (13) and (14):

$$\Gamma(\lambda, \theta) = ae^{-b(\lambda+\theta)} \text{ and } \gamma(\lambda) = ce^{-b\lambda},\ \lambda, \theta > 0 \quad (17)$$

with constants $a, b, c > 0$. Namely, we add proper functions $f_1(\lambda, \theta)$ and $f_2(\lambda)$ to (13) and (14) so that the manufactured solution (17) solves the modified equations:

$$0 = -(\theta + \lambda)\Gamma(\theta, \lambda) - \frac{1}{w}\int_0^{+\infty} \Gamma(\theta, \omega)\Gamma(\omega, \lambda)\pi(\mathrm{d}\omega) + 1 + f_1(\lambda, \theta),\ \lambda, \theta > 0 \tag{18}$$

and

$$\begin{array}{l} 0 = -\lambda\gamma(\lambda) - \frac{1}{w}\int_0^{+\infty} \Gamma(\lambda, \tau)\gamma(\tau)\pi(\mathrm{d}\tau) \\ +\frac{1}{2}M_1 \int_0^{+\infty} \Gamma(\lambda, \tau)\pi(\mathrm{d}\tau) - \bar{X} + f_2(\lambda) \end{array}, \quad \lambda > 0. \tag{19}$$

Here, we still use (15). By the manufactured solution (17), each integral in (18), (19), (15), and hence H is evaluated analytically owing to using the gamma-type π and the tempered stable-type ν. The system consisting of the equations (18), (19), (15) is different from the original integro-Riccati equation. However, they share the common integral terms, suggesting that computational performance of the proposed numerical scheme can be examined against the manufactured solution (17).

We set $a = 1.0$, $b = 0.2$, $c = 0.5$, $w = 1$, and $\overline{X} = 15$ (m^3/s), leading to $H = 115.6514$ (m^6/s^2). For the discretization, we fix $\overline{\eta} = 0.05$ (1/h). Tables 1 and 2 show the computed H with its relative error (RE) and convergence rate (CR) for $\beta = 0.25$ and $\beta = 0.50$, respectively. The CRs have been computed by the common arithmetic [16]. Similarly, Tables 3 and 4 show the computed Γ and γ with its maximum nodal errors (NEs) and CRs for $\beta = 0.25$ and $\beta = 0.50$, respectively.

The numerical solutions converge to the manufactured solutions, verifying the proposed numerical scheme computationally. The CRs of H, Γ, γ are larger than 2.2 for $\beta = 0.25$ and is larger than 0.7 for $\beta = 0.50$ except for the finest level at which the discretization error of c_i, λ_i dominates. The obtained results suggest that choosing the smaller $\beta = 0.25$ is more efficient in this case possibly because the smaller β better harmonizes the domain truncation and node intervals. Note that numerical solutions did not converge to the manufactured solution if $\beta = 0.75$, suggesting an important remark that using a too large β should be avoided.

Table 1. Computed H with its RE and CR ($\beta = 0.25$).

n	Computed H	RE	CR
10	115.1436	4.39.E–03	2.29.E+00
20	115.5474	3.15.E–02	4.04.E+00
40	115.6451	5.48.E–05	6.88.E+00
80	115.6514	4.64.E–07	3.33.E+00
160	115.6514	4.61.E–08	

Table 2. Computed H with its RE and CR ($\beta = 0.50$).

n	Computed H	RE	CR
10	114.3099	1.16.E–02	7.43.E–01
20	114.8497	6.93.E–03	1.08.E+00
40	115.2710	3.29.E–03	1.57.E+00
80	115.5230	1.11.E–03	2.29.E+00
160	115.6251	2.27.E–04	

Table 3. Computed Γ, γ with their NEs and CRs ($\beta = 0.25$).

n	NE of Γ	NE of γ	CR of Γ	CR of γ
10	1.47.E–01	4.35.E–02	2.22.E+00	2.51.E+00
20	3.15.E–02	7.66.E–03	3.94.E+00	4.58.E+00
40	2.06.E–03	3.21.E–04	6.79.E+00	6.98.E+00
80	1.86.E–05	2.54.E–06	3.21.E+00	−3.51.E–02
160	2.01.E–06	2.60.E–06		

Table 4. Computed Γ, γ with their NEs and CRs ($\beta = 0.50$).

n	NE of Γ	NE of γ	CR of Γ	CR of γ
10	4.89.E–01	1.53.E–01	6.70.E–01	7.29.E–01
20	3.07.E–01	9.20.E–02	8.55.E–01	1.04.E+00
40	1.70.E–01	4.47.E–02	1.24.E+00	1.77.E+00
80	7.21.E–02	1.32.E–02	1.93.E+00	3.08.E+00
160	1.90.E–02	1.56.E–03		

With $n = 100$, $\overline{\eta} = 0.05$ (1/h), and $\beta = 0.25$, we present numerical solutions to the integro-Riccati equation. Figures 1 and 2 show the cases with a small controlling cost $w = 0.5$ and a large cost $w = 5$, respectively. The functional shapes of Γ, γ are common in the two cases, while their magnitudes are significantly different. In both cases, numerical solutions are successfully computed without spurious oscillations.

Finally, Fig. 3 shows the computed optimized objective H for a variety of couple $(\overline{X}, w)$. Figure 3 suggests that the optimized objective H is increasing with respect to w. This observation is theoretically consistent with our formulation because increasing the controlling cost should associate a larger value of the objective. Dependence of H on the target discharge $\overline{X}$ is less significant, but seems to be moderately increasing with respect to $\overline{X}$ for each w. This is considered due to that maintaining a higher level is more costly in general for the computed cases here because the minimum discharge is only 0.1 (m^3/s).

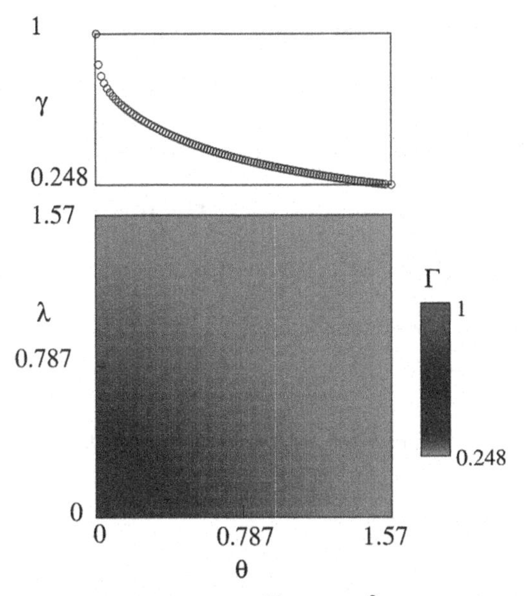

Fig. 1. Computed Γ, γ with $\overline{X} = 10$ (m^3/s) and $w = 0.5$.

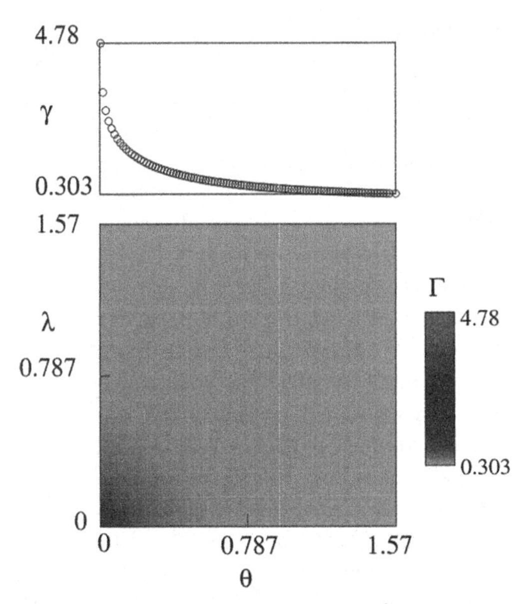

Fig. 2. Computed Γ, γ with $\overline{X} = 10$ (m^3/s) and $w = 5$.

In all the computational cases, the computed Γ are positive semi-definite, suggesting that the optimal controls are stabilizable owing to the formula (11). As demonstrated in this paper, the proposed numerical method suffices for computing the optimal control of the infinite-dimensional SDE under diverse conditions.

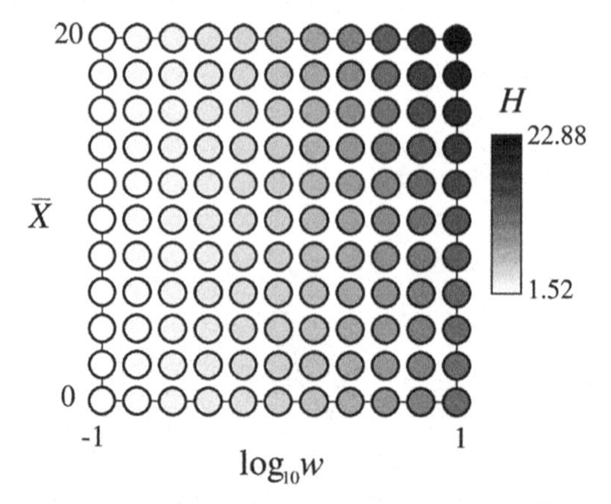

Fig. 3. Computed H for a variety of the couple $(\overline{X}, w)$.

4 Conclusion

We presented a novel control problem of an infinite-dimensional SDE arising in environmental management and discussed that it is a highly non-trivial problem due to the noise irregularity. We derived the integro-Riccati equation as a computable optimality equation. Our numerical scheme sufficed to handle this equation.

Currently, we are dealing with a time-periodic control problem of a supOU process as an optimization problem under uncertainty in long-run. Rather difficult is a mathematical justification of the optimality equation under the non-standard space-time noise. Accumulating knowledge from both theoretical and engineering sides would be necessary for correctly understanding control of infinite-dimensional SDEs. Exploring the theoretically optimal discretization of the proposed scheme is also interesting. Applying the proposed framework to weather derivatives [17] based on river hydrological processes will be another future research direction.

In this paper, we considered a control problem under full information which is a common assumption in most of the stochastic control problems. This means that the observer of the target system has a complete information to construct an optimal control, which is not always technically possible in applications. A possible way to abandon the full-information assumption would be the use of a simpler open-loop control in which the coefficients of optimal controls may be optimized by a gradient descent. Interestingly, for deterministic LQ problems, it has been pointed out that the open-loop problem is often computationally harder than the closed-loop ones [18, 19] because of the non-convex domain of optimization. This finding would apply to stochastic LQ control problems as well. In future, we will compare performance of diverse types of controls including the presented one and the open-loop ones using actual data of environmental management. In particular, modeling and control of coupled hydrological and biochemical dynamics in river environments are of great interest as there exist a huge number of unresolved issues where the proposed stochastic control approach potentially serves as a powerful analysis tool.

We focused on the use of a dynamic programming principle, while the maximum principle can also give an equivalent control formulation based on forward-backward stochastic differential equations. These two principles characterize the same control problem from different viewpoints with each other, naturally leading to different numerical methods for its resolution. Currently, we are investigating an approach from the maximum principle for exploring a more efficient numerical method to compute the LQ and related control problems under uncertainty.

Acknowledgements. JSPS Research Grant (19H03073, 21K01573, 22K14441), Kurita Water and Environment Foundation (Grant 21K008), Environmental Research Projects from the Sumitomo Foundation (203160) support this research. We used ACCMS, Kyoto University, Japan for a part of our computation. The authors thank all the members of mathematical analysis study group in Shimane University, Japan for their valuable comments and suggestions on our research.

References

1. Lü, Q., Zhang, X.: Mathematical Control Theory for Stochastic Partial Differential Equations. Springer, Cham (2021). https://doi.org/10.1007/978-3-030-82331-3
2. Xiong, M., Chen, L., Ming, J., Hou, L.: Cluster-based gradient method for stochastic optimal control problems with elliptic partial differential equation constraint. Numerical Methods for Partial Differential Equations (in press)
3. Oizumi, R., Inaba, H.: Evolution of heterogeneity under constant and variable environments. PLoS ONE **16**(9), e0257377 (2021)
4. Nadtochiy, S., Zariphopoulou, T.: Optimal contract for a fund manager with capital injections and endogenous trading constraints. SIAM J. Financ. Math. **10**(3), 698–722 (2019)
5. Bonnans, J.F., Zidani, H.: Consistency of generalized finite difference schemes for the stochastic HJB equation. SIAM J. Numer. Anal. **41**(3), 1008–1021 (2003)
6. Lima, L., Ruffino, P., Souza, F.: Stochastic near-optimal control: additive, multiplicative, non-Markovian and applications. Eur. Phys. J. Spec. Top. **230**(14–15), 2783–2792 (2021). https://doi.org/10.1140/epjs/s11734-021-00185-y
7. Urbani, P.: Disordered high-dimensional optimal control. J. Phys. A: Math. Theor. **54**(32), 324001 (2021)
8. Yoshioka, H.: Fitting a supOU process to time series of discharge in a perennial river environment, ANZIAM Journal (in press)
9. Moon, W., Hannachi, A.: River Nile discharge, the Pacific Ocean and world climate–a seasonal synchronization perspective. Tellus A: Dyn. Meteorol. Oceanogr. **73**(1), 1–12 (2021)
10. Abi Jaber, E., Miller, E., Pham, H.: Linear-Quadratic control for a class of stochastic Volterra equations: solvability and approximation. Ann. Appl. Probab. **31**(5), 2244–2274 (2021)
11. Barndorff-Nielsen, O.E.: Superposition of ornstein-uhlenbeck type processes. Theor. Probab. Appl. **45**(2), 175–194 (2001)
12. Barndorff-Nielsen, O.E., Stelzer, R.: Multivariate supOU processes. Ann. Appl. Probab. **21**(1), 140–182 (2011)
13. Benth, F.E., Rüdiger, B., Suess, A.: Ornstein-Uhlenbeck processes in Hilbert space with non-Gaussian stochastic volatility. Stochast. Processes Appl. **128**(2), 461–486 (2018)
14. Fabbri, G., Gozzi, F., Swiech, A.: Stochastic Optimal Control in Infinite Dimension. Springer, Cham (2017)
15. Griffiths, M., Riedle, M.: Modelling Lévy space-time white noises. J. Lond. Math. Soc. **104**(3), 1452–1474 (2021)

16. Kwon, Y., Lee, Y.: A second-order finite difference method for option pricing under jump-diffusion models. SIAM J. Numer. Anal. **49**(6), 2598–2617 (2011)
17. Bemš, J., Aydin, C.: Introduction to weather derivatives. Wiley Interdisciplinary Reviews: Energy and Environment, e426 (2021)
18. Fatkhullin, I., Polyak, B.: Optimizing static linear feedback: Gradient method. SIAM J. Control. Optim. **59**(5), 3887–3911 (2021)
19. Polyak, B.T., Khlebnikov, M.V.: Static controller synthesis for peak-to-peak gain minimization as an optimization problem. Autom. Remote. Control. **82**(9), 1530–1553 (2021)

Acceleration of Interval PIES Computations Using Interpolation of Kernels with Uncertainly Defined Boundary Shape

Eugeniusz Zieniuk, Marta Czupryna(✉), and Andrzej Kużelewski

Institute of Computer Science, University of Bialystok, Ciołkowskiego 1M, 15-245 Białystok, Poland
{e.zieniuk,m.czupryna,a.kuzelewski}@uwb.edu.pl

Abstract. In this paper, interval modification of degenerate parametric integral equation system (DPIES) for Laplace's equation was presented. The main purpose of such a modification was to reduce the computational time necessary to solve uncertainly defined boundary problems. Such problems were modeled with the uncertainly defined shape of the boundary using proposed modified directed interval arithmetic. The presented algorithm of mentioned interval modification of DPIES was implemented as a computer program. The reliability and efficiency tests were proceeded based on boundary problems modeled by Laplace's equation. Both, obtained solutions, as well as computational time, were analyzed. As a result, the interval kernels interpolation (using Lagrange polynomial) caused a reduction of necessary interval arithmetic calculations, which also caused accelerations of computations.

Keywords: Boundary value problems · Collocation method · Parametric integral equation system · Interval arithmetic · Uncertainty

1 Introduction

The finite element method (FEM) [1] and boundary element method (BEM) [2] are the most often used methods for solving boundary problems. Nowadays, it is very important to include the uncertainty of measurements in solving such problems. Therefore, a lot of modifications have appeared in the literature, such as stochastic [3], interval [4] or fuzzy [5]. However, the disadvantages of FEM and BEM, mainly the necessity of discretization, caused an increasing amount of calculations and extend the computation time.

Therefore, the parametric integral equation system (PIES) [6,7] (where the classical discretization was eliminated) was proposed. The advantages of this method in solving uncertainly defined problems were presented in [8,9]. These papers focused on the simplest and often used interval arithmetic [10,11] for modeling the uncertainty. Unfortunately, the direct application was time-consuming, and obtained solutions were often overestimated.

D. Groen et al. (Eds.): ICCS 2022, LNCS 13353, pp. 589–596, 2022.
https://doi.org/10.1007/978-3-031-08760-8_48

Results of degenerate PIES (DPIES) application for solving boundary problems [12] was the motivation to test its applicability for uncertainly defined problems. The mathematical formalism of DPIES allowed the development of more efficient algorithms for the numerical solution of PIES.

In this paper, interval modification of DPIES (to include the uncertainty) was presented. Uncertainty modeling is more reliable because it includes e.g. measurement errors. The proposed strategy was implemented as a computer program and tested on examples of problems described by Laplace's equation. Interval DPIES occurs to be faster than the interval PIES [6,8,9]. The additional advantage was the reduction of the number of interval calculations. This reduced not only the computational time but also the unnecessary overestimations.

2 IPIES with Boundary Shape Uncertainty

The application of interval arithmetic [10,11], to model the boundary shape uncertainty, seemed to be sensible and elementary. However, obtained solutions to so-defined boundary problems raised doubts about their reliability and usefulness. Details were discussed in [9] and the interval PIES (IPIES) [8,9] using a modification of the directed interval arithmetic was proposed. The solutions on the boundary (of two-dimensional problems modeled by Laplace's equation with boundary shape uncertainty) can be obtained by solving IPIES [9]:

$$0.5u_l(z) = \sum_{j=1}^{n} \int_{\widehat{s}_{j-1}}^{\widehat{s}_j} \{\boldsymbol{U}_{lj}^*(z,s)p_j(s) - \boldsymbol{P}_{lj}^*(z,s)u_j(s)\}\boldsymbol{J}_j(s)ds, \tag{1}$$

where $\widehat{s}_{l-1} \leq z \leq \widehat{s}_l$ and $\widehat{s}_{j-1} \leq s \leq \widehat{s}_j$ are exactly defined in parametric coordinate system. They correspond to beginning and ending of segment of interval curve $S_m = [\boldsymbol{S}_m^{(1)}, \boldsymbol{S}_m^{(2)}]$ $(m = j, l)$ which model uncertainly defined boundary shape. The function $\boldsymbol{J}_j(s) = [\underline{J}_j(s), \overline{J}_j(s)]$ is the Jacobian of interval curve segment $\boldsymbol{S}_j(s)$. Functions $\boldsymbol{U}_{lj}^*(z,s) = [\underline{U}_{lj}^*(z,s), \overline{U}_{lj}^*(z,s)], \boldsymbol{P}_{lj}^*(z,s) = [\underline{P}_{lj}^*(z,s), \overline{P}_{lj}^*(z,s)]$ are interval kernels defined as:

$$\boldsymbol{U}_{lj}^*(z,s) = \frac{1}{2\pi} ln\left(\frac{1}{[\boldsymbol{\eta}_1^2 + \boldsymbol{\eta}_2^2]^{0.5}}\right), \quad \boldsymbol{P}_{lj}^*(z,s) = \frac{1}{2\pi} \frac{\boldsymbol{\eta}_1 \boldsymbol{n}_1(s) + \boldsymbol{\eta}_2 \boldsymbol{n}_2(s)}{\boldsymbol{\eta}_1^2 + \boldsymbol{\eta}_2^2}, \tag{2}$$

where $\boldsymbol{n}_1(s) = [\underline{n}_1(s), \overline{n}_1(s)], \boldsymbol{n}_2(s) = [\underline{n}_2(s), \overline{n}_1(s)]$ are interval components of normal vector $n = [n_1(s), n_2(s)]^T$ to interval segment $\boldsymbol{S}_j$. Kernels include the boundary shape uncertainty by the relationship between interval segments $S_m (m = l, j = 1, 2, 3, ..., n)$ (defined in the Cartesian coordinate system):

$$\boldsymbol{\eta}_1 = \boldsymbol{S}_l^{(1)}(z) - \boldsymbol{S}_j^{(1)}(s), \quad \boldsymbol{\eta}_2 = \boldsymbol{S}_l^{(2)}(z) - \boldsymbol{S}_j^{(2)}(s), \tag{3}$$

where $S_m (m = j, l)$ are segments of interval closed curves.

Integral functions $p_j(s), u_j(s)$ are parametric boundary functions defined on segments $\boldsymbol{S}_j$. One of these functions is defined by boundary conditions, then the other is searched by numerical solution of IPIES. The boundary conditions were defined exactly (without uncertainty) to unambiguously analyze the influence of interval kernels interpolation.

3 Interval Degenerated Kernels

The direct application of the collocation and Galerkin methods to solve PIES was presented in [7,13]. The collocation method was easily applied but produced less accurate solutions. The Galerkin method was more accurate but needed more computational time. Therefore, in this work, to accelerate the calculations, the interval kernels in IPIES were replaced by degenerate ones. The literature provides methods (without the uncertainty), such as Fourier or Taylor series expansion [14,15] to obtain such kernels. However, in this paper, a new strategy using Lagrange polynomials was proposed.

3.1 Generalized Lagrange Polynomials

The form (1) was used for $l = j$ because of the kernels singularity. Interpolation was used only beyond the main diagonal ($j \neq l$). The application of generalized Lagrange interpolation (for functions of two variables) was very simple, because the kernels $\boldsymbol{X}^*_{lj}(z,s)(\boldsymbol{X} = \boldsymbol{U}, \boldsymbol{P})$ already were defined in a unit square using normalized parameters $0 \leq z, s \leq 1$:

$$\boldsymbol{X}^*_{lj}(z,s) = \sum_{a=0}^{p-1}\sum_{b=0}^{m-1} \boldsymbol{X}^{(ab)}_{lj} L^{(a)}_l(z) L^{(b)}_j(s), \tag{4}$$

$$L^{(a)}_l(z) = \prod_{k=0,k\neq a}^{p-1} \frac{z - z^{(k)}}{z^{(a)} - z^{(k)}}, \quad L^{(b)}_j(s) = \prod_{k=0,k\neq b}^{m-1} \frac{s - s^{(k)}}{s^{(b)} - s^{(k)}}, \tag{5}$$

and interval values of $\boldsymbol{X}^{(ab)}_{lj} = \boldsymbol{X}_{lj}(z^{(a)}, s^{(b)})$ were easily determined from the formulas (2) at the interpolation points $z^{(a)}, s^{(b)}$. These points were defined by roots of Chebyshev polynomials, to avoid Runge's phenomenon. The number of such nodes and their distribution determine the interpolation accuracy. Only $\boldsymbol{X}^{(ab)}_{lj}$ was defined as intervals, so the amount of interval data was significantly reduced. Substituting (4) to the (1) the degenerate interval parametric integral equation system (DIPIES) (for $l \neq j$) was obtained:

$$\begin{aligned} 0.5u_l(z) = \sum_{j=1}^{n} \Bigg\{ &\sum_{a=0}^{p-1}\sum_{b=0}^{m-1} \boldsymbol{U}^{(ab)}_{lj} L^{(a)}_l(z) \int_0^1 L^{(b)}_j(s) p_j(s) \\ -&\sum_{a=0}^{p-1}\sum_{b=0}^{m-1} \boldsymbol{P}^{(ab)}_{lj} L^{(a)}_l(z) \int_0^1 L^{(b)}_j(s) u_j(s) \Bigg\} \boldsymbol{J}_j(s) ds, \end{aligned} \tag{6}$$

where $l = 1, 2, ..., n$ and for $l = j$ general PIES (1) was used.

The separation of the variables made it possible to move the Lagrange polynomials $L^{(a)}_l(z)$ and interval values $\boldsymbol{P}^{(ab)}_{lj}, \boldsymbol{U}^{(ab)}_{lj}$ outside the integral. The uncertainty of the integrand is determined only by Jacobian $\boldsymbol{J}_j(s)$. Other values (polynomials $L^{(b)}_j(s)$ and unknown functions $u_j(s)$ or $p_j(s)$) are exactly defined (without uncertainty).

3.2 Numerical Solution

Solution of (6) is to find unknown functions $p_j(s)$ or $u_j(s)$. They were approximated by $\widetilde{p}_j$ or $\widetilde{u}_j$ series using Chebyshev polynomials as base functions $f_j^{(k)}$:

$$\widetilde{p}_j(s) = \sum_{k=0}^{M-1} p_j^{(k)} f_j^{(k)}(s), \quad \widetilde{u}_j(s) = \sum_{k=0}^{M-1} u_j^{(k)} f_j^{(k)}(s), \qquad j = 1, ..., n. \tag{7}$$

The collocation method was applied to determine the unknown $u_j^{(k)}$ and $p_j^{(k)}$. In this method, the Eq. (6) is written in the so-called collocation points $z^{(c)}$, where $s_{l-1} < z^{(c)} < s_l$. Substituting approximating series (7) to (6) degenerate IPIES (for $l \neq j$) is presented as:

$$0.5 \sum_{k=0}^{M} u_l^{(k)} f_l^{(k)}(z^{(c)}) = \sum_{j=1}^{n} \left\{ \sum_{a=0}^{p-1} \sum_{b=0}^{m-1} \boldsymbol{U}_{lj}^{(ab)} L_l^{(a)}(z^{(c)}) \int_0^1 L_j^{(b)}(s) \sum_{k=0}^{M} p_j^{(k)} f_j^{(k)}(s) \right.$$
$$\left. - \sum_{a=0}^{p-1} \sum_{b=0}^{m-1} \boldsymbol{P}_{lj}^{(ab)} L_l^{(a)}(z^{(c)}) \int_0^1 L_j^{(b)}(s) \sum_{k=0}^{M} u_j^{(k)} f_j^{(k)}(s) \right\} \boldsymbol{J}_j(s) ds, \tag{8}$$

where $l = 1, 2, ..., n$ and for $l = j$ the series (7) were used in IPIES (1).

The DIPIES presented at the collocation points $z^{(c)}$ can be obtained (in an explicit form) for any boundary problem. The values obtained by multiplying the Lagrange polynomial $L_l^{(a)}(z^{(c)})$ by the interval values $\boldsymbol{U}_{lj}^{(ab)}$ or $\boldsymbol{P}_{lj}^{(ab)}$ create a vertical vector, whereas, the values obtained by calculating the integrals create a horizontal vector. Therefore, using k collocation points on all n segments, a two $n \times k$ - dimensional vectors are obtained.

Even theoretically, such a strategy has a significant advantage over IPIES. Till now, each element of the matrix required computing integrals with interval kernels. In proposed DIPIES, only the elements on the diagonal are calculated directly from (1). Outside the main diagonal, they are calculated as the product of two previously obtained vectors. This is a much less time-consuming operation than interval integration. Additionally, the number of interval arithmetic calculations (necessary to obtain the matrix coefficients) was significantly reduced.

4 Tests of the Proposed Strategy

The solutions' obtained using proposed DIPES are compared with those obtained using IPIES. A different number of interpolation nodes (for the $\boldsymbol{U}_{lj}$ and $\boldsymbol{P}_{lj}$ kernels) was considered. Intel Core i5-4590S with 8 GB RAM with MS Visual Studio 2013 (version: 12.0.21005.1 REL) compiler on Windows 8.1 64-bit system was used during tests. Although the average time from 100 runs of the algorithm was presented, the again obtained values (even for the same nodes number) can differ slightly. The exact interpolation of kernel P for adjacent segments occurred to be troublesome, so it was also obtained by classical integration (1). The examples were solved using three collocation points on each segment.

4.1 Elementary Example with an Analytical Solution

The elementary problem (presented in Fig. 1) was considered to confirm the correctness of the obtained solutions. The exactly defined analytical solution is $u = 100y$. Table 1 presents chosen solutions and the average relative error between IPIES and DIPIES solutions (calculated separately for the lower and upper bound of the interval). The average time necessary to calculate the elements of the G matrix (U kernel) and H matrix (P kernel) are also presented. All solutions were obtained at 20 points in the cross-section, where $x = 0.75$ and y changes from 0 to 1 (dashed line in Fig. 1).

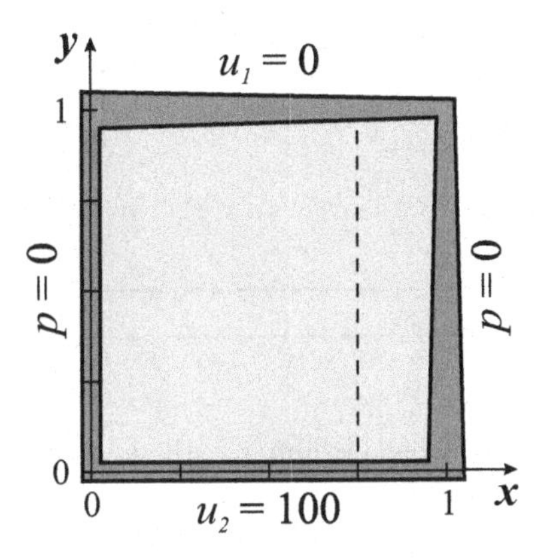

Fig. 1. The problem example with shape uncertainty.

Table 1. Solutions in the domain (computational time and the average relative error).

$100y$	IPIES	DIPIES ($U \times P$)			
		3×2	2×3	3×3	3×5
20	[18.87, 20.92]	[19.07, 21.16]	[18.89, 21.12]	[18.82, 20.89]	[18.83, 20.89]
40	[40.36, 39.70]	[40.27, 39.66]	[40.25, 39.81]	[40.34, 39.70]	[40.34, 39.69]
60	[61.60, 58.65]	[61.44, 58.52]	[61.50, 58.71]	[61.62, 58.67]	[61.62, 58.67]
80	[82.55, 77.78]	[82.46, 77.70]	[82.37, 77.70]	[82.58, 77.82]	[82.58, 77.82]
average relative error [%]		[1.65, 0.98]	[1.02, 0.80]	[0.13, 0.07]	[0.08, 0.06]
time U [ms]	10.91	3.11	2.85	3.19	3.43
time P [ms]	6.64	4.76	4.93	5.17	5.86

The number of interpolation nodes was defined as $U \times P$. For example, 3×2 means the 3 nodes for U kernel and 2 for P kernel. Exact analytical solutions are located inside all of the interval solutions. Despite the elementary example, the DIPIES method occurs to be much faster than IPIES (especially for the U kernel) with the average relative error lower than 0.1%.

4.2 Example of a Problem with a Complex Shape (40 Segments)

The shape of the boundary in the next example was defined using 40 segments. The considered uncertainly defined shape with the boundary conditions is presented in Fig. 2. Solutions of such a problem were obtained in the cross-section (dashed line in Fig. 2), where $y = 3$ and x changes from 1 to 20.

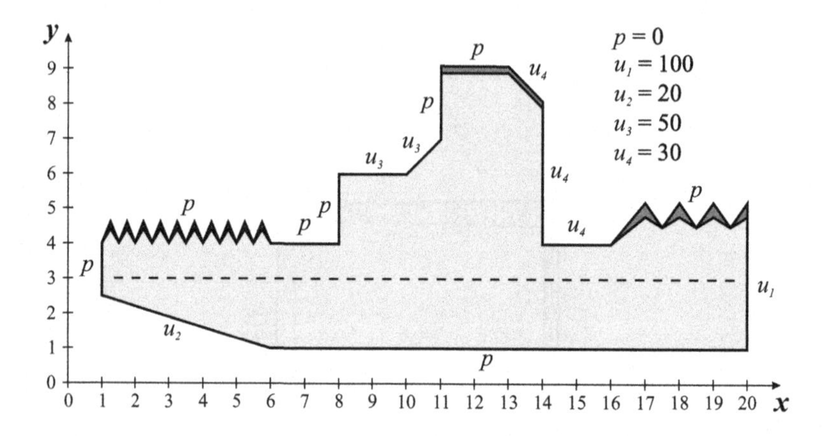

Fig. 2. The problem example with shape uncertainty.

The computational time and the average relative error of the solutions are presented in Table 2. Even three nodes are sufficient, for interpolation of U kernel, to obtain a solution with the error of 0.03%. Unfortunately, despite the exclusion of P kernel interpolation for adjacent segments, similar accuracy was obtained using eight nodes. However, even then, the DIPIES is faster than IPIES.

Table 2. Solutions in the domain (computational time and the average relative error).

	IPIES	DIPIES ($U \times P$)				
		2×8	3×3	3×5	3×8	5×8
time U [ms]	983.08	53.34	78.85	79.56	76.22	151.75
time P [ms]	625.23	491.2	125.78	226.27	502.51	459.12
average relative error [%]		[1.67, 1.68]	[0.64,0.91]	[0.29, 0.35]	[0.04, 0.03]	[0.05, 0.06]

5 Conclusions

The paper presents DIPIES, obtained by replacing kernels in IPIES with degenerate ones. The generalized Lagrange interpolation was used to obtain such kernels. The accuracy of the interpolation is determined by the number of nodes and their arrangement. Moreover, for the interpolation of the U kernel, a smaller number of nodes was sufficient compared to the P kernel. Therefore, the P kernel interpolation was used only for not adjacent segments. The effectiveness of the strategy was tested on the example of problems (modeled by Laplace's equation) with an uncertainly defined boundary shape. Two examples were solved. The results were compared with the analytical and numerical solutions (obtained by IPIES). Obtained results, present a high potential of the method. The application of degenerate kernels in the IPIES reduced the number of interval arithmetic operations and accelerated the calculations with satisfactory solutions accuracy.

References

1. Hughes, T.J.R.: The Finite Element Method: Linear Static and Dynamic Finite Element Analysis. Prentice-Hall, Englewood Cliffs (1987)
2. Brebbia, C.A., Telles, J.C.F., Wrobel, L.C.: Boundary Element Techniques: Theory and Applications in Engineering. Springer-Verlag, New York (1984). https://doi.org/10.1007/978-3-642-48860-3
3. Kaminski, M.: Stochastic boundary element method analysis of the inter face defects in composite materials. Compos. Struct. **94**, 394–402 (2012)
4. Ni, B.Y., Jiang, C.: Interval field model and interval finite element analysis. Comput. Methods Appl. Mech. Eng. **360**, 112713 (2020)
5. Behera, D., Chakraverty, S.: fuzzy finite element based solution of uncertain static problems of structural mechanics. Int. J. Comput. Appl. **69**(15), 6–11 (2013)
6. Kapturczak, M., Zieniuk, E., Kużelewski, A.: NURBS curves in parametric integral equations system for modeling and solving boundary value problems in elasticity. In: Krzhizhanovskaya, V.V., et al. (eds.) ICCS 2020. LNCS, vol. 12138, pp. 116–123. Springer, Cham (2020). https://doi.org/10.1007/978-3-030-50417-5_9
7. Zieniuk, E., Szerszen, K., Kapturczak, M.: A numerical approach to the determination of 3D stokes flow in polygonal domains using PIES. In: Wyrzykowski, R., Dongarra, J., Karczewski, K., Waśniewski, J. (eds.) PPAM 2011. LNCS, vol. 7203, pp. 112–121. Springer, Heidelberg (2012). https://doi.org/10.1007/978-3-642-31464-3_12
8. Zieniuk, E., Czupryna, M.: The strategy of modeling and solving the problems described by Laplace's equation with uncertainly defined boundary shape and boundary conditions. Inf. Sci. **582**, 439–461 (2022)
9. Zieniuk, E., Kapturczak, M.: Interval arithmetic in modeling and solving Laplace's equation problems with uncertainly defined boundary shape. Eng. Anal. Bound. Elem. **125**, 110–123 (2021)
10. Moore, R.E.: Interval Analysis. Prentice-Hall, New York (1966)
11. Markov, S.M.: On directed interval arithmetic and its applications. J. Univers. Comput. Sci. **1**(7), 514–526 (1995)
12. Zieniuk, E., Kapturczak, M., Kużelewski, A.: Degenerate parametric integral equations system for Laplace equation and its effective solving. Lect. Not. Eng. Comput. Sci. **II**, 819–824 (2017)
13. Bołtuć, A., Zieniuk, E.: The application of the Galerkin method to solving PIES for Laplace's equation. AIP Conf. Proc. **1738**, 480097 (2016)

14. Chen, J., Hsiao, C., Leu, S.: A new method for Stokes problems with circular boundaries using degenerate kernel and Fourier series. Int. J. Numer. Meth. Eng. **74**, 1955–1987 (2008)
15. Jozi, M., Karimi, S.: Degenerate kernel approximation method for solving Hammerstein system of Fredholm integral equations of the second kind. J. Math. Model. **4**(2), 117–132 (2016)

Uncertainty Occurrence in Projects and Its Consequences for Project Management

Barbara Gładysz and Dorota Kuchta(✉)

Wroclaw University of Science and Technology, Wybrzeże Wyspiańskiego 27, 50-370 Wrocław, Poland
dorota.kuchta@pwr.edu.pl

Abstract. Based on a survey (with 350 respondents), the occurrence of uncertainty, defined as incomplete or imperfect knowledge in the project planning or preparation stage, was described and quantified. The uncertainty with respect to customer expectations, project result, methods to be used, duration and cost of project stages, and (both human and material) resources was considered, and its consequences for project management and entire organizations were analysed. The results show that the scope of uncertainty in projects cannot be neglected in practice and requires the use, in the project planning or preparation stage, and during the whole project course, of advanced project and uncertainty management methods. The questionnaire used in the document is recommended to be applied to organizations in order to measure and track the scope of uncertainty in the projects being implemented and adopt a tailored approach to uncertainty management. Agile approaches seem to be highly recommendable in this regard. Future research directions are proposed, including the application of a special type of fuzzy numbers to project management.

Keywords: Knowledge about project · Project success · Customer requirements · Project uncertainty · Project planning

1 Introduction

Uncertainty, along with risk, has been the subject of research for many years and has been defined in the literature in various ways, also in the context of project management (see, e.g., [1–3]). This is so because risk and uncertainty (independently of the specific definition adopted) are fundamental issues in project management, addressed both by researchers [4, 5] and by professional project management standards [6]. The relatively high rate of project failures, together with an analysis of their causes [7–9], indicates that project risk and uncertainty management should be the focus of project managers and teams. However, optimism bias [10, 11] often prevents project managers and teams from applying enough effort to this aspect of project management in the project planning (or preparation) stage. As a result, project teams begin to work on projects using unrealistic project plans or an erroneous image of what the project is aiming at. At the same time, they are unaware of their lack of knowledge, which inevitably leads to project failure.

D. Groen et al. (Eds.): ICCS 2022, LNCS 13353, pp. 597–610, 2022.
https://doi.org/10.1007/978-3-031-08760-8_49

Therefore, the objective of this paper is to show that optimism is often not justified because uncertainty in the project occurs fairly often in practice. It is pretty common that in the project planning or preparation stage, the information about the project is uncertain, and therefore the plans and the expected consequences of the projects for the organisations must be treated with the necessary caution. This statement will be proved with respect to various manifestations of project uncertainty. On top of that, the objective is to propose a tool that can be used to measure and quantify the degree of uncertainty (in its various manifestations) in projects that have already been realized in a given organization, which may be helpful to improve the organizational project management process.

Similar research is presented in [3], but in that approach, the uncertainty was measured without considering the differences between the knowledge (about issues that must be known in the project planning or preparation phase) before the project starts and after the project ends. In our opinion, this aspect is an important measure of uncertainty because a piece of information (e.g., on a project parameter) available in the project planning phase can be seen as certain only if it does not change with time in a substantial way.

In this paper, we adopt the approach of [1, 12]: we define uncertainty as the state of not knowing for sure, and place project uncertainty in the framework of seven 'Ws' questions asked with respect to projects in [12]: Who? (who are the parties involved?); Why (what do the parties want to achieve?)? What? (what is the deliverable product the parties are interested in?); Which way (how will the plans in each lifecycle stage deliver what is needed?); Wherewithal (what key resources are needed to achieve execution of the plans?); When? (when do all relevant events have to take place?); Where? (where will the project take place?). Seen in this way, uncertainty means that we do not have full knowledge at the project planning or preparation stage of how to answer the respective questions correctly and precisely. Two cases will be referred to: one when, in the given moment, we are aware of our lack of knowledge, and the other one when we will be aware of it only in the future. Our research refers specifically to the aspects What, Which way, and Wherewithal.

The outline of the remainder of this paper is as follows: In Sect. 2, we explain our research methods. As we chose to measure the knowledge present among the project team in two steps – one referring exclusively to the period before the project started and one referring to the differences between the situation before the project started and the actual outcomes – the results of the questionnaires are analysed in two respective steps. Thus, in Sect. 3, we present the results of basic statistical analyses referring to the knowledge about the project in the period before its start, and in Sect. 4, we discuss the results of basic statistical analyses referring to the differences in knowledge before the project started and after the completion of the project. In Sect. 5, we present a discussion and some conclusions.

2 Research Methods

2.1 Measurement of Knowledge Available to the Project Team

The tool used was a questionnaire measuring the level of knowledge about selected project elements present among the project team. We had to decide how to measure this aspect in a way that would allow us to expect the majority of respondents to give reliable answers. We selected the following measures:

a) Regarding the time before the project start: the degree of agreement among project team members as to the expectations of the customer and the expected project result. If project members were not in agreement as to what the customer expected or what the project result should have been, it is legitimate to conclude that they did not really understand the project and the proper way of implementing it. Therefore, they did not really know what they were supposed to do.
b) Regarding the time before the project start: the assessed level of knowledge of individual project features (like cost, duration, etc.) present among the project team. If the respondents did not think that they knew the respective features of the project well before the project started, it is also legitimate to conclude that they did not really know how to implement the project properly.
c) Regarding the time after project completion: the end-start differences in customer expectations. If, after the project had ended, it turned out that the customer had different expectations than those expressed before the project started, it would be difficult to claim that the project team knew what it meant to terminate the project with success.
d) Regarding the time after project completion: the end-start differences in various project parameters and features. If the actual project realisation was substantially different from the project plan (with respect to time, cost, resources, etc.), it would be difficult to claim that the project team knew what to do to bring the project to success.

2.2 Questionnaire-Based Survey

The primary research method used was interviews structured as presented in Sect. 2.1. The principal goal of the interviews was to find out whether and to what degree, the project team knew before the projects started what the ultimate customer expectations were and what exactly to do to satisfy them. The interviews were conducted in 2020. A total of 350 Polish companies/organisations were surveyed, including 118 micro, 131 small, 71 medium, and 30 large ones (Table 1). The respondents were selected among project managers or heads of the companies/organisations.

Table 1. Size of organisations from the research.

Company/Organisation size	Number	Percentage
Micro (below 10 people)	118	33.7
Small (10–49 people)	131	37.4
Medium (50–249 people)	71	20.3
Large (over 250 people)	30	8.6
Total	350	100.0

The CATI [13] method was used for the interviews.

In an attempt to characterise, at least to a certain degree, the organisations represented by the respondents, one question was asked about the average duration of the projects implemented by the organisation; this could range from several days to several years (Table 2).

Table 2. Average project duration in the organization.

Duration class	A few days	Month	Quarter	Half-year	Year	A few years	Total
Percentage	15.9	13.9	17.6	18.2	19.3	15.2	100

Another question characterised the type of project management applied in the organisations. Three options were proposed [14]:

- traditional (waterfall) project management;
- agile/extreme project management;
- a combination of the two.

It turned out that all three types of project management were used in all the organisations represented in the survey. However, specific types of project management approaches prevailed in individual organisations. In 32% of the organisations, the most frequently used management type was the traditional approach, in 7%, the agile/extreme one, and in 11% a combination of traditional/agile.

In the next stage of the interview, an attempt was made to determine the knowledge present among the project team. As indicated above, we decided to investigate this problem in two steps (one referring to the time before the project start and one referring to the differences between project start and project end) through four types of information we hoped to gather:

I. the indicator (for the time before the project started) of agreement among the project team as to what the customer expected or what the project result should be;
II. the assessed level of knowledge (for the time before the project started) of the project team about individual project features (cost, duration, etc.);

III. project end-start difference in customer expectations;
IV. project end-start difference in individual project features.

Let us now present the details of the questionnaire used in parts I, II, III, and IV, together with the variables and scales used.

Part I (differences of opinion among members of the project team before the project started) consisted of two question groups:

- differences of opinion among project team members as to the expectations of the customer (variable D_ClientExp);
- differences of opinion within the project team as to the outcome of the project (variable D_Results).

The respondents were asked to assess the differences on a Likert scale, with the following meanings: 1 - practical lack of differences; 2 - very small differences; 3 - small differences; 4 - medium differences; 5 - large differences; 6 - very large differences; and 7 - substantial differences.

Part II (level of project team knowledge before the project started) covered the following aspects:

- knowledge of the methods and technologies needed to implement the project (K_Methods);
- knowledge of the time needed for individual project stages (K_Time);
- knowledge of the costs of implementing individual project stages (K_Cost);
- knowledge of the appropriate quantity of human resources (K_PeopleNo);
- knowledge of the needed human resource competencies (K_PeopleComp);
- knowledge of material resources needed (K_MatRes).

The assessments were made on the same 7-level Likert scale, with the following meanings: 1 - almost none; 2 - very small; 3 – small; 4 – medium; 5 – large; 6 - very large; and 7 - complete. In each case, when the answer was four or less (i.e., the level of knowledge was medium or lower), the respondents were asked whether lack of knowledge was the reason for disagreement or conflicts in the project team. There were two possible answers: yes or no.

In part III, the project managers defined the differences between the expectations of the customer at the beginning and the end of the project (D_StartFinish). These differences were assessed on the same 7-point Likert scale as in Part I.

Part IV, devoted to the differences between selected project features in the planning stage of the project and after its completion, was applied only with respect to the projects for which the responses "medium" or more were given in part II. We asked about the following differences:

- between the planned and achieved results (D_StartFinish_Results);
- between the methods or technologies planned for use in the project and those actually used (D_StartFinish_Methods);
- between the planned and actual cost of individual tasks (D_StartFinish_Cost);

- between the planned and actual quantity of human resources needed for project tasks (D_PeopleNo_StartFinish);
- between the planned and actual human resources competencies needed to perform project tasks (D_StartFinish_PeopleComp);
- the planned and actual quantity of material resources needed to perform project tasks (D_StartFinish_MatRes).

These differences were assessed on the same 7-point Likert scale as in Part I and III. To analyse the data retrieved from the surveys, we applied the software package SPSS (https://spss.pl/spss-statistics/).

3 Results of Basic Statistical Analyses Related to Project Start (Parts I and II of the Questionnaire)

3.1 Differences of Opinion Among the Project Team Regarding Customer Expectations and Expected Project Result (Part I)

Respondents were asked about the differences of opinion concerning customer expectations (D_ClientExp) and what the project result (D_Results) should be. The highest values on the Likert scale corresponded to the highest differences and the lowest values to the lowest ones. The distributions of both variables (D_ClientExp, D_Results) are illustrated in Table 3.

Table 3. Descriptive statistics of variables D_ClientExp and D_Results.

Variable	Mean	Median	Mode	Variance	Differences	
					1,2,3	4,5,6
D_ClientExp	2.71	2.00	1.00	2.96	69%	31%
D_Results	2.75	2.00	1.00	2.74	70%	30%

The two probability distributions of the differences of opinion within the project team are right-handed. No outliers were identified. Estimators of the expected values of opinion differences in both analysed dimensions (D_ClientExp, D_Results) are equal to approx. 2.7, and of the variance to approx. 2.8. Median Me = 2, which in the Likert scale means very small differences. So, the level of uncertainty in the aspect considered here may seem not to be very high: in about 70% of the projects, the differences of opinion were almost none, small or very small. However, in about 30% of the projects, the differences of opinion here were medium, large, very large, or substantial. The last statement is very important. In about 30% of projects, the project team members were definitely not in agreement (before the project started) as to what the customer wanted and where the project was heading at. In all of these cases, it would be difficult to claim that the project team had complete knowledge about the project.

Let us now look at the assessment of the level of knowledge about project details prior to project start.

3.2 Level of Project Team Knowledge Concerning Individual Project Parameters (Part II)

The respondents were asked about the level of the project team's knowledge (during project planning) of various aspects of the project necessary for a good project plan. The higher the answer value, the higher the knowledge evaluation. The following variables were used: K_Methods (methods to be used in the project), K_Time (duration of individual project stages), K_Cost (cost of individual project stages), K_PeopleNo (number of human resources needed), K_PeopleComp (competences of human resources needed) and K_MatRes (amount of material resources needed). Descriptive statistics of the probability distributions are given in Table 4.

Table 4. Characteristics of probability distributions of the knowledge of the project team about the project before its start.

Level of knowledge	Mean	Median	Mode	Variance	Level of knowledge (Likert scale)	
					1,2,3,4	5,6,7
K_Methods	5.71	6.00	6	1.49	13%	87%
K_Time	5.64	6.00	6	1.59	13%	87%
K_Cost	5.40	6.00	6	1.77	22%	78%
K_PeopleNo	5.47	6.00	5	1.58	20%	80%
K_PeopleComp	5.65	6.00	5	1.14	12%*	88%
K_MatRes	5.59	6.00	6	1.41	15%	85%

* no "almost no knowledge" level was observed

All probability distributions are left-handed with a median equal to 6 (on the Likert scale - very high level of knowledge). The means are about 5.5, and the modes 5 or 6. So again, the first impression might be that the knowledge level with respect to project plan details was relatively high. It is essential to underline, however, that there is a considerable group of projects (between 13% and 22%) for which the level of knowledge in some aspects was medium or less. This means that, in a relatively large portion of projects, the project team did not really know how to implement the project – for example, in more than 20% of the projects examined, the level of knowledge about the cost of individual project stages could not be described as large, and the same is true for the knowledge about the needed human resources and so forth.

We also observed unusual project teams (outliers) in which the level of knowledge was very small. This is especially true for costs (5% of projects were characterised by a small or very small level of knowledge, and 4% by almost none). In terms of knowledge about methods and technologies, time of implementation of individual project stages, and appropriate quantity of human or material resources, there were several projects with the respective knowledge at the "almost none" level. This means that organisations, on some occasions, deal with projects where the knowledge of how to implement them is extremely low.

The next section is devoted to the differences between the knowledge accessible to the project's team before the project started and after it has finished.

4 Differences Between the Knowledge in the Project Planning Stage and After Project Completion (Parts III and IV of the Questionnaire)

Here we present the results regarding the differences between the knowledge accessible in project planning stage and after project completion. The occurrence of these types of differences would indicate that uncertainty was present in the project planning stage.

4.1 Differences in Customer Expectations Expressed Before the Project Start and After Project Completion (Part III)

Here we present the results regarding the differences in customer expectations as expressed prior to and at the end of the project. If they were significant, the project team would certainly have been unable to know exactly how to implement the project. The distribution of the differences between the expectations of the customer at the beginning and at the end of the projects is presented in Table 5.

Table 5. Characteristics of probability distributions of differences of customer expectations

Area	Mean	Median	Mode	Variance	Differences of opinion (the Likert scale)		
					1, 2	3, 4, 5	6, 7
D_StartFinish	1.97	2.00	1	1.54	79%	18%	3%

The probability distribution is right-handed. The estimator of the expected value of the differences in customer expectations (D_StartFinish) is 1.97, and the variance estimator is 1.54. The median Me = 2, which in the Likert scale indicates very small differences. The largest fraction of projects (79%) includes those in which the differences in the expectations of the customer at the beginning and end of the project were almost none or very small. However, if we distinguish projects in which the differences in customer expectations between the project beginning and project end were not very small, we arrive at a rather high percentage of 21%. In addition, numerous outliers were identified, where the uncertainty was much higher. Indeed, in 3% of projects (extreme outliers), there were very large or substantial differences. This means that, in a certain group of projects, the customer may change their vision (or the way of communicating it) in a substantial way, so it would be difficult to claim the project team was able to know what to do at the project beginning.

4.2 Differences Between Selected Project Plan Details and the Reality (Part IV)

The managers of the projects in which the differences in the expectations of the customer at the beginning and end of the project were not negligible (i.e., medium, large, very large, or substantial) assessed the level of differences between the expected and actual project results as well as the differences between the planned and actual values of the various project plan elements (the same ones were considered in part II of the questionnaire). The respective variables were introduced in Sect. 2.

Table 6 provides estimates of the probability distribution parameters of the respective differences.

Table 6. Statistical characteristics of the distribution of the differences between project plans and the actual outcomes.

	Median	Mode	Variance	Differences (Likert scale)	
				1,2,3	4,5,6,7
D_StartFinish_Results	3.0	2	1.71	57%	43%**
D_StartFinish_Methods	2.5	2	1.40	46%*	54%**
D_StartFinish_Time	3.0	1	1.74	40%*	60%
D_StartFinish_Cost	2.5	1	1.93	52%	48%**
D_PeopleNo_StartFinish	2.5	2	1.86	46%*	54%**
D_StartFinish_PeopleComp	2.5	2	1.72	75%*	25%**
D_StartFinish_MatRes	2.5	2	1.96	64%*	36%**

* the level "almost none," ** the level "substantial differences" was not observed

In all the analysed categories, the expected value of the differences between the different project aspects at the beginning and the end of the project is the same: about 3.6 (on a scale of 1–7), and the variance is equal to about 1.75. In 20% to 60% of the projects, the differences between the plan and the reality were at least medium. This means that, in a considerable group of projects, it was impossible that the project team could have possessed fairly good knowledge about the project.

Outliers were observed in the case of such categories as methods and techniques, time, and people's competencies. They correspond to projects with very large or substantial differences between the project plan and the actual outcomes. In such projects, complete knowledge about the project was certainly not present among the project team before the project started.

5 Discussion

The study started with the investigation of various aspects of uncertainty in the sense of lack or imperfection of knowledge. In each of the investigated aspects of uncertainty, it was shown that, although in most projects the uncertainty was not acute, in each case there was a considerable group of projects where the given uncertainty manifestation had such an intensity that it was bound to considerably reduce project success probability. The cardinality of the "problematic" group of projects depends each time on the thresholds we choose while defining such a set. For example, should it comprise only big and very big differences or medium differences too? Only almost none and a very small level of knowledge, or small and medium too? In our analysis, we took arbitrary decisions in this regard, but in each case, the decision-maker should take their own decision, depending on the level of each uncertainty manifestation they accept and the level they would prefer to avoid by using relevant project management methods. Independently of the thresholds, we identified several extreme cases - projects (outliers) where the respective aspect of uncertainty was present at the highest possible intensity level and which were certainly extremely difficult to manage, whatever the uncertainty acceptance level is chosen.

The first aspect investigated was the differences of opinion among the project team concerning the customer expectations and the project result (the element "What" from [12]). This aspect has not been investigated in the literature in the context of project uncertainty so far. As far as the investigated sample is concerned, in 30% of projects, these differences were medium or more, and in less than 30% of projects, they were judged as "practically nonexistent." The last statement means that there were some differences of opinion about the two aspects among the members of the project team in more than 70% of projects. In such a high percentage of projects, the project team was not completely unanimous as to what the project was aiming at. Such uncertainty may lead to a considerable loss of time, effort and enthusiasm. Respective methods of requirements management [16] and of making imprecise requirements precise [17] should be applied.

The next aspect taken into account was the knowledge the project team had (in their own opinion) about various project parameters and other aspects before the projects started. The following items were considered:

- Methods to be used (item Which way from [12]): here, in 13% of projects, the knowledge was judged to be medium, small, very small, or almost none, and in less than half of the projects the knowledge was seen as very high or complete. This means that in the case of more than 50% of the projects, the methods to be used were not fully known before the project started. Such situations require special, Agile-oriented management approaches [20]. It has to be underlined that two outliers were identified here: projects for which the knowledge of the methods to be used was judged as being very small;
- Duration of individual project stages (item When from [12]): the distribution is here similar to that of methods. This shows that in over 50% of projects, advanced methods of time management should be applied – without them, the work in the project and in the whole organization executing the project is difficult to schedule. Approaches like Agile management [14] or Critical chain [19] are recommended. Four outliers (with very small knowledge) were identified;
- Cost of individual project stages (item Wherewithal from [12]): here, the uncertainty appeared to be the most intense. In 22% of the projects, the knowledge in this aspect was evaluated as being less than medium, in 9% (which constituted 31 projects) as less

than small, and in 1% (which constituted four projects) as very small or practically nonexistent. In less than half of the projects, the knowledge was seen as very high or complete. Uncertainty linked to project cost may lead to serious financial problems in the project or even in the whole organization. It seems that apart from advanced methods of project cost estimation [20, 21] and Agile approaches, advanced methods of uncertainty management should be taken into account in project cost management [12];

- The number of necessary human resources (item Wherewithal from [12]): here, the uncertainty also appeared to be very intense: the knowledge in this aspect was seen as being less than medium in 20% of projects and only in about 40% of projects it was evaluated as being very high or complete. In fact, in many projects, e.g., IT projects, the project cost is closely related to the number of human resources used; thus the similarity of the situation in this and the previous aspect is not surprising. This type of uncertainty may lead to serious delays (due to the lack of necessary personnel) or additional costs (due to outsourcing necessity). Here methods of human capacities and work velocity recording [14] should be applied in order to avoid such a high level of uncertainty in planning the number of human resources. One outlier with very small knowledge was identified in this aspect;
- Necessary human competencies (item Wherewithal from [12]): here, the distribution turned out to be similar to that of methods and time. The consequences of this type of uncertainty will be similar to those identified in the area of the number of human resources. Specialized methods of human resources planning [22] would be a remedy;
- Amount of necessary material resources (item Wherewithal from [12]): in 15% of cases, the respective knowledge was less than medium, with one outlier characterized by very small knowledge. Material resources planning is very important, especially in cases when the lead time between the placement of an order and delivery is high or when time-consuming public procurement procedures are necessary. Reserves or other remedies have to be applied in order, on the one hand, to not run short of materials and on the other, not to be left with unnecessary inventory [23].

The incomplete knowledge in the planning or preparation stage results not only in a shortage or surplus of resources or the necessity to introduce big changes with respect to the project plan but may also have an influence on human relationships and thus on team spirit – which is an important project success factor [24]. In order to evaluate the scope of the negative influence of uncertainty on project spirit, we asked those respondents who indicated that the level of knowledge before the project started was, in at least one aspect, small, very small, or almost none, whether this lack of knowledge led to disagreements in the project team. The results are presented in Table 7.

In about half of the projects with incomplete knowledge, there appeared to be disagreements caused by this uncertainty type. This means that uncertainty (defined as incomplete knowledge), apart from the obvious negative influence on cost and time (due to necessary changes or incorrect amount of resources), may also deteriorate the team spirit. As the intensity of uncertainty is non-negligible, advanced methods of project team spirit evaluation and management have to be used in everyday project management practice [25].

Table 7. Projects in which, due to incomplete knowledge, there were conflicts in the project team.

Incomplete knowledge in terms of	Fraction of projects in which the incomplete knowledge resulted in conflicts
Time	45.2%
Number of human resources	42.4%
Competencies	57.5%
Material resources	56.3%

The next aspect taken into account was uncertainty (lack of knowledge) measured by the "end-start" differences in the aspects considered above. As we mentioned earlier, this problem has not been considered so far in the literature on uncertainty in projects. The project managers in whose projects the expectations of the customer changed during the time of project duration considerably (at least to the medium degree, which happened in 11% of cases) declared that at least medium differences between the final and the planned values or methods had occurred: in 60% of projects for the duration of project stages, in about 50% of projects for methods, cost and number of human resources, and in about 30–40% of projects for the other aspects. This shows that the lack of certainty observed on the customer side results in considerable uncertainty for the project team in the project planning stage. In the case of methods, time, and competencies, several outliers were observed, where the respective differences were very high.

6 Conclusions

To sum up, uncertainty, understood as the lack of knowledge is present in the planning or preparation stage of projects to a non-negligible degree. The frequency values will depend on the choice of the definition of the notion "negligible," but in any case, we will be talking about dozens of percentages of projects with substantial uncertainty. Also, the fact that outliers with extreme uncertainties were observed should not be forgotten. An outlier may seem insignificant in statistical terms but may be the opposite in practical ones: it may be a project of utmost strategic importance or high budget, where the extreme lack of certainty in the planning stage will be reflected in its failure, which in turn may even be devastating for the organization as whole (for its reputation or financial situation).

The questionnaire used in the research presented in this paper (or a similar one, a combination with that from [3] would be advisable) could be used in each organization with respect to already finished projects within a certain project type so that the uncertainty scope linked to various project aspects in the given organization (and its changes) can be determined and described. Other project uncertainty management methods should be also involved. A track of uncertainty management tools that have been used and could be used (according to [12]) should be kept, so that an adequate choice of uncertainty management methods for the organization and for the given project type can be made and continually updated.

As far as research on project uncertainty management is concerned, it would be advisable to introduce Z fuzzy numbers, or its more general version - Z* augmented fuzzy numbers [25], to the project planning process. These numbers combine the features of "traditional" fuzzy numbers [26] (applied to project management for decades [27]), which allow modeling of uncertainty and partial knowledge (e.g., uncertain duration of a project task) with the possibility to quantify the context in which the fuzzy information was given (thus, e.g., the experience and optimism degree of the person who is the information source, the measured uncertainty scope for a given project type in a given organization and the moment when the information was received), in order to obtain an adjusted representation of uncertain information.

Funding. This research was funded by the National Science Center (Poland), grant number 394311, 2017/27/B/HS4/01881 Grant title: Selected methods supporting project management, taking into consideration various stakeholder groups and using type-2 fuzzy numbers.

References

1. Baccarini, D.J.: Project Uncertainty Management. Published by Baccarinì, D.J. (2018) https://www.researchgate.net/profile/David-Baccarini
2. Perminova-Harikoski, O., Gustafsson, M., Wikström, K.: Defining uncertainty in projects–a new perspective. Int. J. Proj. Manag. **26**, 73–79 (2008)
3. Wyrozębski, P.: Risk and Uncertainty in Project Planning Process. In: Trocki, M., Bukłaha, E. (eds.) Project Management - Challenges and Research Results, pp. 73–102. Warsaw School of Economics Press, Warsaw (2016)
4. Acebes, F., Poza, D., González-Varona, J.M., Pajares, J., López-Paredes, A.: On the project risk baseline: Integrating aleatory uncertainty into project scheduling. Comput. Ind. Eng. **160**, 107537 (2021)
5. Barghi, B., Shadrokh Sikari, S.: Qualitative and quantitative project risk assessment using a hybrid PMBOK model developed under uncertainty conditions. Heliyon (2020)
6. Practice Standard for Project Risk Management. Project Management Institute, Chicago (2009)
7. Chaos Report. https://www.standishgroup.com/sample_research_files/
8. CHAOSReport2015-Final.pdf. Accessed 21 Jan 2022
9. Project Management Statistics: Trends and Common Mistakes in 2022 (2022). https://teamstage.io/project-management-statistics/. Accessed 21 Jan 2022
10. Shahriari, M., Sauce, G., Buhe, C., Boileau, H.: Construction project failure due to uncertainties—a case study (2015). https://www.scopus.com/inward/record.uri?eid=2-s2.0-84941629705&partnerID=40&md5=96e09515330cb07291ded73fa898ce7a
11. Prater, J., Kirytopoulos, K., Tony, M.: Optimism bias within the project management context: A systematic quantitative literature review. Int. J. Manag. Proj. Bus. **10**, 370–385 (2017)
12. Wang, J., Zhuang, X.,Yang, J., Sheng, Z.: The effects of optimism bias in teams. Appl. Econ. **46**, 3980-3994 (2014)
13. Chapman, C., Ward, S.: How to Manage Project Opportunity and Risk. John Wiley and Sons Ltd. (2011)
14. Lavrakas, P.: Computer-assisted telephone interviewing (CATI). In: Proceedings of the Encyclopedia of Survey Research Methods (2013)
15. Wysocki, R.K., Kaikini, S., Sneed, R.: Effective project management : Traditional, agile, extreme (2014). http://site.ebrary.com/id/10814451

16. Pinto, J.K., Slevin, D.P.: Project success : Definitions and measurement techniques. Proj. Manag. J. (1988)
17. Coventry, T.: Requirements management – planning for success! techniques to get it right when planning requirements. In: Proceedings of PMI® Global Congress 2015—EMEA. Project Management Institute, London (2015)
18. Kaur, R., Bhardwaj, M., Malhotra, N.: Analyzing Imprecise Requirements Using Fuzzy Logic. Int. J. Eng. Res. Technol. **1**, (2012)
19. Turner, J.R., Cochrane, R.A.: Goals-and-methods matrix: Coping with projects with ill defined goals and/or methods of achieving them. Int. J. Proj. Manag. **11**, 93–102 (1993). https://doi.org/10.1016/0263-7863(93)90017-H
20. Correia, F., Abreu, A.: An overview of critical chain applied to project management. In: Proceedings of the International Conference on Manufacturing Engineering Quality Production Systems MEQAPS, pp. 261–267 (2011)
21. Boehm, B., Abts Ch.: Software cost estimation with COCOMO II, 1st edn. Prentice Hall PTR, Upper Saddle River (2000)
22. Practice Standard for Project Estimating. Project Management Insitutute, Chicago (2011)
23. Project Management Body of Knowledge (PMBOK® Guide). Project Management Insitutute, Chicago (2008)
24. Lock, D.: Project Management, 10th edn. Gower Publishing Limited, Farnham (2013)
25. Aronson, Z., Lechler, T., Reilly, R., Shenhar, A.: Project spirit-a strategic concept. 1, (2001)
26. Carmeli, A., Levi, A., Peccei, R.: Resilience and creative problem-solving capacities in project teams: A relational view. Int. J. Proj. Manag. **39**, 546–556 (2021)
27. Zadeh, L.A.: The concept of a linguistic variable and its application to approximate reasoning-I. Inf. Sci. (Ny). (1975)
28. Shipley, M.F., de Korvin, A., Omer, K.: BIFPET methodology versus PERT in project management: fuzzy probability instead of the beta distribution. J. Eng. Technol. Manag. **14**, 49–65 (1997)

Getting Formal Ontologies Closer to Final Users Through Knowledge Graph Visualization: Interpretation and Misinterpretation

Salvatore Flavio Pileggi(✉)

School of Computer Science, Faculty of Engineering and IT,
University of Technology Sydney, Sydney, NSW, Australia
SalvatoreFlavio.Pileggi@uts.edu.au

Abstract. Knowledge Graphs are extensively adopted in a variety of disciplines to support knowledge integration, visualization, unification, analysis and sharing at different levels. On the other side, Ontology has gained a significant popularity within machine-processable environments, where it is extensively used to formally define knowledge structures. Additionally, the progressive development of the Semantic Web has further contributed to a consolidation at a conceptual level and to the consequent standardisation of languages as part of the Web technology. This work focuses on customizable visualization/interaction, looking at Knowledge Graphs resulting from formal ontologies. While the proposed approach in itself is considered to be scalable via customization, the current implementation of the research prototype assumes detailed visualizations for relatively small data sets with a progressive detail decreasing when the amount of information increases. Finally, issues related to possible misinterpretations of ontology-based knowledge graphs from a final user perspective are briefly discussed.

Keywords: Ontology · Knowledge graph · Data visualization · Uncertainty

1 Introduction

Knowledge Graph (or Semantic Network) has recently gained popularity in a variety of contexts and disciplines where it is ubiquitously used without a commonly accepted well-established definition [9]. An effective use of Knowledge Graphs is considered to be critical as it provides structured data and factual knowledge in scope [23].

In the more specific field of knowledge representation and reasoning [7], a knowledge graph is understood as a graph-structured representation of a given knowledge base composed of interrelated entities and associated semantics. It is considered an effective framework for knowledge integration, visualization, unification, analysis and sharing at different levels. Knowledge graph relevance

D. Groen et al. (Eds.): ICCS 2022, LNCS 13353, pp. 611–622, 2022.
https://doi.org/10.1007/978-3-031-08760-8_50

is evident in literature as it is widely adopted in the context of different disciplines and applications (e.g. recommendation systems [35] or education [6]) along a number of issues and challenges, such as, among others, identification [31], embedding [34], refinement [24], construction [19] and learning [18]. On the other side, Ontology has gained a significant popularity within machine-processable environments [14], where it is extensively used to formally define knowledge structures. Additionally, the progressive development of the Semantic Web [4] has further contributed to a consolidation at a conceptual level and to the consequent standardisation of languages (e.g. RDF/RDFS [22], OWL [1], SPARQL [2]) as part of the Web technology. Knowledge graphs are not necessarily specified by using formal standard languages. However, it becomes a compelling need wherever the target knowledge is dynamically understood as part of interoperable and re-usable environments. Additionally, ontological structures are a de facto requirement in many practical cases (e.g. [5,8]).

This work focuses on customizable visualization/interaction, looking at Knowledge Graphs resulting from formal ontologies. As concisely discussed later on in the paper, the most common approaches for visualization work manly at an ontology level, focusing on the ontology schema. That is normally considered to be effective to meet the needs of ontology developers and, more in general, to overview the knowledge structure. However, Ontology and Knowledge Graph are in general two well different concepts as the former is commonly understood as a formal specification of the knowledge space, while the latter is a representation as a graph of the same knowledge space that should meet visualization requirements from a final user perspective. In other words, ontology works mostly on the machine side and Knowledge Graph provides a domain or application specific interface for final users. Therefore, there is a practical gap between Ontology and Knowledge Graph visualization that quite often forces an ad-hoc approach. This work aims to reduce such a gap through a standard, yet customizable, strategy that presents ontologies as application-level Knowledge Graphs. It ultimately wants to get a given knowledge space as close to final users as possible. Additionally, interaction capabilities are empowered by a smooth integration within the visualization environment of formal query and informal search functionalities. While the proposed approach in itself is considered to be scalable via customization, the current implementation of the research prototype assumes detailed visualizations for relatively small data sets with a progressive detail decreasing when the amount of information increases.

The paper follows with a related work section, which concisely addresses ontology visualization. The approach provided and the implementation of the current prototype is proposed later on, while Sect. 4 discusses the interpretation of ontology-based knowledge graphs from a final user perspective. As usual, the paper ends by summarising the contribution looking at possible future work.

2 Related Work

Ontology visualization methods and tools have been extensively discussed in the recent past [15]. Major difficulties are related to the descriptive nature of ontologies, which normally provide vocabularies to dynamically build knowledge according to a non-prescriptive approach. Moreover, ontologies are normally specified by using rich data models. In the specific case of Web ontologies developed upon OWL, we can distinguish at least three different conceptual sub-sets: the TBox is normally associated with the ontology schema (concepts/classes and relationships), the ABox is commonly related to instances of concepts and, finally the RBox includes inference rules and relational axioms. In addition, metadata can be associated with entities (e.g. through annotation properties in OWL).

An exhaustive overview of those solutions is out of the scope of this paper. However, looking at the different techniques, two main class of solution can be identified as follows:

- *Indented list.* It is a very popular method among developers and, therefore, within ontology editors. For instance, Protege [11], which is the most famous ontology editor by far, adopts such a technique to provide simple and effective browsing of the different components, i.e. classes, properties and individuals ((example in Fig. 1). While such an approach can be useful also to final users in a variety of possible applications and situation, it is in principle designed for ontology developers and it is not suitable as a primary technique to interact with knowledge graphs. The most common approach is currently to integrate indented list with other visualization techniques (e.g. through plug-ins in Protege [33]).

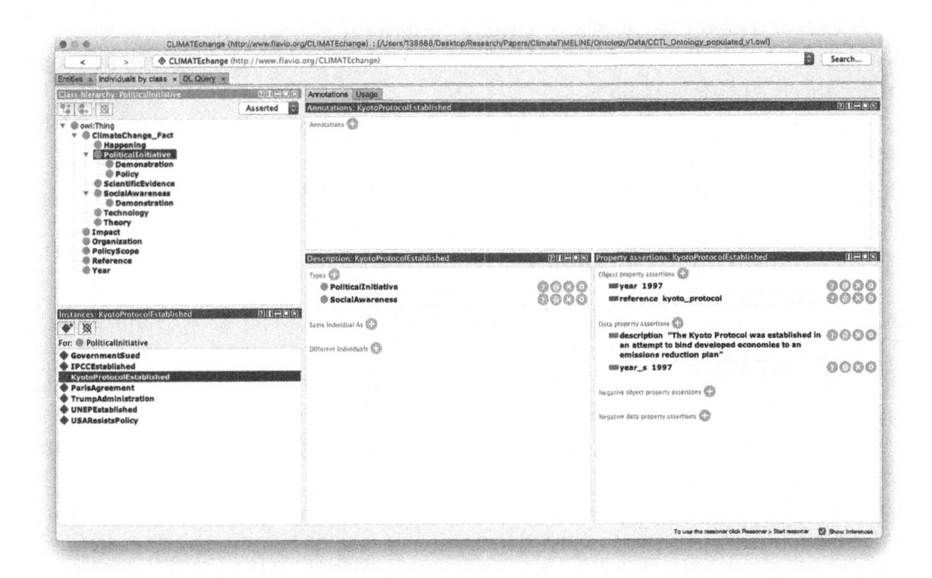

Fig. 1. Visualization of Climate Change TimeLine ontology [30] in Protege.

- *Graph-based visualization.* Methods based on graph visualization [16] are considered to be quite effective and are, therefore, extensively used in a number of different contexts. An interesting comparison of available tools from a practical perspective is proposed in [12]. A well-known visualization tool is VOWL [20] which, like most tools, focuses mainly on the ontology schema to provide an overview of the ontology structure. Populated ontologies are more rarely addressed (e.g. [3]).

The solution proposed in the paper aims to a flexible and customizable visualization of knowledge graphs underpinned by formal ontologies.

3 Ontology-Based Knowledge Graph Visualizer (OB-KGV)

This section aims to provide an overview of the tool developed from a final user perspective. *Ontology-Based Knowledge Graph Visualizer (OB-KGV)* is developed in Java upon the reasoning capabilities provided by PELLET [32] and its query wrapper. Therefore, the tool takes in input an OWL ontology and produce a post-reasoning visualization within a customizable environment to adapt to the different characteristics of the considered data structures as well as to the intent and the extent of the visualization at an application level.

The approach proposed is considered to be scalable via customization and the priority is currently on usability, according to the Technology Acceptance Model [17] and looking at users with a minimum yet existing background in ontology-related fields. Some simplifications and terminology refinement could further increase the user base.

In the next sub-sections, the main features of the tools are discussed by firstly addressing a generic visualization mode, then looking more specifically at the provided views and, finally, at the interaction capabilities.

3.1 Visualization of Semantic Structures

An example of visualization by adopting the most comprehensive available view - i.e. the *ontology view* - is shown in Fig. 2. This is a small dataset defined according to the virtual table model [29] which allows to systematically represent a common dataset in a relational table as an OWL structure.

Such a representation considers all OWL entities, including class, object and data properties and instances. Annotation properties can be accessed though the query/search interface but are not visualised as part of the Knowledge Graph in the current prototype to avoid a further increase of the density for the main knowledge structure. It means that basically entities and relationships of any kind can potentially be considered in the visualization. Such a generic view can be customised in terms of entity filtering, clustering, labelling and visualization area/style.

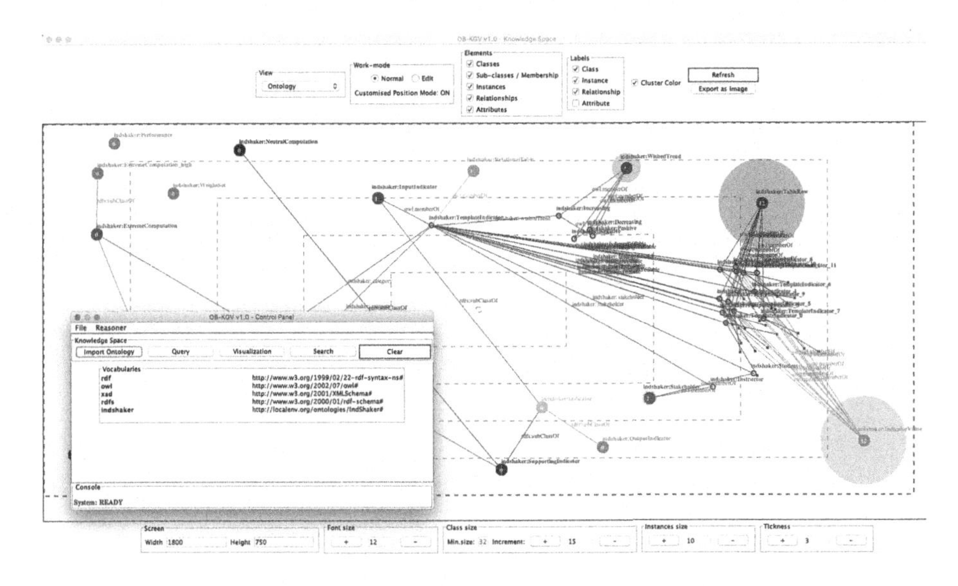

Fig. 2. Visualization of a small ontology.

3.2 Customized Views

In addition, three different pre-defined yet customizable views are provided by the tool:

- *Schema view.* It aims to overview the ontology schema and, therefore, the knowledge structure according to a classic visualization approach enriched in this specific case by the information on the cluster size (namely the number of instances belonging to a given class). Such a view considers OWL classes and the relationships existing among them. It is built by processing the sub-class, domain and range OWL relational axioms. Therefore, it is effective only when such structures are effectively used in the considered ontologies.
- *Data view.* It implements in a way the opposite philosophy by focusing on data - i.e. the ontology population - by considering the instances of classes (but not classes) and the relationships existing among such instances. Also attributes can be considered in order to provide an effective understanding of the available data.
- *Knowledge Graph view.* It can be considered a middle term between the two previous views as it focuses on classes and instances and relationships existing among them avoiding data details (attributes). It is by definition a balanced view which is considered to be suitable in a wide range of cases.

The Climate Change TimeLine [30] is an ontology that organizes Climate Change-related information according to a time-line structure. Such an ontology is visualized according to the different views in Figs. 3 and 4.

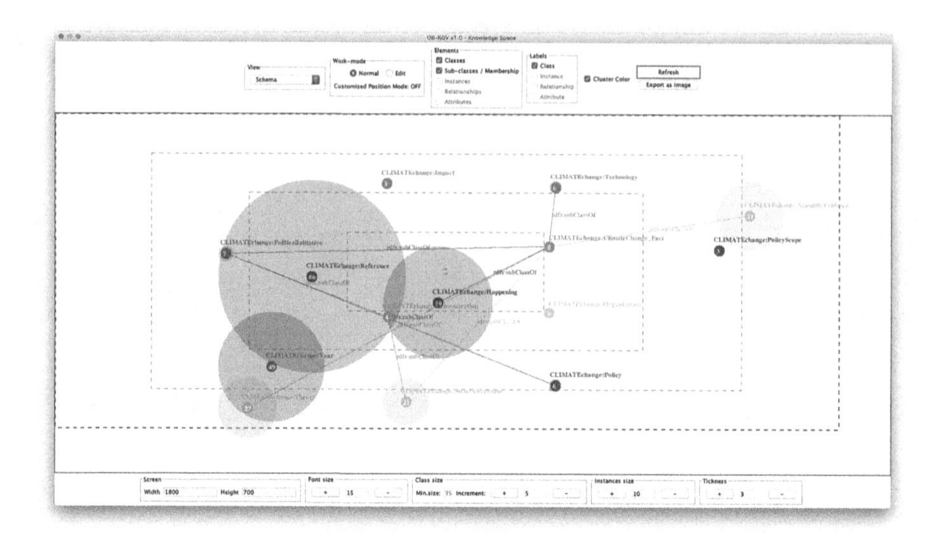

Fig. 3. Schema view.

3.3 Query and Search

The current research prototype enables the interaction with the knowledge structure by providing two integrated mechanisms:

- *Searching (natural language).* The searching tool allows users to browse the semantic network by providing keywords that are matched with available ontological entities like in common search engines. Results are provided as a list, as well as the interested components are highlighted in the graph. While the former mechanism is critical to identify formal concepts in big or dense graphs, the latter provides a visual representation of the result set. This interaction mode is potentially suitable to any user, including also user without any background in ontology-related fields.
- *Formal Query (SPARQL).* The query interface supports formal query in SPARQL [2]. Such an approach enables the power and the complexity of SPARQL [25] but it is evidently suitable only to users with specific technical skills in the field.

The two interfaces previously described are designed to work as part of an integrated framework. An example is reported in Fig. 5: first, the search interface is used to search the keyword "theory" within the Climate Change TimeLine ontology (Fig. 5(a); then, the formal concept identified ("CLIMATE-change:Theory") is adopted to retrieve the associated cluster by formal query (Fig. 5(b)).

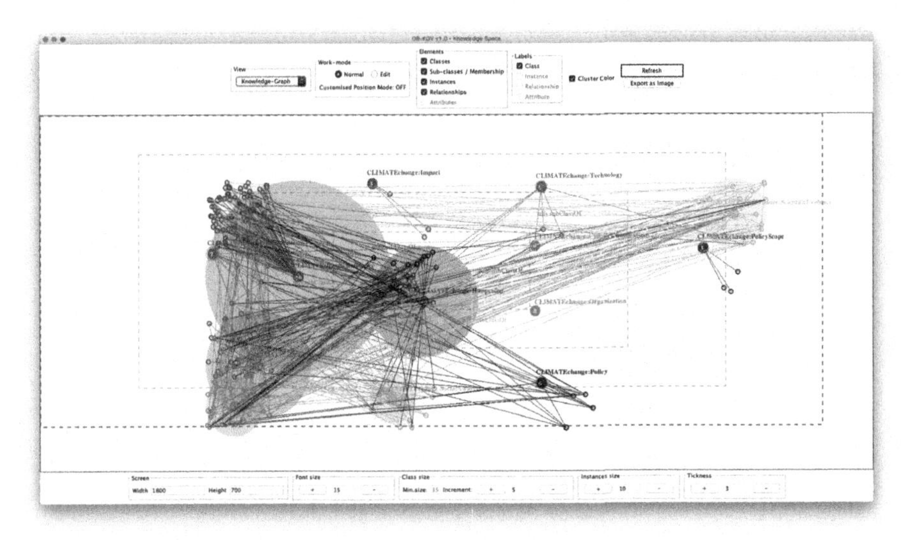

(a) Knowledge Graph view.

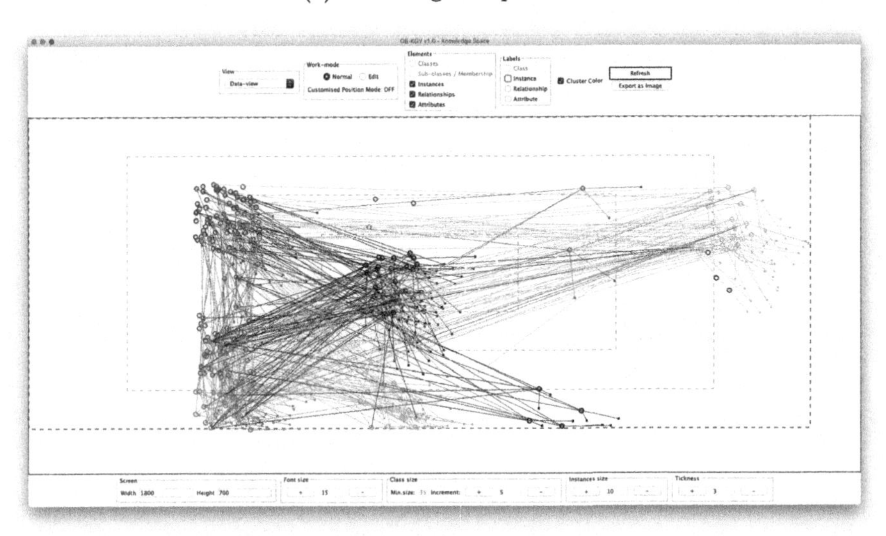

(b) Data view.

Fig. 4. Knowledge graph vs Data view.

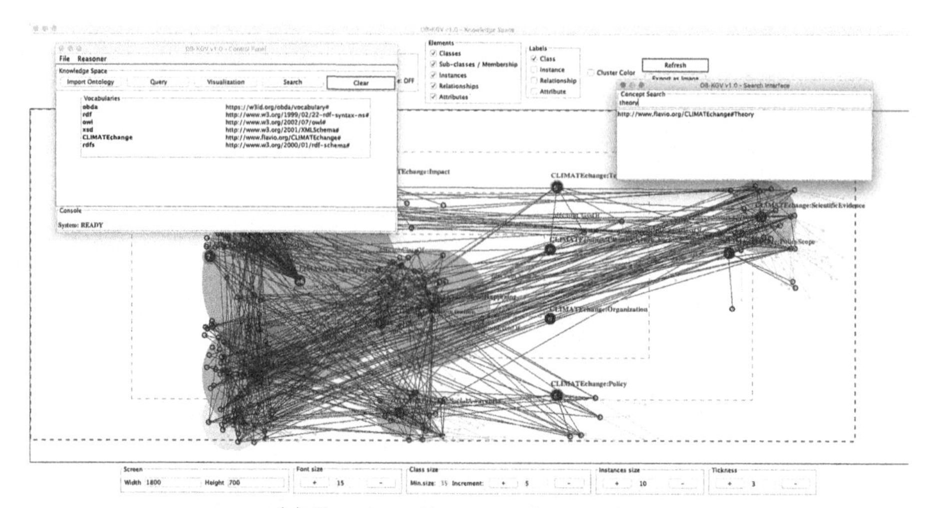

(a) Search on the semantic structure.

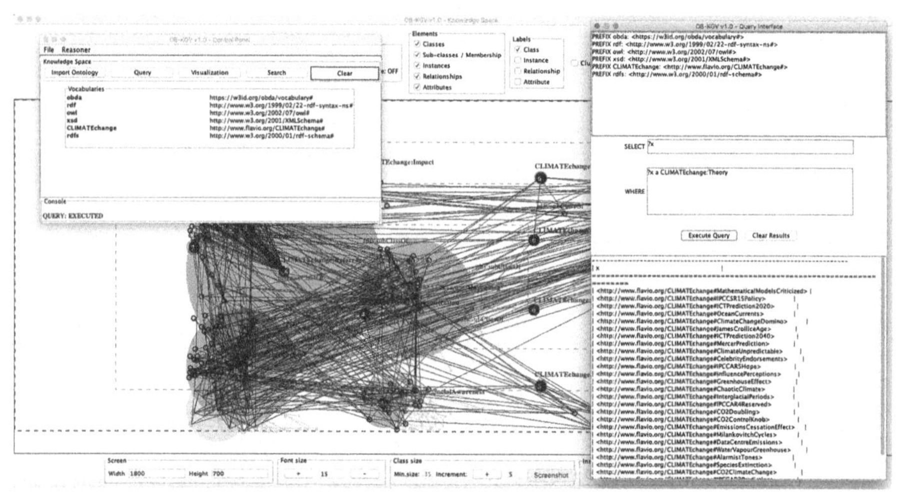

(b) Formal query (SPARQL) on the semantic structure.

Fig. 5. An example of interaction by using the searching and query interfaces.

4 Ontology-Based Knowledge Graph Interpretation: A User Perspective

Formal ontologies in computer systems implies a certain complexity as an attempt to represent human knowledge within a machine World. Even the formalization of a very simple domain (e.g. [26]) relies on the contextual understanding of the information. Such a complexity is normally function of the adopted data model capability and may increase significantly when advanced features are adopted (e.g. [27]).

By adopting Ontology-based Knowledge Graphs, final users are expected to directly interact with the knowledge space with a proper level of abstraction. That is normally enabled by automatic reasoners that provide a user level knowledge infrastructure by computing inference rules in the background of the system. While at a theoretical level the resulting knowledge space represented as a graph is supposed to be a reasonable interface for final users, there are a number of practical aspects to be considered.

Indeed, more than one factor may impact an effective interaction with the knowledge space. The usability of tools [21] and of underlying ontologies [10] definitely play a critical role, although the understanding of "quality" could be in some case domain or application-specific. From a user perspective, the risk of misinterpretation of the knowledge space [13] can be addressed according to two different dimensions: the expertise in a given domain and the technical skills. These dimensions are briefly discussed in the following sub-sections.

4.1 Expertise in the Domain

In general terms, from a domain knowledge point of view, three different classes of final users may be defined as follows:

- *Internal User.* By having a full knowledge and unambiguous understanding of the vocabulary adopted - i.e. the underlying ontology - it is the ideal user, regardless of technical skills.
- *Expert User.* As an expert in the target domain, such a kind of user is supposed to be able to correctly understand conceptualizations. It differs from the previous because has no knowledge of the specific ontology adopted but it is expected to have the background to understand it at an application level.
- *Generic User.* This is a user without a specific expertise in the considered domain.

For internal users the risk of misinterpretation of the target knowledge is in theory very low, if not negligible. As per definition, a detailed knowledge of the vocabulary is a guarantee of correct interpretation of concepts and, therefore, of associated data. Additionally, the comprehensive understanding of the vocabulary facilitates an holistic view as well as more specific and fine-grained, yet contextual, analysis.

An expert user as previously described is expected to be by far the most common user. That is the typical case of a professional or a researcher in a given domain who is using a third-party ontology-based tool. Ontology-based Knowledge Graph should be a suitable asset as long as proper annotations are associated to main concepts and entities. Such a metadata normally contributes to fill the gap between domain knowledge and concrete implementations. However, due to the intrinsic complexity of certain domains (e.g. [28]), an unified view is not always easy to achieve and requires a detailed knowledge of the adopted vocabulary.

A generic user could take advantage of KGs, which are expected to enable in fact the interaction with the knowledge space. However, the lack of specific knowledge in the domain increases significantly the risk of misinterpretation.

4.2 Technical Skills

Looking at a user classification from a more technical perspective, we can reasonable define the following categories:

- *Ontology-specific background.* A user with a specific ontology background is supposed to fully understand and have a concrete expertise in the technology adopted (e.g. able to use formal query languages).
- *Technical background.* User with a significant technical background other than ontology-specific. It could be a data analyst for instance.
- *Non-technical user.* Users without a clear technical background.

An ontology-specific background allows a direct and effective interaction with the knowledge space by using formal query languages without brokers. That is evidently the ideal situation which is, however, unrealistic in most cases. A technical background as previously described includes a wide range of possible users and doesn't necessarily guarantee effective interaction. However, a technical background could allow an understanding of metrics for graph analysis and could facilitate a relatively good quality of experience. A non-technical user mainly relies on abstraction mechanisms at different levels (e.g. interfaces based on a natural language) which could be application-specific.

Generally speaking, the ability to properly interact with the knowledge space reduces the risk of misinterpretation.

5 Conclusions and Future Work

OB-KGV aims at an exhaustive and highly customizable solution for the visualization of knowledge graphs specified as formal OWL ontologies. Developed in Java upon PELLET [32], the research prototype provides an interactive post-reasoning visualization environment in which final users can browse the knowledge structure by adopting both formal query and natural language.

The approach proposed is considered to be scalable via customization. However, the current implementation of the research prototype assumes detailed visualizations for relatively small data sets with a progressive detail decreasing when the amount of information increases. Such a prototype is also supposed to evolve in the next future to incorporate functionalities aimed at improving the user experience as well as to integrate advanced functionalities in terms of graph analysis, data integration and learning.

The most evident limitation is the current lack of user validation that could be object of future work.

References

1. W3C - OWL 2 Web Ontology Language Document Overview (Second Edition). https://www.w3.org/TR/owl2-overview/. Accessed 18 Sept 2020
2. W3C - SPARQL 1.1 Overview. https://www.w3.org/TR/sparql11-overview/. Accessed 18 Sept 2020
3. Bach, B., Pietriga, E., Liccardi, I., Legostaev, G.: OntoTrix: a hybrid visualization for populated ontologies. In: Proceedings of the 20th International Conference Companion on World Wide Web, pp. 177–180 (2011)
4. Berners-Lee, T., Hendler, J., Lassila, O.: The semantic web. Sci. Am. **284**(5), 34–43 (2001)
5. Chen, H., Luo, X.: An automatic literature knowledge graph and reasoning network modeling framework based on ontology and natural language processing. Adv. Eng. Inform. **42**, 100959 (2019)
6. Chen, P., Lu, Y., Zheng, V.W., Chen, X., Yang, B.: KnowEdu: a system to construct knowledge graph for education. IEEE Access **6**, 31553–31563 (2018)
7. Chen, X., Jia, S., Xiang, Y.: A review: knowledge reasoning over knowledge graph. Expert Syst. Appl. **141**, 112948 (2020)
8. Fang, W., Ma, L., Love, P.E., Luo, H., Ding, L., Zhou, A.: Knowledge graph for identifying hazards on construction sites: integrating computer vision with ontology. Autom. Constr. **119**, 103310 (2020)
9. Fensel, D., et al.: Introduction: what is a knowledge graph? In: Knowledge Graphs, pp. 1–10. Springer, Cham (2020). https://doi.org/10.1007/978-3-030-37439-6_1
10. Gangemi, A., Catenacci, C., Ciaramita, M., Lehmann, J.: Modelling ontology evaluation and validation. In: Sure, Y., Domingue, J. (eds.) ESWC 2006. LNCS, vol. 4011, pp. 140–154. Springer, Heidelberg (2006). https://doi.org/10.1007/11762256_13
11. Gennari, J.H., et al.: The evolution of protégé: an environment for knowledge-based systems development. Int. J. Hum. Comput. Stud. **58**(1), 89–123 (2003)
12. Ghorbel, F., Ellouze, N., Métais, E., Hamdi, F., Gargouri, F., Herradi, N.: Memo graph: an ontology visualization tool for everyone. Procedia Comput. Sci. **96**, 265–274 (2016)
13. Glazer, N.: Challenges with graph interpretation: a review of the literature. Stud. Sci. Educ. **47**(2), 183–210 (2011)
14. Guarino, N.: Formal ontology, conceptual analysis and knowledge representation. Int. J. Hum. Comput. Stud. **43**(5–6), 625–640 (1995)
15. Katifori, A., Halatsis, C., Lepouras, G., Vassilakis, C., Giannopoulou, E.: Ontology visualization methods - a survey. ACM Comput. Surv. (CSUR) **39**(4), 10 (2007)
16. Lanzenberger, M., Sampson, J., Rester, M.: Visualization in ontology tools. In: International Conference on Complex, Intelligent and Software Intensive Systems, 2009. CISIS 2009, pp. 705–711. IEEE (2009)
17. Lee, Y., Kozar, K.A., Larsen, K.R.: The technology acceptance model: past, present, and future. Commun. Assoc. Inf. Syst. **12**(1), 50 (2003)
18. Lin, Y., Liu, Z., Sun, M., Liu, Y., Zhu, X.: Learning entity and relation embeddings for knowledge graph completion. In: Twenty-Ninth AAAI Conference on Artificial Intelligence (2015)
19. LiuQiao, L., DuanHong, L., et al.: Knowledge graph construction techniques. J. Comput. Res. Dev. **53**(3), 582 (2016)
20. Lohmann, S., Negru, S., Haag, F., Ertl, T.: Visualizing ontologies with VOWL. Semant. Web **7**(4), 399–419 (2016)

21. Marangunić, N., Granić, A.: Technology acceptance model: a literature review from 1986 to 2013. Univers. Access Inf. Soc. **14**(1), 81–95 (2015)
22. McBride, B.: The resource description framework (RDF) and its vocabulary description language RDFS. In: Staab, S., Studer, R. (eds.) Handbook on Ontologies. International Handbooks on Information Systems, pp. 51–65. Springer, Heidelberg (2004). https://doi.org/10.1007/978-3-540-24750-0_3
23. Noy, N., Gao, Y., Jain, A., Narayanan, A., Patterson, A., Taylor, J.: Industry-scale knowledge graphs: lessons and challenges: five diverse technology companies show how it's done. Queue **17**(2), 48–75 (2019)
24. Paulheim, H.: Knowledge graph refinement: a survey of approaches and evaluation methods. Semant. Web **8**(3), 489–508 (2017)
25. Pérez, J., Arenas, M., Gutierrez, C.: Semantics and complexity of SPARQL. In: Cruz, I., et al. (eds.) ISWC 2006. LNCS, vol. 4273, pp. 30–43. Springer, Heidelberg (2006). https://doi.org/10.1007/11926078_3
26. Pileggi, S.F.: A novel domain ontology for sensor networks. In: 2010 Second International Conference on Computational Intelligence, Modelling and Simulation, pp. 443–447. IEEE (2010)
27. Pileggi, S.F.: Probabilistic semantics. Procedia Comput. Sci. **80**, 1834–1845 (2016)
28. Pileggi, S.F., Indorf, M., Nagi, A., Kersten, W.: CoRiMaS-an ontological approach to cooperative risk management in seaports. Sustainability **12**(11), 4767 (2020)
29. Pileggi, S.F., Crain, H., Yahia, S.B.: An ontological approach to knowledge building by data integration. In: Krzhizhanovskaya, V.V., et al. (eds.) ICCS 2020. LNCS, vol. 12143, pp. 479–493. Springer, Cham (2020). https://doi.org/10.1007/978-3-030-50436-6_35
30. Pileggi, S.F., Lamia, S.A.: Climate change timeline: an ontology to tell the story so far. IEEE Access **8**, 65294–65312 (2020)
31. Pujara, J., Miao, H., Getoor, L., Cohen, W.: Knowledge graph identification. In: Alani, H., et al. (eds.) ISWC 2013. LNCS, vol. 8218, pp. 542–557. Springer, Heidelberg (2013). https://doi.org/10.1007/978-3-642-41335-3_34
32. Sirin, E., Parsia, B., Grau, B.C., Kalyanpur, A., Katz, Y.: PELLET: a practical OWL-DL reasoner. Web Semant. Sci. Serv. Agents World Wide Web **5**(2), 51–53 (2007)
33. Sivakumar, R., Arivoli, P.: Ontology visualization protégé tools-a review. Int. J. Adv. Inf. Technol. (IJAIT) **1** (2011)
34. Wang, Q., Mao, Z., Wang, B., Guo, L.: Knowledge graph embedding: a survey of approaches and applications. IEEE Trans. Knowl. Data Eng. **29**(12), 2724–2743 (2017)
35. Wang, X., He, X., Cao, Y., Liu, M., Chua, T.S.: KGAT: knowledge graph attention network for recommendation. In: Proceedings of the 25th ACM SIGKDD International Conference on Knowledge Discovery and Data Mining, pp. 950–958 (2019)

Towards Mitigating the Eye Gaze Tracking Uncertainty in Virtual Reality

Konstantin Ryabinin[1,2](✉) and Svetlana Chuprina[2]

[1] Saint Petersburg State University, 7/9 Universitetskaya Emb., Saint Petersburg 199034, Russia
kostya.ryabinin@gmail.com

[2] Perm State University, 15 Bukireva Str., Perm 614068, Russia
chuprinas@inbox.ru

Abstract. We propose a novel algorithm to evaluate and mitigate the uncertainty of data reported by eye gaze tracking devices embedded in virtual reality head-mounted displays. Our algorithm first is calibrated by leveraging unit quaternions to encode angular differences between reported and ground-truth gaze directions, then interpolates these quaternions for each gaze sample, and finally corrects gaze directions by rotating them using interpolated quaternions. The real part of the interpolated quaternion is used as the certainty factor for the corresponding gaze direction sample. The proposed algorithm is implemented in the VRSciVi Workbench within the ontology-driven SciVi visual analytics platform and can be used to improve the eye gaze tracking quality in different virtual reality applications including the ones for Digital Humanities research. The tests of the proposed algorithm revealed its capability of increasing eye tracking accuracy by 25% and precision by 32% compared with the raw output of the Tobii tracker embedded in the Vive Pro Eye head-mounted display. In addition, the certainty factors calculated help to acknowledge the quality of reported gaze directions in the subsequent data analysis stages. Due to the ontology-driven software generation, the proposed approach enables high-level adaptation to the specifics of the experiments in virtual reality.

Keywords: Eye tracking · Virtual reality · Uncertainty mitigation · Ontology-driven software generation · Quaternion-based model

1 Introduction

Modern Virtual Reality (VR) head-mounted displays (HMDs) enable not only the user's head orientation and position tracking, but also eye gaze tracking capabilities. These new capabilities were included in HMDs not long ago and have a lot of interesting applications, which make eye tracking a very promising extension for traditional VR technologies. First, eye tracking enables the so-called

This study is supported by the research grant No. ID75288744 from Saint Petersburg State University.

D. Groen et al. (Eds.): ICCS 2022, LNCS 13353, pp. 623–636, 2022.
https://doi.org/10.1007/978-3-031-08760-8_51

foveated rendering, which is an optimization technique that reduces rendering workload by lowering the image quality in the user's peripheral vision [19]. Second, eye tracking provides new ways of interaction with VR scene objects [15]. For example, a "hands-free" gaze-based selection of objects can be implemented, which may be useful for instantly getting the context information about the scene contents. Alternatively, eye tracking in multiplayer VR games/meetings allows making the users' avatars more vivid by animating their eyes in accordance with actual users' gaze. Last but not least, eye tracking has opened up novel ways to study the user's behavior. In simulators, eye tracking can be used for estimating the correctness of the user's behavior by checking if the user looks in the expected direction or not. For example, in the car driving simulator, the user can be alerted when looking off the road [20]. In special test stands, eye tracking can help to study the information perception mechanisms by providing data about the order of objects the user focuses on [14] because the direction of eye gaze provides a strong cue to the person's intentions and future actions [6].

While traditional eye tracking with remote stationary tracking devices is a well-defined methodology [9,14,26], VR brings some new challenges dealing with the accuracy and precision issues [17,27,30]. In eye tracking, accuracy is defined as "average angular error between the measured and the actual location of the intended fixation target", and precision "is the spread of the gaze positions when a user is fixating on a known location in space" [27]. Modern VR HMDs still have quite a low angular resolution, sampling rate, and peak signal-noise ratio of embedded eye tracking sensors compared with remote stationary eye trackers [17,30]. The tasks like foveated rendering require just an approximate gaze direction and their performance is not much affected by tracking inaccuracy. In contrast, for studying human behavior, eye tracking accuracy and precision are crucial. Inaccurate/noisy raw tracking data lead to an increase in the uncertainty of the analysis stage results and thereby can ruin the entire study.

At the same time, VR as a set of immersive visualization technologies provides new potential to organize advanced experiments for human behavior study within a field of Digital Humanities research [12,28,29]. Proper handling of the eye gaze tracking uncertainty is badly needed to make the results of those experiments reliable.

In this work, we propose a unified approach to evaluate and mitigate the eye gaze tracking data uncertainty in VR. The proposed approach is based on our experience of using Vive Pro Eye HMD with embedded Tobii eye tracking sensor as an immersion and measurement hardware within the Digital Humanities research. The uncertainty mitigation algorithm is an essential part of the so-called VRSciVi Workbench that is a set of tools within the SciVi visual analytics platform[1] aimed to automate the conducting of the experiments in VR and handling the results of these experiments by means of scientific visualization and visual analytics tools.

[1] https://scivi.tools/.

2 Key Contributions

We propose ontology-driven tools to automatically generate the software for conducting eye-tracking-based experiments within the immersive VR environments mitigating the uncertainty of data reported by embedded eye trackers. The following hitpoints of the conducted research can be highlighted:

1. Novel quaternion-based model of the eye gaze tracking uncertainty.
2. Novel algorithm to mitigate the eye gaze tracking uncertainty at runtime that enables to increase the eye tracking accuracy and precision by ca. 30%.
3. Ontology-driven high-level software development tools to generate the interface for parametrization of the proposed algorithm and to adapt it for integration in a software to meet the specifics of different VR-based experiments.

3 Related Work

Eye gaze tracking is a research methodology with more than a hundred years of history that is nowadays accessible and intensively used in a wide range of scientific domains [9]. The related hardware has progressed over several decades and its evolution is methodically outlined in the review by Shehu et al. [26]. The particular protocols of conducting experiments and algorithms for processing the collected data are elaborated and well-documented [9,14]. The corresponding software provides implementations of these algorithms along with the needed visualization and analytics techniques [3,25]. But the convergence of eye gaze tracking with VR brings new challenges in both of these fields demanding the development of new tools for handling gaze tracks in close relation to the VR scene objects [10] and immersion features [5,16].

Huge problems of eye tracking in VR are low resolution, accuracy, and precision of the trackers embedded in HMDs, which hinder the use of these trackers in special cases like medical or humanities research. The temporal resolution of embedded eye trackers is normally capped at the VR scene refresh rate, which is 90 Hz (corresponding to the refresh rate of the modern VR HMDs) [17,30]. In contrast, the modern stationary eye trackers operate on up to 2000 Hz [1]. As found by D. Lohr et al. on the example of Vive Pro Eye HMD, embedded trackers can have internal non-toggleable low-pass filters rejecting fast saccades even within the available sampling rate [17]. That means, it is fundamentally impossible to use such devices to study phenomena like ocular microtremor, etc.

A. Sipatchin et al. conducted a very elaborate case study of Vive Pro Eye's usability for clinical ophthalmology and concluded that although this device has "limitations of the eye-tracker capabilities as a perimetry assessment tool", it has a "good potential usage as a ready-to-go online assistance tool for visual loss" and the "upcoming VR headsets with embedded eye-tracking" can be introduced "into patients' homes for low-vision clinical usability" [27]. Another important contribution of this research group is the accuracy and precision measurement of the Vive Pro Eye's embedded Tobii eye tracker. They found out that the accuracy deteriorates dramatically as the gaze location moves away from the display center and the spatial accuracy of the tracker is more than 3 times lower than the manufacturer claims. The same phenomenon was shown by K. Binaee et al. for "binocular eye tracker, built into the Oculus Rift DK2 HMD" [2]. This suggests that a lot of embedded eye trackers suffer from this problem. Despite the obvious limitations, Vive Pro Eye HMD is recognized as an apparatus with high ergonomic characteristics suitable for the eye-tracking-based study of human behavior and information perception [11,17,18]. According to [11,12,17,18,28, 29], it can be considered as a consensus that VR research on improving the HMD user experience and embedded eye trackers quality is worth continuing because VR-based technologies provide great possibilities to conduct human-centered studies.

Along with the hardware improvements constantly provided by manufacturers, software improvements are possible as well by applying post-processing filters to the raw data of embedded eye trackers. For example, K. Binaee et al. propose a method for embedded eye tracker post-hoc calibration based on homography calculated via a random sample consensus and a dynamic singular value decomposition [2]. This method allows to increase the eye tracking accuracy up to 5 times compared with the built-in HMD calibration technique, but the disadvantage is the post-hoc nature of the method which means the calculations are off-line.

S. Tripathi et al. propose a self-calibrating eye gaze tracking scheme based on Gaussian Process Regression models [31]. This scheme avoids the explicit calibration of tracking devices while maintaining competitive accuracy. The disadvantage is that this scheme is only applicable for scenes with moving objects and cannot be used for static scenes.

A. Hassoumi et al. propose improving eye tracking accuracy by using a so-called symbolic regression, which seeks an optimal model of transforming the eye pupil coordinates to the gaze location within "a set of different types of predefined functions and their combinations" using a genetic algorithm [8]. This approach allows a 30% accuracy increase compared with the standard calibration procedures. This method was used for monocular eye tracking systems, but modern VR HMDs use binocular systems, so the method cannot be applied to VR directly.

H. Drewes et al. propose a calibration method based on smooth pursuit eye movement that is 9 times faster than the traditional calibration process (which uses 9 stationary points aligned by the regular grid as fiducial gaze targets with known coordinates), but the resulting accuracy is slightly lower [7].

A. Jogeshwar et al. designed the cone model "to acknowledge and incorporate the uncertainty" of eye gaze detected by embedded trackers in VR [12]. But the way to evaluate and mitigate this uncertainty is still an open question, especially for cases with small objects of interest. Our work contributes to solving this problem aiming to mitigate the eye gaze tracking uncertainty within Digital Humanities research.

4 Facing the Eye Gaze Tracking Uncertainty in Virtual Reality

We faced the problem of uncertain eye gaze tracking data during the work on the research project "Text processing in L1 and L2: Experimental study with eye-tracking, visual analytics and virtual reality technologies"[2] (supported by the research grant No. ID75288744 from Saint Petersburg State University). One of the goals of this project is to study the reading process of humans within a VR environment using eye gaze tracking as a measurement technique.

The experiments are conducted using a specific VR setup including Vive Pro Eye HMD connected to the VR rendering station based on the AMD Ryzen 9 CPU and NVidia Titan RTX GPU. The rendering is performed by Unreal Engine 4. The eye tracking data are collected using the SRanipal SDK[3] plugin for Unreal Engine. The experiment control and the data analysis are performed by the VRSciVi Workbench within the SciVi visual analytics platform [22].

Considering the case of reading a relatively large poster ($40° \times 26°$ of vision area in size, see Fig. 1), we found out that the data retrieved from the embedded eye tracker is far from being suitable to trace the reading process on the level of individual letters/syllables or at least of individual words (the width of each letter is approx. 0.63°). Figure 1 demonstrates the virtual scene (rendered by Unreal Engine 4) as viewed by the informant with the 183 words long text displayed on the wall poster. Figure 2 shows the map of the gaze fixations measured by the Vive Pro Eye HMD embedded Tobii eye tracker. Circle size depicts fixation duration. Fixations are detected according to the method suggested in [16]; dwell time threshold is set to 250 ms, angular threshold is set to 1°.

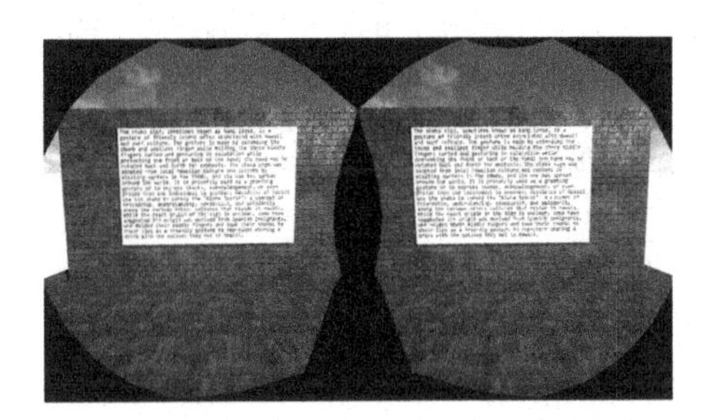

Fig. 1. The virtual scene as viewed by the informant

The shaka sign, sometimes known as hang loose, is a
gesture of friendly intent often associated with Hawaii
and surf culture. The gesture is made by extending the
thumb and smallest finger while holding the three middle
fingers curled and gesturing in salutation while
presenting the front or back of the hand; the hand may be
rotated back and forth for emphasis. The shaka sign was
adopted from local Hawaiian culture and customs by
visiting surfers in the 1960s, and its use has spread
around the world. It is primarily used as a greeting
gesture or to express thanks, acknowledgement, or even
praise from one individual to another. Residents of Hawaii
use the shaka to convey the "Aloha Spirit": a concept of
friendship, understanding, compassion, and solidarity
among the various ethnic cultures that reside in Hawaii.
While the exact origin of the sign is unclear, some have
suggested its origin was derived from Spanish immigrants,
who folded their middle fingers and took their thumbs to
their lips as a friendly gesture to represent sharing a
drink with the natives they met in Hawaii.

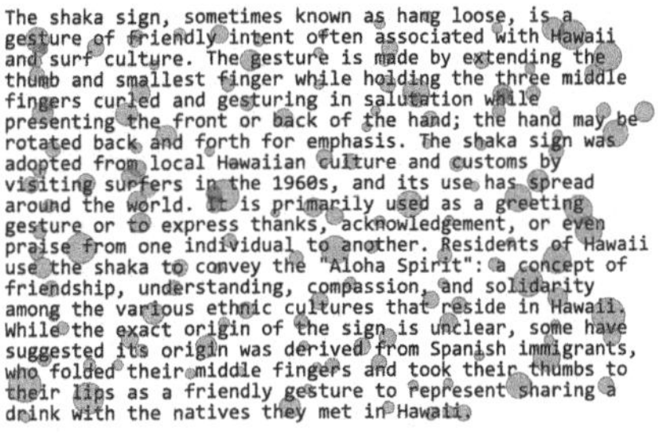

Fig. 2. The map of obtained gaze fixations

The fixations heatmap in Fig. 2 reveals obvious distortions: fixations miss the actual words and seem to be shifted. Such distortions hinder any subsequent analysis of the obtained data. As stated by A. Jogeshwar et al. to get reliable results of experiments the gaze uncertainty should be properly acknowledged [12].

[2] L1 and L2 stand for the native and foreign languages respectively.

[3] https://developer-express.vive.com/resources/vive-sense/eye-and-facial-tracking-sdk/.

As mentioned above, the VRSciVi Workbench is the central software element of our experimental setup. It controls the VR scene content, manages data extraction, transformation, and loading processes, and provides tools to visualize and analyze the collected data. This platform is driven by ontologies, which enable its extensibility and adaptability [23]. So, we have introduced the new components to SciVi for handling the eye gaze tracking uncertainty without changing the source code of any other components. These components are publicly available in the SciVi open-source repository and can be used in any SciVi-based VR project (see Sect. 6).

5 Evaluating and Mitigating the Uncertainty of the Vive Pro Eye Sensor Data

5.1 Evaluating the Uncertainty

To evaluate the eye gaze tracking uncertainty of the Vive Pro Eye sensor, we use a 5×5 points pattern similar to the one described in [27]. We display the pattern points on the billboard in the virtual scene (a white poster on the wall, see Fig. 1). Points are shown one by one, each for 2000 ms, and the informant is asked to stare at them. For each point, the first 500 ms of gaze data are discarded as suggested in [7] to trim the initial saccade. The gaze locations are measured and averaged during the subsequent 1500 ms. The result of the comparison of ground-truth and reported gaze locations based on 10 trials is demonstrated in Fig. 3. The blue points depict ground-truth gaze locations (the points of the pattern the persons were looking at) and the red ellipses bound the averaged gaze locations reported by the Vive Pro Eye sensor individually calibrated for each person using a built-in 5-points calibration procedure. It must be noted that the gaze locations are represented in the 2D texture space of the billboard the pattern is shown on. The reported gaze coordinates are calculated by obtaining the hit points of the gaze ray with this billboard and transforming these coordinates from the virtual scene coordinates to the billboard texture space. The gaze ray is reported by the SRanipal SDK plugin for Unreal Engine, which is the default (and the only official) way to interact with the embedded Tobii eye tracker of Vive Pro Eye HMD.

Figure 3 clearly shows the non-uniform nature of eye gaze tracking uncertainty, which aligns with the results reported in [2,27]. We confirm the accuracy loss in the display periphery: the smallest angular error of 0.02° is located near the center and the biggest one of 2.4° is located near the border of the billboard; the average angular error is 0.77° and the standard deviation is 0.43°. The baseline accuracy in our case is about 1.7 times higher than reported in [27] because according to the setup of our experiment we are inspecting a narrower field of view. Moreover, the billboard is not tied to the user's head, so the user can slightly rotate their head reducing the angular distance between the target point and the vision area center. Still, this accuracy does not suit our needs related to the study of the reading process in VR because individual letters of the texts

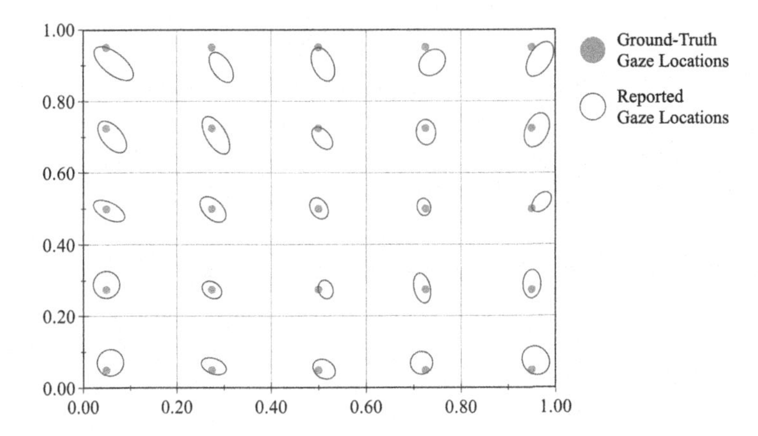

Fig. 3. Comparison of ground-truth and reported gaze locations (Color figure online)

considered are ca. 0.63° in size. Therefore the accuracy should be improved by proper accounting of the gaze tracking uncertainty.

5.2 VR HMD Frame of Reference

Figure 4 shows the frame of reference used in our calculations by uncertainty evaluation and mitigation. The point O and the vectors $\boldsymbol{x}$, $\boldsymbol{y}$, $\boldsymbol{z}$ build up the left-handed coordinate system $\mathcal{H}$ tied to the user's head. These elements are constructed by Unreal Engine based on the HMD positioning system and represented in the global coordinate system $\mathcal{G}$ of the virtual scene. The $\boldsymbol{r}$ vector represents gaze direction represented in $\mathcal{H}$ as reported by the SRanipal SDK plugin and the vector $\boldsymbol{g}$ represents true gaze direction. The angle ϕ determines how far the user looks to the side from their forward direction according to the reported by SRanipal SDK, and the angle ψ represents the angular difference between reported and true gaze directions.

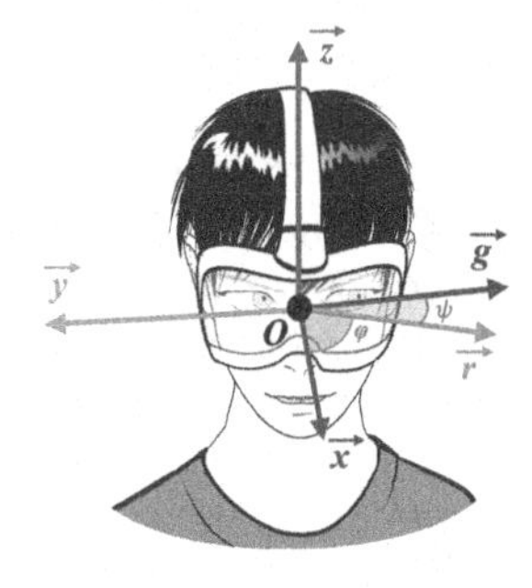

Fig. 4. VR HMD frame of reference

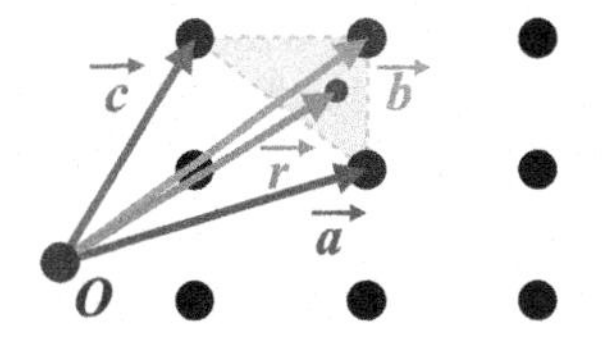

Fig. 5. Basis for reported gaze vector

5.3 Quaternion-Based Uncertainty Representation and Mitigation

To represent eye gaze tracking uncertainty we propose using a quaternion-based model [13]. Having a ground-truth gaze unit vector $\boldsymbol{g}$ and reported gaze unit vector $\boldsymbol{r}$, we can encode the tracking error as a quaternion q that represents the shortest arc rotation from $\boldsymbol{g}$ to $\boldsymbol{r}$:

$$q = \begin{cases} \{\boldsymbol{r} \cdot \boldsymbol{g} + 1, \boldsymbol{r} \times \boldsymbol{g}\}, & \boldsymbol{r} \cdot \boldsymbol{g} \neq -1 \\ \{0, \quad \boldsymbol{u}\}, & \boldsymbol{u} \cdot \boldsymbol{r} = 0, \ \boldsymbol{r} \cdot \boldsymbol{g} = -1 \end{cases}. \quad (1)$$

This quaternion should be normalized to be a versor (unit quaternion):

$$q_{r \to g} = \frac{q}{|q|}. \quad (2)$$

For this versor,

$$\boldsymbol{g} = q_{r \to g} \ \boldsymbol{r} \ q^*_{r \to g}, \quad (3)$$

where $q^*_{r \to g}$ is a complex conjugate of $q_{r \to g}$, $q^*_{r \to g} = \{\mathrm{Re}(q_{r \to g}), -\mathrm{Im}(q_{r \to g})\}$.

The real part of $q_{r \to g}$ represents a cosine of half angle between $\boldsymbol{g}$ and $\boldsymbol{r}$, which means $\mathrm{Re}(q_{r \to g}) = 1$ if $\boldsymbol{r} = \boldsymbol{g}$ and $\mathrm{Re}(q_{r \to g}) = 0$ if $\boldsymbol{r} = -\boldsymbol{g}$. In this regard, we propose to interpret the real part of $q_{r \to g}$ as a certainty factor (CF) of reported gaze direction:

$$\mathrm{CF}(\boldsymbol{r}) = \mathrm{Re}(q_{r \to g}). \quad (4)$$

The predicted angular error ψ_r of $\boldsymbol{r}$ can be extracted from CF as

$$\psi_r = 2 \arccos\left(\mathrm{CF}(\boldsymbol{r})\right). \quad (5)$$

Consequently, if we could find a corresponding versor $q_{r \to g}$ for an arbitrary reported gaze vector $\boldsymbol{r}$, we can then evaluate, how certain is the reported gaze direction using the formula (4), predict its angular error using the formula (5) and even correct the uncertain direction to the true one using the formula (3).

We propose tackling this problem by spatial interpolation of a discrete set of versors obtained during the custom calibration process. The general description of the calibration algorithm is as follows:

1. Place the white billboard in the VR scene, tied to the user's head. The size of this billboard is a matter of experimenting, for now we end up with $56° \times 37°$ This billboard will be a canvas for displaying target points. Since it is tied to the user's head, the user has to only rotate the eyes to look at the points, and not the head, which allows accounting the eye gaze direction only.
2. Choose the calibration pattern. This is still a matter of experimenting, but to start with, we use a traditional 3×3 regular grid of points, which center matches the center of the billboard.
3. For each point $P_i = \{x_i, y_i, z_i\}$ (coordinates are represented in the $\mathcal{H}$ reference frame mentioned in Sect. 5.2) from the pattern, $i = \overline{1, n}$, $n = 9$:
   3.1. Display P_i as a filled circle of radius $0.38°$ related to the vision area.

3.2. During 1000 ms:
   3.2.1. Decrease the circle radius down to 0.18°. As stated in [21], the size reduction helps the user to concentrate on the point's center.
   3.2.2. Discard the first 500 ms of gaze data to trim the initial saccade [7].
   3.2.3. Use the subsequent data to calculate the $\boldsymbol{r}_i$ vector averaging the reported gaze vectors.

3.3 Calculate the ground-truth gaze vector as $\boldsymbol{g}_i = -P_i/|P_i|$. This formula is valid because P_i is represented in $\mathcal{H}$.

3.4 Calculate the versor $q_{\boldsymbol{r}_i \to \boldsymbol{g}_i}$ using the formulas (1) and (2). Let us denote this versor as q_i for brevity.

3.5 Store the calibration tuple $\langle \boldsymbol{r}_i, q_i \rangle$ for future reference.

After this custom calibration procedure, each time the gaze direction $\boldsymbol{r}$ is reported and its uncertainty is evaluated and mitigated using the following algorithm:

1. For the vector $\boldsymbol{r}$, obtain the following 3 reference vectors (see Fig. 5):
   $\boldsymbol{a} = \boldsymbol{r}_{n/2+1}$,
   $\boldsymbol{b} = \boldsymbol{r}_j | \boldsymbol{r} \cdot \boldsymbol{r}_j \to \max,\ j = \overline{1,n},\ j \neq n/2+1$,
   $\boldsymbol{c} = \boldsymbol{r}_k | \boldsymbol{r} \cdot \boldsymbol{r}_k \to \max,\ k = \overline{1,n},\ k \neq n/2+1,\ k \neq j$.
2. Let the matrix M be composed from the coordinates of $\boldsymbol{a}$, $\boldsymbol{b}$, and $\boldsymbol{c}$ represented in the basis of $\mathcal{H}$:
   $$M = \begin{pmatrix} a_x & b_x & c_x \\ a_y & b_y & c_y \\ a_z & b_z & c_z \end{pmatrix}.$$
3. If $|M| = 0$ (that means, $\boldsymbol{a}$, $\boldsymbol{b}$, and $\boldsymbol{c}$ are linearly dependent), take the versor associated with the nearest vector among the calibration tuples:
   $q_{\boldsymbol{r} \to \boldsymbol{g}} = q_l | \boldsymbol{r} \cdot \boldsymbol{r}_l \to \max,\ l = \overline{1,n}$.
4. If $|M| \neq 0$,
   4.1 Find the coordinates $\{t_a, t_b, t_c\}$ of $\boldsymbol{r}$ in the basis $\{\boldsymbol{a}, \boldsymbol{b}, \boldsymbol{c}\}$:
   $(t_a\ t_b\ t_c)^\top = M^{-1}(r_x\ r_y\ r_z)^\top$.
   4.2 Use these coordinates to calculate the desired versor interpolating the versors from calibration tuples, which correspond to the vectors $\boldsymbol{a}$, $\boldsymbol{b}$, and $\boldsymbol{c}$: $q_{\boldsymbol{r} \to \boldsymbol{g}} = t_a q_a + t_b q_b + t_c q_c = t_a q_{n/2+1} + t_b q_j + t_c q_k$.
5. Calculate the gaze CF by the formula (4) and the corrected gaze vector $\boldsymbol{g}$ by the formula (3).

The results of applying the described algorithms are discussed in Sect. 7.

6 Implementation of Uncertainty Mitigation Algorithm in VRSciVi

VRSciVi Workbench within the SciVi platform is suited for controlling immersive VR environments. VRSciVi has the client-server architecture. VRSciVi Server is responsible for displaying the VR scene using Unreal Engine 4 as a graphics rendering system. VRSciVi Client is a set of SciVi platform plugins to build and control virtual scenes, as well as to collect and analyze related data about human activities within VR. The communication between Client and Server relies on the WebSocket protocol, which is detailed in [22].

The algorithm of uncertainty mitigation is implemented within VRSciVi Server and controlled by VRSciVi Client. The main plugin provided by VRSciVi Client is currently a so-called VRBoard, which enables placing different visual stimuli on the billboard located in the virtual world (see. Fig. 1). VRBoard provides needed settings for the uncertainty mitigation algorithm and a command to start the custom calibration procedure. Like any other SciVi plugin, VRBoard is specified by a light-weight application ontology that is used for four main purposes. First, it documents the plugin. Second, it drives the automatic generation of a graphical user interface code for the plugin to allow the users to set up the needed parameters. Third, it drives the automatic generation of execution code to properly run the plugin in the given software environment. Fourth, it allows SciVi to organize proper communication of this plugin with the others, taking into account the computing nodes the plugins run on. Thanks to underlying ontology and built-in reasoning mechanism, all of these goals are achieved in a uniform way. The fragment of this ontology is demonstrated in Fig. 6.

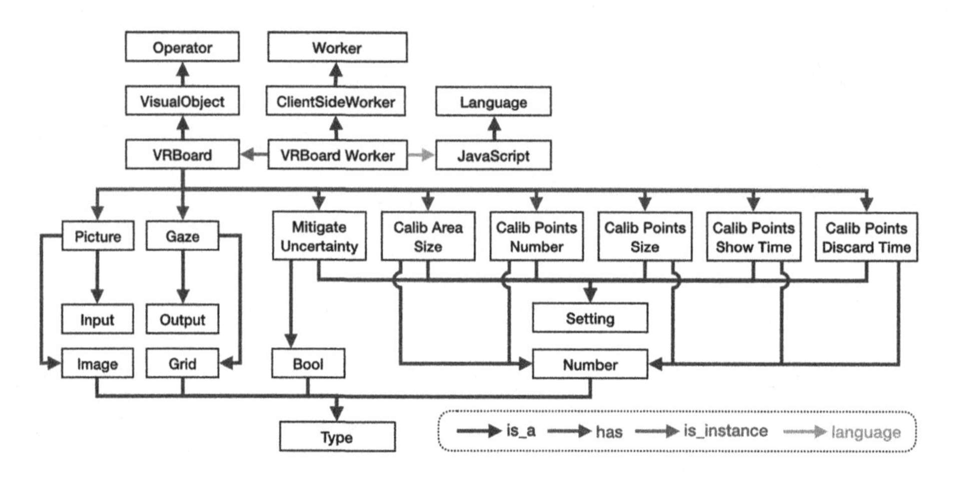

Fig. 6. Fragment of VRBoard ontology used in VRSciVi

Ontologies are designed within the ONTOLIS visual editor and have a proprietary JSON-based format .ONT [4]. As a rule, we use only a restricted set of basic relationship types to maintain the unified ontology model to specify different application ontologies for solving different tasks and reusing the built-in SciVi reasoning mechanism. This basic relationship set enables to improve the efficiency of code generation [24].

As can be seen in Fig. 6, VRBoard has the input "Picture" of the "Image" type that denotes a visual stimulus to be placed in VR and provides "Gaze" of the "Grid" type as an output. Also, there are several numerical settings related to the calibration ("Calib Area Size" – the angular size of the calibration grid, "Calib Points Number" – the number of target points in the calibration pattern, "Calib Point Size" – the angular size of the target point, "Calib Point Show Time" – the time to show each point, "Calib Point Discard Time" – the time to discard for trimming the initial saccade). The settings mentioned above help us in experimenting to find the optimal calibration strategy.

The "Mitigate Uncertainty" Boolean flag toggles the error correction. If it is set to True, reported gaze vectors are rotated by the calculated versors, otherwise just CF values are calculated and transmitted along with unchanged reported gaze vectors. "VRBoard Worker" provides a client-side JavaScript-implementation of VRBoard. If needed, this implementation makes the VRBoard plugin along with the uncertainty mitigation algorithm available in any software generated by SciVi.

Like the entire SciVi platform, VRSciVi Workbench is an open-source project publicly available on the Web: https://scivi.tools/vrscivi.

7 Discussion

To estimate the quality of uncertainty mitigation, we run the accuracy assessment described in Sect. 5.1 considering not only the reported gaze vectors, but also the corrected ones. Figure 7 sums up the results of 10 trials. In each trial, first, the built-in calibration was performed, then the custom calibration, and after that, the gaze data of sequential looking at 25 target points (aligned by 5 grid) were collected. Red ellipses bound reported gaze targets (just like in Fig. 3) and green ellipses bound corrected gaze targets. Blue points denote ground-truth target locations. Each sample's background highlights whether the correction algorithm improved both accuracy and precision (green), improved either accuracy or precision (yellow), impaired accuracy and precision (red).

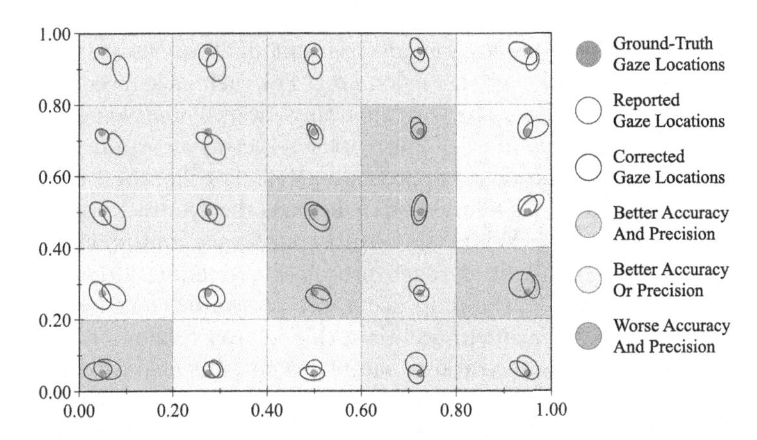

Fig. 7. Comparison of ground-truth, reported, and corrected gaze locations (Color figure online)

The comparison of accuracy and precision of reported and corrected gaze data collected in this experiment is given in Table 1.

Table 1. Comparison of reported and corrected gaze data accuracy and precision

	Min. error	Max. error	Av. error	Std. dev. of error
Reported gaze data	0.03°	2.15°	0.71°	0.41°
Corrected gaze data	0°	1.59°	0.53°	0.28°
Uncertainty Mitigation	100%	26%	25%	32%

As shown in the table, the proposed uncertainty mitigation algorithm increases the eye tracking accuracy by 25% and precision by 32%. The maximal registered error (1.59°), average error (0.53°) and standard deviation (0.28°) are still too large to reliably distinguish individual letters in the texts during the reading study that requires the error to be no more than 0.63°. Nevertheless, the minimal error of 0° proves the potential of the proposed approach.

To further improve the accuracy, other calibration patterns should be examined [8] along with more sophisticated versors' interpolation strategies. At the same time, it is important to keep the calibration process as simple as possible so as not to exhaust the user [8,31]. Currently, the built-in and custom calibration routines take ca. 30 s and 15 s respectively, which is fairly fast.

8 Conclusion

We propose a quaternion-based model suitable to evaluate and mitigate eye gaze tracking uncertainty in VR applications. This model requires a single-pass custom calibration of an eye tracker performed right after a built-in calibration procedure. During the custom calibration, the differences between reported and ground-truth gaze directions are encoded as unit quaternions (versors), which are afterward interpolated over the vision area. For each gaze direction sampled from the eye tracking device, the corresponding interpolated versor is used for correction, whereby the real part of this versor serves as a certainty factor.

We implemented the proposed approach in VRSciVi Workbench to use with Vive Pro Eye VR HMD. This allowed us to increase the eye tracking accuracy by 25% and precision by 32%. While the resulting accuracy and precision are still not high enough to reliably study such processes as reading large texts (more than 100 words long) in VR, the proposed approach can be considered as having potential for future refinement. In addition, this approach allows evaluating the uncertainty for each eye gaze tracking sample to consider it during the upcoming analysis stage. Ontology-driven software development tools of SciVi provide high-level means to generate the interface for parametrization of the proposed algorithm and to adapt it for integration in a software to meet the specifics of different VR-based experiments.

For future work, we plan to try overcoming the limitations of the currently implemented algorithm by using machine learning methods for versor interpolation, as well as to experiment with more sophisticated custom calibration strategies.

References

1. Andersson, R., Nyström, M., Holmqvist, K.: Sampling frequency and eye-tracking measures: how speed affects durations, latencies, and more. J. Eye Mov. Res. **3**(3) (2010). https://doi.org/10.16910/jemr.3.3.6
2. Binaee, K., Diaz, G., Pelz, J., Phillips, F.: Binocular eye tracking calibration during a virtual ball catching task using head mounted display. In: Proceedings of the ACM Symposium on Applied Perception, pp. 15–18 (2016). https://doi.org/10.1145/2931002.2931020
3. Blascheck, T., Kurzhals, K., Raschke, M., Burch, M., Weiskopf, D., Ertl, T.: State-of-the-art of visualization for eye tracking data. In: EuroVis - STARs, pp. 63–82 (2014). https://doi.org/10.2312/eurovisstar.20141173
4. Chuprina, S., Nasraoui, O.: Using ontology-based adaptable scientific visualization and cognitive graphics tools to transform traditional information systems into intelligent systems. Sci. Vis. **8**(1), 23–44 (2016)
5. Clay, V., König, P., König, S.U.: Eye tracking in virtual reality. J. Eye Mov. Res. **12**(1) (2019). https://doi.org/10.16910/jemr.12.1.3
6. Clifford, C.W.G., Palmer, C.J.: Adaptation to the direction of others' gaze: a review. Frontiers Psychol. **9** (2018). https://doi.org/10.3389/fpsyg.2018.02165
7. Drewes, H., Pfeuffer, K., Alt, F.: Time- and space-efficient eye tracker calibration. In: Proceedings of the 11th ACM Symposium on Eye Tracking Research & Applications, pp. 1–8 (2019). https://doi.org/10.1145/3314111.3319818
8. Hassoumi, A., Peysakhovich, V., Hurter, C.: Improving eye-tracking calibration accuracy using symbolic regression. PLoS ONE **14**(3) (2019). https://doi.org/10.1371/journal.pone.0213675
9. Holmqvist, K., Andersson, R.: Eye-Tracking: A Comprehensive Guide to Methods. Paradigms and Measures. Lund Eye-Tracking Research Institute, Lund (2017)
10. Iacobi, J.C.E.: Software for analyzing user experiences in virtual reality using eye tracking. KTH Royal Institute of Technology (2018). https://kth.diva-portal.org/smash/get/diva2:1231972/FULLTEXT01.pdf. Visited 16 Feb 2022
11. Imaoka, Y., Flury, A., de Bruin, E.D.: Assessing saccadic eye movements with head-mounted display virtual reality technology. Frontiers Psychiatry **11** (2020). https://doi.org/10.3389/fpsyt.2020.572938
12. Jogeshwar, A., Diaz, G., Farnand, S., Pelz, J.: The cone model: recognizing gaze uncertainty in virtual environments. In: IS&T International Symposium on Electronic Imaging (2020). https://doi.org/10.2352/ISSN.2470-1173.2020.9.IQSP-288
13. Kuipers, J.B.: Quaternions and Rotation Sequences: A Primer with Applications to Orbits. Aerospace and Virtual Reality. Princeton University Press, Princeton (1999)
14. Kuperman, V., et al.: Text reading in English as a second language: evidence from the multilingual eye-movements corpus (MECO). Stud. Second Lang. Acquisition (2020)
15. Lang, B.: Eye-tracking is a game changer for vr that goes far beyond foveated rendering (2013). https://www.roadtovr.com/. https://www.roadtovr.com/why-eye-tracking-is-a-game-changer-for-vr-headsets-virtual-reality/. Visited 16 Feb 2022

16. Llanes-Jurado, J., Marín-Morales, J., Guixeres, J., Alcañiz, M.: Development and calibration of an eye-tracking fixation identification algorithm for immersive virtual reality. Sensors **20**(17) (2020). https://doi.org/10.3390/s20174956
17. Lohr, D.J., Friedman, L., Komogortsev, O.V.: Evaluating the data quality of eye tracking signals from a virtual reality system: case study using SMI's eye-tracking HTC Vive. arXiv (2019). https://arxiv.org/abs/1912.02083. Visited 16 Feb 2022
18. Mirault, J., Guerre-Genton, A., Dufau, S., Grainger, J.: Using virtual reality to study reading: an eye-tracking investigation of transposed-word effects. Methods Psychol. **3** (2020). https://doi.org/10.1016/j.metip.2020.100029
19. Mohanto, B., Islam, A.B.M.T., Gobbetti, E., Staadt, O.: An integrative view of foveated rendering. Comput. Graph. (2021). https://doi.org/10.1016/j.cag.2021.10.010
20. Popov, A., Shakhuro, V., Konushin, A.: Operator's gaze direction recognition. In: Proceedings of the 31st International Conference on Computer Graphics and Vision (GraphiCon 2021), pp. 475–487 (2021). https://doi.org/10.20948/graphicon-2021-3027-475-487
21. Qian, K., et al.: An eye tracking based virtual reality system for use inside magnetic resonance imaging systems. Sci. Rep. **11**, 1–7 (2021)
22. Ryabinin, K., Belousov, K.: Visual analytics of gaze tracks in virtual reality environment. Sci. Vis. **13**(2), 50–66 (2021). https://doi.org/10.26583/sv.13.2.04
23. Ryabinin, K., Chuprina, S.: Development of ontology-based multiplatform adaptive scientific visualization system. J. Comput. Sci. **10**, 370–381 (2015). https://doi.org/10.1016/j.jocs.2015.03.003
24. Ryabinin, K., Chuprina, S.: Ontology-driven edge computing. In: Krzhizhanovskaya, V.V., et al. (eds.) ICCS 2020. LNCS, vol. 12143, pp. 312–325. Springer, Cham (2020). https://doi.org/10.1007/978-3-030-50436-6_23
25. Sharafi, Z., Sharif, B., Guéhéneuc, Y.G., Begel, A., Bednarik, R., Crosby, M.: A practical guide on conducting eye tracking studies in software engineering. Empir. Softw. Eng. **25**, 3128–3174 (2020)
26. Shehu, I.S., Wang, Y., Athuman, A.M., Fu, X.: Remote eye gaze tracking research: a comparative evaluation on past and recent progress. Electronics **10**(24) (2021). https://doi.org/10.3390/electronics10243165
27. Sipatchin, A., Wahl, S., Rifai, K.: Eye-tracking for clinical ophthalmology with virtual reality (VR): a case study of the HTC Vive Pro eye's usability. Healthcare **9**(2) (2021). https://doi.org/10.3390/healthcare9020180
28. Skulmowski, A., Bunge, A., Kaspar, K., Pipa, G.: Forced-choice decision-making in modified trolley dilemma situations: a virtual reality and eye tracking study. Frontiers Behav. Neurosci. **8** (2014). https://doi.org/10.3389/fnbeh.2014.00426
29. Sonntag, D., et al.: Cognitive monitoring via eye tracking in virtual reality pedestrian environments. In: Proceedings of the 4th International Symposium on Pervasive Displays, pp. 269–270 (2015). https://doi.org/10.1145/2757710.2776816
30. Stein, N., et al.: A comparison of eye tracking latencies among several commercial head-mounted displays. i-Perception **12**(1), 1–16 (2021). https://doi.org/10.1177/2041669520983338
31. Tripathi, S., Guenter, B.: A statistical approach to continuous self-calibrating eye gaze tracking for head-mounted virtual reality systems. In: 2017 IEEE Winter Conference on Applications of Computer Vision (WACV), pp. 862–870 (2017). https://doi.org/10.1109/WACV.2017.101

On a Nonlinear Approach to Uncertainty Quantification on the Ensemble of Numerical Solutions

Aleksey K. Alekseev, Alexander E. Bondarev(✉), and Artem E. Kuvshinnikov

Keldysh Institute of Applied Mathematics RAS, Moscow, Russia
bond@keldysh.ru

Abstract. The estimation of the approximation errors using the ensemble of numerical solutions is considered in the Inverse Problem statement. The Inverse Problem is set in the nonlinear variational formulation that provides additional opportunities. The ensemble of numerical results is analyzed, which is obtained by the OpenFOAM solvers for the inviscid compressible flow containing an oblique shock wave. The numerical tests demonstrated feasibility to compute the approximation errors without any regularization. The refined solution, corresponding the mean of numerical solutions with the approximation error correction, is also computed and compared with the analytic one.

Keywords: Uncertainty quantification · Ensemble of numerical solutions · Euler equations · Oblique shock · OpenFOAM · Inverse Problem

1 Introduction

The standards, regulating the applications of Computational Fluid Dynamics [1, 2], recommend the verification of solutions and codes that implies the estimation of the approximation error. By this reason the computationally inexpensive methods for *a posteriori* estimation of the pointwise approximation error are of current interest. We consider herein the computationally inexpensive approach to *a posteriori* error estimation based on the ensemble of numerical solutions obtained by different algorithms [3–5]. This approach may be implemented in a non-intrusive manner using certain postprocessor. The refined solution and the approximation errors are estimated using the combination of solutions at every grid node, which is processed by the Inverse Ill-posed Problem in the variational formulation. The results of the numerical tests for compressible Euler equations are provided. The ensemble of solutions is computed by four OpenFOAM solvers for the inviscid compressible flow. We compare the refined solution and approximation errors computed by Inverse problem with the analytic solution and corresponding error estimates.

D. Groen et al. (Eds.): ICCS 2022, LNCS 13353, pp. 637–645, 2022.
https://doi.org/10.1007/978-3-031-08760-8_52

2 The Linear Statement for Approximation Error Estimation Using the Distances Between Numerical Solutions

We consider an ensemble of numerical solutions (grid functions) $u_m^{(i)}$ $(i = 1 \ldots n)$, computed on the same grid by n different solvers similarly to [3–6]. Papers [3–5] address the estimation of the approximation errors from the differences between solutions and Inverse Problem posed in the variational statement form with the zero order Tikhonov regularization.

A local pointwise analysis of the flowfield is used. The projection of the exact solution $\tilde{u}$ onto the grid is noted as $\tilde{u}_m$, the approximation error for i-th solution is noted as $\Delta u_m^{(i)}$. Here m is the grid point number ($m = 1,\ldots,L$).

Then $N = n.(n-1)/2$ equations that relate the approximation errors and the computable differences of the numerical solutions $d_{ij,m} = u_m^{(i)} - u_m^{(j)} = \tilde{u}_{h,m} + \Delta u_m^{(i)} - \tilde{u}_{h,m} - \Delta u_m^{(j)} = \Delta u_m^{(i)} - \Delta u_m^{(j)}$. The following relations hold for a vectorized form of the differences $f_{k,m}$:

$$D_{kj} \Delta u_m^{(j)} = f_{k,m}. \tag{1}$$

Herein, D_{kj} is a rectangular $N \times n$ matrix, the summation over a repeating index is implied (no summation over repeating index m).

The Eq. (1) has the form:

$$\begin{pmatrix} 1 & -1 & 0 \\ 1 & 0 & -1 \\ 0 & 1 & -1 \end{pmatrix} \begin{pmatrix} \Delta u_m^{(1)} \\ \Delta u_m^{(2)} \\ \Delta u_m^{(3)} \end{pmatrix} = \begin{pmatrix} f_{1,m} \\ f_{2,m} \\ f_{3,m} \end{pmatrix} = \begin{pmatrix} d_{12,m} \\ d_{13,m} \\ d_{23,m} \end{pmatrix} = \begin{pmatrix} u_m^{(1)} - u_m^{(2)} \\ u_m^{(1)} - u_m^{(3)} \\ u_m^{(2)} - u_m^{(3)} \end{pmatrix}. \tag{2}$$

in the simplest case of three numerical solutions.

The solution of system (1) is invariant to a shift transformation $\Delta u_m^{(j)} = \Delta \tilde{u}_m^{(j)} + b$ (for any $b \in (-\infty, \infty)$). By this reason, this problem of approximation error estimation is underdetermined and ill-posed (unstable). A regularization ([7, 8]) is usually necessary in order to obtain the steady and bounded solution of the ill-posed problems. The zero order Tikhonov regularization is used in works [3–5]. It enables to obtain solutions with the minimum shift error $|b|$, since the condition of $\|\Delta u_m\|_{L_2}$ minimum restricts the absolute value of b:

$$\min_{b_m}(\delta(b_m)) = \min_{b_m} \sum_j^n (\Delta u_m^{(j)})^2/2 = \min_{b_m} \sum_j^n (\Delta \tilde{u}_m^{(j)} + b_m)^2/2. \tag{3}$$

One may see that:

$$\Delta\delta(b_m) = \sum_j^n (\Delta \tilde{u}_m^{(j)} + b_m)\, \Delta b_m, \tag{4}$$

and the minimum of the norm occurs at b_m that equals the mean error (with the opposite sign):

$$b_m = -\frac{1}{n} \sum_j^n \Delta \tilde{u}_m^{(j)} = -\Delta \bar{u}_m. \tag{5}$$

Thus, the error $\left|\Delta u_m^{(j)}\right|$ cannot be less than $|b_m|$ (the mean error value) in the linear approach. By this reason, the error estimates contain unremovable errors.

The variational statement of the Inverse Problem implies the minimization of the functional:

$$\varepsilon_m(\Delta \vec{u}) = 1/2(D_{ij}\Delta u_m^{(j)} - f_{i,m}) \cdot (D_{ik}\Delta u_m^{(k)} - f_{i,m}) + \alpha/2(\Delta u_m^{(j)} \Delta u_m^{(k)}). \quad (6)$$

The functional (6) consists of the discrepancy of the predictions and observations and the zero order Tikhonov regularization term (α is the regularization parameter). The efficiency of the linear statement was confirmed by the set of numerical tests presented in papers [3–5].

3 The Nonlinear Statement for the Estimation of the Refined Solution and the Approximation Error on the Ensemble of Numerical Solutions

The shift invariance discussed above causes an instability and the presence of unremovable error. Fortunately, this invariance is purely linear effect. So, some transition to the nonlinear statement allows eliminating the shift invariance. Let's consider the corresponding opportunities.

We consider three formulations with the different number of nonlinear terms. The simplest one is presented by Eq. (7) (and marked as Variant 1):

$$\begin{pmatrix} 2\tilde{u}_m + \Delta u_m^{(1)} & 0 & 0 & \tilde{u}_m \\ 1 & 0 & -1 & 0 \\ 0 & 1 & -1 & 0 \\ 1/3 & 1/3 & 1/3 & 1 \end{pmatrix} \cdot \begin{pmatrix} \Delta u_m^{(1)} \\ \Delta u_m^{(2)} \\ \Delta u_m^{(3)} \\ \tilde{u}_m \end{pmatrix} = \begin{pmatrix} (u_m^{(1)})^2 \\ u_m^{(1)} - u_m^{(3)} \\ u_m^{(2)} - u_m^{(3)} \\ (u_m^{(1)} + u_m^{(2)} + u_m^{(3)})/3 \end{pmatrix}. \quad (7)$$

It contains nonlinear term $\left(u_m^{(1)}\right)^2$ at right hand side that prohibits the shift invariance $u_m^{(j)} = \tilde{u}_m^{(j)} + b$. Additionally, this statement enables a direct estimation of refined solution $\tilde{u}_m$, which corresponds to the average of numerical solutions with account of approximation errors.

The solution of Eq. (7) corresponds to the minimum of the following functional (without any regularization and with the dependence of the matrix D_{ij}^m on local parameters):

$$\varepsilon_m(\Delta \vec{u}) = 1/2(D_{ij}^m \Delta u_m^{(j)} - f_{i,m}) \cdot (D_{ik}^m \Delta u_m^{(k)} - f_{i,m}). \quad (8)$$

The nonlinear system (7) is degenerated that can be directly checked. Fortunately, the determinant of matrix operator D_{ij}^m is far from zero. By this reason the linearization of the following form $(D_{ij}^m)^n (\Delta u_m^{(j)})^{n+1} = f_{i,m}$ (n — number of iteration) can be used at the iterative minimization of the functional (8).

The second formulation (variant 2 presented by Eq. 9) contains two nonlinear terms:

$$\begin{pmatrix} 2\tilde{u}_m + \Delta u_m^{(1)} & 0 & 0 & \tilde{u}_m \\ 0 & 2\tilde{u}_m + \Delta u_m^{(2)} & 0 & \tilde{u}_m \\ 0 & 1 & -1 & 0 \\ 1/3 & 1/3 & 1/3 & 1 \end{pmatrix} \cdot \begin{pmatrix} \Delta u_m^{(1)} \\ \Delta u_m^{(2)} \\ \Delta u_m^{(3)} \\ \tilde{u}_m \end{pmatrix} = \begin{pmatrix} (u_m^{(1)})^2 \\ (u_m^{(2)})^2 \\ u_m^{(2)} - u_m^{(3)} \\ (u_m^{(1)} + u_m^{(2)} + u_m^{(3)})/3 \end{pmatrix}. \tag{9}$$

The third formulation (variant 3 presented by Eq. 10) contains all linear terms and all squares of the numerical solutions:

$$\begin{pmatrix} 2\tilde{u}_m + \Delta u_m^{(1)} & 0 & 0 & \tilde{u}_m \\ 0 & 2\tilde{u}_m + \Delta u_m^{(2)} & 0 & \tilde{u}_m \\ 0 & 0 & 2\tilde{u}_m + \Delta u_m^{(3)} & \tilde{u}_m \\ 1 & -1 & 0 & 0 \\ 1 & 0 & -1 & 0 \\ 0 & 1 & -1 & 0 \\ 1/3 & 1/3 & 1/3 & 1 \end{pmatrix} \cdot \begin{pmatrix} \Delta u_m^{(1)} \\ \Delta u_m^{(2)} \\ \Delta u_m^{(3)} \\ \tilde{u}_m \end{pmatrix} = \begin{pmatrix} (u_m^{(1)})^2 \\ (u_m^{(2)})^2 \\ (u_m^{(3)})^2 \\ u_m^{(1)} - u_m^{(2)} \\ u_m^{(1)} - u_m^{(3)} \\ u_m^{(2)} - u_m^{(3)} \\ (u_m^{(1)} + u_m^{(2)} + u_m^{(3)})/3 \end{pmatrix}. \tag{10}$$

The considered nonlinear statement enables the estimation of approximation errors without any regularization. In addition, the refined solution $\tilde{u}_m$ can be computed.

4 The Minimization Algorithm

We apply the steepest descent iterations (k is the number of the iteration) for the minimization of the functionals (6) and (8):

$$\Delta u_m^{(j),k+1} = \Delta u_m^{(j),k} - \tau \nabla \varepsilon_m. \tag{11}$$

The gradient is obtained by the numerical differentiation. The iterations terminate at a small value of the functional $\varepsilon \leq \varepsilon_* \ (\varepsilon_* = 10^{-8}$ was used in these tests).

The solutions depend on the regularization parameter α in the linear statement. At $\alpha = 0$ the magnitude of $\left|\Delta u^{(j)}(\alpha)\right|$ is not bounded. The limit $\alpha \to \infty$ is not acceptable also, since $\left|\Delta u^{(j)}(\alpha)\right| \to 0$. A range of the regularization parameter α exists where the weak dependence of the solution on α is manifested (numerical tests determine it as $\alpha \in \left(10^{-6}, 10^{-1}\right)$). The solution $\Delta u^{(j)}(\alpha)$ is close to the exact one in this range and may be considered as the regularized solution [8].

5 The Test Problem

The solutions and the errors considered in the paper correspond to the oblique shock waves, which are chosen due to the availability of analytic solutions and the relatively high level of the approximation errors. The test problem is governed by the two dimensional compressible Euler equations.

The shocked flowfield is engendered by a plate at the angle of attack $\alpha = 20°$ in the uniform supersonic flow ($M = 4$). The results are compared with the projection of the analytic solution on the computational grid.

The inflow parameters are set on the left boundary ("*inlet*") and on the upper boundary ("*top*"). The zero gradient condition for the gas dynamic functions is specified on the right boundary ("*outlet*"). The condition of zero normal gradient is posed for the pressure and the temperature, and the condition "*slip*" is posed for the speed, that ensures the non-penetration on the plate surface.

The following solvers from the OpenFOAM software package [9] are used:

- *rhoCentralFoam* (rCF), which is based on a central-upwind scheme [10, 11].
- *sonicFoam* (sF), which is based on the PISO algorithm [12].
- *pisoCentralFoam* (pCF) [13], which combines the Kurganov-Tadmor scheme [10] and the PISO algorithm [12].
- *QGDFoam* (QGDF), which implements the quasi-gas dynamic equations [14].

It is important for the problem considered in this paper that these solvers use the quite different algorithms.

We estimate the refined solution and approximation errors by minimizing the functional (8) using Expressions (7),(9),(10). The analytic solution corresponds to the oblique shock wave (Rankine-Hugoniot relations).

6 Numerical Results

The Figs. 1 and 3 present the pieces of vectorized grid function of the density obtained by the Inverse Problem in comparison with the exact error. The Figs. 2 and 4 present the pieces of the vectorized density error in comparison with the exact error. The index along the abscissa axis $i = N_x(k_x - 1) + m_y$ is defined by indexes along $X(k_x)$ and $Y(m_y)$. The periodical jump of variables corresponds to the transition through the shock wave. Results for $\alpha = 20°$, $M = 4$ are provided.

The Fig. 1 demonstrates close results for nonlinear estimations of the refined density (Var1, Var2 and Var3) for numerical solutions computed using rCF, pCF and sF algorithms. Var1, Var2 and Var3 correspond to nonlinear estimation of $\tilde{u}_m$ by Eqs. (7), (9) and (10). Figure 2 shows that the numerical density errors for linear and nonlinear estimates are close. The error estimates are shifted with respect to the exact error.

The Figs. 3 and 4 provide the density and its errors for the set of numerical solutions, which are computed by rCF, pCF and QGDF solvers. Var1 corresponds to nonlinear estimation of $\tilde{u}_m$ by Eqs. (7). The replacement of sF by QGDF weakly affects on the behaviour of variables.

Both linear and nonlinear approaches provide close results for the error estimates. However, nonlinear options (in contrast to linear one) provide the refined solution $\tilde{u}_m$ and do not use any regularization that presents the main advantage of this approach.

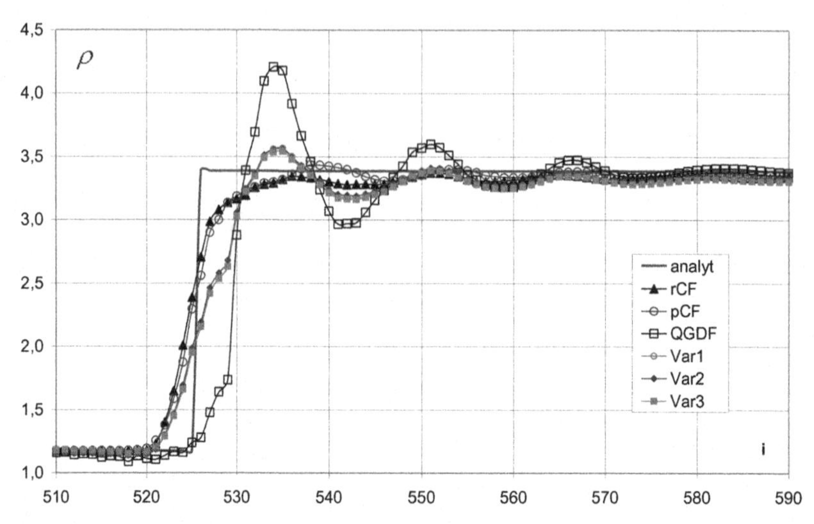

Fig. 1. The vectorized values of the density computed by rCF, pCF, sF, refined solutions obtained by the Inverse Problem (Var1, Var2 and Var3) and the analytic solution.

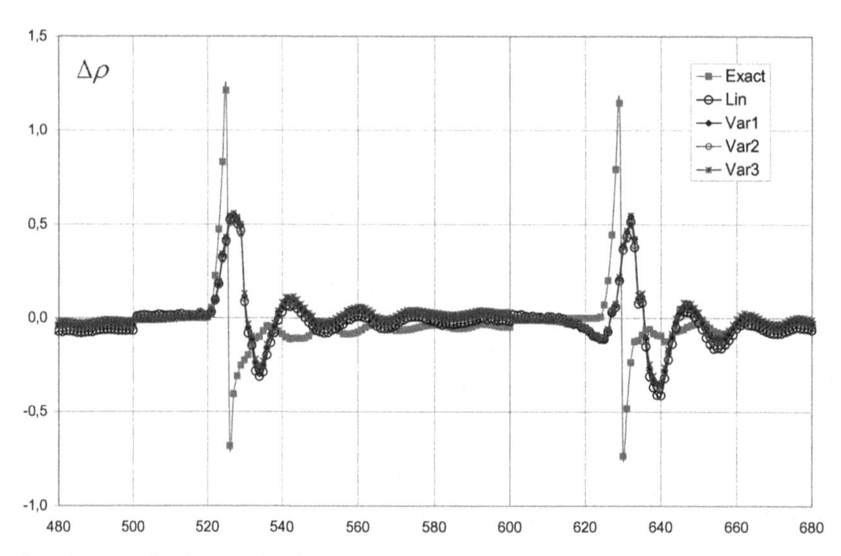

Fig. 2. The vectorized exact density error for solution computed by rCF algorithm, the linear error estimates, and the nonlinear estimates of error (Var1, Var2 and Var3).

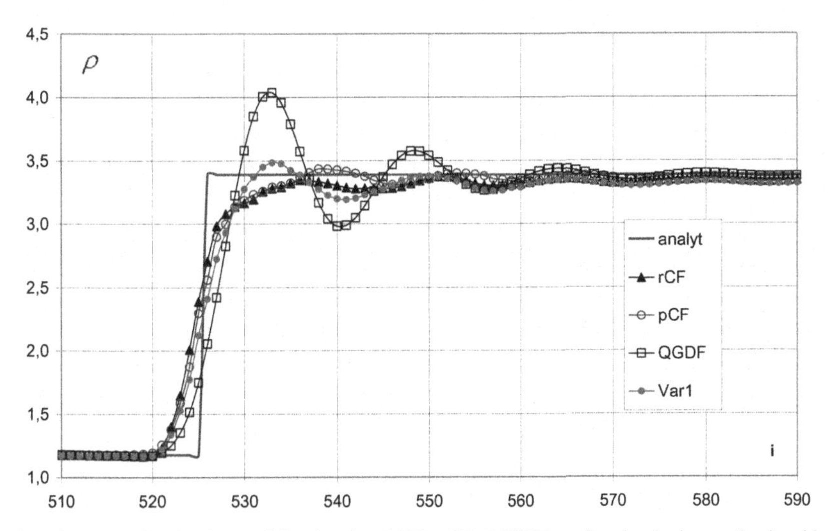

Fig. 3. The vectorized values of the density (rCF, pCF, QGDF), refined solutions, obtained by the Inverse Problem (Var1), and the analytic solution.

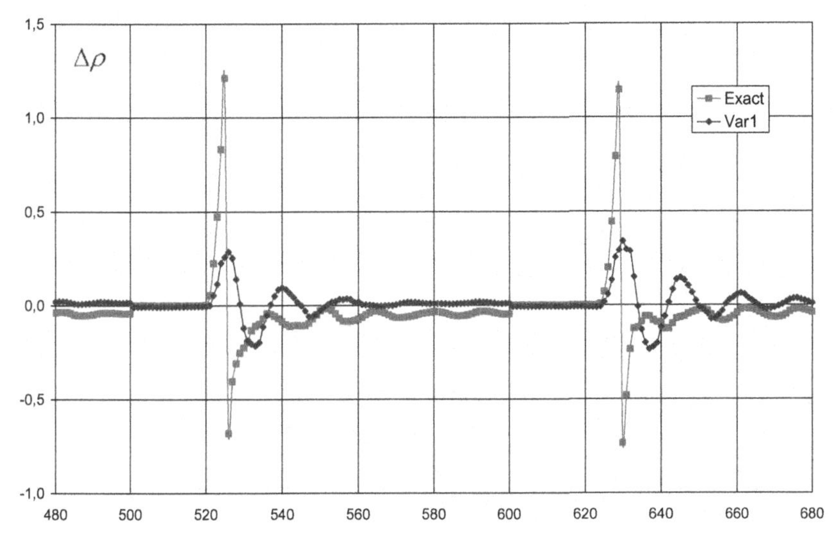

Fig. 4. The vectorized exact density error for solution computed by rCF solver and the nonlinear estimates of error (Var1).

7 Discussion

The Inverse Problem (IP) based approach was used in the linear statement by [3] for the supersonic axisymmetric flows around cones, for the crossing shocks (Edney-I and Edney-VI patterns by [4] and for the oblique shock flow pattern by [5].

Here, we consider a nonlinear extension of IP-based method [3–5] for the oblique shock flow pattern. The analytic solutions and exact errors (obtained by the comparison with the analytic solutions) are used as the etalons.

No regularization is used at the nonlinear case in contrast to the linear statement [3–5]. This circumstance is caused by the nonlinearity that prevents from the linear shift invariance that is specific for the linear statement.

Additionally, the nonlinear option enables the explicit estimation of the refined solution $\tilde{u}_m$. At first glance, it seems to be impossible to obtain four independent variables $\tilde{u}_m$, $\Delta u_m^{(1)}$, $\Delta u_m^{(2)}$, $\Delta u_m^{(3)}$ using three input parameters $u_m^{(1)}$, $u_m^{(2)}$, $u_m^{(3)}$. The unexpected success of the nonlinear approach may be caused by implicit correlations between $\tilde{u}_m$ and $\Delta u_m^{(i)}$ that are contained in the input data $u_m^{(i)}$. The correlations between $\Delta u_m^{(i)}$ really exist and are demonstrated by [6]. One of the reasons for these correlations is the similar behaviour of the approximation errors at discontinuities, see Fig. 5.

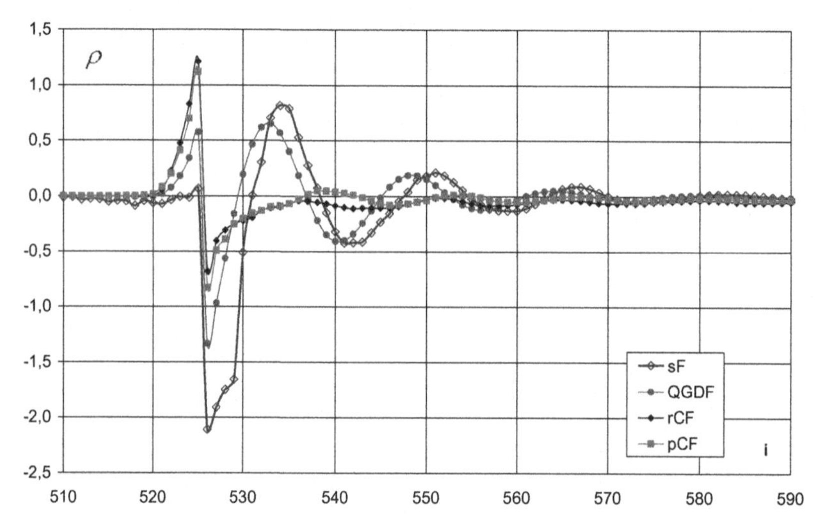

Fig. 5. The correlated nature of approximation errors in the vicinity of shock wave.

However, the additional analysis of the advantages and limitations of the nonlinear version is necessary.

8 Conclusion

An ensemble of numerical solutions obtained by different methods is considered for estimation of the pointwise refined solution and the approximation errors. The post-processing is applied, which is based on the Inverse Problem. The linear and nonlinear statements are analyzed and compared.

The nonlinear approach (in contrast to linear one) provides the estimation of the refined solution $\tilde{u}_m$, does not use any regularization, and, formally, has no limitations on the accuracy of error estimation that is specific for linear approach [3–5].

The four solvers from the OpenFOAM software package are used in numerical experiments. The two-dimensional inviscid flow pattern engendered by the oblique shock wave is considered as a test case.

The numerical tests demonstrate the successful estimation of the refined solution and the pointwise approximation error in the nonlinear statements.

References

1. Guide for the Verification and Validation of Computational Fluid Dynamics Simulations, American Institute of Aeronautics and Astronautics, AIAA-G-077–1998 (1998)
2. Standard for Verification and Validation in Computational Fluid Dynamics and Heat Transfer, ASME V&V 20–2009 (2009)
3. Alekseev, A.K., Bondarev, A.E., Kuvshinnikov, A.E.: A posteriori error estimation via differences of numerical solutions. In: Krzhizhanovskaya, V.V., et al. (eds.) ICCS 2020. LNCS, vol. 12143, pp. 508–519. Springer, Cham (2020). https://doi.org/10.1007/978-3-030-50436-6_37
4. Alekseev, A.K., Bondarev, A.E.: The estimation of approximation error using inverse problem and a set of numerical solutions. Inverse Problems Sci. Eng. **29**(13), 3360–3376 (2021). https://doi.org/10.1080/17415977.2021.2000604
5. Alekseev, A.K., Bondarev, A.E., Kuvshinnikov, A.E.: A comparison of the Richardson extrapolation and the approximation error estimation on the ensemble of numerical solutions. In: Paszynski, M., Kranzlmüller, D., Krzhizhanovskaya, V.V., Dongarra, J.J., Sloot, P.M.A. (eds.) ICCS 2021. LNCS, vol. 12747, pp. 554–566. Springer, Cham (2021). https://doi.org/10.1007/978-3-030-77980-1_42
6. Alekseev, A.K., Bondarev, A.E.: On a posteriori error estimation using distances between numerical solutions and angles between truncation errors. Math. Comput. Simul. **190**, 892–904 (2021). https://doi.org/10.1016/j.matcom.2021.06.014
7. Tikhonov, A.N., Arsenin, V.Y.: Solutions of Ill-Posed Problems. Winston and Sons, Washington DC (1977)
8. Alifanov, O.M., Artyukhin, E.A., Rumyantsev S.V.: Extreme Methods for Solving Ill-Posed Problems with Applications to Inverse Heat Transfer Problems. Begell House (1995)
9. OpenFOAM, http://www.openfoam.org. Accessed 20 Jan 2022
10. Kurganov, A., Tadmor, E.: New high-resolution central schemes for nonlinear conservation laws and convection-diffusion equations. J. Comput. Phys. **160**(1), 241–282 (2000). https://doi.org/10.1006/jcph.2000.6459
11. Greenshields, C., Wellerr, H., Gasparini, L., Reese, J.: Implementation of semi-discrete, non-staggered central schemes in a colocated, polyhedral, finite volume framework, for high-speed viscous flows. Int. J. Numer. Meth. Fluids **63**(1), 1–21 (2010). https://doi.org/10.1002/fld.2069
12. Issa, R.: Solution of the implicit discretized fluid flow equations by operator splitting. J. Comput. Phys. **62**(1), 40–65 (1986). https://doi.org/10.1016/0021-9991(86)90099-9
13. Kraposhin, M., Bovtrikova, A., Strijhak, S.: Adaptation of Kurganov-Tadmor numerical scheme for applying in combination with the PISO method in numerical simulation of flows in a wide range of Mach numbers. Procedia Comput. Sci. **66**, 43–52 (2015). https://doi.org/10.1016/j.procs.2015.11.007
14. Kraposhin, M.V., Smirnova, E.V., Elizarova, T.G., Istomina, M.A.: Development of a new OpenFOAM solver using regularized gas dynamic equations. Comput. Fluids **166**, 163–175 (2018). https://doi.org/10.1016/j.compfluid.2018.02.010

On the Use of Sobol' Sequence for High Dimensional Simulation

Emanouil Atanassov(✉) and Sofiya Ivanovska

Institute of Information and Communication Technologies,
Bulgarian Academy of Sciences, Sofia, Bulgaria
{emanouil,sofia}@parallel.bas.bg

Abstract. When used in simulations, the quasi-Monte Carlo methods utilize specially constructed sequences in order to improve on the respective Monte Carlo methods in terms of accuracy mainly. Their advantage comes from the possibility to devise sequences of numbers that are better distributed in the corresponding high-dimensional unit cube, compared to the randomly sampled points of the typical Monte Carlo method. Perhaps the most widely used family of sequences are the Sobol' sequences, due to their excellent equidistribution properties. These sequences are determined by sets of so-called direction numbers, where researches have significant freedom to tailor the set being used to the problem at hand. The advancements in scientific computing lead to ever increasing dimensionality of the problems under consideration. Due to the increased computational cost of the simulations, the number of trajectories that can be used is limited. In this work we concentrate on optimising the direction numbers of the Sobol' sequences in such situations, when the constructive dimension of the algorithm is relatively high, compared to the number of points of the sequence being used. We propose an algorithm that provides us with such sets of numbers, suitable for a range of problems. We then show how the resulting sequences perform in numerical experiments, compared with other well known sets of direction numbers. The algorithm has been efficiently implemented on servers equipped with powerful GPUs and is applicable for a wide range of problems.

Keywords: quasi-Monte Carlo · High-dimensional simulation · Low-discrepancy sequences

1 Introduction

The quasi-Monte Carlo methods are built upon the idea to replace the random numbers, typically produced by pseudorandom number generators, by specially crafted deterministic sequences. Most of the Monte Carlo algorithms can

This work has been financed in part by a grant from CAF America. We acknowledge the provided access to the e-infrastructure of the Centre for Advanced Computing and Data Processing, with the financial support by the Grant No. BG05M2OP001-1.001-0003, financed by the Science and Education for Smart Growth Operational Program (2014–2020) and co-financed by the European Union through the European structural and investment funds.

D. Groen et al. (Eds.): ICCS 2022, LNCS 13353, pp. 646–652, 2022.
https://doi.org/10.1007/978-3-031-08760-8_53

be thought of multi-dimensional integral approximations by average of random samples

$$\int\limits_{E^s} f(x)\,dx \approx \frac{1}{N}\sum_{i=0}^{N-1} f(x_i),$$

therefore the corresponding quasi-Monte Carlo method would use a special s-dimensional sequence to compute the same integral. In this setting the dimension s is called the "constructive dimension" of the algorithm. One justification for the use of low-discrepancy sequences is the Koksma-Hlawka inequality (see, e.g., [3]), which connects the accuracy of the algorithm with one measure of equidistribution of the sequence, called star-discrepancy. For a fixed dimension s, the best possible order of the star-discrepancy is believed to be $N^{-1}\log^s N$, and sequences that achieve this order are called low-discrepancy sequences. Other measures of the quality of distribution of sequences are also used, for example the diaphony [10] or the dyadic diaphony [1].

The Sobol' sequences are one of the oldest known families of low-discrepancy sequences [9], but they are still the most widely used, partially because of their close relation to the binary number system, but mostly because of their good results in high-dimensional settings, see, e.g., [8]. As an example, the price of a financial asset can be modelled by sampling its trajectories along time, where each time step requires one or more random numbers in order to advance. This means that the constructive dimension of the algorithm is a multiple of the number of time steps. Even though the dimension rises fast in such situations, the Sobol' sequences retain their advantage compared to Monte Carlo. Other notions of dimension have been proposed, in order to explain and quantify such effects, e.g., the effective dimension or the average dimension, see [4]. Although these quantities can explain the observed advantage of a quasi-Monte Carlo method based on the Sobol' sequences, they do not change the necessary dimension.

Although initially the low-discrepancy sequences used in quasi-Monte Carlo methods were fully deterministic, there are many theoretical and practical reasons to add some randomness to a quasi-Monte Carlo algorithm. Such a procedure, which modifies a low-discrepancy sequence in a random way, while retaining its equidistribution properties, is usually called "scrambling". In this work we are going to use only the simplest and computationally inexpensive procedure of Matoušek [5], which is also one of the most widely used and provides equal grounds for comparison. Essentially, it consist of performing a bitwise **xor** operation with a fixed random vector for each dimension.

Even though there are more complex scrambling schemes, e.g., the scrambling proposed by Owen in [6], they do not solve one important problem that arises when the constructive dimension of the algorithm becomes high compared with the number of points. Consider a problem where the constructive dimension is s and the number of points is $N = 2^n$ and $s > N/2$. The first bit of the term of the Sobol' sequence in each dimension is determined by the index and the first column of the matrix of direction numbers $A_i = a^i_{jk}$. Since we have s such columns and the index is between 0 and $2^s - 1$ inclusively, only the first n

bits of the column are important. Moreover, the matrix of direction numbers is triangular with ones over the main diagonal, so we have only $n-1$ bits that can take values of either 0 or 1, or 2^{n-1} possibilities. Therefore the Dirichlet principle ensures that if $s > N/2$, there will be two dimensions where the first bits of the sequence will be exactly equal. Even if we add scrambling, the situation does not improve, because the bits will be either exactly equal for all terms of the sequence, or exactly inverted. Thus the correlation between these two dimensions will be much higher in absolute value, compared with Monte Carlo sampling. If the algorithm uses only the first bit of the terms of the sequence, e.g., for making some binary choice, choosing subsets, etc., then such correlation is obviously problematic. There are also many algorithms where a discrete value is to be sampled. In such case, the subsequent bits reveal even worse problems, since the available choices become even less. Thus, even if $s < N/2$, we can still have dimensions where certain bits always coincide in the original Sobol' sequence. Owen scrambling would alleviate such a problem, but only to an extent. We should also mention that following the typical definition of direction numbers related to primitive polynomials, the available choices are even more restricted, especially in the first dimensions.

In the next section we describe our approach to quantify this problem and define measures that have to be optimised. Then we describe our optimisation algorithm. In the numerical experiments we show how these sets perform on typical integration problems.

2 Optimisation Framework

We define measures that should lead to optimal directional numbers in view of the problems described in the previous section. Because of computational costs, we try to obtain numbers that are reusable in many settings. Thus we to cover all dimensions up to some maximal dimension and number of points in a range from 2^m up to and including 2^n. We define several stages. The first stage performs filtering. Since we must avoid coincidences in the columns of the matrices A_i, for a given column of order j in dimension i, we define a measure of coincidences as follows:

$$P(a) = \sum_{k=m}^{n} \sum_{r=1}^{i-1} 2^k c_k(a, a_j^r),$$

where $c_k(a, a_j^r)$ is one if the columns a and $a_j^r = \{a_{ij}^r\}$ coincide in their upper k bits, otherwise zero.

We are going to search for direction numbers column by column, filtering the possible choices to only those, which attain minimum of the function above. Optionally, we can restrict those choices to only columns that follow the usual construction of the Sobol' sequences using primitive polynomials. This seems to improve the results for lower number of dimensions, but is very restrictive when dimensions are higher than 64, so in our experiments we are not using this option. In the next stage we minimise a measure of the equidistribution properties of the sequence, related to Walsh functions. It is very similar to the notion of the diadic diaphony, adjusted to be easily computable for the Sobol' sequence. Consider

$$R(\sigma, n) = \sum_{(m_1,\ldots,m_s)\in L} w(m_1,\ldots,m_s)\, r(\sigma, n; m_1,\ldots,m_s),$$

where $w(m_1,\ldots,m_s) = \prod_{i=1}^{s} 2^{-\alpha k_i}$ if $2^{k_i-1}\ m_i \leq 2^{k_i}$. The weight of $m_i = 0$ is taken to be 1. The set L consists of those integer tuples $(m_1,\ldots,m_s)$, that have at least one non-zero value and $0 \leq m_i < 2^n$. The quantity $r(\sigma, n, m_1,\ldots,m_s)$ depends on the Sobol sequence and the multidimensional Walsh function corresponding to $(m_1,\ldots,m_s)$ and is 0 or 1 depending whether the integral computes exactly to 0 when using the sequence σ with 2^n terms or not. We obtained good results when limiting the set L to only contain tuples with up to 4 non-zero integers, since quasi-Monte Carlo methods have difficulty outperforming Monte Carlo when the effective dimension is higher than 4.

When the number of dimensions s gets higher, the value of this measure grows, especially if the power α is lower. Rounding errors accumulate too. Thus for reasonable values of the dimension s and range of number of points between 2^m and 2^n, we obtain multiple values of the column of direction numbers with measures that either equal or close. Thus the outcome of the optimisation is usually a set of acceptable numbers for the given column. In order to cover multiple possibilities for the number of points we take the measure R computed for different values of the number of points, between m and n, weighted by the number of points used. We usually set $\alpha = 1$, although the choice $\alpha = 2$ is also logical due to the connection with the diadic diaphony. With $\alpha = 1$ we get reproducible results, while choices like $\alpha = 1.5$ may produce different results for different runs due to accumulation of rounding errors and non-deterministic order of summation when computing in parallel. The last stage is a validation stage, where we test the obtained direction numbers on a suitably chosen integral and select those that gave the best results. As these computations are expensive, we do not test all possibilities. Our approach to the validation stage is explained more in details in the next section. The outcome of the optimisation is a set of direction numbers that can be used for all dimensions less than some maximal dimension and for number of points that is less than some fixed power-of-two.

3 Optimisation Algorithm

We proceed to compute direction numbers column by column, initially computing all first columns dimension by dimension and then proceeding to the second columns and so on. In this way the choice of a column from one dimension is only dependent on the columns that are at the same position or ahead in previous dimensions. The filtering stage is straightforward - computing the measure F for each possible value of the column. For a column at position $j \geq 1$ in the dimension $i \geq 1$ we have 2^{n-j} possibilities since the main diagonal consists of ones only. We form an array M of all columns for which F attains its minimum and proceed to the next stage, where we compute the value of R for all columns $x \in M$, selected at the previous stage. If the value of n is too high and this stage becomes prohibitively expensive, we can select random subset of the array of possible columns. In order to take into account possible rounding errors, we select those values for the column, for which $R(\sigma) \leq (1+\varepsilon) \min R(\sigma)$, forming a set Q, and proceed to the next stage. As we only know the first few columns, we only compute terms in R that correspond to these columns, i.e., m_i have only as many bits as is the number of columns. This computation, when done on GPUs, is relatively fast, provided we maintain an array of weights and use dinamic programming. The set Q can be large, but we choose randomly only a small subset for validation. In our experiments we used up to 10 values. For these values of the column we compute the etalon integral of

$$f(x) = \left(\frac{1}{\sqrt{s}} \sum_{i=1}^{s} (2x_i - 1) \right)^2$$

which quantifies the amount of correlation between dimensions. We select the column that produces the best result (performing multiple computations, e.g., 40, randomly filling the yet unfilled positions in the binary matrices). The validation step allows to obtain direction numbers with smoother behaviour.

4 Numerical Results

We tested direction numbers obtained by using our algorithm in comparison with the direction numbers, provided at [7], using criterion D_7, following [2]. In all cases we consider integration problems in a high-dimensional unit cube E^s and we apply the Matoušek scrambling, obtaining 100 different estimates for the integral, which assures us that the RMS error computed is representative. The first subintegral function is used regularly for such kinds of tests:

$$f_1(x) = \left(\frac{1}{\sqrt{s}} \sum_{i=1}^{s} \Phi^{-1}(x_i) \right)^2,$$

where Φ^{-1} is the inverse of the c.d.f. for the normal distribution,

$$f_2(x) = \max\left(0, S \exp\left(\frac{1}{\sqrt{s}} \sum_{i=1}^{s} \Phi^{-1}(x_i) - \frac{\sigma^2}{2} \right) - K \right).$$

The function f_2 corresponds to an approximation of the value of an European call option, such that the exact value of the integral can be computed from the Black-Scholes formula. We used $S = 1, K = 1.05$, $T = 1$, $\sigma = 0.10, r = 0$. In order to better compare the results across the dimensions, we always normalise by dividing by the exact value of the integral. On Fig. 1, 2, 3 and 4 we see how the direction numbers produced by the algorithm compare with the fixed direction numbers. The number of points of the sequence is 2^{10} and 2^{12} and the maximum number of dimensions is 1024 or 16384 respectively. One can see how the accuracy evolves with the increase in the constructive dimension. Our algorithm seems to produce suboptimal results for small dimensions, up to 64. However, it starts to outperform when the number of dimensions becomes larger. It is possible to use fixed direction numbers for the first dimensions and extend the set following the algorithm, but we observe worse results for the larger dimensions in this way. For brevity we do not present results for other integrals, but our experience is that for such integrals where the constructive dimension is high and there are many interactions between dimensions without clear domination of a small number of variables the direction numbers produced by our algorithm are promising. This is a typical situation for all the integrals that we have tested. When the number of dimensions becomes bigger than 64, the new direction numbers show clear advantage.

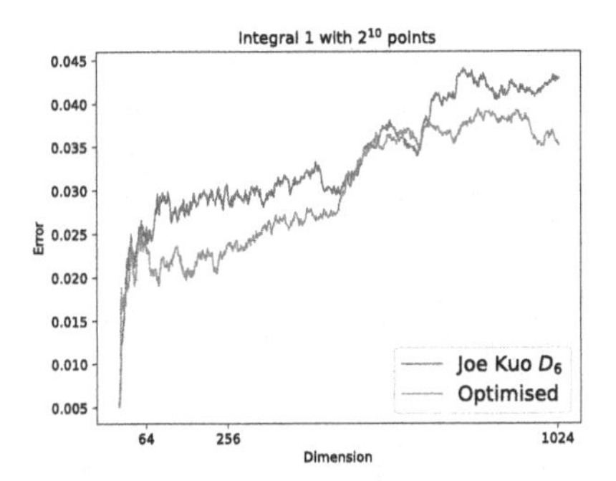

Fig. 1. Estimate the integral 1 for different dimensions using 2^{10} points

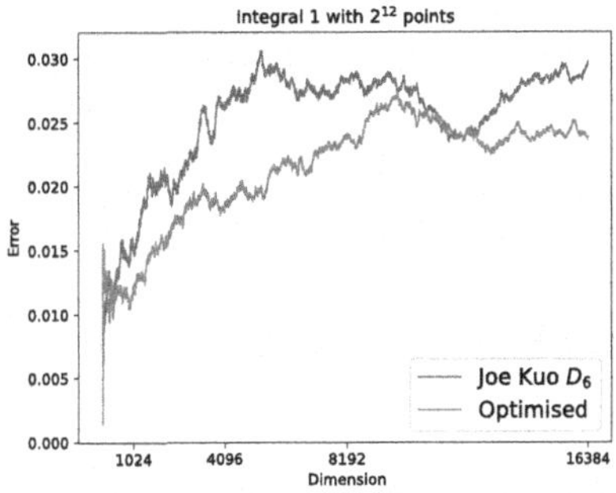

Fig. 2. Estimate the integral 1 for different dimensions using 2^{12} points

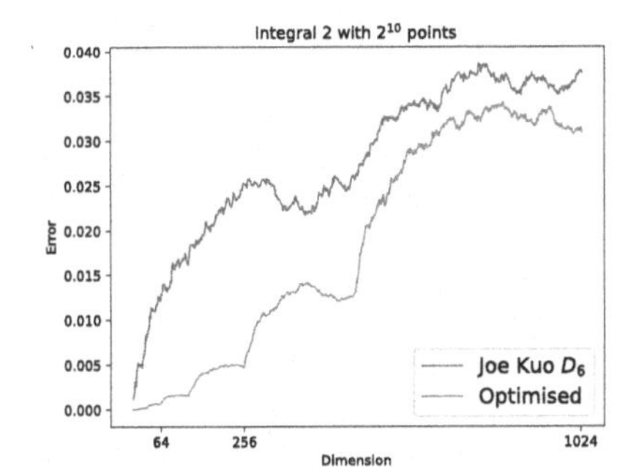

Fig. 3. Estimate the integral 2 for different dimensions using 2^{10} points

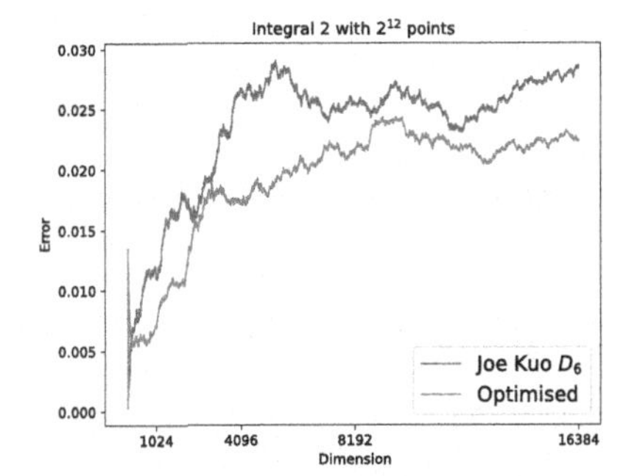

Fig. 4. Estimate the integral 2 for different dimensions using 2^{12} points

The algorithm yields results relatively fast for small values of the number of points, like 2^{10} or 2^{12}. For larger values of the number of points, the number of computations should be reduced to make it run faster. This can be accomplished, e.g., by first selecting a suitable set of direction numbers for smaller number of points and then extending it, thus drastically limiting the number of possibilities to consider. Validation can also be limited or even skipped entirely if needed.

5 Conclusions and Directions for Future Work

Our algorithm produces direction numbers that are suitable for use in algorithms with high constructive dimension, especially in situations where the number of dimensions are comparable with the number of trajectories/points. When the same computations are to be performed using different parameters, it is feasible to compute a set of direction numbers specific for the problem, especially when servers with powerful GPUs are available. We aim to make publicly available sets of direction numbers that are optimised and validated for some combinations of dimensions and number of points that are widely used. In this work we did not explore the possibility to set different weights for the different dimensions, thus allowing for, e.g., declining importance of the dimensions, but our framework is well suited to that problem and we hope that even better results will be obtained for simulations such.

References

1. Hellekalek, P., Leeb, H.: Dyadic diaphony. Acta Arithmetica **80**, 187–196 (1997)
2. Joe, S., Kuo, F.Y.: Constructing Sobol sequences with better two-dimensional projections. SIAM J. Sci. Comput. **30**, 2635–2654 (2008)
3. Kuipers, L., Niederreiter, H.: Uniform Distribution of Sequences. John Wiley (reprint edition published by Dover Publications Inc., Mineola, New York in 2006) (1974)
4. Liu, R., Owen, A.B.: Estimating mean dimensionality of analysis of variance decompositions. J. Am. Stat. Assoc. **101**(474), 712–721 (2006). http://www.jstor.org/stable/27590729
5. Matousek, J.: On the L2-discrepancy for anchored boxes. J. Complex. **14**, 527–556 (1998)
6. Owen, A.B.: Scrambling Sobol' and Niederreiter-Xing points. J. Complex. **14**, 466–489 (1998)
7. Sobol sequence generator. https://web.maths.unsw.edu.au/~fkuo/sobol/. Accessed 15 Feb 2022
8. Sobol, I.M., Asotsky, D.I., Kreinin, A., Kucherenko, S.S.: Construction and comparison of high-dimensional Sobol' generators. Wilmott **2011**, 64–79 (2011)
9. Sobol, I.M.: On the distribution of points in a cube and the approximate evaluation of integrals. Ussr Comput. Math. Math. Phys. **7**, 86–112 (1967)
10. Zinterhof, P.: Uber einige Abschätzungen bei der approximation von Funktionen mit Gleichverteilungsmethoden. Sitzungsber. Osterr. Akad. Wiss. Math. Natur. Kl. **II**(185), 121–132 (1976)

MiDaS: Extract Golden Results from Knowledge Discovery Even over Incomplete Databases

Lucas S. Rodrigues(✉), Thiago G. Vespa, Igor A. R. Eleutério, Willian D. Oliveira, Agma J. M. Traina, and Caetano Traina Jr.

Institute of Mathematics and Computer Sciences, University of São Paulo (USP), São Carlos, Brazil
lucas_rodrigues@usp.br, caetano@icmc.usp.br

Abstract. The continuous growth in data collection requires effective and efficient capabilities to support Knowledge Discovery in Databases (KDD) over large amounts of complex data. However, as activities such as data acquisition, cleaning, preparation, and recording may lead to incompleteness, impairing the KDD processes, specially because most analysis methods do not adequately handle missing data. To analyze complex data, such as performing similarity search or classification tasks, KDD processes require similarity assessment. However, incompleteness can disrupt the assessment evaluation, making the system unable to compare incomplete tuples. Therefore, incompleteness can render databases useless for knowledge extraction or, at best, dramatically reducing their usefulness. In this paper, we propose `MiDaS`, a framework based on a RDBMS system that offers tools to deal with missing data employing several strategies, making it possible to assess similarity over complex data, even in the presence of missing data at KDD scenarios. We show experimental results of analyses using `MiDaS` for similarity retrieval, classification, and clustering tasks over publicly available complex datasets, evaluating the quality and performance of several missing data treatments. The results highlight that `MiDaS` is well-suited for dealing with incompleteness enhancing data analysis in several KDD scenarios.

Keywords: Knowledge discovery in databases · Missing data · RDBMS framework · Similarity queries

1 Introduction

The growing advances in data collection and organization demand effective and efficient capabilities to support Knowledge Discovery in Databases (KDD) processes over large amounts of complex data, such as data preprocessing, mining, and result evaluation. As more and more data are available for analysis, missingness increases as well. However, dealing with missing data is a challenging problem for KDD, as data preparation and analysis tools do not properly handle missingness.

D. Groen et al. (Eds.): ICCS 2022, LNCS 13353, pp. 653–667, 2022.
https://doi.org/10.1007/978-3-031-08760-8_54

Similarity queries are well-suited to retrieve complex data since comparisons based either on identity or ordering are mostly inappropriate for complex data. Recently, new techniques have been widely explored for information retrieval, such as performing similarity comparisons over image collections to analyze and visualize Content-Based Image Retrieval (CBIR) results [18]. They are also being employed for classification, such as in Instance-Based Learning (IBL), to perform predictions based on data, mostly of them stored in Relational Database Management Systems (RDBMSs) [5,16]. Moreover, being able to express similarity queries in RDBMSs is becoming even more relevant for diverse applications that take into account the results of data mining in the KDD process.

Incompleteness problems can occur at any KDD step, including during data acquisition, data integration from different sources and as results of failures in phenomena observation and measurement [12]. Moreover, missingness can hamper data mining since similarity queries and other information retrieval techniques do not handle attributes with missing values, reducing the data available and thus the effectiveness of the KDD process.

Similarity queries over incomplete databases usually ignore attributes with missing values because distance functions cannot measure dissimilarity among incomplete objects. Figure 1 illustrates this problem with a dataset composed of medical exams of the brain, heart, and stomach. In this example, the heart and stomach exams with NULL for patient P_3 are missing data. The similarity between Patients P_1 and P_3 cannot be evaluated because most distance functions, such as Euclidean, cannot assess the similarity among those patients' records.

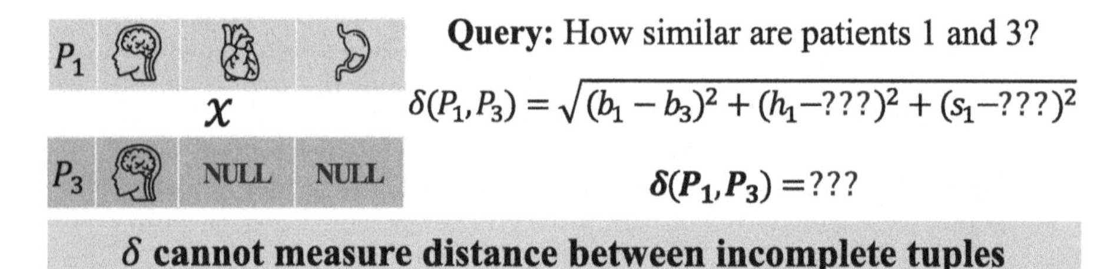

Fig. 1. Similarity searches over data with missing values.

Existing techniques to handle missing data include deleting incomplete tuples or imputing probable values for each missing attribute. Over the years, imputation methods have explored diverse heuristics to infer values, including the global mean, an average of the neighborhood, and regression models [4,13,17]. Recently, an approach dealing with this problem took advantage of the intrinsic information embedded in the data, using correlations among the similarity of near neighbors to weight similarity queries over incomplete databases, well-fitting the data distribution, and improving the results [18].

In recent years, several works have explored functionalities to support KDD tasks for Content Based Retrieval Systems (CBRSs) [1,14,16]. However, they do not consider the missing data problem in the RDBMS environment, thus rendering data mining impracticable over incomplete data collections. Here we aim at including additional functionalities into an RDBMSs to handle missing data for KDD scenarios, to support information retrieval scenarios, especially for executing similarity queries and to help classification or clustering tasks.

We introduce `MiDaS`: the ***Mi**ssing **Da**ta in **S**imilarity* retrieval framework. It is an RDBMS-centered framework that provides tools based on well-suited strategies to deal with missing data for similarity comparisons and KDD tasks. `MiDaS` comprises two layers of functionalities to handle missing data and support similarity search. The missing data treatment layer exports functionalities to handle incompleteness based on six reliable, long-used approaches proposed in the literature. The query engine layer provides similarity retrieval operators suitable for queries over complete and incomplete databases. Therefore, now that information is the "new gold" for companies, `MiDaS` incorporates tools for data preparation and analysis into the RDBMS environment, allowing the use of all available data to mine the "gold" right there where the data is stored. Besides, we intend to contribute to KDD issues, providing an efficient way to deal with incompleteness in a single environment and support information retrieval tasks with quality and performance. The main contributions of this work are as follows:

- Provide a "missing data engine", including strategies for tuple deletion, several relevant imputation heuristics, and a recent approach based on data correlation that neither discard nor infer values.
- Support k-Nearest Neighbors (k-NN) and Range similarity retrieval operators even over incomplete data scenarios.
- Conduct a performance analysis of the `MiDaS` Framework, to reveal the efficiency when treating several missing data rates in complex databases.
- Present a thorough experimental analysis of the `MiDaS` strategies to show the advantages and applicabilities when the framework handles missing data for KDD scenarios, including similarity retrieval, classification based on neighborhood, and clustering analysis.

The remainder of this paper is organized as follows. Section 2 presents the relevant background and works that explore similarity searches and missing data. Section 3 presents `MiDaS` and Sect. 4 shows its experimental analyses. Finally, Sect. 5 concludes the paper.

2 Background and Related Work

Comparing by similarity is the usual way to query and retrieve complex data. Here we briefly discuss related concepts concerning this work.

We call "complex" an attribute that may be compared under varied "aspects", so it must be previously defined, for example, using a distance function over certain features of the objects. Similarity queries perform comparisons between complex objects using low-level representations called feature vectors, obtained by Feature Extraction Methods (FEMs), and stored in RDBMSs. For example, there are several suitable FEMs to process images, based on features such as color, texture, shape, or on learning approaches, such as Histogram Oriented Gradients (HOG), Haralick, and VGG16 [7,19]. We assume that $\mathbb{S}$ is a domain of features already extracted and stored in attribute `S` of a RDBMS relation and S as the active domain of that attribute. Thus, S is the set of values stored, which are compared by distance functions in the similarity queries. A distance functions (δ) assess the dissimilarity between two complex objects (s_i, s_j) and return a real value from $\mathbb{R}^+$. There are several distance functions, such as Euclidean, Manhattan, Chebyshev, and others [9], and some are called a "metric" when it meets the Metric Spaces properties.

Each similarity query, either a Range or a k-Nearest Neighbors query (k-NN), is posed over a complex dataset $S \subseteq \mathbb{S}$ and requires a central object $s_q \in \mathbb{S}$ called the query center, a distance function δ, and a similarity limit. A Range Query (R_q) retrieves every tuple where $s_i \in S$ whose distance to s_q is less or equal than the limit given by a similarity radius (ξ), measured as $\delta(s_q, s_i) \leqslant \xi$. A k-NN Query (Knn_q) retrieves the tuples where s_i is one of the k nearest to s_q, where the limit is the amount k of tuples retrieved.

2.1 Missing Data and Treatments

The missing data problem in RDBMS occurs when a complex value or some of its components is `NULL` (e.g., when s_i is an array). Incompleteness negatively affects the quality of the data and, consequently, the experts' analyses.

Several works explore viable solutions to the problem. **Deletion methods** drop tuples or attributes with missing values. However, these methods are widely questionable, as they reduce the data available for analyses and may lose relevant information [18]. **Imputation methods** explore statistical heuristics, such as imputation based on the mean value of an attribute or exploring regression models to predict probable values [13]. Machine Learning heuristics perform imputation based on the element neighborhood, searching for the k nearest neighbors to infer missing values [4]. Imputation methods based on decision trees identify natural partitions on the data or use supervised classification algorithms to infer probable values [17].

A recent alternative for both deletion and imputation, called **SOLID** [18], exploits correlations among attributes to handle incompleteness. It uses existing values to identify pairwise attribute correlations to weigh their contribution to the similarity assessment. Hence, **SOLID** takes advantage of correlations to execute weighted queries over an incomplete dataset, employing all the data available, thus avoiding either discarding tuples or imputing values.

2.2 Content-Based Retrieval Systems

Several Content-Based Retrieval Systems (CBRSs) based on RDBMS were proposed to perform information retrieval at diverse scenarios and applications [3,5,11,14,16]. Table 1 summarizes the most relevant related works, regarding the following aspects:

- **Open-Source RDBMS:** whether it is based on Open-Source RDBMS, which brings flexibility and eases incorporating new techniques and algorithms;

- **Similarity-Retrieval Tasks:** whether it provides mechanisms for information retrieval over complex datasets (range and k-NN comparisons);
- **Handle Missing Data:** whether it supports or applies strategies to deal with incompleteness, including data cleaning and preparation aiming at KDD;
- **Self-Contained Functions:** whether it seamlessly integrates with the RDBMS architecture without requiring external interactions or add-ons tools, easing user customization.

Table 1. Related works and our proposed `MiDaS` according to relevant aspects.

Work	Year	Open-source RDBMS	Similarity retrieval	Missing data	Self-contained
SIREN [3]	2006	✗	✔	✗	✗
FMI-SiR [11]	2010	✗	✔	✗	✗
SimbA [5]	2014	✔	✔	✗	✗
Kiara [16]	2016	✔	✔	✗	✗
SimAOP [1]	2016	✔	✔	✗	✗
MSQL [14]	2017	✔	✔	✗	✗
`MiDaS`	2022	✔	✔	✔	✔

Open-Source RDBMS. Most of the related works are based on open-source architectures, mainly on PostgreSQL [1,5,14,16]. Open-Source RDBMS increases availability and makes it easier for the user to tune its functionalities or to include new algorithms and procedures. However, [3] and [11] were developed using Oracle and its specific tools. Both suggest it can be extended to PostgreSQL but do not describe how. We choose PostgreSQL because it enhances transparency and applicability for missing data management and similarity query execution.

Similarity-Retrieval. Many of the works support mechanisms to execute similarity queries in an RDBMS, highlighting the relevancy and suitability of similarity retrieval over complex data. They aim at extending existing retrieval operators with new abilities, either modifying the RDBMS core and other modules [11] or creating other relational operators using plain SQL [3,5,14,16], to allow posing both range and k-NN queries. We followed the Kiara concept [16] as the underpinning for the similarity operations in our framework, extending it to handle missing data. Our solution is adaptable to other similarity retrieval frameworks, as it does not depend on a specific RDBMS architecture.

Handle Missing Data. None of the previous works that explore similarity retrieval on RDBMSs provide mechanisms to handle missing data. In fact, they just discard incomplete tuples, losing potentially helpful information and knowledge from the databases. `MiDaS` provides suitable strategies to infer missing data to enable similarity queries over incomplete data, mainly using the **SOLID** approach, which improves flexibility and the quality of the analysis.

Self-contained Functions. The previous works only provide functionality for the RDBMS environment to execute similarity queries, but none of them deals with incompleteness. `MiDaS` can help with both missing-aware data preparation and similarity retrieval, providing a novel environment that the user can customize using strategies to deal with missing data in KDD processes.

3 The `MiDaS` Framework

The `MiDaS` **Framework** incorporates tools to handle incompleteness in complex data databases and support knowledge retrieval based on similarity queries. It aids KDD processes, particularly data preparation and data analysis combined with the functionalities of an RDBMS. Currently, `MiDaS` is implemented in PostgreSQL[1]. As shown in Fig. 2, the `MiDaS` architecture is composed of two layers:

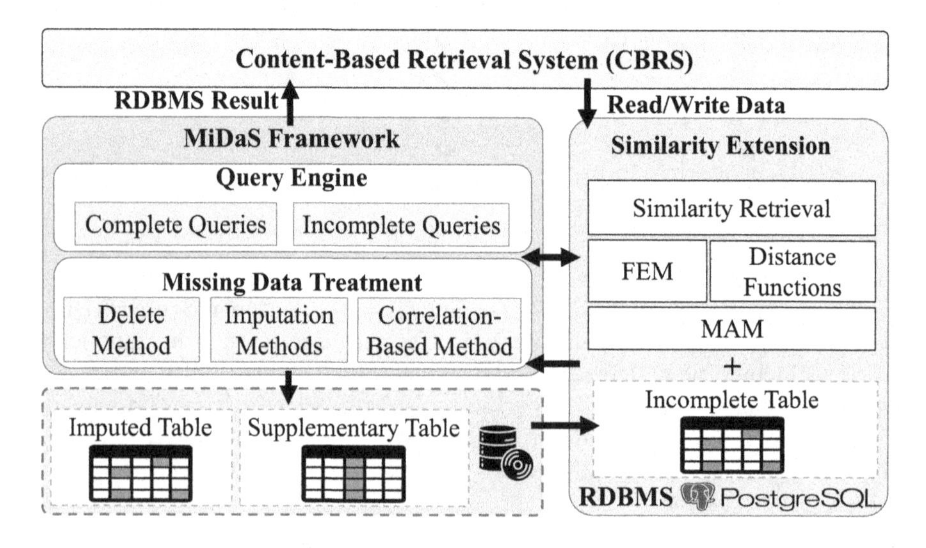

Fig. 2. Overview of `MiDaS` framework.

- **Missing Data Treatment Layer:** provides six reliable approaches for data preparation, with emphasis on complex datasets and applications for data mining scenarios.
- **Query Engine Layer:** implements the k-NN and Range query operators, enabling similarity queries over complete and incomplete data, providing tools for information retrieval at varying applications.

3.1 The Missing Data Treatment Layer

This layer is able to handle missing data through Missing Data Treatment (MDT) modules, which can be helpful for a variety of applications. `MiDaS`

[1] PostgreSQL: Available at: https://www.postgresql.org/. Access date: 09 Fev. 2022.

includes six MDT, covering the main approaches available from the literature: one based on tuple deletion, four on imputation following varying heuristics, and one on attribute correlation, described as follows.

The **(i) Deletion** MDT just ignores the tuples with missing values without changing the original table. The imputation MDTs assign a value for each missing value. There are four options, as follows: **(ii) Mean Imputation (MI):** fills the missing values using the mean of the existing values of the attribute in the entire table. **(iii)** ***k*****-NN Imputation (kNNI):** fills the missing values using the average value of the k-Nearest Neighbors of each tuple, where k is defined by the user. **(iv) Regression Imputation (RI):** predicts missing values based on linear regression using other attributes and the observed values of the attribute. **(v) Decision-Tree Imputation (DTI):** identifies data partitions based on the similarity between elements and applies a classification algorithm to infer a probable value for each missing case. Each imputation MDT receives an incomplete table and the required user-defined parameters and returns the imputed table. The user chooses how to process the output further, either updating the original table or storing the output as a temporary table.

Finally, **(vi) SOLID MDT** evaluates pairwise correlations among selected attributes. SOLID receives the incomplete table and the *correlation threshold* as a real-valued parameter and discards the attributes with low correlations just for each pairwise comparison. Hence, SOLID generates a supplementary table (**"weights_tableName"**) with two attributes: the compatible attribute pair and the corresponding weights, as explained in Sect. 2. As with the other MDTs, SOLID does not update the input table. Every MDT can be activated using either PL/Python or SQL statements with storing functions/procedures into the RDBMS.

`MiDaS` executes the MDTs in either `Pre-Computation` or `Query Run-time` mode, bringing flexibility to the user when choosing how to treat and prepare the incomplete datasets. In `Pre-Computation` mode, an MDT is processed, and the result is saved as a new table in the RDBMS as the imputed or supplementary table. When similarity queries are posed, this table is used instead of the original. On the other hand, execution in `Query Run-time` mode re-executes the MDT method at each query execution. `MiDaS` covers most of the missing data handling needs in various scenarios, producing data more suitable for KDD processes improving the quality of data analysis. Moreover, as the `MiDaS` is Open-Source and based on an open RDBMS, other MDTs can be included.

3.2 The Query Engine Layer

Similarity queries over either complete or incomplete datasets require appropriate retrieval operators, and this topic is especially relevant for RDBMSs. Current RDBMSs do not support those operators, although similarity queries may be expressed in SQL combining existing relational operators and functions. However, considering all the variations involved, especially when missing data must be handled, expressing similarity queries can be troublesome. Thus, the `MiDaS` was also developed to provide resources to express similarity queries, implementing the Query Engine layer to provide similarity retrieval operators. It executes both Range and k-NN queries over complete and incomplete data using any of the provided MDTs.

We provide the similarity retrieval operators based on the concept that the result of a (sub-)query over a database relation is also a relation. Hence, we implement each operator as an SQL function that returns a table, callable at the FROM clause of a SELECT command. Statement 1.1 presents the general syntax of the command to pose a similarity query over complete or incomplete tables. It can be integrated with the elements of the SQL language, such as referring to other tables and clauses of the select statement and other RDBMS commands. As the result, the MiDaS is a robust and practical similarity engine for complex data retrieval, able to handle complete and incomplete data.

```
SELECT queryResult.* FROM simQueryWithMissing(
    table_name     anyelement,     -- input table
    mdt_method     VARCHAR,        -- MDT method
    sim_operator   VARCHAR,        -- knn or range
    sim_criterion  NUMERIC,        -- k or radius value
    obj_query      COMPLEX_DATA,   -- query center
    dist_func      VARCHAR         -- distance function
    ) as queryResult;
```

Statement 1.1. SIMILARITY QUERY WITH MIDAS FRAMEWORK.

The `simQueryWithMissing` function requires the following parameters:

- **`table_name`:** the table where the similarity search must be performed;
- **`mdt_method`:** the MDT method, choosing `deletion`, `mean`, `knn`, `regression`, `dt` or `solid` to handle missing data, or `NULL` for complete data;
- **`sim_operator`:** the similarity comparison operator: specify `knn` or `range`;
- **`sim_criterion`:** the similarity limit: the k for `knn` queries or radius for `range`;
- **`obj_query`:** the query center;
- **`dist_function`** the distance function. `MiDaS` Framework provides `L1` to Manhattan, `L2` to Euclidean, or `Linf` to Chebyshev. Also, the user can define other distance functions as stored functions.

Execution in `Pre-Computation` or `Query Run-time` mode is transparent for the query layer, as when the table for a Pre-Computation MDT exists, the query is executed using the pre-processed table to speed up execution. Figure 3 illustrates the query execution that retrieves the five patients closest to patient `pat110886`, using the SOLID approach as the MDT method over the table of MRI chest exams stored in table `chestMRI`, using the `L2` distance function. When the weights table for SOLID is not already prepared, `MiDaS` automatically chooses the `Query Run-time` mode, calls the MDT to generate the table with the default parameters, and stores it in the RDBMS, to speed up further queries. When the table is already computed, then the query is executed in `Pre-Computation` mode.

4 Experimental Analysis

This section describes the datasets used in the experiments, how the experiments were evaluated, the analysis of their performances, and how `MiDaS` can be applied for the KDD process, including information retrieval, classification, and clustering tasks, highlighting the framework applicability.

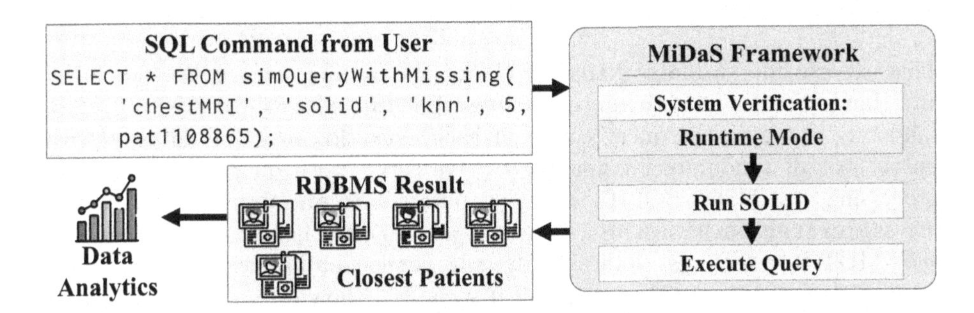

Fig. 3. Query execution example with `MiDaS` framework.

4.1 Datasets and Implementation Specifications

The experiments were performed using four datasets of images, including medical images, objects, and a font types collection. Table 2 shows their number of tuples (n), attributes (d), and a brief description.

Table 2. Image datasets used in this work.

Dataset	n	d	Description
ds-Spine [6]	54	10	Lumbar muscles and vertebral bodies MRI
ds-Coil [15]	100	72	Objects posing over 360 rotation, with 5 interval degrees
ds-Letters [10]	152	52	Font types of alphabetic in lower and upper cases letters
ds-Brats [2]	1251	5	Glioma MRI scans with pathologically confirmed diagnosis

Data collected and processed in January 25, 2022.

Every tuple in a relation may have missing values. For example, each attribute of the dataset *ds-Letters* is the image of a letter in a font, each font having $d = 52$ letters, but some letters may be missing for a specific font. We extracted features from the four image datasets using five FEMs: Local Binary Patterns (LBP), Zernike, Haralick, Histogram of Oriented Gradients (HOG), and VGG16 (a CNN model) [7,19]

`MiDaS` was implemented using PostgreSQL 13.2, incorporating PL/Python scripts and the well-known open libraries Pandas, Numpy, Scipy, Sklearn, and others, for missing data treatment. We used the C++ language for operations of bulk-loading features from complex collections and SQL scripts for distances functions and similarity retrieval operations. We create the scripts as MiDaS RDBMS, developing the imputations heuristics based on models from Sklearn and using SOLID scripts from the repository of [18]. Scripts and more details about data management are available in a Git repository[2], which also provides the extracted features.

4.2 MiDaS Performance

This experiment evaluates MiDaS performance running each MDT for k-NN and Range queries in both runtime modes. Each query was executed 50 times, randomly changing the query center at each execution over the datasets with up to 50% of randomly missing values. We setup both queries with no pre-processing and MDT=SOLID, changing parameters `sim_operator` and defining `sim_criterion` with `k` = 21 and `range` = 0.55. Changing each query for each MDT just requires modifying the `mdt_method` parameter. Tables 3 and 4 present the average wall-clock time (in seconds), showing the `Query Run Time` (QT columns) and the MDT execution time, concerning the `Pre-Computation` (PCM columns).

Table 3. k-NN queries elapsed time (in seconds) of each MDT in 50% missingness.

Dataset	Deletion		SOLID		MI		kNNI		RI		DTI	
	QT	PCM	QT	PCM	QT	PCM	QT	PCM	QT	PCM	QT	PCM
ds-Spine	0.04	–	0.11	0.70	0.24	2.10	0.22	4.56	0.22	9.06	0.20	42.54
ds-Coil	0.18	–	0.43	18.66	0.76	0.58	0.84	0.74	0.66	6.32	0.74	15.74
ds-Brats	0.16	–	0.14	6.14	0.64	0.44	0.62	0.58	0.60	1.10	0.62	2.46
ds-Letters	0.12	–	0.42	14.72	0.76	0.52	0.78	0.84	0.78	5.46	0.76	12.84

Table 4. Range queries elapsed time (in seconds) of each MDT in 50% missingness.

Dataset	Deletion		SOLID		MI		kNNI		RI		DTI	
	QT	PCM	QT	PCM	QT	PCM	QT	PCM	QT	PCM	QT	PCM
ds-Spine	0.04	–	0.11	0.70	0.18	2.10	0.24	4.56	0.20	9.06	0.26	42.48
ds-Coil	0.18	–	0.43	18.66	0.72	0.58	0.74	0.84	0.68	6.32	0.43	15.70
ds-Brats	0.08	–	0.15	6.14	0.60	0.44	0.64	0.58	0.60	1.10	0.68	2.46
ds-Letters	0.18	–	0.72	14.72	0.74	0.52	0.76	0.84	0.78	5.46	0.72	12.84

Notice that only the QT time is spent in `Pre-Computation` mode (PCM), whereas both (QT+PCM values) times are spent in `Query Runtime` mode (QT). Overall, the query time of the `Pre-Computation` mode varies from 0.04 to 0.78

[2] Git repository of MiDaS: https://github.com/lsrusp/MiDaS.

seconds for k-NN and Range queries. The Query runtime mode adds the time to execute SOLID or an imputation MDT, leading to a total time-varying from 0.44 up to 42.54 seconds (PCM columns). For other missing rates, the queries give similar times.

These experiments show that `MiDaS` achieves high performance in both modes, allowing a quick comparison between the various MDT approaches. We highlight that `Pre-Computation` mode is better suited for scenarios where frequent queries are posed over datasets that undergo few or no updates because they only require the MDT method to be executed once. The `Query Runtime` mode is better suited for fast-changing datasets because it allows the MDT to track data evolution.

4.3 `MiDaS` for Knowledge Discovery on Databases

We show the applicability of `MiDaS` MDT approaches for distinguished scenarios where missingness can cause hardness or even make tasks impracticable for KDD. Therefore, we present MDTs analysis, highlighting the quality and advantages for similarity retrieval, classification, and clustering scenarios.

Similarity Retrieval. We analyzed SOLID MDT results to deal with incompleteness using dataset *ds-Spine*, exploiting the advantages of weighted queries based on correlation data. Figure 4 shows its tuples distribution using the MDS Projection [8], making it possible to visualize the multidimensional space distribution of elements using the distance matrix. The distance matrix is based on the Euclidean distances between tuples over (a) complete data, (b) 10% missingness, (c) posing SOLID for 10%, (d) 50% missing rate, and (e) SOLID execution for 50% missing, respectively for each case. Blue points are complete tuples, red are incomplete tuples, and black points are the SOLID application. We highlight that the SOLID takes advantage of the correlation between complex attributes to any missingness scenarios, well-fitting the data distribution to allow posing queries that achieve high-quality results even at large amounts of missingness.

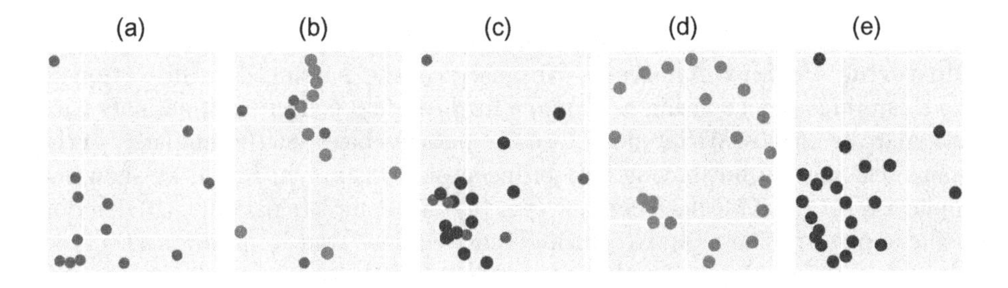

Fig. 4. Similarity retrieval over *ds-Spine* in scenarios of (a) complete data, (b) 10% missingness, (c) SOLID for 10%, (d) 50% missing rate and (e) SOLID execution for 50% missing. Blue points are tuples complete, red are incomplete ones, and black points are evaluation results from SOLID. (Color figure online)

Classification Scenarios. We analyzed the instance classification based on a k-NN Classifier model using MiDaS imputations based on Mean, k-NN, and Regression for cases of 50% missing rate. The k-NN Classifier employs Euclidean distance to measure dissimilarity. This experiment shows how a treated dataset can be better suited to employing our framework for supervised tasks since untreated missingness can bias classification. In this task, we classified the elements from the *ds-Letters* dataset using the font type as the "class" of the tuple, such as "Times New Roman", "Verdana", etc. We split the train/test sets at 50% each, changing k between 3 to 21, and executed the evaluation using F1-Measure, Precision, and Recall, as shown in Fig. 5. The results show that the employed imputation methods were well-suited for classification scenarios, resulting in high values up to 0.89, 0.89, and 0.90 for F1-Measure, Precision, and Recall, respectively. Therefore, the MiDaS can be applied and is fast enough in practice to support missingness treatment for labeled datasets, such as *ds-Letters* (see Sect. 4.2).

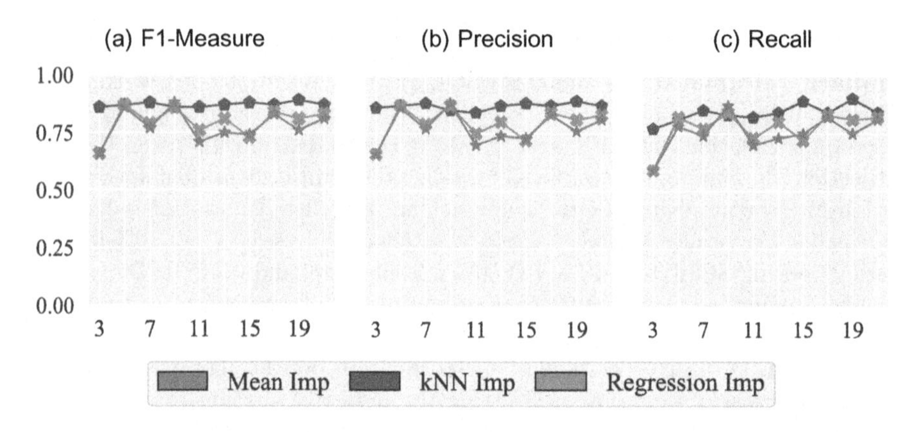

Fig. 5. Classification evaluation for *ds-Letters* dataset using F1-measure, precision, and Recall of k-NN classifier.

Clustering Patient Analysis. We analyzed the missing data distortion in the clustering scenario using *ds-Brats*, a medical data collection of patients with several exam images. We intend to discover patients based on the similarity of the exams distribution but having 50% of incomplete exams. In Fig. 6, we show the application of the KMeans technique over (a) only complete patients, (b) deletion of incomplete patients, and (c) using the Decision Tree Imputation (DTI) to impute values for the tuples that were deleted in (b) case. As seen in (b) with only complete tuples, missingness can be highly harmful to clustering because the clusters' distortions blur the patient group's discovery. (c) DTI takes advantage of the data partitions based on highly similar objects and aid in reducing the clusters distortions, spending just a few seconds to deal with missingness (see Tables 3 and 4). Each distinguished color in Fig. 6 represents a patients' cluster.

We evaluate the clustering distribution in (a), (b), and (c) using the silhouette metric (the closer to one, the better the clusters), resulting in 0.473, 0.204, and 0.571, respectively, highlighting that missingness treatments are advantageous and improve clustering tasks. Hence, the `MiDaS` makes it possible to discover patient groups and enables, for example, the behavior analyses of each patient group to support the physician's decisions using the most helpful information.

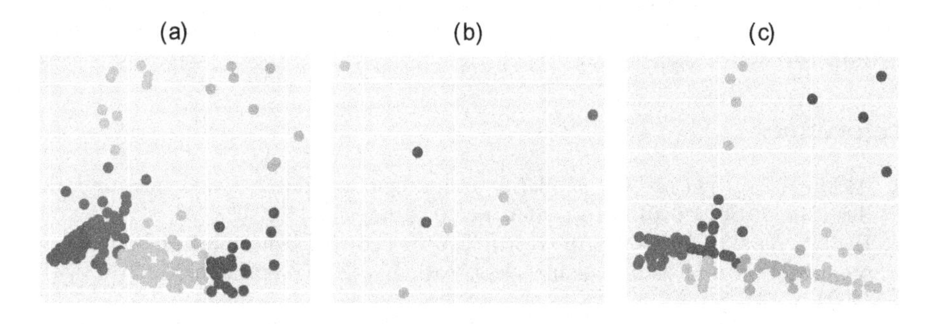

Fig. 6. Clustering analysis for KMeans application over *ds-Brats* in scenarios over (a) only complete patient data, (b) deletion of 50% incomplete patient records, and (c) use Decision Tree Imputation (DTI), focusing on the missing impact and the DTI advantages to impute values for tuples deleted case in (b).

As shown in the varied scenarios evaluated in this section, we highlight that the `MiDaS` provides many methods well-suited to handle data incompleteness and that it is very useful for several KDD scenarios. Therefore, this Section supports our assertion that the `MiDaS` provides a simple but effective approach to dealing with incompleteness, leading to obtain high-quality results improving analytics by extracting golden insights from the data in a timely manner.

5 Conclusions

In this work, we presented `MiDaS`: the ***Mi****ssing* ***Da****ta in* ***S****imilarity* retrieval framework, implemented as an extension of the PostgreSQL RDBMS that is able to provide well-suited treatments for missing data, enabling analysis tasks for Knowledge Discovery in Databases. `MiDaS` aims to meet the requirements of KDD tasks by embedding proper tools into an RDBMS and bringing data processing directly to where the data is stored. It provides six MDT methods well-suited for dealing with missing data and provides features for similarity retrieval that do not discard nor impute values and for performing classification and clustering tasks. For similarity retrieval, it provides query operators that are able to handle both complete and incomplete datasets, seamlessly integrated with the other features of the RDBMS. Therefore, we claim that `MiDaS` is a novel and practical environment for users to handle missingness using the provided MDTs, which allows performing KDD analyses integrated to execute similarity queries over complete or incomplete data. The experimental evaluation shows that MiDaS provides strong resources to deal with incompleteness in large databases, increasing the quality of results from various KDD scenarios and the efficiency of similarity query execution.

Acknowledgements. This research was supported in part by the Coordenação de Aperfeiçoamento de Pessoal de Nível Superior - Brasil - Finance Code 001, by the São Paulo Research Foundation (FAPESP, grants No. 2016/17078-0, 2020/07200-9, 2020/10902-5), and the National Council for Scientific and Technological Development (CNPq).

References

1. Al Marri, W.J., et al.: The similarity-aware relational database set operators. Inf. Syst. **59**, 79–93 (2016). https://doi.org/10.1016/j.is.2015.10.008
2. Baid, U., et al.: The RSNA-ASNR-MICCAI BraTS 2021 benchmark on brain tumor segmentation and radiogenomic classification. arXiv preprint arXiv:2107.02314 (2021)
3. Barioni, M.C.N., et al.: SIREN: a similarity retrieval engine for complex data. In: The 32nd International Conference on VLDB, pp. 1155–1158. ACM (2006)
4. Batista, G., et al.: A study of k-nearest neighbour as an imputation method. In: Soft Computing Systems - Design, Management and Applications. Frontiers in Artificial Intelligence and Applications, vol. 87, pp. 251–260. IOS Press (2002)
5. Bedo, M.V.N., Traina, A.J.M., Traina-Jr., C.: Seamless integration of distance functions and feature vectors for similarity-queries processing. JIDM **5**(3), 308–320 (2014). https://periodicos.ufmg.br/index.php/jidm/article/view/276
6. Burian, E., et al.: Lumbar muscle and vertebral bodies segmentation of chemical shift encoding-based water-fat MRI. BMC Musculoskelet. Disord. **20**(1), 1–7 (2019). https://doi.org/10.1186/s12891-019-2528-x
7. Coelho, L.P.: Mahotas: open source software for scriptable computer vision. arXiv preprint arXiv:1211.4907 (2012)
8. Cox, M.A., Cox, T.F.: Multidimensional scaling. In: Handbook of Data Visualization, pp. 315–347. Springer, Heidelberg (2008). https://doi.org/10.1007/978-3-540-33037-0_14
9. Deza, M.M., Deza, E.: Encyclopedia of distances. In: Encyclopedia of Distances, pp. 1–583. Springer, Berlin (2009). https://doi.org/10.1007/978-3-642-30958-8
10. Hajder, S.: Letters organized by typefaces - standard windows fonts with each letters organized in classes by typeface (2020). https://www.kaggle.com/killen/bw-font-typefaces
11. Kaster, D.S., et al.: FMI-SiR: a flexible and efficient module for similarity searching on oracle database. J. Inf. Data Manag. **1**(2), 229–244 (2010). https://periodicos.ufmg.br/index.php/jidm/article/view/36
12. Lei, C., et al.: Expanding query answers on medical knowledge bases. In: The 23rd International Conference on Extending Database Technology, pp. 567–578. OpenProceedings.org (2020). https://doi.org/10.5441/002/edbt.2020.67
13. Little, R.J., Rubin, D.B.: Statistical Analysis with Missing Data, vol. 793. Wiley (2019). https://doi.org/10.1002/9781119482260
14. Lu, W., Hou, J., Yan, Y., Zhang, M., Du, X., Moscibroda, T.: MSQL: efficient similarity search in metric spaces using SQL. VLDB J. **26**(6), 829–854 (2017). https://doi.org/10.1007/s00778-017-0481-6

15. Nene, S.A., et al.: Columbia object image library (coil-100). Technical report, Department of Computer Science, Columbia University, New York, USA (1996)
16. Oliveira, P.H., et al.: On the support of a similarity-enabled relational database management system in civilian crisis situations. In: ICEIS 2016 - Proceedings of the 18th International Conference on Enterprise Information Systems, pp. 119–126. SciTePress (2016). https://doi.org/10.5220/0005816701190126
17. Rahman, M.G., Islam, M.Z.: Missing value imputation using decision trees and decision forests by splitting and merging records: two novel techniques. Knowl. Based Syst. **53** (2013). https://doi.org/10.1016/j.knosys.2013.08.023
18. Rodrigues, L.S., Cazzolato, M.T., Traina, A.J.M., Traina, C.: Taking advantage of highly-correlated attributes in similarity queries with missing values. In: Satoh, S., et al. (eds.) SISAP 2020. LNCS, vol. 12440, pp. 168–176. Springer, Cham (2020). https://doi.org/10.1007/978-3-030-60936-8_13
19. Simonyan, K., Zisserman, A.: Very deep convolutional networks for large-scale image recognition. arXiv preprint arXiv:1409.1556 (2014)

Teaching Computational Science

Computational Science 101 - Towards a Computationally Informed Citizenry

Victor Winter(✉)

University of Nebraska at Omaha, Omaha, NE 68182, USA
vwinter@unomaha.edu

Abstract. This article gives an overview of CSCI 1280, an introductory course in computational science being developed at the University of Nebraska at Omaha. The course is intended for all students, regardless of major, and is delivered in a fully asynchronous format that makes extensive use of ed tech and virtual technologies.

In CSCI 1280, students write programs whose excution produce graphical block-based artifacts. This visual domain is well-suited for the study of programming fundamentals and also aligns well with scientific simulations based on cellular-automata. An overview of CSCI 1280s simulation framework for percolation theory and the spread of infectious diseases is given.

Keywords: Functional programming · Patterns · Computational science

1 Overview of CSCI 1280

At the University of Nebraska at Omaha (UNO), *CSCI 1280 - Introduction to Computational Science* is a freshman-level gen ed science course that embraces the domain of computational science as a means to provide a gentle and engaging introduction to programming.

All programs in CSCI 1280 are written in a freely available programming environment called *Bricklayer* [8,9] which extends the functional programming language SML with a collection of graphics and computational science libraries. Using these libraries, programs can be written to construct block-based artifacts (AKA pixel art) like the ones shown in Fig. 1.

In addition to enabling the creation of art, Bricklayer's block-based infrastructure is also suitable for scientific exploration. A variety of scientific models and simulations are based on cellular automata [3]. For example, diffusion models and simulations can be used to study heat-diffusion, spreading of fire, ant behavior, and biofilms [6]. Cellular automata have also been used to model and study the formation of biological patterns (e.g., molds), Turing patterns, as well as adhesive interactions of cells [1].

D. Groen et al. (Eds.): ICCS 2022, LNCS 13353, pp. 671–677, 2022.
https://doi.org/10.1007/978-3-031-08760-8_55

(a) A Sierpinski triangle.

(b) An artist's rendition of Saturn.

Fig. 1. Examples of 2D and 3D Bricklayer artifacts.

1.1 Contribution

CSCI 1280s curriculum provides a gentle introduction to programming as well as computational science. The visual domain used allows concepts to be covered in a fashion that is not math intensive and therefore suitable for all majors (STEM majors as well as non-STEM majors). This visual domain is also enables instructional design to heavily leverage interactive technology to achieve educational goals.

2 Patterns

CSCI 1280 begins with an informal study of patterns, a concept the National Science Teachers Association (NSTA) has classified as crosscutting all the sciences [5].

The first assignment in CSCI 1280 asks students to use a Bricklayer web app, called the *Grid*, to create a pattern. A link to the Grid is provided below.

https://bricklayer.org/apps/CSCI_1280/Grid_CSCI_1280/grid.html

A key question in the *create-a-pattern* assignment is whether students understand the patterns created by their peers. A thought experiment students are asked to perform is the following:

Ask yourself whether you could continue working on (i.e., increase the size of) a pattern created by another student?

To enable students to engage in such thought experiments, the patterns created by students are shared online in a 3D interactive art gallery. A link to such a gallery is provided below.

https://mygame.page/csci1280-create-a-pattern-art-gallery-02

These art galleries employ PlayFab leaderboards allowing players (i.e., students) to rate the patterns of their peers using a Likert scale. A painting can be rated by "shooting" (i.e., positioning the crosshair and pressing the left mouse button) the desired circle below the painting. Players can report their ratings and visit a ratings summary page to view how the paintings in a gallery have been rated by the class as a whole.

To strengthen students' understanding of pattern the create-a-pattern assignment is followed by a discussion of several forms of symmetry that a grid-based artifact can possess. Specifically, vertical and horizontal reflection symmetry is discussed as well as 2-fold and 4-fold rotational symmetry. To help develop a students understanding of these symmetries, a web app called *Mystique* was developed. A link to Mystique is provided below.

http://mystique.bricklayer.org

(a) A Celtic forest with a blue will-o-the-wisp off to the left.

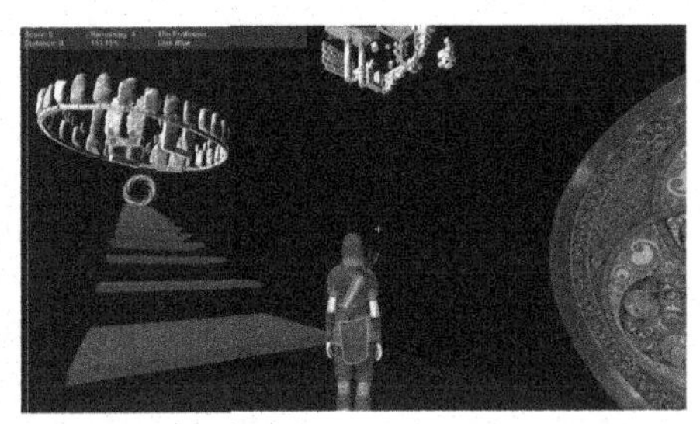

(b) The Upside Down - Portals to various lands.

Fig. 2. Triskelion.

To encourage a more sustained and hopefully enthusiastic engagement with the study of symmetry, a third-person game, called Triskelion, was developed. Triskelion interleaves gameplay with solving Mystique puzzles. The goal of gameplay is to cast magic spells in order to collect will-o-the-wisps which are moving orbs randomly spawned at various locations on a map. When a player collects a will-o-the-wisp they are teleported to a puzzle realm where they must solve a symmetry puzzle in order to return to gameplay. When all the will-o-the-wisps in the first map have been collected, the player is teleported to *The Upside Down* where they encounter two parkour-based pathways. Each pathway ends in a portal that transports the player to a distinct final map. Thus, when playing a game of Triskelion only two of the three maps are visited. The hope here is that curiosity alone will result in students exploring all maps. Screenshots highlighting Triskelion are shown in Fig. 2.

Triskelion supports clan-based competition in which the score a player receives is based on the speed and accuracy with which they solve puzzles. During a match, the score for a clan is the average score of the players in the clan. Educators have access to leaderboards that track a variety of engagements metrics such as: (1) how many games an individual has played, (2) the total time devoted to solving puzzles, and (3) the total time devoted to gameplay.

3 Bricklayer's Percolation Library

Percolation theory [2,7] involves the study of clustering in graphs. In practice, percolation theory has applications in numerous fields of study including: chemistry, biology, statistical physics, epidemiology, and materials science.

An example of a question fundamental in percolation theory is the following: *If a liquid is poured on top of some porous material, will it be able to make its way through the pores in the material and reach the bottom?* If so, the material is said to *percolate.*

When studying percolation, a density known as a *percolation threshold* is of central importance. This threshold represents the material density above which the material almost never percolates and below which the material almost always percolates. While approximate values for percolation thresholds can be determined through simulation, no mathematical solution has yet been discovered for calculating such values. For this reason, percolation theory provides an interesting domain highlighting an important role that computers can play in scientific inquiry.

Bricklayer's Percolation library provides a variety of functions for exploring the basics of percolation. One function can be used to create rectangles having a desired size and density. Another function allows "water" to be "poured on the top" of such a rectangle. Figure 3 shows 2 Bricklayer squares having similar density. One square percolates and the other does not.

Using these two functions students can explore the roles played by rectangle size and density in percolation. Such experimentation is labor intensive and motivates the need for another function, also provided by the Percolation library, which is able to perform numerous tests automatically and summarize the results (e.g., 6540 rectangles out of 10000 rectangles percolated). Through automated testing of this kind it becomes possible to obtain information that could not realistically have been obtained through manual experimentation. This provides students with a concrete and compelling example of the importance of computational power in scientific inquiry.

By varying the density of a fixed size rectangle, automated testing can be used to manually search for the percolation threshold. However, in order to estimate the percolation threshold in a manner that is fairly accurate, a large number of automated tests will need to be performed. To support the search for accurate percolation thresholds, Bricklayer provides a function that performs a Monte Carlo search. This significantly increases the computational complexity of the simulations being performed, which now can take minutes or even hours

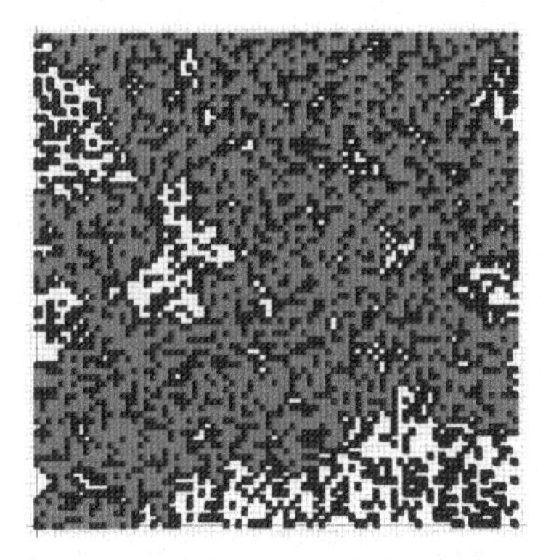

(a) A randomly generated Bricklayer 2D cell structure that percolates.

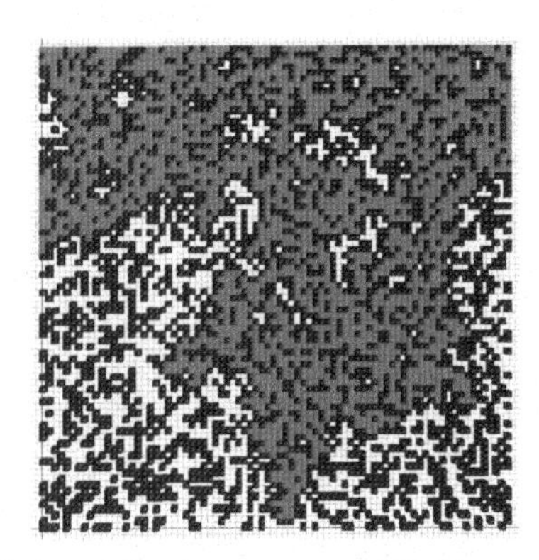

(b) A randomly generated Bricklayer 2D cell structure that does not percolate.

Fig. 3. Pouring water on 80 × 80 squares whose density is approximately 59%.

to complete. However, the information obtainable through Monte Carlo search significantly exceeds the information that can be obtained through manual search based on automated testing. The hope is that through this sequence of ever more computationally demanding experiments and simulations students will gain an understanding of how computational power can meaningfully contribute to scientific exploration.

4 Bricklayer's Infectious Disease Library

Bricklayer's infectious disease library (IDL) implements a fairly straightforward variation of a susceptible, infected, recovered (SIR) transmission model. Though not of medical grade, for questions with known (i.e., non-politically contentious and scientifically agreed upon) answers, Bricklayer's simulations align closely with scientific findings. An example of such a question is: "Given the R_0 value for the measles, what percentage of the population needs to be vaccinated in order to achieve heard immunity?"

Building upon abstractions introduced in the Percolation library, the IDL models a population as a square whose cells represent individuals. In contrast to percolation models, where cells can be in one of two states (open/closed), cells in IDL models represent individuals that can be in one of four states: infectable, uninfectable, infected, and recovered. These four states are modeled via programmatically assignable colors (e.g., individuals that are infectable can be represented as aqua colored cells).

Both uniform as well as Gaussian[1] (i.e., normal) random number generators can be used to model R_0. An additional variable is also provided to

[1] Research has suggested that Gaussian random number generators produce more accurate results than uniform random number generators [4] for simulations of this type.

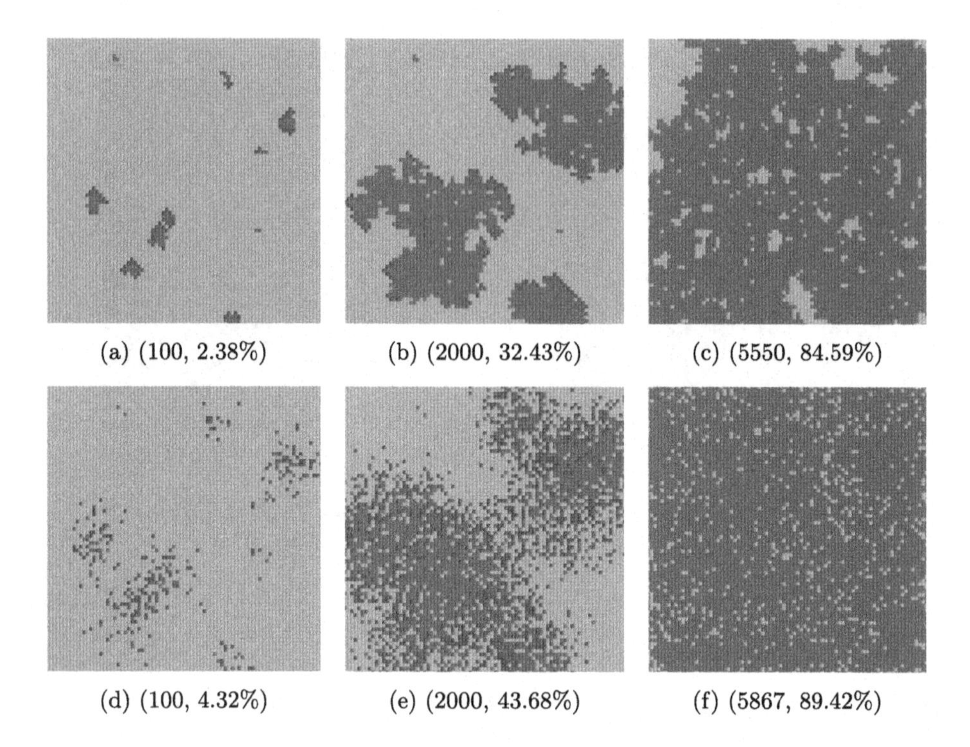

(a) (100, 2.38%) (b) (2000, 32.43%) (c) (5550, 84.59%)

(d) (100, 4.32%) (e) (2000, 43.68%) (f) (5867, 89.42%)

Fig. 4. The impact of social radius on the spread of an infectious disease.

control the probability of transmission (i.e., infection) when two cells interact. A variable is also introduced to model *social distancing* – the region (i.e., the population) where interaction with an infected person is possible. This integer variable denotes a radius defining a neighborhood around a cell of interest (i.e., an infected person). A radius of 0 corresponds to a Von Neumann neighborhood, a radius of 1 corresponds to a Moore neighborhood, and a radius, r, larger than 1 defines a square whose sides are $2 \times r + 1$.

The specification of an IDL simulation consists of: (1) the dimensions of the square modeling a population, (2) a color assignment denoting the state of an individual, (3) the size of the initial infected population, (4) a model of R_0, (5) the probability of infection upon contact, (6) a social radius, and (7) the number of transmission steps to be performed. A simulation stops when either (1) the number of transmission steps is reached, or (2) when the transmission opportunities have been exhausted (e.g., herd immunity is reached). Bricklayer's IDL library assumes that a person can be infected at most one time.

Figure 4 shows the results of social distancing on two otherwise equivalent simulations using a Gaussian number generator with $R_0 = 2.5$. The tuples in the captions denote (number of transmission steps, percent of the population infected or recovered). The three images in the top row were generated using a social radius of 0 while the images in the bottom row were generated using a social radius of 5.

The final assignment in this module is open ended and asks students to use the Bricklayer's IDL to explore the transmission of COVID-19 (e.g., what constitutes herd immunity for COVID-19). Students need to determine what value of R_0 they are willing to consider as well as other aspects like social distancing. The values used should be justified by citing reputable sources. The interpretation of the simulation results, should then be compared to reputable sources.

5 Conclusion

Technology is becoming increasingly intertwined in all walks of life. It's role is so significant that it is essential for all citizens to gain a deeper understanding of what technology can do as well as what its limitations and weakness might be.

A widely acknowledged goal of higher education in general (and general education requirements in particular) is to create a well-informed citizenry able to meaningfully engage in society. As a consequence of the astonishing advances in technology, the prerequisites for such engagement are changing rapidly. By targeting broad student populations having diverse backgrounds, CSCI 1280 seeks to positively contribute to the creation of a *computationally informed citizenry* - a citizenry having sufficient knowledge to participate in the technologically advanced societies of the 21st century.

References

1. Deutsch, A., Dormann, S.: Cellular Automaton Modeling of Biological Pattern Formation. Birkhauser, 2nd edn. (2017). https://doi.org/10.1007/b138451
2. Grimmett, G.: What is percolation? In: Percolation, pp. 1–31. Springer, Heidelberg (1999). https://doi.org/10.1007/978-3-662-03981-6_1
3. Ilachinski, A.: Cellular Automata - A Discrete Universe. World Scientific (2001). https://doi.org/10.1142/4702
4. Lachiany, M., Louzoun, Y.: Effects of distribution of infection rate on epidemic models. Phys. Rev. E. **94**(2) (2016)
5. National Research Council. A Framework for K-12 Science Education - Practices, Crosscutting Concepts, and Core Ideas. The National Academies Press (2012)
6. Shiflet, A.B., Shiflet, G.W.: Introduction to Computational Science - Modeling and Simulation for the Sciences, 2nd edn. Princeton University Press (2014)
7. Stauffer, D., Aharony, A.: Introduction to Percolation Theory. CRC Press (1994)
8. Winter, V., Hickman, H., Winter, I.: A Bricklayer-tech report. In: World Scientific and Engineering Academy and Society (WSEAS) Transactions on Computers, pp. 92–107 (2021)
9. Winter, V., Love, B., Corritore, C.: The bricklayer ecosystem - art, math, and code. In: Electronic Proceedings in Theoretical Computer Science (EPTCS) (2016)

Cloud as a Platform for W-Machine Didactic Computer Simulator

Piotr Paczuła[1], Robert Tutajewicz[1], Robert Brzeski[1], Hafedh Zghidi[1], Alina Momot[1], Ewa Płuciennik[1], Adam Duszeńko[1], Stanisław Kozielski[1], and Dariusz Mrozek[1,2(✉)]

[1] Department of Applied Informatics, Silesian University of Technology, ul. Akademicka 16, 44-100 Gliwice, Poland
dariusz.mrozek@polsl.pl

[2] Institute of Biomedical Informatics, National Yang Ming Chiao Tung University, Taipei City, Taiwan (R.O.C.)

Abstract. Effective teaching of how computers work is essential for future computer engineers and requires fairly simple computer simulators used in regular students' education. This article shows the evaluation of alternative architectures (platform-based and serverless) for cloud computing as a working platform for a simple didactic computer simulator called W-Machine. The model of this didactic computer is presented at the microarchitecture level, emphasizing the design of the control unit of the computer. The W-Machine computer simulator allows students to create both new instructions and whole assembly language programs.

Keywords: Teaching · Computational science · Computer model · Cloud computing · Serverless computing · Platform as a service (PaaS)

1 Introduction

The complexity of computer organization and construction causes that they are usually presented at the model level. The type of model depends on the purpose of the presentation. A commonly used model is the instruction set level description, also known as the Instruction Set Architecture [12,14]. It defines the architecture of the computer and includes a description of registers, a description of memory, and a description of the execution of all instructions performed by the computer. Such a model is sufficient for compiler developers, as well as for assembly language programmers. A more detailed description of the computer organization at the level of the so-called microarchitecture takes into account the registers and functional units of the processor (the central processing unit of the computer) and the main connections between them. The microarchitecture model of the computer allows defining the instruction cycle of the processor, i.e., the successive steps of instruction execution. To present the development of computer organization, the literature proposes didactic computer models exposing selected details of the concepts being explained and simplifying other elements of

D. Groen et al. (Eds.): ICCS 2022, LNCS 13353, pp. 678–692, 2022.
https://doi.org/10.1007/978-3-031-08760-8_56

computer design. Many basic textbooks, e.g. [5,12,14], use such models. While the initial model is usually very simplified, subsequent versions of the model are extended to enable the presentation of increasingly complex real-world processor problems. For example, models Mic-1 and Mic-2 are used in [14] to explain microprogramming control of the processor, while Mic-3 and Mic-4 are used to present the pipelined execution of the instruction cycle. The model using the MIPS processor architecture, presented in the book [5] and its earlier editions, is close to the actual processor organization. This model provides a simple and very clear way to explain the concept of pipelined processor organization, including the transition from non-pipelined to the pipelined organization. It also allows illustrating the problems associated with the pipelined organization and how to solve them. These advantages have determined that the MIPS model is used as a teaching model in computer architecture lectures at many universities around the world.

The presented W-Machine, created by S. Wegrzyn and next developed by S. Kozielski as a hardware device, is a didactic model of a simple, fully-functional computer described at the instruction set level and the microarchitecture level [16]. The specific feature of the microarchitecture model is that it distinguishes all signals controlling the transmission of data and addresses during the execution of the instruction cycle of this computer. This model shows in detail the design of the control unit that generates the mentioned signals that control the buses and functional units. The model includes two variants of the control unit structure: hardwired and microprogrammed. The open structure of the W-Machine simulator allows the users to independently define new machine instructions by creating microprograms that perform the functions of these instructions. Simultaneous visualization of active processor elements during the instruction execution and the possibility to use priorly implemented instructions in the creation of own assembly language program allows and simplifies the student's understanding and learning of the idea of computer operation in the commonly used von Neumann architecture [4]. This brings to the curriculum structure elements that connect the hardware and software. In this way, the course becomes more creative and effective for both the teacher and the student by using the W-Machine simulator, which can be a very good tool to aid in teaching and learning. However, teaching the computer organization and operation on such a low level now faces many challenges. First of all, the pandemic situation changed the conditions in which we live [2], including teaching opportunities [11]. Remote work is not only becoming more and more popular but is even often required [1]. Restrictions in movement mean that the work is more often done from home. Additionally, taking into account the internationalization of universities and frequent difficulties in crossing borders by foreign students, the possibility of remote access to educational tools becomes even obligatory. Cloud computing, due to the shared resources, the range of remote access, and global availability of the platform and services, can be one of the very good solutions to deploy and distribute the educational software [3,7,15,17]. In this paper, we test two alternative approaches for disclosing the W-Machine simulator in the Cloud environment - the Platform-based (i.e., PaaS) and serverless-based.

2 Background

The W-Machine was designed in the seventies at the Silesian University of Technology. The project was first implemented in the form of an electronic device. This device was used for many years to teach students the basics of the design and operation of computers. Currently, the W-Machine simulators are used for this purpose. The W-Machine is a didactic computer designed according to the von Neumann architecture. However, its basic version is devoid of IO devices and consists of the following components: arithmetic-logic unit, main memory, and control unit. These elements are connected by a data bus and an address bus. Figure 1 shows the architecture of the W-Machine. Words representing the instructions and the data are usually sent on the data bus, and the instructions' addresses and data addresses are sent on the address bus.

The main memory unit (RAM) includes two registers: a data register and an address register. The address register contains the address of the memory location for which read and write operations are performed. The data register stores data that is written to and read from memory. Memory stores words that describe the instructions being executed and the data used by those instructions. The data is written as binary signed numbers in the two's complement system. The instruction description word is divided into two parts: the instruction code field and the instruction argument field. The argument of the instruction is most often the address in the memory. The arithmetic logic unit (ALU) performs calculations on the data delivered from memory. The result of the calculations is stored in a register called the accumulator.

The operation of the W-Machine is controlled by a control unit that generates appropriate control signals based on the current state of the machine. Control signals are binary signals that activate relevant operations performed in the processor. Each control signal is responsible for performing one elementary action. Table 1 shows the control signals and the actions they perform. The control unit also includes two registers: an instruction register and a program counter. The instruction register stores the description of the currently executed instruction. Based on the content of this register, the control unit knows which instruction is currently being executed. The program counter contains the memory address

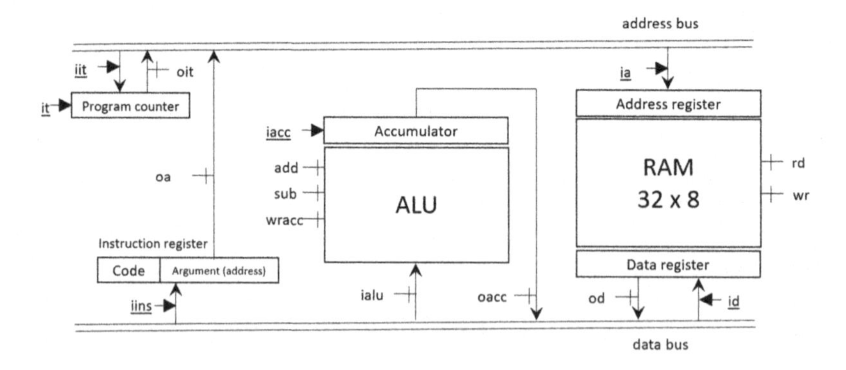

Fig. 1. W-Machine architecture.

Table 1. Control signals in the W-Machine

Signal name	Action
rd	Reads a memory location addressed by the address register
wr	Writes the content of the data register to the memory
ia	The content of the address bus are written to the memory address register
od	Outputs the content of the data register to the data bus
id	Writes the content of the data bus to the data register
ialu	The content of the data bus is fed to the ALU
add	Computes the sum of the accumulator content and the ALU input
sub	Subtracts the content of the ALU input from the accumulator
wracc	Rewrites the content of the ALU input to the ALU output
iacc	Writes the result of the operation in ALU to the accumulator
oacc	Outputs the content of the accumulator to the data bus
iins	Writes the content of the data bus to the instruction register
oa	Outputs the address (the content of the argument field) to the address bus
oit	Outputs the content of the program counter to the address bus
iit	Writes the content of the address bus to the program counter
it	Increments the content of the program counter

Table 2. Default W-Machine instruction list

Instruction name	Action
ADD	Adds the memory word to accumulator
SUB	Subtracts the memory word from accumulator
LOAD	Loads the memory word to accumulator
STOR	Stores the content of the accumulator to memory
JMP	Jumps to the given address
BLZ	(Branch on less than zero) branches when a negative number in accumulator
BEZ	(Branch on equal to zero) branches when the accumulator is zero

where the currently executed instruction is located. However, due to the adopted concept of instruction execution, the program counter stores the address of the next instruction to be executed for most of the instruction execution time.

The execution of each instruction is divided into several successive cycles. The end of each cycle is indicated by a clock signal. Students should try to execute the instruction in as few cycles as possible. In each cycle, the corresponding control signals are activated, which causes the execution of the related actions. These actions constitute a single step of the instruction execution. Therefore, designing the instruction covers determining which control signals are to be active in each of the clock cycles. For example, the ADD instruction consists of three consecutive cycles. The signals *rd*, *od*, *iins* and *it* are active in the first (fetch and decode) cycle. The signals *oa* and *ia* are active in the second cycle. And finally, in the third cycle, they are active *rd*, *od*, *ialu*, *add*, *iacc*, *oit* and *ia*.

Table 2 shows the default list of instructions. Since the control unit is implemented as a microprogrammed one, we can modify existing instructions and add new ones. Moreover, students can use the designed instructions to write programs in a assembly language. With the W-Machine on the cloud, both functionalities can be practiced online, bearing in mind that it requires an active subscription for hosting the simulator operable and covering the hosting costs of the public cloud.

3 Related Works

Cloud computing is an architecture that works effectively in business. Although education has different priorities than business, these two areas have a lot in common. Just like in business, it is essential to reduce costs with high ease of use and flexibility of access. Of course, cloud computing is not intended for all applications. The research presented in [13] proves that in educational applications, high availability and quick adaptability to the changing demand for access are sometimes more important than performance.

The main goal of computer model simulators is to provide students with the opportunity to combine theory with practice. In [10], Prasad et al. analyzed over a dozen simulators for teaching computer organization and architecture, taking into account, i.a., the criteria, like scope, complexity, type of instruction set, possibility to write assembly code, user interface, support for distance learning and free availability. In conclusion, the authors stated that two simulators, MARIE and DEEDS[1] had met defined criteria. MARIE (Machine Architecture that is Really Intuitive and Easy) [9] is a simulator that has two versions: Java program and JavaScript version, and its main advantages are intuitive interface and assembly program execution visualization. DEEDS (Digital Electronics Education and Design Suite) covers combinational and sequential circuits, finite state machine design, computer architecture, and Deeds-McE (a computer emulator). The authors have found DEEDS to be less intuitive than MARIE, but its main advantage is the support for the generation of chip/PCB layout. Other tools analyzed in the article include Qucs (Quite Universal Circuit Simulator), CircuitLogix, ISE Design Suite - Xilinx, Quartus II, and HADES. In [6], Imai et al. present VisuSim used for assembly programming exercises in e-learning cooperative (built-in email communication) mode. Another worth mentioning simulator is the gem5 simulator - a very powerful, universal, and popular tool [8]. Nevertheless, gem5 is too complicated to use for first-year students with little knowledge of how computers work. The same problem applies to the Sim4edu website[2]. On the other hand, there exists a very simple online simulator[3], but it only allows for running assembly programs. In conclusion, none of the mentioned simulators allow students for simple, control signals-based definition of new instructions and utilization of the instructions in developed programs. Moreover, none of the tools is available for the cloud environment, like the W-Machine presented here.

[1] https://www.digitalelectronicsdeeds.com/deeds.html.
[2] https://sim4edu.com/.
[3] https://schweigi.github.io/assembler-simulator/.

4 W-Machine on the Cloud Environment

The previously used W-Machine simulator was redeveloped for the Azure cloud. Cloud provides a hosting environment for scalable web applications, so it seems to be a good alternative for local web servers in the case of many (e.g., thousands) users. Moreover, the cloud environment lowers the entry barrier for hosting and ensures the high and global availability of the deployed applications, like the W-Machine. The tool has been divided into several individual cloud modules (Fig. 2). The bottom App Service is responsible for delivering the W-Machine application, which serves the website files using the HTTP protocol. It is the least loaded element of the system, sending files only the first time the W-Machine service is launched. Subsequent actions on the website are handled by servers with domain logic. The architecture includes two alternative server solutions responsible for the operation of the W-Machine. The first one is deployed using the Azure App Service (PaaS solution) with domain logic implementing the simulator's functionality. The second twin service runs on Azure Functions (serverless solution) with the same functionality. Both service approaches were chosen for their capability to scale an application easily (vertically and horizontally).

Both solutions use the same execution code supporting the logic of the W-Machine. The separation allows changing the front-end that cooperates with the simulator at any time of its operation. Additionally, each back-end service has its own Application Insights service responsible for monitoring its work. The Application Insights services are used to test the response time to requests sent in load tests presented in Sect. 5. The non-relational Redis database is responsible for storing user data, such as the state and settings of the W-Machine for each created session. The database features high efficiency thanks to storing data in a cache, which significantly speeds up the time of writing and reading data at the expense of data persistence and the maximum amount of information stored. The nature of the application does not require collecting a lot of data in a permanent form. Such architecture allows for effective separation of the website responsible for the user interface (which is unnecessary when testing the solution in terms of performance) from the computing services (implementing the teaching application logic), monitoring services, and databases.

The front-end of the W-Machine on the Cloud is shown in Fig. 3. The left panel allows observing automatic or manual execution of the program or instruction prepared by students. The right panel is used for developing the program (visible) or developing the instruction composed of control signals.

5 Experimental Evaluation of Alternative Architectures

For each alternative architecture of the execution environment, we carried out different types of performance tests to check the profitability of each architecture with possible scaling: 1) baseline, 2) load, 3) stress, 4) peak, and 5) soak testing. Each test was performed with the same test procedure. In the first step, we sent a request to get the state of the W-Machine, which simulated loading the

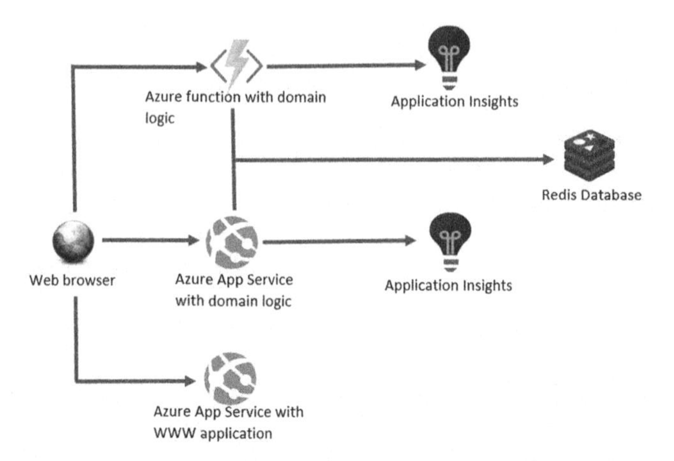

Fig. 2. Architectural alternatives of the W-Machine on the Cloud.

application for the first time. Then, we made a request to execute a simple program that summed two arrays and saved the result to the third array. The program used in all testing cases is the same as the one shown in Fig. 3. Tests were executed by repeating the simple program simultaneously and many times by multiple users. This was done by using K6 open source testing tool that simulates many users and sends execution requests to the tested environment. In our experiments, particular requests were sent with a 500 ms pause between one another since users usually do not act continuously, and a slight time delay accompanies each action they perform.

To test both investigated architectures (PaaS and serverless), the expected maximum load limit was set to 1,000 simultaneous users of the W-Machine. The successive load levels used in the tests were the percentage of the maximum load:

- baseline testing - 200 users (20% of the maximum load),
- load and soak testing - 700 users (70% of the maximum load),
- stress testing - $\leq$1,000 users most of the time, up to 1,500 users,
- peak testing - up to 1,500 users (150% of the maximum load).

The characteristics of performed tests are presented in Fig. 4. As can be observed, the baseline and load tests had three phases taking 1 min, 3 min, and again 1 min. In the first 1-min phase, the number of users increased to the top value, appropriate to the test type. In the second phase, the number of users was constant, and in the third phase, it decreased again from the top value to zero. The difference between the tests lies in the top value. The nature of the soak test was similar except that it was much longer with 2-, 56-, and 2-min periods and 700 users. The stress test relied on the load constantly increasing to a level exceeding the maximum level that the system could achieve without

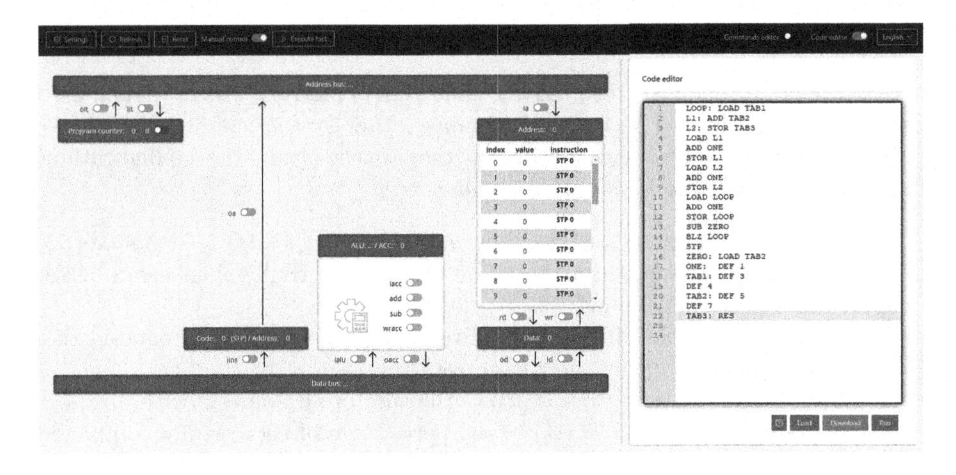

Fig. 3. The W-Machine on the Cloud: left panel for observing or manual testing of the program execution, right panel with a program or instruction editor.

generating a large number of errors. The expected place for a system breakdown was a level greater than 1,000. The peak test had a long period of the normal load of 200 users, followed by a sudden increase in the number of users to a critical number of 1,500 concurrent users, sustained for a short period. After that, traffic decreased drastically and went back to the previous level.

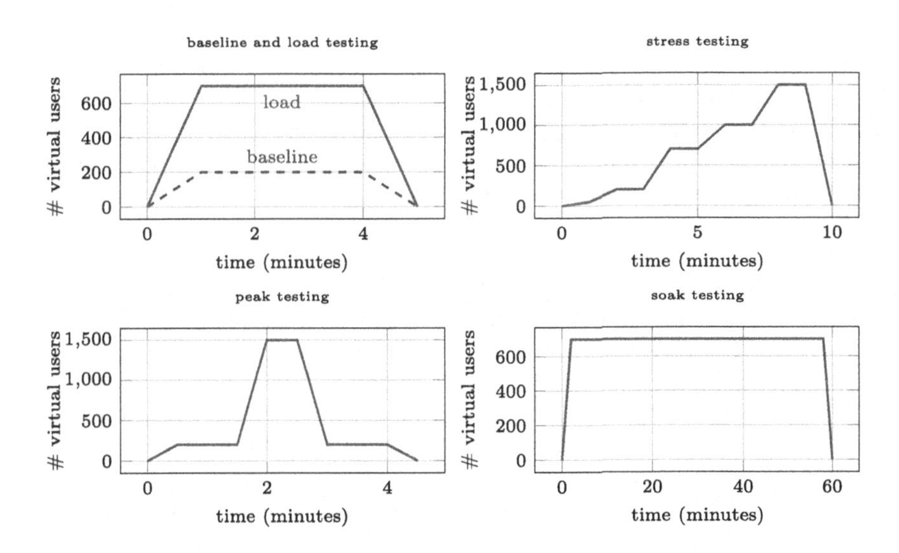

Fig. 4. Characteristics of the performed types of tests.

5.1 Performance Metrics

Each of the performed tests resulted in a collection of metrics, which we then used to analyze and compare performance against other system configurations. We used the following performance metrics to compare the alternative configurations and architectures of the W-Machine simulator:

- The number of requests per second (marked as $requests/s$) - the average number of requests sent throughout the test period. High values mean better application performance under heavy loads.
- The number of successfully completed requests - the sum of all requests successfully completed during the entire test.
- Number of requests with error status - the sum of all requests with errors.
- Success rate (marked as $OK(\%)$) - the percentage of successfully completed requests to all requests.
- Response time metrics: well-known average (marked as Avg), minimum (Min), maximum (Max), and median (Med), i.e., 50th percentile, as well as 90th percentile ($P(90)$) and 95th percentile ($P(95)$) which reflects the behavior of the system and its stability for sudden jumps in the application load (e.g., the value $P(90) = 1000$ ms means that 90% of all sent requests were completed in less than 1000 ms, with 10% of all requests exceeding this time).

5.2 Results

For the W-Machine running on App Services (PaaS), the experiments were conducted separately on several pricing tiers with the varying number of instances of virtual machines (VMs) - 1, 3, up to 10, if it was possible for the particular tier. Such tests allowed us to select the best offer in terms of the target system performance and the costs of keeping it working. We started with the B1 tier, which provides the weakest virtual machines in terms of performance, and does not allow automatic scaling of the number of instances. Additionally, it has a maximum number of parallel applications of 3 at the same time. This tier is used mainly for development and testing. Results of various types of tests are shown in Table 3. As can be observed, the average response times (Avg) in various tests usually reach several seconds even when scaling to 3 instances. This shows that this tier does not guarantee appropriate performance.

The B3 pricing tier, compared to B1, offers 4 times more computing power, which translates into much better results (Table 3). The median (Med) and average response times for baseline testing are well below 100 ms. The obtained results are almost 10 times better compared to the B1 pricing tier. Bandwidth has also increased drastically ($requests/s$). Adding another 2 instances seems to be necessary only for more heavy loads. For example, for the stress test we can observe that the value for P(95) decreased two-fold to a satisfactory level after scaling from 1 to 3 instances. The cost of one B3 instance is 30% higher than that generated by three B1 instances, but the results obtained are many

Table 3. Performance tests for W-Machine working as an App Service for different pricing tiers and variable instance count.

Pricing tier	Testing type	Instances	Cost	Requests/s	OK(%)	Response time to request (ms)					
						Avg	Min	Max	Med	P(90)	P(95)
B1	Baseline	1	54.75	94.14	100	1210	43	5210	1110	2470	2790
		3	164.25	193.82	100	324.33	40.83	3780	158.07	845.85	1150
	Load	1	54.75	90.43	99.85	5890	42.52	22850	6180	9060	10090
		3	164.25	223.88	100	2030	41.23	21530	900.5	5970	7830
	Stress	1	54.75	84.2	98.53	8010	42.41	60000	6640	16650	20050
		3	164.25	220.42	100	2740	40.65	29280	788.54	8970	11270
	Peak	1	54.75	88.92	100	5850	43	45170	2090	19260	26560
		3	164.25	193.48	100	2440	41.09	45920	314.43	5290	17460
B3	Baseline	1	219	271	100	86.82	40.44	839.8	71.1	152.92	192.61
		3	657	289.79	100	50.78	40.51	442.5	45.39	73.49	76.37
	Load	1	219	447.44	100	754.29	40.96	3550	697.09	1430	1740
		3	657	868.12	100	142	40.22	1930	75.54	340.37	496.8
	Stress	1	219	397.84	100	1240	41.04	6780	1060	2590	3280
		3	657	773.42	100	389.27	40.45	6650	111	1030	1720
	Peak	1	219	340.84	100	925.12	40.7	5390	243.4	2710	3130
		3	657	516.76	100	422.23	40	5840	69.7	1150	2300
S1	Baseline	1	73	102	100	1060	42	6540	971	2030	2370
		3	219	202	100	289	41	3800	132.75	736	1110
		10	730	277	100	75.86	40.53	1470	51.28	116.22	159.85
	Load	1	73	105	100	4940	42	17060	5270	7640	8540
		3	219	228	100	1980	40.8	24100	886	5960	9270
		10	730	670.92	100	333.1	40.37	9580	128	744.91	1290
	Stress	1	73	93	98.66	7080	42	60000	6110	14580	16910
		3	219	209	99.74	2910	41	38890	973.32	8910	11940
		10	730	537.67	100	722.53	40	22210	132.4	1840	3860
	Peak	1	73	85	96.71	5940	43	60000	2110	13180	16700
		3	219	206	100	2230	40.98	35770	346.6	5010	15970
		10	730	444.12	100	608.86	39.74	12620	88.22	1540	4100
S3	Baseline	1	292	258.16	100	115.84	40.46	2790	86.12	229.55	291.29
		3	876	288.91	100	52.44	40.39	1540	46.21	76.29	80.52
		10	2920	289.48	100	51.22	40.6	437.34	45.73	74.68	78.01
	Load	1	292	427	100	812	40	4120	797	1510	1710
		3	876	810.14	100	188.09	40.22	2650	80.43	481.87	737.51
		10	2920	1012.85	100	51.76	40.21	477.4	45.88	75.15	79.3
	Stress	1	292	377	100	1330	40	6370	1120	2850	3190
		3	876	718.28	100	459.4	40.31	6600	113.57	1390	2180
		10	2920	1229.77	100	56.7	40.27	1310	48.19	80.61	89.02
	Peak	1	292	333	100	967	40	5630	235	2830	3300
		3	876	515.59	100	430.79	40.32	8200	71.2	1170	2210
		10	2920	837.77	100	59.95	40.03	813.87	48.43	83.49	98.2
	Soak	3	876	970.88	100	214.31	39.16	3600	94.99	563.65	793.94
P3V3	Baseline	1	957.76	291.5	100	46.92	40.17	1150	44.46	55.27	56.95
	Load	1	957.76	1007.66	100	53.63	39.85	1380	46.3	68.77	82.63
	Stress	1	957.76	1012.8	100	177.03	39.9	2470	94.92	439.7	556.21
	Peak	1	957.76	717.69	100	157.71	40.08	1980	87.23	376.39	472.05

times better. The instances in the S1 tier have comparable parameters to those offered in the B1 plan. However, in the S1 tier, we can scale the application up to 10 instances, which translates into much greater available performance. In fact, single instances in this plan are slightly better than those available in the B1 plan. Increasing the number of instances to 10 brings noticeable performance gain for more demanding tests - for the load and stress tests the average response time decreased at least 10 times and the throughput increased at least 6 times.

The S3 tier offers performance identical to that available in the B3 plan. The main difference lies in the maximum number of instances that can be allocated to serve the application. Given the large amount of resources offered in this tier, the differences in the baseline test results for the varying instance numbers are negligible. The first differences are visible when the number of users is increased to 700 (S3, load testing). For 10 instances, the application works visibly below its maximum capabilities and still has a computational reserve, which can be seen in a slight increase in bandwidth ($requests/s$) compared to the configuration using 3 instances. Median and average response times are much lower than for the S1 tier and comparable to those obtained with the B3 tier for the same number of instances. The performed stress and peak tests also showed a large reserve of computing power available with the S3 tier and 10 instances. This can be noticed by observing the P(95) value, which remains at a level lower than 100 ms. For this pricing tier (S3), we also carried out the soak tests using 3 application instances (S3, soak). After a 4-h test, there were no errors in the application ($OK(\%) = 100$), and its operation during this time was stable (the number of requests per second was equal to 970.88 with the average response time 214.31 ms). The achieved throughput was similar to that obtained in the load test for the same configuration (3 instances). The relatively low P(95) value of less than 1 s also proves that the load is distributed fairly evenly over the available application instances. Due to the high costs of the P3V3 pricing tier, we only tested the application working on one active instance. This layer provides extensive computing resources (8 cores and 32 GB memory per instance). Comparing the results obtained for the P3V3-based deployment of W-Machine with those obtained for the configuration with the use of the S3 plan, we can notice how important the performance of a single machine is. The results obtained for the load test are almost identical when comparing them with the S3-hosted application with 10 instances, while the costs generated by both solutions are entirely different. The P3V3 pricing tier is about 3 times cheaper here, offering similar performance. Table 3 (P3V3) shows that the W-Machine simulator provides perfect performance results for all tests, keeping the median response time below the 100 ms limit.

The application responsible for the main logic of the W-Machine uses the .NET 5 platform, which allows it to be implemented as an App Service using three different environments: Windows and Linux servers and the Docker containerization tool. The monthly costs generated by each of them are \$730, \$693.5, and \$693.5 respectively for Windows, Linux, and Docker (with the S1 tier configured for 10 application instances). To compare their actual performance, we performed the same tests for each OS platform. When testing the four different scenarios shown in Table 4, no significant differences in the results obtained were observed. The differences are within the limits of the measurement error and are not a sufficient basis for determining the superiority of a given OS platform, taking into account only the differences in performance. However, costs may vary significantly, especially for premium tiers, like P3V3 ($957.76 per month for Windows, $490.56 per month for Linux).

The second of the tested architectures for the W-Machine relied on serverless Azure Function service. This approach differs in operation from those using dedicated virtual machines. The first difference is the capability to scale active instances to zero. This results in a cold start of the application, which occurs when the website is inactive for a long time. Lack of network traffic deactivates all active instances executing the W-Machine code, so the first requests after resuming traffic take much longer. The second difference is the inability to set a fixed number of active instances, as we can do in App Service (PaaS). Automatic scaling of the application, which is an immanent feature of the approach, works very quickly, as indicated by very low P(95) values for the performed tests in Table 5 (we performed the same tests as for the various App Service configurations).

We could observe that in baseline testing, after just the first 30 s, 4 instances of the application were created, which remained until the end of the test. In load testing, 9 instances were allocated almost immediately that served 700 simulated users very efficiently. However, the maximum response time for the test exceeded 10 s due to the aforementioned cold start. Requests sent before creating the instances responding to the traffic adequately often resulted in a long delay in execution. A similar effect could be seen for stress testing and peak testing, where the increase in traffic is sudden and requires the creation of

Table 4. Performance tests for App Service on S1 tier for different environments.

Testing type	Environment	Requests/s	OK(%)	Response time to request (ms)					
				Avg	Min	Max	Med	P(90)	P(95)
Baseline	Windows	277	100	75.86	40.53	1470	51.28	116.22	159.85
	Linux	277.18	100	76.2	49.22	2060	51.16	118.55	155.25
	Docker	278.28	100	73.62	40.44	1630	50.28	114.77	156.01
Load	Windows	670.92	100	333.1	40.37	9580	128	744.91	1290
	Linux	642.57	100	369.93	40.31	1319	115.69	918.4	1560
	Docker	631.29	100	385.66	40.49	10920	114.96	1000	1710
Stress	Windows	537.67	100	722.53	40	22210	132.4	1840	3860
	Linux	539.2	100	784.23	40.27	19980	102.47	1230	5410
	Docker	563.4	100	727.91	40.54	18000	137.21	1730	3220
Peak	Windows	444.12	100	608.86	39.74	12620	88.22	1540	4100
	Linux	435.82	100	621.29	40.35	15500	89.49	1430	2850
	Docker	452.92	100	568.02	40.47	12040	86.62	1650	3260

Table 5. Performance tests for W-Machine working on serverless Azure Function.

Testing types	Requests/s	OK(%)	Response time to request (ms)					
			Avg	Min	Max	Med	P(90)	P(95)
Baseline	283.04	100	63.24	43.21	346	56.81	81.44	92.96
Load	949.21	100	88.04	42.6	10150	69.68	140.71	186.73
Stress	1063.51	99.99	144.06	42.97	56110	113.53	236.19	290.47
Peak	665.99	99.95	185.24	43.38	19400	88.5	434.46	567.57

new instances of the application. The load spikes were much more pronounced here, and the first errors appeared due to the lack of computing power of the available instances. The stress testing started with 4 instances at the level of 200 simulated users. As the workload increased, the number of instances increased to 6 for 700 users, 10 instances for 1,000 users, and up to 14 instances for 1,500 users. The dynamic addition of resources allowed the application to maintain a stable response time. The values of P(90) and P(95) did not exceed the limit of 300 ms, which proves the very good responsiveness of the system. The biggest challenge for the application was the peak testing, which subjected the service to a very rapid increase in load. It took a full minute for the traffic to stabilize, during which the number of active application instances increased from 4 to 14.

6 Results Summary and Concluding Remarks

Providing computer simulators that can be quickly deployed, globally available in many regions worldwide, and fast responding to users' requests is crucial for globally implemented distance learning. In terms of functionality, our W-Machine simulator complements the solutions mentioned in Sect. 3 by providing a low entry barrier for observing and planning transfers of data and addresses, managing particular steps of the instruction execution (on the control signals level), and development and debugging of created instructions and programs implemented in assembly language. When surveying the usability of the tool among 71 first-year students, we obtained 73% satisfaction in terms of usability, 71% in terms of creativity and 51% in terms of the attractiveness of the tool. Moreover, to our best knowledge, it is the first cloud-based computer simulator. Thus, it is very flexible in the context of worldwide accessibility for a large group of users at the same time. The results obtained during our experiments allow formulating conclusions regarding the impact of individual parameters of the offered approaches on the actual system efficiency.

When starting the research on alternative approaches to the application deployment (App Services and Azure Functions), the primary assumption was to find the best solution in terms of performance and generated costs. To qualify the configuration as meeting the stable and quick operation criteria, we assumed the maximum median and average response time limit not exceeding 200 ms. Among all the tested versions, we can distinguish three candidate solutions. The first is a 3-instance B3 pricing tier setup, which translates into a monthly cost of $657.00. The price is for the Windows version; therefore, the final price may be reduced to $153.30 when using the Linux version. Unfortunately, this tier has limited scaling capabilities. Another configuration worth a closer look at is the one based on the S3 tier, which is 33% more expensive than B3 but also more scalable. The application running on the Azure Function service is also a promising alternative. In some cases, the best request throughput and average response times may also prove to be the best budgetary solution. For example, in the case of applications with short but intense periods of high load, a much better alternative is to choose the serverless service (Azure Function). However, with the assumed constant load, the situation may also change in favor of App Services with an appropriate pricing tier for the Application Service Plan.

Future works will cover further development of the W-Machine on the cloud covering the implementation of other architectures of the simulator itself toward the inclusion of the interrupt and I/O modules.

Acknowledgments. The research was supported by Statutory Research funds of the Department of Applied Informatics, Silesian University of Technology, Gliwice, Poland (grant no. 02/100/BK_22/0017).

References

1. Al-Mawee, W., Morgan-Kwayu, K., Gharaibeh, T.: Student's perspective on distance learning during COVID-19 pandemic: a case study of Western Michigan University. Int. J. Educ. Res. Open (2021). ISSN 2666–3740
2. Alashhab, Z.R., Anbar, M., Singh, M.M., Leau, Y.B., Al-Sai, Z.A., Alhayja'a, S.A.: Impact of coronavirus pandemic crisis on technologies and cloud computing applications. J. Electron. Sci. Technol. **19**(1), 100059 (2021)
3. Encalada, W.L., Sequera, J.L.C.: Model to implement virtual computing labs via cloud computing services. Symmetry **9**(7), 117 (2017)
4. Goldstine, H.H.: The Computer from Pascal to von Neumann. Princeton University Press, Princeton (2008)
5. Hennessy, J.L., Patterson, D.A.: Computer Architecture: A Quantitative Approach, 6th edn. Elsevier, Amsterdam (2019)
6. Imai, Y., Imai, M., Moritoh, Y.: Evaluation of visual computer simulator for computer architecture education. In: IADIS International Conference e-Learning, pp. 17–24 (2013)
7. Jararweh, Y., Alshara, Z., Jarrah, M., Kharbutli, M., Alsaleh, M.N.: Teachcloud: a cloud computing educational toolkit. Int. J. Cloud Comput. **2**(2–3), 237–257 (2013)
8. Lowe-Power, J., Ahmad, A., Armejach, A., Herrera, A., et al.: The gem5 simulator: Version 20.0+ (2021). https://hal.inria.fr/hal-03100818/file/main.pdf
9. Nyugen, J., Joshi, S., Jiang, E.: Introduction to MARIE, A Basic CPU Simulator. The MIT (2016)
10. Prasad, P., Alsadoon, A., Beg, A., Chan, A.: Using simulators for teaching computer organization and architecture. Comput. Appl. Eng. Educ. **24**(2), 215–224 (2016)

11. Soni, V.D.: Global impact of e-learning during COVID 19 (2020). https://dx.doi.org/10.2139/ssrn.3630073. Available at SSRN 3630073
12. Stallings, W.: Computer Organization and Architecture. Designing for Performance, 10th edn. Pearson, Hoboken (2016)
13. Tamara Almarabeh, Y.K.M.: Cloud computing of e-learning. Modern Appl. Sci., 11–18 (2018). ISSN 1913-1844
14. Tanenbaum, A.: Structured Computer Organization. Pearson, Uppersaddle River (2013)
15. Wang, B., Xing, H.: The application of cloud computing in education informatization. In: 2011 International Conference on Computer Science and Service System (CSSS), pp. 2673–2676. IEEE (2011)
16. Węgrzyn, S.: Foundations of Computer Science. PWN, Warsaw (1982)
17. Yadav, K.: Role of cloud computing in education. Int. J. Innov. Res. Comput. Commun. Eng. **2**(2), 3108–3112 (2014)

Snowflake Generation

Valerie Maxville(✉)

Curtin University, Perth, WA, Australia
v.maxville@curtin.edu.au
http://www.curtin.edu.au/

Abstract. For over fifty years we have worked to improve the teaching of computer science and coding. Teaching computational science extends on these challenges as students may be less inclined towards coding given they have chosen a different discipline which may have only recently become computational. Introductory coding education could be considered a checklist of skills, however that does not prepare students for tackling innovative projects. To apply coding to a domain, students need to take their skills and venture into the unknown, persevering through various levels of errors and misunderstanding. In this paper we reflect on programming assignments in Curtin Computing's Fundamentals of Programming unit. In the recent Summer School, students were challenged to simulate the generation and movement of snowflakes, experiencing frustration and elation as they achieved varying levels of success in the assignment. Although these assignments are resource-intensive in design, student effort and assessment, we see them as the most effective way to prepare students for future computational science projects.

Keywords: Education · Computational science · Programming · STEM

1 Introduction

There has been a movement in recent years pushing for everyone to learn how to code. A recent post by the Forbes Technology Council [1] cited twelve of fifteen members supporting coding for everyone. It's a great idea, however, teaching coding isn't always easy. Huggard [2] writes of negative attitudes and "programming trauma" as barriers to learning. Initiatives such as CoderDojo seek to raise enthusiasm through coding clubs [3,4], aiming to take students through the levels of Digital Proficiency (Fig. 1). These clubs are aimed at school children to create and/or extend an interest in coding, aiming prepare students for future employment opportunities. There is a question on the long-term impact of programming and STEM interventions, and further study is required to gauge what duration and regularity is required for students to maintain interest [5].

In the past, university level coding was primarily confined to Computer Science and Information Technology courses. These introductory programming classes have a long history of low pass-rates along with the challenge of assessing programming "ability" [6]. At Curtin University, we are seeing rapid growth in

D. Groen et al. (Eds.): ICCS 2022, LNCS 13353, pp. 693–706, 2022.
https://doi.org/10.1007/978-3-031-08760-8_57

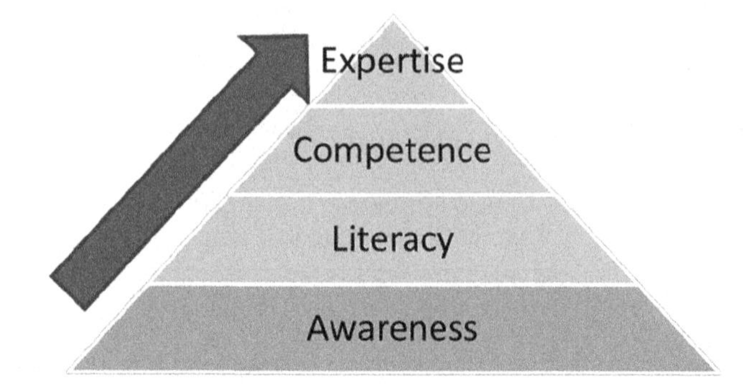

Fig. 1. Digital Proficiency Levels (adapted from ECOL Foundation, 2011) [4]

the number of students taking programming courses, whether through choosing computing degrees; finding their courses now include programming; or taking the units as electives. Engineering students need at least one full unit of programming for their degree to be accredited, and science students are also required to take programming in their first year. So, how do we adapt our courses to be suitable for a wider range of backgrounds, and even more diverse fields of application? There have been initiatives in the computational science area, including [7] and [8], which highlight key skills and concepts required for science students and researchers.

This increase in students does not occur in a vacuum, and we have additional pressures to accommodate. Directives to reduce expenses can have a detrimental impact on staff: student ratios at exactly the time that we find ourselves struggling to scale teaching. In addition, a mandate to reduce the number of assessments forces changes that are not based on improving outcomes or learning, and may indeed have the opposite effect. The media will have us believe that the students are part of a "snowflake generation", unable to cope with the stresses of life - which would include study. However, that stigmatises the challenges many are having with mental health in an unpredictable world [9]. This is compounded by the challenges of a worldwide pandemic, affecting the health, families and livelihoods of students. And those teaching through these time, requiring agility to move quickly between teaching modes and the loss of the 1-1, face to face interaction that often makes all the difference.

In this paper, we discuss approaching these challenges through a snowflake-related assessment, part of an introductory programming unit taken by science and engineering students. As the students persevered through the assignment and the unit, each of their journeys was unique, forming a delicate network of connections of ideas: snowflakes, in a good way.

> *"Snowflakes fascinate me... Millions of them falling gently to the ground... And they say that no two of them are alike! Each one completely different*

from all the others... The last of the rugged individualists!" - Charles M. Schulz[1]

2 Fundamentals of Programming

The unit, Fundamentals of Programming, was developed in 2017 for new courses in Predictive Analytics and Data Science at Curtin University. The predicted number of students in its first run was 20. By the time semester started there were 120 students enrolled. Since becoming part of the first year core for both science and engineering course, the unit is now attracting over 700 students at the main campus, and is delivered in Malaysia, Mauritius, Dubai and Sri Lanka.

As a new course, that author was given Unit Learning Objectives as a guide to what was required in the unit. The language was to be Python were clearly looking for a "vanilla" introductory programming unit. After eight years as the Education Program Leader at the Pawsey Supercomputing Centre, the author had observed the skills and tools used by a range of computational scientists, and the key issue in data science raised by leaders in eResearch and digital humanities. The science students would need to learn about scripting and automation; notebooks and reproducibility; software carpentry; and would need to work in a Linux environment.

In terms of assessment, a typical 50% exam, 20% assignment, 15% mid-semester test and 15% practical test was in place. Students needed to submit a reasonable attempt for the assignment, and receive 40% on the exam to pass the unit. From the start, the "practical test" was reworked to be assessed as a series of 5 practical tests worth 3% each. The tests were to be competency-based, with students allowed to resubmit until they could demonstrate the required skills. Student response to this approach was overwhelmingly positive. The assignment required the demonstration of the work, aiding Academic Integrity.

We were then given a directive from the university that no unit could have more than three assessments. An initial exemption was approved, but eventually the mid-semester test was dropped and the assignment and practical test weightings increased to 30% and 20% respectively. With the pandemic, the exam could no longer be held in a large, controlled examination venue. In response, the exam was replaced with a Final Assessment: 24-h open book, practical and delivered in Jupyter Notebook. In the current offering of the unit, the assignment weighting has been increased and the Final Assessment decreased to both be 40% to better reflect the effort required. This is in response to students questioning the weighting with respect to the time required.

The unit is both easy and hard. It is hard in that you cannot pass without doing the work. It is easy, in that it is very scaffolded, and if you do the work, you will pass. With Python, plotting is trivially easy, and is used from week three to provide a visual representation of coding outcomes - a huge advantage

[1] Snowflake quotes sourced from: https://kidadl.com/articles/best-snowflake-quotes-that-are-truly-unique and https://routinelynomadic.com/snowflake-quotes-sayings-captions/.

compared to Java and other languages where graphics sits beyond a barricade of pre-requisite knowledge. We have quite a few experienced coders taking the unit, but even more of the students have no coding experience. Given this range of backgrounds, we work hard to keep the unit challenging so that we can develop and extend all of the students.

3 The Assignment

Each semester we have an assignment scenario where students are asked to develop a simulation for a "real world" problem. Previous topics have included the spread of disease; parking cars; brine shrimp life cycles; and the game of cats. The assignment is assessed on three major views of the student's work: 30% demonstration of code, 30% implementation of code, and 40% for the reflective report.

Challenge assignments are not new in programming units - they have been a staple from the earliest courses. In [10] the patterns of student performance in coding challenges is explored. These are smaller task than those undertaken in this unit - where we also challenge the student in terms of planning, time management and persistence. Just as the assignment takes around four weeks for the student, they take time to assess. We find each assignment takes 20–30 minutes to mark. With a large class, that may mean 300 h of marking. Some automation of assessment is possible, which tends to lead to a very defined project where student solutions can be put through a test suite. We believe that there is much to gain from a "loose" specification, allowing for creativity and choices, which are then justified in the submitted report.

3.1 Generating Snowflakes

For the 2022 Summer School, we asked students to simulate the growth of a snowflake (such as in Fig. 2); and then generate multiple snowflakes and simulate their movement in space. The inspiration was taken from a video of a scientist who had discovered the secret to creating custom snow flakes [11]. With some additional source material provided [12], students were ready to start thinking about the approach to the problem.

"No two snowflakes are alike." - Wilson Bentley

Table 1 gives the outline of the problem given to the students. It is intentionally open to interpretation, allowing students to take their work in an individual direction. Within the simulation, there are five features that are expected to be implemented and expanded on (Table 2): object behaviour, movement, terrain and boundaries, interaction and visualisation/results. Bonus marks can be gained for utilising a game engine to implement the problem, and for other exceptional approaches e.g. complex interactions, entertaining visualisations, or complex terrain rendering.

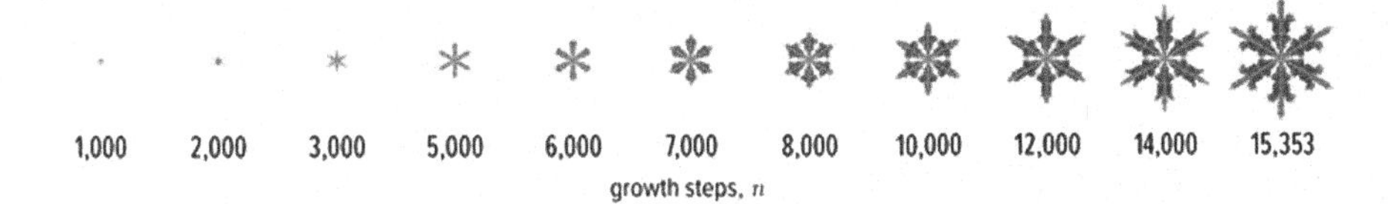

Fig. 2. Snowflake growth [13]

Table 1. Assignment: problem definition

The Problem
We will be simulating the generation of snowflakes. Each snowflake will have data as to its shape, then be drawn into a sky/world where it will progress down in a simulated drifting movement
You will be given some sample code, showing a range of approaches to similar problems. For the assignment, you will develop code to model the snowflakes using objects, and to add features to the simulation (e.g. more functionality, statistics, graphics). Your task is to extend the code and then showcase your simulation, varying the input parameters, to show how they impact the overall simulation

Much of the work for the assignment is to consider the different facets of the problem, and where they might take them. Sample code was given for the sub-problem of taking a quarter of a snowflake and duplicating and flipping it to make the whole snowflake.

At this point in the semester, we have just covered object-orientation (OO), which is a challenging concept for most students. In practicals and tests, the students have seen a range of array-based simulations. Many choose to ignore OO in the assignment, and stay in the comfort zone of arrays. Very high marks are still possible.

The reminder: *'Think before you code!'* is included in the assignment specification. Invariably, students reflect on their assignments and their surprise at the amount of thinking that was required, and how often a challenging idea took a small amount of code (and vice-versa). They also get to see large-scale repetition and the need for functions.

3.2 Scaffolding

Each semester there is an air of bewilderment about where to start with the assignment. To help with the initial steps, we provide some starter code (Table 3). This has typically been a simplistic model of a simulation, with x, y positions or population counts held in arrays. These approaches can be extended for those who find coding difficult, while stronger students will be able to convert to objects to represent each entity. High marks are possible with either approach. For the snowflake assignment, students were given code to take a quarter of a snowflake and flip it three ways to form a symmetrical snowflake twice as high

Table 2. Assignment: required extensions

The required extensions are:
1. **Object Behaviour:** Extend to have the snowflakes represented as objects. Each snowflake will have a position (in the overview plot) and its data/shape. Prompts: What makes the snowflake unique in its growth? How will you capture/simulate those factors?
2. **Movement:** Snowflakes should have a movement based on gravity and (simulated) wind. Prompts: Is the "wind" constant or does it vary? Does the "wind" affect all snowflakes uniformly?
3. **Terrain and Boundaries:** You could have a 2d terrain read from a file, including some representation of the height of the terrain. There should be boundaries to stop the snowflakes going beyond the grid. Prompts: How do you stop the snowflakes moving to invalid spaces? What happens when they hit a boundary?
4. **Interaction:** Your code should recognise when snowflakes meet/overlap and respond accordingly. Decide what this does to their movement and to the visualisation. Prompts: Do the snowflakes affect each other's path? [How] will overlaps affect the visualisation?
5. **Visualisation/Results:** Enhance the display for the simulation to show multiple snowflakes moving in a world/sky. You should display a log of events and/or statistics for the simulation and also be able to save the simulation state as a plot image or a csv file. Prompts: How will you differentiate the snowflakes? What useful information could you display?
6. **BONUS:** Utilise a game engine or plotting package to allow interaction with the movement and generation of the snowflakes. Prompts: What interaction could you have? Discuss this in future work if you don't attempt it. Note: Your program should allow command line arguments to control the parameters of the experiment/simulation

and twice as wide. This was also used within one of the Practical Tests to ensure students were familiar with the code as preparation for the assignment.

In previous semesters, only specific starter code was given. Now we also make a collection of starter code from earlier semesters available. This might seem to be giving too much help, however there is learning in reading other people's code and evaluating its applicability. Students frequently talk about the "gremlin approach" or the "shrimp approach" when they have taken inspiration from the sample code.

This semester we gave more of a walkthrough of the approach to developing the sample code. The code wasn't actually given - it appeared in a screenshot in an announcement. Students could then follow the steps towards the final version of the supplied. We are moving towards providing coding case studies as we feel providing an exemplar of the process of coding is more valuable than just giving access to code that is beyond their level.

Table 3. Help given to students as a starting point

The assignment is now available on the Assessments page. You will have until Friday 18th March to complete it (just over 4 weeks)
There is starter code from previous assignments on the page with the assignment specification. For this assignment, I'm including an approach and some starter code for a snowflake that may be useful:
Making Snowflakes
As it's a very visual problem, you should start with making sure you can plot what you generate. The versions I went through to get to this are:
1. Generate and plot a 2d array of random integers. Have a variable "size" to make it easy to update the array size
2. Make an array of zeros 2x the #rows and 2x # cols, then slot the array from (1) into the top left of the array
3. Now slot duplicates of the quarter into the other three spaces: top right, bottom left, bottom right. (I multiplied the values by 2,3,4 to make it easier to check it was correct)
4. Modify the slotting code in (3) to flip rows/cols when copying to give symmetry
5. Modify (4) to put the building of the full snowflake into a function - it will be easier to use and make the code more readable. The function should stand alone - so don't reference variables that are outside the function (I changed their names to avoid confusion)
I've put some screenshots in below to show the steps. Hopefully this will give something of a starting point for the assignment, noting that snowflake symmetry is a little different to this - but you only need to generate a quarter, then hand off the duplication to the function
You are welcome to take a different approach! Four or eight pointed snowflakes are OK (and easier), but six is more realistic

3.3 Reflection

The journey through the assignment, and the coding experience, is different for each student. 40% of the marks for the assignment are allocated to the report, to allow reflection and showcasing of their work - giving maximum chance for letting the marker know what they did and why. A perfect program would only achieve 60% in the unit without the report. The report structure is given in Table 4.

In coding assignments, we have noted some patterns in the students' performance, which can be related to the Digital Proficiency Levels in Fig. 1:

1. **I don't want to talk about it** - code doesn't go much beyond the sample code combined with snippets from lectures and practicals. Usually accompanied by an apology *(Proficiency: Awareness)*.

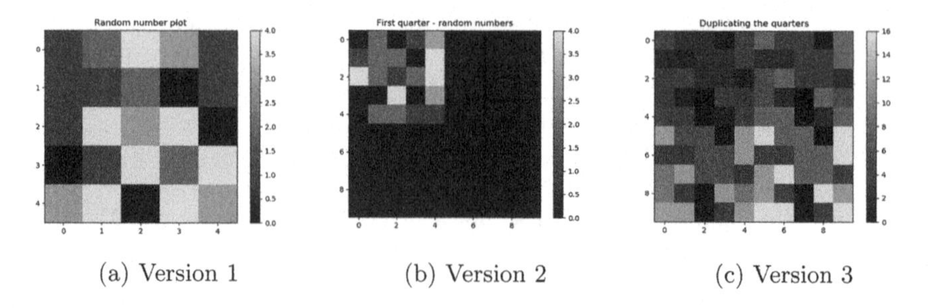

(a) Version 1 (b) Version 2 (c) Version 3

Fig. 3. Example output of first three versions of snowflake code

2. **I think I'm getting it!** - Has a go at each of the required extensions, may still struggle with object-orientation, so will often just have the x, y position in the class and still be doing most of the work with arrays *(Proficiency: Literacy)* (Fig. 3).
3. **Hits the marks** - works to the rubric to have a solid attempt at all requirements. Could get full marks... given more time *(Proficiency: Competence).*
4. **Exceeds expectations** - brings functionality, complexity and artistry that surprises and excites the markers *(Proficiency: Expertise).*

In attempting the assignment, students not only need to have some coding skill, they then need to apply that to a simulation. This simulation came in two parts: generating the snowflake(s) and moving the snowflakes through the scene (plot window).

Students in the *Awareness* level would tend to address only the second part, taking the advice that many of the extensions could be attempted without a fancy snowflake generator. These students generated matplotlib.pyplot `scatterplot( )` with circles to represent the snowflakes. Working from the "gremlins" code-base, the snowflakes were generated at random points at the top of the scene, then worked with one of: a speed in the x and y directions; a random movement in the x and y directions or a random choice from a list of possible movements in the x and y directions.

With a bit more understanding (*Literacy*) we began to see the basic simulations be based on simple objects with positions. Plots were still `scatterplot( )` circles moving down the scene, but now there were colormaps and background colours creating a more "snow-like" scene. One other extension would be attempted, usually wind or terrain. With the wind, the snowflakes would have a breeze come through, either uniformly or randomly affecting the snowflakes. The terrain would be represented as blocks, with tests for each snowflake to see if it hit the ground (and disappeared). These students were excited about their work as they had achieved more than they expected!

Students who had gained enough knowledge for *Competence* attempted all extensions and spoke with excitement of the additional work they could have done. Their snowflakes were pixelated images, often based on the supplied code. This required a shift from scatterplots to displaying a compiled image with

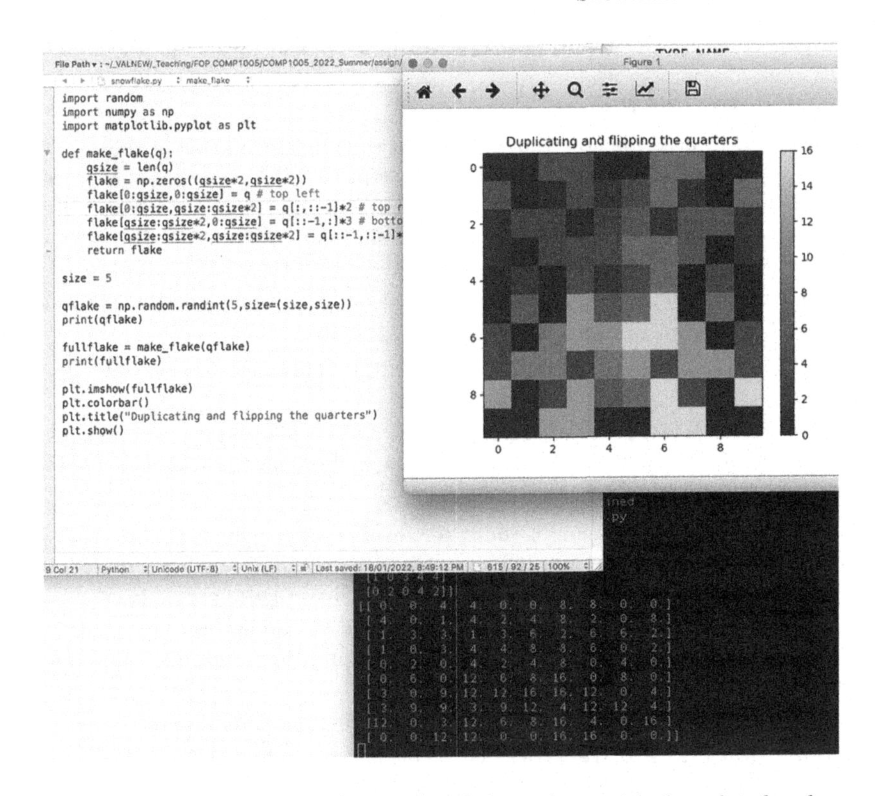

Fig. 4. Output of versions (4 and 5) is below, along with the related code

`imshow( )`, with the 2-D array representation of the snowflake being held in an object, along with its position. These simulations often tested for the terrain and would create a layer of snow when the snowflake landed. Alternatively, they identified overlapping snowflakes, with print statements to report these "interactions" (Fig. 4).

The assignments where the students had gained some *Expertise* were delightful. They had taken the time to not only convert to objects, but also to utilise a game engine (which was not covered in the unit). They had a variety of snowflakes, including variation in size and colour. Wind was controlled though interaction with the game, throwing gusts through the snowflake field. Overlapping snowflakes would change colour or change direction.

Living in Western Australia, most of the class has never seen snow, so that may have added another level of curiosity (and challenge) to the assignment.

3.4 Assessment

With many years experience in setting and assessing assignments for Computer Science students, changing the common student attitude that a fully-functional program is all-important has been a bit of a crusade. We want the students

Table 4. Report structure to guide student reflection

Simulation Project Report
You will need to describe how you approached the implementation of the simulation, and explain to users how to run the program. You will then showcase the application(s) you have developed, and use them to explore the simulation outputs. This exploration would include changing parameters, simulation time and perhaps comparing outcomes if you switch various features on/off
Your Documentation will be around 6–10 pages and should include the following:
1. **Overview** of your program's purpose and features
2. **User Guide** on how to use your simulation (and parameter sweep code, if applicable)
3. **Traceability Matrix** for each feature, give a short description, state where it is implemented, and how you've tested it (if unfinished, say "not implemented")
4. **Showcase** of your code output, including:
– Introduction: A discussion of your code, explaining how it works, any additional features and how you implemented them. Explain the features you are showcasing, modifications and parameters you are investigating
– Methodology: Describe how you have chosen to set up and compare the simulations for the showcase. Include commands, input files, outputs - anything needed to reproduce your results
– Results: Present the results of at least three different simulations
5. **Conclusion and Future Work:** Give conclusions and what further investigations and/or model extensions could follow

to write with good style, to make considered decisions in their approach and to reflect on their creation and be able to critique their own work. The marks awarded for the submissions are as follows:

- **Code Features.** (30 marks) Based on implementation and documentation.
- **Demonstration.** (30 marks) Students demonstrate their code and respond to questions from the markers.
- **Report.** (40 marks) As described in Table 4.

The bonus of this is the students who are less strong on coding, are often quite comfortable writing a report about their experience. This is also very important for science and engineering students, who's future programs will not be an end in themselves - they will be the basis for decision-making and analysis. For those situations, the ability to discuss their code, their decisions, assumptions and limitations are vital.

3.5 Academic Integrity

The easy accessibility of information and answers has aided programmers incredibly. Unfortunately, that also increases the potential for students to find or source answers which may range from how to use colormaps, through testing for the intersection of objects, to being able to pay someone to do their assignment for them. The more students in the class, the more difficult identifying these issues becomes.

> *"We're like snowflakes. Each of us is unique, but it's still pretty hard to tell us apart."* - Tony Vigorito.

Our most time-consuming, but also most effective, mechanism for assessing academic integrity is the assignment demonstrations. Unless a cheating student has done some exceptional preparation, they will find it difficult to describe code they haven't written themselves. We also see very different style in the code from students in our course, to code they may have accessed or procured.

> *"They say that there can never be two snowflakes that are exactly alike, but has anyone checked lately?"* - Terry Pratchett.

In a more formal approach, we also have an in-house tool, Tokendiff [14], for comparing student assignments for similarity. This tool provides a report on each pair of assignments to indicate the code that is too similar. It is able to account for rearranged code, and for renamed variables, which is about as far as a student goes if they "need" to copy.

By providing quite obscure problems, following a family of simulations and scripting that is fairly unique, students have indicated there is *nothing* on the Internet to help with their assignments. The challenge is to continue to come up with new ideas!

4 Results

Students took an impressive range of paths with the same starting point for the assignment. Typically we see strains of similar assignments, particularly at the lower end of scores. That was not noticed in this smaller class group, where, beyond the supplied scaffolding code, they went in different directions.

4.1 Variations on a Theme

In line with the proficiency categories, the images below provide a indication of the output of simulations in the students' assignment work.

4.2 Reflections and Reactions

> *"Every avalanche begins with the movement of a single snowflake, and my hope is to move a snowflake."* - Thomas Frey.

Getting that first snowflake movement was cause for celebration. That was soon followed by a drive to have more or different snowflakes, to adjust their movement or change the colours to be more relatable. Students expressed relief

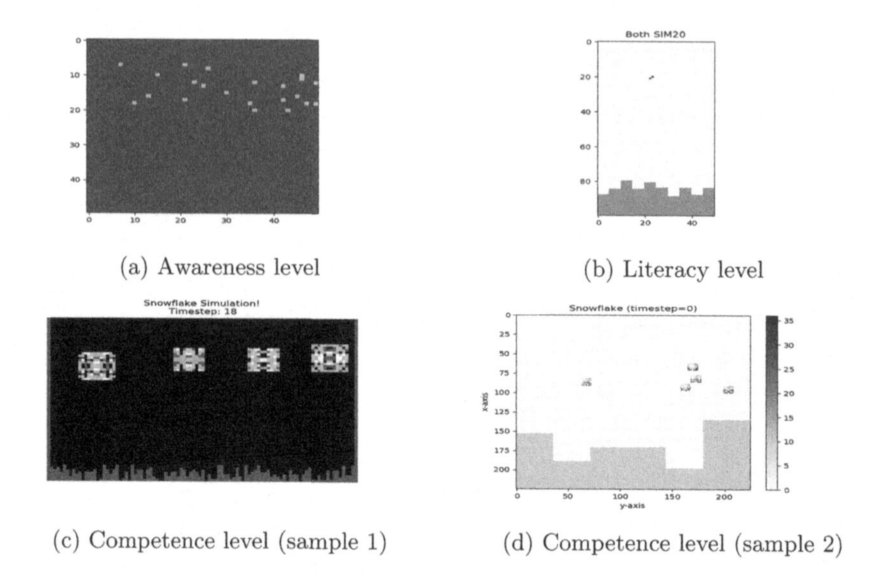

(a) Awareness level

(b) Literacy level

(c) Competence level (sample 1)

(d) Competence level (sample 2)

Fig. 5. Example output of submitted assignments at competency levels 1–3

Fig. 6. Output at expertise level

and pride in their coding outcomes, at all levels of proficiency. The subject for the assignment was so approachable, it was possible to show family members their work and share the excitement of the activity (not always possible with university assignments) (Fig. 5).

As the students moved through the unit, and the assignment, the snowflake took on a greater meaning. Where once it may have been an insult or slur against a generation, they were being creative and taking it as a challenge. The analogy extends further in an education environment, where we might see the students' knowledge and skills growing from a small core. As they extend themselves to new concepts, tentatively at first, then with more strength, their snowflake of understanding grows. It spans from the simple concepts of variables and control structures, then reinforces and extends on those with collections and functions. And this is a viable snowflake, but it can go further: into object-orientation, scripting and automation and beyond. Eventually they follow their unique path of connections and concepts to make their snowflake - similar but different to that of any other student. We, as educators, need to create the right conditions for that growth to happen (Fig. 6).

> *"Advice is like snow - the softer it falls, the longer it dwells upon, and the deeper it sinks into the mind."* -Jeremiah Seed.

5 Conclusion

When teaching Fundamentals of Programming, we aim to impart far more than how to put lines of code together. We realise many students find coding daunting and confronting: and programming languages are not gentle when they find an error! So we work to provide a safe environment and find interesting, relevant and often quirky scenarios to help students overcome their fears and go boldly into the code. We also aim to spark their imaginations to see the potential of simulations, analysis, visualisation and automation.

The assignments are a necessary leap into the unknown, with some scaffolding provided. Students are free to explore their own line of implementation - there is no correct or model solution. For many students, this in itself is unsettling, however we seem to win most of them over by the end of semester. Regardless of the proficiency of the student, they find the assignment challenging, and will usually present something they are proud of, or be able to reflect on what they would have liked to do - which is clearer once they done things the "wrong" way.

The students in the course, reputably part of a "snowflake generation", are initially hesitant. They may be expert users of technology, but coding strips that away to the essence of problems solving without wizards and templates. To move beyond the unforgiving nature of the interpreter, through the bewildering pedantic-ness of code and then confidently put forward a solution to a not-yet-resolved challenge, takes grit and perseverance. Many would back away from this uncomfortable space, but these students push on. They hopefully have become more confident when encountering the unknown, and will take this strength on with them in their future study and careers.

Acknowledgements. The author is grateful for the efforts of all students who have undertaken Fundamentals of Programming at Curtin. In particular, the Summer School students of 2022 were brave enough to try the new format for the unit, and to produce the astonishing work discussed in this paper.

References

1. Forbes Technology Council: Should Everyone Learn To Code? 15 Tech Pros Weigh In On Why Or Why Not. Forbes, 20 May 2020. https://www.forbes.com/sites/forbestechcouncil/2020/03/20/should-everyone-learn-to-code-15-tech-pros-weigh-in-on-why-or-why-not/?sh=492945a7693e. Accessed 10 Feb 2022
2. Huggard, M.: Programming Trauma: Can it be Avoided? (2004)
3. CoderDojo. https://coderdojo.com/. Accessed 10 Feb 2022
4. Sheridan, I., Goggin D., Sullivan, L.: Exploration of Learning Gained Through CoderDojo Coding Activities (2016). https://doi.org/10.21125/inted.2016.0545
5. Fletcher, A., Mason, R., Cooper, G.: Helping students get IT: investigating the longitudinal impacts of IT school outreach in Australia. In: Australasian Computing Education Conference (ACE 2021), pp. 115–124. Association for Computing Machinery, New York (2021). https://doi.org/10.1145/3441636.3442312
6. Pirie, I.: The measurement of programming ability. University of Glasgow (United Kingdom). ProQuest Dissertations Publishing (1975). 10778059
7. Wilson, G.: Software Carpentry: lessons learned [version 2; peer review: 3 approved]. F1000Research 2016 3:62. https://doi.org/10.12688/f1000research.3-62.v2
8. SHODOR: A National Resource for Computational Science Education. http://www.shodor.org/. Accessed 4 Feb 2020
9. Haslam-Ormerod, S.: 'Snowflake millennial' label is inaccurate and reverses progress to destigmatise mental health. The COnversation. https://theconversation.com/snowflake-millennial-label-is-inaccurate-and-reverses-progress-to-destigmatise-mental-health-109667. Accessed 20 Feb 2022
10. Kadar, R., Wahab, N.A., Othman, J., Shamsuddin, M., Mahlan, S.B.: A study of difficulties in teaching and learning programming: a systematic literature review. Int. J. Acad. Res. Progress. Educ. Dev. **10**(3), 591–605 (2021)
11. Veritasium: The Snowflake Mystery - YouTube video. https://www.youtube.com/watch?v=ao2Jfm35XeE. Accessed 10 Feb 2022
12. weather.gov Snowflake Science. https://www.weather.gov/apx/snowflakescience. Accessed 10 Feb 2022
13. Krzywinski, M., Lever, J.: In Silico Flurries - Computing a world of snowflakes, 23 December 2017. https://blogs.scientificamerican.com/sa-visual/in-silico-flurries/. Accessed 19 Feb 2022
14. Cooper, D.: TokenDiff - a source code comparison tool to support the detection of collusion and plagiarism in coding assignments (2022). https://bitbucket.org/cooperdja/tokendiff. Accessed 6 Nov 2022

Extensible, Cloud-Based Open Environment for Remote Classes Reuse

Jakub Perżyło, Przemysław Wątroba, Sebastian Kuśnierz, Jakub Myśliwiec, Sławomir Zieliński(✉), and Marek Konieczny

AGH University of Science and Technology, Kraków, Poland
{slawek,marekko}@agh.edu.pl

Abstract. The purpose of the presented work was to foster the reuse of whole remote classes, including the workflows involving multiple tools used during the class. Such a functionality is especially useful for designing new classes and introducing new topics into curricula, which happens quite often in laboratory-based courses in the Information Technologies area. The proposed approach allows professors to shorten the time it takes to create remote collaboration environments designed and prepared by their peers, by integrating multiple tools in topic-oriented educational collaboration environments. To achieve that, we split the system, code-named 'sozisel', into two parts - one of them responsible for facilitating creation of educational environments templates (models) describing the online class workflows composed of multiple tools, the other being responsible for executing the actual classes. For the environment to be extensible, we decided to use open, standards-based technologies only. This stands in contrast with most of commercially available environments, which are frequently based on proprietary technologies. The functional evaluation carried out on the sozisel prototype proved that the system does not require significant resources to be used in classes of medium size.

Keywords: Cloud-based remote education · Collaborative education · Model driven architecture

1 Introduction

The proliferation of broadband Internet access, as well as the breakout of the COVID-19 pandemic, resulted in significant growth in usage of remote collaboration technologies in recent years. In face of the necessity to conduct remote classes, many teachers, including university professors, began to use environments capable of organizing remote meetings much more frequently than they used to do before. Commercial companies such as Zoom, Cisco, Google, and

The research presented in this paper has been partially supported by the funds of Polish Ministry of Science and Higher Education assigned to AGH University of Science and Technology.

D. Groen et al. (Eds.): ICCS 2022, LNCS 13353, pp. 707–718, 2022.
https://doi.org/10.1007/978-3-031-08760-8_58

Microsoft are major representatives of remote meeting environments vendors, and deliver mature products facilitating the remote meetings. The default set of tools used during the meetings can be enhanced with many plugins dedicated for the platforms. Due to business reasons, the vendors are rather reluctant to integrating their environments with external tools. Although it can be possible to call into the video conference from an external tool, the more sophisticated features are typically not exposed to the outside world. Free versions of the products are frequently limited in terms of functionality and number of session participants. Open source video conferencing solutions, such as Jitsi Meet or Jami, provide an interesting alternative at least from a researcher point of view by being open to modifications and extensions.

The work presented in this paper was motivated by the specific needs of IT education area. In AGH University practice it is quite common to update courses by developing new content for laboratory classes. Such an update is typically conducted by a small group of people, who lead the classes initially, and later share them with others. By sharing we mean not only passing on the static content (written instructions, etc.), but also the way of presenting the topics to a group of students, instructing about best practices, typical caveats, etc. Note also that subsequent occurrences of the newly developed labs may be distant in time, which impedes class sharing. We also observed similar needs among high school teachers, who cooperate with universities in the Małopolska Educational Cloud project (refer to [7] for an overview of the project and collaboration patterns).

Our idea was to build an environment that would facilitate two aspects of IT education: (1) sharing both the curricula, and sets of educational tools between professors, and (2) making the classes more repeatable by preparing, storing, and executing a workflow for a lab class. To make it possible we decoupled the processes of describing the class workflow and executing it. As a result of the definition phase, a model for the class is created, including the tools to be used during the class, the input data (files, links, etc.) for the activities to be performed during the class, and the tentative schedule of events (e.g. start of a survey). The model is converted into an executable workflow when a class execution phase starts.

The structure of the paper is as follows. Section 2 surveys the important developments in the subject area. Section 3 analyzes the requirements for sharing and reusing online sessions and describes the approach proposed in this article, which results in creation of reusable workflows. Section 4 discusses the results of evaluating a proof of concept implementation of the system code-named ‘sozisel’, which implements the discussed functionality. Section 5 concludes the paper and points out the directions of future work.

2 Related Work

Many universities and commercial companies provide trainings to online communities. There are many motivations behind that: to reduce costs, to improve curricula, or to increase their reach [4–6]. In the latter case, there are two common options – either to support self-service learning (e.g., by following the Massive Open Online Courses (MOOCs) paradigm[1,2]) or to follow a train-the-trainer approach.[3] In either case, the processes conducted in the learning environment need to be repeatable and predictable. Wide adoption of online teaching by educational institutions makes the requirement even stronger.

The predictability and repeatability requirements which apply to online environments consisting of various tools are especially important when teaching technical courses. Arguably, practical verification and experimentation in teaching engineering and computational sciences is crucial [1,3] for the students to succeed. Therefore, teaching technical courses online involves not only collaboration-related tools, but also virtualized lab environments [2]. As a common approach, before an environment is made available to a wide community, it is tested with regard to its functionality and appropriate resources needed by the environment are provided, so that the self-service users' experience is satisfactory.

In comparison with the self-service case, little attention is paid to systematically prepare teacher-led remote classes. Platforms focused on technology-oriented trainings provide extensive curricula, powerful tools[4,5] and flexible virtualized lab environments[6,7] but they lack of online session templates. Courses offered according to the train-the-trainer approach are supported with tools which commonly do not form an integrated environment. Reuse of the already completed class sessions also suffers from little support. In this paper we propose an approach which leverages open tools to provide an added value by allowing the trainers to share complete environments for leading classes instead of sharing class instructions only.

3 Sharing and Reusing Online Sessions

In this section we describe our approach to supporting educational communities by providing means for structured composition of online class workflows. We assume that the communities are composed of teachers who provide similar, or at least related, courses to their respective students, and are willing to share their how-to knowledge within the community. We start from analyzing their requirements and follow with describing and justifying the main architectural elements of the proposed online class reuse system.

[1] edX, https://www.edx.org/.
[2] Coursera, https://www.coursera.org/.
[3] Cisco Networking Academy Program, www.netacad.com.
[4] Cisco Packet Tracer, https://www.netacad.com/courses/packet-tracer.
[5] Mininet, http://mininet.org.
[6] Cisco DevNet https://developer.cisco.com.
[7] P4 behavioral model version 2, https://github.com/p4lang/behavioral-model.

3.1 Actors and Requirements

When a new content in the form of laboratory classes is introduced into a course, typically a single person or a small group of people are responsible for design and initial evaluation of the new content. After that, the content is shared with a larger group of teachers who conduct the classes. In case of classes led remotely, the sharing process can be supported by a structured feedback, e.g., by recordings of past sessions and written (or recorded) notes. The process is depicted in Fig. 1.

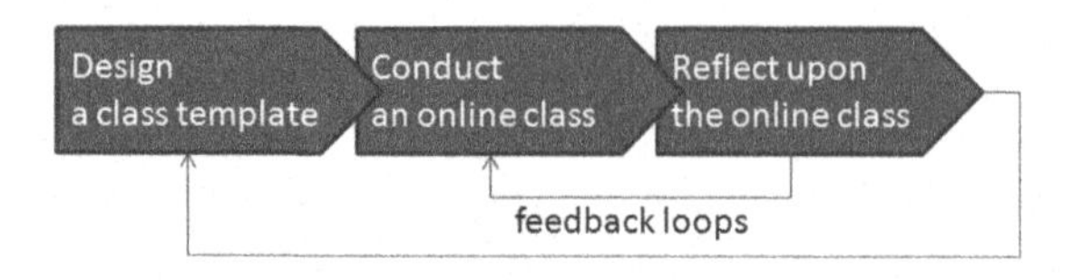

Fig. 1. Online class experience sharing idea.

Analyzing raw recordings is a time consuming task. Moreover, the person who attempts to lead a new class based on such materials only, needs to create a new environment practically from scratch, by instantiating and configuring the tools, etc. The presented system provides a remedy for the mentioned obstacles by focusing on sharing not only static materials, but the whole teaching expertise, including support for:

1. reproducing online sessions,
2. customizing the sessions to the needs of the teacher,
3. reflecting upon past sessions.

In the proposed approach, users acting as 'Designers', who produce the reference templates for classes, are aware of how to teach the respective subjects, what tools should be used and in what sequence, etc. They are also able to allot time for specific activities and include the assignments in the templates. After a template for a specific class is designed and shared, the users acting as 'Teachers' use it for instantiating actual classes. Typically, the people who design and develop the class, are teaching a few groups of students, and later share their experience with others. Therefore, the roles of 'Designer' and 'Teacher' are not disjoint. The template sharing and class instantiation processes are depicted in Fig. 2.

Note that in the presented process a Teacher is able to re-produce a class based either on a template, or on a previously stored instance. That directly reflects the observation of the behaviours of the teachers involved in the Małopolska Educational Cloud – they prefer to reuse the sets of tools and resources they used before over templates prepared by others. Simply put, they feel confident when using the services and content again.

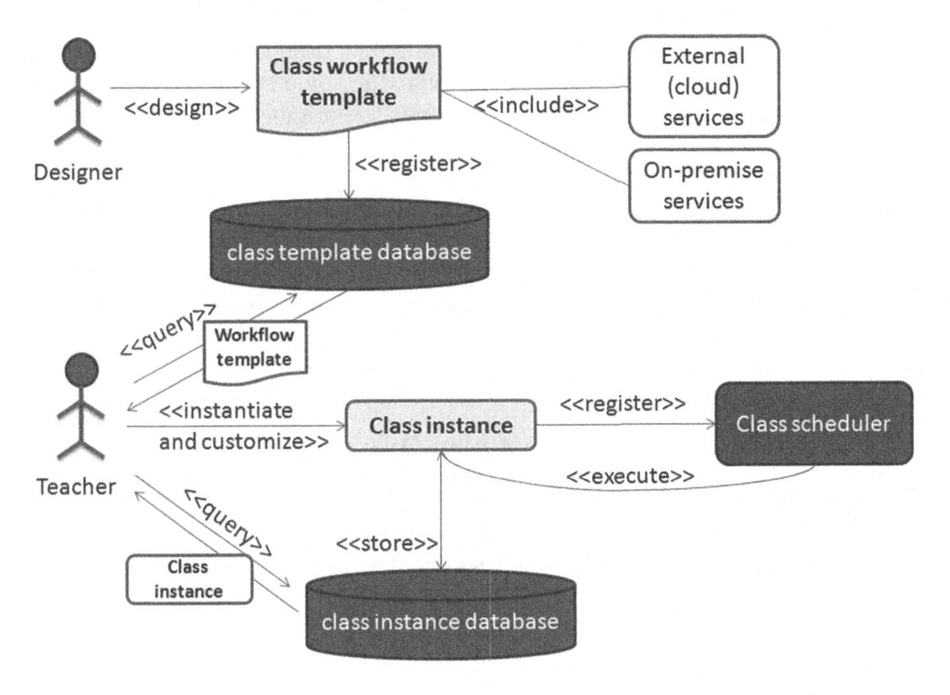

Fig. 2. Cooperation between users sharing class workflows.

3.2 Analysis of Requirements

The requirements related to sharing expertise can be analyzed with regard to their statics and dynamics. In short, the former reflects the traditional way of sharing course materials, while the latter involves the use of environment-managed workflows.

Support for reproducing the sessions in what we call 'static aspect' involves re-creation of the sets of tools used during the sessions themselves, without requiring detailed knowledge of the tools' configuration, requirements, etc. That could be achieved by incorporating tools into an integrated environment that is able to instantiate them on demand. Customization options include the ability to provide different input (e.g., questions for a quiz) each time the online session is re-created. Reflection includes the ability to provide comments after completing the class. Both customization and reflection need to be addressed by introducing appropriate records (which refer to respective resources) in the instance database. We call such records 'session resources'.

Regarding the dynamics of the class, the respective requirements can be elaborated as the following characteristics of remote class workflows. First, the difference between reproducing and simply replaying an online class should be pointed out. By reproducing we mean using the same tools (possibly in the same order), but not necessarily with the same time frames and using the same resources (e.g., displaying the same content). In this way the environment would adapt to a particular person's style of teaching. Note that with regard to customization, the class workflow should be completely modifiable, it should allow introducing new tools, reordering of events, and even omitting some steps. Of course, after modification the workflow should not overwrite the original one, but be kept as a separate entry in the class instance database. Finally, with regard to reflection – it is safe to assume that in reality that activity is rarely performed just after the class is finished. Nonetheless, given the sequence of events that occurred during the class, the environment should be able to perform some automatic operations, including: (1) synchronizing the recording to be saved with the log of events contained in the class instance workflow (2) automatic post-processing of service usage results in a service-dependent way (e.g. blurring faces, etc.). Such functionality will make a later reflection easier.

3.3 Architecure and Implementation Overview

The basic building blocks of the proposed environment is depicted in Fig. 3. The environment integrates cloud based and on premise services and exposes the session design, creation and management interfaces in the form of web applications. A database storing session templates and recordings serves as a backend.

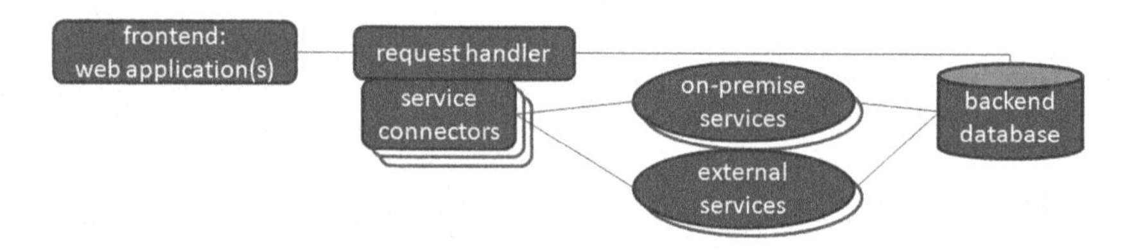

Fig. 3. Basic building blocks of the proposed environment.

Cloud services are referenced using software connectors that leverage their respective APIs. For consistency, we assume the same model of integration with local (on premise) services. Such a modular design provides for extensibility, which is needed to incorporate new, topic-specific services. Note that the usage of the backend storage is coordinated by the request handler, which is responsible for storing the events that are reported by multiple services in a consistent, synchronized way. That provides for easier review and annotation of past classes.

The prototype described in this paper integrates the following services: (1) audiovisual connectivity service (2) quizzes and surveys service, and (3) shared whiteboard service. These correspond to the respective audiovisual service, virtual desktop service, and virtual whiteboard service described in [7], which describes the functionality of an environment built for MEC using commercial products. Regarding the most important implementation technologies for the prototype of the environment, we chose the following ones:

- Jitsi Meet[8] for audivisual connectivity between participants,

[8] Jitsi Meet, https://jitsi.org/jitsi-meet.

- Nginx[9] as a reverse proxy for frontend calls,
- PostgreSQL[10] for the (backend) instance database,
- Phoenix Framework[11] for server-side model-view-controller pattern implementation,
- GraphQL[12] as a basis for our API calls,
- React[13] for building frontend web applications.

Figure 4 illustrates how the chosen technologies fit into the environment architecture. The main criteria for the enumerated choices included their openness, popularity and versatility.

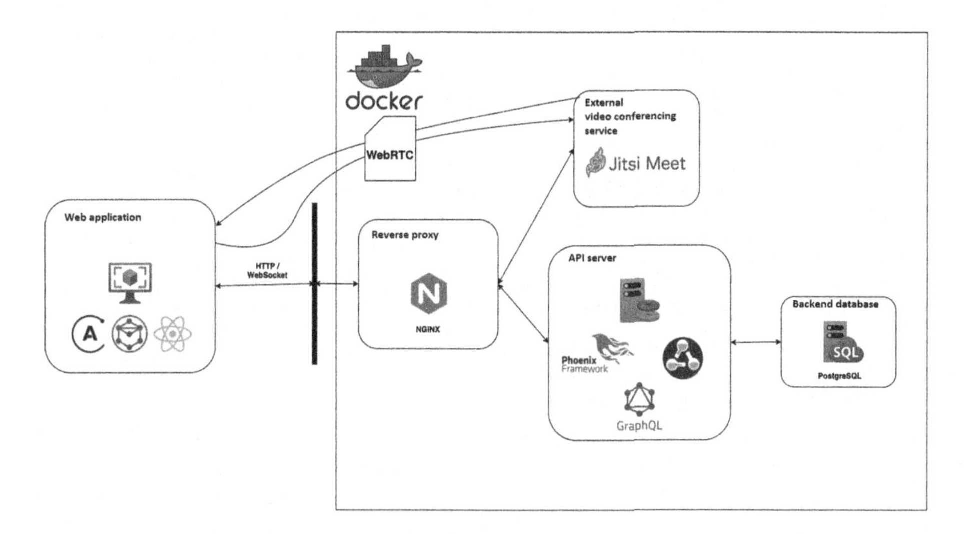

Fig. 4. Technologies used for prototype implementation.

The prototype implementation does not ideally fit the architecture. For the sake of implementation simplicity, we decided not to store the video session recordings inside the database, but rather hold references to their locations. The logical model of the database (simplified) is illustrated in Fig. 5.

The database is logically split into two parts, related to class templates and instances, respectively. A user who plays the Designer role, creates templates for online classes and provides the required session resources, references to which are held in the 'session templates' and 'session resources' tables, respectively. As mentioned in Sect. 3.2, they include, e.g., documents to be presented, questions for quizzes or surveys. Additionally, the Designer provides an agenda for the

[9] Nginx, https://nginx.org.
[10] PostgreSQL, https://www.postgresql.org.
[11] Phoenix Framework, https://www.phoenixframework.org.
[12] GraphQL, https://www.graphql.org.
[13] React, https://reactjs.org.

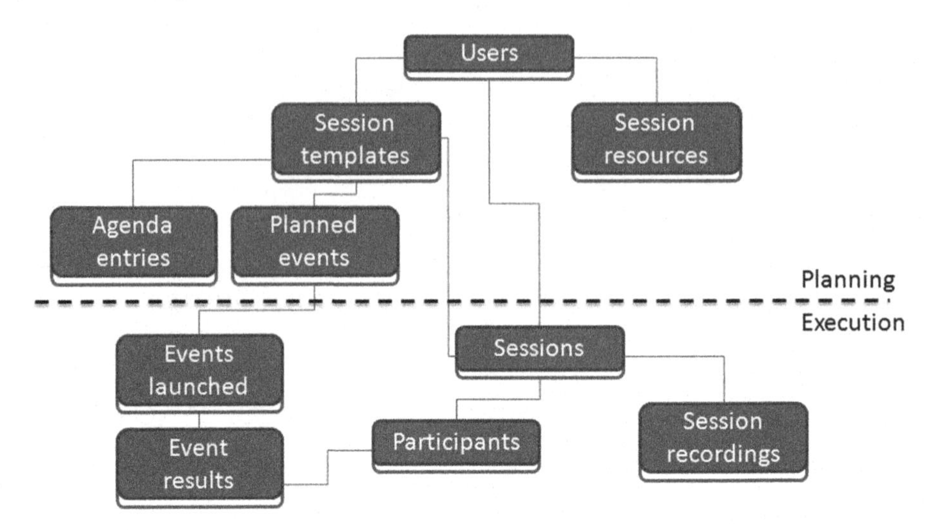

Fig. 5. Main tables of the backend database.

session in a textual form (table: 'agenda entries') and tentative timeline (table: 'planned events'). Based on that information, a user playing the Teacher role is able either to schedule a planned session directly, or to clone a session definition and customize it before scheduling (table: 'sessions').

When a scheduled class is conducted, the resources and tools are ready to be used by the session participants at the times defined by the schedule. However, it is up to the session leader (role: Teacher) if the timeline is followed strictly. The Teacher has an option either to delay, speed up or omit some planned events, depending on the students' learning paces or any other factors. The actual sequence of events is recorded in the 'events launched' table so that the users have an automatically recorded feedback, which can be used to modify the designs of future classes. Moreover, the video stream of the sessions are recorded (table: 'session recordings'), so the event log can be reviewed in synchronization with the recording, which can be useful from didactic point of view. Finally, the results of using the respective tools are kept in the 'event results' table.

4 Proof of Concept Evaluation

The sozisel prototype that was implemented used Jitsi Meet for video conferences. Additionally, it was capable of providing quizzes, surveys, and whiteboards. The services were hosted on premise. Figure 6 contains a screenshot of the project welcome page. The prototype is available for experiments at https://sozisel.pl. At the time of writing the article there was no English version of the system. However, for the sake of readability, the screenshots include automatically translated text.

Fig. 6. Sozisel.pl welcome page.

During the tests we focused on both the design and instantiation phases of the online sessions. We started from template design (Fig. 7). In this use case, the Designer first creates a template, names it, and provides initial agenda. Then, tools to be used are selected (the screenshot contains forms that refer to a quiz, a questionnaire, and a whiteboard). Additionally, a tentative workflow is designed by setting tool usage start times as offsets from the start of the class, as well as respective durations. There is also an option to provide initial contents to be displayed by the tools.

The session scheduling form is depicted in Fig. 8. In the presented use case, the Teacher instantiates a session based on a template created and shared (made public) by a Designer. The teacher is able either to schedule the session in accordance to the designed workflow, or to copy the design and modify it before the session is executed.

Figure 9 shows a Teacher view of an ongoing session. In the screenshot, a quiz has just ended and therefore the grades are displayed by the Teacher's application. Because the quiz was the last planned activity, the application displays also an 'end session' button at the bottom of the screen.

The prototype was tested on a PC equipped with 6-core Intel Core i5-9600K CPU@3.70GHz with sessions for up to 25 users. During the sessions, all of the implemented tools were used. In the largest sessions the CPU load (average, calculated for a single core) was 3% for backend, 0% for frontend, and 200% for the video session (Jitsi). The users did not suffer from any lags resulting from the CPU load, and based on the numbers it is safe to assume that the same machine could host another session for a second 25-person class.

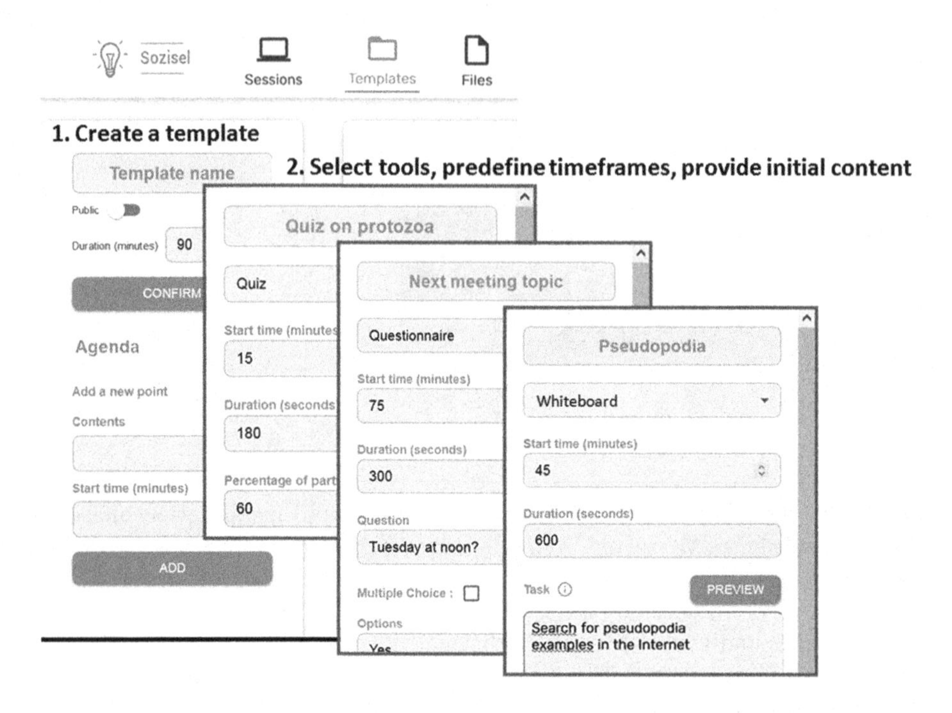

Fig. 7. Sozisel event planning phase.

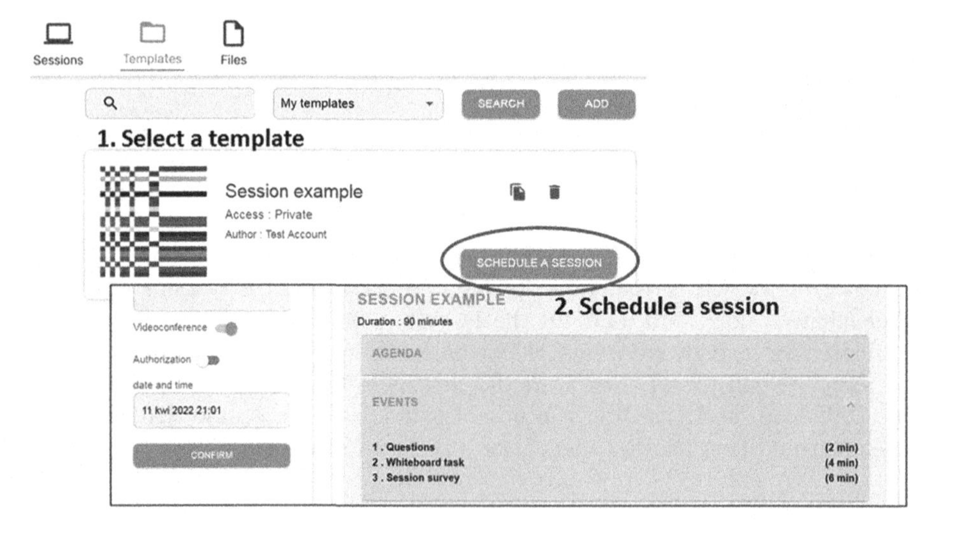

Fig. 8. Session scheduling

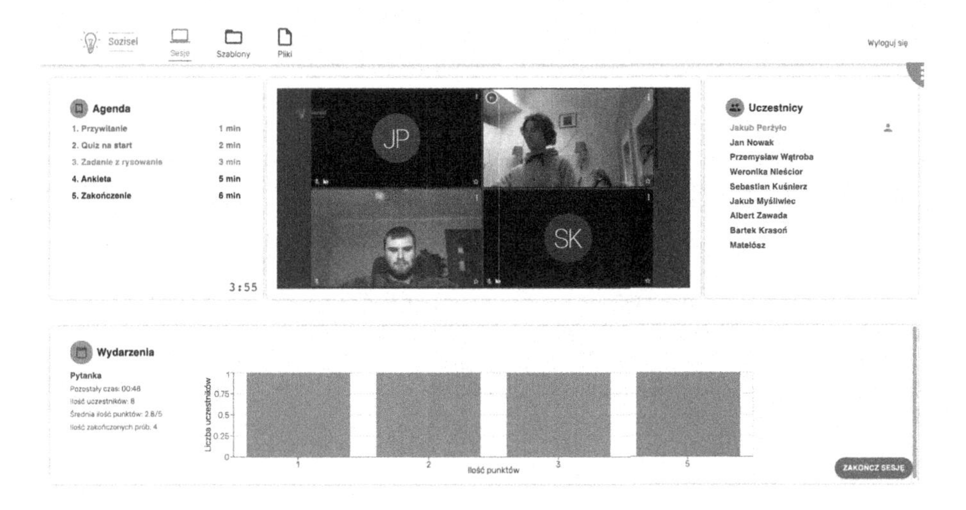

Fig. 9. Teacher view (just after a quiz was finished).

5 Conclusions and Future Work

Sozisel prototype proved to be a useful tool for medium-size online classes. It is based on open technologies, and designed in a modular way. The first direction of its further development is to leverage its modularity to incorporate more tools into the system or to integrate it with external services. Short-term goals include integrating game-based quizzes (developed at AGH university in another project), and computer networking specific services (we plan to use Kathara), and using it in real-world scenarios, for remotely led IT classes. As a long-term goal, we want to expose sozisel as a platform for EEDS, a model-based deployment environment for educational services, developed earlier at AGH [2]. In that way, multi-platform model-based automatic orchestration functionality would be added to the scheduling phase.

References

1. Feisel, L.D., Rosa, A.J.: The role of the laboratory in undergraduate engineering education. J. Eng. Educ. **94**(1), 121–130 (2005)
2. Gandia, R.L., Zieliński, S., Konieczny, M.: Model-based approach to automated provisioning of collaborative educational services. In: Paszynski, M., Kranzlmüller, D., Krzhizhanovskaya, V.V., Dongarra, J.J., Sloot, P.M.A. (eds.) ICCS 2021. LNCS, vol. 12747, pp. 640–653. Springer, Cham (2021). https://doi.org/10.1007/978-3-030-77980-1_48
3. Lee, H.S., Kim, Y., Thomas, E.: Integrated educational project of theoretical, experimental, and computational analyses. In: ASEE Gulf-Southwest Section Annual Meeting 2018 Papers. American Society for Engineering Education (2019)

4. Lynch, T., Ghergulescu, I.: Review of virtual labs as the emerging technologies for teaching STEM subjects. In: INTED2017 Proceedings 11th International Technology, Education and Development Conference Valencia, Spain. 6-8 March, pp. 6082–6091 (2017)
5. Morales-Menendez, R., Ramírez-Mendoza, R.A., et al.: Virtual/remote labs for automation teaching: a cost effective approach. IFAC-PapersOnLine **52**(9), 266–271 (2019)
6. Perales, M., Pedraza, L., Moreno-Ger, P.: Work-in-progress: Improving online higher education with virtual and remote labs. In: 2019 IEEE Global Engineering Education Conference, EDUCON, pp. 1136–1139. IEEE (2019)
7. Zielinski, K., Czekierda, L., Malawski, F., Stras, R., Zielinski, S.: Recognizing value of educational collaboration between high schools and universities facilitated by modern ICT. J. Comput. Assist. Learn. **33**(6), 633–648 (2017). https://doi.org/10.1111/jcal.12207

Supporting Education in Advanced Mathematical Analysis Problems by Symbolic Computational Techniques Using Computer Algebra System

Włodzimierz Wojas and Jan Krupa(✉)

Department of Applied Mathematics, Warsaw University of Life Sciences (SGGW), ul. Nowoursynowska 159, 02-776 Warsaw, Poland
{wlodzimierz_wojas,jan_krupa}@sggw.edu.pl

Abstract. In this paper we present two didactic examples of the use of Mathematica's symbolic calculations in problems of mathematical analysis which we prepared for students of Warsaw University of Life Sciences. In Example 1 we solve the problem of convex optimization and next in Example 2 we calculate the complex integrals. We also describe a didactic experiment for students of the Informatics and Econometric Faculty of Warsaw University of Life Sciences.

Keywords: Symbolic calculations · Computational mathematics · Mathematical didactics · CAS · Convex optimization · Complex integrals

Subject Classifications: 97B40 · 97R20 · 90C30 · 97I60

1 Introduction

Computer Algebra Systems (CAS) such as WxMaxima, Mathematica, Maple, Sage and others are often used to support calculations and visualization in teaching mathematical subjects [2,3,9–11,13]. This paper is intended as a follow-up of the article [12] on supporting higher mathematics education by dynamic visualization using CAS. The didactic value of using software supporting visualization in teaching higher mathematics is generally beyond doubt. For example, the ability to visualize 3D objects with options of dynamic plot and animation, with various options for colour and transparency selection, appears to be very useful didactic tool in teaching students more advanced mathematics topics. In paper [12] we presented 3D example (Example 1) of dynamic visualization primal simplex algorithm steps containing 3D feasible region, corner points, simplex path and level sets using Mathematica [6,14]. Obtaining a comparable didactic effect without any specialized computer programs such as CAS, would be difficult if possible. In our opinion, it is much more difficult to assess the didactic value of the use of symbolic computing using CAS in teaching higher mathematics. Symbolic computation in CAS allow us to solve mathematics problems symbolically,

D. Groen et al. (Eds.): ICCS 2022, LNCS 13353, pp. 719–731, 2022.
https://doi.org/10.1007/978-3-031-08760-8_59

check and trace our calculations done by hand and present the results in an attractive graphic form. However, it would be difficult to unequivocally answer the question of whether, from the didactic point of view, it is better to solve and present mathematical problems symbolically using CAS or present manual calculations on the board. It would probably be valuable to learn both approaches. In the 1990s, there was a discussion between Steven Krantz and Jerry Uhl about the relevance of teaching Calculus students with Mathematica as part of the Calculus & Mathematica project [4,8]. In this discussion, the views of both sides were divided and there was no common consensus as to the advantage of one approach over the other. The discussion shows that the problem of evaluating an approach with or without CAS in teaching higher mathematics is not simple or unequivocal. It seems that the ability to use symbolic CAS calculations can be particularly useful in more advanced research problems carried out by students, for example as part of a diploma thesis. In this paper we present and discuss two didactic examples of advanced mathematical analysis problems in which we use symbolic calculations. These examples were prepared by us for students of Warsaw University of Life Science. We also present a didactic experiment with the participation of the Informatics and Econometric Faculty students in their first contact with CAS.

2 Example 1: Convex Programming

Convex programming is part of mathematical programming. This example was prepared for students of the Informatics and Econometric Faculty of Warsaw University of Life Sciences within the course of Mathematical Programming. We solved the following convex optimization problem with Mathematica:
find the global minimum value of the function $f(x_1, x_2, x_3) = (x_1^2 + x_2^2 + x_3)e^{x_3}$ over the greatest set D, over which f is convex. Determine several examples of level sets of function f.

This approach based on the fact that for convex function defined on convex set, local minimum is a global minimum.

Theorem 1 (See [1]**).** *Let S be a nonempty convex set in $\mathbb{R}^n$, and let $f : S \to \mathbb{R}$ be convex on S. Consider the problem to minimize $f(x)$ subject to $x \in S$. Suppose that $x_0 \in S$ is a local optimal solution to the problem.*

1. *Then x_0 is a global optimal solution.*
2. *If either x_0 is a strict local minimum or f i s strictly convex, x_0 is the unique global optimal solution and is also a strong local minimum.*

Theorem 2 (See [1]**).** *Let S be a nonempty open convex set in $\mathbb{R}^n$, and let $f : S \to \mathbb{R}$ be twice differentiable on S. If the Hessian matrix is positive definite at each point in S, f is strictly convex. Conversely, if f is strictly convex, the Hessian matrix is positive semidefinite at each point in S.*

Theorem 3 (See [1]**).** *Let D be a nonempty convex set in $\mathbb{R}^n$, and let $f : D \to \mathbb{R}$. Then f is convex if and only if* $\operatorname{epi} f = \{(x, x_{n+1}) \in \mathbb{R}^{n+1} : x \in D, x_{n+1} \in \mathbb{R}, x_{n+1} \geq f(x)\}$ *is a convex set.*

Theorem 4 *Let D be a nonempty closed convex set in $\mathbb{R}^n$, $S = \operatorname{int}D \neq \emptyset$, and let $f : D \to \mathbb{R}$ be continuous on D and convex on S. Then f is convex on D.*

Using Mathematica we can find the stationary points of f and the greatest set D, over which f is convex.

Listing 2.1. Calculation of the stationary points of f in Mathematica:

```
In[1]:= f = (x1^2 + x2^2 + x3)Exp[x3];
In[2]:= {f1 = D[f, x1], f2 = D[f, x2], f3 = D[f, x3]}//Simplify
Out[2]= {2e^x3 x1, 2e^x3 x2, e^x3 (1 + x1^2 + x2^2 + x3)}
In[3]:= r = Solve[{f1 == 0, f2 == 0, f3 == 0}, {x1, x2, x3}, Reals]
Out[3]= {{x1-> 0, x2-> 0, x3-> -1}}
In[4]:= f/.r[[1]]
Out[4]= -1/e
```

From the listing 2.1 we have that the only stationary point of f is $P_0 = (0, 0, -1)$. $f(P_0) = f(0, 0, -1) = -\frac{1}{e}$

Listing 2.2. Calculation of Hessian Matrix of f in Mathematica:

```
In[5]:= H = {{D[f, {x1, 2}], D[f, x1, x2], D[f, x1, x3]}, {D[f, x1, x2], D[f, {x2, 2}],
D[f, x2, x3]}, {D[f, x1, x3], D[f, {x2, x3}], D[f, {x3, 2}]}};
{MatrixForm[H], Det[H]}//Simplify
Out[5]= { ( 2e^x3       0           2e^x3 x1
            0           2e^x3       2e^x3 x2
            2e^x3 x1    2e^x3 x2    e^x3 (2 + x1^2 + x2^2 + x3) ) , 4e^(3x3) (2 - x1^2 - x2^2 + x3) }
In[6]:= H0 = H/.r[[1]]; MatrixForm[H0]
Out[6]= ( 2/e  0    0
          0    2/e  0
          0    0    1/e )
```

From listing 2.2 we have:

$$H(x_1, x_2, x_3) = \begin{pmatrix} 2e^{x_3} & 0 & 2e^{x_3}x_1 \\ 0 & 2e^{x_3} & 2e^{x_3}x_2 \\ 2e^{x_3}x_1 & 2e^{x_3}x_2 & e^{x_3}(2 + x_1^2 + x_2^2 + x_3) \end{pmatrix}$$

and

$$H(P_0) = H(0, 0, -1) = \begin{pmatrix} \frac{2}{e} & 0 & 0 \\ 0 & \frac{2}{e} & 0 \\ 0 & 0 & \frac{1}{e} \end{pmatrix}.$$

$H(P_0)$ is positive definite, so in P_0 f attains local strict minimum.

Now we will find all $P = (x_1, x_2, x_3) \in \mathbb{R}^3$ such that Hessian Matrix of f $H(x_1, x_2, x_3)$ is positive definite.

From Listing 2.2 we also have:

$$\left.\begin{cases} W_1(P) = 2e^{x_3} > 0 \\ W_2(P) = 4e^{2x_3} > 0 \\ W_3(P) = 4e^{3x_3}(2 - x_1^2 - x_2^2 + x_3) > 0 \end{cases}\right\} \Leftrightarrow 2 - x_1^2 - x_2^2 + x_3 > 0.$$

Let define function $g(x_1, x_2) = x_1^2 + x_2^2 - 2$ on $\mathbb{R}^2$.

We can prove that g is strictly concave on $\mathbb{R}^2$ directly from the definition of concave function.

Listing 2.3. Checking concavity of function g on $\mathbb{R}^2$ in Mathematica :

```
In[1]:= g[{x1_, x2_}] = x1^2 + x2^2 - 2;
In[2]:= {x = {x1, x2}, y = {y1, y2}};
In[3]:=Assuming[0 < λ < 1 && Element[{λ, x1, x2, y1, y2}, Reals] &&
Or[x1 ≠ y1, x2 ≠ y2], g[λx + (1 - λ)y] < λg[x] + (1 - λ)g[y]//FullSimplify
Out[3] = True
```

Listing 2.4. Checking concavity of function g on $\mathbb{R}^2$ in Mathematica:

```
In[4]:= λg[x] + (1 - λ)g[y] - g[λx + (1 - λ)y]//FullSimplify
Out[4] = -((x1 - y1)^2 + (x2 - y2)^2)(-1 + λ)λ
```

From Listing 2.4 or 2.3 we have that:
$g(\lambda x + (1 - \lambda)y) < \lambda g(x) + (1 - \lambda)g(y)$
for each $x, y \in \mathbb{R}^2$ such that $x \neq y$ and for each $\lambda \in (0, 1)$ ($\mathbb{R}^2$ is convex set)

So the sets $D = \text{epi } g = \{(x_1, x_2, x_3) \in \mathbb{R}^3 : x_3 \geq x_1^2 + x_2^2 - 2\}$ and $S = \text{int } D = \{(x_1, x_2, x_3) \in \mathbb{R}^3 : x_3 > x_1^2 + x_2^2 - 2\}$ are convex (see Theorem 3). From Fig. 1 we can also conclude that the set D is concave.

Hence on the set $S = \{(x_1, x_2, x_3) \in \mathbb{R}^3 : x_3 > x_1^2 + x_2^2 - 2\}$ Hessian of f is positive definite, hence the function f is strictly convex on S and from Theorem 4 we conclude that f is convex on the set $D = \{(x_1, x_2, x_3) \in \mathbb{R}^3 : x_3 \geq x_1^2 + x_2^2 - 2\}$. Hence from the theorem 1, local strict minimum of f is also global strict minimum (the least value) of f on the set D.

Of course the calculations and simplifications in Listing 2.4 we can do by hand. Calculations in Mathematica or other CAS are only one option for checking our hand calculations.

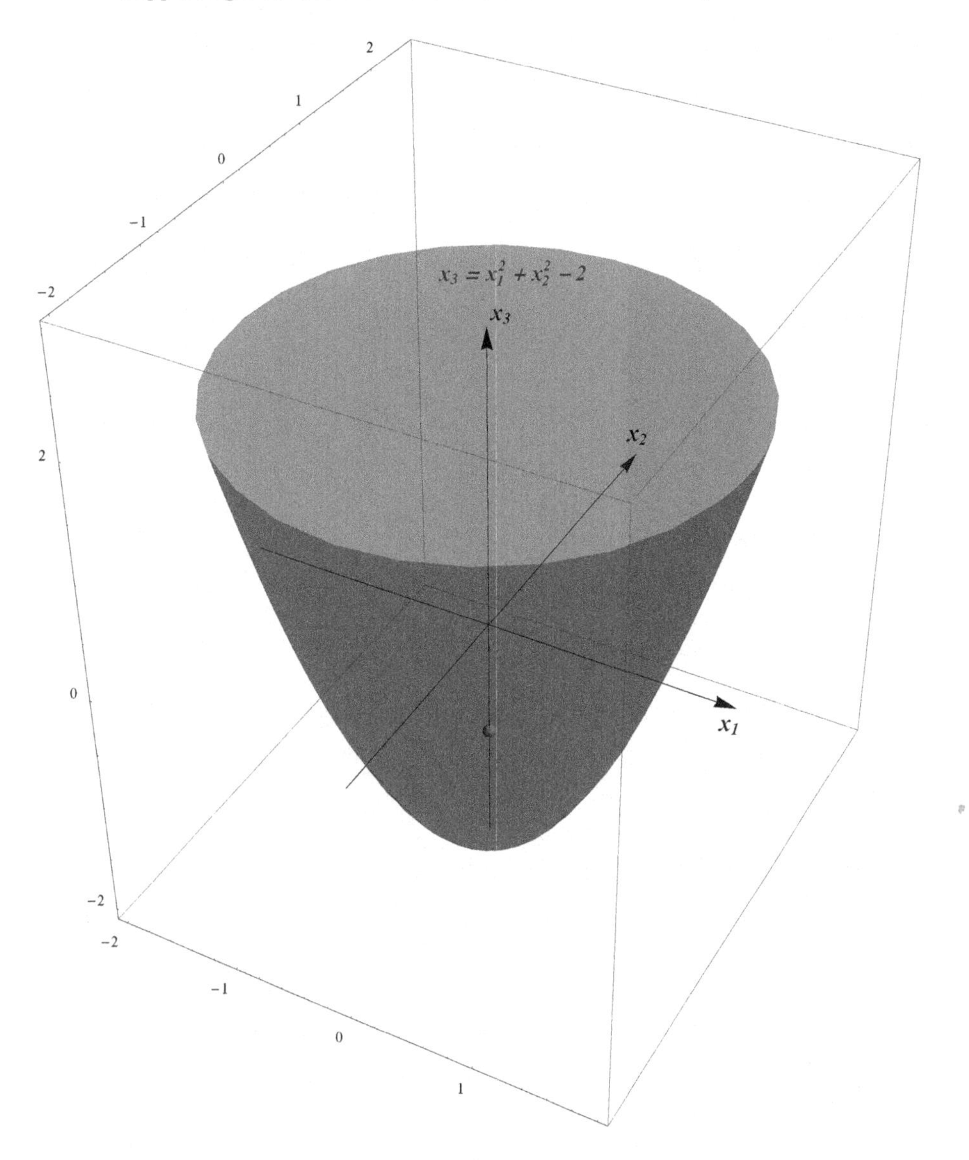

Fig. 1. The set D

So we get the finale answer: $f_{\min} = f(0, 0, -1) = -\dfrac{1}{e}$.
CAS allows us to solve problem symbolically and also visualise it. Since symbolic computations are performed in CAS, so visualisations are performed there as well.

We show below dynamic plots presenting level sets of function f.

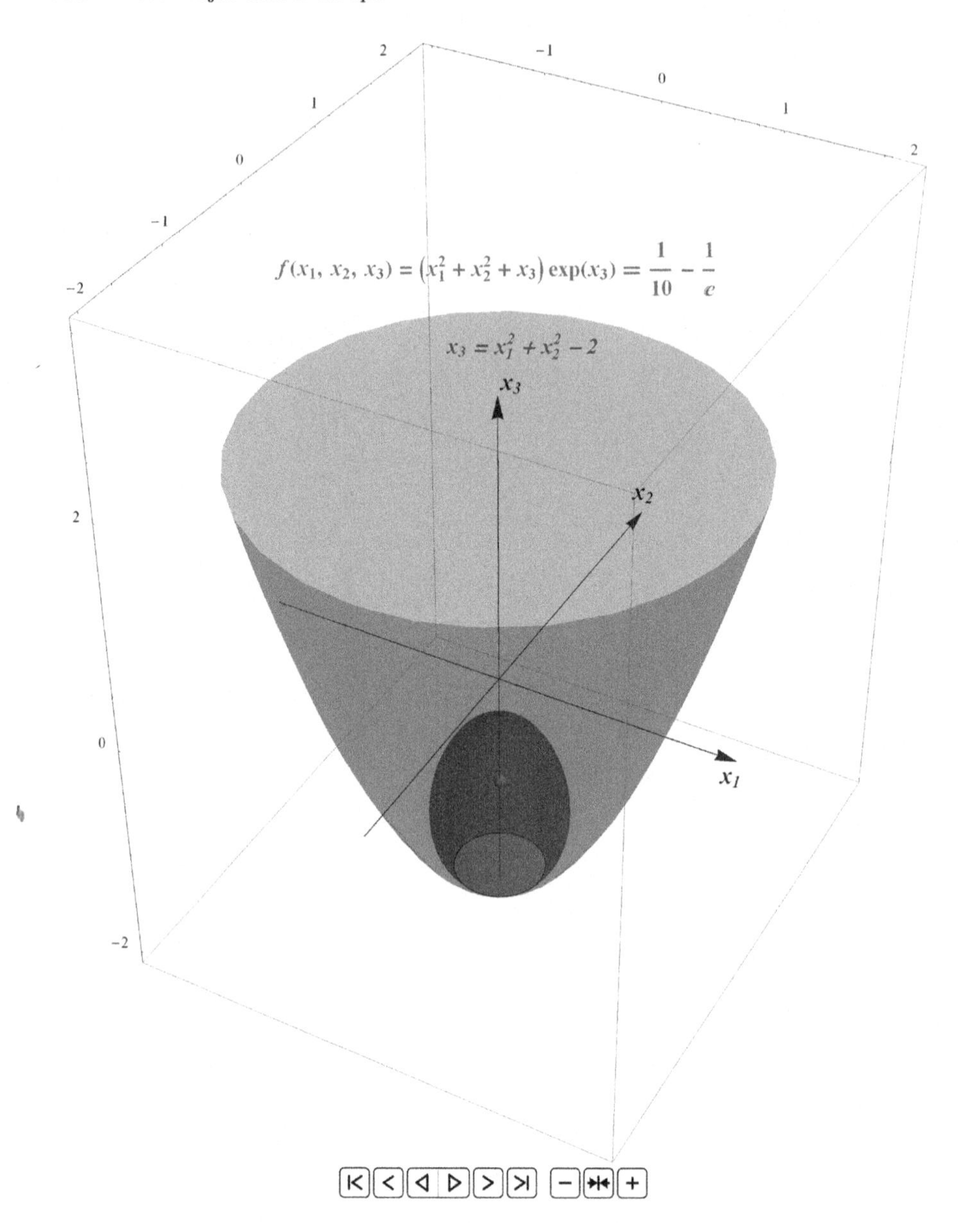

Fig. 2. Dynamic plots of the level surfaces of convex function f on the convex set D

Dynamic version of the Figs. 3, 2 can be found at:
https://drive.google.com/file/d/1F7xOQaU7YSN3ar-z12rDgZJAFj06aDdU/view?usp=sharing

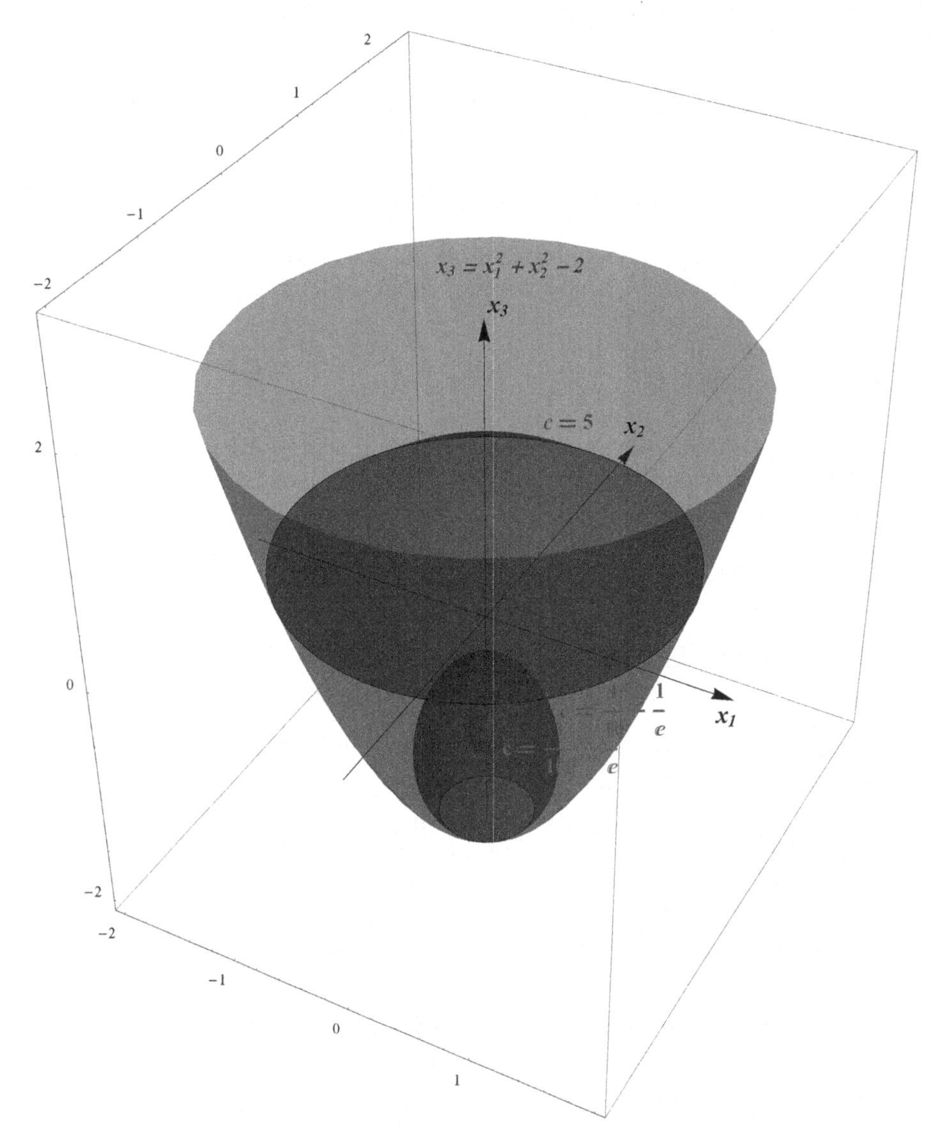

Fig. 3. Plots of the level surfaces $\{(x_1, x_2, x_3) \in \mathbb{R}^3 : f(x_1, x_2, x_3) = c\}$ of convex function f on the convex set D: for $c = 5, -1/e + 1/10, -1/e + 1/100$

3 The Didactic Experiment with Example 1

This experiment was carried out on in two independent groups of second-year students of the Informatics and Econometric Faculty of Warsaw University of Life Sciences within the course of Mathematical Programming. Students were after courses of one and multivariable analysis and linear algebra. As part of the mathematical programming course, students were acquainted with the basic definitions and theorems in the field of convex programming. The students declared that they had not had contact with the CAS symbolic calculations before. In the first group of 36 students the Example 1 (only in CAS version) was demonstrated to the students and discussed in detail by the lecture. The presentation together with the discussion lasted about 20 min. After the presentation, the students answered two questions.

1. Were the presented symbolic calculations in Mathematica helpful in understanding the example? They chose one of four options:
   a) they were not helpful,
   b) they were a bit helpful,
   c) they were helpful,
   d) they were very helpful.
2. Did the symbolic calculations demonstrated in this presentation broaden your knowledge of computational techniques? They chose one of four options:
   a) did not broaden
   b) did broaden a little,
   c) did broaden,
   d) did greatly broaden.

We received the following results. For the question 1.: 0% of the students chose the answer a), 44% answer b), 53% answer c) and 3% answer d). For the question 2.: 0% of the students chose the answer a), 61% answer b), 39% answer c) and 0% answer d).

In the second group of 25 students the Example 1 was presented in two versions. First, in a traditional way - with manual calculation without CAS, and then with the use of CAS. The presentation of each version together with the discussion lasted about 20 min. After these two presentations, the students were to refer to the following statement by selecting one of the sub-items.

Comparing the two presented versions of the solution of the Example 1, I think that:

a) it should first of all get acquainted with the first version of the solution (without CAS) and getting acquainted only with the second version (with CAS) is not enough,
b) it should first of all get acquainted with the second version of the solution (with CAS) and getting acquainted only with the first version (without CAS) is not enough,
c) each of the presented versions of the solution is equally good and you only need to get acquainted with one of them,
d) it should get acquainted with both versions of the solution and getting acquainted only with one of them is not enough,
e) I cannot judge which version of the solution is good enough, I have no opinion on this.

We received the following result: 32% of the students chose sub-item a), 52% sub-item b), 0% sub-item c), 16% sub-item d) and 0% sub-item e).

Analysing the obtained percentages of answers in the first group of students, it is worth emphasising that in both questions (1 and 2), none of the students chose the answer a). All students found that the presented CAS example was in some way helpful in understanding the problem of convex programming as well as it broaden their knowledge about computational techniques to some extent. Most of the students (53%) found that the presentation with the use of CAS was significantly helpful in understanding the problem of convex programming. 39% of the students of the first group found that the presentation significantly broadened their knowledge of computational techniques. In the second group, students compared two approaches to solving the convex programming problem: without CAS and with CAS. Analysing the percentage of responses in this group of students, we can see that the majority of students (52%) decided that it should first of all get acquainted with the second version of the solution (with CAS) and getting acquainted only with the first version (without CAS) is not enough. It also seems important that none of the students of this group considered both versions to be equally good and it is enough to get acquainted with one of them (0% for answer c)). Given that this was the first contact of students of both groups with CAS methodology, the results of the experiment would suggest that supporting the teaching of higher mathematics through CAS programs with the use of symbolic computing may have significant educational value.

4 Example 2: Calculating Complex Integrals with Mathematica

In this example we present some our experiences in teaching elements of complex analysis students of Environmental Engineering Faculty of Warsaw University of Life Science. Complex analysis in this faculty was one of the parts of higher mathematics course. In the framework of this course complex potential fluid flow model in two dimensions was presented. To understand this model and to be able to solve connected with it tasks, the ability to calculate complex integrals along a curve is required, so in this course we spent some time practicing computing complex integrals ([5,7]). Let us solve the following complex integral problem using Mathematica.

Is the following equation true? Justify the answer.

$$\int_{K_1} \bar{z}\,\mathrm{Re}\,z\,\mathrm{d}z = \int_{K_2} \bar{z}\,\mathrm{Re}\,z\,\mathrm{d}z,$$

where K_1 is a directed segments from point A to point B and K_2 is a directed broken line ABC presented in Fig. 4.

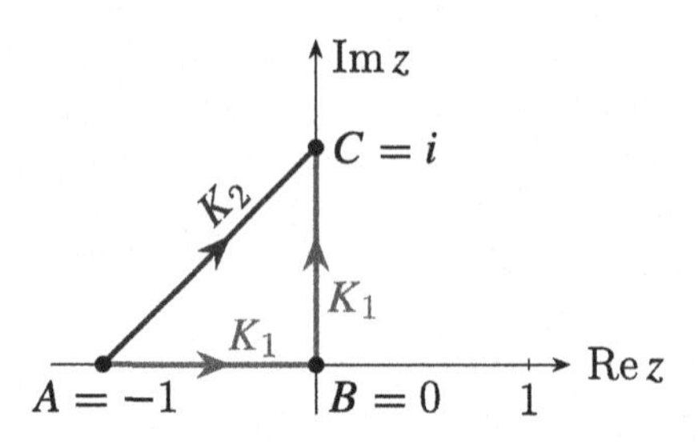

Fig. 4. Directet segments K_1 and K_2

Let's note that the integrands in both integrals are the same. If the function $f(z) = \bar{z}\,\mathrm{Re}\, z$ is holomorphic in simply connected region containing K_1 and K_2, then equation is true. Let's see if Cauchy-Riemann conditions are satisfied for function f. We have:

$$f(z) = f(x+iy) = (x-iy)x = x^2 - ixy = u(x,y) + iv(x,y),$$

where $u(x,y) = x^2$ and $v(x,y) = -xy$ are real and imaginary parts of function f respectively. Determining partial derivatives of functions $u(x,y)$ and $v(x,y)$ we get: $\dfrac{\partial u}{\partial x} = 2x$, $\dfrac{\partial u}{\partial y} = 0$, $\dfrac{\partial v}{\partial x} = -y$, $\dfrac{\partial u}{\partial y} = -x$. Both Cauchy-Riemann conditions: $\dfrac{\partial u}{\partial x} = \dfrac{\partial v}{\partial y}$, $\dfrac{\partial u}{\partial y} = -\dfrac{\partial v}{\partial x}$ are satisfied only in point B. So, function f is not holomorphic in any simple connected region containing K_1 and K_2. Let us calculate both integrals. Let's use the following parametrizations of segments AB, AC and BC.
For AC: $z_1(t) = t + i(t+1)$, $t \in [-1, 0]$.
For AB: $z_2(t) = t$, $t \in [-1, 0]$.
For BC: $z_3(t) = it$, $t \in [0, 1]$.

So, we get:
$\displaystyle\int_{K_1} \bar{z}\,\mathrm{Re}\, z\, \mathrm{d}z = \int_{K_1} \bar{z}_1\,\mathrm{Re}\, z_1\, \mathrm{d}z_1 = \int_{-1}^{0} [t - i(t+1)]t(1+i)\, \mathrm{d}t = (1+i)\int_{-1}^{0} [t^2 - i(t^2+t)]\, \mathrm{d}t = (1+i)\left[\frac{1-i}{3}t^3 - \frac{i}{2}t^2\right]_{-1}^{0} = \frac{1}{6} + \frac{1}{2}i.$
$\displaystyle\int_{K_2} \bar{z}\,\mathrm{Re}\, z\, \mathrm{d}z = \int_{AB} \bar{z}\,\mathrm{Re}\, z\, \mathrm{d}z + \int_{BC} \bar{z}\,\mathrm{Re}\, z\, \mathrm{d}z = \int_{AB} \bar{z}_2\,\mathrm{Re}\, z_2\, \mathrm{d}z_2 + \int_{BC} \bar{z}_3\,\mathrm{Re}\, z_3\, \mathrm{d}z_3 = \int_{-1}^{0} t^2\, \mathrm{d}t + \int_{0}^{1} 0\, \mathrm{d}t = \frac{1}{3}.$
Finally: $\displaystyle\int_{K_1} \bar{z}\,\mathrm{Re}\, z\, \mathrm{d}z = \frac{1}{6} + \frac{1}{2}i \neq \frac{1}{3} = \int_{K_2} \bar{z}\,\mathrm{Re}\, z\, \mathrm{d}z$. Equation is not true.

Listing 4.1. Calculation of the integral $\int_{K_1} f(z)\,\mathrm{d}z$ in Mathematica using parametrisation:

```
In [1] := f[z_] = Re[z] Conjugate[z];
In [2] := z1[t_] = t + I(t + 1);
In [3] := f1t = f[z1[t]]z1'[t]//ComplexExpand
Out[3] = (1 − I)t + 2t^2
In [4] := Integrate[f1t, {t, −1, 0}]
Out[4] = 1/6 + I/2
```

Listing 4.2. Calculation of the integral $\int_{K_2} f(z)\,\mathrm{d}z$ in Mathematica using parametrisation:

```
In [5] := z2[t_] = t;
In [6] := f2t = f[z2[t]]z2'[t]//ComplexExpand
Out[6] = t^2
In [7] := i1 = Integrate[f2t, {t, −1, 0}]
Out[7] = 1/3
In [8] := z3[t_] = It;
In [9] := f3t = f[z3[t]]z3'[t]//ComplexExpand
Out[9] = 0
In [10] := i2 = Integrate[f3t, {t, 0, 1}]
Out[10]= 0
In [11] := i1 + i2
Out[11]= 1/3
```

Listing 4.3. Calculation of the indefinite complex integral of $f(z)$ in Mathematica:

```
In [12] := Integrate[f[z], z, Assumptions− > Element[z, Complexes]]
Out[12] = Integrate[Re z Conjugate[z], z, Assumptions− > Element[z, Complexes]]
```

Listing 4.4. Calculation of the complex integrals $\int_{K_1} f(z)\,\mathrm{d}z$, $\int_{K_2} f(z)\,\mathrm{d}z$ in Mathematica:

```
In [13] := Integrate[f[z], {z, −1, I}]
Out[13] = 1/6 + I/2
In [14] := Integrate[f[z], {z, −1, 0, I}]
Out[14] = 1/3
```

Note that the Mathematica program did not compute the indefinite complex integral of the function $f(z)$. By examining the holomorphism of the function $f(z)$, we showed that the result of the integration can depend on the curve along which we integrate, so it was necessary to calculate the integrals on both sides of the equality. Both of the integrals we calculated using Mathematica in two ways: parameterizing K_1 and K_2 and also directly - using the Integrate procedure for two points: -1 and I or for three points: -1, 0 and I.

5 Summary and Conclusions

This paper analyses the problem of supporting the teaching of higher mathematics with the help of symbolic computation using CAS. The article presents two original examples of supporting teaching advanced problems of mathematical analysis with the use of symbolic computation in CAS. It also includes a didactic experiment with the participation of the Informatics and Econometric Faculty students in their first contact with CAS. We have not encountered a similar approach to the analysis of this issue - based on didactic experiments, in the available literature.

Supporting the teaching of higher mathematics with the help of symbolic calculations using CAS programs is an alternative approach to the traditional method of calculations by hand for solving mathematical tasks and presenting calculations to students. The Example 1 presents a convex optimization problem solved step by step using Mathematica procedures. This example was presented to two groups of students as part of a didactic experiment carried out in Mathematical Programming class. In the first group, most of the students (53%) rated the presented symbolic calculations in Mathematica as helpful in understanding the example; 3% of students found the presentation very helpful and 44% a little helpful in understanding the example. All the students of the first group stated the presentation to some extent broadened their knowledge of computational computing techniques. In the second group, more than half of the students (52%) decided that it should first of all get acquainted with the version of the Example 1 using CAS and getting acquainted only with the version without CAS is not enough. The obtained percentage results of the student questionnaires seem to suggest that the use of symbolic calculations in CAS to solve the Example 1 was significantly helpful in understanding this example. The advantage of using the integrated CAS program environment for a symbolic solution of the task is the possibility of graphical illustration of the individual steps of symbolic calculations using graphical procedures of CAS programs. Nevertheless, independent use of the symbolic computational procedures of CAS programs by students to solve more advanced tasks in the field of higher mathematics requires not only a good knowledge of these programs but also a deep understanding of mathematical issues in the field of a given subject. The Example 2 shows that the "automatic" use of symbolic computational procedures in Mathematica does not make sure that this usage is fully correct and that the result obtained is correct. For example, without checking the holomorphicity in Example 2, we cannot know whether computing the complex integral from point to point in Mathematica, using the procedure Integrate depends or not on the curve along which we are integrating. The support of traditional methods of teaching mathematics with CAS programs is rather unobjectionable and is quite widely regarded as a useful didactic tool for teaching mathematics in various fields. Nevertheless, replacing traditional methods of teaching higher mathematics with CAS-based teaching is debatable, as shown by the discussion between Steven Krantz and Jerry Uhl [4,8]. However, it seems that such controversy will not arise when students are taught both approaches independently with a collaborative orientation - supporting one approach over the other.

References

1. Bazaraa, S.M., Shetty, C.M.: Nonlinear programming. In: Theory and Algorithms, 2nd edn. Wiley (1979)
2. Guyer, T.: Computer algebra systems as the mathematics teaching tool. World Appl. Sci. J. **3**(1), 132–139 (2008)
3. Kramarski, B., Hirsch, C.: Using computer algebra systems in mathematical classrooms. J. Comput. Assist. Learn. **19**, 35–45 (2003)
4. Krantz, S.: How to Teach Mathematics. AMS (1993)
5. Lu, J.K., Zhong, S.G., Liu, S.Q.: Introduction to the Theory of Complex Functions. Series of Pure Mathematics. World Scientific (2002)
6. Ruskeepaa, H.: Mathematica Navigator: Graphics and Methods of Applied Mathematics. Academic Press, Boston (2005)
7. Shaw, W.: Complex Analysis with Mathematica. Cambridge University Press (2006)
8. Uhl, J.: Steven Krantz versus calculus & mathematica. Notices Am. Math. Soc. **43**(3), 285–286 (1996)
9. Wojas, W., Krupa, J.: Familiarizing students with definition of Lebesgue integral: examples of calculation directly from its definition using mathematica. Math. Comput. Sci. 363–381 (2017). https://doi.org/10.1007/s11786-017-0321-5
10. Wojas, W., Krupa, J.: Some remarks on Taylor's polynomials visualization using mathematica in context of function approximation. In: Kotsireas, I.S., Martínez-Moro, E. (eds.) ACA 2015. SPMS, vol. 198, pp. 487–498. Springer, Cham (2017). https://doi.org/10.1007/978-3-319-56932-1_31
11. Wojas, W., Krupa, J.: Teaching students nonlinear programming with computer algebra system. Math. Comput. Sci. 297–309 (2018). https://doi.org/10.1007/s11786-018-0374-0
12. Wojas, W., Krupa, J.: Supporting education in algorithms of computational mathematics by dynamic visualizations using computer algebra system. In: Computational Science - ICCS, pp. 634–647 (2020). https://doi.org/10.1007/978-3-030-50436-6_47
13. Wojas, W., Krupa, J., Bojarski, J.: Familiarizing students with definition of Lebesgue outer measure using mathematica: some examples of calculation directly from its definition. Math. Comput. Sci. **14**, 253–270 (2020). https://doi.org/10.1007/s11786-019-00435-2
14. Wolfram, S.: The Mathematica Book. Wolfram Media Cambridge University Press (1996)

TEDA: A Computational Toolbox for Teaching Ensemble Based Data Assimilation

Elias D. Nino-Ruiz(✉) and Sebastian Racedo Valbuena

Applied Math and Computer Science Laboratory, Department of Computer Science, Universidad del Norte, Barranquilla 080001, Colombia
{enino,racedo}@uninorte.edu.co

Abstract. This paper presents an intuitive Python toolbox for Teaching Ensemble-based Data Assimilation (DA), TEDA. This toolbox responds to the necessity of having software for teaching and learning topics related to ensemble-based DA; this process can be critical to motivate undergraduate and graduate students towards scientific topics such as meteorological anomalies and climate change. Most DA toolboxes are related to operational software wherein the learning process of concepts and methods is not the main focus. TEDA facilitates the teaching and learning process of DA concepts via multiple plots of error statistics and by providing different perspectives of numerical results such as model errors, observational errors, error distributions, the time evolution of errors, ensemble-based DA, covariance matrix inflation, precision matrix estimation, and covariance localization methods, among others. By default, the toolbox is released with five well-known ensemble-based DA methods: the stochastic ensemble Kalman filter (EnKF), the dual EnKF formulation, the EnKF via Cholesky decomposition, the EnKF based on a modified Cholesky decomposition, and the EnKF based on B-localization. Besides, TEDA comes with three toy models: the Duffing equation (2 variables), the Lorenz 63 model (3 variables), and the Lorenz 96 model (40 variables), all of which exhibit chaotic behavior for some parameter configurations, which makes them attractive for testing DA methods. We develop the toolbox using the Object-Oriented Programming (OOP) paradigm, which makes incorporating new techniques and models into the toolbox easy. We can simulate several DA scenarios for different configurations of models and methods to better understand how ensemble-based DA methods work.

Keywords: Data Assimilation · Ensemble Kalman Filter · Education · Python

1 Introduction

Computational tools are widely applied in different contexts of science to develop curiosity and interest in undergraduate and graduate students towards some specific discipline. This seems to be an extraordinary strategy in which instructors

GitHub Package Repository https://github.com/enino84/TEDA.git.

D. Groen et al. (Eds.): ICCS 2022, LNCS 13353, pp. 732–745, 2022.
https://doi.org/10.1007/978-3-031-08760-8_60

and students feel comfortable boarding topics from different branches of science. Indeed, it has helped in the context of (applied) math (and in general STEM courses), wherein students offer some sort of resistance to learning [10]. Nowadays, Data Assimilation (DA) has become a relevant field of research, learning, and teaching owing to recent meteorological anomalies and climate change. In sequential DA, an imperfect numerical forecast $\mathbf{x}^b \in \mathbb{R}^{n\times 1}$ is adjusted according to a vector of real noisy observations $\mathbf{y} \in \mathbb{R}^{m\times 1}$, where $\mathbf{x}^b \in \mathbb{R}^{n\times 1}$ and $\mathbf{y} \in \mathbb{R}^{m\times 1}$ are the background state and the observation vector, respectively, n is the model size, and m denotes the number of observations. To estimate prior parameters, typically, ensembles of model realizations are used. Since each ensemble member has a high computational cost, ensemble sizes are commonly bounded by the hundreds while model resolutions are in the millions. This provides a set of difficulties that are of high interest to the data assimilation community, researchers in general, and instructors. For instance, if seen from a researcher's perspective, the impact of sampling errors is something of high relevance, especially since these methods can be employed, for instance, to predict hurricane initialization with a fairly high percentage of certainty [2]. Another application of DA that is highly active is the combination of Global Positioning System (GPS) and Meteorology (MET) with DA methods to improve weather predictions [12]. However, from an instructor or student perspective, who is trying to explain or learn these topics, then the availability of software, toolboxes, or other tools to aid in teaching is still very limited. There are several DA-oriented toolboxes but usually they are designed for investigative use, and most of them are not flexible when wanting to introduce new models or methods. In [9], authors implemented an open interface standard for a quickly implement of DA for numerical models. This implementation has a few well known methods and focuses on parameter estimation. In [1] authors present a tutorial in Python that shows the implementation of common sequential methods and the idea behind them. These implementations have good approaches to introduce researchers, professors or students to the area of DA, but have some minor inconveniences, for example, by the time this paper is written [9] is available in Java, C++, and FORTRAN which are well know programming languages but are not that simple to understand, also the toolbox needs to be downloaded. We propose a Python-based toolbox to facilitate the comprehension of DA topics that can be used offline or online in Google Colab. This toolbox follows the Object-Oriented Programming (OOP) paradigm, which brings flexibility; it counts with different ensemble-based methods, numerical models, and covariance and precision matrix estimators. Our implementation is education-oriented, and therefore, it is more detailed than other operational DA implementations.

The structure of this paper is as follows: Sect. 2 briefly discuss ensemble-based methods, well-known issues in the context of ensemble DA, and numerical toy models; the topics are restricted to all functionalities in our educational toolbox, in Sect. 3 presents the toolbox and how to employ it to simulate and test ensemble methods. Section 4 demonstrates some results of the use of the toolbox with different methods and models. Conclusions and future directions of TEDA are stated in Sect. 5.

2 Preliminaries

Learning or teaching STEM (Science, Technology, Engineering and Mathematics) topics can be challenged; instructors often apply motivational strategies to keep students focused on their course content [5]. DA is not the exception: this subject combines information from different branches of science: Statistics, Applied Math, and Computer Science. DA can be seen as a challenging field to study, do research, and teach at a first look. However, we believe that by using a proper toolbox wherein numerical experiments are run, and analyses are provided; students can quickly assimilate concepts from different fields of science. We consider DA as a field of extreme importance given current meteorological and climate conditions. Climate change is a real issue that all people must be aware of. It impacts different branches of our lives, starting with the planet we live on. Waterfloods, forest fires, and tsunamis are just consequences of everything we have done to our world so far. Despite how we got to this point, the main question remains: what can we do to mitigate the impact of climate change in our lives, economies, societies, and planet? We can answer this question in many manners, one of them is by providing scientists, instructors, and students with a context such as DA.

Our toolbox provides filter implementations, numerical models, and multiple plots to enrich learning and teaching. We release our learning toolbox with ensemble-based filter implementations, numerical models, and DA scenarios to efficiently perform the teaching process; we briefly discuss all of them in this Section.

2.1 Ensemble-Based Data Assimilation

The Ensemble Kalman Filter (EnKF) is a sequential ensemble-based method which employs an ensemble of model realizations to estimate moments of prior error distributions:

$$\mathbf{X}^b = \left[\mathbf{x}^{b[1]}, \mathbf{x}^{b[2]}, \dots, \mathbf{x}^{b[N]}\right] \in \mathbb{R}^{n \times N}, \tag{1a}$$

where $\mathbf{x}^{b[e]} \in \mathbb{R}^{n \times 1}$ denotes the e-th ensemble member, for $1 \leq e \leq N$, at time k, for $0 \leq k \leq M$. The empirical moments of (1a) are employed to estimate the forecast state $\mathbf{x}^b$:

$$b = \frac{1}{N} \sum_{e=1}^{N} \mathbf{x}^{b[e]} \in \mathbb{R}^{n \times 1},$$

and the background error covariance matrix $\mathbf{B}$:

$$\mathbf{P}^b = \frac{1}{N-1} \mathbf{\Delta X}^b \left[\mathbf{\Delta X}^b\right]^T \in \mathbb{R}^{n \times n},$$

where the matrix of member deviations reads:

$$\boldsymbol{\Delta}\mathbf{X}^b = \mathbf{X}^b - \bar{b}\mathbf{1}^T \in \mathbb{R}^{n \times N}\,. \tag{1b}$$

The posterior ensemble can then be built by using synthetic observations:

$$\mathbf{X}^a = \mathbf{X}^b + \boldsymbol{\Delta}\mathbf{X}^a\,,$$

where the analysis updates can be obtained by solving the following linear system in the stochastic EnKF [4]:

$$\left[\mathbf{H}\mathbf{P}^b\mathbf{H}^T + \mathbf{R}\right]\boldsymbol{\Delta}\mathbf{X}^a = \mathbf{D}^s \in \mathbb{R}^{m \times N}\,,$$

or in the dual EnKF formulation [8]:

$$\left[\left[\mathbf{P}^b\right]^{-1} + \mathbf{H}^T\mathbf{R}^{-1}\mathbf{H}\right]\boldsymbol{\Delta}\mathbf{X}^a = \mathbf{H}^T\mathbf{R}^{-1}\mathbf{D}^s \in \mathbb{R}^{n \times N}\,,$$

the innovation matrix reads $\mathbf{D}^s \in \mathbb{R}^{m \times N}$ whose e-th column $\mathbf{y} - \mathbf{H}\mathbf{x}^{b[e]} + \boldsymbol{\varepsilon}^{[e]} \in \mathbb{R}^{m \times 1}$ is a synthetic observation with $\boldsymbol{\varepsilon}^{[e]} \sim \mathcal{N}\left(\mathbf{0}_m,\, \mathbf{R}\right)$. Note that, the synthetic observations are sampled about the actual ones $\mathbf{y}$. Yet another possible implementation is via the space spanned by the ensemble of anomalies (1b), this is

$$\mathbf{X}^a = \mathbf{X}^b + \underbrace{\boldsymbol{\Delta}\mathbf{X}^b\mathbf{W}^*}_{\boldsymbol{\Delta}\mathbf{X}^a} \tag{2a}$$

where the analysis innovations $\boldsymbol{\Delta}\mathbf{X}^a$ can be computed by solving the following linear system onto the ensemble space:

$$[(N-1)\mathbf{I} + \mathbf{Q}^T\mathbf{R}^{-1}\mathbf{Q}]\mathbf{W}^* = \mathbf{Q}^T\mathbf{R}^{-1}\left[\mathbf{Y}^S - \mathbf{H}\mathbf{X}^b\right]$$

where $\mathbf{Q} = \sqrt[-1]{N-1}\mathbf{H}\boldsymbol{\Delta}\mathbf{X}^b \in \mathbb{R}^{m \times N}$. The implementation (2a) is well-known as the EnKF based on Cholesky decomposition.

In operational DA scenarios, the ensemble size can be lesser than model dimensions by order of magnitudes, and as a direct consequence, sampling errors impact the quality of analysis increments. To counteract the effects of sampling noise, localization methods are commonly employed. In the context of covariance tapering, the use of the *spatial-predecessors* concept can be employed to obtain sparse estimators of precision matrices [6]. The predecessors of model component i, from now on $P(i,\, r)$, for $1 \le i \le n$ and a radius of influence $r \in \mathbb{Z}^+$, are given by the set of components whose labels are lesser than that of the i-th one.

In the EnKF based on a modified Cholesky decomposition (EnKF-MC) [8] the following estimator is employed to approximate the precision covariance matrix of the background error distribution:

$$\widehat{\mathbf{B}}^{-1} = \widehat{\mathbf{L}}^T\widehat{\mathbf{D}}^{-1}\widehat{\mathbf{L}} \in \mathbb{R}^{n \times n}\,,$$

where the Cholesky factor $\mathbf{L} \in \mathbb{R}^{n \times n}$ is a lower triangular matrix,

$$\left\{\widehat{\mathbf{L}}\right\}_{i,v} = \begin{cases} -\beta_{i,v,k} & , v \in P(i,r) \\ 1 & , i = v \\ 0 & , otherwise \end{cases},$$

whose non-zero sub-diagonal elements $\beta_{i,v,k}$ are obtained by fitting models of the form,

$$\mathbf{x}_{[i]}^T = \sum_{v \in P(i,r)} \beta_{i,v,k} \mathbf{x}_{[v]}^T + \gamma_i \in \mathbb{R}^{N \times 1}, 1 \leq i \leq n,$$

where $\mathbf{x}_{[i]}^T \in \mathbb{R}^{N \times 1}$ denotes the i-th row (model component) of the ensemble (1a), components of vector $\gamma_i \in \mathbb{R}^{N \times 1}$ are samples from a zero-mean Normal distribution with unknown variance σ^2, and $\mathbf{D} \in \mathbb{R}^{n \times n}$ is a diagonal matrix whose diagonal elements read,

$$\begin{aligned} \{\mathbf{D}\}_{i,i} &= \widehat{\mathbf{var}} \left(\mathbf{x}_{[i]}^T - \sum_{v \in P(i,r)} \beta_{i,v,k} \mathbf{x}_{[j]}^T \right)^{-1} \\ &\approx \mathbf{var}\left(\gamma_i\right)^{-1} = \frac{1}{\sigma^2} > 0, \text{ with } \{\mathbf{D}\}_{1,1} = \widehat{\mathbf{var}} \left(\mathbf{x}_{[1]}^T\right)^{-1}, \end{aligned}$$

where $\mathbf{var}(\bullet)$ and $\widehat{\mathbf{var}}(\bullet)$ denote the actual and the empirical variances, respectively. The analysis equations can then be written as discussed in [8].

2.2 Numerical Models

The Lorenz 96 model is a dynamical system formulated by Edward Lorenz in [7]; it mimics the behaviour of particles fluctuacting in the atmosphere. The dynamics of particles $x_i \in \mathbb{R}$, for $1 \leq i \leq n$, are described by the following set of Ordinaty Differential Equations:

$$\frac{dx_i}{dt} = (x_{i+1} - x_{i-2})x_{i-1} - x_i + F,$$

wherein periodical boundary conditions are assumed $x_{-1} = x_{N-1}, x_0 = x_N, x_{N+1} = x_1$. This is one of the most widely employed toy models in the context of DA given its chaotic nature when the external force F equals 8.

The Lorenz 63 model is a dynamic system of three Ordinary Differential Equations introduced by Edward Lorenz, again. The equations are described as:

$$\frac{\mathrm{d}x}{\mathrm{d}t} = \sigma(y - x), \frac{\mathrm{d}y}{\mathrm{d}t} = x(\rho - z) - y, \frac{\mathrm{d}z}{\mathrm{d}t} = xy - \beta z,$$

where $\sigma, \rho,$ and β are parameters corresponding to the Prandtl number (usually 10), Rayleigh number , and some physical dimension of the layer itself (usually 8/3), respectively. The system exhibits a chaotic behavior when $\rho = 28$. This model can be seen as a simplified mathematical model for atmospheric convection.

The Duffing Equation is a non-linear second order differential equation presented by Georg Duffing in [3]. This equation is used to model damped and driven oscillators. This initial value problem reads:

$$\ddot{x} + \delta\dot{x} + \alpha x + \beta x^3 = \gamma \cos(\omega t), \text{ with } x(0) = 0, \text{ and } \dot{x}(0) = 1,$$

where δ controls the amount of damping, α denotes the linear stiffness, β stands for the amount of non-linearity in the restoring force, and γ is the amplitude of the periodic driving force.

3 Teaching Data Assimilation (TEDA) Python Toolbox

We provide a simple manner to learn and teach DA-related concepts: a computational toolbox named TEDA (Teaching Ensemble-based Data Assimilation). Our toolbox is written using Python, and we employ the Object-Oriented-Programming (OOP) paradigm to make easier the merging of our toolbox into any other educational or research program. The github repository of our package reads https://github.com/enino84/TEDA.git. We allow the users to analyze results from different perspectives. For instance, our visualizations allow users to understand error correlations, the structure of correlations, model trajectories, ensemble uncertainties, the effects of assimilation on model trajectories and correlations, the results of applying covariance tapering on precision matrices or localization methods on covariance ones, the error evolution of small perturbations, prior and posterior estimation of errors, among others. We employ some useful metrics to provide a wide spectrum of the forecast and the assimilation processes through specialized metrics, for instance, the ℓ_2-norm of errors or the Root-Mean-Square-Errors, both of them well-known in the DA community. We release TEDA with five well-known sequential data assimilation methods: the EnKF, the stochastic EnKF, the EnKF via Cholesky, the EnKF via a modified Cholesky decomposition, and the EnKF via B-Localization. Besides, three numerical models are available for testing and creating DA benchmarks: Lorenz 96, Lorenz 63, and Duffing's Equation. We briefly discuss all methods and models in our toolbox in Sect. 2. We release TEDA in two different manners: as a Python package and a Jupyter notebook.

As a Python package, TEDA consists of five major classes: the *Model* and the *Analysis* ones, both of them abstract classes; these ones define the methods that must be implemented for the definition of new models and ensemble methods. The *Background* class has the methods to create the initial ensemble, the initial forecast, and other required methods for simulations. The *Observation* class defines methods to generate synthetic observations jointly their error distributions (Gaussian by default). The *Simulation* class triggers simulations in DA scenarios given objects of the previous four classes; from here, we can run simulations by calling the "run" method, which will generate outputs such as error plots and statistics. We can modify all parameters for classes and methods in TEDA as needed. We document classes and methods by using Docstring; a convention described within PEP 257 [11]. The purpose of these is to provide to final users brief overviews of objects, attributes, and methods. We show the class diagram of TEDA in Fig. 1. A representation of how folders are arranged can be seen in Fig. 2.

The TEDA notebook is an implementation of our toolbox on Jupyter notebooks. This allows students and instructors to simulate, test, and study sequential data assimilation methods easily. This notebook can be deployed in any web-based interactive development environment for notebooks, code, and data. For instance, we can run on-line the TEDA notebook by using Google Colab, a well-known free cloud service to host and execute Jupyter notebooks. This aspect is relevant since computer power is not needed from students to run our toolbox (which can be seen as a social inclusion initiative); a conventional laptop or low-memory computer with a web browser is sufficient for TEDA. This also can support breaking the social gap regarding students and computational resources; this is a well-known issue in developing countries. Since TEDA is based in OOP, instructors and students can develop their own methods and integrate them into our toolbox as required.

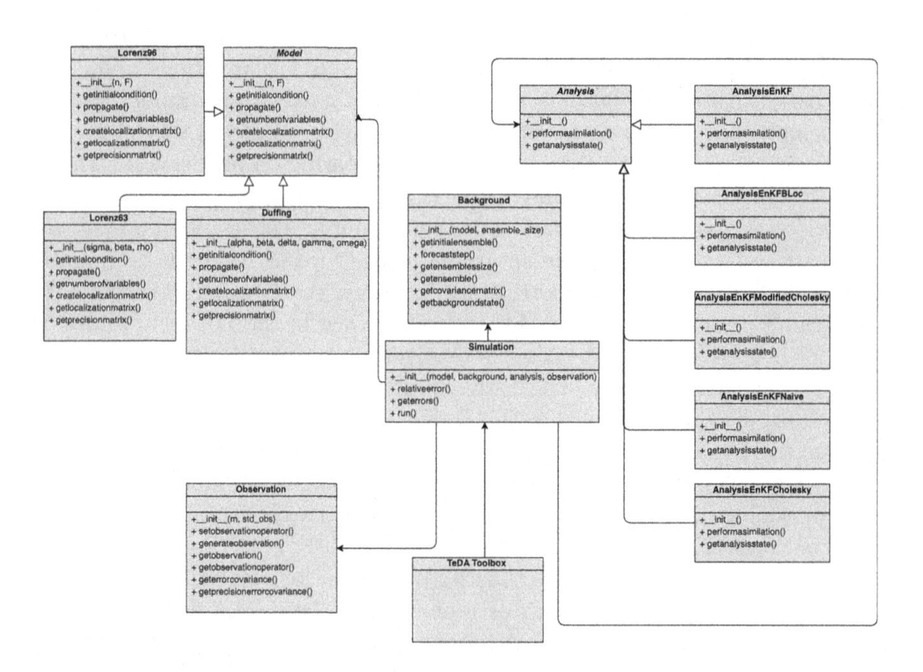

Fig. 1. Class diagram of the TEDA learning toolbox.

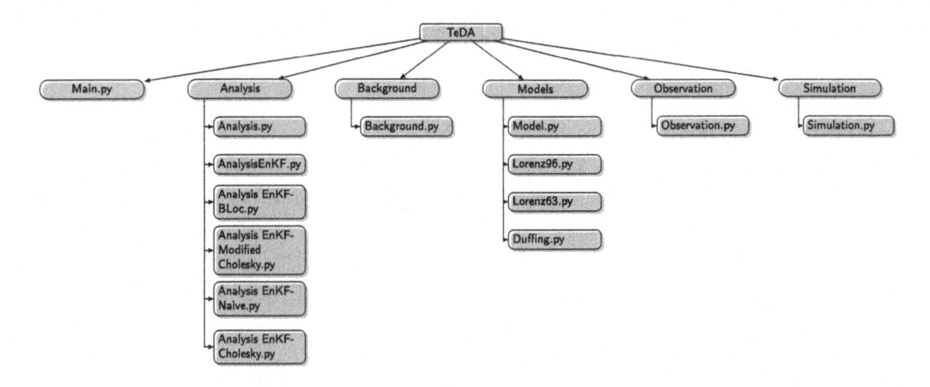

Fig. 2. Tree of the TEDA folder. The necessary files to run (or use) the assimilation toolbox.

4 Demonstration of the TEDA Toolbox

In this Section, we briefly show the capabilities of our package and all potential analyses that students and instructors can conduct via TEDA. Our toolbox attacks four important points during learning ensemble DA: initial ensemble generation, assimilation step, effects of sampling errors, and covariance matrix estimation. Our toolbox provides visual reports of all previous concepts jointly quantitative (statistical) analyses. All plots in this Section can be generated from the TEDA visualization toolkit.

In the context of initial ensemble generation, TEDA allows instructors and students to set up initial perturbation errors to create initial ensembles of model realizations. In Fig. 3a and 3b, we show a visual report of generating an initial ensemble for the Duffing equation via TEDA. In the first row, we show how the initial random perturbations are adjusted according to Duffing dynamics. For this model, by adding Normal white noise to the initial condition ($x(0) = 0$ and $x'(0) = 1$), we can see how the small initial perturbations are amplified by the non-linear dynamics encapsulated in the numerical model. Besides, the chaotic behavior of this model is clear; small perturbations ($t = 0$) tend to completely different conditions in time ($t = 18$). This can help students to understand how non-linear dynamics can drive forecasts to completely different states in future steps. The second row shows that initial Gaussian perturbations (in blue) are mapped to non-Gaussian kernels after numerical model integration (in red). For instance, the ensemble distribution is non-Gaussian (and even more, it is multi-modal). This makes clear the fact that non-Gaussian distributions are frequently found in real-life contexts. This is a well-known issue in the DA community; despite non-Gaussian errors being present in background distributions, scientists rely on Gaussian assumption on prior errors mainly to computational resources: for Gaussian assumption on prior and observational errors, closed-form expression can be obtained for computing analysis members. In the third row, the first column shows the different trajectories for all initial perturbations. Trajectories are identified by colors; again, it is evident the chaotic behavior of the Duffing model; the second column allows us to identify zones from which initial perturbations start (blue dots) and where they end (red dots) after model integration. Despite the fact that initial solutions are close (in space), we can see how the non-linear nature of the Duffing equation makes them follow different paths. In the last row, the first column denotes a three-dimensional plot wherein we can see how ensemble members take different trajectories as time evolves. The Duffing equation is time-dependent, and therefore, we can see the multi-modal spiral behavior for each solution. Besides, the second column shows where the initial conditions start and where they end after the model propagation. The last two rows can be employed to show how non-linear dynamics affect model trajectories even for small synthetic perturbations. As can be seen, many explanations can be obtained by just analyzing plots or their combination. This provides a wide vision of how spatial paths relate to small perturbations in initial conditions and how Normal initial perturbations are non-linearly mapped to non-Gaussian shapes after model propagation.

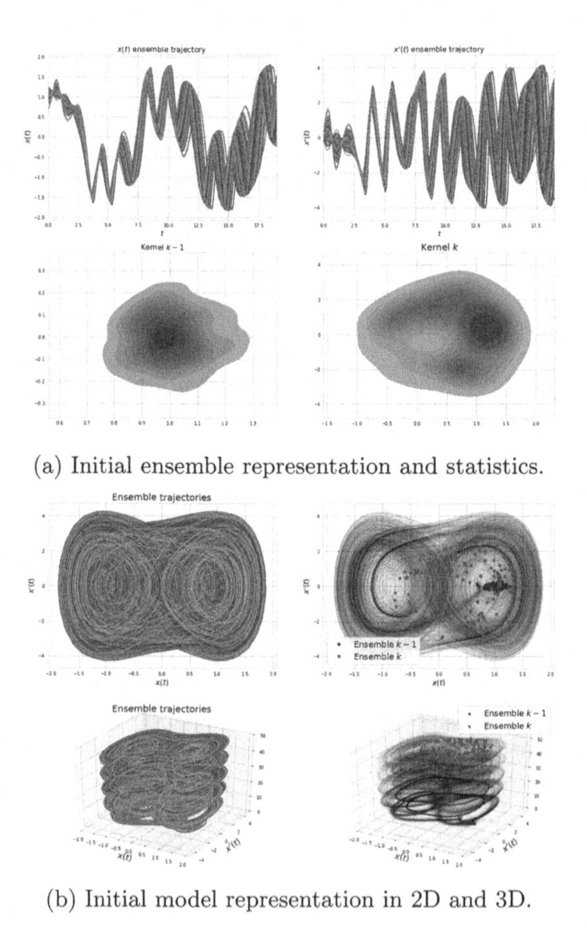

(a) Initial ensemble representation and statistics.

(b) Initial model representation in 2D and 3D.

Fig. 3. Some TEDA plots for data assimilation using the duffing equation

In Fig. 4, we show the effects of ensemble-based DA for different cases: No Data Assimilation (NODA), a single assimilation step, and two assimilation steps. Students will get to this point after creating an initial ensemble, as we did before, and even more, by providing the error distribution of observations jointly their time frequencies. In Fig. 4a, the solid black line denotes the actual state of the system while blue lines represent ensemble member trajectories. Note that uncertainty increases as no DA is performed, and ensemble trajectories are distant from actual states. However, as can be seen in Fig. 4b, a single assimilation step can reduce uncertainty in ensemble trajectories, and even more, it can route ensemble trajectories to actual system states. Uncertainties in blue forecasts are more significant than those in red ones. Of course, uncertainty increases as time evolves (i.e., ensemble trajectories in red). Figure 4c shows that we can reduce uncertainties, in time, by frequently injecting observations into an imperfect numerical forecast. This plot can support the understanding of uncertainties in errors as a function of time. Students can realize that as no actual information is injected into the numerical forecast, model errors will make forecasts diverge from real-life scenarios.

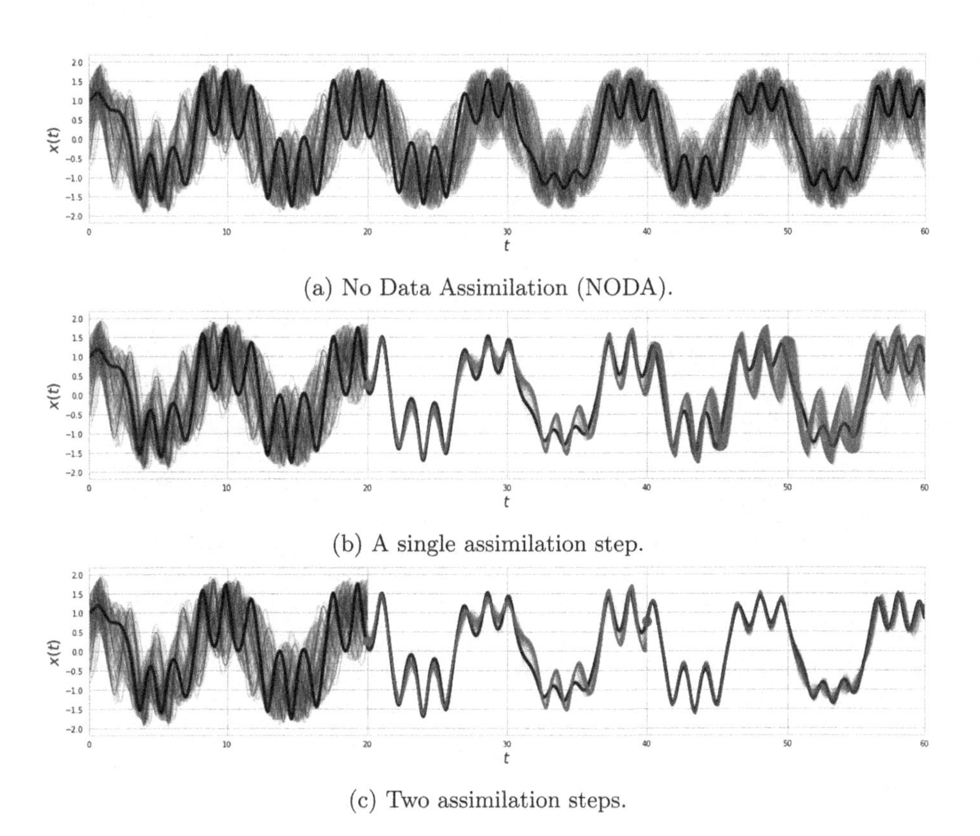

(a) No Data Assimilation (NODA).

(b) A single assimilation step.

(c) Two assimilation steps.

Fig. 4. The effects of DA in the Duffing equation.

In Fig. 5, we can see how error correlations are developed in time (forecast) and how these are updated after DA (analysis). We show in blue and green how observations (red dots) drive the states of the numerical forecast. Besides, we can see how correlation matrices evolve after assimilation steps (i.e., from positive to negative correlations); we want observed components from the model space to get closer to observations. Similarly, Fig. 6 shows plots for the x_2 variable of the Lorenz-96 model. We can see how uncertainties increase and even more, how correlations are developed in time for model variables in Figs. 6c and 6d. These scenarios can show students how error correlations are developed in model components after forecast steps and analysis ones.

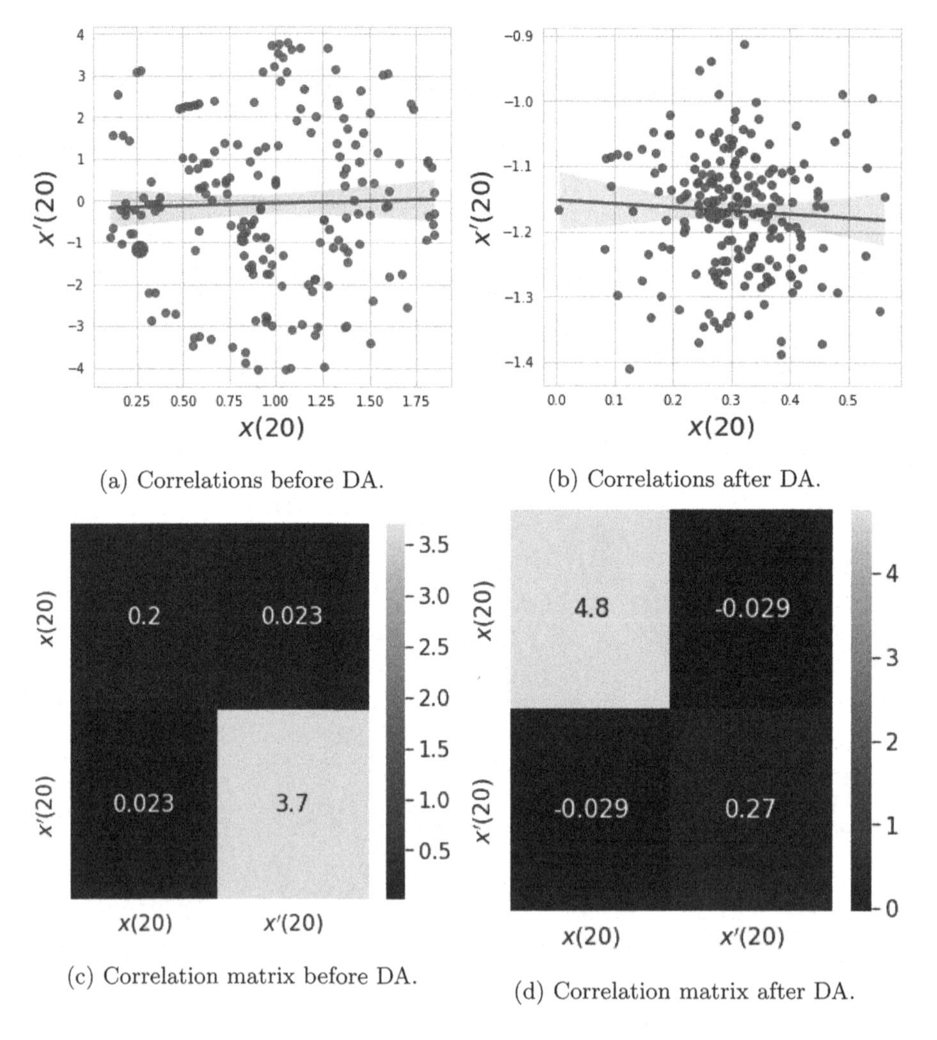

(a) Correlations before DA.

(b) Correlations after DA.

(c) Correlation matrix before DA.

(d) Correlation matrix after DA.

Fig. 5. Initial pool of background members for the experiments. Two dimensional projections based on the two leading directions are shown.

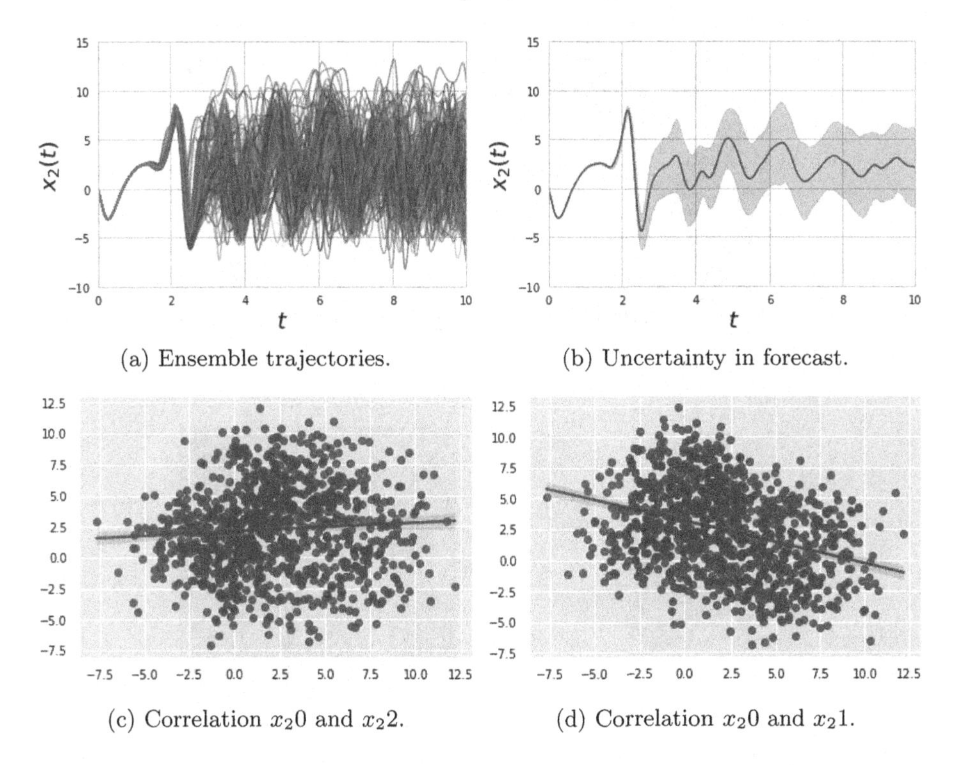

(a) Ensemble trajectories.

(b) Uncertainty in forecast.

(c) Correlation x_20 and x_22.

(d) Correlation x_20 and x_21.

Fig. 6. Some visualizations of TEDA for the Lorenz-96 model.

Figure 7 shows examples of some TEDA plots about localization aspects for the Lorenz-96 model. In Figs. 7a and 7b, we show prior structures (pre-defined student parameters) of background errors for covariance matrix localization via Schur product and precision matrix estimation, respectively. TEDA also enriches analyses by providing additional plots. For instance, a 3d representation of the decorrelation matrix can be seen in Fig. 7c. This figure clearly shows how we expect error correlations to decrease in space. This can provide a visual explanation of error decay regarding a given radius length. Figure 7d shows the ensemble covariance before localization is applied. As can be seen, sampling errors develop correlation in spatially distant model components. For instance, there is no clear structure of error relationships regarding model dynamics. We can then consider applying any pre-defined structures on the samples to mitigate the impact of sampling errors. Results can be seen in Figs. 7e and 7f for the Schur product and modified Cholesky estimators, respectively.

After all experiment are run, TEDA generates a static HTML file where step by step plots are shown, explanations are given, and comparisons are made in selected DA methods for experiments. Images are stored in sub-folders, these can be utilized as needed.

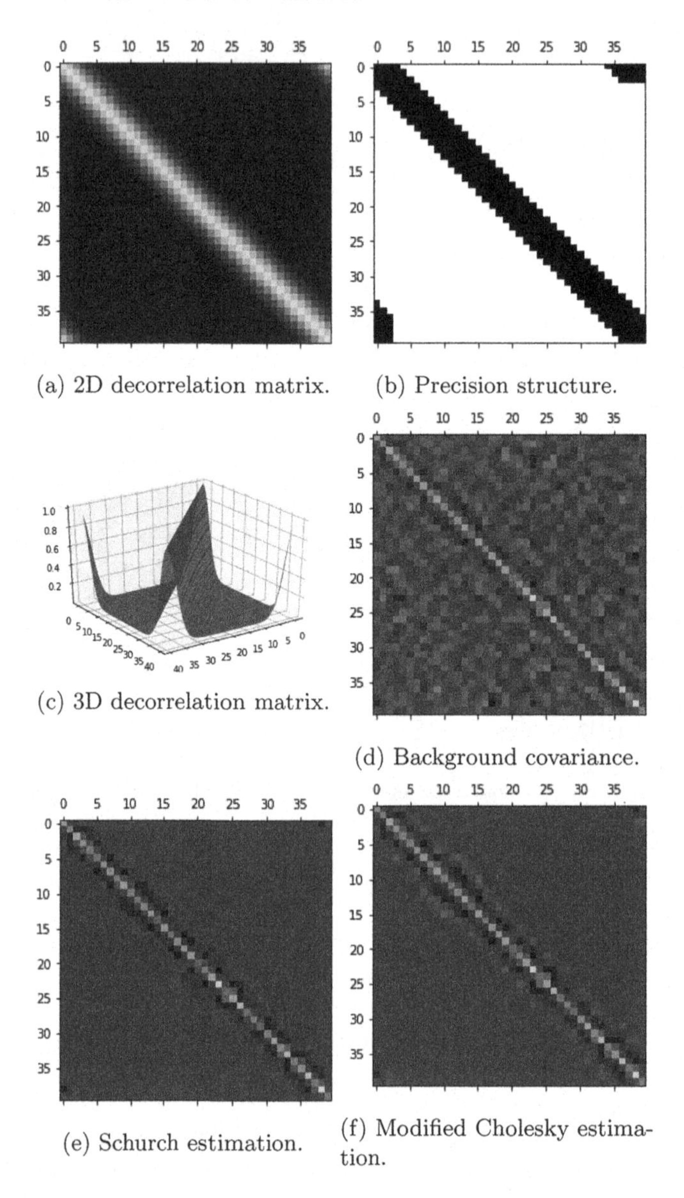

(a) 2D decorrelation matrix. (b) Precision structure. (c) 3D decorrelation matrix. (d) Background covariance. (e) Schurch estimation. (f) Modified Cholesky estimation.

Fig. 7. Some TEDA plots for covariance matrix and precision matrix estimation.

5 Conclusions and Future Research

The TEDA toolbox is an educational software to support the learning and teaching process of DA aspects. TEDA is released with five ensemble-based methods, three numerical models, and a powerful visualization toolbox to facilitate analyses. Since it is OOP (and written in Python following the standard PEP 257), we can easily add new methods and models. Our toolbox is released jointly with a Jupyter notebook to show a practical case of TEDA and, even more, to exploit cloud resources (i.e., Google Colab by just uploading the Jupyter notebook). TEDA can create multiple DA scenarios with different parameters, for instance, various observational networks, ensemble sizes, inflation factors, among many others. This flexibility makes TEDA exceptional to compare results from multiple methods and simulations. This can be exploited in academic contexts to support the learning process. Our toolbox generates a static HTML file for each simulation wherein information about ensemble generation, analysis steps, and other relevant aspects are discussed. We expect to increase the number of models and methods in future releases of TEDA.

References

1. Ahmed, S.E., Pawar, S., San, O.: PYDA: a hands-on introduction to dynamical data assimilation with python. Fluids **5**(4) (2020). https://doi.org/10.3390/fluids5040225
2. Asch, M., Bocquet, M., Nodet, M.: Data assimilation: methods, algorithms, and applications. SIAM (2016)
3. Duffing, G.: Erzwungene Schwingungen bei veränderlicher Eigenfrequenz und ihre technische Bedeutung. No. 41–42, Vieweg (1918)
4. Evensen, G.: The ensemble kalman filter: theoretical formulation and practical implementation. Ocean Dyn. **53**, 343–367 (2003). https://doi.org/10.1007/s10236-003-0036-9
5. Freeman, K.E., Alston, S.T., Winborne, D.G.: Do learning communities enhance the quality of students' learning and motivation in stem? J. Negro Educ. 227–240 (2008)
6. Levina, E., Rothman, A., Zhu, J., et al.: Sparse estimation of large covariance matrices via a nested lasso penalty. Ann. Appl. Statist. **2**(1), 245–263 (2008)
7. Lorenz, E.N.: Predictability: a problem partly solved. In: Proceedings of the Seminar on Predictability, vol. 1 (1996)
8. Nino-Ruiz, E.D., Sandu, A., Deng, X.: An ensemble Kalman filter implementation based on modified Cholesky decomposition for inverse covariance matrix estimation. SIAM J. Sci. Comput. **40**(2), A867–A886 (2018)
9. OpenDA-Association. Openda (2021). https://github.com/OpenDA-Association/OpenDA
10. Tharayil, S., et al.: Strategies to mitigate student resistance to active learning. Int. J STEM Educ. **5**(1), 1–16 (2018). https://doi.org/10.1186/s40594-018-0102-y
11. Van Rossum, G., Warsaw, B., Coghlan, N.: Pep 8-style guide for python code. Python. org 1565 (2001)
12. Wang, B., Zou, X., Zhu, J.: Data assimilation and its applications. Proc. Natl. Acad. Sci. **97**(21), 11143–11144 (2000)

Uncertainty Quantification for Computational Models

Forward Uncertainty Quantification and Sensitivity Analysis of the Holzapfel-Ogden Model for the Left Ventricular Passive Mechanics

Berilo de Oliveira Santos, Rafael Moreira Guedes, Luis Paulo da Silva Barra, Raphael Fortes Marcomini, Rodrigo Weber dos Santos, and Bernardo Martins Rocha(✉)

Universidade Federal de Juiz de Fora, Juiz de Fora, Brazil
{berilo.santos,rafael.guedes,raphael.marcomini}@engenharia.ufjf.br,
{bernardomartinsrocha,rodrigo.weber}@ice.ufjf.br,
luis.barra@ufjf.edu.br

Abstract. Cardiovascular diseases are still responsible for many deaths worldwide, and computational models are essential tools for a better understanding of the behavior of cardiac tissue under normal and pathological situations. The microstructure of cardiac tissue is complex and formed by the preferential alignment of myocytes along their main axis with end-to-end coupling. Mathematical models of cardiac mechanics require the process of parameter estimation to produce a response consistent with experimental data and the physiological phenomenon in question. This work presents a polynomial chaos-based emulator for forward uncertainty quantification and sensitivity analysis of the Holzapfel-Ogden orthotropic constitutive model during the passive filling stage. The fiber orientation field is treated as a random field through the usage of the Karhunen-Loève (KL) expansion. The response and uncertainty of the constitutive parameters of the model considered here are also investigated. Our results show the propagated uncertainties for the end-diastolic volume and fiber strain. A global sensitivity analysis of the constitutive parameters of the complete model is also presented, evidencing the model's key parameters.

Keywords: Cardiac mechanics · Karhunen-Loève expansion · Polynomial chaos

1 Introduction

Cardiovascular diseases are the leading cause of death in the world, however, many of them can be avoided if there is a previous diagnosis. Computational models are essential tools to understand better the cardiac system and, above

Supported by UFJF, CAPES, CNPq (Grants 310722/2021-7, 315267/2020-8), and FAPEMIG (Grants APQ-01340-18, APQ 02489/21).

D. Groen et al. (Eds.): ICCS 2022, LNCS 13353, pp. 749–761, 2022.
https://doi.org/10.1007/978-3-031-08760-8_61

all, how it is affected by pathologies or disorders [13]. Cardiac tissue is made up of fibers that are fundamental in various aspects of the heart, and therefore, changes in its typical orientation can result in improper functioning, as is the case with hypertrophic cardiomyopathy (HCM) [12] that affects the heart muscle. HCM is responsible for the thickening of the cardiac muscle (especially the ventricles or lower heart chambers), increased left ventricular stiffness, fiber disarray, and cellular changes [9].

During the last years, computational models of the cardiovascular system have evolved significantly. In particular, the construction of patient-specific models that could be applied in the clinical setting is a high goal. However, the construction of patient-specific models involves several sources of uncertainty, ranging from personalized geometries based on medical images to parametric uncertainty inherent in the underlying mathematical model.

Recently, many studies on uncertainty quantification for cardiac electromechanics have been performed [3,4,10,14,15]. The work of [14] was one of the first to apply uncertainty quantification (UQ) techniques for the problem of passive filling of the left ventricle (LV). They considered as inputs the constitutive parameters of a transversely isotropic constitutive model [6]. In [15] similar analyses were carried out, but now considering the fiber orientation field as a random field through the KL expansion. In [4] another source of uncertainty was added in the LV model, where geometry was also considered as uncertain through a parametrized strategy for mesh generation. The previous studies focused on the passive filling phase of the LV only. A sensitivity analysis and forward uncertainty quantification study of the complete cardiac cycle was presented in [3]. In [10] a sensitivity analysis of a detailed human fully-coupled ventricular electromechanical model was conducted using the HO model. However, due to the usage of a more complex coupled electromechanical model with many parameters, only one parameter of the HO model was evaluated.

In this work, we focus on forward uncertainty quantification and sensitivity analysis of the passive filling phase of the left ventricular mechanics. We considered as uncertain input parameters the parameters from the Holzapfel-Ogden (HO) constitutive model [7] and the fiber field as a random field using the truncated Karhunen-Loève (KL) expansion. The analyses were performed using surrogate models (emulators) based on the Polynomial Chaos Expansion (PCE), as in previous works on cardiac mechanics [3,4].

The remaining of this manuscript is organized as follows: in Sect. 2 the mathematical models, the numerical methods, the techniques used for uncertainty quantification, and sensitivity analysis are presented. Next, Sect. 3 describes the computer implementation and computational experiments; while the numerical results are presented on Sect. 4. Section 5 ends this work with conclusions, limitations, and possible future works.

2 Models and Methods

2.1 Cardiac Mechanics

The focus of this study is to model the phenomenon of cardiac mechanics that corresponds to the (passive) filling of the left ventricle (LV) by the blood during the diastolic phase. At this stage, blood fills the LV cavity and exerts pressure on the endocardial surface. The following problem describes the passive filling of the LV:

$$\nabla \cdot (\mathbf{FS}) = \mathbf{0}, \qquad \text{in} \quad \Omega_0, \tag{1}$$

$$\mathbf{u} = \mathbf{0}, \qquad \text{on} \quad \partial\Omega_{\text{base}}, \tag{2}$$

$$(\mathbf{FS})\mathbf{N} = p_{\text{endo}}\mathbf{F}^{-T}\mathbf{N}, \qquad \text{on} \quad \partial\Omega_{\text{endo}}, \tag{3}$$

where $\mathbf{F}$ is the deformation gradient tensor, $\mathbf{S}$ is the second Piola-Kirchhoff stress tensor, $\mathbf{u}$ is the displacement field, $\mathbf{N}$ is the unit normal vector of the endocardium surface, and p_{endo} is the applied pressure on the endocardium. For simplicity, we considered zero displacement boundary conditions at the base of the endocardium.

The Holzapfel-Ogden (HO) constitutive model was used to describe the LV tissue stress-strain relationship. The strain energy function of the HO model for the incompressible case is described by:

$$\Psi = \frac{a}{2b}\left[\exp\left\{b\left(I_1 - 3\right)\right\} - 1\right] + \frac{a_f}{2b_f}\left[\exp\left\{b_f\left(\max\left(I_{4f}, 1\right) - 1\right)^2\right\} - 1\right] \tag{4}$$

where a, b, a_f, b_f, a_s, b_s, a_{fs}, and b_{fs} are the material parameters, and I_1, I_{4f}, and I_{8fs} are invariants given by: $I_1 = \text{tr}(\mathbf{C})$, $I_{4\text{f}} = \mathbf{f}_0 \cdot (\mathbf{C}\mathbf{f}_0)$, $I_{4\text{s}} = \mathbf{s}_0 \cdot (\mathbf{C}\mathbf{s}_0)$, $I_{8\text{fs}} = \mathbf{m}_0 \cdot (\mathbf{C}\mathbf{s}_0)$ where $\mathbf{C} = \mathbf{F}^T\mathbf{F}$ is the right Cauchy-Green tensor, $\mathbf{f}_0$ and $\mathbf{s}_0$ are the fiber and sheet directions in the reference configuration, respectively. The second Piola-Kirchhoff stress tensor $\mathbf{S}$ of Eq. (1) is given by $\mathbf{S} = 2\frac{\partial\Psi(\mathbf{C})}{\partial\mathbf{C}} - p\mathbf{C}^{-1}$, where p is the pressure.

2.2 Left Ventricular Geometry and Fiber Orientation

A simplified geometric model of the left ventricle, generated from the equations of a family of a truncated ellipsoid, where the wall thickness is homogeneous, was considered. Figure 1 (left) shows its dimensions with measurements that typically represent the human LV. The finite element mesh generated from this geometry, as shown in Fig. 1 (middle), is composed of a total of 1786 nodes and 6395 tetrahedral elements.

A typical fiber orientation field for the LV is illustrated on Fig. 1 (right). The microstructure of the cardiac tissue is represented as a constant function per element, where each element has unit vectors $\mathbf{f}_0$, $\mathbf{s}_0$, and $\mathbf{n}_0$ that describes the fiber, sheet, and normal directions in the reference configuration. To generate the fiber orientation, the Laplace-Dirichlet Rule-Based (LDRB) rule-based algorithm

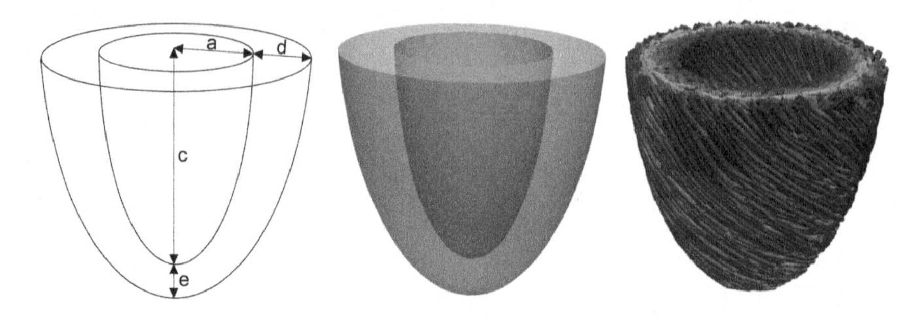

Fig. 1. The left ventricular geometry and its corresponding finite element mesh are shown on the left and middle panels; whereas a typical visualization of the fiber field orientation is shown on the right. The following dimensions were used: a = 2.0 cm, c = 6.0 cm, d = 1.3 cm, and e = 1.0 cm, which results in the approximately 50 mL of cavity volume.

developed by [1] was used. The baseline fiber orientation used in this study has a helical angle of 60° on the endocardium surface and varies linearly throughout the myocardium (transmural direction) up to the value of -60° on the epicardium surface.

2.3 Numerical Solution

Numerical solutions of the problem given in Eq. (1) were obtained by the finite element method (FEM) implemented in the open-source library FEniCS. For discretization, a mixed method of the Taylor-Hood type was used to approximate $(\mathbf{u}, p)$ with approximations $\mathbb{P}_2 \times \mathbb{P}_1$, that is, of degree 2 for the displacement field and degree 1 for the pressure field, respectively. The non-linear LV filling problem is solved with Newton's method. In addition, an incremental procedure was used and the total pressure to be applied, of $p_{endo} = 2.7\,\text{kPa}$, was divided into 50 steps as performed in [8].

2.4 Polynomial Chaos Expansion Surrogate Models

An emulator or surrogate model is an approach that aims to solve a complex problem in a simplified and, consequently, faster way. In the specific case of this work, an emulator based on polynomial chaos expansion (PCE) was adopted, which has been successfully used in other works [3,4,14] on cardiac mechanics to perform forward uncertainty quantification and sensitivity analyses.

PCE is a technique for generating low-cost computational approximations for a quantity of interest Y, usually obtained after solving the governing equations (the forward problem). Let $f(\mathbf{Z})$ be the simulator of this quantity of interest Y, where $\mathbf{Z}$ are the input parameters. This quantity is expanded into a series of orthogonal polynomials Ψ_j with random entries. The PCE for Y is given by:

$$Y = \sum_{j=0}^{\infty} b_j \Psi_j \left(\{Z_n\}_{n=1}^{\infty}\right) = f_{PCE}(\mathbf{Z}) \tag{5}$$

where b_j are the coefficients to be determined and Ψ_j are the orthogonal polynomials with their respective random variables $Z_1, Z_2, \ldots, Z_n$. In practice, the expansion is truncated at a finite number of terms, and the approximation of Y, obtained by the emulator f_{PCE} and denoted by $\hat{Y}$, can be expressed by:

$$f(\mathbf{Z}) = Y \approx \hat{Y} = \sum_{j=0}^{N_p-1} c_j \Psi_j(\mathbf{Z}) = f_{PCE}(\mathbf{Z}) \tag{6}$$

where c_j are the coefficients of the expansion to be determined and $\Psi_j(\mathbf{Z})$ are the orthogonal polynomials. The polynomial expansion of Eq. (6) has degree p for D input parameters. The number of terms is given by: $N_p = \frac{(D+p)!}{D!p!}$.

For an improved accuracy of the surrogate model, it is recommended to use a number of terms (samples) greater than N_p to create it. In general, a multiplicative factor m is adopted. That is, the number of samples is given by $N_s = mN_p$. There are different ways to determine the coefficients c_j that determine the emulator of a quantity. In this work, the stochastic collocation method [14] was adopted, which together with the choice of N_s for $m > 1$ results in a least-squares problem. More details on this procedure can be found in [4,14].

At this point, it is important to note that the emulators for the mechanical problem, defined in Eq. 1, are polynomials that approximate the outputs of the simulator. Furthermore, once the emulators are built, due to their polynomial nature, these values can be calculated cheaply by evaluating the polynomial (emulator) for a given set of parameters which makes them appropriate for forward uncertainty quantification and sensitivity analysis.

2.5 Sensitivity Analysis via Sobol

Sensitivity analysis measures the impact of input parameters on some output data of a problem. In this work, the first order and total Sobol [16] indices were adopted.

Let Y be a scalar quantity of interest for which we want to assess the impact of the input parameters $\mathbf{X} = \{X_1, X_2, \ldots, X_D\}$. The first order Sobol index expresses the direct influence of a parameter X_i on the variance of the quantity of interest Y. This index is given by:

$$Si = \frac{\mathbb{V}[\mathbb{E}(Y|X_i)]}{\mathbb{V}(Y)} \tag{7}$$

where $\mathbb{E}$ denotes the expected value, $\mathbb{V}$ represents the variance, and $\mathbb{E}(Y|X_i)$ denotes the expected value of the output Y when the parameter X_i is fixed. The first order Sobol sensitivity index represents the expected reduction in the variance of the analyzed quantity when the parameter X_i is fixed.

The total Sobol index represents possible interactions between the input parameters and their effects on the Y quantity. For the input X_i it is denoted by S_{Ti}, and is given by:

$$S_{Ti} = \frac{\mathbb{E}[\mathbb{V}(Y|X_{\sim i})]}{\mathbb{V}(Y)} = 1 - \frac{\mathbb{V}[\mathbb{E}(Y|X_{\sim i})]}{\mathbb{V}(Y)} \quad (8)$$

where $X_{\sim i}$ represents all input parameters except the X_i parameter.

In this work the Sobol sensitivity indices were calculated from the PCE surrogate models using the implementation of the `ChaosPy` [5] library.

2.6 Random Fiber Field

Uncertainty in the fiber field of the cardiac microstructure was considered so far, with two approaches employed in previous works [3,15]. One way [3,4,10, 14] is to consider the parameters α_{endo} and α_{epi} used in the rule-based fiber generation algorithm as uncertain input parameters. The other approach treats the fiber orientation as a random field and is generated via the Karhunen-Loève expansion [15]. This approach has the advantage of considering local variations in the cardiac microstructure, in contrast to the parametric approach where such variation is not present and is only represented through the α_{endo} and α_{epi} parameters.

We consider that the orientation of the fibers (only) will be represented as a random field. Fiber orientation is considered as the sum of a random field representing a perturbation $\mathbf{F}$ and the original fiber orientation field $\mathbf{f}_{micro}$ that follows the properties of the microstructure of cardiac tissue. Then, the orientation of the fibers is given by

$$\mathbf{f}(\mathbf{x}, \theta) = \mathbf{f}_{micro}(\mathbf{x}) + \mathbf{F}(\mathbf{x}, \theta), \quad (9)$$

where θ represents the dependence of the perturbed field $\mathbf{f}$ on some random property. The perturbation $\mathbf{F}$ is represented as a random field using the truncated KL expansion as follows:

$$\mathbf{F}(\mathbf{x}, \theta) = \bar{\mathbf{F}}(\mathbf{x}) + \sum_{k=1}^{n_{kl}} \eta_k(\theta)\sqrt{\lambda_k}\boldsymbol{\phi}_k(\mathbf{x}), \quad (10)$$

where $\bar{\mathbf{F}}(\mathbf{x})$ is the expected value of the stochastic field in $\mathbf{x}$ and $\{\eta_k(\theta)\}$ represents a set of independent Gaussian random variables and $(\lambda_k, \phi_k(\mathbf{x}))$ are the eigenvalues and eigenfunctions of the following integral:

$$\int_D C(\boldsymbol{y}, \boldsymbol{x})\phi_i(\boldsymbol{y})\mathrm{d}\boldsymbol{y} = \lambda_i \boldsymbol{\phi}_i(\boldsymbol{x}), \quad (11)$$

where D is the domain of the cardiac tissue of interest and $C(\mathbf{y}, \mathbf{x})$ the covariance function. Without loss of generality, it is assumed that $\bar{\mathbf{F}}(\mathbf{x}) = 0$ and that the covariance function has the following exponential form given by:

$$C(\boldsymbol{x}, \boldsymbol{y}) = \sigma_{\mathrm{KL}}^2 \exp\left(-\frac{|\boldsymbol{x} - \boldsymbol{y}|^2}{2l_{KL}^2}\right) \quad \forall \boldsymbol{x}, \boldsymbol{y} \in D, \quad (12)$$

where σ_{KL}^2 is the variance of the field and l_{KL} is the correlation size that defines the spatial scale over which the field exhibits significant correlation [15].

In practical terms, to compute the KL expansion given in Eq. (9), the following generalized eigenvalue problem needs to be solved:

$$\mathbf{T}\boldsymbol{\phi}_k = \lambda_k \mathbf{M}\boldsymbol{\phi}_k, \quad \text{with} \quad \mathbf{T} = \mathbf{M}^T\mathbf{C}\mathbf{M}, \tag{13}$$

where $\mathbf{M}$ is the mass matrix calculated using the finite element method.

The truncated KL expansion reduces the dimensionality of the stochastic space from infinity to n_{KL} and provides a parametric representation of the random field $\mathbf{F}(\mathbf{x}, \theta)$ through n_{KL} random variables. The uncertainty (or randomness) in the fiber orientation field comes from n_{KL} independent random variables $\eta_1, \ldots, \eta_{n_{KL}}$, which follow normal distributions with mean zero and unit standard deviation, i.e. $\eta_i \sim \mathcal{N}(0, 1)$.

2.7 Quantities of Interest

To evaluate the parametric uncertainty in the response of the model given by Eq. 1 we considered the following outputs or quantities of interest (QoI): cavity volume (as a function of applied pressure), the end-diastolic volume (EDV), and an average fiber strain measured at the end of diastole [3]. Fiber strain was computed as:

$$\varepsilon_{fiber} = \mathbf{f}_0^T \mathbf{E} \mathbf{f}_0, \tag{14}$$

where $\mathbf{E}$ is the Green-Lagrange strain tensor and $\mathbf{f}_0$ is the fiber orientation in the undeformed configuration. The average fiber strain is computed using a set of 20 points uniformly distributed in the LV domain and Eq. (14) to compute their values.

3 Implementation and Experiments

3.1 Implementation

The computational experiments of this work were all carried out in a code implemented in the Python programming language with support for scientific computing through the `NumPy` and `SciPy` libraries. The library `ChaosPy` [5] was used for uncertainty quantification, sensitivity analysis, and the construction of the KL expansion. The library `FEniCS` [11] was used for solving the forward problem (passive filling of the LV) using finite elements. The library `ldrb` [1] was used to generate the fiber field in finite element meshes LDRB.

3.2 Numerical Experiments

Two experiments were carried out for UQ and SA. The first one considered only the 4 input parameters of the HO model (transversely isotropic case) and no uncertainties in the fiber field. We assumed uniform distributions for the uncertain parameters of the HO model, allowing them to vary from half its reference value to a two-fold increase in the reference value. Uniform distributions were chosen due to the lack of data for parameters [4,14,15], and also to represent the population average values [2]. Table 1 summarizes the baseline parameters values and the corresponding distributions used in the experiments for UQ and SA.

Table 1. Baseline parameter values of the HO model and the distributions used to construct the PCE surrogate models. The uniform distributions are bounded by below and above with half and twice of the baseline value, respectively.

Parameter	a	b	a_f	b_f
Baseline value	228	7.78	116.85	11.83
Distribution	$\mathcal{U}(114, 456)$	$\mathcal{U}(3.89, 15.56)$	$\mathcal{U}(58.42, 233.7)$	$\mathcal{U}(5.92, 23.66)$

The second experiment extends the previous one, where uncertainty in the fiber field is now included in the analysis via the KL technique. The settings for the HO parameters were the same of the first experiment. For the KL expansion a total of $n_{KL} = 8$ terms was used, and the following parameters were considered: $\sigma_{KL} = 0.5$ radians and $l_{KL} = 1.0$ cm, as previously used in [15]. We considered $p = 2$ and $m = 2$, which for the first experiment with $D = 4$ resulted in $N_s = 30$ samples, whereas for the second experiment with $D = 12$ resulted in a total of $N_s = 182$ samples (train data) that were generated for the construction of the surrogate models. Finally, the accuracy of the surrogate models were carried out via a new set of $N_s^{test} = 100$ simulations (test data).

4 Results

In the following, we present the numerical results obtained in this work. First, a preliminary study to define the number of terms in the KL expansion was carried out, followed by a study to show the prediction capabilities of the PCE surrogate models employed for further analyses. Then, results of the forward UQ and SA of the passive filling LV problem using the HO model are presented.

4.1 Karhunen-Loeve Expansion

First, to define the number of terms in the truncated KL expansion for representing the random fiber field, we computed the eigenvalues of the generalized problem defined in Eq. (13). Figure 2 shows the first 128 eigenvalues, where it is clear their fast decay. To avoid a large number of input parameters, we adopted $n_{KL} = 8$ for further studies with random fiber fields.

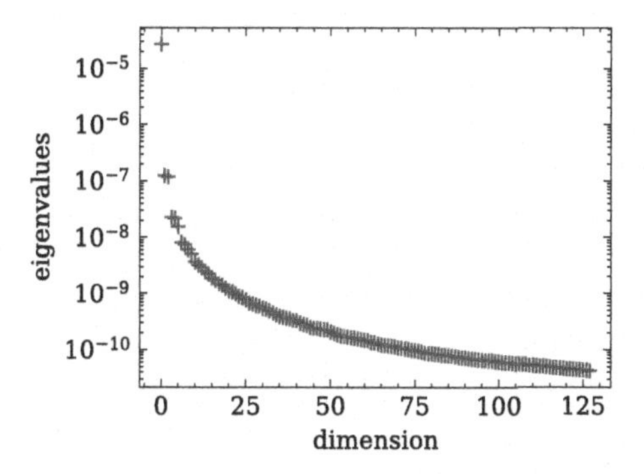

Fig. 2. Eigenvalues of the KL expansion for the LV mesh considered in this work.

4.2 Polynomial Expansion Emulator

After creating the emulator using N_s samples, a new set of N_s^{test} new samples was generated with the FEM simulator. This study aims to assess the prediction capabilities of the surrogate models on test data.

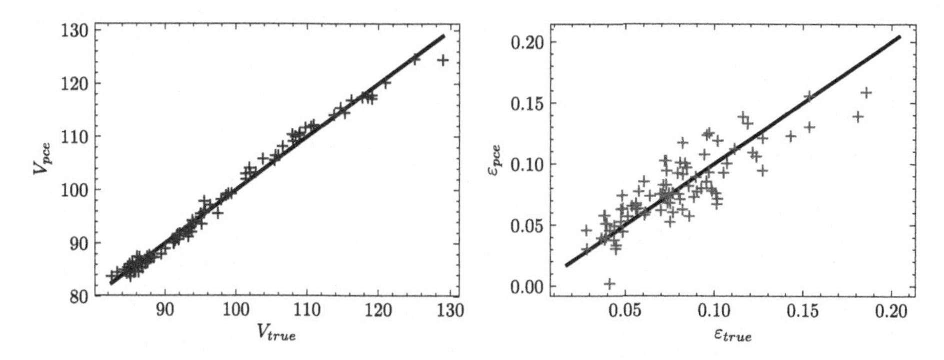

Fig. 3. Prediction results of the end-diastolic volume (left) and fiber strain (right) using the PCE surrogate model on the set of test samples.

Figure 3 shows the true values of the outputs (EDV and average fiber strain) versus the predicted values obtained by the PCE surrogate models. The black line suggests the exact predictions of the true values, while the points are the predictions. The closer the points to the solid black line, the better predictability is for the surrogate models. One can observe that the PCE surrogate models can predict the outputs very well in general.

4.3 Forward Uncertainty Quantification

Forward uncertainty quantification for experiments 1 (HO parameters only) and 2 (HO parameters and random fiber field via KL) were carried out to evaluate the uncertainties on the QoIs. Figure 4 (left panel) shows the propagated uncertainty to the cavity volume as a function of pressure for experiments 1 and 2, labeled as HO and HO+KL, which consider as input the HO parameters only and the HO parameters and random fiber field, respectively. The solid line represents the mean response, whereas the shaded region represents the mean ± standard deviation. One can observe more variations for the volume on the upper limit for the second experiment, as expected since the inclusion of a random fiber field increases uncertainty. The right panel of Fig. 4 shows the distributions of the end-diastolic volume obtained for the two cases, where one can observe its asymmetry (for both cases) and the fact that the HO+KL case presents more spread towards larger volume values.

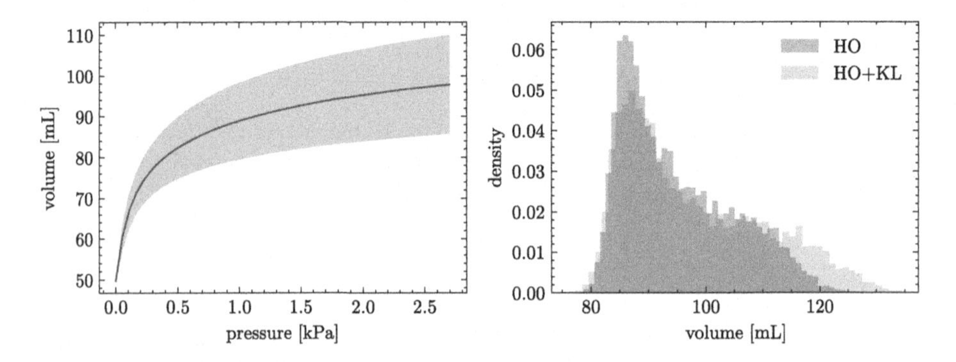

Fig. 4. (Left) Forward uncertainty quantification for the pressure-volume curve during the passive filling stage of the left ventricle dynamics for the parameters of the HO model (only) and HO+KL for the case with random fiber field. The solid line represents the mean value, and the shaded region the confidence interval within one standard deviation. (Right) Distributions of the end-diastolic volume.

Table 2 shows some statistics estimated from the PCE surrogate models for the end-diastolic volume and average fiber strain. In general, local variations in the fiber orientation can significantly impact the outputs. The expected EDV volume value in both cases is very similar, with a slight increase in the coefficient of variation in experiment 2 concerning experiment 1. The average fiber strain ε_{fiber} resulted in a smaller expected value for the second experiment. This reduced fiber strain is a result of the inclusion of uncertainties in the fiber field combined with the fact that it consists of an averaged quantity over a set of elements in the LV.

4.4 Sensitivity Analysis

Finally, we present a global sensitivity analysis based on Sobol indices for experiment 1 dealing with the four parameters of the transversely isotropic HO model. The computation of the main and total Sobol indices was carried out using the PCE surrogate models for EDV and ε.

Table 2. Propagated uncertainties on the quantities of interest (EDV and ε_{fiber}) for experiments 1 (HO) and 2 (HO+KL). The coefficient of variation denotes the ratio between standard deviation and expected value.

Experiment	Exp. 1		Exp. 2	
Statistics	EDV	ε_{fiber}	EDV	ε_{fiber}
Expected value	94.992	0.138	97.890	0.079
Standard deviation	9.456	0.035	12.188	0.041
Coefficient of Variation	0.099	0.254	0.125	0.518

Table 3. Sensitivity analysis via main and total Sobol indices for the end-diastolic volume and average fiber strain quantities of interest.

Parameters	End-diastolic volume		Average fiber strain	
	Main	Total	Main	Total
a	0.04109086	0.04786979	0.04350967	0.05388241
b	0.91614816	0.93681466	0.92107288	0.93461802
a_f	0.00537262	0.01050244	0.00269112	0.01456773
b_f	0.01411364	0.02808783	0.01202287	0.01763529

Table 3 presents the main and total Sobol indices for the a, b, a_f, and b_f parameters with respect to end-diastolic volume and average fiber strain. The results show that the b parameter present in Eq. (4) is the one that clearly has the most impact on the QoIs analyzed. It is also worth noting through the total Sobol indices that some level of interaction between the parameters is also present.

5 Conclusions

In this work, we presented a forward uncertainty quantification and sensitivity analysis of a constitutive model usually employed in finite element analysis of cardiac mechanics. In particular, we focused on the passive filling problem of the left ventricle and its derived quantities of interest, such as cavity volume and strain. The analyses explored as uncertain inputs the constitutive parameters of the Holzapfel-Ogden model and the fiber orientation field that defines the cardiac microstructure as a random field using the Karhunen-Loève expansion. Due to the high computational cost of solving the passive filling problem of the LV, the UQ and SA studies were carried out with the surrogate models based on polynomial chaos expansions.

The UQ and SA results can be summarized in the following findings. The model outputs analyzed in this work were highly sensitive to local variations and uncertainties in the fiber orientation. The inclusion of local variations in the fiber field increased the upper limit of the range of uncertainty for the end-diastolic volume. The parameter appearing in the exponent of the isotropic term of the transversely isotropic Holzapfel-Ogden constitutive model is the most sensitive parameter in the material model considered in this work.

5.1 Limitations and Future Works

Some limitations of this work are worthy of discussion for improvements in future works. One of the main limitations of this work is the fact that only the fiber orientation was treated as a random field, in spite of the fact the cardiac tissue is usually modeled as an orthotropic material. This limitation comes from the fact that perturbing the fiber, sheet, and normal directions simultaneously would demand a more complex approach for applying the KL expansion and keeping the orthogonality between the vectors defining these directions. As a consequence of this first limitation, the Holzapfel-Ogden model was limited to the transversely isotropic case. Future works should overcome these limitations by exploring all the parameters in the HO model, including uncertainty in the entire local microstructural vectors of cardiac tissue, including other quantities of interest, and also studying the entire cardiac cycle. Another limitation of this work is the choice of the distributions of the uncertain model parameters. Although it is based on a set of reference values of the literature, further studies should consider an inverse uncertainty quantification approach to better characterize parameter distributions.

Acknowledgments. Supported by UFJF, CAPES, CNPq (Grants **423278/2021-5**, 310722/2021-7, 315267/2020-8), and FAPEMIG (Grants APQ-01340-18, APQ 02489/21).

References

1. Bayer, J.D., Blake, R.C., Plank, G., Trayanova, N.A.: A novel rule-based algorithm for assigning myocardial fiber orientation to computational heart models. Ann. Biomed. Eng. **40**(10), 2243–2254 (2012)
2. Cai, L., Ren, L., Wang, Y., Xie, W., Zhu, G., Gao, H.: Surrogate models based on machine learning methods for parameter estimation of left ventricular myocardium. Roy. Soc. Open Sci. **8**(1), 201121 (2021)
3. Campos, J., Sundnes, J., Dos Santos, R., Rocha, B.: Uncertainty quantification and sensitivity analysis of left ventricular function during the full cardiac cycle. Phil. Trans. R. Soc. A **378**(2173), 20190381 (2020)
4. Campos, J.O., Sundnes, J., dos Santos, R.W., Rocha, B.M.: Effects of left ventricle wall thickness uncertainties on cardiac mechanics. Biomech. Model. Mechanobiol. **18**(5), 1415–1427 (2019). https://doi.org/10.1007/s10237-019-01153-1
5. Feinberg, J., Langtangen, H.P.: Chaospy: an open source tool for designing methods of uncertainty quantification. J. Comput. Sci. **11**, 46–57 (2015)
6. Guccione, J.M., Costa, K.D., McCulloch, A.D.: Finite element stress analysis of left ventricular mechanics in the beating dog heart. J. Biomech. **28**(10), 1167–1177 (1995)

7. Holzapfel, G.A., Ogden, R.W.: Constitutive modelling of passive myocardium: a structurally based framework for material characterization. Philos. Trans. Roy. Soc. Math. Phys. Eng. Sci. **367**(1902), 3445–3475 (2009)
8. Karlsen, K.S.: Effects of inertia in modeling of left ventricular mechanics. Master's thesis (2017)
9. Kovacheva, E., Gerach, T., Schuler, S., Ochs, M., Dössel, O., Loewe, A.: Causes of altered ventricular mechanics in hypertrophic cardiomyopathy-an in-silico study (2021)
10. Levrero-Florencio, F., et al.: Sensitivity analysis of a strongly-coupled human-based electromechanical cardiac model: effect of mechanical parameters on physiologically relevant biomarkers. Comput. Methods Appl. Mech. Eng. **361**, 112762 (2020)
11. Logg, A., Mardal, K.A., Wells, G.: Automated Solution Of Differential Equations By The Finite Element Method: The FEniCS book, vol. 84. Springer Science & Business Media (2012)
12. Mosqueira, D., Smith, J.G., Bhagwan, J.R., Denning, C.: Modeling hypertrophic cardiomyopathy: mechanistic insights and pharmacological intervention. Trends Mol. Med. **25**(9), 775–790 (2019)
13. Oliveira, R.S., et al.: Ectopic beats arise from micro-reentries near infarct regions in simulations of a patient-specific heart model. Sci. Rep. **8**(1), 1–14 (2018)
14. Osnes, H., Sundnes, J.: Uncertainty analysis of ventricular mechanics using the probabilistic collocation method. IEEE Trans. Biomed. Eng. **59**(8), 2171–2179 (2012)
15. Rodríguez-Cantano, R., Sundnes, J., Rognes, M.E.: Uncertainty in cardiac myofiber orientation and stiffnesses dominate the variability of left ventricle deformation response. Int. J. Numer. Methods Biomed. Eng. **35**(5), e3178 (2019)
16. Saltelli, A., et al.: Global Sensitivity Analysis: The Primer. John Wiley & Sons (2008)

Automated Variance-Based Sensitivity Analysis of a Heterogeneous Atomistic-Continuum System

Kevin Bronik[1](✉), Werner Muller Roa[1], Maxime Vassaux[1], Wouter Edeling[2], and Peter Coveney[1]

[1] Centre for Computational Science, Department of Chemistry, UCL, London, UK
k.bronik@ucl.ac.uk

[2] Department of Scientific Computing, Centrum Wiskunde and Informatica, Amsterdam, Netherlands

Abstract. A fully automated computational tool for the study of the uncertainty in a mathematical-computational model of a heterogeneous multi-scale atomistic-continuum coupling system is implemented and tested in this project. This tool can facilitate quantitative assessments of the model's overall uncertainty for a given specific range of variables. The computational approach here is based on the polynomial chaos expansion using projection variance, a pseudo-spectral method. It also supports regression variance, a point collocation method with nested quadrature point where the random sampling method takes a dictionary of the names of the parameters which are manually defined to vary with corresponding distributions. The tool in conjunction with an existing platform for verification, validation, and uncertainty quantification offers a scientific simulation environment and data processing workflows that enables the execution of simulation and analysis tasks on a cluster or supercomputing platform with remote submission capabilities.

Keywords: Heterogeneous multi-scale model · Sensitivity analysis · Computational science · Atomistic and continuum coupling · Molecular dynamics simulation · Finite element method · Coupling simulation · High performance computing

1 Introduction

Evidently, many scientific and research fields use sensitivity analysis for their critical application and computational programs, where an automated method which can lead to highly accurate diagnosis and analysis of the level of uncertainty affected by either mathematical modelling or programming aspect can be used to reduce uncertainties quicker and more cost effective. To monitor and challenge such problem that has attracted increased attention in recent years, this study aimed to construct a general automated platform that not only can remove some unnecessary steps to solve the problem through automating the process but also it can accelerate the analysis by enabling the execution of simulation and analysis tasks on a cluster or supercomputing platform.

D. Groen et al. (Eds.): ICCS 2022, LNCS 13353, pp. 762–766, 2022.
https://doi.org/10.1007/978-3-031-08760-8_62

With particular attention to the problems, this project's special focus was analysing a publicly available open-source code SCEMa (see Fig. 1) (https://GitHub.com/UCL-CCS/SCEMa). The selected code here [6] was a heterogeneous multi-scale scientific simulation model. It's highly non-linear stress/strain behaviour which is the result of a coupling of an atomistic and a continuum system made this code an attractive candidate for characterizing, tracing, and managing statistical uncertainty analysis. Here, to facilitate uncertainty quantification, the analysis and sampling methods that was applied to the problem was based on polynomial chaos expansion. Although other existing approaches such as Metropolis-Hastings, quasi-Monte Carlo sampler and stochastic collocation sampler could be also applied to the problem of uncertainty quantification. To measure sensitivity across the whole varying input space, the known direct variance-based method of sensitivity, so called Sobol first-order sensitivity index and total-effect index (see Fig. 2 and Fig. 3) were considered.

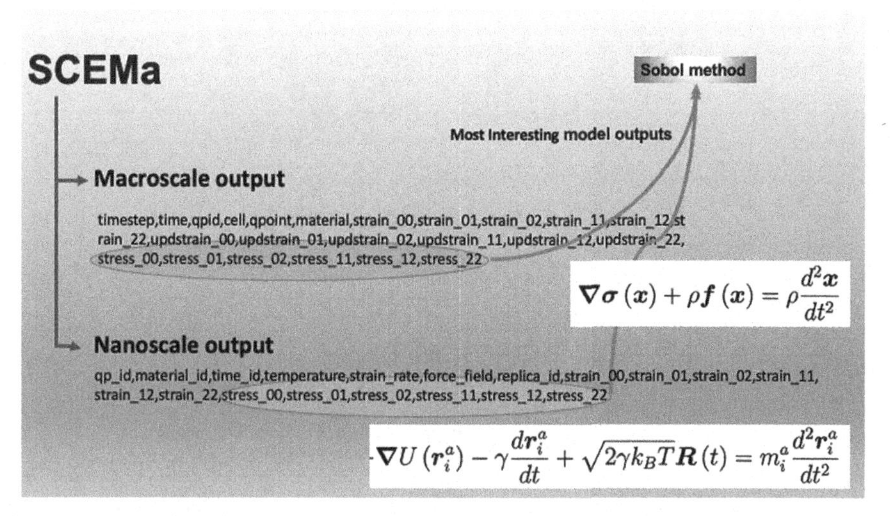

Fig. 1. SCEMa (Simulation Coupling Environment for Materials), a heterogeneous multiscale method implementation

2 Computational Approach

The implementation of the uncertainty quantification tool FabSCEMa (https://github.com/UCL-CCS/FabSCEMa) proceeded with regards to Fabsim, an automation toolkit for complex simulation tasks (https://github.com/djgroen/FabSim3), EasyVVUQ, a python library designed to facilitate verification, validation and uncertainty quantification (VVUQ) for a wide variety of simulations (https://github.com/UCL-CCS/EasyVVUQ) and also other existing toolkits under VECMA Toolkit (https://www.vecma-toolkit.eu/) [1, 2, 7]. To facilitate uncertainty quantification of the Heterogeneous Multiscale scientific simulation models, SCEMa code, it was necessary to execute the programs on heterogeneous computing resources such as traditional cluster or supercomputing platform (this work used extensively the ARCHER2 UK National Supercomputing

Service). The main reason for this limitation lies in the fact that the sampling space was very large and execution of such multiple simulations almost impossible on any local development machine.

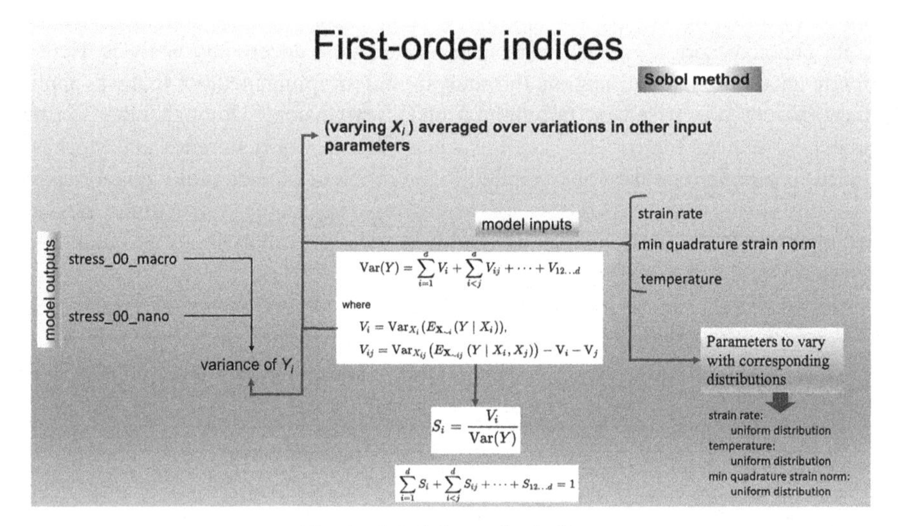

Fig. 2. Sobol first-order indices

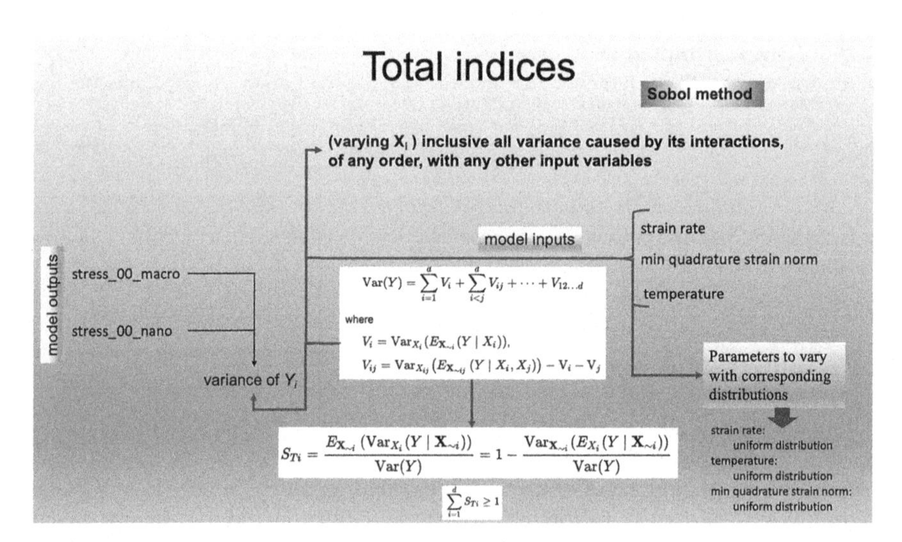

Fig. 3. Sobol total indices

3 Results

Figures 4 and 5 show a visual comparison between first order and total order sensitivity indices based on polynomial chaos expansion, polynomial order 2 and polynomial order 3. Here the tool, we have implemented in this study, after simulating SCEMa code (running several samples with different topologies) uses the normal stress predictions (stress_00_macro and stress_00_nano), which are the result of macro and nano (the

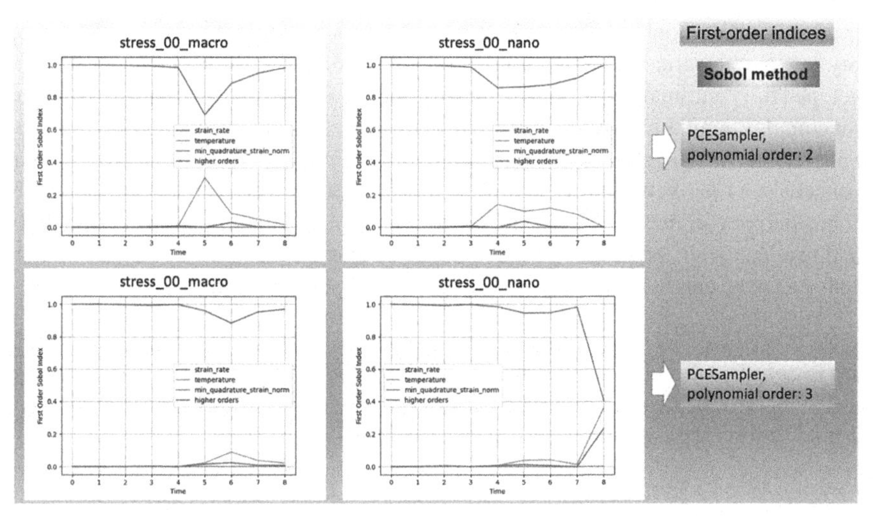

Fig. 4. Sobol first-order sensitivity indices based on polynomial chaos expansion, polynomial order 2 and polynomial order 3

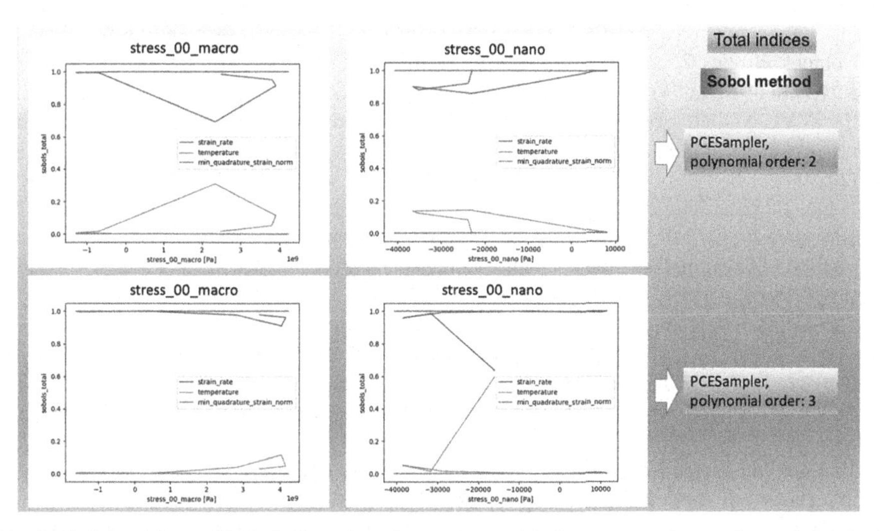

Fig. 5. Sobol total sensitivity indices based on polynomial chaos expansion, polynomial order 2 and polynomial order 3

atomistic and a continuum) simulation outputs, as especially selected multiple outputs which are analysed through independent sensitivity analyses [3–5]. The model multiple inputs (a tensor of uncertain model inputs) are strain rate, min quadrature strain norm and temperature (see Fig. 2 and Fig. 3) where the corresponding distributions are based on uniform distribution.

4 Conclusion

In this study, we demonstrated an automated computational tool for the study of the uncertainty in a mathematical-computational model of a complex coupling system. Here the developed fully automated analysis method tool can be easily modified and applied to any other complex simulation system where it can provide an accurate detailed knowledge of the uncertainty with regards to analysis and sampling methods such as polynomial chaos expansion in a relatively short period of time. This tool is not dependent on the number of input-output parameters that are used in the uncertainty quantification algorithm. The visual results, the visual picture of the importance of each variable in comparison to the output variance, can help researchers/experts easily interpret wide and varying data.

Acknowledgments. We thank EPSRC for funding the Software Environment for Actionable and VVUQ-evaluated Exascale Applications (SEAVEA) (grant no. EP/W007711/1), which provided access to the ARCHER2 UK National Supercomputing Service (https://www.archer2.ac.uk) on which we performed many of the calculations reported here.

References

1. Groen, D., Bhati A.P., Suter, J., Hetherington, J., Zasada, S.J., Coveney, P.V.: FabSim: Facilitating computational research through automation on large-scale and distributed e-infrastructures. Comput. Physics Commun. **207**, 375-385 (2016)
2. Richardson, R.A., Wright, D.W., Edeling, W., Jancauskas, V., Lakhlili, J., Coveney, P.V.: EasyVVUQ: a library for verification, validation and uncertainty quantification in high performance computing. J. Open Research Softw. 8(1), 11. https://doi.org/10.5334/jors.303 (2020)
3. Sobol, I.: Sensitivity estimates for nonlinear mathematical models. Matematicheskoe Modelirovanie **2**, 112–118. In: Russian, translated in English in Sobol' , I. (1993). Sensitivity analysis for non-linear mathematical models. Mathematical Modeling & Computational Experiment (Engl. Transl.), 1993, 1, 407–414 (1990)
4. Sobol, I.M.: Uniformly distributed sequences with an additional uniform property. Zh. Vych. Mat. Mat. Fiz. **16**, 1332–1337 (1976) (in Russian); U.S.S.R. Comput. Maths. Math. Phys. **16**, 236–242 (1976) (in English).
5. Sudret, B.: Global sensitivity analysis using polynomial chaos expansions. Reliability Eng. Syst. Safety **93**(7), 964-979 (2008)
6. Vassaux, M., Richardson, R.A., Coveney, P.V.: The heterogeneous multiscale method applied to inelastic polymer mechanics. Phil. Trans. R. Soc. A **377**(2142), 20180150 (2019)
7. Wright, D.W., et al.: Building Confidence in Simulation: Applications of EasyVVUQ. Adv. Theory Simul., **3**, 1900246 (2020) https://doi.org/10.1002/adts.201900246

Author Index

Printed in the United States
by Baker & Taylor Publisher Services